Training to Fly
Military Flight Training 1907–1945

Rebecca Hancock Cameron

AIR FORCE
History and Museums
PROGRAM
1999

Foreword

The volume at hand, *Training to Fly: Military Flight Training, 1907–1945*, is an institutional history of flight training by the predecessor organizations of the United States Air Force. The U.S. Army purchased its first airplane, built and successfully flown by Orville and Wilbur Wright, in 1909, and placed both lighter- and heavier-than-air aeronautics in the Division of Military Aeronautics of the Signal Corps. As pilots and observers in the Air Service of the American Expeditionary Forces, Americans flew combat missions in France during the Great War. In the first postwar decade, airmen achieved a measure of recognition with the establishment of the Air Corps and, during World War II, the Army Air Forces attained equal status with the Army Ground Forces.

During this first era of military aviation, as described by Rebecca Cameron in *Training to Fly*, the groundwork was laid for the independent United States Air Force. Those were extraordinarily fertile years of invention and innovation in aircraft, engine, and avionics technologies. It was a period in which an air force culture was created, one that was a product of individual personalities, of the demands of a technologically oriented officer corps who served as the fighting force, and of patterns of professional development and identity unique to airmen. Most critical, a flight training system was established on firm footing, whose effective test came in combat in World War II, and whose organization and methods continue virtually intact to the present day.

This volume is based primarily on official documents that are housed in the National Archives and Records Administration. Some, dating from World War II, remained unconsulted and languishing in dust-covered boxes until the author's research required that they be declassified. She has relied upon memoirs and other first-person accounts to give a human face to training policies as found in those dry, official records.

Training to Fly is the first definitive study of this important subject. Training is often overlooked because operations, especially descriptions of aerial combat, have attracted the greatest attention of scholars and the popular press. Yet the success of any military action, as we have learned over and over, is inevitably based upon the quality of training. That training is further enhanced by an understanding of its history, of what has failed, and what has worked.

RICHARD P. HALLION
Air Force Historian

Contents

	page
Foreword	iii
Introduction	1

PART I

The First Decade, 1907–1917 7

CHAPTER ONE

Beginnings: Men and Machines 11
Institutional and Intellectual Underpinnings of Military Aviation 11
Airplane Trials 18
Training the Army to Fly 21
A One-man, One-plane Training Air Force 26
New Airplanes, New Men 28
First Tactical Organization 39

CHAPTER TWO

The Signal Corps Aviation School 43
College Park, Maryland 44
Augusta, Georgia 47
Diversification 50
North Island, California 56
Growing Pains 67

CHAPTER THREE

Prelude to War: Reform, Operational Training, Preparedness 71
The Case before Congress 72
Training Excursions into the Field 79
Struggling Out of Isolation 87
Breakout 92
On the Brink of War 96

Contents

PART II

The End of Illusions 101

CHAPTER FOUR

Training at Home for War Overseas 107
Ground Schools 112
Primary Flying Training 121
Advanced Flying Training 131
    Pursuit 133
    Observation 135
    Bombardment 141
Too Little, Too Late 143

CHAPTER FIVE

Air Service, American Expeditionary Forces 147
Primary Training 149
Advanced Training 160
Specialized Training 165
    Pursuit 166
    Observation 173
    Bombardment 181
    Aerial Gunnery 189
Unit Training 191
Looking Back 194

PART III

Peace 203

CHAPTER SIX

Postwar Retrenchment 213
Organization 213
Flight Training 222
    Primary Flying School 224
    Advanced Flying School 226
Specialized Training 227
    Observation 227
    Pursuit 229
    Attack 233
    Bombardment 235

Tactical Unit Training 236
Early Recovery 239

CHAPTER SEVEN

Boom and Bust: The Air Corps Years 241
Planning and Organization 242
The Air Corps Training Center 246
    Primary Flying School (Including Basic Training) 252
    Advanced Flying School 257
Tactical Unit Training 261
Instrument Flying 264
The Beginning and End of "Normalcy" 270

CHAPTER EIGHT

Training an Air Force: The GHQ Era 273
The GHQ Air Force Perspective 276
Training a Wing 283
Training the Specialties 286
    Fighters (Pursuit) 287
    Attack 290
    Bombardment 291
    Observation and Reconnaissance 295
The End of an Era 299

PART IV

Rearming, 1939–1941 305

CHAPTER NINE

Individual Pilot and Aircrew Training 313
Individual Pilot Training 316
    Primary 320
    Basic 324
    Advanced 328
Aircrew Training 334
    Navigator 337
    Bombardier 342
    Gunnery 347

Contents

CHAPTER TEN

Operational Training 351
Specialties 358
    Dive Bombing 358
    Light Bombardment (Attack) 359
    Heavy Bombardment 361
    Pursuit (Fighters) 365
    Observation and Reconnaissance 367
On the Cusp 370

PART V

Training for War: Planning, Procuring, Organizing .. 375

CHAPTER ELEVEN

Picking the Men, Training the Pilots 383
Manpower Procurement and Classification 384
Pilot Training 388
    Primary 390
    Basic 395
    Advanced 400
        Single-engine 400
        Twin-engine 405
        Observation 407
    Transition Training 408
        Bombers 410
        Fighters 413
        Observation and Reconnaissance 415
Advance and Retreat 415

CHAPTER TWELVE

Not Just a Pilot's War: Individual Training of Navigators,
    Bombardiers, and Gunners 421
Navigator Training 422
Bombardier Training 428
Flexible Gunnery Training 438
Summation 448

Contents

CHAPTER THIRTEEN

Crew and Unit Training: Organization, Technology, and Doctrine 451
Organizing and Administering Operational Training 454
    Heavy Bombardment 458
    Very Heavy Bombardment 460
    Medium and Light Bombardment 461
    Fighters 463
Combat "Readiness" 464
Training, Doctrine, and Tactics 467

CHAPTER FOURTEEN

Training for Strategic Bombardment 481
Heavy Bombardment 484
    Training Curriculum 487
Very Heavy Bombardment 503
Radar in Strategic Operations 509
Assessments 515

CHAPTER FIFTEEN

Crew and Unit Training for the Tactical Air Forces 521
Medium Bombardment 525
Light Bombardment 535
Fighters 538
Assessments 552

An End and a Beginning 557

Appendices
Accident Statistics of the Interwar Period 564
Army Air Forces Training Command and Its Predecessor Flying Training Commands 565
Major Changes in Undergraduate Pilot Training, July 1939–January 1943 566
Flying Training Graduates, July 1939–August 1945 567
Location and Supervision of Pilot and Bombardier Training, July 1940 568
Bombardier Requirements in Relation to the Group and Pilot Programs 569

Contents

Notes 571
Abbreviations Used 637
Selected Bibliography 641
Index 657

MAPS

U.S. Air Service AEF Training Centers in France, 1917–1918 199
Army Air Forces Flying Training Command Sites, July 1944 418–419

TABLE

Flying Fields of the Aviation Section of the U.S. Air Service, November 11, 1918 145

CHARTS

Air Corps Aviation School 74
Stages of Flying Training, World War I 109
Flying Schools, World War I 111
Gunnery Branch, World War I 129
Assistant Chief of Air Staff, Training, Directory Chart, December 30, 1944 455

PHOTOGRAPHS

Brig. Gen. Adolphus Greely; Brig. Gen. James Allen 12
Capt. Charles DeForest Chandler with Balloon Detachment at Fort Omaha 13
Thomas E. Selfridge and Alexander Graham Bell 17
Orville Wright with his flyer at Fort Myer 18
Lt. B. D. Foulois, Lt. F. P. Lahm, Lt. G. C. Sweet, Maj. C. McK. Saltzman, Maj. G. O. Squier, Capt. C. DeF. Chandler, and Lt. F. E. Humphreys 20
The Wright flyer endurance and speed tests 21
Foulois and Lahm at College Park 23
The Wright Model B and instructor Parmelee with Foulois 29

Contents

Wright instructor Coffyn	30
Henry H. Arnold and Thomas DeW. Milling with John Rodgers	31
Arnold in the Wright B	32
Glenn Curtiss with his aircraft	35
Lts. P. W. Beck, G. E. M. Kelly, and J. C. Walker, Jr.	36
The Curtiss Model D	38
Maneuver Camp, Fort Sam Houston	40
Lt. Roy C. Kirtland and a Wright B	45
Training camp at Augusta	49
Glenn Curtiss and the *June Bug*	51
Lts. L. E. Goodier, J. E. Carberry, and W. R. Taliaferro with a Curtiss pusher	52
The Burgess Model H	53
Hangars at North Island with Curtiss Models E and G	57
Capt. Arthur C. Cowan; 1st Lt. Walter R. Taliaferro	60
Civilian flying instructor Oscar Brindley	63
A JN–4	64
Martin T trainers	66
North Island workshop	75
1912 Maneuvers, Connecticut	80
S.C. No. 11	82
Lts. Moss L. Love, Carleton G. Chapman, and Herbert A. Dargue	85
Lahm's crash in Manila Bay	86
The Breese Penguin	90
1st Lt. Thomas DeW. Milling with S.C. No. 26	91
Maj. Frank Lahm and a JN–4	94
Hiram Bingham with Air Service trainees	113
Col. Clinton G. Edgar	122
Maj. Hugh Knerr and the North Island site	123
Training aircraft on Scott and Brooks Fields	124
Curtiss JN–4A and a Boeing trainer	125
Maj. Gen. William L. Kenly with two of his associates	133
Thomas-Morse S–4 Scout	134
Balloon observation and aerial gunnery	136
Brig. Gen. Benjamin D. Foulois and Secretary of War Newton D. Baker	138
Col. Raynal Bolling and Issoudun, France	148
General Pershing at Tours	150
The field at Issoudun	151
Beaumont Barracks at Tours	152
Nieuport 17 and Breese Penguin	154
Sopwith F–1 Camel and Morane-Saulnier	156
Farman trainer and Caproni triplane cabin	158

Contents

French Breguet 14 and Spad XIII 163
Engineering Department at Issoudun 164
Pursuit pilot instruction and a Spad XIII 167
Repairing a Nieuport 27 168
Maneuver instructions diagram 170
Issoudun Field No. 7 and Nieuport 28s 171
S.E. 5, Bristol Fighter, and Avro 504 173
Gunnery firing device 174
Tours observation tower 175
Panel signaling, observers' classroom, and camera gun repair 176
Camera instruction, stripping practice, and code class 178
Qualification and medical tests for airmen 179
DH-4 180
Salmsons 181
Military Aeronautics certificate, Caproni bomber, and a bomb
  of the type used for training 183
The Clermont-Ferrand training site and a Breguet-Renault 185
Caproni biplane bomber, S.C. 40070 187
Handley-Page bomber 188
The St. Jean-de-Monts training site 191
Salmson 2A2 194
Maj. Gen. Charles T. Menoher 215
Maj. Gen. Mason M. Patrick and Maj. Herbert A. Dargue 217
Training activities at Kelly Field 223
Brooks Field training site 225
DH-4B 228
Lt. Col. Harold E. Hartney 230
SE-5s and an MB-3A 232
2d Lt. Truman Landon and an O-2 234
An MB-2 235
Aerial bomb training 238
Maj. Gen. James E. Fechet; Brig. Gen. Frank P. Lahm 243
Duncan Field 247
The Ruggles Orientator 248
Brig. Gen. James E. Chaney 249
PT-11s 254
A BT-2 255
Keystone B-6A and a B-3A instrument board 259
Boeing P-12B, Thomas-Morse O-19, and Douglas O-25 260
Langley Field 263
Instrument flying training with the Ocker Box and sextant 266
The Link trainer 267
A Keystone B-6A used in the Air Corps mail operation 268

Contents

Douglas O–46A	269
Lt. James H. Doolittle	271
Maj. Gen. Oscar Westover	275
The Air Corps Tactical School graduating class of 1928	276
Brig. Gen. Frank Andrews and his staff	278
Lt. Col. George Brett	282
Curtiss Condor B–2 and Martin B–12	284
March Field	285
Capts. Frank O'D. Hunter and Hugh M. Elmendorf	288
Capt. Ira C. Eaker and formation flying instruction	289
A–12 Shrike	291
Aerial photography and instruction in darkroom technique	297
Air Corps Training Center aircraft, Kelly Field	315
Student pilots, Randolph Field	316
Pilot trainees	317
JN–4 training crash	322
Proper parachute adjustment	323
Primary flying training graduates	324
Blind flying trainer	326
RAF advanced flying graduates	330
Advanced training in flexible gunnery and engines	332
AT–6 and AT–11	333
Navigator ground school instrument training	340
Bombardier ground school firing practice	344
Bombardier and navigator training	347
Col. Walter R. Weaver	348
P–36	353
B–18	354
Ground school instruction in aerial manuevers	357
A–20As	360
B–17 and crew	362
P–40	367
O–47	369
Pilot trainees	387
Ryan School	391
PT–22	394
BT–13	397
Flying cadets and a Vultee Valiant	399
Pilot trainee and an AT–6D	401
Advanced training in formation and instrument flying	403
Link gunnery trainer	404
AT–7	406
B–17	409

Contents

B–26 413
P–38 414
Lt. Gen. Barton K. Yount 416
Instruction in celestial navigation and map and chart reading 424
Brig. Gen. Robert W. Harper 427
Bombardier and navigator in training 429
Night training of bomber crew and AT–11 spot camera 431
D–8, Sperry S–1, and Norden bombsights 433
Dual trained bombardier and dead-reckoning navigator 435
Flexible gunnery training and the Casper range 439
Ground school gunnery training and enemy aircraft recognition 442
Moving target practice 443
Air-to-air and air-to-ground gunnery practice 445
Combat aircrew members 453
Heavy bombardment crew and B–24 459
Very heavy bombardment crew and B–29 461
Medium and light bombardment crews in the B–26 and B–25 462
Flying sergeants and P–39s 464
High-altitude formation flying 469
B–25 cockpit mockup used for instrument training 471
Aircraft recognition training 474
A crew training with the Norden bombsight 476
B–29 483
Gunnery ground schools in England 489
B–17 and gunner locations 491
Camera bombing with simulated targets 496
Gunnery practice against towed targets 497
B–24 502
Tail and remote gunners' compartments 509
Maj. Gen. Barton K. Yount with his staff and Training Command
    headquarters building 518
A–20 522
Nose and tail gunners in B–26 526
Low-altitude skip bombing practice 528
Medium bombardment trainees 532
Light-bombardment strafing practice 536
School of Applied Tactics training session 541
Low-level bombing training 546
P–61 night fighter 549

The photographs on pages 276 and 282 are courtesy of Lt. Gen. Devol Brett, USAF (Ret.). The art reproduced on page 123 is courtesy of the U.S. Air Force Academy Library.

Introduction

I "prayed that I might not be posted to a Training Squadron" wrote British author C. S. Lewis in his memoir of flying during the Great War. Another war later, airmen expressed the same sentiment, with many inventing any excuse to avoid a training assignment. Compared to the excitement of combat operations, most pilots felt that they had been insulted, passed over, effectively put out to pasture, when they became part of the training establishment. Moreover, training considerations often have been stepchildren in the U.S. air service's systems acquisition and budgetary processes. The lack of enthusiasm about training is indicated further by the scanty attention paid to the subject by historians and diarists. It appears that only "official histories" such as the one at hand address, in any but the most anecdotal fashion, what many find to be a dull topic dealing with a support function.

But, at a cost in blood, training prepares men (and now women) to fight, and also initiates them into the warrior culture that pervades military life. Training is an all-pervasive phenomenon in an air force. It is a constant. It takes place all of the time, during war and peace. In the global conflicts of the first half of the twentieth century, even front-line squadrons trained as well as fought.

As it happens, training is not a colorless endeavor. The first part of this narrative, for example, describes an individualistic, dangerous, and innovative era in the history of flight. The airplane was new and military men had to invent a definition for it and for themselves as airmen. At the same time, they succeeded in laying down fundamental guidelines for air training that have lasted to the present day. Admittedly, however, institution-building and professionalization (themes central to this book's chapters covering the interwar years) and curriculum and program development (a large part of the training story during the buildup and World War II period) intrinsically lack the drama of individual exploits, technological discovery, or combat.

The history of flight can be seen as the secular, technological equivalent of man's religious grasp for a world beyond himself, but this study is not intended as metaphor. Although it will touch upon the ways in which airmen and the public considered flying to be a transcendent experience, and upon the

Introduction

messianic element among aviation proponents, it is mostly an earthbound institutional and policy history of the organizational structures developed for flight training, and of the methods used to teach military men to fly airplanes. One cannot discuss flight training without describing some of the technological developments that dictated the skills and tactics airmen had to master, or the doctrine that either grew out of training experience or that drove training practices. Certainly doctrine is the synapse between training and operations.

There are a number of topics not covered in this history of flight training. In many cases, they are worth at least one book of their own. Except at the beginning, when "flight" was synonymous with lighter-than-air craft, this volume only addresses heavier-than-air training. Helicopters made their appearance in the Army during the period under discussion also, but they are not described here. Maneuvers, arguably an important aspect of training, are mentioned but not discussed in detail.

The history of the Air Force would be enriched by biographies of its leading players, but to date not many have been written. Because institutions not only grow and change in reaction to external events but also because of individual decisions and personalities (although there have been few definitive biographies to draw upon), I have tried to be mindful of the human dimension along with budgetary considerations, curriculum and program planning, administrative apparatus, and training techniques. I have relied heavily upon memoirs and oral histories to flavor the official record with personal experience.

During the first forty years of its existence, before the creation of the independent Air Force, U.S. military aviation professionalized and created a culture that set it apart from its parent, the ground-based United States Army. The air arm demonstrated a marked technocratic bias, and it evolved training practices empirically, showing relatively little interest in theory or military traditions and hierarchy.[1] More than the ground combat arms, it glorified the individual, specifically the warrior-pilot who flew against extremely dangerous odds, in training as well as in combat. The flight training program formulated rules and regulations in part to defy air force culture, in that training procedures atempted to curb the eccentric, the dangerously individualistic, the tendency for airmen to rely only on themselves and each other. James Gould Cozzens served in the Army Air Forces during World War II. His wartime diaries provide a fascinating glimpse into the human side of decision-making and, in his well-regarded novel *Guard of Honor*, he described the high-wire act that became integral to air force culture:

> Flying in those days was a business set apart by its unexampled dangers; and those who flew were joined in the bond of their undefined, informal co-operative effort to shut their minds to the plain fact that if the war continued they were all going to die — perhaps by enemy action, perhaps by accident; perhaps this week, certainly next month. They supported each

Introduction

other in fending off the normal animal despair; now by braving it with cumbersome and elaborate humor — *take the piston rods out of my kidneys and assemble the engine again*; now by a solemn deprecatory indirection which did not blush to use such euphemisms as "grounded for good."

For those who survived it was a bond.[2]

As long as technology remained relatively simple, flight training centered on the individual's mastery of nature and machine. During the interwar years, the air service accommodated to the development of larger, faster aircraft by instituting specialized functions, and wrote its training directives around those specialties. Separate training programs for each pilot specialty evolved and, reluctantly, the Air Corps added them for nonpilot aircrew members. As World War II loomed, crew training and teamwork were emphasized. In general, air training became more collective, more corporate, less a story of persons than of institutionalized programs and processes. At the same time, each year a new crop of airmen earned their wings, and each of them rediscovered the sensation of flight. Because men's enthusiasms, mistakes, judgments and fears — just as much as technological imperatives — determined the evolutionary direction of manned flight, even a history of training cannot be reduced to a study only of mechanization.

From the experiences of the first generation of military men learning to fly, and two world wars, a number of questions emerge. How closely did air and ground training converge? How well did training replicate operational practices? Was training patterned on doctrine or the other way around? Was training policy a carefully considered construct with predictive value, or was it principally reactive; in other words, was it personnel or crisis management? How important were standardized training techniques? In what ways did peacetime and wartime training differ? How did the relationship between Army aviation and the aircraft industry evolve? What feedback did training officials receive from field commanders, and how did they respond? To what degree did the military rely upon civilians for training and facilities to substitute for or supplement its own activities? How were theory and practice, quantity and quality, balanced? Was air training affected more by manpower or equipment shortages, and how did it respond to the rare surfeit of either? Upon what standards were airmen selected, promoted, and assigned to specialties? Which pilot and aircrew specialties dominated at different times, and which managed to implement the most successful training programs? Were the lessons learned from one war applied to the next?

All of those matters arose during the era covered in this volume. Sometimes airmen devoted intense scrutiny to one or another of them, sometimes they were oblivious, sometimes they had more pressing concerns. No checklist of answers to the questions above can be supplied at the conclusion of this narrative. The answers varied according to time and place, circumstance and

Introduction

personalities. Historical answers are found in the specifics, embedded in the dailiness of human events.

I hope that this study sheds light on the ways the Army's air arm dealt with those and other considerations as it defined itself and its mission over time, the manner in which it translated professional and doctrinal concepts into a training program, and how well airmen accomplished or failed to resolve the issues before them. Presumably, this history will help to illuminate why and when airmen advanced or retreated as they did, and how their beliefs and actions, given external historical circumstance and technological change, created an Air Force.

Part I
The First Decade 1907–1917

I

On October 5th, we moved in, built a shed for the machine, set up the pylon and track, and Wilbur began our pilot training. At the end of about three hours' dual, we were turned loose and made our first solo flights. A few days later I was even considered qualified to carry passengers and did so, taking Lieutenant Sweet of the Navy as my victim for a flight around the field.
— Colonel F. P. Lahm[1]

So Frank Purdy Lahm described the Army's first aviation training given to him and his fellow lieutenant, Frederic E. Humphreys, in 1909. Their instructor, endlessly patient and kindly Wilbur Wright, with the sober eye and stiff collar, looked to be the religious midwesterner he was. Every day except Sunday, as long as the weather was clear, he took his pupils up in the new Army airplane, helping them to learn the sense of balance and steering he brought to flying from his experiments with gliders. Every evening, they mulled over the future of aviation, discussed the fine points of airplane control, and analyzed the dangerous effect of winds. The students progressed from straight flights to gentle turns and, most important, they learned how to return safely to the ground. In less than a month's time, Lahm had taken up his "victim" Lieutenant Sweet, training was considered complete, and the young officers pronounced to be qualified aviators.

His instruction of the two lieutenants those October mornings in 1909 fulfilled Wilbur Wright's final requirement in the terms of his contract with the Army. Nearly two years earlier, the Chief Signal Officer of the U.S. Army had issued invitations for bids on the production of a "gasless flyer." After meeting specifications for airplane speed, weight, and power, and completing a series of flight tests, the winning bidder would be obligated to instruct two Army officers in the operation of the airplane.

1907-1917

The little Wright biplane, the only successful competitor for the Army contract, flew ten miles on its distance test, established a world record flight of one hour, twelve minutes, and forty seconds on the endurance test, and overcame the required 40 mile-per-hour speed to reach a maximum of 47.431 miles per hour on part of the speed test. An eager public and press, as well as the Aeronautical Board of the Signal Corps appointed to observe the trials, stood watching as the Wrights accomplished their breathtaking feats.

Even though the Army bought the first airplane for the U.S. government in 1909, the promise of military aviation was not then readily apparent. The Army had acknowledged aeronautics two years earlier when it established an Aeronautical Division within the Signal Corps. That tiny office took on the responsibility of maintenance and flight training for its new dirigible (rigid airship) and its new airplane. But, within the Army, no great stir followed the purchase of either of the two astonishing inventions.

Oddly enough, the Army paid minuscule attention to a phenomenon enjoying tremendous popular appeal. Exhibition flying attracted huge crowds of ticket-buying customers. (It also cost dearly in loss of life.) Stunt fliers rivaled vaudeville stars in their celebrity. Man's newfound ability to fly, the culmination of centuries of dreaming and ill-fated experimentation, commanded widespread enthusiasm and support among the public and the engineering-minded in the scientific community. Yet the military seemed generally unfazed by it and disinclined to capitalize upon its possibilities. In his yearly reports, the Chief Signal Officer pointed to the great strides in aviation and aerostation (the term used for ballooning) made by European nations, but could say little on behalf of American military efforts. Army officers conducted some firing tests at captive balloons and improvised with bombsights, but had too little time and too few people to permit experimentation in depth. With one training dirigible, one Wright airplane, three small balloons, one lieutenant on aviation duty, one officer licensed as a balloon pilot, and nine enlisted men, it was entirely appropriate in 1910 for the Chief Signal Officer to describe military aeronautics as at a standstill.[2]

Though nearly stillborn, aviation expanded suddenly when the Congress appropriated $125,000 for aeronautics for fiscal year 1912, with $25,000 available on March 3, 1911. As a result, the Signal Corps purchased new equipment, inaugurated a broader and more vital training program, established the Signal Corps Aviation School at College Park, Maryland, and explored other training sites. Most significant, it shifted the emphasis away from aerostation toward heavier-than-air flight. The three short years between 1911 and the outbreak of war in Europe were critical for aviation, not because the U.S. Army developed the doctrine, tactics, or aircraft to take it effectively into combat — it did not — but because it articulated concepts of training and professionalization that would launch it effectively into the postwar era.

Part I

As was inevitable during peacetime, between 1909 and 1914 military aviation concentrated upon training. For the most part, training consisted of teaching the basic skills of taking off, turning, and landing a low-powered airplane. Training correlated only incidentally with established rules of engagement on the battlefield because the U.S. Army had only the vaguest sense about the possible uses of the airplane in war. Specialization arose only insofar as exhibition flying could be distinguished from military flying. And although the aviation program comprised both research-and-development and training elements, engineering and experimentation with aircraft types and engines remained secondary to teaching men to fly.

Congressional hearings in 1913 led to the act of July 18, 1914, that created the Aviation Section and gave Army aeronautics official standing and credibility. Passage of the act culminated several years of advocacy from outside and within the Signal Corps to secure institutional stability and funding for Army aviation. It also redressed some personnel and organizational problems. Regrettably, however, it failed to authorize an expanded training establishment, the particular focus of Signal Corps special pleading.

Nonetheless, by 1914 it could no longer be said that military aviation was synonymous with training alone. Specifications for new aircraft, and modifications to the older machines, differentiated between airframes and engines intended for training and for field operations. Diversification also took place at the Signal Corps Aviation School with the segregation of training and experimentation–repair into two departments. Moreover, early in 1915 the newly created National Advisory Committee for Aeronautics assumed the role of shepherding civil and military aviation research, which clarified the school's training function and diminished its oversight of scientific and technological innovation.

Although training remained a high priority, the Aviation Section now had separate training and operational units. The 1st Aero Squadron, operating with American ground forces on the Texas border in 1915, had been garrisoned at San Diego with the Signal Corps Aviation School until it moved to San Antonio. There it joined Army forces monitoring the Mexican insurrection.

Only a few short weeks after passage of the legislation creating the Aviation Section, Americans felt the reverberations from the first clashes of the cataclysmic war that would engulf much of Western Europe, Russia, and the United States. Military reform, underway since the turn of the century, assumed a new guise as America launched pell-mell into the "preparedness" movement, a drive to strengthen her defenses against vaguely defined but threatening outside forces. Closer to home, American military intervention in the civil strife in Mexico aggravated the national awareness of a potentially immediate threat

1907-1917

to U.S. borders and created the opportunity for the first operational assignment of an aviation unit. As the preparedness movement gathered momentum, the Aviation Section re-emphasized training in accordance with the growing national sentiment calling for the training and maintenance of both a military reserve and a larger force-in-being.

ONE

Beginnings: Men and Machines

Every flight that I have made down here has been more or less hazardous.
— Lt. Benjamin D. Foulois[1]

In the United States, the possibilities of observation from the air led to the use of balloons as early as the Civil War. While balloons remained in the Army inventory, they were never assigned a very active operational role, and the military took little initiative in exploring their potential. Sufficient interest in aeronautics persisted nonetheless, particularly in the minds of the first Chief Signal Officer of the U.S. Army, Brig. Gen. Adolphus Greely, and of his successor, Brig. Gen. James Allen, that by August 1908 the government had purchased the Baldwin airship (dirigible).

The best thinking of the time held that aircraft would be used in observation and reconnaissance and in courier activities. The Signal Corps, whose mission it was to relay information, therefore assumed ownership of these mobile, piloted vehicles of observation and communication. Its Aeronautical Division was established on August 1, 1907, to "have charge of all matters pertaining to military ballooning, air machines, and all kindred subjects."[2] Even though the Signal Corps lacked a well-articulated concept of military employment for the balloon, and had almost no notion about the possible uses of the heavier-than-air machine, the Aeronautical Division began with high hopes for the future of flight in the Army.

Institutional and Intellectual Underpinnings of Military Aviation

Chief Signal Officer Allen appointed Capt. Charles DeForest Chandler to be the first chief of the U.S. Army's Aeronautical Division. Having visited British aerostation facilities while in Europe in 1905, Chandler's fitness for his new assignment came from his presumed familiarity with balloons. The following year, he and Maj. Samuel Reber of the Signal Corps represented the War

1907-1917

Brig. Gen. Adolphus Greely, the first Chief Signal Officer of the U.S. Army

Brig. Gen. James Allen, General Greeley's successor

Department in a free balloon ascent, and in 1907 Chandler qualified for balloon certificate No. 8 issued by the Federation Aeronautique Internationale (FAI), the only agency at that time to have established flight performance standards in any type of aircraft.[3]

Army men participated in events that drew audiences eager to see the old-fashioned free balloons and the newer dirigibles that, by virtue of being motor-driven, operated with greater flexibility and mobility. Balloon races pitted teams that were often supported by foreign as well as American aeronautical organizations, entrepreneurs, and government interests. Many of the developers and participants not only enjoyed the sporting element but were eager to explore the scientific and mechanical principles of manned flight. The Army permitted some of its officers to take part in balloon meets and later in airplane races, albeit on their own time, because their activities brought recognition to Army aeronautics. Probably, too, beneath the sober professional face, senior Army men were indulgent of sporting events as appropriate to the traditional roles of officer and gentleman.

Chandler's Aeronautical Division staff consisted of Cpl. Edward Ward and First-Class Pvt. Joseph E. Barrett.[4] Despite its insignificant size, the fledgling division received encouragement from a handful of older Signal Corps officers who perceived the potential of aeronautics as more than its showcase appeal. Among these were balloonist Samuel Reber and Maj. George O. Squier, the latter having earned a doctorate in electrical engineering from Johns Hopkins University. Keenly interested in the application of scientific discoveries to the

Beginnings

Capt. Charles DeForest Chandler, the first chief of the U.S. Army's Aeronautical Division, originally gained fame as a balloonist. He is shown here with the gas house built to supply the Balloon Detachment stationed at Fort Omaha, Nebraska. Cpl. Edward Ward, one of Chandler's staff members stationed at Fort Omaha, is seated in a balloon's concentrating ring and is surrounded by fellow Signal Corps personnel.

military, Squier promoted publication in technical fields, pushed research-and-development projects, and pioneered in the use of radio and photography in the Signal Corps. As commander of the Signal School at Fort Leavenworth, Kansas, he also introduced aeronautics into the Army educational system. He would continue to be an active supporter of aviation when, in 1917, he became Chief Signal Officer.[5]

Another scientifically minded officer who furthered Army aviation in its earliest days, Lt. Col. William A. Glassford, had begun writing about aeronautics, military history, and meteorology in the 1890s and later visited European countries to report on aeronautical experimentation abroad. He not only remained interested in theoretical and intellectual matters but, like Reber and Squier, helped to professionalize aeronautics as a viable career path within the military.[6] Other officers soon to associate themselves with aviation would link the old Signal Corps interest in balloon reconnaissance to the new civilian enthusiasm for sport ballooning and exhibition flying and to the American scientific community's fascination with invention and applied science.

The Army introduced training in aerostation as early as 1902 at Fort Myer, Virginia, when it organized a balloon detachment.[7] In 1905 the unit was stationed at the large cavalry and field artillery drill ground at Fort Omaha, Nebraska. By 1906, dictated by War Department General Order No. 145,[8] balloon instruction was supposedly in the curriculum at the Army schools at Fort Leavenworth, but lack of equipment there precluded a thoroughgoing practical course.[9] After further consideration, the Signal Corps decided against concentrating its aeronautical instruction at Fort Leavenworth.

That decision may have been fateful. Because of an Army reorganization in 1903, the Leavenworth schools were prospering as a result of a rigorous curriculum and improved leadership and instructional staff. Officers attending the Leavenworth schools became conversant with military doctrine and the application of new technologies to battlefield situations.[10] The Signal School was a relatively late arrival among the schools, having been organized in 1905 and placed under the directorship of Major Squier. It provided training in electrical communications (and theoretically in aeronautics) within the context of combined arms operations. The curriculum addressed tactical as well as technical matters.

In spite of Squier's forceful advocacy, Leavenworth was not selected as the primary site for training in aerostation. The Signal Corps thereby lost a valuable opportunity to integrate aeronautics into the Army's most forward-thinking curriculum, and the experience would be repeated when officers began training to fly airplanes. The fact that the theory and practice of military aeronautics was only a footnote in the Army's educational system undoubtedly contributed to the isolation of aviation within the service and to the mutual suspicion that grew up between airmen and their fellow ground officers. Moreover, it prevented many high-ranking and leadership-bound Army officers in the combat arms from considering the possible uses of airplanes in wartime.

At the time the Board of the Academic Schools decided to move aeronautics out of Leavenworth except for a token presence, it took note of a young assistant instructor: Lt. Frank Purdy Lahm "is undoubtedly the best equipped [aero]naut in the army, if not in the United States."[11] It must be noted that Lahm's recognition derived as much from his singularity as from his accomplishments — almost no other Army officers were active in aeronautics before 1907. Lahm made his first balloon flight in July 1904 while visiting his father in Paris. The younger Lahm was then teaching French at West Point, and when he returned to France in 1905 he finished instruction and received his balloon pilot's license. "That," he later commented, "is what got me into the game."[12] Lahm went on the following year to win the first Gordon Bennett International Balloon Race, and thereby called himself to the attention of senior Signal Corps officers. Lahm's involvement in balloon meets brought him into contact with balloon and engine inventors and also introduced him to those experimenting with heavier-than-air machines. Again through his father, an active promoter of aeronautics, Lieutenant Lahm met the Wright brothers during the summer of 1907 when he and they were visiting Paris.[13] Lahm wrote enthusiastically to the Chief Signal Officer about the meeting, and as a result of the letter, Lahm's relative celebrity, and possibly the Academic Board's report, General Allen requested that Lahm be assigned to duty with the Signal Corps.[14]

Lighter-than-air activities occupied the small cadre of officers and men who came into aeronautics between the creation of the Aeronautical Division in August 1907 and the acceptance of the Wright airplane in the summer of 1909. In 1907, Chief Signal Officer Allen saw little future for heavier-than-air machines. The Wright brothers wrote to the Board of Ordnance and Fortifications in June: "We believe that the principal use of a flyer at present is for military purposes."[15] On the contrary, according to Allen, the Wright brothers' "flying machine is not suitable for military purposes, and an appropriation from Congress with a view of purchasing one or more of these flying machines is not recommended."[16]

Given the Army leadership's vague disinterest in aeronautics, officers eager to explore the new field turned to research and opinion reported in nonmilitary publications. Even later, when the service began to generate its own instructional materials, it lacked a theoretical or doctrinal basis and depended upon technical information provided largely from civilian sources. Magazines such as *Scientific American*, *Aeronautics*, and *Flying* published detailed articles and drawings of airplanes, balloons, motors, and control systems. They covered foreign aeronautical developments and discussed theory as well as practice. Thus, despite the narrow confines of the Army, quizzical Signal Corps officers became familiar with aeronautical happenings.

Occasionally, too, officers posted overseas took the opportunity to report on foreign military aeronautics. From these firsthand investigations and the burgeoning scientific literature, the Aeronautical Division tracked European developments in dirigibles, gliders, and heavier-than-air devices. Compiled into Signal Corps Bulletins, the reports were often quite encompassing, outlining the design and construction of free balloons and dirigibles, the means for obtaining hydrogen, glossaries of aeronautical terms, and other technical data.[17]

Although Army officers themselves had limited opportunity to test published theories and speculations, they encountered spirited debate about the future and utility of flight in public forums, and in the pages of newspapers, magazines, and journals. They became part of informal and professional networks for the exchange of opinion and information. At the International Aeronautical Congress held in New York in October 1907, for example, Chief Signal Officer General Allen explained the plans and ongoing activities of the Signal Corps. Squier came from Fort Leavenworth to talk about "the advantages of an aeronautic division in active operations." Glassford, then Chief Signal Officer of one of the Army's departments, presented a paper titled "Our Army and Aerial Warfare."

On that occasion, it was the congress President, Willis L. Moore, Chief of the U.S. Weather Bureau, who delivered the loudest call to improve prospects for military aviation. Given Moore's field of study, it must have been surprising to hear him contend that the future of aeronautics lay in military, not in commercial, ventures: "It is evident that the first application in aerial navigation will be the art of war, and it is clear that its main usefulness will be in reconnaissance, for the [bomb]loads which can be carried will be small." Moore was convinced, along with the leadership of the Signal Corps at the time, that the dirigible possessed advantages over the "flying machine." It could potentially carry more surplus weight and could rise more quickly than the airplane, conferring "great advantage both for attack and defense, as evidenced by all contests between birds." Again, Moore reinforced the general belief that "the chief use in war, . . . both of the dirigible balloon and of the flying machine will be in scouting and in directing artillery fire by use of wireless telegraphy. . . . Their offensive operations will be limited, although occasional lucky shots may prove decisive."[18]

Those within and outside the Signal Corps called upon the U.S. Congress and the Army General Staff to acknowledge the unused potential of military aviation. The Aero Club of America, for example, which numbered Army airmen among its early members, consistently advocated larger appropriations of public funds. By 1910, the influential club had spawned several splinter organizations that joined in pressing for more government funding.[19] In his speech before the International Aeronautical Congress in 1907, Moore lambasted the U.S. government for its failure to properly assess and finance military aeronautics. "In the United States," he charged, "the Government has done practically nothing toward building dirigible balloons. This has been left to private initiative."[20] On the heels of those rousing challenges, the Aeronautical Convention passed a resolution asking President Theodore Roosevelt to bring to the attention of Congress "the advisability of providing the departments of the Government charged with these duties, funds sufficient to establish aeronautical plants commensurate with those of other nations."[21]

Despite lobbying and the Chief Signal Officer's yearly pleas, Congress spent parsimoniously on aviation for some time. Nonetheless, enthusiastic Signal Corps officers remained hopeful. Glassford, for instance, commented to the press that "there is always the consolation that when Uncle Sam really sees

Beginnings

the situation he takes no halfway measures, and so I look upon the present agitation and attention to aerial machines of warfare as fortunate in that Congress will be convinced of the necessity for radical action to enable us to 'catch on' and catch up."[22]

In the meantime, by the summer of 1908, two new officers had volunteered for aeronautical duty. Lt. Thomas E. Selfridge was the most experienced of the would-be military pilots. He had designed flying machines built by Professor Alexander Graham Bell's Aerial Experiment Association, an organization initially established to promote tests of Dr. Bell's tetrahedral kite. Selfridge also piloted the White Wing, a design of dirigible manufacturer Thomas Baldwin. Other flights in Aerial Experiment Association airplanes followed. Selfridge accepted a detail to the Signal Corps on August 3, 1907, and received his FAI airship license in August 1908. His interest in experimentation continued; he drew up a comprehensive plan for a flying field (then called an aerodrome) and an experimental plant where motors, propellers, and other equipment could be tested.[23] His highly promising career would soon be cut short when he became the first Army man to die in a plane crash during the Wright airplane trials that September.

About the same time that Selfridge joined the Aeronautical Division, infantryman 2d Lt. Benjamin D. Foulois attended the Signal School, where he became intrigued by military aviation. The school required each student to write a thesis on some aspect of Signal Corps activity. Foulois's interest, piqued by the paucity of information on military aeronautics, led him to choose as his topic "The Tactical and Strategical Value of Dirigible Balloons and Aerodynamical Flying Machines." His academic interest and the fact, he surmised, that he weighed 126 pounds and "didn't displace quite as much as some of the others did when it came to flying," resulted in his assignment to the Office of the Chief Signal Officer.[24] In July he assumed command of the balloon detachment at Fort Myer.

Thomas E. Selfridge (*left*) was not only an early airplane pilot, he also collaborated with Alexander Graham Bell (*right*) in aircraft design.

1907-1917

Airplane Trials

While the Aeronautical Division practiced ascents and cross-country trips from the drill ground in the Army's first motor-powered balloon, Dirigible No. 1, unobtrusive Orville Wright arrived at Fort Myer to begin assembly of his "flyer" for government trials. At the beginning of the year, the Board of Ordnance and Fortifications officially authorized the Chief Signal Officer, who had earlier expressed reservations about heavier-than-air flight, to accept bids that would lead to the purchase of an airplane.[25] Considerable negotiation between the government and putative airplane manufacturers ended in an agreement with Orville and Wilbur Wright, the only bidders able to meet the Army's specifications. The Wrights signed the contract on February 10 and Orville, the pilot for the preliminary tests, delivered the airplane to Fort Myer on August 20, 1908.

A festival atmosphere surrounded the Wright airplane performance trials. Even President William Howard Taft milled among the curious onlookers. "That precedent," wrote Chandler and Lahm in their memoir, "was soon followed by many of his cabinet members and of course by everyone of consequence in the social set of the Capitol City. Some of the socialites brought beverages and sandwiches in their autos, which they shared with their friends, thus making quite an unusual social event of the official trials."[26] Midway, the

Orville Wright walks toward the camera in this photograph taken as his flyer is prepared for demonstration at Fort Myer in 1908.

Beginnings

trials ceased. On September 17 Orville Wright went aloft with Selfridge as Army observer. A crack in the airplane's right propeller fouled a rudder guy wire, causing both to break. The plane crashed. Wright was severely hurt, and Selfridge died from his injuries a few hours later.

Selfridge's death was the first in flight training's bloody trail. Professionally, too, this early loss landed a particular blow to the new enterprise. The Army had no other officer with Selfridge's experience in heavier-than-air flight. Chandler and Lahm were balloonists; Foulois had only glancing familiarity. Some of the senior Signal officers indulged in scientific and professional inquiry but lacked any firsthand experience. Only Selfridge had both designed and flown heavier-than-air craft. His death was the first of a great many but, coming so soon, it brought a changed and more sober mien to the trials, which would not resume for another nine months.

The following year, near the end of June 1909, Orville and Wilbur Wright brought the rebuilt biplane back to Fort Myer, where they began assembling and adjusting it in the airplane shed on the drill ground. Renewed public curiosity grew so great that every day people thronged the field to watch the anticipated official trials. All were primed for the opening event on June 28. The catapult used to hurl the airplane forward and into the air had been erected, the motor had been repaired, and the airplane appeared ready to fly. "At last," an *Aeronautics* reporter opined, "all was ready but the wind, and that refused absolutely to abate, even though 'Uncle Joe' Cannon and Chairman Tawny of the committee on appropriations, a large number of Senators, Representatives, officers and just plain misters, misses and mistresses of high degree fretted and fumed in the sweltering sun. The machine was new and untried and it was not deemed advisable to attempt a flight in the wind that was blowing, so all and each of those assembled disassembled themselves and traveled home."[27]

The Wright brothers, not to be pushed or rushed, patiently tested their engineering changes in practice flights and delayed the official trials until they were confident of both the airplane and the weather. Lahm recalled that "flights were made only in light winds, and while large crowds and high officials were often disappointed, the Wrights were adamant in their decision not to fly unless conditions were just right."[28]

Foulois, a member of the Army Oversight Board, sat reading a book about flight as he waited for the trials to start. As the Wrights tinkered with the machine, he peppered them with questions about why and what they were doing. Foulois later recalled that Wilbur finally remarked with some asperity that one could only learn about flight by fixing and flying an airplane, not by talking about it. The lieutenant thus became an assistant to the inventors: "I donned my coveralls, stuck a pair of pliers, a screw driver, cotton waste, and a bar of soap in my pockets... and got to work."[29]

The first official trial took place late on the afternoon of July 27. With

1907-1917

Lahm as passenger observer, Orville Wright kept the plane in the air for one hour, twelve minutes, and forty seconds, more than fulfilling the endurance requirement. The speed test, put off because of uncertain weather until July 30, was the final trial. This time Foulois accompanied as the observer. The plane carrying the two men averaged forty-two miles an hour, again surpassing the Army requirement.[30] After the horrible disaster of the previous year, the trials of 1909 — attended by wellwishers, the Army Oversight Board, President Taft, and the Secretaries of the War and Navy Departments — demonstrated that heavier-than-air flight was a reality with potential, if yet uncertain, military use. The Wrights were now obliged to provide flight training to two officers.

Also serving with Lt. B. D. Foulois (*second from right*) as members of the Army Oversight Board were (*left to right*) Lt. Frank P. Lahm, Lt. G. C. Sweet, USN, Maj. C. McK. Saltzman, Maj. George O. Squier, Capt. Charles DeF. Chandler, and 2d Lt. F. E. Humphreys.

Beginnings

The Wright flyer (*above*) is ready for the first official trial held July 27, 1909, which set the world record for time in flight. The picture to the *right* shows the aircraft just before it took off for the final trial, this one for speed, held on July 30.

Training the Army to Fly

By late summer of 1909, the U.S. Army owned its first and only airplane but had no place to train officers to fly it. As he looked toward the imminent airplane trials, General Allen had worried that "there is not a suitable [training ground] that I know of in this country today.... One thing that has kept back both dirigible balloons and aeroplanes is the fact that there has been no market for them and also that there has been no place where aviators could practice their art."[31] Flight training required considerable open space, uninterrupted by buildings and trees, and large enough to house an airplane hangar.

The Signal Corps eliminated from consideration Fort Omaha, the site for ballooning, because of its harsh winter weather and geographical isolation. A location near the capital city made sense for securing military and commercial supplies.[32] Moreover, the climate appeared to be suited to the needs of aviation. Nobody could fly safely in high winds or in cold and rain. Not only was the airplane, with its low-powered motor, dangerously unstable in turbulent air, but pilots, who sat exposed on the wing with feet extended on a crossbar, typically wore light clothing and no helmets or goggles. Bulky gear obscured visibility and restricted ease of movement, yet it was equally difficult to operate the controls with cold hands and feet, or to see when dust and rain blew in one's face.

Weather, in other words, was key. James H. Doolittle, the famous Army Air Corps test pilot, recalled the first time he witnessed flying at an air meet in California during the winter of 1909–1910: "I can remember some of the pilots putting their fingers in their mouths and holding [a finger] up to see if one side cooled faster than the other, and if it did there was too much breeze to fly."[33] The commercial flying schools and airplane manufacturers that sprang up around the country began relocating where genial climate permitted year-round flying. The Wrights themselves opened a winter training camp near Montgomery, Alabama.[34] Glenn Curtiss, who would become a principal supplier of Army and Navy aircraft, moved his flying operations from Hammondsport, New York, to San Diego, California.

The Army too soon would look south and eventually settle in Southern California, but initially it selected an area near Washington, D.C. The airplane trials had been held at Fort Myer because of its proximity to Washington, headquarters of the Army and its Aeronautical Division, but the commandant at Fort Myer (a cavalry and field artillery post) refused to relinquish the parade ground for further flight training. He was already disgruntled because the trials had disrupted his summer training schedule of mounted drills. Moreover, the Wrights expressed reluctance to teach beginners to fly on the small, enclosed area.

Frank Lahm set out to find another spot. In balloon ascensions, he had had a bird's-eye view of much of the countryside around Washington, and recommended that the Army lease a 160-acre tract near College Park, Maryland, about eight miles northeast of Washington, D.C. Few changes would be required besides cutting down some trees to allow a straight course nearly two-thirds of a mile long to be laid out diagonally across the soft, sandy field.[35] The Signal Corps agreed to lease the site, signed the contract in September, and began clearing the land, constructing a hangar, and laying the monorail starting track. By the end of the month, General Allen had notified Wilbur Wright, who was to be the instructor, that the Signal Corps was ready to begin training at College Park.[36]

The Chief Signal Officer selected Lahm and Foulois as the two officer trainees. It was not a difficult choice, given the fact that hardly anyone else was available. Even so, at the last minute General Allen decided to send Foulois to the International Congress of Aeronautics at Nancy, France, and to replace him with 2d Lt. Frederic E. Humphreys, a new volunteer detailed from the Corps of Engineers. Wilbur Wright and the two lieutenants rented rooms in College Park.

Flight training began the morning of October 8. Trainees usually flew early

Beginnings

A tract near College Park, Maryland (*top*) was chosen for development as the Signal Corps's first training facility. Its proximity to Washington, D.C., and its spacious setting seemed conducive to the task at hand. Chief Signal Officer General Allen originally selected Benjamin Foulois (*above left*) and Frank Lahm (*above right*) to train there. However, Allen decided to send Foulois to an aeronautical congress in France and assigned another man to replace him.

in the morning or late in the afternoon, when the winds slackened. Wilbur Wright strictly observed the Sabbath, so nobody went aloft on Sundays, or on days when it rained or the winds blew up. Following Wright's three flights to check out the airplane and the field, Lahm, who had been in the airplane twice during the trials, took to the air along with Wright as the first student pilot. The flight lasted five minutes and eight seconds. Next, Humphreys, up for the first time, accompanied Wright for a flight lasting four minutes, fifteen seconds. Training ended for the day.

The airplane, Signal Corps (S. C.) No. 1, was a pusher-type, so-called because its two propellers were located in the rear. The Wrights taught by what came to be called the "dual method," whereby an instructor accompanied a student from the beginning. The student sat in the middle of the leading edge of the lower wing and the instructor sat to his left, where his weight compensated for the weight of the engine on the right. The Wrights' invention, a "warping" mechanism, affected right- and left-side lift, which stabilized lateral or rolling motion. Two sticks, one outside each seat, controlled the elevator (vertical motion). A single stick between the two men operated the rudder (lateral motion). The rudder lever was hinged to the warp control and could be used with warping, or it could be used independently. A foot control worked the spark advance, which regulated the speed of the engine.[37]

The student pilot first learned how to take off. A catapult (formed of a weight, ropes, and pulleys) launched the airplane down a monorail track and into the air as the pilot pulled on the elevator, and the motor carried the plane forward and up. Fairly quickly, Wright amended the procedure, adding engine power to send the plane down the track.

Initial flights were short hops, made at a low altitude of twenty to thirty feet. As the sensation of being airborne became more familiar and the new pilots gained confidence in handling the airplane, flights lasted longer, sometimes as much as half an hour in optimal wind and weather conditions. Additionally, with more complicated maneuvers, Wright took the student pilots to higher altitudes, where greater airspace provided a margin of safety. There, in learning how to gauge a turn, the student checked to see that a piece of string hanging from the landing skids' crosspiece remained parallel to the skids.

Landing, made with a dead motor, could be the most difficult task to accomplish. The pilot brought the airplane down at the proper angle, cutting the motor and cruising along the bumpy ground on the skids that served as landing gear. The balky engine occasionally stopped in midair, so Wright also demonstrated powerless landings from a higher altitude, shutting off the motor in midflight and showing his pupils how to glide safely to earth.[38]

Airmen who knew and learned from the Wrights were convinced that it was the inventors' experience with gliders that perfected their sense of balance while in motion.[39] Even more than their technical knowledge, born of construction and repair of the airplane, and their greater amount of flying time, the Wrights passed on an awareness of the "feel" of flying, an instinct for gauging the wind, and a sense of balance and glide crucial in flying the early airplanes that were so susceptible to motor or equipment failure. Personally, the

Beginnings

Wrights were methodical and patient teachers. Wilbur not only demonstrated how to fly his machine, but he also talked at length with his two students. Lahm remembered that "no question, . . . however unimportant it might seem, failed to have careful consideration and a well thought out answer; and you may be sure we asked many questions."[40]

Lieutenants Humphreys and Lahm soloed for the first time on October 26. A newspaper reporter complimented Humphreys on his graceful return to earth: "Neither Wilbur nor Orville ever made a gentler landing. The enlisted men of the Signal Corps broke into a cheer as the youthful lieutenant squirmed out from beneath the wire trusses."[41] Thereafter the two new pilots flew alone or together, with Wright mostly observing from the ground. On November 3 an uninitiated passenger, Lt. George Sweet, who had been the Navy observer at the airplane trials, accompanied Lahm. The two Army lieutenants also practiced flying together so that each could learn to operate the controls from the instructor's position.[42]

Benjamin Foulois returned from France, eager to be included. Although Wright had no contractual obligation to teach a third man, he stayed on a few days, flying with Foulois about an hour and a half.[43] Afterward, Humphreys gave another hour and a half's instruction. Foulois neither soloed nor practiced taking off and landing by himself.

November in Washington turned cold and rainy and a cutting wind curtailed flying time. Since Wilbur Wright had been quoted in the newspapers as saying that a man could be taught to fly in not much more than an hour's actual flying time, Lahm and Humphreys apparently had obtained sufficient experience to be considered fully trained. Foulois remained very much a novice, but it was he who soon would be the only active Army pilot. As the more senior officer, Chandler was primarily an administrator and aeronaut; Selfridge was dead; and three new lieutenants had been relieved from aeronautical duty not long after they arrived. Humphreys, just assigned to aviation on June 11, was recalled to his corps as soon as he finished flight training. Lahm, too, after four years with the Signal Corps, rejoined the branch in which he was commissioned because of a limitation in the time officers could remain on detached duty.[44] Foulois — with little more than three hours' instruction, no experience in taking off and landing, and never having flown an airplane alone — found himself to be the single Army pilot in charge of instructing others (and himself) in the art of flying.

Since Washington appeared to be unsuitable for year-round training, the Army looked elsewhere. Among the original sites under consideration, San Antonio seemed promising because of its mild weather and the presence of a large military reservation at Fort Sam Houston.[45] After repairs to the airplane, which suffered damage on November 5 when Lahm and Humphreys crashed, fortunately without injury to either, Foulois and the enlisted detachment left for Texas with the single Army aircraft, S. C. No. 1. That November, as Foulois wryly put it, "the entire combat arm of the United States... consisted of one badly damaged airplane, eight attached enlisted mechanics, one civilian mechanic, and one untrained pilot — me."[46]

Before he left Fort Myer, having completed his contractual obligation to the Army, Wilbur Wright supervised repairs to the airplane and gave Foulois a final opportunity to glean last-minute advice. Wright assured Foulois that he would be happy to answer any questions put to him by mail, leaving Foulois, as he commented in his well-known remark, to become the first pilot in history taught by correspondence course. When Foulois expressed his anxieties to General Allen, proposing that the government hire the Wrights for continued instruction, Allen replied that the Signal Corps had no more money. As Foulois later reconstructed the conversation, Allen reassured him: "You'll learn those techniques [soloing, landing and taking off] as you go along.... Just take plenty of spare parts — and teach yourself to fly."[47]

A One-man, One-plane Training Air Force

No money. That was the watchword in American aeronautics almost until the outbreak of World War I, although Chief Signal Officer Allen encouraged aviation, as did his successors Brig. Gen. George P. Scriven and Maj. Gen. George O. Squier. For his part, Allen seemed resigned to the fact that Congress might continue to turn a deaf ear to appropriation requests. At one point he opined that legislators were not opposed to aeronautics, but "it is merely a question of money, Congress feeling that just at this time the country is too poor to do anything in the matter."[48] Whatever he truly believed to be the case, Allen continued in private correspondence and in his annual reports to lobby diplomatically for an appropriation so that the United States, initially the leader in development of heavier-than-air flight, might now become competitive with European nations, particularly Germany and France.[49]

In sending his one airman to San Antonio, Allen explained Foulois's mandate more grandly in public than he had in private to Foulois. Not only would Foulois teach himself to fly, Allen proclaimed, but Fort Sam Houston would "be used for the training of officers and men during the winter."[50] The San Antonio newspapers trumpeted the arrival of the young Army lieutenant and the first Army airplane, repeating the notion that Foulois had come to teach other Signal Corps officers to fly.[51] Foulois himself realized that he had been given neither the money nor the commitment to undertake a large training program, and under any circumstances, he was hardly the man to assume such a task. He was forced to do exactly as Allen had directed — take plenty of spare parts and teach (only) himself to fly.

Very quickly it became apparent that, although San Antonio had been selected for its mild climate and calm air, the prevalent wind velocity in the area frequently exceeded all expectations. Foulois found Weather Bureau data on wind patterns to be virtually useless. It was misleading to measure the force of winds in towns blocked by buildings and other obstructions, when the Army airplane shed was located on an open plateau buffeted by the "unbroken force of the wind from all directions."[52] The gusty winds caused "the old girl" — No. 1 — "to buck like an unbroken cow pony."[53] Having received no instruction in taking off or landing, Foulois nonetheless adequately managed to take off and make straight flights in the choppy air, but his landings were consistently terrifying. Effectively, most of his descents to the earth were crash landings. He continually broke pieces of the plane, so his store of spare parts served him well.

Between the gully-washing rainstorms, the turbulence, and the mesquite-covered, uneven terrain, on-the-job training proved to be very trying, as Foulois reported relatively good-humoredly to the Aeronautical Division. He wrote the Wrights frequently, asking how to perfect basic maneuvers and, after accidents, what might have caused them. Responses usually came in the return mail. Under the circumstances, Foulois made remarkable progress in learning to fly. He gradually gained sufficient confidence to request permission for longer and longer flights. His senior officers were reluctant to let him range too far, since they were not unmindful of the difficult task he had been set and worried about the safety of the Army's only pilot and the constant damage to the Army's only airplane. Foulois was gratified by the solicitousness of Capt. Arthur S. Cowan, who relieved Chandler as chief of the Aeronautical Division on July 1, in allowing him to make cross-country flying part of his self-imposed curriculum:

> Every flight that I have made down here has been more or less hazardous. If I waited for perfect weather conditions, I would average about one short flight per month. However, every flight has taught me something new. I have been worrying a lot for the past few months with the thought that the Office might think that I have not been doing much with the machine, but the wind blows almost continuously the whole year round in this country, and it would be absolutely foolish to attempt flights in very high winds with this low power machine. I sincerely hope that the Office will let me go ahead with the plans, for remodeling the machine and equipping it with a powerful engine.[54]

As Foulois indicated, flying with the low-powered engine caused both frustration and danger. A twenty-five- to thirty-horsepower motor could barely lift an airplane into the air and, with no reserve power, the pilot made all his maneuvers with the throttle wide open. Thus, as pioneer airman and future Commanding General of the Army Air Forces Henry H. Arnold described it, "the performances secured were the results of skillful piloting.... There was such a small margin between a successful flight and disaster."[55] The high-powered engine that Foulois hoped to mount on his plane would indeed permit increased maneuverability, but greater speed also meant greater danger.

1907-1917

Foulois contrived another useful modification. That summer he drew up blueprints for a wheel assembly to replace the skids used on the bottom of the airplane.[56] About the same time, the Wrights, too, began using wheels, as they notified Foulois. At an air meet in May they attached wheels to their airplane, and shortly thereafter the Signal Corps purchased a set for the Army airplane.

Foulois's air force was a humble affair, particularly when compared to the grand role for aviation that the Army had now drafted. The 1910 Field Service Regulations authorized fully equipped aeronautical companies for service with mobile forces. Aero companies would be part of battalions, thus fulfilling the airplane's projected mission in aerial communications. Everybody recognized, however, how far the regulations strayed from reality, since they made no provision for money or personnel.[57] As the Chief Signal Officer stated, "At present not even a model of such a company could possibly be organized, nor will it be possible to do so until the Signal Corps is increased by suitable legislation."[58]

New Airplanes, New Men

The following year, 1911, was a banner one. Congress acted, specifically appropriating money for Army aeronautics for the first time. On March 3, 1911, the Signal Corps received $25,000 of the $125,000 earmarked for fiscal year 1912. With its newly appropriated funds, the Signal Corps immediately ordered five airplanes: three Wrights (one manufactured by W. Starling Burgess under the Wright patents), and two built by the Curtiss Company. One Wright and one Curtiss were intended for Fort Sam Houston.[59] Until they arrived, Robert F. Collier, publisher of *Collier's Magazine* and a zealous aviation enthusiast, rented his own Wright B to the Army for $1.00 per month.[60] Since the Wright Company made changes in the control mechanism of the B, one of its pilots, Phillip O. Parmelee, accompanied the new plane to Texas in order to instruct Benjamin Foulois in its use. Much-used, battered No. 1 went on "tactical reserve status," as Foulois called it,[61] from which it was retired eventually to the Smithsonian Institution.

Foulois was thrilled with the new airplane, believing it to be the "best military machine in the world today," and he was grateful for some hands-on instruction.[62] He would later claim that civilians Parmelee and Frank Coffyn, the Wright instructor who replaced Parmelee, taught him to fly strictly "for military tactical purposes and not for information."[63] Since nobody in the Army at that time had a clear sense of how one would employ an airplane for "military tactical purposes," presumably Foulois meant that he participated in aviation's first military operation, scouting duty on the eastern end of the Texas-Mexican border (an action of short duration and of little significance militarily).

It also appears that Foulois needed more help in the fundamentals of flying

Beginnings

The Wright airplane used as a trainer for use at Fort Sam Houston was the Model B with wheels attached, as Foulois suggested. The Wright pilot Phil Parmelee accompanied the craft to Texas and instructed Benjamin Foulois in its use.

than he admitted. Upon his return to Dayton, Parmelee reported that Foulois handled the new control mechanism awkwardly. Although the Wright Company suggested that Foulois come to Dayton for further training, the Chief Signal Officer refused to spare him from duty in San Antonio.[64] The Wrights did not abandon Foulois, however; they sent another company-trained pilot, Frank Coffyn, to Texas. Coffyn came to the same conclusion as Parmelee regarding Foulois's piloting abilities, taking particular note of the lieutenant's crash landings. Except during rainstorms and high winds and on Sundays, Coffyn and Foulois flew together every day for nearly a month. Foulois felt that at last he was capable of training others and properly maintaining the airplane.[65] Coffyn agreed, with reservations, urging Foulois to come to Dayton for a special course that included "many details about the mechanical construction, and theory of the aeroplane which can only be obtained at the Wright factories."[66]

If Foulois was unable to take advantage of the opportunity, the Signal Corps capitalized upon it for others, sending the next group of men for more extended training with the manufacturers from whom the Army purchased airplanes. Now, with the purse strings loosened, for the first time the Aeronautical Division could anticipate having a small inventory of airplanes. Even so, it had almost no competent pilots, no clear criteria by which to judge candidates, and no means at hand to train them within the Army. But with offers for

1907-1917

Wright pilot Frank Coffyn went to Texas to instruct Foulois in the Model B and suggested that Foulois come to Dayton to train there too.

training in hand from the airplane manufacturers, the Signal Corps could now address personnel procurement.

The Chief Signal Officer scoured the ranks for volunteers, ideally captains of not more than five years of service in that grade, or unmarried lieutenants, of medium weight. Clearly, not all those who met the requirements and signed on to fly were fit for the job. General Allen wanted a long list of candidates, "as none of the officers to be detailed have had any experience in this work [and] it will probably be necessary to try out quite a number before suitable ones are found."[67] He selected a few and, as a result of civilian manufacturers' willingness to train military aviators and the Army's inability to provide instruction, ordered them to Dayton, Ohio, and San Diego, California, for training.

Lieutenants Thomas DeWitt Milling and Henry H. Arnold reported to the Wright Company in Dayton. Neither man was a Signal officer and neither had any background in aviation. The Chief Signal Officer tapped both for reasons obscure to themselves. A dashing Cavalry officer and one of the Army's best polo players, Milling fit the developing profile of the military pilot — the independent young man who embraced excitement and an element of risk. Milling was stationed at Fort Leavenworth when he received a telegram from the Aeronautical Division: "Will detail with Signal Corps for aeronautical duty be agreeable to you?"[68] Similarly, high-spirited Arnold recalled that "out of the blue an official letter arrived from the War Department. Would I be willing to volunteer for training with the Wright Brothers at Dayton as an airplane pilot?"[69] Both lieutenants answered the summons in the affirmative.

Early in May 1911 the two officers arrived in Dayton. A newly built brick building on the outskirts of the city housed the Wright airplane factory, where what later would be termed ground training took place. A cow pasture called Simms Station, a few miles outside town, served as the flying field. The class consisted of three military officers, including Navy Lt. John Rodgers, and four civilians. Each was assigned an instructor from the Wright team. Cliff Turpin taught Milling and Al Welsh coached Arnold.

Beginnings

The Army sent Henry H. Arnold (*left*) and Thomas DeWitt Milling (*right*) to the Wright School in Dayton, Ohio, in 1911. Future Navy flier John Rodgers (*center*) also trained there with them.

The experience of the officers in Dayton much more resembled formal training than the instruction Wilbur Wright had given Humphreys and Lahm only a couple of years earlier. Not only had the Wrights' knowledge and experience broadened in that short time, but their business had expanded, bringing with it the benefits of greater depth and scale. More systematically than had been possible previously, at the factory the Army officers learned about the construction, maintenance, and assembly of the motor and the airframe. Next came ground training — aeronautical theory and techniques of flight — followed by flight instruction.[70] From that sequential approach arose the system of flight training that the Air Service and its successors — the Air Corps and the Army Air Forces — would employ thereafter.

Trainees became familiar with the airplane by sitting in a primitive simulator consisting of an old airplane, without landing gear or tail assembly, balanced on sawhorses. The motor was unattached but the wings were movable, which allowed the student to operate the warping lever. The Wright training technique called for dual instruction in the air, but on the balancing machine the student practiced alone. The sawhorse trainer was reminiscent of Foulois's unruly flights in San Antonio when his "unbroken cow pony" bucked the Texas winds.[71] As Arnold later described that first simulator:

> The lateral controls were connected with small clutches at the wingtips, and grabbed a moving belt running over a pulley. A forward motion, and

31

1907–1917

Seated in a Wright B two-seat trainer at the Wright Flying School in Dayton is novice pilot "Hap" Arnold.

the clutch would snatch the belt, and down would go the left wing. A backward pull, and the reverse would happen. The jolts and teetering were so violent that the student was kept busy just moving the lever back and forth to keep on an even keel. That was primary training, and it lasted for several days.[72]

The two lieutenants turned in weekly progress reports. They spent much time, they explained to the Chief Signal Officer, on the balancing machine trying to get the feel for the unnatural motions of the warping lever. When not at the factory or in the air, they often went out to the field in the afternoons to watch Wright pilots land and take off. The students made several flights with an instructor before they began to handle the controls. On Arnold's third lesson he was allowed to put his hand on the elevator. He tried the warping lever on his ninth lesson. Welsh began teaching him to land on his twelfth lesson, and on his nineteenth session, Arnold landed without assistance.[73] Ten days after he first touched the controls, Arnold reported: "During the week I have made twelve flight[s] by myself. My instruction under the personal supervision of the instructor in the machine is finished and from now on all my flights will be made alone for experience."[74]

At about the same time, Milling, usually more discursive than Arnold, wrote that he continued working in the mornings at the factory. "My afternoons have been spent on the aviation field and I have made eight flights during the week, and I am now able to fly the machine alone, having made the last flight by myself. I made the flight alone after a total of one hour and fifty five minutes spent in the air."[75] As the student pilots became more experienced, they were permitted to fly under more threatening conditions, attaining altitudes up to 1,100 feet in fifteen-mile-per-hour winds. They would have preferred more practice, but several men shared each training airplane, and the much-used machinery was frequently out of commission for repairs.

Orville and Wilbur Wright did not relinquish their students entirely to company instructors. They invited the young airmen for Sunday dinner and conversation, and visited the field and hangar for discussion and observation. Arnold remembered "the best of all in that Simms Field 'hangar,' of course, were when the Wright brothers themselves joined us." As Humphreys and Lahm had discovered, the Wrights answered many questions in their "courteous, almost diffident" manner. "Their presence in the hangar always made the sessions different. Despite their mild, retiring way of listening until everyone else had made his speech about this incident or that phenomenon, or what the exact future of an air development would be, you always felt them there."[76] Orville Wright flew by himself nearly every day, mostly to test a new skid or elevator or some aspect of the control. He also flew with the students. Wright observed Milling's first solo flight and his spiraling descent from an altitude of 600 feet. "He was sufficiently impressed," Milling believed, that "he gave me personal instruction many times thereafter ... I have always felt that it was the instruction I received from Orville Wright that carried me through my flying career without being killed."[77]

In mid-May, Lieutenants Milling and Arnold completed basic flight training. Wright Company manager F. H. Russell cautioned that they still required considerable practice to become thoroughly at ease in the airplane. Since the Army officers were not the only students receiving instruction at the factory, totally personalized attention was not possible, and each student had to build on the fundamentals he had learned from the company. "It might be advisable," Russell suggested, "to get them on to their own machines as soon as it is practicable." He penned a note at the bottom to the effect that Milling "is flying remarkably."[78]

From the beginning of his flying days, Tommy Milling's associates considered him to be one of the best Army pilots. Grover C. Loening, who joined the Wright Company as an aeronautical engineer in 1914, met Milling at the Nassau Boulevard air meet soon after Milling began flying and immediately recognized the young officer as "one of the first real natural-born flyers. He had cool daring, a fine hand, and a very keen set of senses. His boyish smile made his way easy, and he soon had become the leading and most skillful pilot at that time in the government service."[79] He also became the only pilot to fly equally comfortably in airplanes with entirely different control mechanisms.

The long-time balloon pilot, Chandler, also went to the Wright Company at summer's end 1911 to learn to fly heavier-than-air machines. He too exhibited a natural gift, according to Orville Wright, although as a more senior officer he would spend more time in administrative duties than in the pilot's seat. By the time he arrived in Dayton, Arnold had given him some instruction at College Park. Chandler's training at Dayton was a courtesy on the part of the Wrights, who normally provided training along with each purchase of a Wright airplane.[80] Chandler spent two weeks in Dayton, visiting the shops and learning airplane construction in the mornings and flying at Simms Station in the afternoon. Orville Wright, normally a man of gentlemanly restraint, applauded Chandler's dexterity. In describing Chandler's accomplishments at the completion of his training, he also drew a general picture of the qualities of temperament and judgment that he considered essential in a good pilot:

> His handling of the machine showed excellent early instruction in the fundamental principles of flying and skill far ahead of what we expect in one of his experience. But the feature of his flying that impressed me most was his quick perception of the beginning of disturbances of equilibrium which enabled him to make correction before the disturbances had time to assume a more serious nature. The ability to detect disturbances in their infancy is of the utmost importance to the skillful operation of a flying machine, but it is a talent not possessed by many beginners... Captain Chandler possesses a combination of good judgement and nerve that especially fits him for this work. In fact, I have never flown with any one of equal experience who has shown better promise for becoming an expert.[81]

Both Wrights put enormous stock in the ability to gauge the effects of wind on airplane performance. Grover Loening recalled Wilbur Wright's unusual ground training: "One of the most interesting things about Wilbur... was the hours of practice he would put in at the controls of the plane, sitting in the seat, hangar doors all closed, no one around, quietly sitting there imagining air disturbances and maneuvers and correcting the rudder and warping wings and elevator to suit."[82] Loening too stressed the importance of understanding the aerodynamic response of an airplane under various wind conditions. In training, the pilot had to get the feel of the air in order to master the technical skills of landing and taking off. Yet Loening felt that "unless this is accompanied by an intelligent understanding of the actions of aeroplanes in the air, the pilot is little more than a somewhat instinctive automaton."[83]

The Army's other training venture took place in San Diego, California, with the Curtiss Company. Of the $25,000 available for aviation as of March 3, 1911, the Signal Corps earmarked approximately $6,000 for the purchase of a Type IV Military airplane from Glenn Curtiss. This aircraft was to have a long and successful career as the primary Army training airplane, almost until the American entry into World War I. The Curtiss airplane used the principles of construction developed by the Wrights,[84] but it was a single-seater, which meant that a trainee had to fly unaccompanied by an instructor. More important, the control mechanism differed from that of the Wright machine.

For some time Curtiss had been sending a flurry of letters to the War

Beginnings

Glenn Curtiss is seated at the controls in one of his aircraft (*left*). His eight-cylinder training machine is pictured above.

Department, offering his services. He invited the military to view his airplane experiments and tests, which he claimed would result in partial credit for the results going to the Signal Corps.[85] By the winter of 1910, Curtiss planned a training school to be located on North Island in San Diego. There he offered to instruct officers at no expense to the government. A number of officers applied for the program and, although General Allen was receptive to the Curtiss offer, he was then unprepared to accept it. But by the time Curtiss opened his aviation school on January 21, 1911, Lt. Theodore Ellyson, USN, was one of the students, and the Army had begun sending its people.

The Chief Signal Officer had been courteously noncommittal to Curtiss's correspondence until he had evidence that Curtiss's airplanes would be useful to the military and that money was available for their purchase. Curtiss's offer then became more attractive. Once the Signal Corps ordered Curtiss airplanes, it authorized Brig. Gen. Tasker Bliss, commanding the Department of California, to detail men to San Diego for company training. Curtiss benefited from training the military for free because the officers suggested design changes that resulted in increased airplane sales to the Army. Curtiss began producing passenger-carrying airplanes, for example, at the insistence of the Army officers enrolled in the Curtiss classes, who explained that normally two men (a pilot and observer) would be engaged in any Army mission.[86] This early relationship between Curtiss and the Signal Corps pointed the way toward what would become a necessary and mutually beneficial research-and-development partnership between the Air Force and the aeronautical industry.

The Aeronautical Division chose three volunteers — 1st Lt. Paul W. Beck

1907-1917

and 2d Lts. G. E. M. Kelly and John C. Walker, Jr., all Infantry officers — to join Lieutenant Ellyson in Curtiss's first class. Arnold, who would train Beck thereafter in San Antonio, thought him a "fireball of enthusiasm. He had been infected with the air bug at the first aviation meet he had witnessed at Los Angeles, thereafter plaguing his superiors for detail. . . . As a flyer he was said to be not only eager, but to possess the kind of 'pilot's luck' that impressed even such phlegmatic pioneers as our instructors."[87] Walker had been stationed at Fort Sam Houston. There he had worked with Foulois on modification and maintenance of the Wright plane.[88] His familiarity with the construction and operation of the Wright and Curtiss machines probably stood him in good stead, since the two airplanes shared few common control features.

Besides learning to fly, the students at San Diego observed and aided Curtiss's experiments, particularly overwater flying. They began constructing a seaplane, a modified standard biplane with an eight-cylinder, fifty-horsepower engine, and with wheels and landing gear replaced by floats. Among the Army students, Walker was especially enthusiastic about this venture.[89] All those en-

In Curtiss's first class (at North Island) were Lieutenants Paul W. Beck (*right*), G. E. M. Kelly (*lower right*), and John C. Walker, Jr. (*below*).

36

rolled in the Curtiss School of Aviation spent hours daily in the shops (as they did in training at the Wright factory) learning the assembly, repair, and maintenance of airplanes and engines. They inspected every part of the airframe and engine. Each student made alterations and repairs to his own training airplane and was responsible for deciding when it was ready to fly. Lieutenant Kelly showed a special mechanical aptitude.[90]

The method of flight instruction consisted of several stages that progressed from running the airplane along the ground to turning and banking at relatively high altitudes. Initially the student skimmed an airplane at reduced power in a straight line along the ground. He gradually increased speed and practiced balancing the ailerons (the Curtiss airplane substituted ailerons for the warping mechanism). Short hops in the air of twenty-five to one hundred yards followed. Then came straightaway flights of about three-fourths of a mile in a full-powered machine. Finally, the trainee flew half-circles and circles, glides, and made powerless landings from higher altitudes.[91]

The primary difference between the Curtiss and Wright instructional methods came from the fact that Curtiss, in an effort to reduce head resistance, built only a single seat into his airplane. As a result, a student pilot learned from lectures and demonstrations, and although an instructor observed him from the ground, he was always alone in the air. The training airplane was sufficiently low powered that the trainee was not likely to harm himself in high-flying maneuvers he was not competent to execute. In fact, it took some while before the student was allowed to take to the air at all. Only when he was judged able to manipulate the controls and to understand fully the aerodynamic principles was he given an airplane sufficiently powered to get him off the ground. The initial training machine was equipped with a four-cylinder, twenty-five-horsepower engine. New controls and an eight-cylinder engine replaced it in order to attain the higher altitudes necessary for making turns.[92]

The Curtiss method of instruction was, on balance, less satisfactory than the Wright method. Curtiss students never had the opportunity to learn what the Wrights considered so important — the responses of the airplane to various wind conditions — at the side of an experienced teacher. A consensus, certainly among the Wright-trained pilots, held that the solo method of instruction failed adequately to prepare new pilots. Beck's eventual catastrophes seemed to bear this out. The Curtiss Company apparently drew the same conclusion, so that by 1912 they were building two-seat, dual-control training airplanes.

If the self-training technique was inferior to dual instruction, the Curtiss control mechanism was much easier to learn and operate than the Wright. To change altitude in a Curtiss machine, the pilot pulled a wheel forward or backward; he turned the wheel to operate the rudder. A frame across his shoulders controlled lateral movement: he leaned right to drop the right wing and leaned left to drop the left. These motions were considered, even by

1907-1917

Wright-trained men, much more natural and thus less difficult to master than the counterintuitive Wright controls.[93]

The three Army lieutenants were still in training when they were ordered to report to San Antonio immediately. Curtiss objected that the young officers were not yet competent to fly independently, and Beck, as the most senior of the three, needed a great deal more help if he was expected to instruct others. The three officers left for Fort Sam Houston nonetheless, probably because the airplanes on order had begun to arrive and only Foulois was on duty.

Frank Coffyn accompanied the first Wright B to San Antonio in April 1911. Wright-trained pilots at Fort Sam Houston and at College Park were then able to fly an airplane whose essential features they already knew. Instructor Eugene Ely arrived to assist the partially trained Curtiss pilots. No. 2, the new Curtiss plane, was a Type IV Military, or Curtiss Model D. Although still a one-seater, some space behind the pilot allowed room for a passenger to perch precariously. But its engine was much more powerful than those on which Beck, Walker, and Kelly had trained.

Before leaving San Diego, Beck learned that the Signal Corps planned to buy a high-powered Curtiss airplane, and he argued forcefully for the purchase of a smaller, four-cylinder machine instead. Curtiss instruction always began in the four-cylinder model, and the trainees felt strongly that it was too dangerous to learn on a plane that required relatively high speed in order to take off. The four-cylinder training airplane lifted at about twenty-eight miles per hour, whereas an eight-cylinder machine required a speed of approximately thirty-four miles per hour. Also, landings in the heavier machine, Beck explained to Allen, "are more difficult owing to the increased weight and

The Curtiss Model D, S.C. No. 2

Beginnings

speed."[94] But the new Model D came equipped with an eight-cylinder, sixty-horsepower engine.[95] Its arrival in San Antonio posed a dilemma that Army men would confront thereafter as they hailed the availability of the newest in aviation technology but then struggled with the requirements for training and the frequent hazards that accompanied its use.

First Tactical Organization

The Army Field Service Regulations of 1910 allowed for the formation of an aeronautical unit, but one man and one airplane hardly constituted a company. Now, with three new Curtiss pilots, two additional Wright-trained officers, additional trainees coming into the program, and several airplanes, some tactical organization seemed possible. Major Squier, the Division Signal Officer, ordered the formation of what became the Provisional Aero Company. It would be commanded by Lieutenant Beck who, although a junior member in experience, was senior officer in military rank. Squier also directed Lieutenant Foulois to draft a set of regulations for the air-ground training of officers and mechanics. In compliance, Foulois prepared "Provisional Airplane Regulations for The Signal Corps, United States Army, 1911."

Until then Foulois had given informal orders and worked alongside his ten men. "There had been," he remarked in an understatement, "no need to issue written instructions on the care and maintenance of air machines, since we had only one."[96] Now, with an air fleet in the making, "Provisional Airplane Regulations" specified responsibilities of pilots and ground crew, types of repair and maintenance needed for airplanes, flying rules, organization of the aero company, and qualifications for aviators; it also included an analysis of meteorology. All this seemed relatively straightforward, if somewhat grandiose and even superfluous, given the magnitude of day-to-day operations.

But the exercise of applying traditional Army methods to flight training revealed a hilarious aspect that Foulois himself recognized even at the time. He couched the rules in terms of the drill-ground procedures of Army field regulations, in a by-the-numbers approach. For example, he stated where each of the men (numbers 1 through 9) who would assist in take-off should stand. This was followed by marching instructions, including commands by the pilot such as "move out" and "prepare for flight, march." At this latter command, enlisted men numbers 1 and 2 were to move at double time around their respective ends of the machine and halt behind the propellers. Man number 3 was to follow number 2 and halt three paces to the rear of the center of the elevator. The mechanic (a civilian) was to "move promptly by the right flank of the aeroplane, halting in front of the seat braces, facing the machine."[97] So it continued. That kind of direction to aviation ground crews was preposterous on its face. Foulois admitted that the men had attempted the procedure a few times, but rather quickly it went by the board. Even so, Foulois incorporated in

39

1907-1917

Maneuver camp, 1911, Fort Sam Houston

his regulation safety rules that he considered to be valid decades later.

The Curtiss pilots of the Provisional Aero Company were not fully proficient when they reported for duty, and their predicament worsened. Fort Sam Houston, then occupied for maneuvers, offered very little room compared to the extensive flying fields at North Island, and the high-powered Curtiss Model D made taking off and landing hazardous for the novice airmen. Walker sustained a frightening scare, almost crashing after his airplane stalled and sideslipped in midair. He landed unhurt but was so unnerved that he asked to be relieved from aviation duty. In short order he was transferred, having spent less than six months and a good bit of government money in aeronautics. Kelly had arrived in San Diego five days later than his fellow officers, so he was even less prepared than they. On May 10 he went up alone in No. 2 for his qualification flight. It was probably not an opportune time for flying as the winds had been capricious for some days. Kelly took off and remained in the air for about five minutes, but as he came in for a landing, one side of the airplane hit the ground and bounced back up, whereupon he climbed again to return for another attempt. The second time he made a dangerously sharp turn and the airplane then dived into the ground. Kelly was thrown from the plane, his skull fractured. He died a few hours later.

The airmen bitterly debated the reasons for Kelly's death, unresolved despite the findings of the official report. Foulois blamed Beck for failing to make proper repairs to No. 2 after his own accident. The Board, however, determined that Kelly had deliberately swerved, possibly to avoid hitting an encampment. This judgment confirmed newly developing attitudes among pilots. Arnold said that, during his own training, accidents were always attributed to pilot error: "It was seldom the plane, or an unknown quantity in the air, but almost always the pilot, who was blamed for being in error. You *had* to believe that to keep up your morale."[98]

Beginnings

No matter the uncertainties about its cause, Kelly's death precipitated the cancellation of air training at Fort Sam Houston. Most were not sorry to leave the field. Wright instructor Frank Coffyn considered San Antonio unacceptable for training new men because of atmospheric conditions, particularly under the circumstances pertaining at Fort Sam Houston: small, crowded spaces that prohibited an adequately long take off or landing or long, straightaway flying.[99] The Commanding General of the Maneuver Division ardently agreed, questioning in fact the point of teaching soldiers to fly.[100] Although the Army did not give up flying, it disbanded the small Provisional Aero Company, and shipped all men and equipment to College Park, Maryland.

TWO

The Signal Corps Aviation School

"...fairly good alighting grounds always beneath"
— Anthony Jannus, "College Park Flying Grounds"[1]

The first appropriation for aviation not only brought orders for new airplanes, it also permitted the Signal Corps to train more officers to fly. That meant, in turn, that the Chief Signal Officer could now make plans for a permanent training station. Greater organizational size and dedicated physical facilities, as any bureaucrat understood, did more to ensure permanence than even the most carefully articulated mission statement. The death of a second Army airman, possibly attributable to inappropriate terrain and unsuitable facilities, reinforced the conviction that the new, dangerous venture of flight training could not be conducted on most Army posts.

The Wrights had taught Lahm and Humphreys at College Park, Maryland, scarcely a year and a half before, and civilian fliers continued to use the field. The Signal Corps decided to return to the area and leased a tract owned by the National Aviation Company. The plat was large enough for circular practice flights more than six miles in diameter, with "fairly good alighting grounds always beneath."[2]

Bids for the construction of hangars went out in May. By early July the facility was operating, and the Wright- and remaining Curtiss-trained pilots reported to College Park from Dayton and San Antonio. The Chief Signal Officer issued a memorandum on July 3 announcing the inauguration of the Signal Corps Aviation School.[3]

The Army designated no other tactical squadron after it dissolved the Provisional Aero Company. It established the Signal Corps Aviation School to provide flight training and encompass all the remaining operational elements, including fifteen enlisted men who were successors of the original balloon detachment of 1902. The Aviation School, according to the Chief Signal

Officer, would "form the headquarters or home station of an aeronautical company, and all men in the United States on aeronautical duty will be assigned to this company and records [will be] kept at company headquarters of the ability and the qualifications of officers and enlisted men on this duty."[4]

College Park, Maryland

Lieutenant Roy C. Kirtland had been selected as a Wright pilot at the same time as Milling and Arnold, but his training was delayed while he supervised procurement and construction at College Park. Kirtland served as secretary of the Aviation School for the nearly two years of its existence in Maryland and also became an instructor after his own flight training. Beck was the only remaining pilot from the disbanded Provisional Aero Company to be recalled to the school, now that Kelly was dead, Walker had fled aviation, and Foulois was assigned to duty in the Division of Militia Affairs in the War Department. In June the school hired Henry S. Molineau as a civilian machinist to oversee all repairs and maintenance of aircraft engines.[5] That same month, Capt. Charles Chandler transferred from Fort Leavenworth, where he had attended the service schools, to take charge of the new installation and resume duties as chief of the Aeronautical Division in Washington. Besides teaching pilots and the airplane mechanics, Arnold became quartermaster of the school during its tenure at College Park and at Augusta, Georgia, where it relocated for winter training.[6]

With a small staff, a training facility, and a complement of airplanes, training activities coalesced by July 1911. The partially repaired Curtiss airplane, all but demolished in Kelly's deadly crash, was shipped back to College Park for further reconstruction. A Curtiss Model E, similar to No. 2 but with a four-cylinder, forty-horsepower engine, arrived in late July. Aggrieved over the price they had paid for the high-powered motor of No. 2, the school put the bigger engine on the new Curtiss E, and the rebuilt No. 2, revamped with the smaller engine, served for beginner instruction.

Late that summer Lt. Frank M. Kennedy began flight instruction on the now safer No 2, an airplane he first encountered by assisting Beck in rebuilding it at Fort Sam Houston after Kelly was killed. In the prescribed Curtiss method, Kennedy began flying short hops in a single-seater. He was one of the last pilots to learn this way, because the school soon implemented the double-seat, dual-control technique for all pilot training. The single-pilot approach would be eliminated in favor of dual training, and that training method became linked to, and therefore justified by, evolving doctrine. The Chief Signal Officer stated that "for military purposes it has been conclusively shown that the two-place machine is necessary for reconnaissance purposes."[7] Officially, aerial reconnaissance was the only mission recognized by the Army.

The first Wright B delivered to College Park was ready to fly by July 1.

Signal Corps Aviation School

Lt. Roy C. Kirtland in a Wright B training airplane, 1911, the type shown parked above, outside the hangars at College Park, Maryland

The similar Burgess-Wright B arrived shortly afterward. Arnold and Milling began instructional duties; Arnold gave Chandler follow-up training, and Milling taught Kirtland. Although the Wright pilots enjoyed greater continuity in transitioning to their new airplane than Curtiss pilots did, they too encountered training difficulties. Two seats on the Wright airplane permitted two men to fly, one as the instructor. But the side-by-side seats lacked a full set of duplicate controls. Both Arnold and Milling learned to fly while sitting in the left seat, using their right hand to control the warping and rudder lever and their left hand to manipulate upward and downward motion. Their students, then, sat in the right seat, the elevator lever in their right hand and the warping-rudder control in their left. This method produced alternating generations of pilots who operated the controls from the right or the left seat. Moreover, the right-seat pilot needed compensating weight on the left, so he had to carry a passenger or a weight whenever he flew. The drawbacks of such a system were evident, but a design solution was not yet possible. Arnold, Milling, and Chandler decided that officers learning to fly Wright airplanes be assigned right- or left-handedness to divide right- and left-seat pilots equally.[8] In 1912 a set of duplicate controls for each seat became standard on Wright airplanes, and in time the Service implemented a single control system for all Army airplanes.

Until controls were standardized, flight training was unnecessarily complicated by the substantial differences between the two airplane types, such that most pilots who qualified on one never flew the other. This resulted in two camps at the Aviation School, each convinced of the superiority of "their" airplane and training methods. No doubt the intense dislike of the Wrights and Curtiss for one another, fueled by acrimonious lawsuits between the two parties, exacerbated the partisanship. Since the Wrights and Curtiss personally trained the first group of Army airmen, some of their rancor rubbed off on the officers at the Aviation School.

Offsetting that factionalism, pilots composed a small, somewhat marginalized, group outside the mainstream of Army life. Although not inclined to think and act as one, they drew together as representatives of a new endeavor yet to receive the wholehearted blessing of the military hierarchy. They believed themselves to be in the vanguard of what later came to be seen as a revolution in the art of warfare. Moreover, to a flight-crazed public, they were a band of celebrities. Practically speaking, less than a decade after the invention of the airplane, there were too few military pilots to sustain hardened cliques — not until Kennedy began training in August 1911 did the Army have even two Curtiss pilots. The initially good-natured rivalry persisted for a time, but would shortly become venomous, adversely affecting the administration of the training program and, in some quarters, tarnishing the reputation of Army pilots.

Through the summer and winter of 1911, the officers of the Signal Corps Aviation School worked at College Park, but because no housing had been built they made daily trips, usually by streetcar, to and from Washington. Their workday began around 7:00 A.M. They flew until at least the middle of the morning, quit when the wind picked up, then began flying again in the late afternoon and often stayed until dark during the long summer evenings. Sometimes one or more of the men would return to desk duty in the Office of the Chief Signal Officer to prepare various reports and schedules. Chandler in particular juggled duties because he was in charge of both the Aviation School and the Aeronautical Division.

Airmen emphasized the hands-on, technical aspects of training, believing that little could be learned from the traditional study of the principles of warfare. No textbooks or case studies could provide grist for discussion about aerial rules of engagement. Yet, even from a practical perspective, the small school staff had received minimal instruction and understood relatively little about the mechanics and construction of engines. They were concerned about their ability to conduct in-depth training and impart thorough technical expertise to what they hoped would be large groups of new pilots. Chandler, Arnold, and Milling therefore urged the Chief Signal Officer to hire an instructor from one of the manufacturers for each beginning pilot. Clearly, such a plan was not feasible, given budgetary realities.

To become a confident, safe pilot required countless hours in the air under varying conditions. A shortage of airplanes meant less flying time and diminished competence and safety. Thus the school staff also requested that each of the next fifteen officers detailed for pilot training be given his own airplane, whenever possible of the type of his choice. Thereafter, each of those officers would instruct one new pilot trainee.[9]

Whether the Chief Signal Officer agreed with his airmen's recommendations mattered little because the Army was unwilling to channel much money toward an expensive and as yet unproven enterprise. The practice of purchasing aircraft one or two at a time inhibited standardization of aircraft design and training methods, and kept accident rates high and pilot competence low, but providing one airplane for every pilot was a luxury well beyond the Chief Signal Officer's means. Nevertheless, he looked toward the time when the ratio would be one airplane for every two officers and six enlisted men. As he stated publicly, flight training "involves a certain amount of danger." Therefore, "the details of officers and men of a particular aeroplane should, as much as possible, be permanent in order that the officers and enlisted men may thoroughly know the particular characteristics of that particular machine."[10]

Considerations of safety also suggested that the Army train more of its senior enlisted men as airplane mechanics. At the time, the Signal Corps employed only one experienced civilian technician, so the pilots usually trained the enlisted troops. Milling and Arnold felt that they had benefited greatly from their time at the factory at Dayton, because the training of mechanics fell to them. They photographed the Wright and Curtiss airplanes, labeling every part from nuts and bolts to wing coverings. That useful exercise identified the components and established the nomenclature for airplanes thereafter.[11] Molineau remained the chief technician throughout the College Park term and, because of his supervision and the Milling–Arnold system, a fairly well-trained enlisted crew headed by knowledgeable noncommissioned officers evolved.

Augusta, Georgia

Faced with the typically windy, cold Washington winter, the Army began looking south for a winter training camp. Chandler set off on a shopping trip during the fall of 1911, traveling through the Tidewater region of the Carolinas and Georgia to inspect areas that Weather Bureau maps showed as having wind velocities and temperatures conducive to flying. Local chambers of commerce and congressional boosters enthusiastically touted the virtues of their cities. After his survey, Chandler recommended a spot near Augusta, Georgia, on a farm just east of the city line. Known as the Hay Farms, it was about three miles long and a mile wide. A road bisected the tract lengthwise, leaving enough space along the sides for landings and takeoffs down the entire field.[12] General Allen approved leasing the property for the winter.

The Signal Corps Aviation School left College Park on the afternoon of November 28, 1911, arriving in Georgia about midnight on the twenty-ninth in a special train of nine cars. Personal circumstances delayed the arrival of Beck and Kennedy until after the turn of the year, but the other four pilots, the medical officer Lt. J. P. Kelly, and nineteen enlisted men set up camp, ready, as Chandler announced optimistically, to "receive additional officers of the army for instruction."[13] Tents sheltered the airplanes; outbuildings nearby served for repair and maintenance. A nine-room house on the property quartered the enlisted, and the officers stayed at a hotel in town.

The group reassembled and checked out the Wright, Burgess-Wright, and Curtiss eight-cylinder airplanes, and training began. Milling, besides flying the Wright airplanes, had begun to learn the Curtiss type under Beck's direction at College Park, and by mid-December he was flying on his own. Chandler established a schedule for what came to be called ground training. Every day except Sundays and holidays the officers assembled for instruction in "telegraphy, wireless telegraphy, gas engines, general principles and structural features of various aeroplanes, and the design of field equipment and appurtenances for aviation service of the United States Army."[14]

The upbeat Captain Chandler reported that winter training got off to an auspicious start. "Since the arrival of the Aviation School at Augusta the weather has been particular[ly] fine with clear skies, and the local residents assert that it is the normal winter conditions."[15] But within the week the wide blue skies began to dump rain, and soon the Savannah River reached flood stage. The enlisted men prepared to tow the equipment to high ground, but fortunately the waters failed to swamp the camp. Nonetheless, the rainy season lasted well into January and prevented any extended stretch of flying. Once the rain and high winds abated, the thermometer dropped to fourteen degrees, and several inches of snow whitened Augusta. By mid-January, ice covered the hangars. Again in February it snowed heavily, and in March, flood waters flowed through the city.

Regardless of the fact that the gentle southern climate became the butt of jokes, the pilots took advantage of the short periods of good weather to log air time. The additional officers of the Army who had been expected in early December showed up desultorily thereafter. Lieutenant C. Sherman of the Corps of Engineers (who, after the war, would draft some of the Air Service's earliest tactical doctrine) spent his leave at the Aviation School, taking lessons from Arnold in the Wright airplane. The next spring, Arnold and Milling demonstrated their wares to Cavalry officers Capt. Robert E. Wood and Lt. C. P. Chandler.[16] Kirtland instructed Lt. Col. C. B. Winder of the Ohio National Guard in the Wright.

Colonel Winder's instruction became a sore subject at the Aviation School. Pilots disliked instructing members of the National Guard because the school was low on instructors, airplanes, and spare parts, and they resented spending their meager resources on the militia. Moreover, when a state purchased an airplane, the manufacturer would train a National Guard pilot just as it did for the Regular Army. But what especially incensed the military men was that Winder, having learned to fly, then advertised his services commercially as an

Signal Corps Aviation School

Augusta in fine weather, during the rain, and in snow

exhibition pilot. The Army had just provided free training for a highly lucrative civilian occupation! Capt. Frederick B. Hennessy posted a newspaper clipping to the Adjutant General about Winder's training, which, Hennessy seethed, "occupied the time and attention of instructors of this school from March until May, just in order to permit him to become a professional exhibition flier, all of which is bound to arouse the hostility of the civilian aeroplane training schools." Whether the civilian schools cared is hard to know, but certainly Hennessy spoke for the Army pilots. Needless to say, the Chief of the Division of Militia Affairs saw the situation differently.[17] Because Winder had already received War Department approval for instruction, the issue was moot for the moment. But it continued to surface, and each time Army pilots fumed and complained.[18]

More constructive, from the perspective of the Aviation School, a Regular Army man, Lt. Leighton W. Hazelhurst, was ordered for aviation duty on March 1 and immediately began instruction on the Wright machine. The Wright B had become the workhorse training airplane, but because Milling could fly both the Curtiss and Wright planes, the two types achieved some parity in their use. Chandler, Kirtland, Arnold, Milling, and Hazelhurst flew the Wright craft; Beck, Kennedy, and Milling flew the Curtiss biplane. Beck also planned to fly both types, so began learning the Wright controls.[19]

At the end of January 1912, Wilbur Wright favored the Aviation School with a visit. He stopped on his way from New York to Dayton, spent two afternoons with the Army pilots, and dined with them Saturday night at the Augusta Country Club. "It is," Captain Chandler enthused, "the first time that either of the famous brothers [has] visited the Army Aviation School since its establishment last summer at College Park." Wright examined the standard Type B machine with his characteristic thoroughness. "Until the time of his departure the next afternoon he cheerfully answered a steady fire of questions from the officers concerning pro[s]pective improvements in military aeroplanes and a wide range of questions covering aviation in general."

With the abysmal weather and what was becoming a pattern of frequent accidents, the first experiment in winter training had to be judged as only marginally successful. Shortly after rejoining the school, in his first flight in the high-powered Curtiss airplane, Beck, who had been warned of the danger, encountered wind currents caused by a large hay barn in front of him. A draft sucked the airplane down, and the wing hit a tree, hurling the airplane to the ground. Fortunately, Beck was unhurt, but before he and the others left Augusta, Beck crashed twice more in the Curtiss D. Miraculously he walked away both times. Kennedy survived a more injurious accident. He had already qualified for his civilian FAI license using the eight-cylinder Curtiss. While practicing landings on February 19, he crashed and was thrown from the airplane. It was reported that his head plowed a six-inch trench into the ground, but, although hospitalized, he recovered. Undoubtedly Kennedy would have died but for the protection afforded by his leather helmet.[20]

At the end of March 1912 the Aviation School packed up and departed for College Park. Subsequently, the school made brief forays into other locales, including another winter in Augusta and, although the staff discussed other training sites, College Park remained the primary campus until the Aviation School made North Island, California, its home in 1913.

Diversification

Grover Loening, who would be hired as the Army's first civilian aeronautical engineer, recalled watching Glenn Curtiss's early-morning flight at Mineola, New York, on July 4, 1908, when Curtiss won the *Scientific American* prize:

> I . . . learned that turns had to be most carefully negotiated because the excess power was so low that the plane would often sink dangerously near the tree-tops on a turn. . . . The stability was nil — flying them felt like sitting on the top of an inverted pendulum ready to fall off on either side at any moment. The speed range was nothing at all. High speed, landing

Signal Corps Aviation School

Glenn Curtiss in his *Scientific American* award-winning aircraft, the *June Bug*

speed, climbing speed were all within one or two miles an hour, because the planes got off into the air with no reserve whatever, and only because of the effect of the ground banking up of air which was not then at all understood.[21]

A couple of years later, the hot Texas winds buffeted Foulois's little plane about the sky when he too flew with low power at open throttle. On the other hand, death also stalked beginning pilots who tried to fly a high-powered airplane. One of the earliest lessons to become codified in the American system of pilot training was a progressive training sequence employing increasingly higher-powered aircraft.

By February 1912 the Aviation School had set aside four machines with low-horsepower engines for beginners.[22] But since the school had by then defined its mission beyond the level of primary training, it also wanted airplanes with a more powerful engines for advanced training and for field service. Therefore, in 1911, the Signal Corps drew up requirements for weight-carrying military airplanes with dual controls, and sent the specifications out for bid early in 1912.[23]

The new tactical aircraft were intended for reconnaissance. The weight-carrying military machine therefore transported observation equipment, not bombs. The Signal Corps ordered two types of airplanes. The first, the Scout, carried a two-man crew: a pilot, and an observer, who operated photographic

1907-1917

Lieutenants Lewis E. Goodier, Joseph E. Carberry, and Walter R. Taliaferro (*left to right*) pose with a Curtiss-built pusher-type aircraft.

and radio equipment. This airplane was to reconnoiter "when hostile armies are in contact or approaching contact." It should be capable of remaining airborne for three hours, attain a speed of at least forty-five miles an hour with a maximum of sixty, and ascend carrying a load of 450 pounds to at least 2,000 feet in ten minutes. The other type of airplane, the Speed Scout, designed for "strategical reconnaissance," carried a pilot who was to "locate and report large bodies of troops." The Speed Scout was to be fast, short-range, and maneuverable.[24] In January 1912, the War Department ordered five airplanes that fit the specifications: one Wright Speed Scout and four Scouts (two Wright-built, one Curtiss, and one Burgess-Wright with a seventy-horsepower Renault engine).[25]

The Curtiss Scout Model E that began testing in March 1912 paved the way for what would become the standard training aircraft. It had the Curtiss control system of a single wheel on a column that could pivot from one person to the other, but it allowed for the Wright method of dual instruction that required two seats.[26] Shortly after the Curtiss E came into the inventory as S.C. No. 8, the Wright Company delivered its first Scout, a similar but slightly larger and stronger version of the standard Type B.[27]

The Burgess Company, of Marblehead, Massachusetts, with the assistance of Greely S. Curtis (no relation to Glenn H. Curtiss), was the new entrant into the competition for Army contracts. Yachtsman and yacht designer W. Starling Burgess, who purchased the use of the Wright patent on a royalty basis, brought to his new enterprise a gift for airplane construction using fine materials and elegant craftsmanship. The first airplane delivered to the Signal Corps, a Model F that became S.C. No. 5, essentially duplicated the Wright Type B. More significant was the follow-on Burgess H, a weight-carrying Scout. Although it kept the soon-to-be outdated Wright controls, it incorporated some elements of European design and, in a major change in configuration, placed the propeller in front of the pilot. All previous Army airplanes were of the pusher type that propelled the airplane from behind. The Burgess tractor design "pulled" the airplane, initiating a change that would become standard.

By the spring of 1912 all three manufacturers were also constructing airplanes with pontoons. Members of the Aviation School trained on seaplanes, then called hydroplanes, flying over the Anacostia River at Washington Barracks near Washington. Hennessy took charge of the equipment and enlisted troops quartered there.[28]

The first congressional appropriations permitted the Signal Corps much greater diversity in its growing but still small inventory. For the first time, aircraft were earmarked specifically for training or operations. Some of the airplanes intended for field service instead became trainers and frequently tactical aircraft were demoted to trainers as they aged. All training planes saw continuous hard use. Chandler reported that the four airplanes used exclusively for training in early 1912 quickly wore out owing to "continuous use and they are not now suitable for service at maneuvers or in time of war."[29] Between 1908 and December 1913 the Signal Corps purchased twenty-four airplanes, fifteen of which were still employed at the beginning of the new year.[30]

Expansion meant more pilots and mechanics as well as airplanes. With additional manpower, the Aeronautical Division could begin to implement

The Burgess H tractor, a weight-carrying Scout, with Wright lever controls

training standards and establish an organization, two goals impossible to attain barely two years earlier. The FAI license, the only form of certification then generally recognized, did not call for proficiency in certain maneuvers peculiar to military aviation. The Signal Corps therefore formulated the rating of Military Aviator, which required a certain distance for cross-country reconnaissance flights, established a ceiling that a pilot must be able to maintain while flying with a passenger, and set altitude and duration-of-flight requirements.[31]

Even with more funds, the Aeronautical Division found that securing officers in sufficient number and of appropriate physical characteristics and mental temperament posed a problem. A board of three officers reviewed the qualifications of all potential candidates.[32] The screening procedure eliminated those obviously unfit for aviation duty, but it did not necessarily produce well-qualified pilots, as Chandler pointed out:

> In the early days of aviation it was supposed that the only requirement for a[n] aviator was sufficient courage, but it has since been proven that one of the most important requisites is physical skill and a natural aptitude for balancing and judging distances properly. A person who does not possess these attributes naturally, can, in the course of time, be trained as an aviator, but will never become really proficient.[33]

The school soon set up an oversight board composed of the commandant and all air officers to judge the proficiency of pilots in training, but quality review mechanisms did nothing to increase numbers. Lacking legislative authority to do more, the Chief of Staff allowed the Signal Corps only ten officers to be detached from line organizations for aviation duty.

The Chief Signal Officer sympathized with airmen's requests for more people. To Glenn Curtiss's offer outstanding to train Army officers without cost to the service, General Allen responded that "the whole difficulty in aviation in the Army is due to the fact that we have not been able to secure additional officers. If this had been done I would have sent some to your school in California."[34] The addition of three officers that summer was offset by the fact that Beck, who was considered the service expert on the Curtiss machine, had to return to his branch in May. Kennedy's accident kept him out of the cockpit for a time, and in October a back injury suffered in a second crash caused him to be relieved permanently from aviation duty. The summer of 1912 brought more devastating losses. On June 11 Hazelhurst flew as the Army observer in a test of a new Wright airplane piloted by A. L. Welsh, who had been Arnold's instructor. The airplane crashed, killing both men.[35]

On the one hand, the Aviation School could not conduct formal classes because the ten officers authorized to aviation trickled in one at a time. On the other hand, each trainee enjoyed intense, personalized instruction. Upon arrival, a student elected the type of airplane control he wished to learn. He was either assigned an instructor or sent to a civilian school for preliminary training.

Students worked at their own pace until they qualified for the FAI test. Thereafter, each was on his own. As Chandler stated, "further proficiency in flying depends entirely upon the enthusiasm and skill of the individual."[36] When ready, the student qualified for the military rating, which made him a graduate of the Aviation School.[37] The new pilot could then devote himself to the study of telegraphy, photography, meteorology, and the construction and operation of mechanical devices and engines.

During the winter of 1912–1913, the Wright pilots, with the exception of Arnold, who assumed duties in the Aeronautical Division, returned to Augusta with five aircraft. Each officer was responsible for a specific airplane.[38] Chandler practiced the right-hand warping control; Harry Graham and Sherman rehearsed for their Military Aviator test; and Milling had become the chief instructor, supervising other airmen and undertaking some experimentation. He also qualified on the Burgess tractor, an airplane he considered "very superior to anything we had received up to that time."[39] Kirtland came later than the others and began work on right-hand warping; he completed the reconnaissance and altitude tests for Military Aviator.[40] When the new military airplanes arrived, the staff put them through their paces. Only Milling could fly the two Scout airplanes readied for field service. Chandler, Graham, Kirtland, and Sherman all flew the training machines.[41]

Although the airmen contended with damaging winds, their 1913 winter stint in Georgia was much more productive than the previous year. Newly appointed Chief Signal Officer Brig. Gen. George P. Scriven informed the Chief of Staff that those in Augusta "unanimously report that this place is excellently well-fitted for an aviation station."[42] Despite the collective enthusiasm for making the Augusta camp a permanent training facility, that winter would be the last time that the Aviation School would be split between two sites. On April 1 the school returned for its final season at College Park.

Throughout this period the Army continued its practice of sending airmen to the factories when it purchased airplanes. In the spring of 1912 Kirtland went to Dayton to learn about the Wright weight-carrying airplane, and Kennedy and Lt. Harold Geiger went to Curtiss's school at Hammondsport, New York, the former to take a course in engine mechanics and the latter to learn to fly. Curtiss also accepted enlisted mechanics, so in June Sgt. James F. Hartman also reported to Hammondsport.[43] Since Beck could not provide adequate instruction on Curtiss airplanes before he was relieved of aviation duty, Lieutenants Samuel H. McLeary, Joseph D. Park, Lewis E. Goodier, Jr., and Lewis H. Brereton concentrated on seaplane flying at the Curtiss factory.[44] Lieutenants Loren H. Call and Eric L. Ellington trained at Marblehead, home of the Burgess Company.

The military presence sometimes resulted in design modifications of prototype aircraft before they were built and delivered. In November 1912, for example, experiments indicated that the Sturtevant four-cylinder, forty-horsepower engine slated for the Signal Corps would not adequately power the Burgess seaplane, so the company increased the horsepower.[45] Based on his training experience with the Curtiss people, Lieutenant Geiger concluded that the Army would benefit from a close working relationship with the manufacturers in the research and development phase: "I believe that an officer (Curtiss flier) should be sent to the factory to assist in developing the machine. I believe that we can accomplish more by working with the manufacturer in developing a military type of aeroplane than in simply depending upon bare fulfilling of our specifications to do this."[46] The give-and-take during the research and development and procurement phases continued, and the Army established the practice of sending one of its officers as an inspector to the factories, but a more intimate partnership of the sort Geiger recommended never developed.

For two winters, the Aviation School split into two groups: the Wright pilots and airplanes moved to Augusta, and the Curtiss pilots and mechanics trained in San Diego. Curtiss had insistently offered to train Army officers free of charge and, in 1912, had proposed to the Chief Signal Officer that more officers *with* their machines report to San Diego, whereupon he could "turn over to you in the spring a squad of men and machines ready for actual service."[47] The Army preferred to continue sending officers as a quid pro quo for machines purchased.[48] The Signal Corps had become enamored of North Island, however, and decided to accept Curtiss's offer to rent adjacent property.

The school did not commit to the move before exhausting efforts to remain permanently near Washington. Bills introduced in both the House and Senate during early 1913 to acquire the airfield at College Park came to nothing. Both military and civilians had used the installation, but the government program had no assurance of stability. The lease on the facility was set to expire on June 30, 1913, and the Signal Corps decided not to renew it. Airmen had already discovered that the climate disallowed much winter training, and no doubt the waste of time and money caused by shifting the Aviation School back and forth influenced the decision. A commission investigating potential locales throughout the United States that might provide year-round flying decided on a site next to Glenn Curtiss's operation. By June 14 the Signal Corps had shipped all its property from College Park to the West Coast.[49]

North Island, California

Glenn Curtiss's tenancy on the peninsula of North Island began in January 1911 when he decided to make a large acreage owned by the Spreckels Company his winter headquarters and site of his experimental and instructional operations. Of North Island, lying in San Diego Bay, Curtiss wrote:

> It is a flat, sandy island, about four miles long and two miles wide with a number of good fields for land flights. The beaches on both the ocean and the bay sides are good, affording level stretches for starting or landing an

Signal Corps Aviation School

The hangars at North Island with two Curtisses parked outside. The pusher Model E is at the left, and the tractor Model G is to the right.

airplane.... North Island is uninhabited except by hundreds of jackrabbits, cottontails, snipe and quail.[50]

An agreement, facilitated by the Aero Club of San Diego, specifically stipulated that Curtiss might invite the War and Navy Departments to share the site.[51] At the end of December, the Signal Corps Aviation School agreed to lease some of the land for $25 a month. The following June Curtiss waived the fee. The first Army school detachment arrived in November 1912 and, as noted, by June of 1913 the school itself had assembled in California. Some uncertainty about the permanency of the arrangement remained, however, because the owners of the property specified that, if requested, the Army would have to move. Nonetheless, airmen hoped to establish a permanent training center in the area.[52] The Navy would, in fact, do so.

Although the school in California would not remain the permanent headquarters for Army air training, the scattering of Aviation School groups among duty stations, and the fleeting deployment of tactical squadrons, had not dissolved the thread of organizational unity that the school had and would represent. In 1911, the Chief Signal Officer ordered that all records of men and officers on aviation duty be kept at the Aviation School. The enlisted detachment at any locale constituted part of an aero squadron, and its personnel records were held with those of the officer corps at the Aviation School.[53] The formal recognition of the school's institutional role took place in December 1913, when, by General Orders 79, the San Diego installation for "theoretical and practical instruction in aviation" was designated the Signal Corps Aviation School, officially taking its place among the Service Schools of the U.S. Army military educational system. Thus it was exempted from control by the department commander in all matters of instruction, administration, and organization.[54]

Throughout the period of transition, the Army school constantly shifted between too few men and too few machines.[55] In late 1912, when part of the school contingency had arrived in California for winter training, the Chief Signal Officer declared that until the authorized personnel ceiling was reached, airplanes would remain out of commission.[56] This delay did not last long, fortunately, and the small number of airmen already on duty continued to fly while the commandant searched for more officers.

The newly official designation failed to fill the ranks with pilots. Furthermore, the more experienced personnel of the Aviation School had changed repeatedly through the first half of 1913. Chandler, Lahm, Arnold, and Sherman returned to troop duty and Milling went to France to investigate European aviation. Geiger, who specialized in the Curtiss seaplane, became commandant of the school until he was replaced by Capt. Arnold S. Cowan, who was not a pilot.[57] New pilots joined the instructional staff after completing manufacturer training. Lieutenant Brereton, for instance, became chiefly responsible for demonstrating seaplaning. By midsummer 1913, Milling had returned from France. At that time, with the entire group gathered in San Diego, the school enrolled fourteen student officers. Forty-eight enlisted men serviced the seven airplanes.[58] In December, a total of twenty officers were on duty at the school.[59]

Virtually any growth in the active air arm could be measured by the size of the Aviation School. With only the rarest exceptions, graduate pilots never joined a tactical unit where they might continue training, ideally in combined arms operations. Instead, most remained at the school unless recalled from aviation altogether. In 1912, the Signal Corps had attempted and failed to organize a tactical squadron, owing to lack of fiscal and personnel resources.[60] The next year, the Chief Signal Officer proposed to establish an organization that would integrate air assets into Army field commands.[61] That year, 1913, the 1st Aero Squadron (Provisional) was organized to support the 2d Division, stationed at Texas City, Texas. These forces assembled as a result of the turmoil in Mexico that spilled across the border. But the period of the squadron's activation was brief, for by the middle of the summer the unit transferred to San Diego where it was incorporated into the school. The pattern repeated itself because, as long as operational and training units shared the same infrastructure and funding, they inevitably borrowed from one to pay for the other.

Airmen were fortunate, however, in their selection of North Island. Occasional puffy winds blowing in from the ocean interfered with flying, but

Signal Corps Aviation School

even in midsummer, the worst time of the year, pilots could expect to log at least two hours in the air every day. Commandant Cowan expressed his satisfaction to the Chief Signal Officer:

> The only objection that we have to find to North Island is its inaccessibility. This, coupled with the fact that the soil is sandy, and when the wind blows the sand gets into everything, makes the Island rather an unsatisfactory place to live. As far as being a suitable place for flying, it would be hard to imagine a better place. There is all the room that could be desired for a training station, with plenty of landing ground and nothing whatever to interfere with us.[62]

Furthermore, the school found the relationship with the Curtiss operation to be advantageous, especially at the beginning when Curtiss instructors provided training and equipment to the Army. At the end of March 1913, the school had eight Curtiss pilots on duty or in training. Among them were Lts. Walter R. Taliaferro and Joseph E. Carberry, who used the Curtiss machines under the supervision of an instructor of the Curtiss school, a practice that continued until the Signal Corps Aviation School acquired additional Curtiss equipment.[63] By the middle of July, the Army school was sufficiently self-supporting to offer training at its own camp. Curtiss still generously allowed the Army to use his machine shop, mechanics, and instructors.[64]

The school made a leap forward when it began bringing in outside experts. As the Chief Signal Officer pointed out, the Aviation School had given "little instruction in anything except the practical side of the art of flight." With pride, he went on to describe the school's meaty technical course that ranged from meteorology to the tensile strength of airplane construction materials. Dr. F. R. Hutton of the American Society of Mechanical Engineers, for example, lectured on internal combustion engines. Col. Samuel Reber from the Chief's Office advised him that only fourteen officers would be in attendance during the winter 1914 course, although they were "a very intelligent set of young men, 6 of whom are graduates of the Military Academy." He also had to admit that "none of these officers have had any technical engineering training, and may be a little bit rusty in mathematics."[65]

The academic program was in the formative stage. Scriven noted that the Aviation School could find no textbooks and "no recognized courses on these subjects in any of the colleges or technical schools of the country, to [serve] as guides."[66] Roy Kirtland recalled a few years later that "the pioneer flyer could learn practically nothing of the theory of flight before he entered upon his training. The practical flyers of that time wrote little or nothing of what they had learned, and the students of aeronautics who did publish such books on aviation as were available had little or no practical experience in flying."[67] Nor could the evolving aviation program look to the Leavenworth model of instruction that used the *applicatory* method — the application of past or

1907-1917

Capt. Arthur C. Cowan (*left*) and 1st Lt. Walter R. Taliaferro (*right*), both posing before the venerable S. C. 1 during good times. Taliaferro would be killed in a crash at North Island in 1915, and the same year Cowan would be the focus of an investigation and eventually be removed from duty at the Aviation School.

hypothetical cases to new situations. Rather, Aviation School students plowed their way through fairly raw technical and engineering data as a complement to their practical flight training. They, unlike their fellow ground officers, had few historical case studies to apply as they considered tactics and military strategy.

The inclusion of a broader tactical perspective, all the more important given the current definition of military aircraft as tools of reconnaissance, would have enriched the Aviation School curriculum. Since the Leavenworth schools failed to provide the intellectual leadership, the Aviation School was left alone to explore the operational possibilities of the new technology, but it in turn tended to ignore conceptualization in favor of technocratic concerns. Once again, had aviation instruction been conducted within the existing military school system, it might have broadened the consciousness of Army planners and tacticians and laid a firmer doctrinal floor beneath military aviation. Considering the time and personnel constraints and the single opportunity for aviation education and training at one place, the Aviation School probably succeeded professionally as well as was possible under the circumstances. As it was, the curriculum demanded much student time and

Signal Corps Aviation School

attention because the lectures were given after flying, which often began at daylight and might continue until dark. *Aeronautics* informed its readers that the Signal Corps required "from nine months to a year, with a lot of experience in cross-country work, before a man can really be said to be an aviator."[68] That nine months to a year was spent training a man to fly, not necessarily educating him in the military arts.

As late as February 1914, the Office of the Chief Signal Officer was unable to provide a set of school rules and regulations "owing to the newness of the work."[69] A year later the Army published the first regulations governing the administration of the Signal Corps Aviation School.[70] By that time, according to Arnold, "the actual training of pilots had advanced to such a state that it was possible to prescribe rules and regulations for their instruction and in addition it was possible to notify prospective students as to the conditions they would meet upon arriving at the school."[71]

The Aviation School moved toward a more structured program via its departmental arrangement. The Experimental and Repair Department undertook major repairs and overhauls of airplanes and engines. According to the Chief Signal Officer's 1915 annual report, the Training Department:

> is devoted to the training of student officers for junior military aviators, the instruction of enlisted men in flying, and the training of suitable enlisted men for aviation mechanicians. The officers are given theoretical and practical courses in the art of flying; in the construction, operation, and repair of aeroplanes and aeronautical motors; in meteorology, and in the navigation of the air. Enlisted men on flying duty are instructed in the art of flying and in the operation and care of aeroplanes and motors. Aviation [mechanics] are trained to repair aeroplanes and motors by a thorough shop course.[72]

Officers received both practical and theoretical instruction in the technical aspects of aviation, whereas the enlisted "branch" of the school "is almost entirely practical, being shop training in the operation and repair of motors and aeroplanes and flying."[73] That arrangement would eventually become institutionalized into two tracks — technical and flight training — with enlisted men engaged in the former and officers (primarily) in the latter.

Besides maintenance, the Experimental and Repair Department handled the vital work of aircraft modification.[74] Colonel Reber, who at the time oversaw aviation in the Office of the Chief Signal Officer, ostensibly supervised the experimental work, but this activity lapsed after the creation of the National Advisory Committee for Aeronautics (NACA) in 1915. Initially intended as a scientifically oriented body, NACA was in fact highly engineering-minded in its approach.[75] The establishment of NACA effectively diminished the Signal Corps Aviation School's guidance of technical innovation and furthered its emphasis on training to the exclusion of more tactical and doctrinal experimentation. The school existed primarily to teach men to fly and to maintain and repair the airplanes.

Furthering the effort to develop technical expertise in its pilots, the school employed its first aeronautical engineer, partly as a result of a 1914 Inspector General's report urging that a pilot's license be granted for knowledge of design and construction of engines as well as flying ability. But Howard C. Davidson claimed that when he attended the Aviation School as a lieutenant in 1916, the instructors "didn't teach us anything but just flying...; we had to learn how to overhaul motors."[76] Thus, during his nonflying hours Davidson and another student pilot "decided we were going to learn something about engines. So we went and got a job in a garage."[77]

As another indication that this narrow perspective regarding engineering theory and practice percolated down from above, in April 1915 General Scriven refused Lieutenant Arnold's request to study aeronautics and engineering at the Massachusetts Institute of Technology or Cornell. The Chief ruled that "It is not advisable to recommend the detail of officers in the Aviation Section of the Signal Corps unless they shall take up flying. As I understand it, you have in view a detail in which you would engage in the study of aviation engineering matters only."[78]

Passage of the 1914 act did little to increase the numbers of officers detailed to aviation, which, among other reasons, caused the school to turn to civilian expertise. For the relatively high salary of $3,600 a year, Grover Loening, appointed through the Civil Service Commission as the Aviation Section's first aeronautical engineer, took charge of the Experimental and Repair Department. Francis "Doc" Wildman left the Curtiss Company to become chief instructor in overwater flying in the Training Department, joining Oscar Brindley, formerly of the Wright Company, who supervised land-based training. Captain Cowan explained to the Chief Signal Officer another reason for the reliance on a civilian staff: "There are now a number of expert aviators in the service, but expert aviators are not necessarily competent instructors. Instructors must have special qualifications in addition to being expert aviators. Teaching men to fly is probably the most dangerous occupation in the world. Men who can do this work and do it well are very rare, and their services are cheap at almost any price."[79]

Finding and keeping competent instructors would remain a perennial problem in flight training. In these early days, whoever was available, and showed some promise in a given area, took up the task. Nor did the school demarcate clearly between enlisted men and officers who taught, or who became pilots. In general, it paid relatively little deference to the hierarchy of rank. For instance, pilot trainees listened especially carefully to the advice of Sgt. Harry B. Ocker (who subsequently became an officer). According to Howard Davidson, "He would get us off to one side and say, 'Now, Lieutenant, I saw Lieutenant What's-his-name do the same thing you did, and we picked him up in a basket.' Well, that scared the hell out of us, so whatever Ocker told us we believed."[80]

The school eliminated two of its most bedeviling problems in 1914. The Office of the Chief Signal Officer directed the commanding officer of the Signal Corps Aviation School to convene a board of experienced airmen to "report upon the suitability and safety of the type B and type C [Wright] machine[s]."[81] As a result of that investigation, the school phased out the Wright airplanes. Ending the initial Wright monopoly, then the acquisition of Wright airplanes altogether, indicated that the American aircraft industry and the military flight training program were coming of age. Concluding its purchases from the Wright Company also brought an end to the considerable friction between the Curtiss and Wright camps that had marred smooth operations and collegiality at the school.

Oscar A. Brindley was one of the civilian flying instructors hired by the Aviation School's Training Division to augment the number of men detailed to aviation.

In the other significant effort to enlarge and improve the aircraft inventory, in early 1914 all the pusher-type airplanes were condemned as unsafe to fly. The old Curtiss and Wright pushers tended to stall and, in an accident, the rear-located engine was likely to be thrown forward onto the pilot. It took some time, however, before the Signal Corps could afford to replace its defunct pushers with newer models.

In fact, the school was almost without training airplanes for the better part of 1914, and it lacked a standard trainer aircraft for some time afterward.[82] It supposedly taught pilots to fly seaplanes, but the Army had no flying boat until the second Curtiss Model F was delivered in September 1914; it was then used in intermediate training. Most of the airplanes used as trainers were modified tactical aircraft. The Burgess H, manufactured under Wright patents, was pulled into use as a training airplane, although it was ideal neither for training nor for field service — too fast for beginners with its seventy-horsepower Renault engine and too heavy for service loads. Loening and his people completely remodeled it for, briefly, advanced training.[83] The school also rejected as unworkable a new Burgess tractor training plane.

To convert to the new controls, Wright pilots practiced on S.C. No. 22, a

1907-1917

Curtiss tractor not intended for use as a trainer (it was classed with the high-powered Curtiss service machines). Although a reliable airplane, Loening thought it so heavy that it "was sort of wished off the ground if there was enough wind."[84] But it was the only machine officially assigned to the school through the summer of 1914. The new Curtiss Model J, ordered as a reconnaissance machine, also found use in training. The Army purchased two of these airplanes in 1914. It removed the first from service in early 1915 after Lt. Frederick J. Gerstner drowned after crashing into San Diego Bay. An accident in October demolished the second, No. 30, killing Lieutenant Taliaferro. The Curtiss Model N, a close relative of the J, was used briefly as a side-by-side trainer. The Signal Corps ordered the first of the JN series, which would evolve into the illustrious Curtiss JN-4, the most successful trainer in World War I. Eight new JN-2s arrived during the summer of 1915; to overcome many of its faults, the JN-2 was converted to a JN-3. The 1st Aero Squadron trained on the JNs at Fort Sill, Oklahoma, and in Texas, and took the "Jennies" — the nickname that stuck — into Mexico with the Pershing Expedition in 1916.[85]

The Aviation School adapted tactical machines for training purposes, but after condemning the pusher it quickly published specifications for airplanes dedicated for training. On March 14, 1914, tentative specifications for a military tractor training biplane went to the Chief Signal Officer. The anticipated aircraft was intended solely for use by beginners, not for cross-country or reconnaissance purposes. "Hence," Captain Cowan explained, "qualities of weight carrying ability, a great range of speed, and a wide radius

A Jenny that served in the Punitive Expedition to Mexico in 1916

of action are not of importance. What is desired is a light weight, strongly built machine possessing as its chief requisite this quality: *ease of maneuver in the air at all speeds*, particularly at its slow speed."[86] The Inspector General's report for 1914 recorded "a general consensus of opinion that [the Curtiss] machine is the best of the American machines for service conditions. There are some who think the Martin machine...is the better machine for training purposes."[87]

In fact, a newcomer to the aircraft industry, Glenn Martin, became the primary supplier of training airplanes between 1914 and 1916. The first Martin trainer began trials August 24. The Army bought it without a motor, installing a Curtiss engine removed from another airplane. An *Aeronautics* reporter admiringly described the two Martin trainers brought for testing in 1915: "The new Martin machines are distinctive by the round, graceful sweep of the wings; the long, tapering, torpedo-shaped body, with a round-nose radiator and a four-wheel chassis.... The metal hood is enameled an olive green, and all surfaces are of an olive drab color."[88] The Army ultimately bought seventeen Martin T and TT trainers.[89] But with only the Martin airplanes on order specifically designed as trainers, flight training remained severely handicapped throughout 1914–1915. "It has been demonstrated beyond question," asserted Cowan, "that satisfactory results in training can be obtained only by using the proper kind of training machine. Attempts to make over service machines into training machines have not been successful, and it is my opinion that this makeshift policy is one that ought not to be followed in this work."[90]

Procuring suitable equipment for preliminary instruction remained one of the principal needs of the training program throughout the mid-teens. The fact that the aircraft inventory at the school expanded dramatically at this time, but that few airplanes were alike, made flight training easy only compared to the job faced by the maintenance and repair people.[91] Moreover, resources were spread too thinly to both instruct and equip the 1st Aero Squadron and to fully staff and equip the school.

In a more positive vein, standardization improved in late September 1915 when the dep control[92] was installed in one of the airplanes, and trainees began to use it. This system had become the standard in all American airplanes by the time the United States entered the European war. Elimination of the differing Curtiss and Wright controls and pusher-type airplanes went far to synchronize training methods, break down the conspicuous rivalry at the school, and reduce the number of training accidents. The introduction of higher-powered, more strongly constructed machines and a standardized control system allowed for greater precision in advanced training maneuvers. The Aviation School now had begun to distinguish phases of training — which would come to be called Primary, Basic, and Advanced — and to identify the types of aircraft appropriate for each.

Even improvements in technology, however, could not eliminate other

1907-1917

Two types of Martin T trainers were produced for the Air Corps in 1914. One featured side-by-side arrangement in the cockpit to permit the instructor to sit at the side of the pilot (S.C. 33, *top left*); the other had a tandem arrangement with the instuructor sitting behind (S.C. 31, *lower right*).

vitiating morale problems. Typically, the military pilot did not see himself as an engineer or administrator. At the time of the creation of the Aviation Section in July 1914, all flying officers had volunteered from the combat arms rather than from the technical branches, and presumably all were drawn by the personal skill, ingenuity, and daring required by their chosen calling.[93] Henry Arnold, writing in the mid-twenties, observed:

> Any of the early aviators still living will remember the difficulties encountered when it was proposed that all planes be equipped with a tachometer for the engine. They all knew that they could tell by ear whether the engine was tuning up properly. They would have scoffed at the pitch and bank indicators that are now included and would have considered any one who used the large number of instruments now installed in a plane far below their own standard as aviators.[94]

To fly pre–World War I aircraft, particularly the Wright-designed planes that sacrificed stability in favor of maneuverability, pilots relied heavily upon their own instincts and the feel of the air. Airmen adopted an empirical rather than a theoretical approach to flight, i.e., experience, gained in the air on a daily basis, served to test the soundness of equipment and training. As expressed by Arnold, pilots were suspicious of artificial or purely technical methods as well as of conclusions not derived from direct experience, including those advanced by senior ground officers or administrators.

Growing Pains

During its short life, the Aeronautical Division operated on a shoestring, even after predictions of future generosity greeted the first congressional appropriation in 1911. A $125,000 budget remained constant in the face of a $1 million request by the Chief Signal Officer for fiscal year 1913. Calls from interested parties to spend on military aeronautics to match European funding levels again filled the press,[95] but by the end of 1913 the Chief Signal Officer was still cautioning his airmen to be ever mindful of cost, given the scarcity of money.[96]

Fiscal realities left the training program in a constant state of readjustment to the requirements of a motley assortment of training aircraft and equipment, brought about by a haphazard procurement process. Without doubt, flight training suffered from a chronic shortage of manpower that was as crippling as the financial shortfall. The division never reached the thirty men authorized; at the end of 1913 the Signal Corps had eleven pilots. In military aviation's first five years, eleven other officers and one enlisted man had died in airplane crashes.[97]

Besides severe manpower and aircraft procurement problems, the advent of a formalized training program at the Aviation School brought to the surface an increasingly vicious competition among pilots trained in Curtiss and Wright methods and aircraft. For example, many years afterward, Kennedy, a Curtiss pilot, accused Chandler, a student of the Wrights, of bias stemming from the time of his and Beck's accidents in 1912:

> Charlie Chandler at first was a great advocate for the Wright type of plane and very carefully ... reported every accident that happened to a Curtiss plane but he very carefully forgot to note the continuous service troubles we had with the Wrights, such as catching on fire when landing about 50% of the time etc.[98]

From the other perspective, Grover Loening, originally a Wright man, considered Beck, trained by Curtiss, to be unreliable.

The rivalry remained muted while the group was small and intact, but it blossomed after two distinct cadres trained at different duty stations. When they all came together at North Island, as Arnold remembered, the contest reached a fierce level:

> Competition between these two camps became very keen and finally

> developed to such an extent that it became dangerous. The exponents of each camp tried to demonstrate in the air the advantages of his particular method of instruction and that the airplane he was flying was so much superior than the other type of airplane. As new students arrived at the school each camp endeavored to persuade the student that his particular method of instruction and machine was far better than the other.[99]

Pilots' highly competitive personalities reinforced the professional rivalry caused by competing methods, equipment, and mentoring. Responding to a letter of complaint about the extreme nervous strain and exhaustion pilots suffered, the Secretary of War pointed out that although thorough training could modify the stress of a dangerous occupation, the underlying problem lay with the kind of men attracted to aviation: "I am afraid the pride of aviators, in their desire to appear always to be ready to go into the air, has been responsible for some of the accidents, although we have no direct evidence to this end."[100]

In mid-1913, the Chief Signal Officer asked the commanding officers of the 1st Aero Squadron and the Signal Corps Aviation School to submit a formal recommendation for a "standard universal control for installation on all aeroplanes."[101] Aircraft standardization underway the following year did much to mute the factionalism even though it did little to curb the competitive, somewhat undisciplined behavior of the pilots, a complaint the Army would continue to levy against its airmen for years to come. In the meantime, the school tried to maintain parity between the two sides by balancing the numbers of airplanes and men assigned to each. But that task was further complicated by the fact that, in the Wright system, a student attempting to learn right-hand warping had to await the time and attention of an instructor able to give him lessons or else change the warping lever from one side to the other.

The schism between pilots lessened but did not disappear for some while. The much deeper and long-lasting fault line between ground and air officers was exposed in airmen's testimony during the 1913 congressional hearings, and an ugly incident at the school a couple of years later. Captain Cowan, a desk-flier, had been appointed head of the Aviation School in 1913. At one point he decided not to remove Wright planes from the inventory despite their high accident record because on "thinking the matter over carefully, I figured out that we really had not given this machine a fair trial, because our officers — all of them who were flying those machines — were really nothing but amateurs."[102] Such an offensively dismissive comment from a man who himself seldom flew undoubtedly enraged those whose professionalism was bought at the expense of life-threatening personal risk, acquired, with few other compensations, day by day.

The antagonism between airmen and administrators flared openly in 1915 when charges were brought against Cowan and the secretary of the school, Capt. William Patterson, for alleged favoritism and insensitivity to safety considerations, and for the fact that both officers received flight pay when they were barely qualified to fly. In its investigation, the Judge Advocate General's office found that although the two men were assigned administrative duties, they were prepared to take up flying if required and thus were not culpable of misusing their positions. The episode led nonetheless to a General Staff investigation of the Aviation Section and testimony before the House of Representatives during hearings on the 1917 appropriations bill. Secretary of War Newton D. Baker censured Chief Signal Officer Scriven and Colonel Reber, who was in charge of the Aviation Section. Though exonerated from wrongdoing, both Reber and Cowan were ultimately relieved of their commands and assigned elsewhere.[103] Capt. William "Billy" Mitchell from the General Staff — who, according to Benjamin Foulois in an astonishing judgment, in view of Mitchell's soon to be well-known iconoclastic style, was expected "to instill old fashioned military discipline among the so-called prima donna pilots then on active duty" — replaced Reber.[104]

Foulois never liked Mitchell and thus might be said to have been biased in his view of War Department motives, but further evidence indicates that competing notions of discipline and control were played out at the school. For example, in January 1917 Colonel William A. Glassford, having replaced Cowan as commandant of the Signal Corps Aviation School, removed Lt. Herbert A. Dargue from his position as the school's officer in charge of training "for lack of cooperation." Glassford, a dedicated Signal officer with an extensive technical background, and an active proponent of military aviation from its beginnings, considered Dargue to be "a strong and capable young officer," but one who needed the "environment of a strict military system to teach him subordination and unquestioning obedience."[105]

Clearly, concern for efficient administration and attendance to duty, honed by generations of Army officers, conflicted with the airman's individualism that often appeared destructive of life, property, and military discipline. Perceptions about conformity versus recklessness, and competing assumptions about authority and rank, warred with one another. The dominance of one or the other value system at any point in time determined much about the decision-making process and organizational control. The deep-seated distrust between pilots and non-flying officers, reflecting the stratified Army hierarchy and an essentially unintegrated flying corps, remained intact throughout the period of Signal Corps — and even Army — control of aviation.

Despite rents in the fabric, military aviation's uncertain and dangerous beginning proved productive. Considering the minimal financial or professional inducement to enter the flying ranks, aeronautics attracted a cadre of committed officers. The Signal Corps Aviation School made a place for itself within the Army structure, if in a limited fashion, and established training facilities and techniques that would prove long-lasting. Its officers began to develop special technical expertise as well as an understanding of the requirements of a flight training program, and of their own unique identity within the parent institution.

1907–1917

Professionalization and an air force culture were in the making. Legislation enacted in 1914 created the Aviation Section and substantiated the combat mission, giving official life to a heretofore nebulous Army function. More important, airmen would soon be forced to look beyond their narrow fascination with a new technology to assess its efficacy in time of war.

THREE

Prelude to War: Reform, Operational Training, Preparedness

> We had used up the word "Armageddon" to describe a war of words made in 1912 by the Progressive Party in behalf of social ideals. We had come to think of war as a primitive rudeness like a backwoods feud, or an outmoded standard of honor, like duelling, as something outlandish, unlikely to occur again, except in small and unstable countries such as Latin America and the Balkans. This confidence, on June 28, 1914!
>
> Mark Sullivan, *Our Times*[1]

Such was American innocence at the time of the assassination at Sarajevo of Archduke Franz Ferdinand, the heir apparent to the Austro-Hungarian throne. The event that precipitated the European war barely caused a ripple in the still rural and small-town life of native-born Americans. For the most part, Americans and their leaders failed to see, in the late summer of 1914, how the carefully crafted war plans and the mobilization of troops in Europe related to them. Moreover, the American military establishment still resembled a constabulary force, the product of a long national policy of isolationism and a conception of armies as drawn from the citizenry in time of need. With few exceptions, the will to create a sizable standing army and navy did not exist. In fairness to American naiveté in 1914, the political and military leadership of the belligerent powers, in a stunning failure of vision, did not themselves anticipate the long, horrible bloodletting that would decimate an entire generation.

Parochial though it was during the first ten years of existence, military aviation partook of the movement for Army reform that had been underway since the early years of the century. The terms of the debate over the practice of military art and administration reflected the temper of the Progressive Era, ushered in by the presidency and typified in the person of Theodore Roosevelt, and which dominated the subsequent administrations of William Howard Taft and Woodrow Wilson. Between 1910 and 1920, military Progressives such as Army Chief of Staff Maj. Gen. Leonard Wood and Secretary of War Henry L. Stimson opposed traditionalists in the Army and in Congress, including Democrat James Hay of Virginia, the powerful chairman of the House Military Affairs Committee. Reformers' views appeared in War Department reports and were aired at congressional hearings that, in turn, resulted in institutional reorganizations and legislation. Most significant for aeronautics, on July 18, 1914, Congress established the Aviation Section in the Signal Corps.

Elsewhere, the world stood at Armageddon, no longer engaged in a war of words. The impact of the European conflict percolated slowly through the layers of U.S. government bureaucracy and the War Department, where the pace and issues of foreign and domestic military policies played out in microcosm in the Aviation Section. If airmen initially appeared relatively unperturbed by events abroad, their thinking followed the evolutionary shift in the national posture toward preparedness for defense. American military aviation groped its way toward maturity, just as the American public and its leaders came to terms with the ramifications of the war on the continent. Plans for expanding flight training became an integral part of Signal Corps proposals brought before Congress and argued within the Signal Corps itself. Yet they were subsumed under the broader issues of military reform and the institutional status of aviation within the Army.

During the years between the birth of Army aviation and the country's entry into war, aeronautics strove to wrest a place for itself in the military system. As evidence of an increase in military preparedness generally and of aviation's insinuation into the structure of the Army's field forces, some of the few military pilots left the Signal Corps Aviation School to set up training units in overseas possessions. Yet the administrative status of aeronautics under the Signal Corps, rather than within or alongside the combat branches, determined the doctrine and tactics by which the air arm defined itself and girded for war.

The Case before Congress

During his 1911–1913 tenure as President Taft's Secretary of War, Henry L. Stimson actively promoted War Department policy planning and supported Chief of Staff General Wood in his battles over Army reorganization. Like other shrewd Progressives, Stimson and Wood made use of journalism and the public forum to press for effective management practices and consolidation of authority within the War Department. More formally, they transmitted ideas for Army organization through the War College Division of the General Staff, and thereafter plans made their way into the public consciousness through the mechanism of congressional hearings.[2]

Prelude to War

Consideration of military aviation became part of this public process. In 1912 Congressman William B. Sharp of Ohio introduced House Resolution 448. In language that suspiciously mirrored aviation interests, the Resolution required that the Secretary of War provide to the Congress results of his investigations into foreign aeronautical developments; the cost of airplanes and the nature of training in the U.S. Army; War Department plans for "increasing the present equipment of aeroplanes, hydro-aeroplanes, and other air craft for the purposes of warfare and national defense"; and recommendations for legislation to increase the number of flying Signal Corps officers and the establishment of additional flying schools.[3] An opportunity firmly in hand, the Office of the Chief Signal Officer quickly compiled documents and correspondence from journals and firsthand reports relating to foreign aviation. The information that the Aeronautical Division had been collecting since 1907, primarily for the education of its officers, could now be aired to political advantage, at the same time avoiding the overt lobbying permitted to interested civilians. It offered greater potential for publicity than the Chief Signal Officer's restrained but forceful advocacy contained in his annual reports to the Secretary of War.

The Signal Corps's report to Congress, titled "Military Aviation," began with a lengthy review of foreign aeronautical training, aircraft, budgets, and doctrine. It then briefly presented a series of proposals to increase the number of Signal Corps officers and aviation squadrons for defending overseas possessions and guarding U.S. coastal defenses. To accomplish this extended operational role, given the fact that the active air arm then had twelve officers and twelve airplanes, the study recommended a considerable expansion of the training program. Five training centers would be located, respectively, on the Atlantic, Pacific, and Gulf coasts, in the Great Lakes region, and elsewhere in the interior. Additionally, hoping to capture the hearts of congressmen, the report specified "as many auxiliary centers as it may be possible to organize, with a view to having a school of instruction in each State."[4] The centers would train pilots for both the Regular Army and Organized Militia; train enlisted mechanics; test aviation devices; and teach ground school subjects such as meteorology, wireless telegraphy, military topography, sketching and reconnaissance, bombing ("the dropping of projectiles from air craft"), the use of small arms from aircraft, and airplane design.[5]

Although modest compared to the considerably larger programs already underway in Europe, the report was highly ambitious, even greedy, given the tiny number of people then employed in American military aviation. The prospectus essentially outlined the existing flight training program and urged that it be duplicated many times over. Where the proposal departed dramatically from previous practice, however, was in its call for formalized coursework and practicum relating to an offensive role for aircraft. At the time, the Army barely recognized any mission besides reconnaissance, and airmen

1907-1917

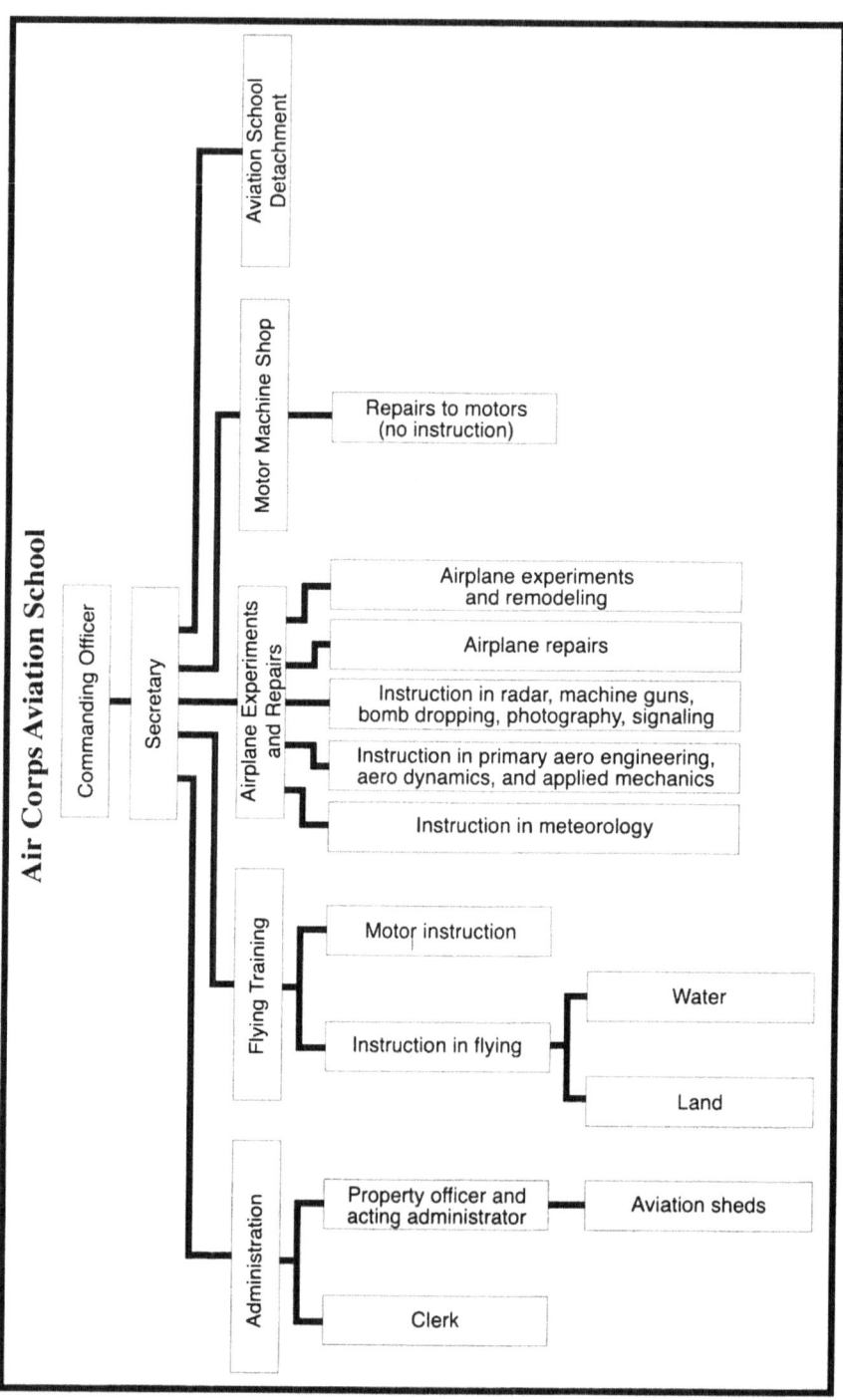

Prelude to War

In the work shop at North Island (*above*), mechanics trained to maintain Army aircraft, and aviation trainees such as Walker, Robinson, and Kelly tested the new aircraft and devices.

experimented with other pursuits only when they had the time or inclination.

Secretary Stimson sent to the House the report as drafted by the Signal Corps, along with an earlier proposed bill to increase the numbers of airmen and provide greater pay and benefits. But Stimson pointedly recommended against the Chief Signal Officer's request for an increase in the number of squadrons and officers until sufficient aircraft justified them and the proposed organization proved effective. He argued instead that training continue to be offered to officers from various branches of the service "that they may keep in mind the special requirements of machines for use in the branch of the service to which they belong."[6]

The Aero Club of America pressed its members and all interested parties to write Congress on behalf of the Signal Corps plans. "If anything is to be done to remove the United States from its present humiliating position in aviation it must be done now."[7] No legislation resulted. The debate had reached public consciousness nonetheless and reemerged in the language of the 1913 appropriations bill and in hearings held that summer. From the time of the 1912 report through the passage of the legislation creating the Aviation Section two years later, momentum grew to give statutory recognition to Army aviation.

Both the 1913 appropriations bill and the 1914 act addressed administrative issues, authorizing an increase in the number of officers of the line on flying duty and providing flight pay. The act of July 1914 determined the basic organizational pattern for Army aviation until 1918 when the Army removed aviation from the Signal Corps. It established a ceiling of 60 officers and 260 enlisted men and cadets, enumerated the age and rank qualifications for air officers, recognized the Military and Junior Military Aviator ratings, and provided for increased pay and benefits.[8]

But the 1914 legislation was in line with congressional enactments passed earlier, and those to come during the interwar years, that failed to provide the means for compliance. In this case, pilots would continue to be volunteers drawn from the line of the Army. Since the Aviation Section was exempted from the Manchu Law, which restricted the time officers could be reassigned from the branch of their commissioning, commanders were reluctant to release many of their young officers to aviation because they could not be replaced. As a result, the Signal Corps failed to attract the number of men authorized for flight duty.

Appropriately, Congress did not step into the evolving battle that pitted the ground Army's chain of command, based on rank, against the airmen's, based on function. One of the reports issued by the Office of the Chief Signal Officer during the hearings stated:

> The pilot of the aeroplane, for whom we all have the highest respect, is the fighting man of the machine. He is the man behind the gun; but from the nature of things, he must be a young, venturesome officer generally without the knowledge of administrative and technical matters which can only come with years of experience and study, and then only to men of a certain type of mentality.[9]

This statement implied that senior nonflying officers of the Signal Corps should retain administrative control at the same time as it acknowledged the special skills and "mentality" required of pilots. But airmen themselves testified that on the working level leadership devolved from technical proficiency rather than from rank — an anomaly in the military system. In his testimony before the House committee, for example, Foulois called Milling the best pilot in the Army. "Personally," he affirmed, "I would be perfectly willing to go under him, although I outrank him by a number of years." Although competence did not officialy overshadow rank, in point of fact, Foulois acknowledged, "the senior officer is perfectly willing to waive his rank when he gets up in the air."[10]

Other men's views echoed Foulois's. In late 1916 Lt. Ralph Royce finished in the Signal Corps Aviation School class just ahead of Lt. Howard Davidson. Although Royce "was junior in Army rank," Davidson recalled, "they didn't know whether he was senior in aviation rank or not. For awhile he was the commanding officer of something or other until they got it all back to the Army rank."[11] Issues of command and control remained tangled and ultimately unresolved as long as the Aviation Section remained part of the Signal Corps and the air arm remained part of the Army.

So, pilots deferred to those they believed had a magic touch, and their senior officers stayed closer to the traditional lines of authority conferred by rank. If a streak of evangelism brought them together, oddly enough it coalesced around training. The thrust of Chief Signal Officer Scriven's

Prelude to War

remarks, and the centerpiece of the reports and proposals submitted in connection with the 1913 hearings, was the outstanding need for an aeronautical training center. Without a structured, systematic program of flight training, aviation could never prove itself within the Army.

The leap from the uncertain present to a vision of the future expressed by air power advocates was considerable, to be sure. The Aviation School at College Park operated on a lease arrangement until June 30, 1913, when it went hat-in-hand to San Diego with no assurance of permanency. Restrictions by lessors and lack of funding precluded extensive construction at either College Park or North Island, and the government could be asked to vacate at any time. Airmen therefore saw no workable alternative to permanent training facilities, government-owned-and-operated, well located, thoughtfully organized and, according to pilots at least, administered by airmen. During the 1913 congressional hearings, the Signal Corps narrowed and refocused its position of the previous year. Whereas the 1912 study proposed five training centers geographically spread over the United States and additional others located in each state, the 1913 plan described a smaller, structured program, based more directly upon experience.

The Signal Corps documents outlined three phases in the process of training military pilots. The instruction of beginners during the first phase, according to Chief Signal Officer Scriven, "depends largely upon climatic conditions," so any school that conducted primary training should be optimally sited to take into account air velocities and currents, and climate. "There are only two or three places in this country that can be so used to advantage," Scriven maintained. Specifically, at San Diego and Augusta, Georgia, "beginner(s) should spend three or four months learning the basics of flying land planes and seaplanes." At the end of this preliminary (primary) phase, a pilot took the test for his FAI certificate.

The trainee then entered the second, the "basic" phase. The Signal Corps hoped to reduce its dependence on civilian operators because it believed military tactics differed from those in commercial aviation. The beginnings of combat flying would be introduced in the intermediate phase. Lieutenant Arnold testified before Congress that when he learned to fly at Dayton, "in 10 days I was flying by myself; the rest of the time I was merely practicing.... It does not take long to teach a man to fly but it takes a long time to make a military aviator." Specifications for new military aircraft were being directed at reconnaissance and cross-country missions. The latter "is very difficult," explained Arnold, "because of the varying currents of air — the wind currents. It is necessary for an aviator to know about the wind currents by the way they strike his machine; he must be able to follow his map and to know where he is; he must learn to know from the sound of his engine whether it is running all right, and many other things." Airmen anticipated that it would take approximately one year for an officer to complete the intermediate (basic) stage and obtain his military license.

The third and final phase of instruction, part of which was not then or ever would be well realized, lay at the heart of the Signal Corps proposal: advanced training in cross-country and reconnaissance missions with a combined arms component. In this respect, the Signal Corps intended a program quite different from ad hoc field operations involving a few solitary men and airplanes on the Texas border. The Signal Corps leadership argued that cooperation with ground troops should be built into flight training in an organized, institutionally based program. The first permanent aeronautics center "should be placed at a locality where the Government already possesses a sufficient amount of land; and first of all...it should be established at a permanent Army station, where the aeronautical personnel can constantly have the advantage of serving with other troops." Although unfavorable atmospheric conditions during parts of the year foreclosed primary training, San Antonio met the requirements for advanced training. There, a man with a military license, according to General Scriven, would become an "aviator of military value under war conditions." Moreover, "I believe it to be a fact," he continued, "that aviation and the training-school establishments should be with the troops; that the aeroplane and its use may be understood by the men who are going to use it, and that the officers of the Army should become familiar with it."

Although the Signal Corps put forth a plan both to centralize and expand the existing elements of flight training and experimentation, the point of departure came from its proposal to train air and ground forces together. The Army defined the air force mission as battlefield support, but it had never formalized training for that role any more than air squadrons had been organically a part of ground units. Had the Signal Corps proposals gone into effect, the system of flight training, the development of doctrine, and the relationship between Army ground and air forces might have evolved more like naval aviation, whose mission was tied closely to the fleet.[12]

Chief of Staff General Wood endorsed the idea of an aeronautical center at San Antonio. But a mild economic depression of late 1913 deepened into 1914, and a slim aviation budget legitimized his caution that "economy and the best interests of the service demand that the construction should be as limited in amount and simple in character as possible, consistent with the needs of the situation."[13] In order to proceed with the center, for the present the War Department would have to restrict expansion in other areas.

Despite all the hopeful indications, the proposal for a permanent aeronautical center did not make its way into law. Congress failed to enact any legislation to establish a government-owned facility until it authorized a California site — again rented — in the War Department appropriations bill for 1917.[14] In the meantime, the Chief Signal Officer continued futilely to beat the drum for aeronautics. And, in his 1914 and 1915 annual reports, the Inspector General opined that the lack of a modern, permanent training facility and the

Prelude to War

scattering of air forces between North Island and the old post in the city of San Diego where the 1st Aero Squadron was stationed seriously hampered aviation development. Although the 1st Aero Squadron in fact moved to San Antonio in 1915, in no way did that garrison constitute an aeronautical center as envisioned by the Signal Corps. San Antonio would become the mecca for air training, but not until 1931.

Ironically, the huge national system of five aviation centers and smaller installations in each state that the Signal Corps proposed in 1912 might have been more easily realized than the more cautious recommendations of 1913 because members of Congress and civic organizations around the country were eager to secure government facilities for local communities. But Congress funded none of the Signal Corps schemes, so most Army aviation remained at North Island — on rented property — until the United States hovered on the precipice of war.

Training Excursions into the Field

The melding of air and ground training occurred sporadically rather than systematically as the Signal Corps had proposed. The first opportunity came in August 1912. That year the Joint Maneuvers of the Regular Army and the Organized Militia, held in Connecticut, added an aviation component. Airmen expected to fly new planes, but manufacturers struggling over the revised specifications for military aircraft were unable to deliver them on time, so the pilots took two of the Aviation School's well-used training machines. The maneuvers consisted of two problems. In the first, the airplanes served for instruction and reconnaissance; in the second, they were attached to one of the two opposing forces in the war game. The aim of the "red" forces was to cut off the water supply of New York City; the "blue" army, with airplanes, defended.[15]

For Problem 1, Army pilots with their old aircraft and a National Guardsman with his Curtiss biplane were scheduled for reconnaissance flights on August 11. High winds in the morning kept the guard pilot from taking off. By the afternoon, the airmen managed some high-altitude flights, then tried to demonstrate aerial communication with the wireless apparatus. This attempt succeeded even less well than the reconnaissance flights. One of the Army pilots, Benjamin Foulois, flew the Burgess-Wright carrying a wireless telegraph, but the weather was so rough that he could not take his attention away from piloting long enough to operate the communications equipment. Also, because he did not carry an observer with field glasses, Foulois could barely distinguish the Artillery from Signal Corps troops.[16] Unfortunately, Problem 1 portended ill for the ensuing performance of the squadron. During the remainder of the exercise, weather continued unreliable for flying; aircraft broke down; and the countryside where they operated, as described by another

1907-1917

The thickly wooded area in which the 1912 Connecticut maneuvers were held

of the pilots, Thomas Milling, was "thickly wooded, densely populated and very hilly and rocky."[17]

After the maneuvers, a blizzard of reports analyzed the experience and offered recommendations. Air officers pointed to a lack of coordination between the several headquarters and the air squadron. They also suggested that because ground commanders had few notions about how to employ airplanes, airmen should be attached to each headquarters and landing fields placed nearby. They complained that their airplanes were woefully underpowered and that much better brakes were needed in order to land in small spaces over difficult terrain. Also, aircraft in the field should be well supplied with spare parts. It was abundantly clear that all military aircraft engaged in reconnaissance should carry two men, because no single individual could pilot, observe, and communicate at the same time. As for support, the pilot should have direct authority over the enlisted troops working on his airplane, and each airplane should have a crew of at least five enlisted men and one senior noncommissioned officer mechanic. The pilots expressed disgust over engine performance and urged their superiors to test engines from companies besides Curtiss and Wright. "*Real* military aviation work," proclaimed one, "must necessarily be cross-country, which needs above all an engine that *never stops* except at the will of the pilot."[18] Apparently the officers found little to criticize in the training that brought them to the maneuvers, but Geiger thought more time should be devoted to the study and practice of military reconnaissance.

The exercises were not without their achievements, and the officers who participated expressed optimism regarding participation in future military operations. Problem 2, according to Colonel Reber, demonstrated that the information obtained by aerial surveillance was much more accurate than that gathered by cavalry patrols used by the other side. He concluded that "to be without an aviation squadron attached to a force which is operating against one that is supplied with aeroplanes is to place an almost insurmountable handicap on the force."[19] In his annual report for 1912, the Chief Signal Officer summed up the utility of the airplane in the instance: "The results obtained by this very inferior equipment proved to be of considerable value, but they should not be accepted by the Army as any criterion for what may be expected of well trained pilots and modern military aeroplanes."[20]

The maneuvers highlighted areas where coordination between air forces and field armies might be better achieved. But reforms in command and control could not eliminate technological barriers, even with "modern military aeroplanes." Given the equipment, it is unlikely that a offensive role for aircraft could have been demonstrated. The Army was buying bigger, stronger airplanes with higher performance engines, but other deterrents to mission success — weather and terrain, for instance — would be difficult to circumvent with the still-primitive equipment. Airmen themselves only obliquely recognized that technical limitations might pose the most serious challenge to their claims.

Because of his assignment to the Division of Militia Affairs, Foulois had remained outside the mainstream of aviation developments until he took part in the 1912 Connecticut maneuvers. Chandler petitioned for his reappointment to the Aviation School, citing the recent Army appropriation bill that exempted airmen from the usual four-year restrictions. Reber joined in pressing for Foulois's reassignment, perhaps exaggeratedly calling him "one of the most if not the most skillful aviator in the United States Army."[21] The Chief of Staff denied the request to station Foulois at North Island, probably on the advice of Signal Corps leaders who wished to diversify aviation. They were then, it may be recalled, arguing before Congress for an increase in the number of training centers and officers authorized for aviation.

> It is believed in this office that it is the present policy of the War Department to establish, so far as is practicable, a reserve in all branches of the Army. If this view is correct, men who have become skilled as aviators should form a reserve, but with duties much wider in scope than is now possible at the College Park station, as, for instance, the establishment of aeronautical centers for the organized militia or at regular Army centers, such as in the Philippines or at Fort Leavenworth. Officers assigned to duty of this nature would naturally be detailed in the Signal Corps, but at present the aviation corps is so small and the number of men and officers engaged in the work necessarily so limited, that any wider application of this plan cannot be considered until another season at least.[22]

This decentralization "policy" effectively returned Foulois to aviation, but at Fort Leavenworth, where he was told to establish an aviation center. The idea of establishing a center among the Army schools was a good one, but Leavenworth did not become an aeronautical center, then or ever. Once more Foulois was the lone airman, with an airplane he was unaccustomed to flying. This time it was a Wright C, and again he wrote to Orville Wright asking for help. Foulois was especially anxious to do well, as on this occasion he was not alone, strictly speaking, because he was surrounded by Army men curious to see what airplanes could do. He was correct in his fear that "the flying that I may do here, will be watched and freely criticized."[23]

In February 1913, the Aviation School contingent in Augusta, Georgia, along with two pilots who had been training in Palm Beach, Florida, transferred to Texas City, Texas, to join ground forces on duty along the border. In so doing the school relinquished its primary training mission to assume operational status as the 1st Aero Squadron. According to Milling, "training took on different objectives. We began concentrating on cross-country orientation, reconnaissance missions with the troops on the ground, and landings and take-offs on difficult terrain."[24] Some beginners clearly had not attained the necessary level of proficiency, so they were released to attend classes at the manufacturing plants. Also, the training airplanes fared badly. Although it could not be said that Army pilots were first blooded in this military operation, several officers secured a rare opportunity for advanced training. They left the border in June.[25]

Also with disappointing results, the Signal Corps attempted to extend operational training by contributing to the War Department's backing of the Taft administration's foreign policy. In the era of "dollar diplomacy," Latin America and the Far East became central to the expansion of American

S.C. No. 11, a Wright Model C, was among the aircraft used by the 1st Aero Squadron when the Aviation School transferred to Texas City, Texas, in 1913.

Prelude to War

commercial enterprise. Though diplomatic and economic successes did not depend on a show of military force, the Navy, in particular, relied heavily on the Monroe Doctrine (Latin America) and Open Door policy (China) to justify its building programs and to promote the efficacy of battle fleets — America's traditional first line of defense — standing against any possible threats to the American sphere of interest. Drawn up in 1911, the Orange Plan consummated the Navy General Board's outline for defending the Philippines and other American possessions in the Pacific in the event of a conflict with Japan. The Army, on the other hand, held firmly to its continental perspective, lacking the philosophical rationale provided by naval theoretician Alfred Thayer Mahan to articulate a global maritime role. President Taft, formerly a High Commissioner in the Philippines, essentially supported the Army viewpoint that war with Japan was unlikely and that a base in the Philippines would be largely indefensible. Throughout this period, therefore, the military presence in the Pacific remained subdued.[26]

Nonetheless, despite the Army's lukewarm enthusiasm for defense of overseas possessions, during 1913 the Signal Corps placed an aviation element in the Philippines and in Hawaii. Both experiments enjoyed brief tenures as, once again, climatic and topographical conditions all but prohibited flying. The Aeronautical Division did not foresee those drawbacks when it set up the operations. Rather, as Chief Signal Officer Scriven informed the Chief of Staff in the summer of 1911, the weather appeared "fairly favorable to this work," and a "good training station for aeroplanes could readily be established near Manila" and another, in time of war, on one of the proximate islands.[27]

Lieutenant Lahm, who was stationed at the Cavalry School at Fort Riley, Kansas, when his regiment was ordered to the Philippines, agreed to take over flight training, made arrangements to ship Wright airplanes, and assumed command of the aviation school at Fort William McKinley in Manila on March 9, 1912.[28] Two enlisted men arrived from the United States, one of whom, Cpl. Vernon L. Burge, had been with the balloon detachment in 1907 and subsequently at College Park. Five men already stationed in the Philippines joined the enlisted detachment. Securing officers for flight training was another matter. When Lahm arrived, none had been assigned. He therefore began teaching Lt. Moss L. Love, stationed at Fort McKinley, and Corporal Burge. Love and Burge both completed the tests for their FAI licenses by June. News of Burge's accomplishment, however, met with a sharp rebuke for infraction against War Department "policy":

> It is not the policy of the War Department to train enlisted men in flying aeroplanes. Their military training is such that very few enlisted men are qualified to observe military operations and render accurate and intelligent reports of what they see from an aeroplane. Another objection is, that very few enlisted men have sufficient knowledge of mechanics to appreciate the stresses to which an aeroplane is subjected during certain maneuvers.[29]

If such a policy existed, no regulation encoded it. Although the War Department had not authorized enlisted pilots, neither had it prohibited them. Burge was only the first of a number of enlisted men who became pilots in the prewar years.[30] The Aviation School encouraged some noncommissioned officers to apply for pilot training and forwarded select applications, which the Chief Signal Officer usually approved.[31]

For a year, although the so-called school had airplanes and instructors and both the Commanding Officer and the Chief Signal Officer of the Philippine Division requested more trainees, nobody volunteered.[32] Finally, in March 1913 Lieutenants C. G. Chapman, Herbert Dargue, and C. Perry Rich began training, and by July they had obtained their Military Aviator ratings. In June, Captain Chandler joined them as the Signal Corps officer in charge.

Because of "new and unfavorable weather and climatic conditions" in the Philippines, the airmen were apparently most comfortable with the old standby Wright B, which the Aeronautical Division expected would "be used in future preferably for experimental and training purposes only."[33] The Philippine Division requested another Wright B and a set of pontoons, plus a higher-powered plane for long-distance reconnaissance. Instead, since the Signal Corps was phasing out the Bs, two Wright Cs arrived over the next few months, despite the fact that Milling, who was not in the Philippines, was considered to be the only one capable of flying the plane.[34]

All the aircraft were equipped with pontoons and flown over the bay from the Pasay beach at the Manila Polo Club. During the rainy season the flying field at Fort McKinley often sat submerged under water and a layer of mud, conditions that had prohibited flying the previous year. Thereafter, between typhoons, the seaplanes took off from Pasay. On August 28 Dargue crashed in the Type B; in September Lahm in a Type C dovetailed into Manila Bay, demolishing the airplane. Both men survived the accidents. But when Lieutenant Rich, in the second Wright C, came in for a water landing on November 14, he was killed and the airplane destroyed when they plunged into the bay.[35]

Although the Signal Corps would remove them from the inventory after nearly everybody agreed they were unsound, the Wright pushers served as primary training airplanes in the Philippines. Curtiss aircraft, which the Philippines contingent also requested, were sent to Hawaii instead. A Burgess seaplane, S.C. No. 17, a high-powered airplane capable of long-distance reconnaissance in windy weather, went, after modifications, to Corregidor Island for service with the Coast Defenses.

By the end of December 1913 training in the Philippines barely limped along. Only Dargue remained among the officers — Rich was dead, Chapman had returned to the United States, and both Lahm and Chandler had been recalled from flight duty. Three training airplanes had been destroyed. A month later Lt. Henry B. Post, who had not been officially part of the training

Prelude to War

Among the flight trainees of the Philippine Division were Lieutenants Moss L. Love (*above*), Carleton G. Chapman (*above right*), and Herbert A. Dargue (*right*).

detachment, crashed into Manila Bay and died. At this point, lacking anybody to train or be trained, it is questionable that a school could be said to exist at all, although a few members of the 1st Aero Squadron and its equipment remained in the Philippine Department. The Defense Board in the Philippine Islands importuned the Signal Corps to continue training, but the Chief's Office maintained that the Aviation School at North Island had been established specifically for the purpose of training and could offer a wide-ranging course of instruction at facilities and in weather appropriate for beginning pilots.[36]

By the end of July 1914, the commanding general of the Philippine Department had one seaplane at Fort Mills as part of the defense of Manila Bay, one Military Aviator, one noncommissioned pilot, and an aviation detachment of eight enlisted men; he spoke of the school at Fort McKinley in the past tense.[37] Dargue continued nevertheless to fly No. 17 through the winter, and participated in tests with the Coast Artillery, until an accident in

1907-1917

Fortunately, when Lt. Frank Lahm's Wright Model C crashed into Manila Bay in the Philippines on September 12, 1913, no one was injured. In another crash a few months later, however, the Philippine Department lost one of its pilot trainees, Lt. Henry B. Post.

January 1915 destroyed the last available airplane. Dargue and his passenger survived unhurt, but Dargue, too, left the Philippines immediately thereafter for the Aviation School at North Island.

In June 1913, the Signal Corps organized a second, even more transitory, Pacific outpost for flight training — Fort Kamehameha in the Hawaiian Islands. Lieutenant Geiger commanded an entourage of twelve enlisted men and one civilian mechanic.[38] He took with him No. 8, a Curtiss E, and a Curtiss G tractor, No. 21.[39] The local population greeted Geiger's arrival as though he were a performer in a one-man circus. Since he had nobody to instruct, he might as well have been an exhibition pilot. Moreover, his two Curtiss airplanes required modifications, which did little to make them acceptable; the harbor at Fort Kamehameha was almost too shallow to use; the winds were treacherous; and two months after he arrived, the enlisted detachment left for San Diego.[40] From the Office of the Chief Signal Officer came the parting word that "it is not the intention of this office to permanently abandon the aviation station at Fort Kamehameha."[41] But the two airplanes were condemned and sold by the end of the year, and General Scriven did not expect to replace them, short of a "grave emergency" in the area. Geiger, too, returned to North Island.[42] As happened in the Philippines, the poor condition of the airplanes precluded any hope of their performing well. By the middle of 1914 the Signal Corps had abandoned training in Hawaii, and only a shred of an aviation presence remained in the Philippines.

Prelude to War

Struggling Out of Isolation

The European war had been building since the Franco-Prussian war of the 1870s, and Americans were not so blinkered as to have no sense of impending conflict. But they utterly failed to foresee the extent of American military involvement. In 1909, Lieutenant Arnold crossed the Indian Ocean aboard ship with several British and German officers with whom he became friendly. He recalled that "everybody knew a war was coming, and both British and German knew [they] couldn't lose it. Every phase of the thing was discussed and hashed over again, except the fact that the United States might be a factor."[43] A full seven years later — two years into the war — another pilot made the astonishing statement that he and his fellows at the Signal Corps Aviation School seldom discussed the role of aircraft in combat: "We knew they were doing these things, but I would say we were not very conscious of it. We were not very conscious of the war itself."[44]

Airmen's preoccupation with world events seems to have been restricted to the growth in foreign aviation and developments in foreign technology. Those matters they conveyed to the American military and political leadership in order to spur spending at home. When the Signal Corps petitioned Congress for $500,000 in 1909, it reasoned that such monies would "shortly place us on at least an equal footing with the European nations, which are devoting so much time and money to this branch of warfare."[45] Congress appropriated no funds for another two years, but the Aeronautical Division continued to watch and record the progress of foreign aeronautics. The Signal Corps's response to House Resolution 448 forced policymakers to recognize that the lack of government funding, or at least a partial subsidy, comparatively straightjacketed Army aviation. With a partnership between government and the nascent industry, it appeared that France, for example, had leaped ahead of the country where heavier-than-air flight was born. Without success, in 1912 air officers asked to procure the most successful French military-type airplanes from the large military aviation contest held in France that year.[46] Although the Army did not approve purchase of European airframes, by the mid-teens it bought foreign engines.

The Aviation Section took note of other aircraft devices manufactured abroad. Yet the Aviation School Experimental Department, the NACA, and briefly the flying field at Mineola (which opened on Long Island in the summer of 1916 and was used for testing the next winter and spring before it was returned to a training station) explored research-and-development projects only superficially. Not surprisingly, the Signal Corps enthused over reports in 1915 about the use of wireless telegraphy on the larger English and German airplanes.[47] By this second year of the war, Americans had identified a great variety of foreign equipment, compared to the limited amount in production or used in training in the United States. Europeans typically classed aircraft into three types — reconnaissance, combat, and pursuit — as opposed to the single reconnaissance airplane in the United States. The U.S. Army had only "taken up the question" of the other types, and by the time another year had passed, the development of "suitable battle machines and speed scouts" was still pending.[48]

If Americans found detailed information regarding foreign technology difficult to obtain even before the war, August 1914 brought nearly an information blackout. That fall, General Scriven requested that a Signal Corps officer be included among the American military observers authorized by the belligerent powers. The Adjutant General replied that neither the German army nor the Allies had any vacancies.[49] Nevertheless, a place must be found for an air officer, Colonel Reber charged: "The importance and magnitude of the operations of the flying corps of these armies are matters of common repute, and it is to be pointed out that none of our military observers of the various armies have had enough experience in aviation to make their reports of value."[50] Fortunately, in early 1915 Col. George O. Squier, while posted as military attaché to London, took note of French and British aviation activities. He compiled at least one report on the British air service and forwarded a copy of the Royal Flying Corps Training Manual to the Signal Corps Aviation School. In February and again at the end of that summer, Captain Chandler went to France to investigate the status of aerostation. He found an increased use of observation balloons by both German and French armies.[51]

Incredibly, military men learned most about the war, including the types and employment of aircraft, not through official channels but from published sources such as magazines and newspapers that were, according to Colonel Reber, of little technical or tactical value. The Aviation Section read about types of combat aircraft; training methods; the locales, numbers, and types of training schools; and the personnel requirements for airmen. An article entitled "Schools and Their Methods" appeared in July 1915 in the London-based *Aeronautics*. "There really appears to us little to choose" in type of machine employed, opined the author. "One instructor may achieve better results with a dual control machine; another may swear by a tractor; a third pins his hopes on a pusher boxkite. The type of machine, indeed, as results have proved, is of much less importance than the quality and method of instruction."[52] American airmen knew well the importance of competent instructors, but considering their own hard-fought battles over the differing control systems and the problems with the pusher airplane, it must have been dispiriting to hear those matters dismissed by those near the battlefield. That same summer the British magazine called upon great numbers of "athletic young men in the country fully possessed of the sporting instinct which helps to make the good aviator" to join the British flying service.[53] At a time when American airmen too hungered for greater numbers but accepted few for pilot training, it must also have been unsettling to realize that in Britain, by the summer of 1915, the urge to fly was close to sufficient qualification.

Prelude to War

Foreign military officers sent to the United States for aeronautical training brought fragmentary information on aviation developments on the continent.[54] Communiqués also arrived from Americans who, during the period of U.S. neutrality, aided in Allied relief services, drove ambulances, and fought in artillery, rifle, and machine gun units, including some of the most distinguished French and British regiments.

Americans also flew in combat. The idea of forming a separate squadron of American airmen fighting with the French was first broached in 1915. In late April 1916 the *Lafayette Escadrille*, comprised of pilots already in service with the French, assembled at the great aviation depot at Le Plessis–Belleville. This squadron saw duty at the front with French forces until it became an American unit in February 1918. Several of its young pilots kept diaries and posted letters and reports back to the United States. They offered invaluable accounts of French training methods that would serve American airmen well in the days to come. Until his death in combat April 16, 1917, for example, Lt. Edmond C. C. Genet flew with the *Lafayette Escadrille*. Through the summer and into the fall of 1916, Genet poured out a stream of material through his American commanding officer. He began collecting "information valuable to U.S. Service about aviation etc.," went on to record data about aircraft flown by both sides, wrote an article on the Vickers machine gun, prepared a report about the Buc school at which he trained, and generally was "taking all the notes" he could and "writing them up in the few spare minutes outside of work hours. They'll all help . . . the U.S.," he noted in his diary, "so I'll do all I can."[55] In the autumn of 1916, another pilot, Lt. James McConnell, wrote a detailed account of his experiences entitled "Flying for France."

In November 1916 *Aviation and Aeronautical Engineering* published "How France Trains Pilot Aviators," written by an unnamed sergeant with the *Layfayette Escadrille* at Verdun.[56] French primary training depended upon the type of machine used; because their many schools employed a multitude of aircraft, they enforced no single method of instruction. The procedure for flying the larger biplanes generally resembled that used at the Signal Corps Aviation School: the *eleve pilote* went up in a dual-control airplane with an instructor, then he flew alone in a less powerful machine while he increased in competence and learned the maneuvers necessary to qualify himself for the military brevet, equivalent to the Military Aviator rating in the United States.

Training for the *avion de chasse*, or fighter airplane, substantially differed. The first phase somewhat resembled the old American Curtiss method. The student began with a sturdy underpowered airplane, nicknamed the "Penguin" because of its short wings and inability to fly. The pilot went up and down the field with an instructor in a dual-control airplane, then in a faster single-seat Penguin. In the next step he "flew" a low-powered monoplane of the Bleriot type that could rise barely three feet off the ground. From short hops he passed on to classes where he learned to go higher, land, and fly evenly. In stages,

1907-1917

The Breese Penguin

flying increasingly higher-powered machines, he became an official *pilote-aviateur* after passing three cross-country flight tests.

Next, at a school of *perfectionnement*, pilots trained on a Morane, Bleriot, or Nieuport. They also practiced with machine guns, fired at targets from the air, and flew in formation. For the first time a pilot attempted "fancy flying" — "how to loop the loop, slide on his wings or tail, go into corkscrews and, more important, to get out of them, and he is encouraged to try new stunts." At that point he joined an *escadrille* at the front where he began "his active service in the war, which, if he survives the course, is the best school of all."[57]

By comparison, the U.S. Army officially trained at one camp for one mission — reconnaissance — with the concept but without the regulations or formalities of a staged training process. Reading about fellow pilots training to fly fighters must have heightened the frustration and sense of isolation among American airmen, whose organizational position within the Signal Corps largely dictated the doctrine by which their mission was defined. In 1913 the Office of the Chief Signal Officer inserted "Report on Progress Made in Aeronautics in the Army Since about March 1, 1913" into the public record of the hearings before the House Committee on Military Affairs. This document stated that "under law the Signal Corps supervises the service of communication, observation, and reconnaissance as effected through wire and wireless telegraph and the telephone. Aeronautics and aviation in military affairs are merely an added means of communication, observation, and reconnaissance and ought to be coordinated with and subordinated to the general service of information."[58]

In their book *How Our Army Grew Wings*, Charles Chandler and Frank Lahm admitted that notwithstanding the "official pronouncements" that aircraft served only for reconnaissance, "the few air officers discussed among themselves the possibilities and probabilities of offensive air warfare when engineering progress would produce improved flight performances — especially weight-carrying."[59] During the 1913 hearings, Capt. Paul W. Beck, the only

Prelude to War

pilot to testify in favor of a bill to create a separate air service, listed four functions for aircraft — reconnaissance, fire control for artillery, offensive action, and transportation. But, he alleged, the Signal Corps mistakenly acknowledged only the first. For those impolitic remarks, the Chief Signal Officer publicly scolded him.

Lieutenant Milling later rued the official unwillingness to experiment with devices such as the Lewis gun that might have been used offensively. "We, who were associated with aviation, felt that the airplane was capable of offensive use if properly designed and armed. Unfortunately the guns available in the Army were neither designed for nor suited to mounting on the airplanes we were flying."[60] When the Lewis gun was tested at College Park, airmen were sufficiently impressed that they requested ten guns for the training program. The Chief of Ordnance refused the request, insisting that Army units could only buy materiel already carried on the inventory. "After we entered the war in 1917," Milling continued, "we found it necessary to purchase these same guns from [Lewis] to equip our planes. This short-sightedness was one of the major reasons why we had neither combat aircraft, nor combat crews in our military structure at the beginning of World War I."[61] As long as the Army offered little support or subsidy for new weaponry (a restriction that crossed all branches and combat arms), and the air arm stayed within the Signal Corps, aviation proponents within the military lacked strategic and tactical doctrine that would have justified a more broadly based testing and training program.

1st Lt. Thomas DeW. Milling stands before S.C. 26, a Burgess Model H.

1907-1917

Breakout

The immaturity of American military aviation in the face of European models of demonstrably greater sophistication could be explained in part by the lack of clear direction in American foreign policy between 1913 and 1916, as a set of Progressive ideals and rhetoric came into collision (but ultimately collusion) with increasing military threats. Within a year of his inauguration in 1913, President Wilson faced not only a European conflict that all expected to be of short duration, but, much closer to home, a depressed U.S. economy and the very real possibility of war with Mexico. Wilson had come into office committed to domestic reform, with little interest in military affairs, but soon he had to take account of not one war but two.

The overt outbreak of hostilities in Europe shook the press, the public, and the War Department from their lethargy regarding the condition of the nation's aerial defenses, as calls for a general buildup of the Army and Navy gathered steam. Even so, among the loud and conflicting clamor for a national defense policy, the seemingly reasonable demands of Army aviation spoke in a tiny voice. In late 1914, Representative A. P. Gardner of Massachusetts lobbed the first volley in the preparedness campaign when he called for an investigation into the state of the nation's military establishment. The following year he dramatized the plight of the Army's pitiful enlisted reserve by assembling *all* of its membership in the dining room of the Willard Hotel in Washington, D.C. — a force sixteen men strong.

Ex-President Theodore Roosevelt rose once more to lead the charge, this time as the foremost preparedness spokesman. Stimson and Wood soon added their voices. In publications and speeches, members of defense-oriented groups, most notably the National Security League, the American Defense Society, and the Navy League, denounced the failure to strengthen America militarily. In 1915, the Aero Club of America (none of whose members, Philip Roosevelt remarked sarcastically to his cousin Theodore, "in fact can fly, or know why a machine does fly") moved into the spotlight, again lobbying forcefully on behalf of military aviation and for preparedness in general. The Club established the National Aeroplane Fund to train and equip airmen for the state militias, and in May 1916 it organized a Preparedness Tournament to demonstrate what the airplane could do. The sinking of the *Lusitania* in the spring of 1915, coupled with political pressure and the burgeoning popularity of preparedness, finally brought the Wilson administration to terms with the movement.[62]

In Congress, throughout 1915 and into the summer of 1916, controversy over military legislation seethed. Should the nation create an aviation reserve and provide systematic air training to the National Guard? Those issues were subsumed under the general debate over increasing all military reserves and federalizing the militia. Although, as noted earlier, airmen held decidedly mixed views about training National Guardsmen, the Chief Signal Officer had supported the practice, stating in 1912 that it would be the Guard "on whom the Government is to depend for the large numbers of aviators that would be required in case of an extended war."[63] The continued advocacy by General Allen and his successor General Scriven on behalf of expanded air training — for both civilians and the Regular Army — translated into a need for vastly increased funding for men, machines, and facilities.

Failure, not reason, brought remedial action. When the 1st Aero Squadron joined the Army strung out along the Mexican border during the Punitive Expedition, airmen were dreadfully ill-equipped and thus made a very poor showing. In January 1915, General Scriven angrily described the decrepit and obsolete aircraft his men were flying. Rectifying the situation with the "ridiculously small" planned appropriation of $300,000, he railed at the Secretary of War, "may result in unnecessary loss of life" since pilots were trapped with "obsolete, old, or patched-up aeroplanes, comparable, some of them, to the Deacon's one-horse shay."[64] But not until the following year, when the pitiful results of America's aerial exploits in Mexico became widely known, did Congress respond.[65] Enacted on March 31, 1916, the Urgent Deficiency Act provided $500,000 for Army aviation in addition to the $300,000 appropriated for 1916. It was followed in August by the stunning 1917 appropriation of $13,281,666 and $600,000 for the purchase of training sites.

For the first time, Army aviation could draw up comprehensive plans with a chance of implementing them. The Deficiency Act allowed the purchase of new training planes and new equipment for the Aero Squadron. The Signal Corps issued specifications for primary and advanced trainers, mostly single-engine tractors with tandem seating and dep control. Virtually all the training aircraft received in the ten or so months before the country went to war were JN–4-series airplanes furnished by the Curtiss Company. Several, equipped with both the Curtiss shoulder yoke and dep control, permitted transition training. In early October 1916, the Signal Corps placed an unprecedentedly large order for thirty-six advanced JN–4B trainers, and on the thirty-first it ordered thirty-six more.

From 1909 onward, but particularly as pressure for war gathered momentum, Signal Corps aviation suffered even more acutely from the shortage of men than the shortage of aircraft. With enough money and American production-line genius, materiel could be produced relatively quickly compared to the slow, drawn-out process of flight training. Throughout 1915 and 1916, the press clamored to increase the number of trained pilots. In June 1915, *Aeronautics* contended that the "pessimism in the press" concerning the few Army airplanes was "not warranted by facts." A reporter quoted a "high military authority" as saying: "We have at the present plenty of machines for our needs and can get machines whenever we want them. Everybody has overlooked the main point. It is not the number of machines that any nation has, but the number of trained fliers."[66] The *Boston Transcript* seconded the point:

1907-1917

Maj. Frank Lahm (*far right*) prepares for a training flight in one of the new JN-4s obtained under the terms of the Urgent Deficiency Act of 1916.

"We might conceivably supply ourselves with aeroplanes, but where would we get the pilots to operate them and the observers capable of rendering them of value? It takes something like six months to turn out a first-class operator, and competent observers could hardly be trained in less time."[67] A year later, in addressing the national convention of the Navy League, Professor Alexander Graham Bell cited the lack of trained pilots as evidence of America's poor state of readiness to defend itself: "But there is one element in relation to the flying machine that we are not producing, that we cannot produce in an emergency, and that is the men. We can produce machines, but not the aviators. That takes time." He went on, "Where are we to get the men, and where are we to train them?"[68] Similarly, the Chief Signal Officer declared: "The war in Europe has emphasized the absolute necessity for an adequate aviation service. The greater need at such a time ... will be for trained men as pilots and observers.... The training of men ... is the crying need of the present time."[69] The same clarion call would echo twenty years later on the eve of World War II.

In loosening the purse strings in yearly appropriations and the Urgent Deficiency Act, and in passing the National Defense Act of June 3, 1916, which unleashed the mobilization process, Congress finally bowed to pressure from the preparedness forces whose calls for action reached a crescendo over the German submarine campaign. By no stretch, however, did the National Defense Act effectively put the country on a wartime footing. It modestly increased the Regular Army strength and expanded and federalized the National Guard. On paper at least, its effect upon the Aviation Section appeared to be more dramatic.

With new authorizations in hand, General Scriven saw the need to act swiftly: "A strong personnel is needed for training and to take advantage of the present wave of enthusiasm which may not last."[70] The act increased the ceiling on officers for aviation and eliminated restrictions of age and rank. It expanded

Prelude to War

the training program as a result of the newly created aviation component in the Signal Officers' Reserve Corps and Enlisted Reserve Corps. To train men coming into these organizations, the Aviation Section signed agreements with civilian manufacturers and aviation schools obligated to comply with requirements pertaining to flying fields, training machines, instructional programs, and competency tests. The government oversaw the primary training of reserve officers and sergeants on flying duty at these schools, paid the tuition for each student who satisfactorily passed his initial flight test, and an additional amount for each man who passed the reserve military aviator test. In July 1916, a new Army facility opened at Mineola, New York, whose purpose was to procure pilots for duty on the border, and to house the New York National Guard's First Aero Company.[71] Shortly thereafter, a similar operation opened in Chicago. Because the weather limited flying hours at these northern schools, in a familiar move, training operations shifted south for the winter. The Chicago school moved to Memphis, Tennessee. Although Mineola did not relocate, it became principally a testing rather than a training station until April 1917.[72]

Other means of producing military pilots grew out of the preparedness movement. After passage of the National Defense Act, the focus of attention turned to universal military training. Privately supported military training associations, several of which trained pilots, sprang up around the country. Another source for securing reserve officers came through the reserve officers training units in colleges. In a widely circulated letter of 1913, General Wood proposed that college presidents institute military training programs in the universities and that summer camps be established to give graduating high school students military training. A number of summer camps were operating by 1915, and an advisory committee of presidents from several prestigious colleges met to consider an air training network. Some of the colleges looked to the government to qualify their young men. While these private efforts and college programs failed to supply a large number of people, they relieved some of the pressure from the still small cadre of Army officers.[73]

Needless to say, the Aviation Section wanted most desperately to enlarge the flight training program of the Regular Army. To seize the advantage offered by the recent legislation, the Signal Corps dredged up its proposal of 1912, putting forth a plan to build a new training facility in the East, another in the central part of the country, and a third on the Pacific coast. When the United States entered the war in April 1917, however, this plan had not come to fruition. The Army still trained only on North Island, where it had been for several years.[74]

The Aviation School did expand insofar as it inaugurated a three-month course for field grade officers, in part to prepare some of them to assume administrative jobs with the Aviation Section. By mid-1916 the course enrolled forty-five officers in addition to the training offered to officers of the Marine

Corps and some enlisted men.[75] Throughout the summer of 1916 the school practiced sending radio communications via airplanes; a course in radio telegraphy became an integral part of the curriculum. More to their liking, officers who passed the test for the Junior Military Aviator rating became eligible for a new night flying course.[76]

By mid-1916, most Army airmen were involved in some type of training. But, asserted Foulois, then commanding the 1st Aero Squadron, not only was the Army's only air force not equipped for training, it "should not be required to do a work that is a proper function of the school."[77] Yet the Signal Corps had long argued that advanced training should take place with ground troops. The Office of the Chief Signal Officer planned therefore for the aviation depot at Fort Sam Houston in San Antonio, Texas, to become an advanced training station for the 1st Aero Squadron and the newly organized 3d Aero Squadron based there.[78] The generous appropriations permitted additional equipment for use by the forces operating in Mexico and for expansion of the advanced training program.[79] Because the squadron had been flying Curtiss reconnaissance airplanes equipped with the Curtiss shoulder yoke, the first new JN–4 trainers sent south included both the yoke and the dep control.[80]

On the Brink of War

In his annual report for 1914, Chief Signal Officer Scriven stated categorically: "If the future shows that attack from the sky is effective and terrible, as may prove to be the case, it is evident that, like the rain, it must fall upon the just and upon the unjust, and it may be supposed will therefore become taboo to all civilized people, and forbidden at least by paper agreements."[81] He stated, even more bluntly, before the House Committee on Military Affairs in December: "As a fighting machine the airplane has not justified its existence."[82]

Officially, from its beginning through the years of neutrality, the Army operated under the premise that aviation was doctrinally and organizationally committed to a single function: reconnaissance. Moreover, insufficient evidence had accumulated to corroborate other capabilities. Airmen could do little more than mutter among themselves about the offensive role of aircraft that was being tried out on the continent, since they could offer little measure of proof of their own.

It lay with outsiders to take a more aggressive stand. In his talk before the National Convention of the Navy League in 1916, Alexander Graham Bell expostulated: "Navies do not protect against aerial attack. This also we know — that heavier-than-air flying machines of the aeroplane type have crossed right over the heads of armies, of millions of men, armed with the most modern weapons of destruction, and have raided places in the rear. Armies do not protect against aerial war."[83] The same summer, Secretary of War Baker conceded that soon the United States might add armed aircraft to its fleet. But

Prelude to War

when the country entered the war less than a year later, there were none.

The preparedness campaign, culminating in the National Defense Act, did not, nor did it intend to, catapult the nation into the mad scramble for mobilization and training for war. As a result, the protagonists of the preparedness movement left the American training program relatively untouched, neither aping continental models nor expanding and deepening the American system of aircraft production and training. External events propelled the country toward the evening of April 2, 1917, when the President of the United States went before a joint session of Congress to ask for a declaration of war. From that point, American aviation was pledged to partnership in an undertaking for which it was unprepared.

Part II
The End of Illusions

II

...and the shells never cease. They alone plunge overhead, tearing away the rotting tree stumps, breaking the plank roads, striking down horses and mules, annihilating, maiming, maddening, they plunge into the grave which is this land; one huge grave, and cast up on it the poor dead. It is unspeakable, godless, hopeless.
— British artist Paul Nash to his wife, November 16, 1917

In August 1914 the German army swept across Luxembourg, pushing the opposing forces into retreat. The Allies retarded the German advance at the River Marne in early September, extinguishing German hopes of any immediate victory and ending the briefly flaring war of movement. Rather than a decisive encounter between the warring nations, the violence of August 1914 preceded three years of appallingly bloody stalemate in the west. Along a line running from the Swiss border to the Channel coast, four million men, soldiers of the Allied and Central Powers, buried themselves in trenches dug into the earth, scarcely hidden behind field fortifications and entanglements of barbed wire. The war of annihilation had become a war of attrition, with both sides resorting to artillery, guns, aircraft, and tanks to support infantry assaults. Pounding of heavy artillery, punctuated by machine gun and rifle fire, beat the dirt and humans into mud and pulp. Counterattack followed charge across the shell-blasted battle front, slaughtering hundreds of thousands of men in attempts to win a few yards of the grisly landscape.

At the time, most Americans found the names of the belligerent countries and their leaders unfamiliar and unpronounceable. They understood neither the tangled causes of the war they too would join, nor the geography of Europe where millions would die. In 1914, the country was still two years away from the recognition, brought about by the "preparedness" campaign, that the United States would become entwined in the incomprehensible, horrific conflict. But in 1917, America joined the combatants, declaring war against Germany and entering the lists at a low point in Allied fortunes.

World War I

Virtually every account of the nation's state of war preparedness in April 1917 enumerates the deficiencies in men, machines, and combat experience. The Navy led the Army in its inventory of the latest technological advances. As for Army aviation, in his recollections after the war, General Pershing lamented:

> The situation at that time ... was such that every American ought to feel mortified to hear it mentioned. Out of the 65 officers and about 1,000 men in the Air Service Section of the Signal Corps, there were 35 officers who could fly. With the exception of five or six officers, none of them could have met the requirements of modern battle conditions and none had any technical experience with aircraft guns, bombs or bombing devices.
>
> We could boast some 55 training planes in various conditions of usefulness, all entirely without war equipment and valueless for service at the front.[1]

To become an employable operational force, the Aviation Section, a tiny component of the Signal Corps, had to secure the support of the War Department, Congress, and public opinion. The endorsements snowballed, resulting in the phenomenal appropriation in July 1917 of $640 million for aviation. Suddenly the impoverished Aviation Section, housed in its one-room rented office, appeared to have been handed a blank check, admittedly to be spent on a near-impossible task — the creation virtually overnight of a widespread and complex network of training schools, curricula, flying fields, training planes, and instructors. All built upon hastily made projections regarding American industrial and manpower capacities, along with the demands from and commitments to the Allies. Experience would demonstrate that, as airmen had been declaring for the past decade, money could not buy time.

A substantial training program depended upon more than an arithmetic increase in the production of pilots and airplanes. Observers, photographers, radio operators, engineers, gunners, and bombardiers for combat crews also had to be trained. And, in the case of pilot training, methods and equipment appropriate to observation, pursuit, and bombardment had to be developed. To handle the vast new administrative apparatus, schools were needed for adjutants, supply officers, and engineers. In other words, aviation planners had to think beyond the mere duplication of earlier facilities and techniques, to institute a fundamental change in the training system: the need for specialization. By the time the war ended, twenty-seven Army flying fields in the United States and sixteen in Europe were training the expanded forces.

To administer the behemoth, the Division of Military Aeronautics and its predecessor organization until April 1918, the Aviation Section of the Signal Corps, formed a Training Section. It was subdivided into branches responsible for ground schooling (known as the Schools of Military Aeronautics), flight training, gunnery instruction, radio, photography, engineering, recordkeeping,

Part II

and technical or enlisted mechanical training. In time, the requirements for specialized advanced training necessitated formation of the aerial observation and bombardment branches.

The administration of training was linked, fatally as it turned out, to the Army's supply system. At the outbreak of war, the War Department was organized into five separate purchasing bureaus that were, according to historian Ronald Schaffer, "several virtually independent systems for buying, financing, storing, and transporting military goods, each serving a particular bureau.... The army's systems were arranged by military functions... while American industries were arranged by commodities."[2] For aircraft procurement, the Bureau of Aircraft Production, not the Division of Military Aeronautics, negotiated with the handful of firms that comprised the American aircraft industry. Under the provisions of the Overman Act of May 20, 1918, the Air Service was created as a distinct line arm with its own procurement organization.[3] Thus, although the 1917 appropriation of $640 million opened the door to massive procurement, Roger Bilstein has pointed out that "few of the companies had ever produced anything more than a training plane, and there were only a handful of designers capable of that job."[4] The ultimate results from the "arsenal of democracy" would be, besides training planes, production of more than 1,200 American-made British deHavillands (DH–4s) and over 13,000 Liberty engines.

Like their comrades at home, air officers in Europe were dependent on the War Department's unwieldy system of supply, and also had to accommodate to the larger scheme of Army mobilization and training. Unlike the Aviation Section in the United States, however, which did not separate from the Signal Corps until May 1918, the Air Service, American Expeditionary Forces (AEF), quickly assumed a status similar to other combat arms when its organization was approved on September 11, 1917. Veteran aviator and now Brig. Gen. Benjamin D. Foulois succeeded Brig. Gen. William L. Kenly as Chief of Air Service in December 1917, and he remained until May 1918, when Maj. Gen. Mason M. Patrick replaced him. Throughout the war, the Chief of Air Service answered directly to the Commanding General, AEF, General John J. Pershing, who held the inflexible view that the function of the General Staff in Washington and its subsidiary departments, such as the Division of Military Aeronautics and Bureau of Airplane Production, was only to supply men and materiel and implement AEF policy. In his memoirs, Pershing confirmed that concept of the chain of command and laid the blame for wartime failures at the doorstep of the Army staff:

> In the absence of any preparation for war beforehand, the principle can hardly be questioned that the commander at the front and not the staff departments in Washington should decide what he needs. The employment of our armies in Europe had been fully covered by general instructions and there were no problems of strategy or questions concerning

103

operations that devolved upon the War Department staff. These were matters for the Commander-in-Chief of the A.E.F. to determine. It remained, then, for the War Department simply and without cavil to support our efforts to the fullest extent by promptly forwarding men and supplies as requested. The Secretary of War was completely in accord with this conception, but it was evident that the staff departments had not grasped it or else the disorganization and confusion were such that it could not be carried out.[5]

By the time the United States entered the war, the exhausted Allied forces yearned for an infusion of fresh American troops. Moreover, aerial warfare was then considered instrumental to success, with both sides committed to the attainment of air superiority. As a result, on May 24, 1917, French Premier Alexandre F. Ribot called for a stupendous increase in air power for the Allied cause. He asked the United States to supply 5,000 pilots, 50,000 mechanics, and 4,500 airplanes to the western front by the spring of 1918. The Aviation Section responded that indeed it could meet and even top those numbers, and in France General Pershing concurred. To fulfill his own ambitions for aviation, in July 1917 Pershing cabled the War Department that he wished to have 260 service, 36 training, and 90 replacement squadrons in France by June 30, 1919.

The American air arm never came close to reaching those figures. In fact, for some time the American military presence as a whole barely registered. The AEF numbered under 62,000 men by the end of September 1917; of the nearly 1,500 airmen sent to Europe that year, virtually all had yet to be trained. Three thousand miles from home, America's would-be army sat stymied in the face of inexperience and massive supply and transportation problems.

In March 1918, nearly a year after the United States entered the war, Germany began another major offensive that shifted action away from the trench warfare of the previous three years. Americans joined in a successful counteroffensive against the ensuing fourth great German drive, and Pershing demanded thereafter that the U.S. Army assume a battle front of its own. Because he believed that Americans would train for open warfare better than the Allies, with their long experience and expectations of trench warfare, he determined that all training should be conducted by Americans. As he later recounted,

... efficiency could be obtained only by adherence to our own doctrines based upon thorough appreciation of the American temperament, qualifications and deficiencies. I recommended the withdrawal of all instruction in the United States from the hands of Allied instructors. This recommendation was promptly approved by the Chief of Staff, who entirely agreed with my views.[6]

Up to that time, through the first year of American wartime involvement, the AEF had been preoccupied with forming a combat organization, separate

Part II

from but dependent upon supplies of domestic personnel, and upon French equipment, rations, and supplies. The sector occupied by the American forces during its year of preparation required railway and port access, plus space, if not ready-built facilities, for billeting and training. As for the Air Service's part in the process, at no time during that first year, or thereafter, did it achieve a surplus of specialists in all the necessary fields. It never obtained the requisite equipment, nor could shipping schedules transport needed men and materiel to France. Because the personnel and aircraft quotas loomed hopelessly out of reach, unrelenting pressure never loosened its stranglehold on the training establishments of the Division of Military Aeronautics in the United States and the Air Service in France.

FOUR

Training at Home for War Overseas

> In my effort to volunteer to serve in [the American air arm], I spent a complete day in Washington looking for its office. By chance, the next day, I found the one-room, rented office of the Aviation Section, Signal Corps of the Army, occupied by Captain Thomas D. Milling and a secretary. I gave them the required information about myself, and one month later I was beginning my training at one of the rapidly organized ground schools in the United States.
> —Captain Douglas Campbell, in Lucien H. Thayer, *America's First Eagles*[1]

The words of Captain Douglas Campbell, America's first World War I ace, hint at the lowly status of American aviation at the outbreak of war and at its subsequent, snowballing growth.[2] Very early, officials decided to implement the pilot training program that they had described earlier to Congress and that was informally in place, namely a series of discrete stages, moving from primary to advanced. A ground school introduction, and specialization that had become common in Europe, would be new additions on each end. Once into flight training itself, individuals would begin on general-purpose training aircraft and, in a final phase, specialize in observation, pursuit, or bombardment. In a system intended to utilize Allied experience and facilities and American manpower, a pilot's training would begin in the United States and end in Europe.

The Air Service kept the phased system of pilot training throughout the war, but the conduct of the different training stages bounced back and forth from the United States to Europe, so that the program never settled into a regular pattern. Few plans were carried out completely because shifting circumstances continually required new responses. The first training formula provided for nonflying training at U.S. ground schools, primary flight training

World War I

at stateside flying fields, and advanced training and assignment to tactical units in Europe. Soon, however, planners abandoned that original idea. Because the Allies offered space and sometimes whole facilities to the U.S. Air Service, many American cadets began taking primary instruction in France, Great Britain, and Italy. And airfields under construction in the United States that were intended for primary quickly expanded to include advanced flying instruction in observation, pursuit, aerial gunnery, and bombing. Training for each specialty became further centralized on certain flying fields.

By spring 1918, planners at home and in France aimed to overturn the original plan so as to give *all* individual pilot training in the United States, and only a short freshening-up course overseas. To implement that policy, as late as October 1918 the Chief of Air Service, AEF, considered it "essential that service equipment be allotted to training in order that all basic training, both preliminary and specialized, be done in the United States."[3] In practice, neither of the basic plans operated systematically during the war and, in fact, the transitional state in which all kinds of training were conducted everywhere, became the norm.

Besides the cooperative arrangements with European allies, flight training at home initially profited by reciprocal agreements with the Canadian arm of the Royal Flying Corps, RFC Canada. Besides providing the model for ground schools and for aerial gunnery instruction, the Canadians taught American pilot recruits in advance of the creation of a U.S. training organization. Lt. Col. (soon Brig. Gen.) Cuthbert G. Hoare, Commander, RFC Canada, met with Chief Signal Officer Squier in May 1917. Hoare spoke about the difficulties in conducting flight training during the harsh Canadian winter, while General Squier worried about the shortage of instructors for the Americans' soon-to-be-built flying schools. Out of their mutual concerns and a subsequent parley came a quid pro quo by which RFC Canada would train three hundred pilots and other ground recruits and organize ten American squadrons to be sent overseas to work with the RFC in France. In return, the Americans would build three airfields in the southern United States and would supply training airplanes for Canadian use.

A nucleus of three American squadrons began training in Canada, and on November 18, 1917, the RFC contingent arrived at Fort Worth, Texas, to train during the winter at Hicks, Everman, and Benbrook Fields, collectively known as Camp Taliaferro.[4] The Canadians were to return home in mid-February, but an extension allowed them to remain in Fort Worth through mid-April; in return, they agreed to train eight American squadrons. When RAF Canada units (the RFC had become the Royal Air Force, the RAF) left the United States, the additional squadrons had not completed training, owing to delays caused by an outbreak of influenza and forced quarantines at the mobilization centers (called "concentration camps"). By mid-April, RAF Canada had successfully graduated 408 pilots from the U.S. facilities, and another 50 had nearly completed

Training at Home

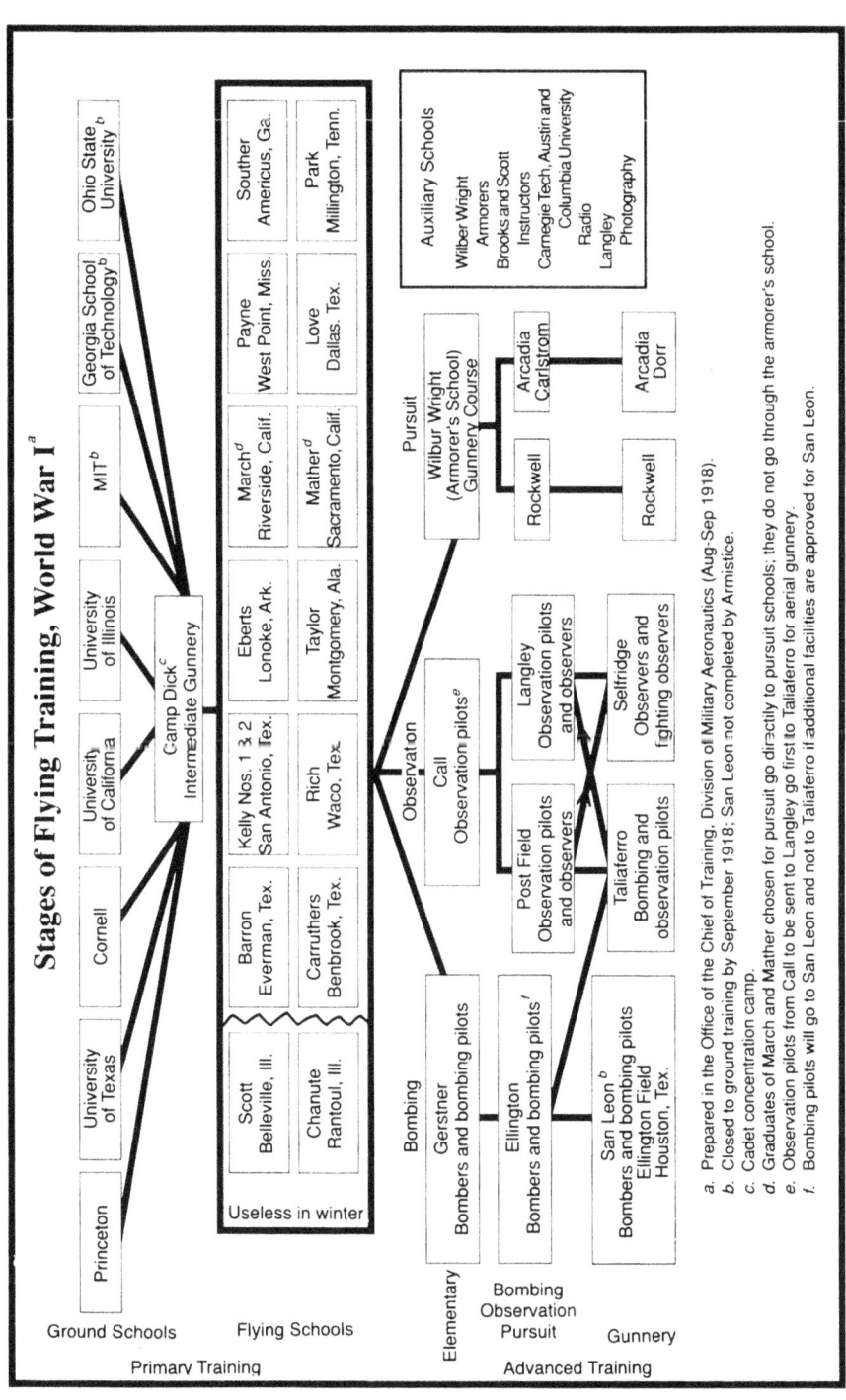

training. Furthermore, 2,500 ground officers and men had been trained, and another 1,600 were undergoing instruction.[5]

Besides pilot training, the advent of specialization brought requirements for crew training. The Flying Branch of the Training Section of the Division of Military Aeronautics (Balloons was the other branch) was therefore subdivided into Primary and Instructors Schools, Observation Schools, Bombing Schools, Pursuit and Gunnery Schools, Photography, Radio, Mechanical Instruction, and Operations. The Training Section and its subordinate offices administered all schools for commissioned and enlisted personnel and the service units throughout the United States (except those specifically attached to other army units), prepared curricula, supervised methods, and distributed supplies.[6]

Throughout the brief war years, the training program reacted principally to aircraft production schedules, availability of spare parts, construction of facilities at home and abroad, and the requirements of the AEF in France. The three thousand miles between the headquarters of the Division of Military Aeronautics[7] and the AEF caused enormous problems in communication. As a result, training plans developed in the United States were constantly altered and usually lagged behind operational needs.

The availability of training airplanes determined the number of trained men that the United States could supply to the combat zone within a year's time. In December 1917, it was estimated that sixty airplanes plus a reserve of two-thirds would be required for each training unit. According to the Training Section, that meant a total of one hundred planes for each single unit, which included a 10 percent loss that "can be expected when flying is constantly being engaged in."[8]

It soon became apparent that the American aircraft industry was incapable of manufacturing the huge numbers of training and combat airplanes that were promised. The end of June 1918, Lt. Col. J. E. Carberry, Chief of Heavier-than-Air, notified his superior that the "most urgent problem in the entire training system was realized to be the co-ordination of the airplane production and the training program." The Bureau of Aircraft Production, he went on to say, was in the process of reevaluating the probable manufacturing capacity, and upon that information the Training Section would revise its programs.[9] The deficit of primary trainers did not remain insurmountable, but shortages of engines, spare parts, and higher-performance aircraft seriously hamstrung specialized advanced flying and technical training. Although the JN–4 proved itself a fine primary trainer, American firms did not build advanced trainers modeled on combat types, which effectively eliminated pursuit and stalled bombardment training in the United States.

The Division of Military Aeronautics enjoyed a surfeit of volunteers for flight training. In fact, pilot surpluses quickly accumulated. But bottlenecks in the flow into and out of primary training, owing to difficulties just mentioned, caused pools to develop and continual shifting of student populations. As a

Training at Home

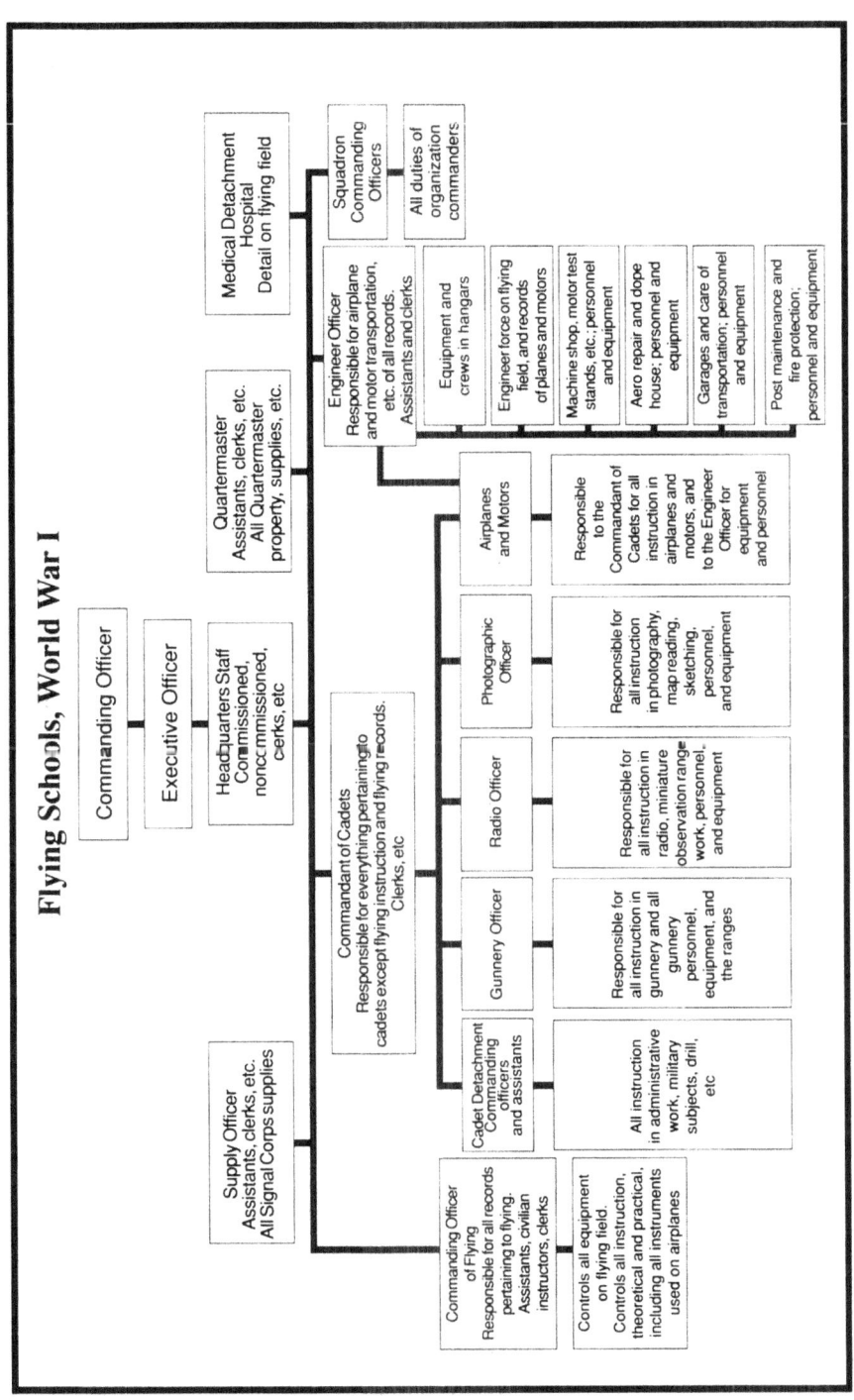

more serious result of the shortages, specialized training conducted in the United States frequently failed to produce men prepared for immediate service over the lines.

Ground Schools

The Air Service instituted the first phase of air training — ground schooling — most quickly because a vast reservoir of eager and qualified young men appeared waiting to be tapped, and because the initial phase required no flying instruction and thus less equipment. As a result of discussions between representatives of the U.S. Army and RFC Canada immediately after the United States declared war, Chief Signal Officer Squier determined to institute the Canadian model of nonflying, or ground school, instruction in American universities.

General Squier called upon Hiram Bingham, who became instrumental in establishing the ground school program and then went on to direct other aspects of flight training at home and overseas. A Yale University history professor and world-traveled explorer, Bingham had volunteered for aviation duty and, though at the age of forty-one was past the pilot's recommended maximum age of thirty, had learned to fly at the Curtiss school in Miami, Florida, in March 1917. Now, as a major in the Signal Officers' Reserve Corps, Bingham led a group to study how Canadian ground schools prepared airmen for service with the RFC.[10] Traveling with him was a contingent representing the Universities of California, Texas, and Illinois; Ohio State; Cornell; and the Massachusetts Institute of Technology — colleges that already had military courses and strong engineering and technical faculties. Those representatives, selected for their familiarity with aeronautics, internal combustion engines, and meteorology, were expected to form the nucleus of a system of instruction in military aeronautics in the United States.[11]

The Americans spent several days, chiefly at the University of Toronto School of Military Aeronautics, where they attended classes, listened to reports from the war zone, and gathered instructional materials and regulations used at the school. The Canadians enrolled a new class every week, graduating students in six weeks' time. Successful ground-school graduates proceeded to flying school. The system served to weed out some unfit or incompetent students early, conserving time and instructional and equipment resources. The American committee decided to adopt the Canadian program in its entirety, only lengthening the course to eight weeks (later extending it to ten weeks, then to twelve). Bingham summed up the purpose and scope of the course:

> Great stress was laid on the importance of developing ability to observe artillery fire and to cooperate with both artillery and infantry. The importance of a thorough knowledge of the machine gun, the internal combustion motor, and wireless telegraphy was emphasized. We decided

Training at Home

Hiram Bingham standing at the center of a group of Air Service men training overseas.

to adopt the British method of dividing the course into two parts: the first, of three weeks, chiefly military studies and infantry drill; the second, of five weeks, technical aeronautics, with particular emphasis on guns and motors.[12]

Upon his return to the United States, Bingham remained in Washington as Director of the U.S. Schools of Military Aeronautics. In May, each of the six colleges represented in Toronto began providing ground training to candidates for commission in the Signal Officers' Reserve Corps. By the end of June, the Schools of Military Aeronautics added Princeton University and the Georgia School of Technology. The eight colleges provided ground schooling until the Air Service deactivated the network near the war's end.[13] The schools accepted pilot candidates from all over the country, not only from among their own student bodies.[14] A cadet stayed an enlisted man throughout ground training; after completing primary training, he received his Reserve Military Aviator (RMA) or Junior Military Aviator rating.[15]

Each school had military staff as well as civilian instructors. The commandant was a Regular Army officer who reported to the Chief Signal Officer; he exercised general supervision of the school and commanded the troops on duty. A civilian president of the Academic Board, who oversaw all technical instruction, was in charge of the teaching staff.[16] At the outset of the

World War I

war, the Army did not have enough officers for instruction at the flying fields, much less in the ground schools. Some of the first men to finish ground schooling went therefore to Toronto for instructor training, then returned as instructors to the schools from which they had graduated.[17]

Civilian faculty members with expertise in various subjects initially developed the curriculum, which soon became standardized in all the schools. The Ground Schools Branch determined the subjects to be taught, number of hours for each, and even suggested examination questions and syllabi.[18] The course included military studies, signaling and radio, gunnery, airplanes, engines, aerial observation, and aids to flight. The last topic encompassed such matters as the theory of flight, the requirements of cross-country and night flying, map reading, and meteorology. "Much of the material in this course is inspirational rather than practical," allowed one description. The number of hours devoted to each subject changed as reports from flying fields indicated the areas in which cadets seemed to be more or less prepared.[19] Added to the curriculum, but never taught by the time of the Armistice, were courses for bombardiers and observers.

The Ground Schools Branch disseminated large amounts of material to the schools. One set of mimeographed stencils, for example, ranged from general information to detailed technical matters, such as diagrams of the Hollocombe-Clift airspeed indicator used by the RFC, and descriptions of various types of bombs employed at the front. The branch also reported on aerial battles taking place in the skies over France, including the conflicting claims of air superiority, and it disseminated booklets such as "Fighting in the Air." The Ground Schools Branch did not generate all the instructional material; articles in aviation magazines and British reference books also circulated. Audiovisual materials included official British motion pictures. Eventually the branch permitted the teaching of French, as long as it "in no way interferes with [cadets'] regular studies."[20] For that purpose, the University of California at Berkeley published *A Handbook of English and French Terms for the Use of Military Aviators*. One of the officers attached to the branch had as his express duty sifting through information from abroad for use in ground-school courses. Bingham thought such materials kept students abreast of the most recent events and therefore was "of great psychological value."[21]

From the beginning, ground schools experienced no shortage of recruits. But instructors quickly realized that careful selection of prospective fliers was crucial for a successful training program, and that not every able-bodied young volunteer was suited to be a pilot. The Air Service wanted "fellows of quick, clear intelligence, mentally acute and physically fit," according to Bingham, so as not to waste "the most expensive education in the world."[22] In midsummer of 1917, General Squier cautioned Harvard University president Lawrence A. Lowell that only the best military students should be steered toward aviation: "Athletes who are quick witted, punctual and reliable. Intelligent men accustomed to making quick decisions are highly desirable. Men who ride well, can sail a fast boat or handle a motorcycle usually make good air pilots."[23]

In its recruiting efforts a few months later, the Office of the Chief Signal Officer requested that Harvard College distribute information and application blanks "to a selected list of the football players in your institution, as men of this class have not only exhibited a pronounced preference for the flying branch of the service, but have proven to be excellent material for training as aviators."[24] Foreign air forces also considered athletically gifted young men to be good candidates. Dr. Graeme Anderson, an RAF flight surgeon, wrote that the "successful flier must be one who has power to coordinate his limb muscles with a beautiful degree of refinement.... It is because of the importance of this delicately coordinated effector response that great importance is attached to a history of sport in the selection of aviators."[25]

Obviously, Air Service pilots were not all college football players, as indicted by the makeup of one squadron in France. According to one of its members, Harold Hartney, "We had one theatre proprietor, 4 salesmen, 3 lawyers, 2 journalists, 5 electrical and gasoline engineers, a concert pianist, a banker, a cotton planter, an automobile race driver, a broker and a mining man." Hartney saw the significance of their peacetime occupations not in their athletic prowess but, as he said:

> To keep one's head above water in any of those professions requires clear, independent constant thinking — not mere mechanical obedience to orders. It was inevitable that these men fought harder because they knew, from their own intelligences, what they were fighting for — an ideal, rather than an immediate military boss. There were also many college students among our flyers. They are harder to analyze because their minds were immature, but some of them became superb pilots and fighters.[26]

Deciding who would become "superb pilots and fighters" remained among the most important tasks of training officials. Even after he was admitted into the program, a man had no assurance of staying. The staff of the Cornell School of Military Aeronautics, according to instructor Lt. Howard Davidson, suspected that some candidates "were just coming into aviation to keep from getting drafted." In response, they devised an informal means of screening out the undesirables. If cadets "didn't look like they were officer material," the school doctor sat them in a chair that turned on its base, and twirled them around till they lost all sense of equilibrium. "He would obligingly spin them until they said they didn't think they cared about being in aviation."[27] The system conspired, albeit in this somewhat comic scenario, to find a medical excuse to reject a man who was physically fit to fly but who did not seem to be officer material. Thereafter, the washout rate remained high. According to one graduate, "a squadron that got through Ground School without suffering fifty per cent 'casualties' was considered so brilliant as to be worthy of suspicion.... No odium attached to the 'busting' process. It was used ruthlessly, often without seeming rhyme or reason, to curb the flood of those who would be flyers."[28]

Although men nearing the end of ground school met examining boards before they were passed on, the issue of selectivity remained among the thorniest in flight training and was never entirely amenable to objective or quantifiable measurement. What were the personal qualities a pilot and (as important) an officer should possess? What were the standards upon which the examining boards should accept or reject ground school applicants for flight training? Which failures still allowed for a second chance, and which meant certain "washout"? How were individuals chosen for a particular specialty?

When cadet Robert Todd reported to the Ohio State School of Military Aeronautics in August 1917, he, along with the other forty or so men in his class, was issued a cot in the armory. That might have been a clue, because he quickly discovered that training consisted of two key elements: the *second* had to do with flying, but of *first* importance — "they were going to make soldiers of us."[29] He and his fellows came to dread memorizing regulations, the hours in drill, practicing salutes, standing guard, and attending lectures on military courtesy and law. Cadet Barney Giles and his classmates at the University of Texas complained about the nearly four hours daily they spent marching around tombstones (an Austin cemetery served as their drill ground).

Bingham acknowledged cadet hostility toward military training but nonetheless kept it at center stage. The Canadians who administered RAF ground schools had explained to their American visitors, before the U.S. training program began, that veterans of the western front "differed radically on the importance of the various subjects of study," but all agreed, according to Bingham, that "undisciplined, unmilitary pilots were extremely undesirable, and that any youth who followed individualistic tendencies to such a degree as to make him appear to be a poor soldier should not be trained as a pilot. They said he would soon come to grief over the lines where team play was so essential."[30]

Once American Army officers set up operations at the front, they made the same point, repeatedly reminding the ground schools to instill in the students a sense of discipline and loyalty. One of the earliest reports from the AEF pertained to ground training in English and French schools. Its author noted that "the smartness of the R.F.C. is evident wherever one comes in contact with its members, also their esprit de corps; and these, together with military courtesy are the main things that they aim to teach at their cadet school."[31] The demeanor of British cadets contrasted with that often reported of American airmen, as another report declared: "The lack of instruction in military courtesy for pilots and observers is very marked in the A.E.F., and consequently the Air Service has been somewhat discredited on this account."[32] In his frequent communiqués, General Pershing minced no words about the importance of rigorous military training for all troops. Pilots were to learn to think and act as officers, and infantry drill was key to that process. This, then, became a central tenet of ground school training, which the expressed objectives of the Schools of Military Aeronautics made clear: "To make soldiers. To eliminate poor material as fast as possible. To discover exceptionally good material — 'Honor men.' To give ground training to future pilots and observers."[33] Technical training appeared last on the list.

An eloquent plea to this effect, running to nearly fourteen closely typed pages, came from Lt. S. M. Clement who, as an observer in France, after considerable discussion with American training officers, reached the conclusion that the "function of the School of Military Aeronautics is primarily a military one and that technical studies should be considered as secondary." Men should be eliminated from ground schools for military rather than scholastic reasons. Cadets' academic grades did not predict whether they would become good pilots. According to Clement:

> The principles upon which [discharge] should be based are not those of ground school achievement, but of potential ability. I appreciate that this is a difficult standard. But aviation is the most highly individualistic branch of the War.... [Becoming a pilot] involves a man's aptitude, his alertness, his leadership, his sense of discipline, his appreciation of his part in the whole scheme, his ability to carry his share; in other words his personality and character. Technical knowledge without these is of little avail in our service.... We are in the position of having no traditions, and we must make them as we go.[34]

The traditional notions of the officer and gentleman, linked to the evolving concepts of professionalism within the air arm, and based as they were on subjective considerations, were never subject to systematic problem solving. The issues were not new, nor would they soon be resolved. Airmen had been attacked before for lacking proper military demeanor and respect for authority. It was a charge that would haunt the air arm for years to come.

Throughout the war, throughout the aviation program, sufficient coordination was lacking among the various administrative agencies in the United States that directed the effort, and between the Division of Military Aeronautics and the AEF in France. To improve communication, officials instituted reporting channels to and from the component schools, made frequent visits and inspections, and continually reevaluated procedures and curricula in light of reports from the western front.

Ground schooling suffered from some unusual internal conflicts that contributed to a lack of clarity in program goals. An implicit friction among civilian organizations, offering what was for most air officers their only basic military training, reflected the uneasy mix of recently activated armchair colonels and Regular Army officers attempting to direct the aviation program from Washington. Bingham, himself newly arrived from the academic world, commented that many ground officers looked upon aviation with incomprehension, if not disdain, and many of these ground officers were not a great deal more comfortable with academe. They were now asked to organize a training program that combined airmen and scholars, itself an unlikely amalgam. As a result, the often conflicting assumptions held by the several groups led to wariness in professional relationships, and some lack of vision in the process of structuring the air training program. Bingham noted anecdotally:

> It may not be out of place to state here that during the first few months of my duty in Washington, the officer who, under General Squier, was in immediate charge of the Aviation Section of the Signal Corps, was not a pilot, had only been up once or twice, was frankly afraid to fly even as an observer, and went so far as to say to me that for the father of seven sons to take flying lessons showed that he did not love his children. I could not help wondering whether the Secretary of War would expect an officer who was afraid of riding horseback to direct the fortunes of the Mounted Service School or even command a cavalry regiment successfully.[35]

The confusion of many in the line Army was exacerbated by the fact that there had been little joint training, so most officers were unfamiliar with aeronautics. The War Department, therefore, had little idea how to teach air doctrine, undertake strategic planning, or construct an operational air force. Nor did it know how to establish adequate coordination between air and ground commanders, or develop tactics for combined arms training.

Not only did it appear to many airmen that some civilians and Army men could not command a flying organization, it also appeared that the schools themselves often competed, based, as likely as not, on old rivalries extraneous to the business at hand. A year after completing his own flight training, then assigned as an instructor at Cornell, Lt. Howard Davidson found, to his amazement, that "Cornell didn't like Yale, and especially didn't like Bingham," a former Yale professor. "They didn't want him to come on the place" when he arrived for inspection.[36]

Despite the inevitable disputes, the Army was fortunate in having the support of a college network that offered a thorough technical introduction to flight. But on both a philosophical and substantive basis as well, the Schools of Military Aeronautics had their critics. Ground school was *too* theoretical, some charged, insofar as it undervalued the "military point of view" of practicality. Moreover, a corollary criticism held that the curriculum packed in too much information, even after the time allowed for the course was lengthened twice. When Lt. Col. C. F. Lee from the Training Division of the RFC examined the newly created American program in September 1917, he considered much in the scheme to be "impracticable and in my opinion of little value." He maintained that "it would take anything up to a year for the Officer to carry away in his head that which is laid down in this curriculum that he should know."[37] Nine months later an American Army officer, Lt. Col. Arthur Woods, reported that in his inspection at Cornell he found a "tendency towards theoretical rather than practical instruction thus increasing a natural tendency of university-trained instructors to use technical language, formulas, and 'curves,' without the preliminary explanation necessary to make them clear to the untechnical minds of the cadets."[38]

On the other hand, some people pointed out, the particular value of ground school lay in the very fact that it captured impatient young men's attention long enough to give them both a theoretical and a practical understanding of aeronautics before they got into the cockpit. Some cadets who had learned to fly previously were said to make especially poor military pilots. Such a man, according to one Signal Corps officer, "is generally captious . . . and much more difficult to deal with in every way. He comes to greatly exaggerate the importance and value of knowing how to fly."[39] Colonel Woods, who had criticized the school at Cornell, believed nonetheless that cadets in ground classes should receive "*all* the instruction in Engines and Airplanes that they will ever need." The flying fields had neither the time nor inclination to teach such subjects. "There the cadets are so nervous and excited," Woods remarked, "and they are so engrossed in learning to fly, that they cannot concentrate so as to listen to verbal instruction, even in the laboratory."[40] The need to impart a substantial amount of information of the sort not generally taught in the airman's subsequent career had to be balanced against the need for practical experience immediately applicable to combat aviation.

Defining relevant subject matter proved a challenge. Military exigency meant constant rethinking of curriculum and approach. "Just as we would get comfortably settled in one course of study," Bingham reflected, "word would come by cable from General Pershing, urging that more stress be laid on something else."[41] To unite theory and practice in the United States at least, school instructors apprised themselves of techniques employed at the training fields. In mid-September 1917, Bingham urged the presidents of the academic boards, and any faculty they selected, to visit the nearest flying school in order to facilitate smooth coordination between these two phases of the training program.[42]

Occasionally, school instructors ventured to the front. In the late summer of 1918, 1st Lt. Stanley T. Williams of the Ground Schools Branch interviewed training officers at Issoudun, France, to solicit their views about the strengths and weaknesses of U.S. ground schools. One, among the reports he compiled, stated that pilots had too little familiarity with airplane engines.[43] This type of complaint would be endlessly repeated throughout the war, and it seldom would be redressed because the ground schools, along with many flying fields, often lacked sufficient — if any — equipment, or the same kind as that used at the front. Although ground schools required few training airplanes, they needed engines and guns for demonstration and practice sessions. The Lewis machine gun and the Curtiss or Hall-Scott motors were the only ones available for training in the early days of the war, but they were by no means the only guns and engines in service in France.

The greatest problem in ground schooling was a lack of equipment that impeded the flow of students through the training sequence. Because there were too few aircraft, many men who had completed ground school could not be sent forward to the flying fields for the first phase of instruction. As a result, fairly early after the United States entered the war, the Ground Schools Branch decided to send some of its graduates to Italy, France, and the RFC schools in Canada, England, and Egypt.

For primary flight training overseas, the schools selected cadets from among the growing pool of those who had "made good" on their examinations.[44] Those selected greeted the chance to learn to fly near the lines with great excitement, and many vied for the coveted positions. Robert Todd, who finished ground school at Ohio State, recalled the assignment process when he went through it: "The names of thirteen men were drawn to go to Canada, while the remaining men were going to Europe. Everyone wanted to go to war and see service before it was over. Some of the men offered money to switch places."[45]

The plan for sending the "best 10 percent" turned into a fiasco. In the event, very little space could be found anywhere for primary training. According to Todd, "the men headed for Europe stayed in a camp on the East Coast and never left the country."[46] Lieutenant Davidson at Cornell recalled that "we had to select the best 10 percent and send them straight to France, and they were going to train them over there. When I got over there, toward the last of 1917, they didn't know what to do with them.... We got them into schools, but I would say the war was almost over when some of them got into flying training."[47] John M. Grider, a cadet who had attended ground school in Chicago, was among those assigned to Italy. But he expressed his disgruntlement when, instead, he and his squadron shipped out to Oxford, England: "We've wasted two weeks studying Italian and two months going to Ground School learning nonsense for now we've got to go thru this British Ground School here. And we hear that everything that we were taught at home is all wrong."[48]

The initial effort to send ground school graduates to Europe for flight training did not satisfactorily solve the problem of student flow. A year into the war, cadets were still graduating from ground school faster than they could be absorbed, in spite of smaller classes and longer courses. By midsummer 1918, the number of ground schools had cut back to five, and students were given the option of volunteering as bombardiers or aerial observers rather than as pilots.

Training at Home

Primary Flying Training

After his round of inspection of the flying schools in mid-November 1917, Inspector General J. L. Chamberlain reported to the Chief of Staff:

> The work of the schools has been seriously handicapped by frequent changes of policy. For instance, at Mt. Clemens, Mich., instructions were received that the work of that school would be advanced flying and that no more beginners would be sent there. Later, instructions were received that this plant would be abandoned for the winter and the personnel sent to a southern camp. Later, this was again changed and instructions were received that there would be no advanced flying and that additional students would be sent there for preliminary training. Later still, instructions were received that flying would be discontinued for the winter and the work of the camp devoted to non-flying instruction.[49]

The primary flying schools faced all of these expedients: shifts from primary to advanced instruction, relocation from northern to southern climates, and abandonment of flying instruction altogether. Until the spring of 1918, Col. Henry H. Arnold, then Assistant Director of Military Aeronautics in charge of the flying schools, considered the situation more a "state of affairs" than a "chain of events." Particularly at the onset of war, the existence of anything recognizable as policy regarding primary flying instruction was highly questionable.

The originally approved program anticipated graduating 540 men monthly from primary during mid-August, September, and October 1917; then by mid-November, the time of the Inspector General's report, the total was to rise to 660 per month.[50] In fact, only 598 had completed primary training by November 30. A shortage of training airplanes on American flying fields, and General Pershing's determination that no cadets could be accommodated for training in Europe, placed the program in immediate and serious jeopardy. Construction of twenty-four training fields, envisioned during the congressional debates in the summer of 1917, became imperative.

Building training facilities was no easy task, although it was accomplished remarkably effectively. When, in 1917, Col. C. G. Edgar assumed command of the division charged with locating and constructing airfields, the Air Service had the school at San Diego on North Island; Camp Kelly in San Antonio; a seaplane camp at Essington, Pennsylvania; and a site at Mineola, New York. Only North Island had been built by the Army specifically to train Regular Army pilots. "When I reported for duty on North Island," one of a class of thirteen, recalled future Air Force general Hugh Knerr, "I found little to match my visions of a military aerodrome — just a few wooden hangars alongside the dusty field and some small airplanes lined up in front of a crude operations office, where men in flying gear were lounging on benches in the shade."[51] Yet North Island in 1917 was the most well established among the flying fields.

World War I

Col. Clinton G. Edgar

Essington and Mineola came into military hands during the height of the preparedness campaign, and both briefly trained Reserve officers. Previously, Essington had been a quarantine station and Mineola, an exposition ground.[52] In contrast, an Army airfield, as Colonel Edgar explained, was a complex of considerable proportion, consisting of "a series of hangars, machine shops, schoolhouse, administration building, garage, one or two aero-repair buildings, barracks for troops, barracks for cadets, officers' mess hall, officers' quarters, commanding officer's house, guardhouse, bakery, quartermaster's stores, aero stores — 54 buildings altogether, I believe."[53]

In May 1917, construction began on Wilbur Wright Field near Dayton, Ohio. Soon afterward, Chanute Field opened at Rantoul, Illinois, as did Selfridge Field near Detroit. By October 31, fourteen facilities had been built, of which nine had begun flight training. During 1917, a number of fields provided primary training: Hazelhurst Field (Mineola, New York), Selfridge Field (Mt. Clemens, Michigan), Wilbur Wright Field (Fairfield, Ohio), Chanute Field (Rantoul, Illinois), Scott Field (Belleville, Illinois), Camp Kelly (San Antonio, Texas), and Rockwell Field (the old North Island site in San Diego). Proposed advanced schools at Houston, Texas, and Lake Charles, Louisiana, were also used for primary training until the necessary equipment could be supplied for specialized instruction.

On December 15, 1917, the five northern schools closed and cadets transferred to the two southern schools. Because of year-round training, southern schools permitted a more even flow of students. Nonetheless, fewer cadets completed primary training during the winter of 1917–1918 than had been hoped. Planners expected that by the spring of 1918 there would be eleven single-unit schools (each unit supposedly composed of 100 airplanes and 144

Training at Home

North Island as Maj. Hugh Knerr, then a new recruit, saw it upon his arrival for primary flying training in September 1917

cadets) and three double-unit schools. In fact, by the time of the Armistice, some thirty training facilities were operating in the United States, the largest number of which were devoted to primary instruction.[54]

Despite the move south and west, several of the finest airfields were geographically or climatically unsuitable for untried new pilots. March Field in California, for example, lay in a valley surrounded by sharp rocky hills that caused considerable air turbulence during the day. And in the case of the flat Texas prairies, although the topography was ideal, the windswept landscape through southern Texas and into Louisiana kicked up summer dust storms, whittling away at engines and clogging propellers. Some fields converted, at least part of the year, to other uses, and at one point, all flying fields closed during the windy middle of the day.

Another major deterrent to progress, the shortage of training planes, lessened by late 1917. Between June and late November, manufacturers met the immediate demand for primary trainers with the delivery of 600 new Curtiss JN–4As — the Jenny, as the airplane became known. On the JN–4A and the revised Model D, a stick replaced earlier Curtiss controls. The famous Jenny remained the ubiquitous primary trainer throughout the war; at some point in

World War I

3d Wing planes on the line at Scott Field in Illinois (*above*) and the flight line consisting of JN-4s at Brooks Field, Texas, an airfield that opened in 1918 to train fliers for World War I.

his career, virtually every U.S.-trained Army pilot learned to fly on the reliable airplane with its eight-cylinder, ninety-horsepower engine and dual stick control.[55] But, well into the first quarter of 1918, although the number of airplanes was adequate, there were too few spare parts and motors. The shortage of parts for the OX water-cooled V-8 engine, the principal engine used on primary trainers, continued through the summer of 1918 and kept many airplanes out of commission.

Depending upon the vagaries of weather, equipment, and individual ability, the aspiring pilot needed six to eight weeks, including forty to fifty hours of flying time, to earn his RMA rating. Ground school had been oriented toward theory, but it also included practice in radio communications, gunnery, engine control, and airplane inspection.[56] Once into the flying phase, a cadet spent his first four to ten hours in dual instruction. An instructor taught a class of four or five students, each of whom he accompanied twice daily for a flight of twenty to thirty minutes. In the air a student had to discern by his instructor's hand signals how the controls worked and how to perform maneuvers. The pupil learned to taxi on the ground, to take off and land, then to turn and execute figure eights, and finally to glide and climb.

A trainee was allowed to solo when the instructor decided he was ready. Flying alone, the student demonstrated his grasp of earlier techniques before he went on to longer, cross-country flights. The cross-country course consisted of three triangular flights of thirty miles or more on each leg, and at least two

Training at Home

A Curtiss JN-4A (*above*) and a Boeing trainer equipped with a Curtiss OX-5 engine (*below*).

straight flights to a destination seventy-five miles or more distant, and a return. The student then began flying acrobatics — recovering from stalls, spiral dives, and loops — once again flying with an instructor. Everybody feared the deadly tailspin, the cause of a great number of fatal accidents. Last, after demonstrating mastery of the required maneuvers, students practiced formation flying in groups of three to six. In all, the solo phase required from sixteen to twenty-one hours in the air.[57]

The final tests determed whether the cadet would wear the silver wings of the RMA. For this test, "there were," according to one successful candidate, "landings over a 'hurdle' and landings for a mark with the motor iding as in ordinary landings, and with the motor 'dead' as in forced landings where the plane comes down with only the power of gravity to keep it under control." Other maneuvers followed — spirals, eights, stalls, and tailspins.[58] Virtually all cadets displayed their abilities in the American Jenny. One airman who had transferred from the RFC where he had been a flight instructor at Gosport, "accustomed to the last word in high powered service machines and to all kinds of training planes," found that qualifying for his RMA in the Jenny with its much narrower range of capability to be a "risky performance."[59] The JN-4B had a climbing speed of 50–55 mph, gliding speed of 60–65 mph, and top speed of not much more.[60] Whatever his previous experience, even if he had made it through ground school and the entire primary phase, if a cadet failed to pass the tests for his RMA rating, he was discharged from the program.

World War I

Training officers distributed written homilies to their students as part of their education. Cadets at Rockwell Field were told, for example, "The one thing that an aviator has to do and to do good is THINK. Learn why it is that a machine does certain things when you move certain controls. Learn just how and why the machine will stay in the air and how the air has to go through the planes to give them an efficient lifting power."[61] One field commander cautioned new pilots: "To summarize all of these precautions follow the instructions laid down by your instructors both verbally and written and annex to this your own good common sense, and you will live longer. It will be some time before your own personal theories in, and knowledge of, Aviation will be worth anything."[62]

For most young men, learning to fly pumped adrenalin into the system and brought a growing sense of self-confidence and self-importance. Yet, looking back at primary training, pilots often assessed their own and the program's maturity more soberly. James P. Hodges recalled that the primary instructors were "little tin gods to us and we thought that every word that came out of them was gospel." Later, however, he and his fellows realized how superficial had been their instuctors' — and their own — expertise for, although the Air Service quickly formed instructors' schools, in many instances the teachers themselves had just learned to fly. (Instructors, therefore, suffered a high mortality rate.) According to Hodges, "when an instructor thought that you could land without cracking up the airplane, he turned you loose. And from then on you were pretty much on your own and taught yourself to fly."[63]

If students at times virtually taught themselves because their instructors were barely one step ahead, many of the senior ground officers had only the vaguest comprehension of flight training. John Macready told an apocryphal, and much repeated, story of primary training at Rockwell Field:

> Of course the principal thing was to teach [cadets] landings, because at that time they had rubber landing gears — just regular cords — and a plane would bounce like a rubber ball . . . . These students would come down and bounce the plane around . . . . Somebody, I guess the adjutant, brought the report in to Colonel Dade [commanding officer], who was an old cavalryman and had a long white moustache . . . . There wouldn't be anybody killed or anything. Somebody'd get a little crushed up now and then. But they would break landing gear or break the end of a wing, loop around. So Colonel Dade looked at the report and he said, "What's the reason for all these broken landing gears? All these broken wings?" The adjutant saluted and said, "Sir, that's due to bad landings." He said, "Take a memorandum. There will be no more bad landings at this field."[64]

The primary phase assumed that large numbers of men would be processed efficiently through a standardized program, even though factors by which to determine success or failure never became quantifiable. Although an instructor's own experience and bias could not help but influence his decision whether to advance a cadet to the next level, speed dominated. In his inspection of flying schools in August 1918, Lt. Col. H. Conger Pratt criticized elements of the "system which takes the initiative away from the flyer and results in a mechanical training.... There is too much attention paid to the time a student remains in a certain stage of instruction and not enough attention paid to the quality of this instruction or the flying ability of the student upon his advance from one stage to another."[65] Standardization, in other words, should not result in lowered standards of proficiency. Unfortunately, in wartime, it usually did.

Although officials of the Division of Military Aeronautics discussed the balance between quality and quantity, they only briefly considered changing the methods that had become established on American flying fields, even though information from abroad indicated that the Allies' practices were quite different. The French *Roleur* system of using nonflying airplanes (Penguins) for primary training resembled the old Curtiss method by which the beginning pilot started on an airplane that was too low-powered to get off the ground. Alone, the cadet taxied before he took off and then graduated to increasingly faster airplanes and more complicated maneuvers. American trainers had abandoned that method in favor of the Wright system of dual instruction to the point when the student soloed. When the pilot transitioned to new aircraft or skills, he returned to an instructor. The Army considered the French method, in that it bought a few Breese Penguins, but the French approach never found acceptance on U.S. flying fields.[66]

An approach developed by the British came closer to displacing American techniques. By October 1918, the War Department announced that all U.S. training fields would use the Gosport system. Named for the School of Special Flying at Gosport, England, it had been developed in 1916 by RFC Col. Robert Smith-Barry. Whereas an American flight instructor trained a class of four or five primary students, then transferred them to new instructors for each ensuing phase, in the Gosport system a student worked with a single instructor throughout his training. In the most critical departure from American methods, the British student began practicing acrobatics almost immediately, since Smith-Barry maintained that training should approximate aerial combat as closely as possible. The intensive, individualized instruction permitted the training of a single pilot to be accomplished in five weeks' time.

By November 1917, the RFC had adopted the Gosport system, but Smith-Barry apparently found himself in bad odor by the early summer of 1918, apparently because of his relentless pestering of senior British officers. He was dispatched to the United States where he continued to proselytize for his methods. Smith-Barry's replacement at Gosport wrote: "Most of the high-ups deserted Smith-Barry, so he had to be removed. First it was suggested that he be posted to far-away Egypt, but then as America was showing great interest in the Gosport system it was decided to exile him with a small staff to the

World War I

U.S.A. and let him expound his theories to the Red Indians."[67] Smith-Barry arrived in the United States with a complement of Avro training planes and requested that he be allowed to demonstrate his methods where senior American Army officials could observe. He was unable to convince the Division of Military Aeronautics to give over Anacostia Field in Washington, D.C., for his purposes, but he was given nine cadets to train at Hazelhurst Field in New York. Thereafter the results of Gosport training would be compared with the American system.

After several months' evaluation, the Director of Military Aeronautics, Maj. Gen. W. L. Kenly, approved conversion to an "amended" Gosport system. But in fact, the American Army adopted very little of the Smith-Barry approach. For example, to apply his method successfully, Smith-Barry argued, the American Air Service must abandon its JNs for the Avro, equipped with a 100-horsepower rotary engine. The RFC, at Smith-Barry's insistence, had replaced their thirteen diffeent types used for primary and intermediate training with the Avro. However, the Jenny was entrenched and widely available in the United States, and replacement of the entire American training fleet was unthinkable. Moreover, Americans were unwilling to tolerate the unavoidably high accident rate that resulted from a system in which pilots flew very few hours and practiced dangerous maneuvers in a demanding airplane.[68] Instead, they retained the American requirement for fifty hours of flying time in primary training. They did, however, obtain new voice-controlled equipment for one-way communication in the air during dual instruction, and a critical element of the Gosport system — a single instructor throughout a student's training — was slated for implementation at all American training fields. But Americans never adequately tested the British system; too few people trained in the Gosport methods before the Armistice, when the Service dismantled most of the wartime training apparatus.[69] The term Gosport remained, but it applied only to the speaking tube that connected two men in flight.

More direct British influence came through cooperative agreements between American and Canadian airmen, hammered out by General Squier and (then) Colonel Hoare in spring of 1917. Among the benefits of the arrangement was the integration of aerial gunnery into the U.S. flight training program. A few Americans who had taken an aerial gunnery course in Canada returned to become instructors at American flying fields. By late 1917, about one-third of Hicks Field, Texas, had been given over to the RFC School of Aerial Gunnery. There, Canadians supplied the planes and equipment to train both Americans and Canadians. In early 1918, the Aviation Section opened its own school at Ellington Field in Houston to train officers and noncommissioned officers as aerial gunnery instructors.

Because of the initial cooperation with the Canadians, the American program closely resembled that of the British. Cadets were introduced to machine guns at the ground school; during primary flight training, they began

Training at Home

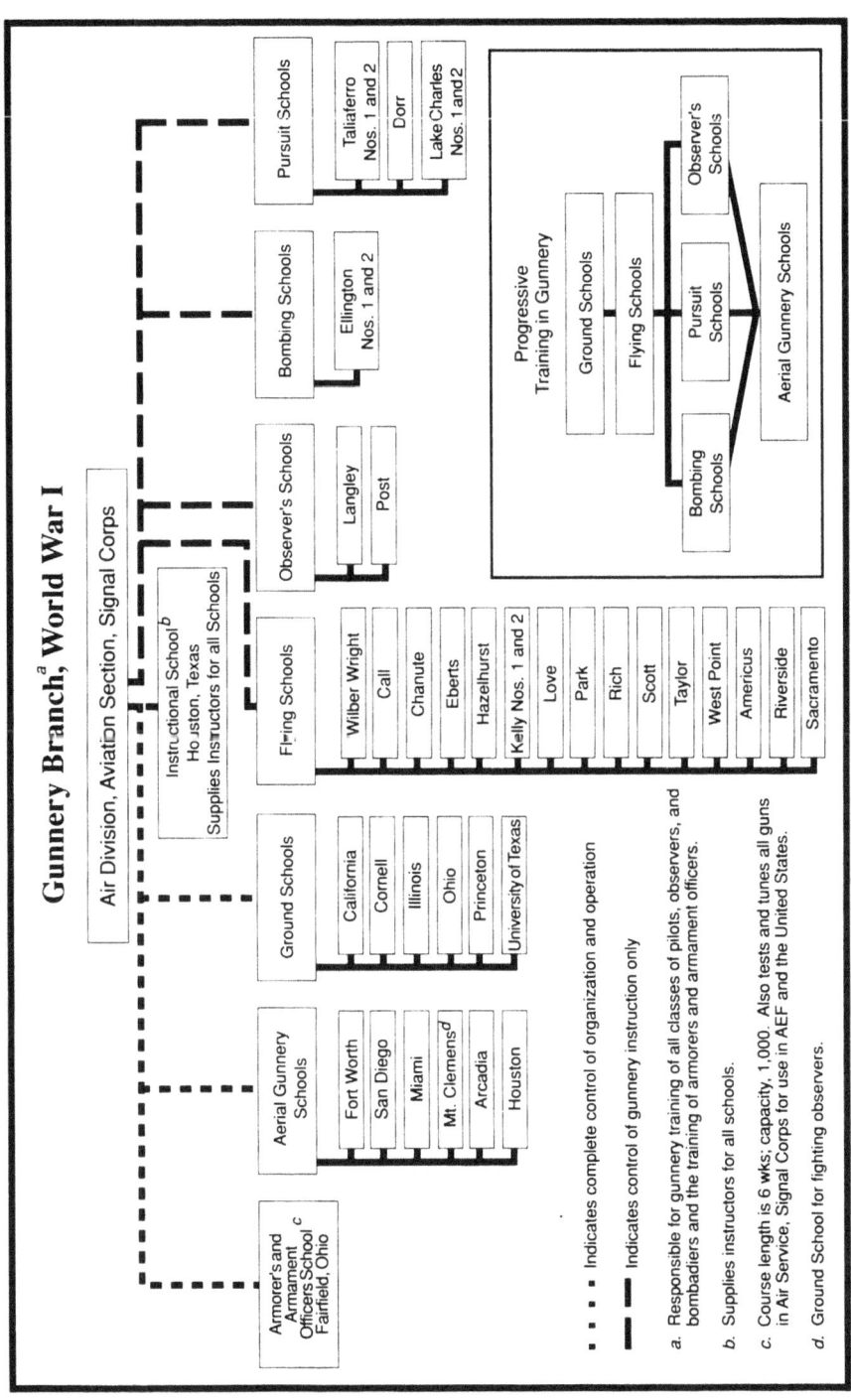

sighting and firing. They started with machine guns on the range, shooting at moving targets. Next came aerial gunnery, firing at targets on the ground and at towed targets and balloons in the air. With a camera gun, the accuracy of fire could be ascertained. All training fields employed the flexible Lewis gun that had first been mounted on aircraft in 1915 and was normally fired by an observer. Americans practiced with a fixed Marlin gun (synchronized through the propeller) instead of the synchronized Vickers, which was unavailable in the United States.[70]

Cadet Briggs Kilburn Adams was among the first group of American students to take gunnery at the RFC School of Aerial Gunnery at Fort Worth. He wrote home enthusiastically, giving a detailed picture of gunnery instruction:

> This gunnery is great fun, for we have so many different sorts of practice. The range work consists in plain target shooting, shooting at silhouettes of machines with aerial sights which allow for the speed of travel, etc. — that is, learning to give the proper deflection of aim so your bullets will cross his line of flight when he is crossing the bullets' line. Then we have surprise targets which pop up at certain intervals here and there, and you load, aim and shoot a burst. It is a training in quickness and precision..... We also have shooting at toy balloons and clay pigeons. Occasionally buzzards fly over and we all pot away at them. In the air we have the camera gun practice, flying the machine and shooting at the same time. Then flying with a pilot while you stand in the rear cockpit with a gun on a swivel and shoot at a target towed by another machine, or silhouettes of machines on the ground, getting practice in diving down again within a few hundred feet, firing a burst and soaring up again.... Then in addition there is the work on the guns, the care and cleaning, and the knowledge of the action and name of parts, etc.[71]

Despite Adams's depiction, in general gunnery instruction tended to be one of the weakest aspects of primary training. The program emphasized learning to fly and, comparatively speaking, other skills merited relatively little attention. Of the possible twelve weeks spent in primary training, perhaps four were allotted to radio, photography, gunnery, and bombing instruction, which were often worked in when bad weather made flying impossible.[72]

Primary training absorbed the lion's share of training resources in the United States. Although planners ultimately decided that the final phase — specialized training — should also take place on American soil, the capacity to build a chain of facilities quickly, the availability of a large pool of pilot candidates, and an adequate supply of primary training equipment determined what Americans would most successfully accomplish. Primary training required quantitative rather than qualitative change from the prewar experience, whereas advanced specialized training posed unforeseen difficulties.

Advanced Flying Training

After finishing the primary course and passing the tests for his RMA rating, the cadet was commissioned and progressed to advanced flight training. AEF manpower and equipment needs on the western front dictated the requirements for this phase. The Air Service program of January 1918 called for sixty service squadrons by June 30 of that year, but the German spring offensive necessitated a commitment of greater numbers of ground troops and a concomitant increase in aviation. The schedule was twice revised, resulting in the Air Service's 202 Squadron Program under which the U.S.-based Training Section would provide specified numbers of pursuit, observation, and day and night bombardment personnel per month. Nearly 146,000 men were scheduled to be in service by June 30, 1919.[73]

Advanced training in the United States adopted the scheme used by tactical squadrons in France of classifying flying personnel (pilots or observers, the latter including all nonpilots) according to mission.[74] Each man was assigned to one of three specialized training programs — pursuit, observation (also called army corps), or bombing. The pursuit pilot, flying a single-seater, usually at high altitude, was the fighter. The observation pilot was accompanied by an aerial observer who gathered information and photographed enemy positions. The bombing pilot and bombardier also flew a double-seat airplane across enemy lines, often at night. Supposedly, all combat airmen had taken some aerial gunnery instruction. Advanced gunnery therefore followed — the pursuit pilot at the pursuit schools and the others at advanced aerial gunnery schools.

Cadets' preferences for specialization were solicited, but for the most part the commanding officer of the primary training school decided which branch of aviation a man joined. After the war, the Director of Air Service, Maj. Gen. Charles Menoher, supposedly clarified the contentious issue of how selection came about: "At no time was the selection of the kind of training a cadet was to take left to his choice. [It] might have been considered, but it was his suitability for this work that decided his classification."[75] In fact, it appears that battlefield requirements, more than aptitude, determined who would go where.

It must be said, however, that the Training Section made every effort to establish the criteria by which individual pilots should be categorized and to outline the tasks required of each mission. In August 1918, the Office of the Director of Military Aeronautics articulated the desirable personality traits for men in each branch:

> Pursuit being purely offensive, a pilot's first qualifications should be aggressiveness and youth. He should be physically quick and alert. Flying should come naturally and easily. He should never be of the heavy, slow-thinking type. He should have initiative and quickness of perception. For Army Corps work, a pilot should be mature, serious, persist[e]nt, pay attention to detail, and be interested in military tactics and man[e]uvers.

World War I

> For Bombing, the older pilots should be chosen. They should be determined, have a good sense of navigation and [be] expert at cross-country flying.[76]

It remained the case that the general profile of pilots chosen for each mission derived from general agreement rather than regulation. The commanding officer of one primary flying field suggested that pursuit pilots be chosen after observing the results of the new acrobatics course. No doubt any number of primary schools employed that informal means of selection.

The burden on the Army, as it tried to build the capability for advanced training in the United States, was ultimately insurmountable. Fields might be used for primary as well as for advanced training, or they might be converted from one type to the other as weather conditions dictated, as equipment became available, or as demand for specialists increased or decreased. Even though curricula for specialized schools had been drawn up by January 1918, the Air Service had not in fact implemented a system with clearly defined stages, nor had it ascertained what constituted completion of a course.

When, in August 1918, the training program had finally pumped too many cadets into primary training and had graduated more RMAs than it could accept in the advanced course, the Training Section curtailed new instruction and centralized existing facilities, establishing groups of schools for pursuit, observation, and bombing. It hoped to improve efficiency so that the number of airmen admitted to pilot training would not drop, but the output of better trained men sent to join operational squadrons would increase. By this time, too, the disarray in aerial gunnery instruction at the flying fields clearly indicated the need for specialized gunnery schools.

Maj. J. R. Moulthrop, executive officer of the Training Section, proposed to General Kenly, Director of Military Aeronautics, that the fields around Ellington become the central locus for bombing training; that Fort Sill, Oklahoma, remain the observation school; and that the three California fields be converted to pursuit training.[77] The concentration of training activities in large schools would centralize logistics, allow standardization of methods of instruction, and eliminate time lost at unsuitable training fields. By the end of May 1918, a bombing school was located at Ellington Field near Houston; a pursuit school at Gerstner Field, Lake Charles, Louisiana, and three other fields to be converted from primary to pursuit; observer schools were at Langley Field, Virginia, and at Post Field, Fort Sill. There were gunnery schools at Selfridge Field, Mt. Clemens, Michigan; at Ellington Field; at Taliaferro Field No. 1, Fort Worth, Texas; and at Wilbur Wright Field, Fairfield, Ohio, which also served as an armorers' and instructors' school.[78]

In April, one unit at Gerstner Field was turned over to pursuit training, including aerial gunnery; the other provided primary instruction. The policy initiated in August of centralizing specialized training resulted in the designation of Dorr and Carlstrom Fields in Florida as a two-unit pursuit and gunnery

Training at Home

Maj. Gen. William L. Kenly (*center*), Director of Military Aeronautics, and two of his associates observe a training flight with pleasure.

school and in the termination of Rockwell Field's primary course to allow pursuit gunnery training. Headquarters was reconsidering the use of Gerstner Field for pursuit, but after a storm destroyed much of the field, it was not reopened during the war. Thus, pursuit training concentrated in Florida and California where weather conditions were most conducive to the demands of fighter tactics.[79]

Pursuit

The approximately nine-week pursuit course encompassed theoretical and practical ground and air training. Ground instruction included study of the organization and employment of pursuit squadrons; the types and assembly of motors, airplanes, and radios; formation flying; German military organization, tactics, and combat aircraft; and fighting methods, maneuvers, attack, and combat, while alone and in formation. Practical ground training included assembly and operation of airplanes, engines, and instruments. Flight instruction consisted principally of acrobatics, formation flying, and gunnery, in which the gun camera photographically recorded the direction and results of fire.[80] Contrary to the popular image of the fighter pilot as lone warrior, the Air Division emphasized the importance of formations:

> As time goes on, Group Flying is becoming more and more advantageous as well as imperative. The day of the individual pilot is past. For a long time the Boche have realized the value of Group Flying as the best means of conserving their pilots and material as well as the most effective way to carry out a mission. Since the Boche always fly in group formation, it stands to reason, that the only way to successfully combat them, is by a group formation of larger size, and better drilled. Hence the importance of group flying in the schools cannot be over emphasized.[81]

World War I

On July 12, 1918, General Pershing requested that the Division of Military Aeronautics supply 125 trained pursuit pilots weekly to the front. The division was unable to comply, principally because of a shortage of advanced trainers. In the original pursuit course at Gerstner Field, for example, much of the equipment was left over from primary training, namely some twenty JN–4Hs and several JN–4Ds and JN–4Cs. Fifty Thomas-Morse Scouts, intended as American-designed advanced trainers, eventually went to Gerstner; approximately 200 were added later at Carlstrom, Dorr, and Rockwell Fields. The Thomas-Morse prototype had been conceived in 1916, but it was rejected until expanded training requirements forced the Army to rely on a broader manufacturing base. The 1917 Scout was powered by the 100-horsepower French Gnome rotary engine and equipped with a Marlin machine gun or camera gun. But even after design and engine changes were effected, the Service still considered it unsatisfactory.[82]

Unfortunately, the U.S. aircraft industry never successfully produced a line of pursuit aircraft, and two-seaters were not diverted to pursuit training. As a result, advanced fighter training could not be completed in the United States. Because the pursuit course was unable to provide as many trained pilots as Pershing wanted, RMAs went overseas to fill the quota in tactical squadrons. Although the pursuit course graduated several hundred men, those sent overseas after completing courses in observation, bombing, and piloting in other two-seat planes were better prepared for combat than most fighter pilots.

Thomas-Morse S–4 Scout

Observation

Many cadets dreamed of becoming one of the daredevil gladiators of the air, one of the famed aces whose victories were counted and lauded and whose names were on the lips of the American public. But the AEF kept up a steady demand for observers and observation pilots. "It is considered," stated one communiqué from Washington to the training fields, "and very wrongly by a good many pilots, that observation work, admittedly the most important work, and one without which the army could not function, occupies a secondary place in aviation, so far as the opportunity of its pilots and observers to distinguish themselves is concerned. This is not the case."[83]

Operational requirements that reflected U.S. military aviation's fledgling history and doctrine dictated that the American training program assign a large number of its people to observation. Compared to pursuit, the training program for observation pilots more closely met its objective, ultimately graduating more pilots than the number mandated by the AEF. By late May 1918, the weekly number of pilots authorized for the two observation pilot schools had been scaled down to 40 men, even though the schools were graduating 60 per week.[84] The Training Section thus contemplated a surplus of observation pilots. But since it was unable to train observers in the same numbers, there developed an ongoing imbalance in production of the two-man crews.

Although the observation mission was familiar, for the first time the Air Service had to craft formal crew training curricula. Besides pilots transitioning to new aircraft, nonpilot observers had to be instructed how to gauge the enemy's activities and to photograph and, frequently, to shoot. Thus, in addition to programs for piloting and gunnery — the two elements of pursuit training — observation required teaching the additional skills of radio communication, photography, and artillery spotting. Because a clear concept of crew training had yet to be formulated, the function of each individual was vaguely articulated, and much of the training for both crew members overlapped; consequently, the observation pilot and aerial observer shared many duties while in the air.

Because observation was the principal arena where ground and air officers' skills converged, officer procurement for the observation training program introduced the greatest friction among the combat branches. Aviation had endured an uneasy isolation within the Army, but observation squadrons now were attached to corps in the AEF, bringing the Air Service under direct command of, and in cooperation with, the ground forces. Moreover, for some time, observers were drawn from the regular line of the Army and were trained at both artillery and aviation schools.

In consultation with the French Aviation Liaison Officer, in early July 1917 American training officials began planning a school for nonpiloting aerial observers. According to the French expert, it would take at least two months to train an observer, and the training should occur at a station housing both

World War I

A balloon observer with radio set in the gondola of a balloon (*left*), and instruction in aerial gunnery at the airplane observers' school (*below*).

airplanes and artillery. In response, Maj. Henry H. Arnold of the Division of Military Aeronautics recommended an immediate transfer of the 3d Aero Squadron from San Antonio to either Fort Sill or Fort Bliss, both artillery posts. He requested that U.S.-trained observers be sent to France for advanced training, and that French officers schooled in the methods of aerial observation used at the front aid the training program in the United States.

The Secretary of War approved the idea of establishing a school at Fort Sill, Oklahoma, in conjunction with the School of Fire for Field Artillery, located about three miles distant.[85] By early February 1918, Langley Field hosted the second school for observers. Commissioned personnel from the Coast Artillery first took a preliminary course at nearby Fort Monroe before their assignment to Langley. In April, the Service created the rating Aerial Observer (airplane or balloon) for graduates of those two special schools, conferring official status to the job previously performed more informally.

Throughout this period, the Signal Corps and Field Artillery skirmished over rules governing chain of command of aerial observers. Because the Aviation Section had decided originally that enlisted cadets lacked the qualifications for training in aerial observation, and that only Artillery and Infantry officers should be assigned as aerial observers, artillerymen and infantrymen had to remain involved in at least the selection process if not in the training. The commandant of the School of Fire for Field Artillery chose from the volunteers of each graduating class a certain number of lieutenants for duty as airplane or balloon observers. Likewise, commanding generals from Artillery and Infantry divisions supplied a list of volunteers from which the Adjutant General chose names of individuals to attend an Air Service School for Aerial Observers.[86]

But division commanders contended that they could neither supply the requisite numbers of officers as volunteer aerial observers nor replace those assigned. For their part, airmen believed that some ground commanders urged their best men not to apply for transfer. At first the Artillery detailed lieutenants from Army regiments only for the period of their training as aerial observers. But the Signal Corps rejected these men and insisted that aerial observers be attached to reconnaissance squadrons rather than on temporary duty from an Artillery regiment.[87] Signal officers argued against the return of observers to their original divisions after they completed the observation course, maintaining that the "eyes" of the Army should remain with the Signal Corps.

Some of the squabbling subsided when, near the close of 1917, General Pershing ordered that twelve Artillery and four Infantry officers be attached to each observation squadron. And the Secretary of War directed in mid-January 1918 that officers on detached service from the Artillery, Infantry, and Staff or Cavalry would, after successful completion of their training, be detailed to the Signal Corps.[88]

In early 1918, the Division of Military Aeronautics Training Section planned that by May 1, U.S. training schools would graduate 260 airplane observers, 60 balloon observers, and 260 observation pilots monthly. In February, 25 observers and 25 pilots were graduating each week from Post Field at Fort Sill, and the figures were expected to double by May 1. Langley began operations later, with a weekly detail of 15 officers from the heavy artillery.[89]

As indicated, quotas for observation pilots continued to be met fairly smoothly; not so with observers. The pilot overage further skewed the balance between pilots and observers such that, by mid-July 1918, the AEF was

World War I

desperate for observers. As one member of the AEF Training Section advised the Division of Military Aeronautics Observation Section:

> We desired 200 artillery observers with aerial gunnery, but stated that the full number called for was desired even if all had not such training. You will have to make every effort to send us fully trained men at the earliest possible date, as the facilities in the AEF will not permit of giving anything more than a refresher course.... If fully trained material is not available, make up the requested number by the best partially trained men available.[90]

Now, as in wars to come, field commanders castigated stateside training staffs for sending poorly trained airmen, but they then went on to demand manpower at any cost. In this instance, the U.S.-based Training Section notified all ground school graduates that, because of the glut of people awaiting pilot training, no cadets would be accepted into the flying schools for several months, but men could volunteer as observers. Otherwise, they would be forced to transfer to other services, face immediate discharge from the Air Service, or wait until such time as they could be trained as pilots. Already enrolled cadets not deemed qualified to be pilots but who were "otherwise desirable officer material" or those who were already qualified as pilots but who were "not at ease in the work" could become bombardiers or artillery observers.[91]

The Air Service was, in other words, forcibly reconsidering its stance that only commissioned officers, not cadets, would be accepted as aerial observers. The dual system of Artillery and Signal Corps observer training had foundered on several

Brig. Gen. Benjamin D. Foulois and Secretary of War Newton D. Baker during the Secretary's visit to the overseas training site at Issoudun

levels, not the least of which was the relative trickle of men from the Field and Coast Artillery. The Signal Corps therefore decided to recruit its own observers from nonpilot cadet volunteers who would receive special training at ground school and additional training with both the Artillery and Air Service. In August 1918, a new policy directed that aerial observers be commissioned in the Air Service rather than the Artillery, Infantry, or Cavalry. Those lacking artillery experience would be given instruction by the Artillery, and all aerial observers would receive training in aviation schools.[92]

The urgent call for more trained observers continued into the fall. With some heat, Lt. Col. Herbert A. Dargue reminded the Director of Military Aeronautics that "the deficiency in observers in France is liable to cause an exceedingly embarrassing situation, unless every effort is to be put forth in the United States to expand observer schools to the absolute limit and train as many observers as possible."[93] In an attempt to boost the morale of those trainees facing a seeming diminution of status and, no doubt, to impress on more men the worthiness of volunteering, the Chief of Training rallied all commanding officers of the flying schools to the view that "there is no question as to the importance of this work or the fact that it is of the same relative importance and dignity as that of the pilot."[94]

By October, the Division of Military Aeronautics had increased authorizations at Langley and Post Fields and considered shortening the observers' course from seven to five weeks.[95] Owing to the different backgrounds of the students — whether commissioned in the Air Service or Artillery, whether cadets or officers — the length of the observer course varied considerably over the relatively short period of its existence. Generally the course matched that offered by the Artillery schools, which were themselves different lengths. In late 1917, the aerial observer course was six weeks long; it later became ten weeks, equal to the School of Fire for Field Artillery. Later, all three schools gave a seven-week course, and finally, to meet the stringent AEF demands for observers, the observer course was reduced to five weeks for commissioned personnel and ten weeks for cadets. Before going overseas, observers spent three additional weeks in the aerial gunnery course at Selfridge Field.[96]

The curriculum changed along with the course length, but the basic format remained consistent and highly practical, built upon skills learned at the Artillery schools. An indoor miniature range, which replicated a battlefield panorama as viewed from the air, allowed students to work over problems in artillery control and reconnaissance. A lantern slide projected a photograph of a sector of the combat zone onto a white screen on the floor. Acting as an aerial observer, one student sat in the middle of an upper gallery with a buzzer or Morse key equivalent to the wireless set on an airplane. Another student on the floor performed the role of the battery on the ground. He placed onto the floor pieces of cardboard coded like the cloth panels used by the Artillery to identify itself and to communicate with the observer. Electric-light flashes representing artillery fire, which in actuality would appear as puffs of smoke, were projected onto the screen. By means of a firing map, the observer could determine coordinates of a target and radio directions for fire to his compatriot on the floor. The other students sitting in the upper galleries and the instructors analyzed the results of the shots.[97]

From its inception, most of the curriculum for observer training had been drafted in consultation with British and French air officers. To maintain close liaison between American observer training and the needs and techniques at the front, a group of French and British officers came to the United States during the summer of 1918. In return, two American observer instructors went overseas to study the training methods at the American and Allied schools, and a combat-experienced American observer followed them and returned to the United States as a training adviser.[98]

From all reports, the American program accrued considerable benefits from having fairly close liaison with Europeans. British and French training methods diverged in some respects, and American officers continued to discuss the advantages of the competing types of signaling, or code system, and whether it should be the pilot or the observer who would shoot or signal. One Signal Corps officer wrote to a friend at the School for Aerial Observers in September 1917 that "at present the tendency abroad... is to have the pilot do all the observing and radio sending — that is, spotting artillery fire, and have the observer on the lookout for the Huns."[99] As late as the end of January 1918, a memorandum to all schools stated that "the exact division of duties between pilot and observer in the United States Air Service has not yet been determined." But it was considered "probable that most of the conduct of artillery shoots will be assigned to the observer."[100] Gunnery was to become the observer's job, and by the summer of 1918 it was also decided that the observer would do all the signaling.[101]

Although the division of duties between pilots and observers was not finalized until the spring of 1918, training officials continued to insist that both pilots and observers be trained in aerial gunnery. They removed gunnery from the jurisdiction of the ground schools and flying branches, although aerial gunnery remained a part of both courses. By mid-May 1918, Taliaferro and Selfridge Fields served as aerial gunnery schools for observers and observation pilots. Since the AEF operated no aerial gunnery school at the time, all advanced aerial gunnery training was to take place in the United States. Another large aerial gunnery school for pilots was planned for Chapman Field, near Miami, but construction had just begun at the signing of the Armistice.

The aerial gunnery curriculum for observation pilots and observers mirrored the English more closely than the French system, although two French instructors from the French aerial gunnery school at Cazaux arrived to demonstrate combat practices. Observation pilots had already received instruction at the observers school in the Marlin gun (the fixed, mounted gun

Training at Home

fired by the pilot) and the camera gun; observers had practiced on the Lewis gun (the flexible, mounted type that allowed a gunner or observer to swivel the gun in several directions) and the camera gun. The aerial gunnery schools gave a three-week course for pilots and observers. Pilots fired the Lewis and Marlin guns; worked with the British synchronizer gear, camera gun, ring sights, model airplanes, and clay pigeons; and practiced range and aerial firing. For observers, the subjects included ground range shooting, aerial shooting at targets on land and water, camera gun practice, ring sight and model airplanes, testing ammunition, and lectures on formation flying and aerial tactics.[102]

Bombardment

Through the years of American neutrality, military officers listened attentively to reports from Europe of aerial dogfights and bombs raining from the skies and slowly instituted a flight training program at home in which gunnery assumed increasing importance. Nonetheless, the U.S. Army officially reckoned the airplane as a nonoffensive tool of war. The initial appropriation for aviation provided only for reconnaissance units. If, nine months into the war, pursuit (the defender) was thought to be the most glamorous, and observation (the "eyes") the most critical, bombardment came in an unpopular third among the specialized branches. Because the Army leadership so tardily and reluctantly developed bombardment as an acknowledged branch of aerial warfare, it was the last section to be organized, and it initially lacked most essential equipment. But it was also subject to less pressure since the AEF requested only two bombing pilots for every five pursuit and three observation pilots.

Once again the Aviation Section set out to convince young men of the critical contribution of nonpursuit pilots and to try to correct the prevailing impression that bombardiers were "Second Grade Men" who had "failed to qualify as flyers, or [were] just passing through the Ground School by a small margin above that required." Such misapprehensions, from the perspective of the Bombing Section at any rate, could be found in reports from ground schools, in the obvious scarcity of volunteers, and in at least one congressional inquiry that prompted a memorandum to all ground schools in defense of the occupation:

> The Bombing Section requires men who possess great skill of accuracy, great brain properties, conservativeness, knowledge of Tactics, and many other qualifications .... [Bombing pilots] penetrat[e] far into the enemy's lines, to points out of range of our own largest guns, the efforts of Cavalry, Infantry, or any other arm of the service, destroying munitions factories, stores and food supplies, crippling transportation, destroying avenues of approach or retreat, demoralizing reinforcements being rushed to the front, reducing heavy fortifications and silencing artillery, harassing infantry on rest periods behind the lines to that extent that when they return to the front again, they have not been sufficiently rested to

World War I

effectively carry on the battle.... This work therefore requires men of great courage, nerve and endurance, who[se] qualifications must not only be equal to the Fighter and Observer but combines [*sic*] many other qualifications not required of the Fighter or Observer.[103]

In spite of the public relations salvo, the notion that the men taken into bombardment, especially as bombardiers, were not the most highly qualified, contained some truth. In one bulletin to the schools, the Office of the Chief Signal Officer tried to put the best face on the situation by observing that bombardiers were ground school volunteers and cadets at flying schools "who failed to qualify as Pilots due only to inability to learn to fly."[104] The statement hardly read as a thumping endorsement.

Planning for bombardment schools began in January 1918, and again the Training Section turned to its allies for advice regarding equipment and techniques. All other aspects of a flight training program on wartime footing had been introduced to some degree in the primary course, but as late as February 1918 no instruction for bombardiers had begun because the service could claim only two experienced instructors, and only two men were then under instruction since no one else had volunteered.[105]

Unable to await a casual influx of candidates, since the commanding general of the AEF had called for 236 bomber pilots and an equal number of bombardiers by May 1, the Aviation Section undertook a recruitment drive for instructors at technically oriented universities, and it ordered 50 ground school graduates to report to Ellington Field toward the last of February for bombardier training. Sufficient cadets were available for instruction as bombing pilots, although they had to train on available equipment until they received "bombing equipped planes."[106]

In fact, training aircraft did not arrive as anticipated, and other equipment remained in similarly short supply. A shipping schedule worked out at the beginning of February called for the delivery of 10,000 dummy bombs per month; as of the third week in March, only 75 had arrived. Although the bombardment school was organized March 1, 1918, and 190 bombardiers and 190 bomber pilots reported for that first one-month course, the school had thus produced no graduates by early April.[107]

By midsummer, conditions improved. Plans were laid for another school at San Leon, Texas, near Ellington Field, but it was not complete before war's end. By then, the broader concern of matching operational requirements in Europe with the type of stateside training program that fed it eclipsed procurement problems.

At least as outlined, in ground school each team of pilot and bombardier was to study theory of bombing and then receive two additional levels of instruction before dropping dummy bombs.[108] Classwork included the study of organizations, map and compass reading, ballistics, photography, tactics, day and night bombing, preparations for raid, miniature range exercises, and study of and then actually dropping dummy and explosive bombs. After commissioning, the bombardier and the pilot held equal rank, and both continued for two more weeks of aerial gunnery training.[109]

At the behest of air officers in France, the bombardment course emphasized formation and night flying. (A third critical element, cross-country, was already integral to reconnaissance training.) To explain how the 7th Aviation Instruction Center in France was teaching formation flying, one American officer wrote home in early December 1917: "It has been found after repeated experiments that it takes about seven hours before a man can keep his position in a group. We are training them to fly in groups of five . . . . In this training formation each man could see the leader easily and the performance of each man in the group could be judged by the observer in the group leader's airplane." AEF squadron members communicated further by a system of previously determined signals.[110] Bombardment training in the United States quickly adopted the same techniques.

Well before the stateside bombing school opened, General Foulois wrote to the Chief Signal Officer regarding the other element that should be stressed in bombardment training: "It is particularly important that plans should be made at once for the training of pilots for night-bombing."[111] Reports from France frequently repeated the necessity for training in nighttime flying and, after the bombing course got under way, it was further subdivided into day and night bombardment. Crews specializing in each received instruction at Ellington Field.

Too Little, Too Late

Between July and September 1917, General Pershing developed a plan that required 260 service squadrons to be on the western front with corresponding backup manpower and equipment by June 30, 1919. Though adjusted periodically, the schedules inevitably required more personnel than the American training program could supply. To meet the demands of the AEF, the Division of Military Aeronautics had begun by the summer of 1918 to redistribute flying activities and centralize advanced training. But if in the last quarter of 1918 the division planned for the future, it found the present conditions, as one officer opined, "woefully inadequate."

The Training Section's task of supplying trained pilots and observers, technicians, and mechanics hinged on the availability of training airplanes, engines, guns, and equipment akin to combat types. Unfortunately, the tactical aircraft authorized for training by the Chief of Staff in the summer of 1918 could not be diverted from the front. Furthermore, American aircraft production proved tortuously problem-ridden, thus dictating success or failure for entire elements of the training program. Despite relatively high manpower reserves, equipment shortages forced training to contract during the summer of 1918.

World War I

Some ground schools closed, and the General Staff capped the number of airmen on the basis of the available equipment "plus [a] moderate margin for safety."

Had the war continued, American training and airplane production might well have reached the quotas set by the AEF, but during the nineteen months of war, by necessity, training policy as crafted by the Department of Military Aeronautics was more often reactive than predictive. Because of geography, the training program in World War I was bifurcated between efforts to mobilize and train air officers and men in the United States, and the special problems of operational training in Europe. Yet the two voices had to sing together, if more often in call-and-response than in duet.

Flying Fields, Aviation Section, U.S. Air Service, November 11, 1918

Field	Location	Special Function	Max. No. Cadets
Baker's	Rochester, N.Y.	Photographic	
Barron[a]	Everman, Tex.	Primary flying	300
Bolling	Anacostia, D.C.	Advanced flying	
Brooks[b]	San Antonio, Tex.	Instructor's school	300
Call[a]	Wichita Falls, Tex.	Observer's school	300
Carlstrom	Arcadia, Fla.	Pursuit flying	400
Carruthers[a]	Benbrook, Tex.	Primary flying	300
Chanute[a]	Rantoul, Ill.	Primary flying	300
Dorr	Arcadia, Fla.	Aerial gunnery	120
Eberts[a]	Lonoke, Ark.	Primary flying	300
Ellington[e]	Houston, Tex.	Bombing school Aerial Gunnery	600
Emerson	Columbia, S.C.	Advanced flying	
1st Reserve Wing:	Mineola, L.I., N.Y.		
Brindley	Commack, L.I., N.Y.	Advanced flying	
Henry J. Damm	Babylon, L.I., N.Y	Advanced flying	
Hazelhurst	Mineola, L.I., N.Y	Advanced flying	
Lufberry	Wautaught, L.I., N.Y	Advanced flying	
Mitchel	Mineola, L.I., N.Y.	Advanced flying	
Roosevelt	Mineola, L.I., N.Y.	Advanced flying	
France	Cocoa Walk, C.Z.	Advanced flying	
Gerstner	Lake Charles, La.	Bombing school	600
Kelly[b]	San Antonio, Tex.	Primary flying	600
Love[b]	Dallas, Tex.	Primary flying	300
March	Riverside, Calif.	Primary flying	300
Mather	Sacramento, Calif.	Primary flying	300
Park Field[b]	Millington, Tenn.	Primary flying	300
Payne Field[b]	West Point, Miss.	Advanced flying	300
Post	Fort Sill, Okla.	Observer's school	315
Rich[d]	Waco, Tex.	Primary flying	300
Rockwell[a]	San Diego, Calif.	Pursuit flying Aerial Gunnery	400
2d Reserve Wing:	Park Place, Houston, Tex.		
Selfridge	Mt. Clemens, Mich.	Aerial gunnery	350
Scott[a]	Belleville, Ill.	Primary flying Instructor's school	300
Souther[a]	Americus, Ga.	Primary flying	300
Taliaferro	Hicks, Tex.	Aerial gunnery	180
Taylor[c]	Montgomery, Ala.	Primary flying	300
Camp Dick	Dallas, Tex.	Cadet gunnery camp	4,500
Wilbur Wright	Fairfield, Ohio	Armorer's school	600
Langley	Hampton, Va.	Observer's school	210

a. One auxiliary field attached.
b. Two auxiliary fields attached.
c. Three auxiliary fields attached.
d. Five auxiliary fields attached.
e. Six auxiliary fields attached.

FIVE

Air Service, American Expeditionary Forces

> I know that I shall meet my fate
> Somewhere among the clouds above
> ....
> Nor law, nor duty bade me fight
> Nor public men, nor cheering crowds,
> A lonely impulse of delight
> Drove to this tumult in the clouds
> I balanced all, brought all to mind,
> The years to come seemed waste of breath
> A waste of breath the years behind
> In balance with this life, this death.
>
> —W. B. Yeats, "An Irish Airman Foresees His Death" (1917)

In drawing up plans for a wartime air arm, the Division of Military Aeronautics in Washington and the Air Service, AEF, extrapolated from the promise, but not the performance, of the prewar aviation program, which had barely struggled to survive. The Aviation Section lacked the administrative structure to mobilize and train an aerial army and had neither articulated nor tested doctrinal concepts. In his memoirs, General Pershing, who had directed the Punitive Expedition and therefore glimpsed something of the possibilities of military aviation, expressed "humiliation" at the primitive state of Army aeronautics on the eve of war. As a result, once into the conflict, much of the guidance for forming a combat force and establishing the training necessary to achieve it came from the Allies.

During the summer of 1917, a group of American officers led by Col. Raynal Bolling sailed to Europe to inspect aviation facilities and consult with Allied leaders. They returned home with information and recommendations

World War I

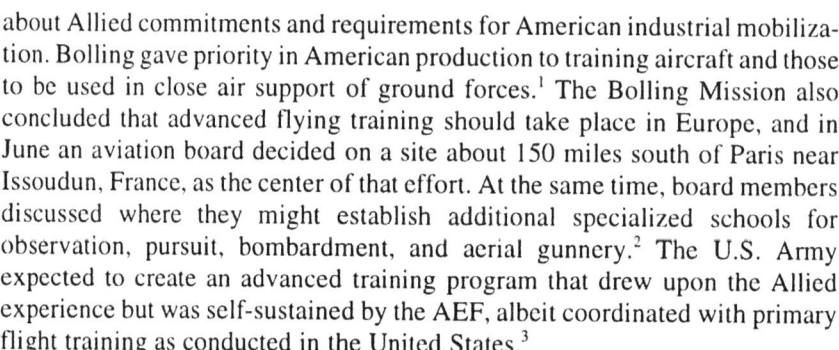

Col. Raynal Bolling and an artist's sketch (J. Andre Smith) of Issoudun, France, as the field appeared during World War I

about Allied commitments and requirements for American industrial mobilization. Bolling gave priority in American production to training aircraft and those to be used in close air support of ground forces.[1] The Bolling Mission also concluded that advanced flying training should take place in Europe, and in June an aviation board decided on a site about 150 miles south of Paris near Issoudun, France, as the center of that effort. At the same time, board members discussed where they might establish additional specialized schools for observation, pursuit, bombardment, and aerial gunnery.[2] The U.S. Army expected to create an advanced training program that drew upon the Allied experience but was self-sustained by the AEF, albeit coordinated with primary flight training as conducted in the United States.[3]

Air officers repeatedly revisited the question of the amount and type of training that should occur in the United States and in Europe. They searched not for the ideal, but for the possible. They ultimately decided that it would be easier to create and coordinate the elements of a program at home rather than near the front, but in fact the American Army did both. Training responsibilities were split between the stateside Aviation Section, renamed the Division of Military Aeronautics in 1918, and the overseas Air Service, the AEF. The former built its entire infrastructure from scratch and was chronically short of equipment. The latter potentially operated at a higher level of chaos since it was subject to greater variables: many different aircraft, competing training methods, and far-flung facilities under the auspices of the French, English, and Italians. Interestingly, however, in his thorough report of November 1918, Colonel Dargue approvingly noted that the "organization of the Air Service in the A.E.F. is such as to make the whole function with very little friction.... This is especially noticeable in the Training Section." He credited an efficient inspection system for much of the success.[4]

Primary Training

Early intentions to conduct only advanced training in Europe immediately went awry. Because the structure for primary flight training had yet to be erected in the United States and because European facilities appeared to have space, Colonel Bolling arranged for several hundred American cadets to be admitted to French training schools, and he contacted the British and Italians to obtain similar commitments. The English accommodated about 200 men, and approximately 500 cadets went to Foggia, Italy, for primary training.[5]

The largest number of men found themselves in France. The Air Service partially gave over the Third Aviation Instruction Center (3d AIC) at Issoudun, France, originally intended for advanced instruction, to primary training. Negotiations for aviation schools at Tours and Clermont-Ferrand to be turned over to the U.S. Army — the former for observation and the latter for bombardment — were also in progress. But Tours, too, was converted to primary training. The old French aero school, located on a plateau across the river at Tours, came into American hands as the 2d AIC in September, and it remained the principal American primary flying school in France until the program's end.

Shifting some primary training to Europe not only disrupted orderly planning at home and abroad, but the arrangement got off to a poor start as faulty communications between France and the United States and a clogged shipping schedule delayed initial arrival of American troops. Frustrated because the French had stated originally that they could accommodate 325 cadets during the flying season ending in mid-September, Colonel Bolling remonstrated with the Chief Signal Officer in mid-August:

> The British, French and Italians have offered to train large numbers of American aviators and mechanics. The offers afford by far the best and surest means of securing the necessary personnel ... for operations early next spring. In fact, they afford almost the only means of securing such personnel.... Yet there have been hopeless delays.... Meanwhile the French have made all arrangements to receive them holding open their schools and seriously interfering with their own training program. Right now the French have a great number of vacancies in their schools awaiting the arrival of American student pilots. In fact, only this week they had one

World War I

General Pershing visiting the 2d Aviation Instruction Center at Tours

school held idle and it is now only partially filled by the arrival of a small detachment of student aviators.[6]

Tardily, cadets streamed into France between October and December of 1917, and the pendulum swung the other direction. Now it became clear that facilities in France had become oversubscribed and airplanes in short supply. A personnel shortage had turned into a glut and it appeared that, even at full capacity, Tours could not expect to handle the load. Moreover, weather conditions during the colder months were inauspicious for teaching novices and also restricted the number of available flying hours. Contrary to the expected production of one hundred pilots a month from Tours, from September through February an average of fewer than thirty-seven men graduated every month.[7] With an increased Allied aviation program, no more room for Americans could be found in the French flying schools. By November the logjam had become sufficiently obvious that General Pershing cabled home that "no pilots should hereafter be sent to Europe until they have taken their military aviation tests in the United States. This is imperative in order to avoid congestion in the European Schools."[8]

The Air Service therefore attempted to reinstate the original policy of providing primary training at home. About the time General Pershing halted cadets' deployment overseas, then Director of Air Service Instruction, Maj. Joseph E. Carberry, commented optimistically to his counterpart on the General Staff that "it will only be a short time before all cadets can be given their preliminary flying training in the U.S., and the necessity of giving preliminary instruction on a large scale in Europe will be discontinued."[9]

But the new year found some thousand cadets in France awaiting primary training. So stunned at the "exceedingly bad" morale and the general condition of training in France — "without exaggeration worse than it can possibly be imagined" — the newly appointed head of the Training Section, Maj. W. A. Robertson, advocated considerable expansion of U.S. training facilities to include all advanced flying.[10] The Air Service ultimately adopted this policy; even so, it took well over a year after the declaration of war for the cadets waiting in France to pass through the system, extending primary flight training there beyond all reasonable hopes. During that interval the French conducted preliminary training at Tours, before turning that facility over to Americans, and at Avord, Chateauroux, and Voves, and briefly, at Vendome.

Many of the first cadets to arrive in France had been specially chosen because of their exceptional records at ground schools. The First English Detachment arrived at Issoudun the middle of October, having been sent first to England for primary training. Finding no space in the English flying schools, the men proceeded to Tours, where again there was no room. On to Issoudun, the cadets met further shortages of facilities and equipment. Since they had no place else to go, they stayed at Issoudun, but instead of learning to fly, they were put to work as laborers in the construction effort then underway. They walked guard duty, cleared land, erected buildings, and engaged in more menial

The field at Issoudun

World War I

tasks. Further compounding their now very low morale, others who had completed flying training in the United States began to arrive in France already commissioned. Many of these men had not been numbered among that initial elite or had enlisted later than those digging ditches in France. Irate inquiries from congressmen and their constituents regarding the unfortunate fate of the supposedly select airmen flooded the Aviation Section. The Service attempted to rectify the unintentional injustice by commissioning all cadets in Europe based on their date of completing ground school. Nonetheless, improving their standing did not assure their entry into the flying schools. This nonflying cadet force, ultimately numbering about a thousand men, each receiving pay of $100 per month, became known as the Million Dollar Guard.

The crisis peaked near the end of 1917. Tours and Issoudun conducted primary training for as many cadets as possible, even though some were left to languish, while other European schools also accepted trainees at overflow levels. Some new arrivals stayed at the Beaumont Barracks in Tours; others lodged at St. Maixent; still others were quartered at AEF headquarters in Paris. In January the Training Section attempted to introduce some order by having all untrained cadets, of whom no more were authorized, removed from the schools and sent to St. Maixent, site of an old French barracks. It was to serve as the concentration point for all aviation troops arriving in the AEF. From

Beaumont Barracks at Tours

there, men were released for training as vacancies occurred. While they waited, some pilots, ground officers, engineers, and adjutants received minimal training. According to Lt. Lucien Thayer, who fought in the war and later chronicled the aviation effort, St. Maixent was "a quiet, contented, prosperous little French town of a few thousand inhabitants, rich in historical traditions." But transformed hurriedly for wartime purposes, the lovely old town did not easily accommodate to the needs of a modernizing combat force. Thayer recalled the remark made by a young lieutenant of the 12th Aero Squadron shot down in combat who, when asked about his treatment as a prisoner of war, replied that it was "a damn sight better than I got in France as a cadet."[11]

The bulk of AEF training took place in France, with the benefit of French expertise and equipment. French officers and Americans from the *Lafayette Escadrille* trained the first American instructors at Tours, but even after the Americans assumed control at Tours, many French instructors remained. Thus, until General Pershing's directive late in the war to employ only American instructors, it was natural for the Air Service to use French training methods. American training schools in France embraced *Roleur* training in which the student pilot learned on the underpowered Breese Penguin, gradually increasing speed on the ground, and then taking flight and practicing more complicated maneuvers in increasingly faster airplanes. In the initial phase, the system differed from dual-control instruction in the United States.

The primary course at Tours required about twenty-five flying hours, whereupon a cadet received his RMA rating and went to Issoudun for advanced training.[12] Despite the enormous pressure of time and space, instructors at Tours sometimes paid more attentive to the quality of a cadet's skill as judged by his landing technique than to the number of flying hours he accrued. Americans held firm to their careful prewar system of flight training, not just for the purpose of evaluating student progress, but because safety considerations remained paramount. The Tours school was already overcrowded, and every available airplane had to be used to the maximum during fair weather. With many beginning pilots aloft simultaneously, their comings and goings were especially dangerous.

Major Howard Davidson, who had been on the instructional staff at the Cornell School of Military Aeronautics before going overseas to assume responsibility for flying instruction at Tours, likened the system developed at Tours to the later Berlin Airlift. The *chef de piste* (supervisor of flying) stood in the middle of the field to control the flow of activity. He recorded the landings and times for each pilot, who was identified by a number painted on the side of his airplane. In the carefully choreographed training ballet, the pattern resembled one gigantic circle, some pilots taking off to join and others leaving to land, all moving counterclockwise in the air. "Anyone who came in and landed across there was in danger," Davidson recalled, "because these cadets were not very quick on perception.... It was worth your life almost

World War I

An American flier of the *Lafayette Escadrille*, Robert Soubiran, standing beside his Nieuport 17 (*upper left*) and the Breese Penguin (*lower right*) on which many young American fliers trained in France.

when you took off, and anything, except the direction that they were going, was just a one-way street."[13]

The French employed a variety of aircraft, in combat and in training. Americans at Avord learned on the Bleriot or the Caudron; promising cadets then passed to the much admired Nieuport for advanced pursuit training.[14] The French could most easily spare the Caudron G–3 for the American primary school at Tours, which was itself modeled directly on the Caudron course at Avord. The Caudron G–3 was a single-engine reconnaissance airplane of 1914 vintage, already outmoded by bomber models developed from it. According to Lucien Thayer, the Caudron was an airplane subsequently "regarded with amusement or derision."[15] To those flying it at the time, its principal, if not single, virtue was its stability. American cadet John Richards, who took instruction on Caudrons at Tours while the school was still in French hands, considered the plane "not a pretty machine, but stable."[16] Davidson agreed, calling it "queer," but with its warping wings (like the old Wright pushers), safe.[17]

Another American cadet who trained in Boulogne unfavorably compared the Caudron to the Bleriot on which he learned. "The Bleriot training is just as good as it is reputed to be — I think — far superior to the Caudron which is like the Curtiss, although Caudron students won't admit it. The difference between the two is about the same as that in learning to swim — the Caudron boys paddle around in shallow water with a pair of floating wings; we are shoved off the dock, almost get drowned, get scared to death and develop a lot of confidence."[18] Although Davidson found the Caudron acceptable, he hardly considered it to be like the Curtiss, and in fact offered to trade two Caudrons to a British training detachment for one Jenny. Oftentimes the British declined to negotiate since good fortune had provided them mostly with Avros, which many Americans and British considered the most desirable training machines.

Because Issoudun had not been equipped for primary training, some cadets went through a hair-raising initial instruction there on outmoded Nieuport combat aircraft. Lt. Col. Walter G. Kilner, then the commanding officer, reported that beginning pilots with some experience on the Curtiss JN–4 (Jenny) did best on the Nieuport 25. The greatest difficulty in this transition arose in learning to work the throttle of the rotary motor. Another problem showed up in the instinctive tendency of U.S.-trained pilots to grab for the stick with their left hand when they found themselves in difficult situations. With the exception of the Bristol Fighter, in European tactical aircraft the stick was to the right, with the other major controls on the left.[19] Harold Hartney, who trained at Issoudun, remarked that "the right hand position on the French Nieuport and Morane planes confused several gallant boys on their first flights in France; the engine quit; they turned back into the field and spun to Eternity." Hartney himself loved flying the Nieuports.

> Fresh from a terrible siege on drab Jennies at home, it was like going from the lumbering trucks of pre-war days to the lightest and liveliest motor cars of 1940. Never have I experienced such a contrast in flying. You guided one of these ships as if it were part of you. They responded almost in exact accordance with your thoughts, instantly, and not like the heavy Jennies in which you would put the stick over and then wait for the wing to get good and ready to come up in response. There was none of that painful old lag.

Nieuports were almost the equal of the Sopwith Camels, about which Hartney also rhapsodized: "those new Sopwith planes bounded into the air like gazelles.... The Camels, instead of circling in wide paths consuming a minute or two, turned the full 360 degrees instantly, almost 'on a dime.'"[20] As he surveyed the differently trained airman at Issoudun, and the competing techniques and aircraft, Colonel Kilner came to the view that the best students

World War I

Sopwith F-1 Camel (*left*)

Morane-Saulnier (*right*)

coming to primary training in France were those with no previous experience at all, because they had the "advantage of having nothing to unlearn."[21]

The Italians agreed to host as many as 500 cadets in a school at Foggia, about 200 miles southeast of Rome.[22] In September 1917, the school, officially the 8th AIC under joint American and Italian jurisdiction, began training the first detachment of forty-six cadets, all honor graduates of American ground schools. The detachment had been sent first to Avord, but when plans for training in Italy crystallized, it was ordered to Foggia. In mid-October, a second detachment arrived.

One of the young pilot trainees, Josiah P. Rowe, Jr., in his letters home and in a series of stories published in his hometown newspaper, *The Daily Star*, of Fredericksburg, Virginia, provided a distinct snapshot of the area and his training. He called Foggia "the backyard of civilization.... Suffice it to say that it is very, very old and very, very dirty and the people are very, very poor." The flying field, which ultimately included two camps, was about a mile and a half from Foggia, situated in largely uncultivated countryside dotted with a few olive orchards and vineyards, mostly given over to sheep-raising. Rowe located it in the "ankle" of Italy, approximately eighty miles from Naples and 120 miles from Rome, "on a perfectly flat area of about fifty square miles between two mountain ranges." Thus, it permitted needed space for takeoffs and landings as well as cross-country navigation. Compared to Issoudun, the facility was luxurious: "Our camp is an excellent one.... We are quartered in large, stone barracks, have real beds with real springs, running water, electric lights, and every reasonable convenience. Our mess hall is fine, for army life, and we have a waiter for every twelve men."

All in all, flight training appeared to be off to an excellent start. The Italians greeted American airmen enthusiastically although, as Rowe recorded, there were some comic elements:

> The Italians are naturally high strung and easily excitable and the stunts which the Americans do don't help in the least to make them more rational and less demonstrative. Every time two planes get within a hundred metres of one another they wave their arms frantically and yell like demons, and when a student looks as if he will surely crash into a building, they tear their hair and jump around distractedly.[23]

The commander of the second American detachment, former New York congressman Capt. Fiorello H. LaGuardia, wrote spiritedly: "Facilities for training are excellent, and there is no reason in the world why we cannot turn out men as quickly and efficiently as the most exacting and fastidious legislator would demand."[24]

Those who went to Italy might have expected to enjoy the benefits of the balmy Mediterranean climate. Unfortunately, neither weather nor topography turned out to be as wholesome as predicted. Malaria afflicted Americans even during the fall and winter months; mountains funneled high winds down the valley toward the airfield. The Italian and American cooperative venture also ran afoul of cultural differences. One cadet painted a different picture from Josiah Rowe of the Italians' reaction to safety concerns, complaining that the Italians refused to cancel scheduled flying during high winds.[25] The American commandant of the school expressed another, typically American, reservation:

> It does not seem good policy to give preliminary training by contract. The time needed for each man varies considerably; some pilots, though able and willing, take necessarily a longer time to master the first elements than others . . . and [a pilot] should not be discontinued just because he has flown more hours than called for by the contract.[26]

Clearly, the Americans were troubled by what they perceived as Italian inflexibility and lack of respect for the individual. Additional difficulties arose, attributable to the language barrier.

Everywhere in Europe, training suffered from logistical and supply problems. In Italy, trainees flew Farman planes powered by 100-horsepower Fiat engines, but there were too few to meet the need. Furthermore, the distance between Foggia and the supply centers aggravated materiel shortages. The Italians were often unable to follow through on their promises, and diplomatic relations probably would have been strained to the breaking point but for the ameliorating presence of a few Italian-American and Italian-speaking officers.

The course in Italy ran from primary through advanced bombardment

World War I

training. It began with dual-control instruction and progressed to solo flying, leading to the brevet of *Pilote Superiore*. Solo and acrobatic flying continued after attainment of the first and second Italian brevets.[27] The advanced level commenced in February with few planes and one Italian officer instructing. But when the huge Caproni bombers arrived in the spring, advanced bombardment training began in earnest. The *Societe Italiano Aviazione* (SIA) aircraft, with their marked tendency to nose-dive, were then discontinued.[28] In spite of the problems encountered at Foggia, between September 28, 1917, and the close of Italian primary training on June 25, 1918, the school graduated 406 pilots, of which 131 went on to complete the bombardment course.[29]

The British trained U.S. cadets in England without administrative oversight by the United States, which produced less confusion than when the two systems attempted to mesh or when one system attempted to convert to another. The British and Americans negotiated a reciprocal agreement that included exchanges with RFC schools in Canada and the training of approximately 200 American cadets in the British Isles.[30] The ship transporting the first group of 53 trainees docked in Liverpool on September 2, 1917. At the School of Military Aeronautics in Oxford, they soon were joined by another 149 men, part of the Italian Detachment that had been ordered first to Italy. Much grumbling by the honor graduates about having to repeat ground school failed to deter the British, who insisted that Americans execute the same training

Pilot trainees while in Italy flew Farman aircraft (*below*) before progressing to the huge Caproni bombers, the triplane model having a two-story cabin (*left*).

course, from the beginning, as British pilots did. At the time, British training squadrons were overtaxed because of bad weather and equipment shortages, so delaying the start of American training may have offered one solution.

By early October, British squadrons began to absorb American cadets. Sixty-six entered a shortened flying course at No. 1 Training Depot at Stamford, Lincolnshire. The rest took a machine-gun course at Grantham. In part, this instruction also repeated ground school, except that the British used mock-ups of the internal structure of the guns — teaching aids "unheard of" in the United States, according to one American officer, and "in some ways the most valuable feature of the entire course."[31] After the men who took flying training completed the course, they reported to night flying squadrons; the others returned to the regular training squadrons.[32]

English as well as Italian-trained pilots learned on the Farman and on DH–6s and JNs similar to the American version. Lt. John Grider, one of the famous American "warbirds" who trained in England, started on DH–6s at Stamford. He found them to be like the Curtiss "except slower and won't spin no matter what you do to them." When his group moved to Thetford, Norfolk, he flew Farmans, or "Rumptys" as they were nicknamed. "These old short-horn Farmans are awful looking buses," he penned in his diary. "I am surprised they fly at all."[33]

English training differed from the slower, cumulative French *Roleur* method. The English Gosport system introduced students almost immediately to combat maneuvers, making it notorious for the high fatality rate in training.[34] Grider commented that "we have the same sort of wild kids here for instructors that we had at Oxford, only more so — wilder and younger. I was told that they kill off more instructors in the R.F.C. than pupils, and from what I've seen, I can well believe it." After he had completed four hours solo on the Rumpty, Grider shifted to the Avro, the generally acknowledged ideal training plane.[35] But, he lamented, "they are entirely different and I have to learn to fly all over again."[36]

Primary training in Europe and the United States shared one insurmountable problem: training airplanes were incompatible with operational types. Furthermore, they were extremely scarce. For example, the fifty or sixty old Caudrons at Tours were constantly cannibalized and repaired so that, by January 1918, when primary training was still in full swing, the school functioned with only some dozen machines, all rebuilt from previous wrecks. The airplanes were said to be, as Thayer remarked, "in such a state of decrepitude that [the students'] teeth chattered as they rolled across the ground."[37] The frequently flown Farmans were unsuited to convoluted maneuvers and so dissimilar to machines in service on the western front that Farman-trained pilots nearly had to start over when they reached France.[38]

This process of constant relearning, or transition training as it would come to be called, was repeated time and again. Students learning to fly JNs, Farmans, Caudrons, DH–6s, and other planes used for primary training in the United States, Canada, Italy, Britain, and France had virtually to retrain when they went onto operational aircraft or (more rarely) to more sophisticated primary training planes. Like any number of officers assigned training duties, Capt. Geoffrey J. Dwyer, in charge of the Flying Training Department in England, underscored the importance of matching training to tactical aircraft. He also urged that students be designated for a specialty at the start of the training sequence. This early selection would avoid duplication and lost time, he claimed, and each man would thus "be trained as a specialist all through rather than given a general training with a specialist finish."[39]

In calling it "general training with a specialist finish," Dwyer neatly summed up the combined training system conducted both in the United States and in Europe. It was neither ideal nor always very workable, and it was certainly not the one planned. It was, rather, the product of wartime necessity: too little training equipment, lack of time occasioned by bad weather through the fall and winter months, and into the first part of 1918, a shortage of facilities. Always, training officers worked against the clock. In judging their results, one had to be mindful, as the Chief of Air Service pointed out, that "a school to turn out 20 pilots a month might be constructed and put in operation in three weeks. A school to turn out 800 pilots a month will take a year before it settles into running order."[40] Not until fifteen months after the Americans entered into the war, and approximately nine months after they began training in France, was the AEF able to relinquish primary training and concentrate on advanced and specialized work.

Advanced Training

> I now wear wings, wings of silver, shoulder bars of gold, a watch of nickel, identification tag of bronze and buttons of brass. If you could see me now you might think I was the display counter of a jewelry store.[41]

The flight training program specified that each RMA, graduate of the primary course with his wings of silver, was selected for advanced training in one of the specialties of pursuit, observation, or bombardment. That approach was not carried out consistently in Europe because of the myriad aircraft flown, the competing systems of primary training, and the relative capriciousness by which Americans were assigned to advanced training in Britain, Italy, or France. By August 1918, 72 airmen had taken advanced training from the French at Avord; 128 from the Italians at Foggia; and 160 had passed through the advanced stage of the British training system.[42] The greatest number went to Issoudun where, to ensure some standardization in the chaotic situation, advanced and specialized training were conducted as separate phases. Many of those completing the advanced course at Issoudun subsequently entered pursuit training, also given at the Issoudun airfields. Observation teams trained principally at Tours and at French artillery schools; bombardment was conducted at Clermont-Ferrand in France, in England, and in Italy. By fits and starts, the AEF implemented this scheme, in spite of the time and resources expended in primary training.

The first sanguine proposals on the part of the French aviation mission to the United States envisioned shipping RMAs to France in July 1917 for a month's advanced training at Issoudun, whereupon combat-ready pilots would go to the front "to engage in the summer battles taking place during August, September and October."[43] But by late fall the American training station at Issoudun, according to Major Robertson, then at AEF headquarters, was "in a hopeless condition of undisciplined chaos."[44] Some cadets arrived with no flight training; others came from French schools where they had started to fly. The airfield was a disaster. Though fairly level, the stony terrain consisted of coarse clay, and when rains descended in the autumn, construction and workers were soon mired in mud. Progress and morale plummeted accordingly. Cadet John Richards, who had already been at Tours and Avord and was at Issoudun to resume training on Nieuports, thought that Avord, which he detested, was "heaven compared to this mud hole."[45] But desperation generated ingenuity. Master Signal Electrician Franklin Perry designed an airplane mudguard that worked fairly well, and was gratefully received at Issoudun and other airfields.

Maj. Carl Spaatz[46] took charge in November, then became the officer in charge of training when Colonel Kilner assumed command. Those two capable officers brought a measure of orderliness and efficiency to Issoudun, and, with the help of French instructors, the program slowly ground into gear. Despite the bad winter weather, advanced and pursuit pilots as well as RMAs began to graduate. By March the worst was past. The base became livable, fields and hangars had been constructed, and Americans assumed firm control of the program. Victor H. Straham went through flying training at Issoudun at that time and wrote home ecstatically to his parents: "You don't really know anything about flying until you take the course that I had at Issoudun France and believe me, it is there on the little planes which go at such speed that you are made a flyer, such is the consensus of opinion of all soldiers and people in France."[47]

Officers conducting advanced training in France were forced to redress the deficiencies in the skills of pilots coming from the United States. Especially by summer 1918, in the face of a concerted effort to integrate U.S. and European training methods more closely, and when a considerable number of trained pilots had arrived to finish any last-minute training before going to the front, disharmonies between the two programs became glaring. Lengthy reports urged corrections in American ground and primary training to minimize retraining in France. Defects in aerial gunnery instruction were usually cited, and because the Air Service was unable to open a school of aerial gunnery until the summer of 1918, it relied entirely on training in the United States. Also, the AEF noted that incoming American airmen were often unfamiliar with the construction and repair of the engines they encountered in France, but once again most of the training equipment used in the United States and in the AEF differed. By the fall of 1918, approximately 75 percent of the engines at the 3d AIC at Issoudun were rotary motors unavailable at American ground schools. AEF officers frequently griped about U.S.-trained pilots' inability to navigate on cross-country flights. Colonel Kilner, for example, while commanding the 3d AIC, complained to the Chief of Air Service: "Pilots received here up to this date have not the slightest conception of cross-country flying; occasionally pilots lose themselves in flying from one field to another, five kilometers apart. It is understood that pilots coming here are supposed to be able to take a cross-country trip without losing their way."[48] Yet the topography and layout of towns and villages, and therefore landmarks for visual naviagation, differed markedly between the U.S. and French countryside. In his report of November 1918, Colonel Dargue made the sensible suggestion that pilots in American schools be given French maps to study.[49] But familiarity with maps did not necessarily solve the problem either, as British pilot C. S. Lewis commented:

> The flat country stretched to the four horizons. To say it looked like a map was a cliché. There was a resemblance, of course ... but the real thing had a bewildering amount of extra detail, a wealth of soft color, of light and shade, that made it, at first, difficult to reconcile with its printed counterpart. Main roads, so importantly marked in red, turned out to be gray, unobtrusive, and hard to distinguish from othe roads. Railways were not clear black lines, but winding threads, even less well defined than the roads....Then there were cloud shadows ... ground mists .....[50]

As described earlier, students coming into advanced training had learned on aircraft of all types. Some at Issoudun were Curtiss-trained from the United States; some had flown Caudrons at Tours and Chateauroux; others had learned on Farmans and Bleriots; a few had flown Nieuports. Some had even begun advanced training on service machines in French or English schools. Issoudun itself had about thirty different types of aircraft that included Spads, Sopwith Camels, Breugets, and Avros. There were seventeen models of the Nieuport.

If confusion reigned for pilots in training, it was a nightmare for supply people and maintenance crews who had to keep airplanes intact — airplanes whose frames, engines, and parts came from England, France, and America. The supply department at Issoudun carried approximately 44,000 separate airframe parts and 20,000 different engine parts.[51] As one officer summed up the situation, the equipment "consisted not of what the school would have preferred but of what it could get."[52]

The 3d AIC tried to untangle the snarl by putting all student pilots through the entire advanced course, regardless of the type or extent of previous instruction. Obviously, the rate of progress varied with the background of the pilot, but at least instruction, if not aircraft, could be standardized. Theoretically, after a three-week course, pilots emerged equally well trained and prepared for specialized instruction in pursuit, observation, or bombardment. The Training Section found, for example, that "after passing through the various machines, in lower fields such as 23 meter, 18 meter, and Avro, there was practically no difference whatever, in the quality of flying between the students trained for 25 hours on Caudron, or 200 hours on Curtis[s]."[53]

Putting all student pilots through the advanced course also provided a screening process for assigning them to specialties. The AEF Training Section maintained that evaluation at the advanced stage in Europe assured greater conformity to battlefront conditions and requirements. Moreover, AEF training officers sometimes questioned the competence of their colleagues on the opposite side of the Atlantic at making assignments, as demonstrated by the remarks of one AEF officer: The "classification of students in the States is not apparently based on any well known factors, as quite frequently, valuable pursuit material has been found in those classified as bombing and observation pilots, and many classified as pursuit pilots have been rank failures as such."[54]

By the time Colonel Bingham (who had been in charge of the Schools of Military Aeronautics in the United States, then became AEF Chief of Personnel) assumed command of the 3d AIC in August 1918, Issoudun had grown to nine major flying fields, approximately two miles apart. Each field provided a particular type or level of training for advanced and specialized

A French Breguet 14 (*right*) and a Spad XIII (*below*)

World War I

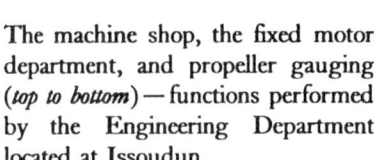

The machine shop, the fixed motor department, and propeller gauging (*top to bottom*) — functions performed by the Engineering Department located at Issoudun

instruction. Each had its own commander and training officer, overseen by the AIC commander and training officer. More than a thousand men had completed advanced training on the Issoudun fields by midsummer.[55]

The curriculum (in idealized form) was an amalgam of American techniques and the inherited French method of successive instruction on increasingly complex aircraft. A typical student might begin advanced instruction on Field 1 with groundwork in aerial gunnery and motors, then go on to the *Roleur* course, frequently on a clipped-wing Morane-Saulnier monoplane with a fifty-horsepower Gnome engine. On Field 2 he flew a dual-control airplane, usually the 23-meter Nieuport, with an instructor. When he satisfied the tester, he went to Field 3 for solo flying, again on the 23-meter Nieuport. He also practiced cross-country flying on whatever aircraft was available and performed some acrobatics in one of the few Avros. The 18-meter Nieuport was flown at Field 9. This segment included a landing class (from ten to thirty landings), a spiral class, and an "airwork" class — all of which involved the student flying solo and instructors observing from the ground through field glasses. At this point, according to Bingham, "if it was found that a student did not readily accustom himself to the delicate and speedier type of ship, he was advised to go in for reconnaissance or bombing rather than to continue the course in pursuit and combat flying."[56] At Field 9, advanced students parted company to take up specialized training.

Specialized Training

By spring 1918, specialized instruction in pursuit, observation, and bombardment had become the AEF's dominant training activity. Not long after the United States entered the war, officials in the Division of Military Aeronautics and the AEF changed their minds about giving all advanced training in Europe and decided that only refresher training would be conducted in France, but it took many months before this approach became possible. Moreover, from late 1917 into the next spring, as the AEF Training Section tried to hammer out an organization, decide how to allocate resources, and determine how to cope with feast or famine in partially trained manpower, it was subject to a series of revised personnel production schedules and reorganizations occasioned by the changing fortunes of war. At the beginning of 1918, the AEF program called for 60 service and 40 training squadrons in France by June 30. This was a far cry from Pershing's audacious demand six months earlier for an eventual full complement of 260 squadrons, yet even the lesser figure, which became fixed at 202 squadrons, loomed only in the distance.

The German offensive begun in March 1918 broke the deadlock along the western front, imposing new demands on air forces on both sides of the lines. Relatively more mobile warfare required more aircraft to provide intelligence regarding shifting enemy lines, prevent enemy reconnaissance, and disrupt enemy resources. Observation, pursuit, and bombardment squadrons had to be mobilized in far greater numbers for support of the more fluid ground war.[57] But at the time that advanced and specialized courses were just beginning to take shape and demand for their graduates was increasing, the aviation program reached its nadir. The escalating need for manpower to feed the Allied defenses gave priority to transportation of ground troops and caused considerable delay in shipping airmen to Europe.

In the spring, newly appointed Chief of Air Service, Brig. Gen. Mason M. Patrick, reshuffled his organization and pressed for an increased aviation program to balance the enlarged ground troop allotments necessitated by the German offensive. Through the summer, as the Germans waged war along the Marne, the Air Service repeatedly cabled home for more airmen trained in the combat specialties. It desperately needed trained pursuit pilots, but if it could not get them, RMAs could be substituted. Additional bombardiers[58] should be sent to team with bomber pilots trained in England. Quotas for observation pilots and observers increased by 75 each per week.[59]

As noted, by late summer, Air Service policy declared that all training except "freshening up" should take place in the United States.[60] The General Staff had agreed to tables of organization by which 202 squadrons would be manned and equipped for frontline duty by the end of June 1919. This figure was considerably lower than the astronomical number projected by Pershing, yet it still placed enormous stress on the AEF and U.S. training programs. In fact, the U.S.-based program could not possibly produce enough fully trained airmen to fill the 60 pursuit, 40 corps observation, 52 army observation, 14 day bombardment, and 27 night bombardment combat squadrons of the 202 Program, so the AEF was forced to assume a stepped-up specialized training effort.[61] General Patrick later wrote that "although I doubted whether [the 202 Program] could be carried out in its entirety, it gave us a 'mark at which to shoot'; it furnished a basis for an estimate of the men and material needed."[62]

The "mark at which to shoot," including projections of the ratio among specialties, varied over time, and military men disagreed about the makeup of a balanced force. In October 1918, the War Department General Staff decided, contrary to Pershing's notion of a 3:2:1.5 ratio, that the air arm should have a ratio of 5:3:1 in pursuit, observation, and bombardment.[63]

Pursuit

Pursuit, or *chasse*, was the specialty most closely tied to the French, both as a legacy of the famous *Lafayette Escadrille* and on an ongoing basis. The policy dictating that all substantial training be carried out in the United States was least easily accomplished with pursuit because the United States lacked the very fast and highly maneuverable fighter airplanes used over the lines. The Allies flew some seventy different single-seat tactical pursuit aircraft, whereas Americans relied mostly on the French Spad XIII with the 220-horsepower Hispano-Suiza motor. But tactical aircraft of any kind were so scarce, even in training in France, that a pursuit pilot often learned on several different kinds of planes, none of which he would fly in combat.

The French provided most of the airplanes and initial pursuit training to the AEF. The first group of men to arrive for *chasse* at Issoudun flew Caudron G–3s during primary training at Tours and then transferred to the Nieuport school at Avord. In late October 1917, the men began flying the fifteen-meter Nieuport at Issoudun. Soon, approximately 40 Curtiss-trained RMAs and other Americans from the French schools at Avord and Tours joined them. Because no independent American pursuit course existed at that time, students learned French methods from instructors at Issoudun. The necessity to give primary instruction to cadets interrupted pursuit training during the last months of 1917 and the first quarter of 1918, but by April this pressure lessened.

To meet increasing demands for pursuit pilots, the Training Section

A student pilot receiving instruction in *chasse* (*right*), and a qualified pilot beside his Hispano-Suiza powered Spad XIII (*below*)

assumed tighter control and restructured the AEF pursuit program. The wide variety of machines in use and the problem of storing the requisite spare parts contributed to the decision to standardize instruction on three aircraft: the Avro, the Nieuport 27, and the Spad. Unfortunately, those airplanes were rarely available, and most training continued on obsolete Nieuports and other less desirable planes. "Their use," according to the Pursuit Division of the Training Section, "was justified because they were the only machines which could be secured."[64] The general scheme, like so many other training plans, was an ideal seldom attained, but, as happened with advanced training at Issoudun, the Training Section at least gained enough experience to define the nature of a successful pursuit program, identify the preferred equipment, and outline an orderly and progressive curriculum.

During the summer, the AEF cabled frantically to the United States for trained pilots and observers, but it received almost none because the U.S. program had become mired in its own attempts to simultaneously mobilize and procure airmen, design a program, obtain training aircraft, and construct training fields. The AEF demanded 125 trained pursuit pilots weekly from the United States, but the Air Service discovered that there were "no pursuit pilots in the States," owing to the fact that there were no acceptable training planes at American fields. As a fallback, RMAs recommended for pursuit ("on what basis of selection," according to the Pursuit Division, "we have never been able to find out") would be accepted. Again, Issoudun had to engage in what

World War I

Repairs are made to the wing and strut of a damaged Nieuport 27.

amounted to complete retraining because the arriving pilots were Curtiss-trained and knew nothing about fighters.[65] Despite the overwhelming obstacles, however, the United States graduated enough pilots to activate twenty pursuit squadrons at the front by the time of the Armistice.

The fighter pilot's mission was to "sweep enemy planes from the air," to outfight other aircraft.[66] He had to climb above, dive below, attack, and twist away from his deadly equal in order to clear the skies for friendly airplanes. So his tactics tended to be hit and run. He normally fired his own gun, which was fixed to shoot in only one direction.[67] The era of the individual ace was passing by the time the United States entered the war. Nonetheless, V-shaped formations typically provided escort at some distance from bombers, which allowed greater offensive action against the enemy. Pursuit training thus aimed to develop agility and quickness in aerial acrobatics.

Americans modified the French and British training methods. The incremental French approach of increasing proficiency by advancing through increasingly high-performance machines seemed to rely on too many different airplanes (although the American program too, not by choice, employed many types). Moreover, under the French system, a pilot never flew with an instructor after his solo flight. Americans preferred the British system of dual–solo–dual in which an instructor accompanied a student, even after he soloed, to correct any faults. What Americans discarded from the British was the short time spent in training. An RFC student was expected to undertake extremely complex and dangerous maneuvers very quickly, leading to "recklessness in pilots," in the minds of Americans, and therefore was "directly responsible for the excessive [*sic*] of deaths in schools, and behind the lines, due to excessive stunting near the ground."[68]

Americans tried to draw upon the best from each system, maintaining instructor control and guidance throughout the process and ensuring that a student pilot flew enough hours in each phase of his training to guarantee a tolerable safety record. Even so, the pursuit course at Issoudun was extremely hazardous — its fatality rate exceeded by more than fivefold that recorded for observation and bombardment training.[69]

By the summer of 1918, the 3d AIC greatly expanded, providing considerable space for each phase of the pursuit course. At Field 8 the newly dubbed pursuit pilot left the eighteen-meter class for his first real fighter airplane, the fifteen-meter Nieuport with an eighty-horsepower engine. First came practice landings, closely observed to confirm a candidate's suitability for pursuit aviation. Instructors knew the tendency of the fifteen-meter to go into a tailspin, so the student was required to take his plane up to an altitude of about 1,200 meters to practice the *vrille*, or tailspin.

Contemplating the tailspin elicited the gallows humor common to military men, as seen in an article published by a little newspaper at Tours entitled "How to Come Out of a Flat Spin, Dead or Crippled for Life." British pilot and later noted author C. S. Lewis recalled that "spinning was the one thing the young pilot fought shy of, the one of two things he hoped he might never do — the other was, catch fire in the air."[70] Instructor Howard Davidson credited Grover Loening, who had been at the Signal Corps Aviation School before the war, with figuring out and teaching pilots how to recover from a tailspin. "Up until that time," Davidson remembered, "we had been prohibited, by instructions from Washington, from doing acrobatics, even making the loop or anything."[71] Recovering from a tailspin thus entered the list of required tests, but it remained fearsomely deadly nevertheless. According to Arthur Sweetser's definitive account of the air war, the tailspin accounted for 30 of the 103 fatalities between January 1 and June 3, 1918. (The second highest number, 23, came from midair collisions.)[72] One young pilot, who trained with the French on Bleriots and then flew with a French *escadrille*, described the supreme test for a pilot as he first attempted that most deadly maneuver:

> Sometimes the plane will hover over the aerodrome, a mere speck up there in the sky. It hovers, circles and hovers again, seemingly for an endless time as the pilot is summoning his courage, screwing it to the point of daring to take that swift dive into terrifying space. It is the first test of nerve, deliberately to fall two or three thousand feet perhaps.... He is fighting the battle with self, summoning his pride, his courage, his determination to do what may mean swift death.... Suddenly there is a

World War I

Instructional aid: drawing of possible maneuvers of an airplane

shout, "There he comes!" He falls, he twists, turns, spins, down, down, down and then — O God! — he comes out of it.[73]

After this harrowing experience, the pilot practiced spirals and then acrobatics at the more distant Fields 4 and 6 to learn techniques for avoiding midair collisions. Bingham wrote of this phase: "Personally, I should have been extremely glad to have been able to avoid the risks due to the necessity for teaching pilots aerial acrobacy in single seater machines, by using more Avros and perfecting the student's acrobacy in that extremely maneuverable dual control machine, but we had to use the planes that we could buy in France."[74]

On Field 7, the student practiced formation flying in a 15-meter Nieuport with a 120-horsepower engine, first in a group of three, then in a group of five. Sometimes groups of fifteen would assemble.[75] A pursuit pilot who flew with the 27th Aero Squadron complained that pilots had not "been taught how to keep in formation.... They have been led to believe that their combat principles involved individual combat principally whereas individual combat is a very rare occurance [sic]."[76] Bingham emphasized: "It was early borne in on us that the aviator who was a grandstand player did not last long against an enemy formation. The successful pursuit pilot must curb his individual daring and his love of taking a sporting chance. Team play, cooperation, and the weight of numbers were all essential."[77]

Field 8 was the combat field where pilots fired machine guns at clay pigeons, flew formations, and used the camera gun in simulated combat with an "enemy." It was here on Field 8, thought Bingham, that "the aggressive spirit of a good polo player or of a first-class football player placed him in the front ranks of the combat pilots."[78] At Field 8, the pursuit student took a course in aerial gunnery.[79] A twenty-day course in ground and aerial machine gunnery at the French school at Cazaux usually followed.

Dogfighting, the form of aerial combat that became glamorized and immortalized during and after the war, was impossibly dangerous in one new type of mission — night flying. All aircraft flown at the time had, certainly by later standards, only the most primitive instruments, and lights on the ground and in the airplane were seldom used. During the 1918 spring offensive, the Germans continued their nighttime bombing raids on London and Paris. The summer's open warfare, which replaced trench fighting, brought greater troop movement, often under cover of night. As a result, nighttime artillery observation also increased.

The 3d AIC inaugurated a pursuit course in night flying on spacious Field 7.[80] Airplanes — either the scarce Sopwith Camel, the valued Avro, or the more common Nieuport 28 — were equipped with navigation lights, and a signaling light that could flash a code to the operator of field searchlights. When scanners detected a night bomber or observation plane, the pursuit pilot went up to pinpoint the intruder and signal its location to the ground. "Immediately," as Bingham dramatically described such a scene, "the searchlights, directed by the listening devices, are turned on the night bomber, who is then held in the powerful rays. The pursuit plane comes up in the blackness behind until he is

Field No. 7 at Issoudun and a line of Nieuport 28s starting their engines

a little below and directly in the rear of his prey, and shoots from a distance of about 20 yards and at an angle of about 10 [degrees] below the night bomber." Obviously, night pursuit was perilous and employed tactics somewhat contrary to those used in daylight, but officials and trainees at Issoudun embraced the mission with enthusiasm. "It was one of our greatest disappointments," recalled Bingham, "that the Armistice was signed just as our night pursuit pilots were receiving the finishing touches of their training in cooperation with the Searchlight Company."[81]

Pursuit pilots also trained in England and, briefly, in Italy. The school at Furbara, Italy, was situated on the seacoast about twenty-six miles from Rome. Furbara, which began operations on April 24, 1918, was intended to be an aerial gunnery range where, additionally, reconnaissance and bombardment pilots were converted to pursuit. The sixteen-day course concentrated on the use of the Lewis, Fiat, and Vickers machine guns and various bombsights.[82] Unfortunately, the field's isolation and a lack of equipment resulted in abandonment of the experiment after only two classes, numbering fifty-two students.[83]

In its totality, the British program was much more extensive than the Italian. Of the American pilots trained by the RAF in England, nearly half went into pursuit aviation, the others into bombardment.[84] General Pershing had argued stubbornly with Allied commanders that American ground forces would not be trained as replacements for Allied units but would join an American army when it formed. Earlier, however, agreements between the United States and Great Britain provided for ten squadrons partially trained in Canada to be distributed to RFC units until the AEF called for them.[85] Because of the time it took for the United States to field its own army, most of the British-trained American pilots served either under British command or in the two British-equipped and British-trained American pursuit squadrons (the 17th and 148th) that fought on the British front.

The greatest number of British-trained pursuit pilots learned on Sopwith Camels. Slightly fewer flew S.E.5s, and a handful, the two-seat Bristol Fighter. The RFC frequently used the two-seat Avro as the transition airplane in which the student pilot sat in the front and the instructor behind. The experience of Lt. John Grider, among the first cadets sent to England to begin training in January 1918, was fairly typical. He went through primary training at Thetford; then, about twenty miles from London, at Colney, he and others took advanced training on Sopwith Pups, Spads, and Avros. From there his squadron went to Turnberry, Scotland, for aerial gunnery training, and finally to Ayr, also in Scotland, for the School of Aerial Fighting. Ayr "is really a beautiful spot," he recorded in his diary, "and I'd like to stay here a while but they kill off pilots too fast for any one to linger very long.... All the flying here is stunting and

Air Service, American Expeditionary Forces

Among the British training aircraft were the S.E. 5 fighter (*right*), the Bristol Fighter (*below*), and the Avro 504 (*bottom right*).

we have service machines. Every time we go up, we are supposed to find another machine and have a dog-fight with it."[86]

Observation

A very significant prewar technological development — machine guns synchronized to shoot through an airplane's propeller arc — allowed airmen to become combatants. Pursuit was an inevitable application of this new technology harnessed to a familiar mission. Now, bombers could be escorted and protected, and observation aircraft could themselves become fighters capable of eliminating an enemy's "eyes." Regardless, however, observation flights in support of ground action remained the most fundamental employment of air forces throughout the war. During the period of trench fighting between 1915 and 1918, artillery became crucial for protecting infantry attackers, and aerial observation aided in directing and adjusting artillery fire. After the siege broke, aerial reconnaissance was essential for pinpointing not only the enemy's

World War I

movements but also the location of friendly troops during the smoke and fire of battle.[87]

Observation required two men — pilot and observer — whose shared duties entailed spotting artillery fire (artillery *reglage* including counterbattery fire direction), infantry contact patrol (observing and reporting infantry activities), reconnaissance, and photography. Observation linked aviation closely to the Artillery, took many of its men from that corps, drew some representation from the Infantry, and required skills beyond piloting and shooting.

The first AEF aerial observers were artillerymen trained at artillery brigade training centers and subsequently instructed in aviation by the French at the Le Valdahon flying field. They received supplemental instruction at the I Corps School at Amanty, France, and completed operational training with French squadrons.[88] As was the case in the United States, AEF artillerymen took the view that all aerial observers trained to work in *reglage* should be commissioned in the Artillery and remain with Artillery rather than Air Service units. From the perspective of airmen, observers should be detailed to the Air Service through training and beyond because aerial observers were used not only for artillery spotting but in reconnaissance, infantry liaison, photography, and aerial gunnery as well. Col. William "Billy" Mitchell, one of the first air officers in Europe, recorded in his journal: "Ground troops kept insisting that their officers be sent up as observers. When they went up in the air, unless they were trained by the Air Service, they were perfectly worthless. Most of them would get air sick. They could not handle their machine guns, had no idea what to do in the face of the enemy and could not tell what they saw on the ground."[89]

A compromise evolved by which the Air Service operated a central observation school for general training of observers and observation pilots. Observer candidates were to be detailed from the Artillery and Infantry as well as from the Air Service, and because air squadrons would be assigned to divisions, corps, and armies at the front with their own commanding generals, advanced training would take place alongside artillery training centers. Accordingly, the AEF placed the Air Service Observers School with the 2d AIC at Tours.

The First Artillery Aerial Observation School (1st AAOS) opened for instruction in April 1918 at Coetquidan; the 2d AAOS located at Souge; the 4th AAOS was

A device allowing the gunner to fire twin Lewis guns

Air Service, American Expeditionary Forces

at Meucon; and the 5th AAOS began at Le Valdahon. The II Corps Aero School at Chatillon-sur-Seine was devoted to artillery fire; an infantry school machine-gun range was also adjacent to the flying field.[90] The Air Service patterned observation training in all these schools after the French since the American army operated on the French front, requiring cooperation between American squadrons and French batteries.[91] Because English observation and communication techniques differed from the French and French-based American approach, the British trained Americans only in pursuit and bombardment.

Construction of the Tours facility began after site inspection in September 1917, but observation training did not commence until January because of delays in the receipt of building materials and the necessity for giving cadets primary instruction at the only field operating at the time. In January, Tours acquired two more fields, and in February, a fourth. In the spring, it added four small fields, known as the St. Avertin group, but they were used mostly for primary instruction. The original instructional staff was split between American and French officers. Of the forty-one students in the first class from the Artillery, Infantry, and the Marines, thirty-three graduated, thus qualifying for advanced aerial gunnery and field training.[92]

During the first five months, airplane shortages limited the school's 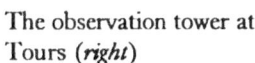 operations, and beginning classes had to rely on the original eight Caudron

The observation tower at Tours (*right*)

Interior of the 2d AIC observation tower at Tours (*left*)

World War I

Panel signaling at the Coetquidan First Artillery Aerial Obervation School (*top left*); observers' classroom in the Second Artillery Aerial Observation School at Souge (*bottom left*); and camera gun repair by a member of the II Corps Aero School at Chatillon-sur-Seine (*far right*).

G–4s. Farmans and Italian SAIs proved unsatisfactory, and the more desirable Sopwith A–2s and B–2s were scarce.[93] By summer, new service planes had begun to arrive, and with less demand for primary training at Tours, the space allocated for it could be reassigned for observation.

The expansion of the installation, which nearly doubled the area, permitted a realignment of the flying fields. Main Field, now No. 1, was used for instruction in aerial photography and for the two-person Caudrons and Sopwiths. No. 2, the erstwhile Farman Field, became home to the new DH–4s and Breguets. The old Spiral Field was renamed No. 3, the combat field. No. 8 and two new fields formed a unit for artillery *reglage* and infantry liaison. The remaining two fields contained ground and aerial gunnery ranges. In September, Tours acquired more property intended for possible expansion of training and for a pool of trained observers awaiting placement at the front.[94] General Pershing visited Tours that August and commended it with his

Air Service, American Expeditionary Forces

characteristic restraint: "The Aviation Instruction Center for Observers gave us a favorable impression in every particular."[95]

The expansion resulted from the increased air program. In August, Chief of Training Colonel Kilner reported that "the importance of Aerial Observation and of the work of the aerial observer is continually emphasized here. The demand for observers and observation pilots at the Front has necessitated the increase of student personnel at the Observers' School, 2d AIC, to the utmost capacity of the field and equipment. The present program contemplates entering every ten days 60 student observers and 35 observation pilots."[96] In a memo drafted the day before, he noted that the increased demand was "to keep up with the ground army. Equipment for it is being delivered by the French and the personnel must be provided by us at all costs."[97]

Through the summer and into the fall, members of the AEF Training Section and the Overseas Training Mission pelted the Department of Military Aeronautics with requests for trained personnel. The Artillery, too, was not meeting its goals, and the Infantry had never done its share, they charged. At one moment, officers bitingly criticized the pace of training at home; in the next, they patiently explained what airmen needed to know, but didn't, about combat tactics. Ultimately, they implored the United States to send whatever men could be produced in order to implement the 202 Squadron Program.

In a further attempt to meet the increased monthly quotas for army and corps observation squadrons at the front, the Chief of Air Service authorized a reduction in the number of teams of observers and pilots in observation squadrons from eighteen to twelve. The Tours complex could now produce 70 observers every ten days as opposed to the earlier 50 every two months. Not counting pilots who had received primary training at Tours, the AEF had graduated 555 observers at the time of the Armistice. But, whereas 950 artillery observers had been requested from the United States before October 31, 1918, only 351 had arrived by war's end; most arrived after October 1, and most were too poorly trained to be sent directly into combat.

Tours had to offer a five-week preliminary course to the U.S.-trained observers. Schools for final instruction in the control of artillery fire had already been set up at Souge, Meucon, Coetquidan, Le Valdahon, and Chatillon-sur-Seine. An agreement between the Chief of Air Service and the Chief of Artillery in early August eliminated inequalities in pay and promotion so that artillery volunteers could transfer without prejudice to the Air Service to become aerial observers. As a result, the artillery school at Saumur, France, provided the bulk of recruits during the fall of 1918.[98]

At the same time, the demand for pilots to man both night- and day-reconnaissance squadrons had increased so much that Issoudun inaugurated a course for observation pilots. It equipped Field 10 with DH–4s newly arrived from American production lines. To meet operational needs, it offered a short, compressed course. In this course, because of the high mortality rate among

World War I

Camera instruction, stripping practice, and a code class, all held at the Tours 2d Aviation Instruction Center

DH-4 pilots (the deHavilland proved to be a flawed, dangerous airplane), ground instruction in artillery spotting and photography received short shrift in favor of flying training. "It was not a satisfactory course," admitted Bingham, "but it was the best we could do under the circumstances, considering the imperative demands from the Front."[99]

Not only did too many DH-4s go down in spectacular flame in combat, they were also considered risky for training. But since the British-designed deHavilland was the only tactical aircraft manufactured in the United States, it satisfied the political need to justify the American airplane production program. Every piece of equipment, all spare parts, and even gas and oil had to come

Air Service, American Expeditionary Forces

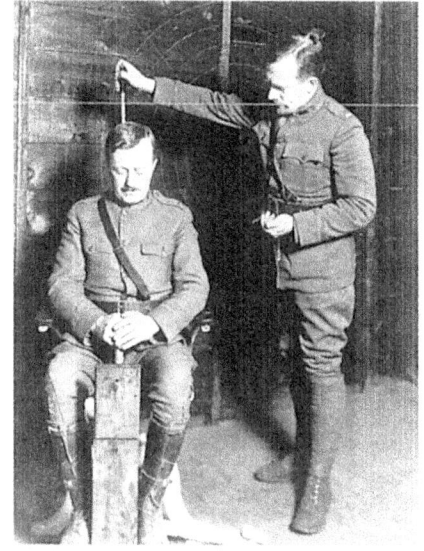

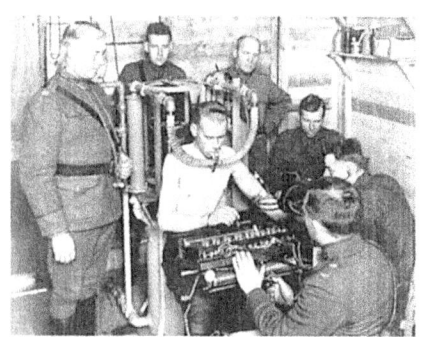

Qualification and medical tests for airmen included equilibrium sensitivity tests (*upper left*), individual reaction times (*upper right*), breathing capacity (*lower left*), and color vision and visual fields tests (*lower right*).

from the United States. It was costly and inefficient to send materiel of this sort to France for training that could be accomplished better at home. Further, training on DH–4s in France did little to advance the plan of offering only refresher training on airfields in the AEF. Finally, to the minds of many airmen, taking into account the poor DH–4 safety record, it was the Salmson or Breguet — the primary tactical aircraft used at the front — rather than the deHavilland that ought to have been built. Nonetheless, the training program could be judged a success numerically in that Issoudun sent 433 observation pilots to the Zone of Advance in a little under the two months that preceded the Armistice.[100]

World War I

The DH–4 multipurpose airplane

The observer performed most of the nonpiloting functions required of the crew, but both men had to learn something about photography, map reading, and radio communication. Furthermore, any pilot had to be practiced in aerial gunnery, even if he was not principally responsible for firing. Tours housed an aerial gunnery school as of January 1918, a Signal Corps radio school in July, and a photography school in August. The Observation Department did not supervise the radio and photography schools, nor were these schools restricted to observation personnel, but the intention was that these schools would help prepare observers for every phase of frontline work except for practical experience in control of artillery fire.[101]

Ultimately, manpower procurement worked fairly well. Pilots generally outnumbered observers, a result of the Air Service's preference for pilots, expanded pilot training programs in the United States and in France, easier refresher training in Europe for pilots than for observers, and the arrival in France of American-made DH–4s that were used in pilot training. Yet the ability to procure officers already familiar with artillery techniques and tactics was a boon to the Air Service in increasing the number of corps observation squadrons at the front.

Furthermore, stodgy observation (from many airmen's perspective) achieved a victory in that, in creating some parity between the branches, it demonstrated that air power could enhance overall combat effectiveness. Inevitably, air and ground forces coordinated better in training than on the battlefield. Regarding the performance of the I Corps Air Service during the battle of Chateau-Thierry, *The Tactical History of the Air Service* contended somewhat self-servingly that "probably the most valuable lesson of the entire

Air Service, American Expeditionary Forces

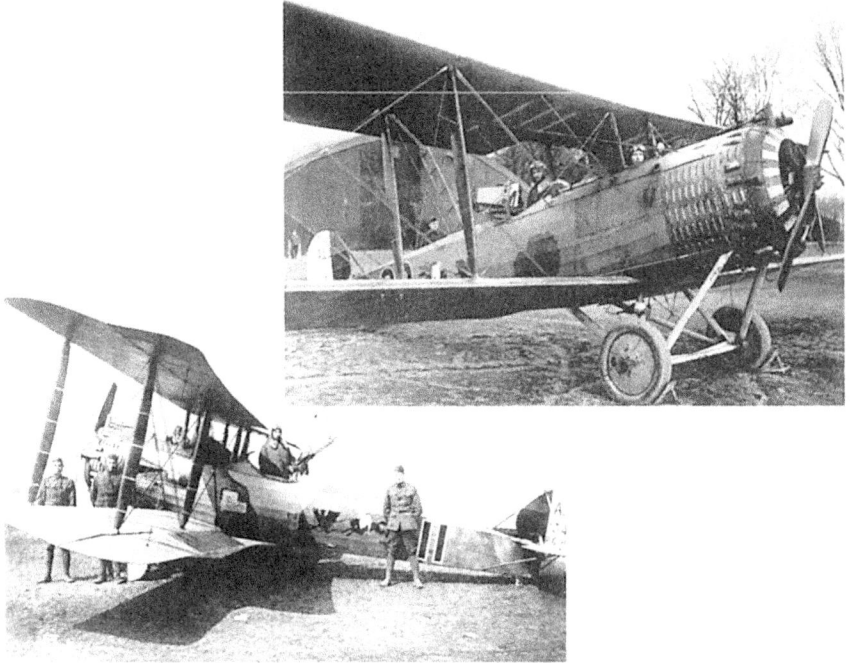

Two French Salmsons, one mounted with an observation camera (*top*), the other outfitted with a gun (*bottom*)

campaign was the knowledge that neither the artillery nor the infantry of the United States divisions here engaged had a sufficient knowledge or experience in the proper use and limitations of the observation Air Service of the corps."[102] At the same time, in his postwar summary of the strengths and weaknesses of observation training, Capt. Harold Wilder emphasized the positive value of close liaison between the Air Service and other army branches. "Complete contact with these branches has been a primary object of the Service and its results have been in proportion to the success of this cooperation. Specialization is the keynote of Observation as witnessed by the fact that the Infantry has contributed some of the best artillery observers and, vice versa, the artillery some of the best photographic observers."[103]

Bombardment

During the first month of engagement in 1914, Germany launched a Zeppelin attack on the Belgian city of Antwerp, the airship *Sachsen* unleashing nearly a ton of small shrapnel bombs. In retaliation, one British pilot bombed the Cologne railroad station and another, the Zeppelin shed at Dusseldorf. Thus encouraged, First Lord of the Admiralty Winston S. Churchill directed the

World War I

Royal Naval Air Service to strike the Zeppelin works at Friedrichshafen, which the pilots of three Avros successfully carried out. The panic among the British population during the ensuing 1915–1916 Zeppelin and 1917 Gotha raids on London was intensified by the realization that the hearthside was as vulnerable as troop convoys, lines of supply, and industrial plants to destruction from the air. Despite heavy losses and marginal effectiveness of Allied bombing forays, by the time the United States entered the war, both sides had activated bomber squadrons, and in the Allied councils, proponents of aerial warfare presented the case for strategic as well as tactical bombardment.[104]

Even before the AEF had established an organizational structure, the Bolling Mission considered arguments favoring an independent strategic bombing force. On arriving in Paris, Pershing was advised by the flamboyant Col. Billy Mitchell, who had become convinced that victory must be sought in the air as well as on the ground, that the Air Service should field a strategic force of thirty bombardment groups and thirty fighter groups as well as a tactical force large enough to balance the ground armies to which it was attached.[105] Not easily bewitched, Pershing steadfastly held to the view that the role of aircraft was support of battlefield activity under the direction of the ground commander. Yet, once the AEF began to organize, the first Chief of Air Service, General Foulois,[106] divided his forces in the Zone of Advance into Tactical and Strategical Aviation sections, each possessing bombers. Tactical operations were to be executed within 25,000 yards (more than 14 miles) of the line, or within reach of long-range artillery; strategic were to occur beyond that perimeter. Foulois put Lt. Col. Edgar S. Gorrell in charge of planning for a bombing campaign as part of the strategic mission.

Bombardment possessed neither the glamour of pursuit nor the doctrinal authority of observation. Thus, it was usually the last choice among recruits. Josiah Rowe went through primary training in Italy, so it was assumed he would continue flying Caproni bombers in advanced instruction. His comment at the prospect: "... my conviction [is] not to fly a bombing machine if I can possibly get another. Bombing is a very important phase of aviation, but it's the last thing I want to do."[107] When Lawrence J. Bauer arrived in France with his silver wings in May 1918, the AEF needed no more pilots, so he volunteered to become a bombardier.[108] Some others were luckier, able to remain pilots even if reduced to flying bombers. About one of his "unfortunate" friends, Charles W. "Chuck" Kerwood, former *Lafayette Escadrille* pilot Edwin C. "Ted" Parsons later wrote: "He passed through the schools at a time when pursuit pilots were a [drag] on the market and there was a crying need for observation and bombing pilots. Hack drivers is what we pursuit pilots were wont to call them. . . . Of all the messy nasty jobs, I believe that of day bombing was the worst."[109]

Another young airman took quite a different view; he thought bombardment was the preferred occupation in the immoral business of war (prefiguring

Air Service, American Expeditionary Forces

Lawrence J. Bauer, whose certificate of graduation upon successful completion of the stateside course in military aeronautics at the University of Illinois is shown here, volunteered for bombardment training while in Europe. The Caproni bomber Type 12 pictured here served to deliver the bombs, both day and night. A typical bomb of the type used for training is demonstrated by the flier at the right.

arguments made for strategic bombardment by the late 1930s). When he was assigned to a "long distance bombing squadron," Briggs Adams wrote to his mother and father: "This bombing is particularly attractive to me, for instead of aiming to kill men, as in fighting on the ground or even in scout [pursuit] fighting, we aim to destroy war manufactories, material things made to kill men. Thus we are striking at the very base of war. And this is most satisfying to me. For I am not in here for the sake of international treaties or patriotism, but to make war on war."[110] The pugilistic Kerwood survived the war; the idealistic Adams and earnest Bauer did not.

In any event, bombardment training fell prey to aircraft production problems and poor facilities — a familiar story — more than to doctrinal debates, politics, or personalities. The production program in the United States failed to produce aircraft that would permit the AEF to field a bomber force. Although at the time he still anticipated a product from the American

World War I

production line, in September 1917 Colonel Bolling placed an order with the Michelin Company to purchase their Breguet bombers. As part of the arrangement for beginning bombardment training in the AEF, the French offered the Americans the Michelin testing field at Aulnat, which was then being used as a bombardment school. The village of Aulnat lay conveniently close to the Michelin works, about 2½ miles east of the city of Clermont-Ferrand. "Nestled away in the mountains of southern France amid the cool haze and the cool damp mists from the hillside," rhapsodized the Air Service history, Americans constructed the 7th AIC.[111] As it turned out, the bucolic countryside was better suited to artistic expression than to large-scale flight training. The school sat in a low, badly drained basin surrounded by the Auvergne mountains. The configuration prevented any expansion of the facility, so the number of bombardment teams that could be trained even when the school ran at full capacity was fewer than the numbers ever demanded for operational purposes. As participant and historian Lucien Thayer summed up the situation:

> Not more than 20 ships could be comfortably kept in the air at one time, formation flying required strict enforcement of discipline, night flying was entirely impractical and there were no auxiliary landing spots beyond the airdrome. These difficulties were recognized from the beginning, but there were also advantages. The school was near to the Michelin factory and the question of supply was at that time and was destined to be for some time in the future a governing factor throughout the Air Service.[112]

Proximity to an aircraft factory boded well, since American forces depended almost entirely upon French aircraft and supplies. The French trained American enlisted mechanics in factories around Paris, and the Michelin and Renault companies agreed to do the same. In the fall, the 33d Aero Squadron detailed fifty-one men to the Michelin and Renault factories. As a result, by the time the first bombardment class got under way in late February, a technical staff was on hand to keep the airplanes and engines intact. But the light bombers on order did not arrive as expected because French and Belgian squadrons had snapped them up. Naturally, new aircraft were shipped first to the front rather than to the American bombardment school, which remained completely without planes until February. When that delivery finally appeared, it turned out to be ten combat-beaten Breguet-Renault planes that had already reached old age. A small but happier technical byproduct of the Michelin/American relationship was a simplified Michelin bombsight. Experiments at the school resulted in an improved sight, the 7th AIC Bomb Sight.[113]

Planning started slowly, and bombardment remained a disappointment in the execution. But much of the initial curriculum development eventually proved valuable. Fortuitously, the training program at Clermont-Ferrand was well served by its commander and the first officer in charge of instruction, who mapped out the proposed requirements and schedules for training in mid-

Air Service, American Expeditionary Forces

The field at Clermont-Ferrand (*top*); the Breguet-Renault (*center*); a group of trainees at Clermont-Ferrand (*bottom*)

November 1917. "The men chosen for bombardment training should, if possible," he informed his superior, "be selected for their qualities of endurance and sport[s]manship, in as much as bombing raids require often 8 hours continuous flying and whereas a *chasse* pilot may refuse an engagement, a bombardment team must complete their mission often under a heavy anti-aircraft fire or during attack by enemy aircraft." Furthermore, the pilot and bombardier should be trained as a team, even to the extent of rooming together.[114]

The program aimed to develop expertise in precision bombing among already advanced pilots and gunners. But, as was the case with other specialties, the course (it extended eventually from four to six weeks) concentrated on transition training for pilots coming from the United States and extra gunnery training for bombardiers. Otherwise, the curriculum included instruction in photography; bombs, sights, and releasing devices; cross-country and formation flying; aerial gunnery; and ground training, with emphasis on map reading. The bombardier also used a simulator — a kind of moving carpet over which he dropped "bombs," calling out varying speeds and headings to the pilot.[115]

Clermont-Ferrand was organized as a day bombardment school, but fairly

World War I

quickly the AEF foresaw the need for night action and urged the U.S.-based establishment to train RMAs in night flying. School officials also insisted adamantly that their instructors have more than theoretical knowledge of combat tactics. Because few American airmen had combat experience when the AEF established its training programs, the French took American instructors into French bombardment squadrons for some training and to give them firsthand experience over the lines. During the spring and summer offensives, it became possible to obtain American instructors who had flown combat sorties.[116]

Unlike the officials at Issoudun, those at Clermont-Ferrand looked forward to receiving American-made deHavillands because the Breguet bomber that was often flown was a slow-moving airplane that proved ineffectual for long-distance bombing raids. Colonel Kilner explained to the Director of Military Aeronautics that "with the coming of the DH–4 and the movement towards developing of fast bimotor planes, it will be possible to penetrate enemy territory for greater distances in the same time as that now taken by Breguet type machine, and with even less risk to equipment and personnel, due to superior speed when attacked. The near future therefore opens the field for long distance bombardment flights and forces the issue on a more thorough training in aerial navigation."[117] In September 1918, the school received its first ten DH–4s, most unequipped with bomb racks. It took until November for the school's inventory to reach forty-eight airplanes.[118]

By the summer of 1918, the AEF was relying on the United States for fully trained bombardment teams as well as for aircraft. In fact, to that point the often repeated but seldom achieved policy of giving only "refresher" or "application" training in the AEF appeared possible. All bombing personnel were then (supposedly) coming completely trained from the United States. Yet the 7th AIC still operated under pressure from the increased aviation program, so that only 15 percent of the requested personnel had arrived by the time of the Armistice.[119] Even had it been possible to attain the personnel, Clermont-Ferrand was considered a poor training site because its location so constrained the size of its facilities. Consequently, bombardment training in France was foiled for the usual reasons: site deficiencies, and shortages of men and equipment.

Nor did the hard work and optimism for bombardment training in Italy come to fruition. The school at Foggia trained 131 American bombardment pilots, 65 of whom served on the Italian front.[120] Of all the training programs undertaken with the Allies — many of which foundered because they lacked tactical aircraft — the program at Foggia seemed particularly promising because students there flew machines identical to the ones used in combat.[121] But the AEF did not field a squadron equipped with the Italian Caproni heavy bombers, even though it was thought to be well suited to long-range attacks. Moreover,

This Caproni bomber, S.C. 40070, was one of the two accepted from a total of five built in America by November 1918.

shortfalls in equipment occurred here too. Foggia lay some distance from the source of supplies, shipments of which were further curtailed by submarine activity in the Mediterranean that exhausted gasoline stores, and by the debacle at the battle of Caporetto that eliminated a great portion of the Caproni inventory.[122]

Clermont-Ferrand trained the instructors and outlined the course to be given at Foggia before the effect of those disasters became evident. The course contemplated inclusion of night flying, impossible in the confined space at Clermont-Ferrand.[123] Bombardment training commenced during the summer of 1918 on a field outfitted with a target and observation tower. The well-regarded Caproni heavy bomber was flown in training, in combat by the *Corpo Aeronautics Militare*, and by the French air service. Including night flying, Caproni training comprised three stages: learning on 200-horsepower, progressing to 450-horsepower, and finally training at the front on 600-horsepower airplanes. Ground instruction consisted of map reading and navigation as well as bombardment subjects. Flight instruction included bomb dropping, cross-country, and night flying. The last phase began in June 1918, when Americans who had trained on 450-horsepower Capronis went to the principal Italian aviation school at Malpensa, a little more than 35 miles northeast of Milan. Seventy-five American officers took the 600-horsepower Caproni course at Malpensa.[124]

Owing to the shortage of aircraft, too few bombardment pilots trained in Italy, and those who completed night bombardment training were found to be insufficiently prepared to fly in AEF night-bombardment squadrons. They were transferred to England for additional training. A review of the situation in September 1918 concluded that the "adoption of Foggia as an American training center was not regarded as a highly successful experiment."[125] Despite its drawbacks, however, the 8th AIC continued in operation until the Armistice.

The German spring offensive coincided with the birth of the British Royal Air

World War I

Force (RAF) and its bombardment arm, the Independent Force, commanded by Maj. Gen. Sir Hugh Trenchard. The RAF as a whole had come to emphasize both tactical and strategic bombardment such that one American officer reported that "every Pilot except the single seater pursuit Pilot is given instruction in bombing. This can be understood when it is realized that at times like the last big drive every machine that can carry bombs is used to bomb mobilized troops."[126] Earlier, the United States and the British Air Ministry had signed an agreement for thirty American Handley-Page squadrons to be equipped and trained in England. For that purpose, the United States would manufacture the component parts of the airplanes that would then be assembled in Britain.[127] The Handley-Page heavy bomber was to be used for tactical purposes and for strategic bombardment of German cities. It could fly for eight hours with a bombload of 1,800 pounds.[128] Some of the British-trained American crews who flew the Handley-Page bombers would serve with the British Independent Force.[129]

Because of the frightening effects of the German nighttime bombing of civilian centers, the British, more than the Italians or French, countered with night bombing attacks, despite the considerable losses and limited efficacy of those raids. By the time of U.S. involvement, the British were conducting an active night bombardment training program. Accordingly, on June 28, 1918, the American Air Service created the Night Bombardment Section headquartered in London, which handled the Handley-Page program in Britain. The Night Bombardment Section also opened in the Paris office of the AEF to coordinate the Handley-Page program with operational activities in France.[130]

In addition to manufacturing facilities, the British made available five airfields along the south coast of England for American mobilization and training. But no Handley-Page parts arrived from the United States until late July, and by November 1, only 160 incomplete sets of parts and 20 Liberty

The Handley-Page bomber was developed to bomb German cities from bases in Britain.

Air Service, American Expeditionary Forces

motors were on hand. The Air Service estimated that 950 sets of parts and 2,375 motors would have been required to equip 30 night-bombardment squadrons.[131] As a result, when the Armistice was declared, no American night bomber squadron had entered combat, although a single squadron was committed on November 9, 1918. That unit had been trained almost entirely in the United States, so it had not benefited from the fairly intensive new instruction in aerial navigation developed by the British.

Aerial Gunnery

Aerial gunnery, as one officer stated, was the sine qua non of aviation. No phase of training excluded it. Yet AEF gunnery training failed to become established on a firm footing because of the daunting deficiencies it shared with nearly every other segment of the wartime training establishment — unsatisfactory facilities and shortages of equipment. An American school did not open until late summer 1918, and until then instruction was scattered throughout the AEF training system. The most consistent instruction was provided at the French school at Cazaux, even though at least one report at war's end claimed that "there was little in the French system that is of value to us."[132]

As soon as Americans arrived on French soil, small groups of men were admitted into the main French gunnery school at Cazaux, located on the lakeshore southwest of Bordeaux and used by the French army and navy primarily for machine-gun practice.[133] The French commandant agreed to take more pilots and observers per month than he had first accepted and to adapt some of the French methods to fit American specialized training.[134] But one American liaison officer inspecting the programs in France and England during the spring of 1918 reported in near disgust to his superiors in Washington: "It is, of course, impossible to train 2300 pilots and observers in such small quantities and the deplorable result is that our observers are being sent to the front to work with French pilots on reconnaissance and artillery *reglage* without having had any aerial gunnery training. I talked personally with one of our observers who fired his first shot in the air at a German machine."[135]

The biggest obstacle to starting an AEF aerial gunnery school lay in finding a site. As the Air Service history later opined, "Conditions in the AEF have by no means been ideal. Suitable sites for Aerial Gunnery Schools are extremely difficult to obtain in a country as thickly populated as France, and lack of ground space, together with sufficient shooting area, has compelled, in a number of cases, the omission of certain features of training."[136] Somewhat unhappily, the Air Service finally settled upon a site at St. Jean-de-Monts near the mouth of the Loire River. The fact that it was underpopulated suited it for gunnery, but also indicated it was a dreary setting close to neither transportation nor sources of supplies. Trees and scrub were cleared and sand banks leveled, but the school did not commence until August 1918. The first class of thirty-six observation and eight pursuit pilots completed training the last week

World War I

in August, but because the war soon ended, most AEF gunnery training was supplied by the French at Cazaux rather than by Americans at St. Jean-de-Monts.

At one time or another, the concentration camp at St. Maixent housed cadets with no flying training and officers with various levels of expertise, all of whom awaited their next phase of training or orders to the front. It was also used for aerial gunnery training. Elsewhere, the small French infantry range at La Planche, used for machine-gun practice after its construction in April, was improved and extended in September to comply with the increased aviation program of the spring and summer. Tours, which conducted most primary flying training and where the observation school eventually located, gave courses administered by the aerial gunnery section of the observers' school and later by a section detached from that school. Four enlisted men who had taken the English gunnery courses at Grantham and Perivale became the first instructors. In a story repeated depressingly often, equipment problems plagued the program; the initial shipment of guns had to be modified for installation on aircraft.

Obtaining additional land to accommodate aerial firing around Tours was difficult because the facility was located in a well-populated area. But in June the Baron de Waldmer donated a site about eleven miles southeast of the airfield at Foret de Larcay. To protect the local inhabitants, guards were posted around the target area and, half an hour before firing commenced, an airplane flew around sounding a loud horn. In October, Larcay added two new ranges.[137] Progress seemed sufficiently hopeful that, by August, Colonel Kilner reported that the 2d AIC gunnery course "corresponds to the one laid out at Selfridge Field," the Army's model program at home.[138]

Advanced and pursuit training at Issoudun also included aerial gunnery. As of the summer of 1918, pilots first went through a course similar to ground instruction in the United States, familiarizing themselves with the Lewis, Vickers, and Marlin guns. They practiced range firing at the same time as they were undergoing *Roleur* flight training. As at Tours, Issoudun adopted the British formula, partially taught by enlisted men trained by the RAF. Upon completion of flight training at Issoudun, the pilot went for advanced gunnery at the French school at Cazaux or at the American school at St. Jean-de-Monts, or for true trial by fire at the front.[139]

The gunnery courses at the AICs were not intended to turn pilots into expert aerial marksmen, since the training centers concentrated on flying. All too often, pilots firing at aerial targets hit the tow plane rather than the target. Gunnery was, however, an essential function for observers and bombardiers, and the Air Service hoped to teach them gunnery away from the flying fields. As part of a reorganization in March 1918, the Training Section proposed an ideal scheme for specialized aerial gunnery training: "The opportunities for training in aerial gunnery will then consist of Furbara in Italy and Cazaux in

Air Service, American Expeditionary Forces

The barracks of the aerial gunnery and ordnance school at St. Jean-de-Monts (*above*) and the aircraft used to train men in gunnery techniques (*below*)

France for pursuit pilots, Vigna-di-Valle in Italy and Cazaux in France for observers, and when established, [St. Jean-de-Monts, France] for pursuit pilots, and such other classes of pilots and observers as conditions require."[140] A strong aerial gunnery course in Italy failed to materialize, but as mentioned, Cazaux and St. Jean-de-Monts figured largely in American gunnery training.

Unit Training

All air training to this point taught men — most of whom were pilots — how to fly, shoot, and bomb. Having finished individual training, young airmen usually looked forward eagerly to winged glory over the battle lines. But their training was not yet over. In late July 1918, a directive from headquarters in the United States proclaimed that each cadet should be disabused of the "mistaken idea that instruction is a necessary evil at present but that he will soon be rid of it... so that he may see that the apparently elementary instruction which he is being given, is to be continued day by day at the Front."[141]

Before going into combat alongside another American or Allied unit, aero squadrons were expected to gain experience working together. In January 1918, Chief of Air Service General Foulois explained this ultimate training stage:

> To prepare squadrons for active service, a certain amount of training as a unit is necessary. The pilots for these squadrons will be taken from those who have completed their individual instruction at the schools in the United States and in this country.... As soon as there is a vacancy at any of the organization training centres, a complete squadron, with its non-flying commissioned officers, will be sent to that place, where it will

receive its flying personnel and equipment, and be ready to work as an organization. The organizational training will cover a period of about six weeks, depending on the weather conditions and instructional facilities. At the end of this time it will be sent to the front for active service.[142]

Some officers urged that operational or unit training be moved away from the front, that is, in the Zone of Supply rather than in the Zone of Advance, where the administrative apparatus of creating a tactical unit could be managed more easily. As it was, squadrons usually arrived, and sometimes lingered, with no service equipment. Nonetheless, the practice of activating squadrons in training centers in the Zone of Advance persisted, in part because the British sent fully organized American squadrons for frontline duty.[143]

On January 16, 1918, the First Pursuit Organization and Training Center opened at Villeneuve-les-Vertus, Marne. The 96th was the first aero squadron to arrive, on February 18. The center and attached aero squadrons moved to Epiez on March 31, when French and British night-bombardment forces temporarily took over Villeneuve. "The purpose of this Center, as I interpret it," explained Maj. B. M. Atkinson who organized the 1st Pursuit Group in May, "is to form pursuit squadrons from completely trained personnel, both commissioned and enlisted, and to coordinate and adjust them to their equipment. At the same time, with the aid of the French here, to break the pilots in over the front. But — as there is no such thing as completely trained personnel and equipment is scarce — problems arise."[144] As evidence of his point, the pilots first had to fly Nieuport 28s without machine guns (then in short supply) even though their mission was to sweep the air of enemy aircraft. Moreover, sixteen of the pilots assigned to the 95th Squadron returned to Cazaux at the end of March because they had been sent to the front without any aerial gunnery training. But at least, when the 94th Squadron reported to the Gengoult airfield for duty, its eighteen pilots had been trained to fire.[145]

Because it was a relatively quiet area until the beginning of the St. Mihiel offensive in September 1918, the Toul sector occupied by the American army allowed for operational training for corps observation squadrons. An airfield and barracks for four squadrons had been constructed for the First Observation Organizational Training Center at Amanty. By April 1918, the squadrons were equipped with twelve airplanes, two with wireless sets but no artillery. According to the commandant "a student was very fortunate if he got even one *reglage* during his course." One inspector reported that "there is no machine gun instruction except a little bit on the ground, and no possibility of any training in this line."[146] Thus, unit training, when it occurred, consisted mostly of exercises and of inconsistent liaison between the Air Service and Artillery and Infantry units.

Pilots of the squadrons in the I Corps observation group had, by and large, not engaged in any effective joint training when the group began operations at the front with the 26th Division. Observers, on the other hand, had trained with

Air Service, American Expeditionary Forces

French squadrons at the front. It appeared that unit training became mostly a matter of introducing ground and air forces to one another. The mission of the I Corps observation group, according to the *Tactical History of the Air Service*, was

> to furnish the means of training our ground troops in the use of the aerial arm.... It was not expected at this time that the work of the 1st Corps group would produce any important tactical results or render any great assistance to the conduct of operation. It was expected, rather, that this period on a quiet sector of the front would serve to complete the schooling of pilots and observers and render them more competent to undertake intensive operations elsewhere on a larger and more complete scale.[147]

The 12th Aero Squadron was ordered to the Baccarat sector, another stabilized, or quiet, area. Besides its visual and photosurveillance functions, and its readiness to assist the Infantry and Artillery if the need arose, it was to train with troops of the line in terrain and panel exercises. Again, according to the *Tactical History*, "great stress was laid upon the matter of exercises." The squadron worked with the Infantry in the proper use of panels and flares as a means of communication between ground batteries and the aerial observer, and with the Artillery in adjusting fire to meet the shifting conditions of a war of movement.

The Chief of Air Service established what was called an "infantry contact school" at Ferme des Greves for III Corps troops working with aero squadrons at the field. This course was intended less for air training than for the instruction of ground troops about the utility of aerial observation, so both air and ground troops participated in terrain maneuvers. Similar schools began at airfields of the I and V Corps groups.

Aerial observation squadrons were also attached to armies. The 91st Aero Squadron arrived at Amanty December 14, 1917, to begin combined arms training with the First Army in the Toul sector. Until April 24, when the Salmson 2A2 airplanes arrived, pilots flew Renaults, taught by instructors from the I Corps Training Center at Gondrecourt, Meuse. The 92d, 24th, and 9th Aero Squadrons formed the 1st Army Observation Group on September 6, 1918, to take part in the St. Mihiel offensive. Unfortunately, confessed an author of the *Tactical History*, "due to the inexperience of the flying personnel, the 24th Squadron was unable to participate to any great extent. The 9th Squadron [Night Reconnaissance] could not function at all because of lack of equipment and training."

Operational training for the third combat specialty, bombardment, fared no better. At Romilly sur-Seine, the location of the training center, "there is nothing there," wryly commented one officer. As of April 1918, pilots at the center were flying Sopwith Strutters, "very poor machines." An inspector took ironic note of the fact that primary training at Clermont-Ferrand possessed

World War I

The Salmson 2A2

more of the excellent 300-horsepower Breguet-Renaults, "machines which are very greatly superior to the machines used at the perfection flying school."[148]

Looking Back

American training and fighting came to an end, tumultuously welcomed, with the Armistice on November 11, 1918. Afterward, the War Department invited pilots, observers, and staff officers to assess their training and operational experiences. Regarding training, their collective evaluation highlighted the lack of proper equipment; failure to develop the combination of military discipline and esprit de corps appropriate to the technical service of aviation; weak liaison between the Air Service and line army and between the Air Service and Division of Military Aeronautics in the United States; and ongoing organizational confusion in the War Department. The latter was owed largely to the passage of the Overman Act in May 1918 that effected a reorganization removing aviation from the Signal Corps. Two separate bureaus were set up — one, the Division of Military Aeronautics, charged with procuring manpower; the other, the Bureau of Aircraft Production, with providing planes, engines, and equipment. This separation of powers posed a dilemma, according to the Director of Military Aeronautics:

> The method of selecting a type to put into production and the final decision whether any plane produced was suitable for its military purpose or not was undetermined. The situation of two sets of officials with equal authority in their respective fields of action, neither responsible to the other, at once demonstrated that neither could be held for the final production of an acceptable plane for the front.[149]

Under the best of circumstances, pilot training was a lengthy process. The

Air Service estimated that a pilot could be completely trained and ready for combat in six months, but few airmen's training followed an ideal timeline. In April 1919 Maj. Gen. Charles T. Menoher, then Director of Air Service, summarized the factors affecting the length of time it took to train airmen: travel time between the various schools; weather conditions; sickness, including the influenza epidemic; lack of materiel; lack of experienced, or inexperienced, instructors; changes in curricula and authority for curriculum development; disciplinary actions and demotions; delays in securing transportation; demands from overseas; the "fact that aviation is a new science"; defects of organization; local conditions at schools; closing of schools and redistribution of students; delays in transition training; and course of instruction at the School of Fire.[150] In sum, it appears that institutional factors, more than any gross misunderstanding of how to teach people to fly, or any procedural inadequacies or timidity, caused the greatest problems for the training program.

The negative results of what became a highly elaborate but imperfect system was, sometimes, a man ill-prepared for his duty, or never given a chance to perform it. For instance, after receiving primary instruction on Jennies in the United States, Lt. George C. Kenney arrived in France in late 1917, where he retrained on Nieuports at Issoudun. As he recalled many years later: "Just as I got assigned to the 15 [meter] where I was supposed to get some acrobatics and some formation work, they sent me to the front. So I went up there without any bothering with acrobatics or gunnery or any of these things."[151] Perhaps Kenney's experience was no more typical than that of a pilot whose training dragged on because of shortages of space or equipment, or the man who completed the entire training sequence but never was called to the front before the war ended.

Contemporary accounts seldom questioned the bravery or dedication of the men who flew in combat. With few exceptions, those men who volunteered as pilots and observers and who possessed the luck, capability, and perseverance to finish their training seemed to be physically and mentally suited for the work. Many airmen judged the civilians and army officers responsible for decisions about personnel and airplane production schedules much more harshly. Delay was endemic. Inequities in pay and promotion, even in petty matters of style, rankled. Cadet John Grider fumed as he waited in Hounslow, England, to go to the front: "I'm an American and I'm proud of it but I'm damned if I can take any pride in the boobs that are running the flying corps. For instance how can we fly when our necks are being choked off by these 1865 model collars? The staff must think they are still in Mexico wearing O[live] D[rab] shirts."[152]

Airmen and ground officers had tangled before the war about their different views of military rank and hierarchy, and airmen were repeatedly chastised during the war for lacking proper discipline and courtesy. Effectively, a new military culture continued to evolve, leaving an uneasy truce between soldier and airman. The technology of the air war required an individualism distinct from the communities that fought together on the ground or at sea. British historian Peter Liddle has pointed out that "the marriage of man and machine allowed for a liberation of the spirit distinctive in the air war"; compared to the other military services it was "something less rigid and structured."[153] Lee Kennett described the formation of "a distinctive caste, a military fraternity like no other."[154]

Between May 21, 1917, and the Armistice, 22,689 cadets entered ground schools and 17,540 graduated.[155] Yet on November 11, 1918, 767 pilots, 481 observers, and 23 aerial gunners were assigned to the 45 American squadrons on the western front.[156] These numbers indicate that the U.S. flight training program in the United States and Europe was a failure for its inability to graduate and send into combat the much greater numbers of airmen required by the several-times revised tables of organization. Many factors, however, lay beyond the control of training officials in either the Air Service or Division of Military Aeronautics, such as delays caused by poor weather and facilities. A further slowdown occurred through the first half of 1918 as a result of the transportation logjam in American ports, where many airmen waited to be sent overseas until after major shipments of ground troops. Throughout that time, pressure on Allied aviation programs to train their own men precluded wholesale training of Americans by other nations.

On the whole, the scarcity of combat aircraft, at every level and almost without reprieve, was most critical. Of the 6,624 combat planes flown by the U.S. Air Service, 4,879 came from the French, 272 from the British, 19 from the Italians, and 1,440 DH–4s from the United States.[157] Likewise, for training, the AEF had to beg, borrow, or build everything it needed. In the rare last case, with the American-built deHavilland airplane for example, the results were puny and wasteful, and by the time the airplane finally arrived on American fields in France, it was already outmoded. Supply problems became less severe by late 1918, and had the war continued as planners anticipated, supply might well have caught up with demand. As it was, American training was most often coping with a shifting set of shortages.

The U.S. Army foresaw none of these factors during the summer of 1917, when it consummated its initial training plans. Fired by the Ribot cable of May 1917, in which the French Premier requested that the Americans build 4,500 aircraft for Allied use, the War Department pledged to produce vast numbers of men and machines. But rather than the 6,000 pilots the Aviation Section thought it could supply, as noted above, 767 pilots and 481 observers were assigned to armies at war's end. From the outset, the goals were hopelessly unattainable, and even the continuously lowered levels remained unreachable, thus fueling the perception of failure among the American public, in Congress, and in military circles.

Airmen did, however, have reason to suspect they would suffer one kind of loss. A naive German airman later wrote that he "applied for a transfer to the Air Force, not from any heroic motive, or for love of adventure, but simply to get away from the mass, from mass-living and mass-dying."[158] The air war did not bring a reprieve from death, and training for it was very costly. High numbers of fatalities always accompanied flight training; accidents on U.S. and European training fields were an everyday occurrence. The definitive number of fatal training accidents is difficult to substantiate. According to Arthur Sweetser, at war's end 8,688 RMAs had graduated from primary training and 204 men had been killed, with 278 total deaths in U.S.-based training programs. A draft report of the Director of Military Aeronautics asserted that during the 1917–1918 fiscal year there were 152 fatalities, of which 86 were caused by stalls that ended in nose dives or tailspins; collisions were responsible for 30; and slide-slips, the other 10.[159] *The Final Report* of the U.S. Air Service records that in the AEF, 218 pilots and observers were killed in training, which amounted to an average of one fatality for every 18 graduates.[160] Colonel Gorrell's invaluable statistical study of the U.S. aeronautical effort gives the aggregate figure of 160 student fatalities in AEF training activities.[161] Under any accounting, the numbers were high, higher than in combat though lower than the casualty figures in air training elsewhere, particularly in Britain.

The test of the training program came with the deployment of air forces over the battle lines. By all reports, individual airmen comported themselves admirably, even nobly, in the face of enormously high loss of life, and despite the fact that many went into combat insufficiently trained. The United States had been effectively unable to train pursuit pilots at home because it had no fighter aircraft; bombardment remained all but untried. As proclaimed, observation and reconnaissance dominated among the specialties. Here, success was measured directly by operations with ground units. Aerial reconnaissance provided useful information to ground commanders, and observation aircraft proved itself in directing artillery fire.

By the time the United States fielded aero squadrons, most combatants recognized that in all missions, aerial combat was a group rather than an individual endeavor, and training emphasized formation flying and cooperation with field armies and other air squadrons. In his reply concerning "lessons learned" from the war, Col. Frank P. Lahm, pioneer aviator and at war's end Chief of Air Service, Second Army, wrote thoughtfully about the need for joint and combined arms training:

> In less than a year we passed from the exploits of individual "aces".... Team work must be the basis of future tactical development.... Bombers and pursuit must know each other and train in the same vicinity. The same applies to pursuit and observation. Moreover it is absolutely essential that corps observation squadrons should train with the line in time of peace.... The Air Service should be concentrated in large units in the vicinity of the training centers for troops.... Perhaps our weakest point has been in the

World War I

lack of understanding between the Air Service (observation in particular) and the line.[162]

Lahm echoed the plaint of 1912, when, without material results, airmen argued before Congress that airfields should be adjacent to schools and posts of the line army. Furthermore, the failure to institute frequent prewar joint exercises also proved to be a dismal portent of how well air and ground officers would understand one another and work together in operational training in the AEF, to say nothing of combat. But in general, the relevance of training to the air combat mission was not well tested during the war because so few squadrons performed frontline duty.

But what was accomplished — all of which took place over a brief nineteen months that began on April 6, 1917, with an air force then numbering only sixty-five officers — was a system of specialized training where there had been none, and the establishment of a combat air arm where none had previously existed. In that plans failed to anticipate events, the training program was a microcosm of the entire blood-drenched four-year war that failed to progress as strategists had mapped. In application, the level of technical advancement during the war did not yet permit a demonstration of military prowess for new instruments of warfare, nor the doctrine for their combined employment that emphasized flexibility and mobility. How American airmen would apply the lessons learned during wartime to the years of peace remained ahead.

U.S. Air Service AEF Training Centers in France, 1917–1918

Part III
Peace

III

"... a broader understanding of the Air Corps' place in the scheme of national defense..."
— Brig. Gen. C. E. Kilbourne[1]

In many respects, World War I ushered in the twentieth century, bringing greater governmental intervention in American industrial affairs and increased efficiency and centralization of production. A more urbanized, industrialized, bureaucratic America emerged from the smoke of the European conflagration to overshadow an older notion of identity based on an agrarian economy and a relatively homogeneous population. Although the processes of modernization had been some time in the making, to many people at the time and to many historians since, the war marked the cataclysmic end of the old social order and the beginning of a new era characterized by a sense of uncertainty, anxiety, and disintegration. Along with millions of young men, traditional values and verities had died.[2]

Labor strikes, racial violence, and political confrontation were among the immediate postwar responses to the anger and fear of "foreign" and disruptive influences. But once the Red Scare of 1919 passed, the country settled hopefully, if naively, into what newly elected President Warren G. Harding called "normalcy." For the first time the United States was a creditor nation, looking toward a decade of unparalleled prosperity. Yet, confounding all prognostications for its postwar role as the preeminent political and economic giant, and despite financial investment in international markets, the United States again adopted a severely isolationist stance in its foreign policy. America, "although clearly the most powerful nation in the world by 1919," according to historian Paul Kennedy, "preferred to retreat from the center of the diplomatic stage."[3]

Anxious to put physical and psychological distance between itself and Europe, America accomplished the process of demobilization and conversion to a peacetime economy in a remarkably speedy, if not orderly, fashion. The

Interwar Years

American Expeditionary Forces (AEF) were dismantled at a rate of nearly 15,000 troops a day, causing President Wilson's Secretary of the Interior Franklin K. Lane to remark that the military structure went "to pieces in a night."[4] It took another year and a half for Congress to articulate a plan and organization for the peacetime armed forces. Unfortunately, that legislation, the National Defense Act of 1920, authorized manpower levels that would not be funded. Thus, military aviation came of age at a time when Americans found war repugnant and when pacifism ran high. Within the Army itself, aeronautics had to prove itself to a leadership unconvinced of its utility in the event of any future conflict.

Organization and administration became the watchwords of the first postwar decade. Through those years the air arm made a determined effort to create a system by which men could be trained to fly and thereafter assigned to tactical units to hone and refine combat skills. Practically speaking, this institutionalization was new to Army aviation. Before the war the handful of airmen remained effectively outside the Army hierarchy. They were seldom attached to army units and, for the most part, stayed at a training facility, practicing their skills and teaching the few other converts to fly. War brought a massive buildup in air forces, and during the nineteen months of U.S. engagement, the Air Service succeeded in putting together an admirable training establishment. Judged in terms of military success, it remained essentially on the sidelines, barely operational and virtually unintegrated into the other combat forces. By the time the United States began to field tactical squadrons, the Armistice was declared.

The frustrating sense that even wartime engagement had not convincingly demonstrated the value of air power lingered through the postwar years. Sadly, as of 1930 at least, the historical records that might have documented specific strengths and weaknesses of the AEF's air operations were lodged with the Air Corps files at the Army War College rather than at the Tactical School, as attested to by Lt. Col. Jacob W. S. Wuest:

> The United States Army Air Corps Operations in the A.E.F. are known in their entirety to no one in the Air Corps or the Army. The orders, operations and reports of air activities have never been thoroughly and systematically studied with an idea of drawing from them sound conclusions regarding the tactics employed, reasons for success or failure, effort expended to secure a given result, etc.[5]

What had not failed of example was the conviction that combat skills included more than piloting; that training in tactical units, not just in schools, was crucial; and that an administrative structure had to be erected to ensure the survival of aeronautics within the Army.

Throughout the war, all phases of the training program had suffered from a critical shortage of airplanes. The inability of American industrial production

Part III

to rectify the crippling shortfall had been the subject of endless wartime memorandums among training officials, and thereafter reverberated through the halls of Congress and filled pages of press copy. Because American industry could not produce the equipment, nearly all of the tactical aircraft flown by Americans during the war were of European manufacture. American airmen returned to a homemade inventory of large numbers of deHavilland DH–4s and JN–4D (Jenny) training airplanes and nearly 12,000 Liberty engines. But what might have been a satisfying postwar sufficiency of equipment became a liability, as stores of soon outdated and decrepit aircraft and engines had to be used until wartime stocks were depleted. Successive models of the DH–4, for example, remained in the active inventory through 1931. Moreover, specialized aircraft still remained to be designed and built.

The Army acknowledged the drawbacks of obsolete weaponry. In 1923, Maj. Gen. William Lassiter headed a committee of officers considering a "War Organization" for the Air Service. At that time, four years after the end of hostilities, his group confirmed that the Air Service was using "deteriorating" war-built aircraft, 80 percent of which were obsolescent training machines or were otherwise unsuitable for combat. The deficit worsened daily and "since it now requires about eighteen months to secure delivery of aircraft after the contract has actually been executed, it is evident that no relief can be expected from the present situation before 1926."[6]

Not only budgetary considerations and wartime surpluses but also lingering uncertainty about the mission of the air forces dictated acquisition decisions. Dissension between airmen and the General Staff and with the Navy over the role of the Air Service in national defense led to lack of clarity in priorities for production of pursuit, attack, and observation aircraft and to a debate, into the 1930s, concerning the development of long-range bombers.

Aeronautical innovation in the United States had lagged behind European efforts throughout the war because there had been only a small industry to sustain it. The postwar years saw America begin to regain ground, owed in part to the growth of civilian aviation that broadened the base of industrial support. The military profited since its technical people worked cooperatively with civilian manufacturers. Also, research and development initiated during the war continued at McCook and Langley Fields. Experimentation with aircraft and engine design, refinement of navigational aids, and studies in aerodynamics added to the combat-oriented work in aerial photography, gunnery, radio, and telegraphy. Unlike the Navy, which closely held its design work, the Air Corps, according to young engineer and pilot Orval Cook, "early in the game adopted a policy of having design work done by civilian industry but, at the same time ... maintaining the capability of feeding into that design work through research that had been done by the Air Corps and also from experience, field experience."[7]

The most significant administrative event during the first postwar decade was the passage of legislation establishing the Air Corps on July 2, 1926. The act authorized an increase in the number of general officers in the Air Corps, one of whom became commandant of the much-heralded Air Corps Training Center. The 1926 legislation also provided for a five-year expansion program that would nearly double the number of commissioned officers, aviation cadets, and enlisted men and substantially increase the aircraft inventory. Unfortunately, the high hopes for a modernized air force would soon, along with the rest of Army planning, run afoul of that perennial shibboleth — lack of money — as times began to change from fiscal and military conservatism to crisis.

President Herbert Hoover left office a beaten and despondent man, having lost the 1932 election to his Democratic opponent, Franklin Delano Roosevelt. As the new President took the oath of office in March 1933, the country foundered in a state of emergency. The economic crisis of 1931–1932 had fueled the fear that the depression might be a terrible, permanent condition. Indeed, a year later the doors of more than a third of all American banks had closed. As one historian so arrestingly put it: "The national economy seemed like a house of cards in a high wind."[8]

Although Roosevelt had been Assistant Secretary of the Navy from 1913 to 1920 and retained some special feeling for the maritime service, the plight of the underfed armed forces caused the new administration less anxiety than did the thousands of unemployed and hungry people and an endangered domestic economy. Amidst the frantic erection and dismantling of federal bureaucracies, the military services and their rivalries warranted slight consideration. Army aviation had fared comparatively well during the previous years of fiscal conservatism but now had little call on national attention and resources. Hoover had not replaced F. Trubee Davison as Assistant Secretary of War for Air. Neither did Roosevelt, who eliminated the position altogether in June 1933. What with the still fresh memory of a war that most Americans bitterly perceived as having failed to ensure international peace, protect democracy, or contribute to American self-interest, the military establishment faced the 1930s with even less promise of public support and growth than it had a decade earlier.

The sense of malaise in national security policymaking deepened during threadbare economic times, but it was rooted in long-held American ideology, values, and traditions. Americans had tended to be disengaged from international concerns, reluctant to support a large standing army, and failed to connect national defense with domestic political matters. Despite the public's lack of interest in military affairs, however, military men themselves fiercely debated national security issues, albeit largely in territorial terms. The Army, Navy, and Air Corps each sought to claim a singular role in national defense, which might then be endorsed by congressional and administrative fiat and rewarded in the budgets. The Air Corps tried vigorously to wrest from the Navy

Part III

a part of the coastal defense mission. It argued that bombardment units should be stationed at critical areas along the coasts to meet any attack on the continental United States. Not surprisingly, the Navy clung tightly to its traditional hegemony over any military action at sea. Army Chief of Staff Maj. Gen. Douglas MacArthur and Chief of Naval Operations Rear Adm. William Pratt reached a seeming detente in 1931. In this arrangement, the Navy essentially ceded some of its responsibility to the Air Corps. But when Pratt retired in 1933, the informal agreement expired.

As a result of the establishment of a combat force, the General Headquarters (GHQ) Air Force, the Air Corps's focus by 1935 shifted back to more familiar terrain, namely the functions and control of air squadrons within the Army. The growing enthusiasm in the Air Corps for long-range strategic bombardment gained headway. Technological advances in airframe and engine design, and in metallurgy, presaged the development of aircraft capable of performing tasks required by the strategic mission. Even though the Air Corps publicly defined the bombardment role in accordance with the defensive posture of national security policies, the emergence of strategic bombardment suggested a more aggressive, autonomous role than support of ground operations and coastal defense.

Air training reflected both the service's publicly espoused and privately held priorities. The Air Corps attempted to integrate relatively unfamiliar overwater piloting and navigation into its mostly land-based training program. In 1932, for example, Maj. Gen. Benjamin D. Foulois ordered the opening of a school at Bolling Field to study navigation and tactics relevant to coastal defense.[9] Senior officers also considered instituting advanced instruction in frontier defense at several airfields. One of the enthusiasts, Lt. Col. Frank M. Andrews, described Selfridge Field in Michigan, which he then commanded, as "essentially a frontier station and we have plenty of [n]avigation problems on these great lakes and over the wide extensive forests in this country."[10] Acting Chief of the Air Corps, Brig. Gen. Oscar Westover, favored the establishment of a center of tactical research to be located at the Air Corps Tactical School, Maxwell Field, Alabama, which would give particular attention to tactics applicable to coast defense. According to Westover:

> The present system of tactical research in our Air Corps is the trying out of ideas in a particular unit and the formulation of a particular doctrine for that unit. As there are many different commanders, the doctrine varies throughout the service, and in the same unit may change when commanders are changed. Much of this doctrine is founded on the particular ideas of an individual and is not based on the research and study from which such doctrines should result. There should be in the Air Corps a clearing house into which ideas can flow, where they can be tried, and where doctrines can be formulated and sent out to the service to be put into practice.[11]

Interwar Years

Except insofar as the Air Corps Tactical School remained a clearinghouse for tactical and doctrinal thinking, more ambitious means for formulating and integrating theory never became a formalized part of the Air Corps training program.

With the establishment of the GHQ Air Force, between 1935 and the outbreak of World War II the service stressed new training elements, particularly navigational competency and more coherent crew training. From GHQ Air Force headquarters at Langley Field, the commander (now) Maj. Gen. Frank M. Andrews let it be known that he expected his pilots to become instrument-rated. Military pilots must break away from their near-total reliance on individual courage and intuition and train using the new instruments under development. In his visits to airfields, Andrews paid special attention to blind and night flying, often taking the controls himself, both to teach and to learn.

As planning for the GHQ Air Force got under way, in 1933 the General Staff asked the Air Corps to cooperate in revising the basic aeronautical doctrinal statement, Army Training Regulation 440–15, and also its field manuals. The regulation had been reworked last in 1926, and the newest version should "accurately present to the service the adopted principles for the utilization of air power and the doctrines that should govern its personnel." This, Assistant Chief of Staff Brig. Gen. C. E. Kilbourne continued, "with a view to a broader understanding of the Air Corps' place in the scheme of national defense and in expectation of doing away with the misconceptions and interbranch prejudices that have prevented the Army from reaching a common understanding and presenting a united front on the subject."[12] But the Air Corps found it difficult to forge doctrine regarding the employment of air forces to which all parties could agree. By the time Kilbourne sent out the Chief of Staff's instructions for the Air Corps studies, Admiral Pratt had retired from the Navy, and the Pratt–MacArthur agreement had become moot. The tug-of-war between the Navy and the Air Corps over the coastal defense mission remained unsettled, and Army air participation in coastal defense remained vague in subsequent employment doctrines. Training instructions, therefore, based upon doctrine, would likewise lack specificity.

The difficulty in writing training materials arose from the fact that the heart of the issue, what Kilbourne called "the Air Corps' place in the scheme of national defense," continued to be debated. All the participants knew the high stakes to be won by formulating training literature, and therefore tried to influence the final results. A board empowered in 1934 to review training methods avoided coming to immediate grips with the Air Corps's mission: "For those organic components of the several arms in which the tactics and techniques are rapidly changing and evolving, due to motorization, mechanization and improvement in means of communication, tentative manuals should issue in limited edition."[13] Not much changed. Four years later, as part of ongoing discussions, GHQ Air Force Chief of Staff Col. W. H. Frank notified the commanding generals of GHQ Air Force wings that "the constant and rapid development in aircraft and equipment makes it advisable to maintain a certain fluidity of tactical doctrine."[14]

Even as the Air Corps tried to articulate new tactics and training based on new doctrine, it struggled to build on the progress it had made already. Yet the 1926 five-year plan dragged on well into the thirties. President Hoover, in one of his last budgetary acts, impounded $2 million of Air Corps appropriations for 1932–1933. When Roosevelt took office, he showed no appetite for change. He, too, withheld funds, this time from the fiscal 1934 budget. Although all tactical air units had finally formed by October 1933, they remained abysmally below strength in manpower and aircraft. As the five-year plan crept toward completion, the initiative for another buildup through a second five-year program was rebuffed. Instead, 1934 witnessed a downward turn in the levels of aircraft procurement, and at midyear, officer strength still remained about 350 shy of the 1,650 authorized by the 1926 act. Both airplane and personnel levels declined further thereafter.[15] Thus, even though the Air Corps replayed its old tune regarding the need for more manpower and equipment, it went largely unheeded by the War Department, the Hoover and Roosevelt administrations, the Bureau of the Budget, and Congress.

Now a seasoned airman familiar with the ebb and flow of manpower and equipment levels during wartime, in mid-1934 Maj. Carl Spaatz, head of the Training and Operations Division, prepared a memo for the Chief of the Air Corps concerning wartime employment of air forces. Spaatz predicted that the "training of pilots and other personnel can be commenced on M[obilization] day in training types, and by utilizing civilian personnel who have already obtained much air experience." But, he warned, the "availability of airplane pilots will greatly exceed the availability of combat equipment for an extensive period of time, probably more than a year."[16] The Air Corps had not fully assimilated the lessons of World War I, but Spaatz reiterated one that it had learned thoroughly: flight training could be ponderously slow and hideously expensive, but its hard-won success was nullified when there was insufficient or obsolete equipment. The Army, and those who funded it, however, paid little attention to Spaatz's implied admonition, and to others like it, as aircraft levels dropped from the mid-1932 high of 1,646 to 855 in June 1936.[17] Thus, when it came to the implementation of plans or training in pursuit of articulated doctrine, training directives were usually ignored because of equipment shortfalls rather than disagreement on principle. As one of many examples, in December 1932 the executive officer at Wright Field explained that compliance with the 1933 training directive was impossible because of a lack of available equipment to carry out the tasks: observation crews could not use camera guns because they had none; pursuit pilots could not fly safely at night without flares. Having a directive did nothing to ensure that the equipment would be stocked or purchased for new airplanes coming

Interwar Years

into the inventory.[18]

Flares and camera guns hardly teetered at the leading edge of aviation technology, so their absence hints at the scarcity of more advanced aircraft and equipment. The late 1920s was a fruitful period for aeronautical research and development. By the late thirties, experimentation in aerodynamics had brought significant advances in engine and airframe design and construction: engine cowling that allowed cooling without reducing drag; metal structure and cantilevered wings; monocoque construction; split wing flaps that enhanced control during takeoff and landing; improved power-to-weight ratio of air-cooled engines that also needed less frequent overhaul; variable-pitch propellers; retractable landing gear; turbo superchargers; high-octane fuel; and much more extensive navigational equipment. Protectively enclosed cockpits had become standard. Monoplanes replaced biplanes.[19]

But owing to depression-era economies, the Air Corps was unable to avail itself fully of the largesse from these improvements. Fortunately, engine and airplane manufacturers found it profitable to produce increasingly sophisticated aircraft and equipment in response to the greatly expanding civil aviation industry.[20] Although unable to consume large amounts, the Air Corps had a taste of the new developments. When the President and Congress finally approved an enormous Air Corps expansion program in early 1939, military men were not unfamiliar with, and would reap the benefits of, a decade of aeronautical progress.

In general, personnel procurement fared better than aircraft acquisition even though policymaking and budget concerns, certainly until the mid-1930s, kept a lid on military "hiring" at a time of frightening national unemployment. Those at the helm of the New Deal labored during the First Hundred Days to put people back to work. The administration used the military as a partner in the work of the Civilian Conservation Corps. Members of the armed forces guided, but gave no military training to, young men employed in reclamation and reforestation projects. The Air Corps joined this endeavor, although some air officers overseeing Conservation Corps activity expressed frustration in having too little time and resources for their normal duties. Opening a wider door to military enlistment and commissioning might have eased the strain on military personnel and offered greater civilian employment, but the time was not conducive to an engorged military. Not until June 1935, to match the anticipated increase in new aircraft procurement resulting from the creation of the GHQ Air Force, did Congress authorize an expansion in the forces. It permitted the Air Corps Reserve to grow to 1,350 men on extended active duty for five years. Once again, however, authority arrived without the money to underwrite it.[21]

As Chief of Staff, General Douglas MacArthur often supported the air arm in its tussles with the General Staff and gave the Air Corps a generous portion of scant fiscal resources. MacArthur knew that the Army's clumsy organizational structure — nine semiautonomous corps areas with attached air units — added to the disarray in War Department strategy and operational efficiency. He attempted to introduce greater coherence in August 1932 with his Four Army Plan by which four field armies answered directly to the Chief of Staff. This scheme did not, however, entail any measure of centralized control of air units by the Air Corps. Furthermore, MacArthur did not share the glorious vision of air power espoused by some of his airmen. He doubted the decisiveness of air warfare in a conflict, as he stated in 1933:

> There is, of course, no question as to the tremendous influence that the airplane will exert upon warfare of the future, but there is as yet only meager experience upon which to base, with any confidence in their accuracy, predictions as to the extent of that influence or the manner in which it will be most effectively utilized. . . . No major battle in or near the United States in which land or sea forces will not constitute the ultimately decisive element can yet be classed as a strong possibility.[22]

This wait-and-see attitude as well as the budget confines wrought by the Great Depression meant that by 1935 equipment inventories had bottomed, personnel levels remained static, and training was stagnant.

Activation of the GHQ Air Force in 1935 as a real rather than paper entity re-energized the Air Corps. It came at a time of greater health in the domestic economy, apparently offering an opportunity to upgrade the pitiful state of the aircraft inventory. Unfortunately, a recession during the late summer of 1937 resulted in further cutbacks in federal spending. Nonetheless, the GHQ Air Force actualized a strike force of concentrated air units, shifted control of combat units to an air commander, and reintroduced a focus on tactical unit training — all critical elements to an independent air mission. Doctrine, organization, and technological advances now dovetailed to permit military aviation a more direct and realizable scope to train a combat air force. In speech after speech in the late 1930s, GHQ Air Force Commander General Andrews trumpeted the importance of air power. Unlike the more flamboyant and well-known air advocate of ten years earlier, Billy Mitchell, Andrews spoke with greater authority because he glimpsed reality rather than dreams.

In October 1938 a radio broadcast of H. G. Wells's *War of the Worlds* caused thousands of horrified listeners to fear that the earth was being invaded from outer space. Although the Martians were not coming, danger was real. The world was about to change, and the Air Corps, too, would never be the same.

SIX

Postwar Retrenchment

[The Air Service] is the newest service in warfare. It has only the experience of one war for a guide. The United States has only had experience in part of this war. There is no precedent for a peace time organization.

—Maj. H. M. Hickam[1]

Between November 1918 and July 1926 when a five-year expansion program was announced in conjunction with the newly authorized Army Air Corps, the Air Service set about formulating policy and doctrine, creating an organization, and establishing its training system. All these institutional steps were vital to the survival of aviation as a component of the Army and to the growth of professionalization within the air arm itself. To build on what they had already achieved and to assure some continuity in methods and management, airmen in the first postwar years set about structuring a permanent peacetime organization.

Organization

In 1919 the War Department petitioned to maintain a force of half a million men, and members of the Air Service envisioned a 239-squadron air arm to be trained at 16 flying schools.[2] To its disappointment, the Regular Army had to content itself with slightly over half the requested number, and aviation anticipated a proportionately reduced share. Were it to return to prewar strength, the Air Service would all but vanish, and the Army, however unconvinced it may have been about the utility of aircraft, had a considerable investment in aeronautics. Thus, the Air Service more than survived, even as budgets and manpower plummeted from their expansive wartime high. All branches of the Army now had to assign priorities among the combat specialties and the reduced number of facilities left open after the war.

The Air Service's uncertainty about its status and authorized strength mirrored the postwar managerial disorganization in the Army and the government as a whole, so the first board appointed in 1919 to explore reorganization came to no decisive conclusions. It seemed clear that the wartime U.S.-based bipartite organization (Department of Military Aeronautics and Bureau of Aircraft Production) had proved unwieldy in the extreme. Those agencies in turn lodged in a different chain of command from the Air Service under the AEF. Subsequent reorganization plans contemplated a more centralized system with all Army air elements under a Director of Air Service. In fact, after the war the Department of Military Aeronautics existed in name only and was abolished in the reorganization of June 1920.[3]

The Air Service instituted a divisional system consisting of Supply, Information, Training and Operations, and Administrative Groups, each headed by First, Second, Third, and Fourth Assistants. The Training and Operations Group was charged with the "operation, supervision, and direction of all flying fields, training schools and organized Air Service units not transferred to the control of Department or other commanders. For such organized Air Service units as are not under the Director of Air Service, but under the control of Department or other commanders, it prescribes the tactical and training methods to be employed."[4] The Training Division included the Primary and Technical Section (primary instruction of individuals in flying, radio, photography, navigation, engineering, and mechanical training) and the Advanced and Tactical Section (advanced training of individuals and units and training in pursuit, bombing, observation, radio, surveillance, attack, photography, navigation, and engineering of tactical units not assigned to the Operations Division).

The Chief of Air Service[5] held responsibility for training but did not control tactical units in the field. Further administrative ambiguity existed within the Office of the Chief of Air Service by the placement of training and operations together. What portion of unit training should be conducted by the Training Division, and what remained to the Operations Division? In June 1919 the Director of Air Service, Maj. Gen. Charles T. Menoher, asked the head of Training and Operations, Brig. Gen. William Mitchell, to prepare a list of activities projected by each of the four groups and indicate priorities, the number of men authorized and the number needed, and the function of each air station. The Chief of Training, Lt. Col. William C. Sherman, who had in the past and would again in 1927 draft training regulations that articulated aviation roles and missions, perceived that an organizational structure influenced doctrine and therefore the nature of a training program. "The system that [is intended to be used] in war should also be employed, as far as practicable, in peace. Otherwise the Army is not a war machine but a peace machine. . . . In time of peace the dividing line between Training and Operations is apparently not so clear." What, during peacetime, should Operations do, he mused. "Who is to be constantly studying new tactical methods and prescribing them for

Postwar Retrenchment

Director of Air Service in 1919, Maj. Gen. Charles T. Menoher

training? Who is to prepare war plans?"[6]

As it developed, the Training Division supervised the training of tactical units, but activities such as maneuvers were not defined as training. The Operations Division prepared war plans and conducted aerial expeditions and races. All divisions helped write training regulations. These overlapping roles and the frequent shift of officers between training and operations activities reduced the likelihood of factionalism that might have occurred in a larger organization with cleanly segregated functions. Moreover, the necessity for personnel exchanges among different offices of the War Department had been identified during the latter part of the war. Yet the functions constituting operations as opposed to training in a peacetime service remained a conundrum. In practice the two were separated by a permeable membrane. In 1929, for example, when flying hours were carefully apportioned according to mission because of tight fuel allocations, one commanding officer remarked that "considerable flying time is credited under operations and miscellaneous, which might have been credited to training. All the work accomplished on Border Patrol missions and Division and Corps maneuvers can also be credited to training as it embraced communications, both radio and visual, formation flying, aerial gunnery, night flying, liaison exercises, field exercises and aerial navigation."[7]

Initially the Training and Operations Group included the Balloon and Airship Section. Col. Charles DeF. Chandler, the veteran balloonist from the earliest days of Army aviation, had recommended in the spring of 1919 that lighter-than-air be administered separately. "The tactical employment of

215

balloon units," he maintained, "is quite a different matter and should be left entirely to the Corps and Army Air Service commanders to correlate their operations with ground troops and airplane units."[8] But in considering this issue, Lt. Col. Oscar Westover reasoned that the Air Service as a whole should be organized along functional lines, as dictated by "the whole trend of modern, efficient organization." The Air Service should expect to cooperate with corps and divisions when in the field, and the "responsibility for complying with the orders should be traceable directly through functional lines of responsibility and not through lines based on materiel."[9]

Although the Balloon and Airship Section separated eventually, Westover's concept won acceptance, and the organization formed along centralized and functional lines. In part this administrative structure was a reaction to the dispersed training activities that characterized the war, when Washington and Europe performed overlapping and often competing tasks, as did the various flying schools. A functional system had the virtue of greater efficiency, but more to the point, it reflected the G system of the Army that the Chief of Staff, General John J. Pershing, carried over from the AEF model: G–1 (Personnel), G–2 (Intelligence), G–3 (Operations and Training), G–4 (Supply), and a War Plans Division. In the postwar adaptation to the General Staff, training was linked with operations rather than controlled by the War Plans Division.[10] In the Office of the Chief of Air Service, the training function mostly stayed on the organizational chart as Training and Operations, but for a time it became Training and War Plans when the Operations Division was discontinued in a December 1921 reorganization.[11]

General Menoher, who had commanded the 42d (Rainbow) Division and VI Army Corps during the war, became the first postwar Director of Air Service. Col. Milton F. Davis, Chief of Training for the Division of Military Aeronautics at war's end, was followed by General Mitchell as Director of Training and Operations. Menoher was a confirmed ground officer at heart, and he and air power advocate Mitchell tangled frequently. No doubt Mitchell was as pleased initially as other airmen when in October 1921 Maj. Gen. Mason M. Patrick took command of the Air Service. Patrick, although also a ground officer and one of Pershing's West Point classmates, lobbied effectively on behalf of aviation within the Army. Moreover, believing his responsibilities to mean that he should understand something about the skills of his young airmen, he took flying lessons from Maj. Herbert A. Dargue at Bolling Field in November 1922. He earned his wings the following June at the age of 59. Dargue commented at the time that "there is probably no one thing that the Chief of Air Service could have done to raise the morale of his officers and men more than to learn to fly himself. Our Air Service is continuously working in what might be called 'The shadow of death.' Accidents have been greatly reduced, but there are still many, a large number of which are fatal."[12]

Despite an uncertain future and in the absence of congressional legislation,

Postwar Retrenchment

In 1922 at the age of 59, Maj. Gen. Mason M. Patrick (*below*), Chief of Air Service, took flight instruction from Maj. Herbert A. Dargue (*left*) at Bolling Field.

in 1919 the Air Service tried to retain as many airmen as possible from the rapidly demobilizing forces and also to recruit new pilots. By war's end all the ground schools in American universities had been discontinued, so this source of manpower disappeared. Reserve Officers Training Corps (ROTC) units in American colleges were still on the drawing boards. In 1919 the Army created the new grade of flying cadet for flight school attendees. At the time, the number of those cadets was not to exceed 1,300,[13] but it rose to an authorized level of 2,500 as a result of the 1920 legislation.[14]

The already trained but increasingly skeletal force was assigned to temporary and permanent stations as members of as yet undetermined numbers of squadrons and groups. Because the government still perceived an immediate threat on the Mexican border, the Army ordered units of the 1st Bombardment Group, 1st Surveillance Group, and 1st Pursuit Group to the Southern Department. Mitchell and Sherman of the Training and Operations Group reported that permanent squadrons were in the process of being organized and relocated, although "some squadrons will be maintained at temporary stations on the Border as long as the present emergency exists."[15]

"In planning the peace time training program," the Director of Operations on the General Staff opined in early 1919, "it will have to be borne in mind that it will take on many features radically different from the war time training. It will be more varied in its scope, the men trained forming the source from which organization commanders, administrative officers and instructors would be drawn for the organization of units and the rapid training of young combat flyers in case of emergency."[16] A consolidated air force would not be trained, commanded, and deployed by the Air Service. Airmen would be trained in a standardized system administered by the Air Service and then be assigned to squadrons under the direction of army corps areas. The Army intended to scrutinize the favored specialty, observation, especially closely. The Adjutant General informed the Director of Air Service in late February 1920 that "the general question of aerial observers (airplane and balloon), the branch or branches of the service in which they are to be commissioned, the scope of their instruction and the duties they will be called upon to perform [is] now under consideration and the decision arrived at will be communicated to you later."[17]

That the nature of command and control of air forces was defined only vaguely should be no surprise, given the lack of unanimity and experience among the several offices promulgating Army doctrine. In 1920 the War Department directed its branches to draft new training regulations and manuals. The first significant result would be a codification of the principles of war in War Department Training Regulation 10–5 of December 23, 1921.[18] A formal statement of air doctrine encoded in training regulations would be issued in January 1926. To draft it, the Air Service farmed out portions of the training literature project to schools and units responsible for particular functions and to the Field Officers School (soon renamed the Air Service Tactical School).[19]

Refinement in air tactics had taken place during the brief time of American engagmenet during the war. Air Service operations had moved away from one-on-one encounters toward the use of larger air elements and an emphasis on unit discipline. But airmen were hard put to articulate rules for air-ground coordination on the battlefield, where the lessons seemed less clear. One young officer addressing the subject of aerial observation in cooperation with the Coast Artillery remarked that "experience may be had before regulations are prescribed. It should be emphasized that...Coast Artillery–Air Service cooperation, even in the matter of conduct of practice fire, is so new and so little known by officers of those branches in the aggregate that it is as yet too early to prescribe definite regulations. Cooperation in war time is yet to be tried."[20]

At the Field Officers School, William Sherman took a similar view. "I don't want you to think that we are crying," he apologized to his friend Bart Yount in the Chief's office, but he and other Air Service strategic thinkers faced an especially difficult task. On the one hand, the "foundation and most of the superstructure of Infantry tactics has existed for many years." On the other, "with the Air Service the case is totally different. We must build from the ground up, on a very limited experience, as compared with the long history of all other arms of the service."[21] As a result, Sherman's solution was to draw heavily upon British doctrine.

Still another senior Air Service officer pointed to the frustration of trying to fit aviation into the already established habits of thought of the Army. In critiquing a draft of the general order governing training, the chief of the

Postwar Retrenchment

Training and Operations Group in 1921, Lt. Col. James E. Fechet, commented that the document was to apply "general statements of doctrine and principles of training" to all branches of the Army. But, he pointed out, "mass psychology, so necessary to be developed in the Infantry and Arms where hand-to-hand combats must take place with enemy troops, does not apply in the same sense to Air Service troops whose enlisted personnel function more as a noncombatant Army except for defense purposes while the actual combat is carried on by the commissioned personnel in the air."[22] Training and recruitment requirements seemed quite different for the Air Service. Pilots had to be possessed of considerable individual initiative and self-reliance, and noncommissioned personnel had to be highly technically oriented.

After considerable deliberation, the Army revised and published Sherman's manuscript on air tactics as Training Regulation 440-15 (TR 440-15), *Fundamental Conceptions of the Air Service*. This document accepted the principle that all air activity supported the ground battle.[23] But Sherman and many other airmen endorsed what many considered too independent a role for aircraft. Regulation writers found the necessity of conforming to senior Army officers' concepts of the employment of aerial forces to be as difficult as attempting to create doctrine from theory alone. Needless to say, General Patrick was especially sensitive to the political dimension. In March 1924 at Fort Leavenworth, he announced that a draft of TR 440-15 had been widely circulated for review and comment. "To say that it did not meet ... unqualified approval is putting the case rather mildly," he admitted. "A study of the comments offered indicates that the text conveyed to the reviewers the erroneous idea that the Air Service intended in the future to fight all wars by itself and that in its opinion the remaining branches of the service could safely stay at home." Such an apostasy was not at all what the regulation writers had intended, Patrick reassured his audience. And although Patrick did not make the point, such an assertion would have constituted hubris utterly unsubstantiated by the Air Service's performance in the war just past. "Now," he continued, "the real meaning was that units of this 'Air Force' might be employed on missions, some of them far removed from the theater of operations, but, nevertheless, that these missions were undertaken absolutely in accord with the general plan of operations of G.H.Q. and were primarily intended to assist all other component parts of the armed forces in carrying out the common mission — victory over the enemy."[24]

And so the final document held. Dated January 26, 1926, TR 440-15 followed the "traditional military view," according to historian Thomas H. Greer. The Army's primary objective was the "destruction of the enemy's armed forces." Greer continues:

> The mission of the Air Service was defined as that of aiding the ground forces to gain decisive success by destroying enemy aviation, attacking surface forces and facilities, and by protecting friendly ground units from

hostile air reconnaissance or attack. In addition, the Air Service was to furnish observation for information and for artillery control, messenger service, and transportation for special personnel.[25]

Although TR 440–15 espoused the "traditional military view," Greer notes that textbooks written and taught at the Tactical School described a more independent function for air power, closer to what Sherman and many other airmen believed. Both the regulation and the somewhat competing Tactical School manual expressed the doctrinal assumptions that took the Army air arm up to the outbreak of World War II.

The difficulty in preparing regulations that outlined the role of aerial forces and thus implicitly prioritized training functions and aircraft procurement lay in the contentiousness over mission between segments of the Army, between the Army and Navy's perceived roles in coastal defense, and ultimately in the fact that national military policy was itself only vaguely articulated. Greer has pointed to the "absence of a clearly defined strategic premise" understood by all. Even the commandant of the Air Corps Tactical School at the close of the decade, for example, later admitted that "he didn't know, or at least could not remember, what strategic assumptions underlay the development of air doctrine at that time. It was surely a question that was much evaded during the entire interval between world wars, and a question which no man, in all truth, could answer with finality."[26]

Nonetheless, by early 1920 the general organizational framework for postwar aviation had become apparent. The Army Reorganization Act of June 4, 1920, amending the National Defense Act of 1916, established the Air Service as one of the combatant branches of the Army along with the Infantry, Cavalry, and Artillery. It authorized the Air Service 1,516 officers, 2,500 flying cadets, and 16,000 enlisted men. The Chief of the Air Service directed training at the special service schools, but the commanders of the nine army corps areas, who were advised by air officers on their staffs, controlled the tactical air units. The 1920 act reaffirmed the concept of a small standing army to be augmented in the event of emergency with a trained Reserve and National Guard.

At the opening of fiscal year 1921 the Air Service had 155 Regular officers; by year's end there were 975. Of that number, only 642 were pilots who had earned their wings.[27] For a time after the war, the commissioned grades of the Air Service held no vacancies. Furthermore, Congress forbade new Army enlistments in early 1921, which effectively curtailed training new cadets. General Menoher made the case that the Air Service should be given special consideration in the allocation of Reserve officers because aviation was subject to special conditions in filling its ranks, namely that because pilots were drawn from young, active duty officers who were detailed to aviation for relatively short periods, more reservists could fill out the lean manpower tables. And, because "there is no such thing as a partially trained flyer," Reserve officer training should be of the same quality and length as that for the Regular Army.[28]

In April Menoher followed with another extensive document outlining the manpower needs of the Air Service. "The problems presented to the Air Service in time of peace are quite unique. No other branch of the Army has the same difficulties to be surmounted." All combat troops in the Air Service were commissioned officers. "The life of the combat personnel is very limited. Their training brings them early to the peak of their efficiency as flyers and their usefulness thereafter rapidly diminishes." The service used its authorized enlisted force of 16,000 men exclusively for repair and maintenance of aircraft, that number was barely sufficient to maintain a minimal combat force of one bombardment group, one attack group, one pursuit group, and the single observation squadron for each corps area. Congress had forbidden further recruitment until the size of the Army dropped toward its allowed strength. That left too few airmen for training and operational roles. "Assuming casualties at the rate of 33⅓ per cent per month," the General stated, and "realizing that the training of a flyer requires at least six months even in war, it is quite apparent that 20,000 flying officers at the outbreak of war should be available."[29] At that time — April 1921 — the Air Service Reserve numbered 5,000 flying officers. Both the quality and quantity of their training had suffered because the Regular Army was too understaffed to provide much assistance. In his annual report, Menoher therefore urged increased compensation and an extension of time on active duty for a fixed number of Reserve officers.

This was Menoher's last hurrah as Chief of the Air Service. He resigned, replaced by Patrick who presided over a further straitened Air Service, cut drastically from $33 million in fiscal year 1921 to $19.2 million for training, operations, procurement, and maintenance the following year.[30] Although Congress reduced military funding, the Air Service still was served, it must be noted, a sizable piece of the military budget pie. Yet the abject picture of an undermanned force that Menoher presented in 1921 remained constant. Between 1923 and 1938 (before the 1939 expansion program), only 10 percent of the qualified aviation cadet applicants graduated from pilot training. Training fatalities added to attrition by disqualification. The Secretary of War had reported in December 1918 that two airmen lost their lives in training for each one killed in battle. Compared to the Army as a whole, deaths of airmen in accidents was forty-nine times as high as that of other officers.[31] Brig. Gen. Noel Parrish, in recalling the postwar years when he was a young pilot, commented that "with the casualty rates we had ... you [were] in combat against nature, ignorance and other factors practically all the time. The weather, gravitation, and so on were your enemy constantly."[32] In sum, drawing from the ranks of aviation cadets and Regular Army officers, the Air Service only managed to train several hundred pilots each year.[33]

Interwar Years

Flight Training

A radio script prepared by the Air Corps in 1931 offered a simplified but fairly straightforward view of the mission of peacetime Army aviation:

> The saying is often heard that "life in the Army is just one school after another." I think that applies most aptly to the Air Corps itself, with its schools for officers, enlisted men, pilots, mechanics, photographers, radio operators or for any other pursuit necessary to the proper education of the personnel that go to make up this [e]ver-growing branch of our country's defense. Now, not only are these schools an important part of the Air Corps training system but there must not be forgotten the follow-up of continued training in these subjects with the Air Corps troops at tactical stations and during maneuvers in the field.[34]

Modestly buried among the enumerated skills for which airmen trained, piloting was in fact the sine qua non of military aviation. Yet the broadcast that day during the depths of the Great Depression, which depicted a busy, prosperous system of flight training, was perhaps too optimistic. At this time, military aviation, especially pilot training, had for over a decade struggled vainly to regain some of its wartime promise.

Immediately after the Armistice, airmen's hopes ran high. The War Department determined to purchase and maintain fifteen flying fields and five balloon schools for training purposes. Of those, the government already owned Rockwell, Langley, Post (at Fort Sill), and Kelly Field No. 1. It leased and expected to buy the others. Early plans anticipated opening several primary schools and separate sites for advanced training in bombardment, observation, pursuit, and gunnery. This extensive operation assumed a somewhat reduced and tightened continuation of the system initiated during the war.

In fact, by 1919 the conduct of primary training had narrowed to March Field in California and Carlstrom Field in Florida. Having only two bases at least aided standardization of methods and, given the shortage of instructors and enlisted mechanics, preserved scarce resources. The pilot school course combined ground school and elementary flight training. Like their wartime predecessors, cadets mostly learned on Jennies. Now, however, training officials considered it "not a satisfactory training plane, but as we have nothing else it is necessary to use them."[35] At March Field cadets flew their last two hours of dual instruction and also soloed on deHavillands. Borrowed from the RAF Gosport system introduced during the war, instructor and student shared a speaking tube for communication in the air, and one instructor taught a student for his entire course.

The first class of cadets at both March and Carlstrom were enlisted men from various Air Service units. Civilians constituted most of the second class.[36] The Training and Operations Group had hoped to graduate thirty men per month, but in the last six months of 1920, Carlstrom had graduated thirty-six

Postwar Retrenchment

One of the earliest sites for Army aviation was Kelly Field, near San Antonio, Texas. In the 1920s when training programs located there, cadets mostly learned on Jennies (*top two images*). Aircraft like the 11th Bombardment Squadron's deHavilland (*lower left*), shown here fitted with a machine gun, returned from the European war and were included in the Air Corps's inventory of training aircraft.

officers and fifty-six cadets; March had graduated twenty-five officers and sixty-seven cadets. A number of the students in the early class, especially at Carlstrom Field, were naval officers not destined for Army squadrons.[37] The course of instruction at the pilot schools during this time lasted four months; its graduates took advanced training in pursuit, attack, bombing, or observation.

Planning aside, in June 1921 only advanced training in observation was given at a specialized school, at Post Field, Fort Sill, Oklahoma. Pursuit and bombardment training took place in the two tactical units at Kelly Field, and the Attack Group provided training for that specialty. Because the Air Service lacked training funds, some students graduating from primary training through the summer of 1921 had to delay advanced work and began transition and night flying at the pilot schools.[38] Even though specialized schools were still mostly paper organizations, training officials decided to lengthen the advanced flying course and shorten the training time spent in a tactical unit.[39]

Clearly, advanced training declined in administrative torpor, but primary training fared not measurably better. The Chief of the Primary and Technical Section reported in early 1920 that neither of the pilot schools had executive officers and that Carlstrom did not even have an Officer in Charge of Flying, one of the most crucial positions in the chain of command at a flight school.[40] The schools keenly felt the critical shortage of manpower. They did not have enough enlisted mechanics to keep the airplanes fully operational. Too few instructors remained in the service to teach students, assuming that students could be recruited in reasonable numbers.

One solution appeared to be a reduction in the number of training fields. By closing March Field at Riverside, California, facilities would be completely centralized and overhead costs further reduced. Next came the removal of the other pilot school from Carlstrom to Brooks Field in San Antonio, Texas. Capt. Hugh Knerr happened to be in pilot training at Carlstrom Field when it closed in June 1922. "The entire garrison," he recalled, "was loaded onto a special train for transportation to Brooks Field ... in one hilarious exodus of men, women, children, cats, canaries, and dogs."[41]

Students graduating from the newly designated but hardly palatial Primary Flying School at Brooks Field went on to the equally primitive facilities at the Advanced Flying School at Kelly Field. By mid-1922 the Air Service, working within the confines of a 125,000-man Army and its own reduced numbers, established its training center for heavier-than-air flight at these two fields in San Antonio, Texas. It modeled school squadrons after tactical units, each consisting of 24 officers, 132 enlisted men, and 16 airplanes.[42]

Primary Flying School

The Primary Flying School was organized as the 11th School Group, consisting of Headquarters 11th School Group, the 46th School Squadron, the 47th School Squadron, and the 62d Service Squadron.[43] The impetus for consolidation had

Postwar Retrenchment

Brooks Field as it appeared in 1920. An unnamed aviator stands beside one of the 96th Aero Squadron's aircraft that returned to Brooks after the war in Europe.

been driven almost entirely by money and manpower shortages, but improved safety and standardization of training methods were positive results. Some Air Service officers had urged, for instance, that primary training be continued throughout stations in the United States and at overseas posts. Yet General Patrick directed that primary flying training be given only at Brooks Field, in spite of the shortage of pilots in overseas departments and the relative ease of recruiting pilot trainees at those posts. The policy of centralization, Patrick informed the Adjutant General in 1923, "is the direct result of a careful consideration of all available statistics on training casualties occurring during the world war period and in view of the success attending the operation of the primary flying school during the past two years, the slightest deviation from this established training policy of the Air Service would unquestionably result in an increased number of casualties."[44] Once again, safety held high priority in the training scheme.

To be eligible to enter flight training, a candidate had to be an unmarried male citizen of the United States between the ages of twenty and twenty-seven and have a high school diploma or its equivalent. Applicants took a physical and educational examination, and those accepted were assigned to a school class. The course covered the theory of flight, regulations pertaining to flying, radio theory, buzzer (code) practice, meteorology, aerial navigation, machine guns and their accessories, instruments, airplane motors, airplanes, personal equipment of the pilot, ground gunnery, bombs and pyrotechnics, theory of photography, primary flying training, cross-country flying, formation flying, night flying, and transition to tactical aircraft.[45] Until 1926, most students learned to fly on the JN–6H. The service estimated the cost for this segment of the training program to be $9,751.23 per student.[46]

Advanced Flying School

By 1924 Kelly Field housed the Advanced Flying School and the 3d Attack Group. The former was organized as the 10th School Group consisting of Headquarters 10th School Group; the 40th, 41st, 42d, and 43d School Squadrons; the 70th Service Squadron; and the 22d Photo Section.[47] Training at Kelly superseded that given in tactical units and at the Observation School. The Advanced, like the Primary Flying School, suffered from a personnel shortage. Since the advanced course encompassed greater variety, it was even more difficult to conduct advanced and specialized training efficiently. Moreover, Kelly housed a mixture of instructional, or school, duties and post activities such as quartermaster, ordnance, and engineering, all of which were partially staffed by Air Service troops and supervised under one command and headquarters. Having a combat group at the field proved useful in attack training, but it was more than offset by having support stretched so thinly.[48]

The first class at the Advanced Flying School entered on July 15, 1922. That course lasted eight months, but thereafter it was shortened to a six-month course with two classes graduating every year. Some students stayed over from one class to the next if they seemed promising as pilots but had been unable to pass their flight tests the first time.

Until 1928 the Advanced Flying School divided its system of instruction into basic and advanced phases. Basic training continued the staged instruction of primary training in which all students worked through increasingly difficult maneuvers. Students went from dual instruction to solo, accuracy, hurdles, figure-eights, 180- and 360-degree turns, performance flights, formations, and cross-country and night flying. These final two activities were especially difficult. Aircraft and landing fields were poorly lit, and the seamless, sparsely populated Texas prairie offered little ambient light besides the moon. The sameness of the topography provided few visual clues to guide the cross-country flyer. Pilots navigated largely by visual means, by reference to geographical and manmade landmarks, because no one had drawn aeronautical maps and navigation aids were primitive. Advanced students spent approximately twelve weeks practicing these skills before beginning specialized training.[49]

Graduates of the advanced course earned ratings as an Airplane Pilot or an Airplane Observer, and cadets received commissions as second lieutenants in the Officers' Reserve Corps. In fiscal year 1923, twenty-seven commissioned officers and forty-eight cadets successfully completed primary training, and twenty-nine officers and forty-five cadets graduated from the advanced course.[50] Specialists coming out of the course were assigned according to the needs of the service and the perceived capabilities of the individual. The first phase of advanced training introduced the students to the specialties so that, whatever function each person eventually performed, each would be familiar with all specialties.[51]

Specialized Training

Observation

Observation occupied a distinctive niche in the Air Service by virtue of its coordination with other branches of the Army. "This cooperation," Colonel Kilner reminded General Menoher, "was not forthcoming from the majority of units in the late war."[52] During 1918 there had been a consistent shortage of observers at the front because the Artillery had to expand rapidly and was unable (and unwilling) to fill the quotas requested by the Air Service, and because too few volunteers meant that officers who otherwise might not have been chosen for the work were detailed to it of necessity. In peacetime, observation training was less pressured, which permitted greater coordination in the training of air and ground forces, and more opportunity for airmen to demonstrate what they could offer.

Although it was only a theoretical construct, Air Service planners had posited an offensive air force of pursuit, bombardment, and attack squadrons that "independently of friendly ground forces, seek[s] to destroy the enemy both in the air and on the ground." Observation, on the other hand, "is primarily an air 'Service' in that it is an adjunct of other services of the Army." Observation "is not offensive and never seeks combat, but will engage in defensive combat if attacked."[53] In 1921, observation comprised two-fifths of the Air Service, an ill-balanced force to the minds of many airmen. Rather, they postulated, 80 percent should be "combat" units.[54] According to General Menoher, however, the highest percentage of Air Service squadrons should be observation, the "only common meeting ground between the Air Service and the other branches."

Observation units, commanded by ground officers, were spread throughout the Army. But as William Sherman wrote to Barton Yount, the esprit de corps required of an air arm could not "be obtained if part of the Air Service belongs to the Artillery, Cavalry, or Infantry, and looks upon the Air Service as a minor

Interwar Years

duty to be performed under exceptional circumstances." Because of the command structure, Sherman continued, "the supervision of the Air Service Commander over these units will be technical — in the widest sense, including the prescription of tactical method.... The supervision by the Air Service over the observation groups will be in the nature of a staff supervision."[55] Air Service officers chafed under a system that restrained their control of and accountability for air forces, and it contributed to their lukewarm enthusiasm for observation aviation.

The Observation School at Post Field, Fort Sill, Oklahoma, opened on July 6, 1920. Cadets and officers studied aerial navigation, artillery, infantry and cavalry liaison, gunnery, maps, photography, meteorology, radio, surveillance, and visual reconnaissance. They used a miniature range for the simulated control of artillery fire. Cadets flew DH–4Bs and tinkered with Liberty motors. Some officers who already knew how to fly received a refresher course; then they, along with others who had gone through the Observation School, took a three-month course at the School of Fire for Field Artillery. The twenty-three officers who had completed this sequence by the spring of 1921 became instructors at the Observation School.[56] The necessity for this allocation of manpower had become evident by the end of the previous year because, according to the Training Division, "the greatest difficulties ... are caused by the shortage of personnel. On account of this, all classes have been small and it has been necessary to retain for duty as instructors a large percentage of the graduates in order to build up the schools and increase the capacity of the output." Personnel in the tactical units therefore, were reduced to an "absolute minimum."[57] Only when graduates of the advanced course could be spared were they sent for further training as communications, photography, armament, or engineering officers.

The two-seater DH–4B

Pursuit

In 1920, when he contemplated Air Service organization, the then Chief of Training Lt. Col. Harold E. Hartney called for creation of separate pursuit and bombardment schools. He offered "academic" reasons for special schools — to standardize training methods, to stimulate research and development, and to institute a curriculum in the tactical units.[58] Moreover, such schools could serve as models for quickly mobilized training in the event of war, a sad lesson learned, Hartney might have added, from the previous conflict. More practically, the Air Service had to consider lower projected manpower levels. At least 3,000 of the 5,000 Reserve officers had no advanced training, and a small number of mobile squadrons could not both train and fight in the event of an emergency. Hartney's thinking echoed that of Sherman and others who saw the necessity for a combat force that focused on operations, including cooperative exercises, not just on the fundamentals of flying.

In any event, specialized schools remained an unaffordable luxury, and tactical units took up the slack. Training in the 1st Pursuit Group during the summer of 1920 included acrobatic instruction, test and practice flights, cross-country flying, border patrols, and reconnaissance flights. Stunting remained important, but along with skills in single-plane maneuvering came an emphasis on formation flying, learned from wartime experience. In the classroom, pilots studied bombing, history of the Air Service, and aerial tactics.[59] Nearly a year later the program remained makeshift, causing training to be "carried on under great difficulties." At the time, approximately 10 percent of primary school graduates were in advanced pursuit training.[60]

By 1923 the Air Service reorganization had been accomplished, and pursuit training took place at the Advanced Flying School rather than in tactical units. Tactics had not changed much, but regulations had been promulgated that were more specific regarding the types of aircraft and armament to be employed, the qualifications to become a pursuit pilot, and the progression in a pilot's training. The fighter pilot, an "aggressive" man, "of good mentality, quick thinking ability, and good physical development" transitioned to combat biplace pursuit aircraft with an instructor, then went to single-seater pursuit planes. The pilot became familiar with fundamentals of flying his airplane and of landing and taking off; then he began to practice acrobatics. Next he flew in three-airplane formations with an instructor in the lead, followed by cross-country flying and instruction in navigation equipment. Combat practices included maneuvering alone and in mock battle with an instructor. Here, pilots also practiced as fighting observers in two-seater aircraft. The number of aircraft in these staged battles increased as the pilot learned how to attack opposing formations. Ground firing preceded aerial gunnery, then came low-altitude bombing, and finally, night flying. Eventually the new pilot repeated all the combat practices with the camera gun so he and his instructor could analyze performance.[61]

Interwar Years

Air Service Chief of Training in 1920, Lt. Col. Harold E. Hartney

During the war, the fighter ace symbolized aerial combat; after the war, pursuit retained its preeminence in the hearts of airmen. Proponents of bombardment somewhat threatened the princely status of the fighter pilot as the decade wore on, and most senior Army officers remained obdurate in believing observation to be the primary if not single legitimate function of military aviation. This internal debate did not entirely relegate the pursuit mission, but its fate, like that of other specialties, was linked to progress in aircraft design.

Major Spaatz, who would become the first Air Force Chief of Staff after World War II, numbered among the active promoters of pursuit aviation within the Air Service. From Mather Field in California and Selfridge Field in Michigan, where he commanded during the 1920s and 1930s, Spaatz wrote training pamphlets on fighter tactics. His interest continued as he went on to attend the Air Corps Tactical School and the Command and General Staff School, and he became Chief of the Training and Operations Division between 1933 and 1935. During those years he stayed in touch with friends also upholding the cause of fighter development. Among them, Capt. Byrne V. Baucom at Kelly Field in the pursuit school squadron wrote Spaatz in frustration about his training problems. In late August 1923 Baucom started training two flights (eight students), but neither he nor they knew whether the trainees were supposed to become fighter pilots or whether they were taking an introductory course in pursuit but would ultimately be assigned to a different specialty.[62] The fact that all advanced specialized training — pursuit, bombardment, observation, and attack — took place at Kelly Field disallowed any great concentration on any one of them. A single command made all the decisions regarding Air Service training, including allocation and use of equipment and training schedules. Those involved in each specialty engaged in lobbying and

Postwar Retrenchment

public relations efforts in behalf of their mission. One officer wrote Spaatz that a "good show put on by your organization" while the Tactical School class was visiting would "undoubtedly make a great impression on all of us and lead the class to believe that Pursuit is the most important part of the Air Force."[63] The low-keyed Spaatz was, however, not given to overblown claims and in fact wrote to General Mitchell that during his long cross-country flight to Kelly Field, "automobiles, freight trains, and practically all means of transportation on the ground passed me."[64]

To the good, completion of the major effort to produce training literature provided classroom materials that specifically addressed specialized military aviation subjects. In the pursuit course, for example, lectures and reading included pamphlets from the TR 440 series such as *The Pursuit Pilot*, *The Pursuit Squadron*, and *Pursuit Aviation*. These were augmented by lectures on pursuit psychology and temperament, organization and administration, functions of adjutants and engineering officers, offensive and defensive tactics during the World War, and technical requirements of pursuit aviation. Lighter but still required reading were memoirs and anecdotal accounts such as *Heroes of Aviation* by Laurence Driggs, *Above the French Lines* by Stuart Walcott, Rickenbacker's *Fighting the Flying Circus*, *The Aviator's Field Book* by Bolcke, German ace Richtohofen's *The Red Air Fighter*, and "The Boys of Twenty," an article on Frank Luke and Joe Wehmet in the *Ladies' Home Journal*. Other entertaining and enlightening materials also looked back to the recent war.[65]

Biplanes remained the standard aircraft for pursuit. They were to be armed with machine guns and bombs, the numbers and capacity depending on the airplane. They carried a minimal arsenal of one .30 caliber machine gun and bomb racks loaded with four 25-pound bombs.[66] In 1924, the 43d Squadron had 20 S.E.5s, 18 Spad XIIIs, and 20 MB–3As, transitioning to those airplanes after "basic" advanced flying. The school anticipated a serious equipment problem. The S.E.5 lived an average of only three or four years, and additional aircraft were always lost in crashes. Soon the school would be out of these aircraft altogether, and no replacements were on order. Similarly, it would shortly be low on Spads.

In 1923 Spaatz suggested that pursuit students go directly from the Curtiss trainer to what he called the T.M. (probably the Thomas Morse S–4, a World War I airplane used subsequently as an advanced trainer). S.E.s and Spads could be eliminated completely, and replacement aircraft, he argued, could be constructed specifically for pursuit. At the time Baucom agreed. "However," he wrote Spaatz, "I feel that I shall be faced with just the reverse policy here."[67] The next year Baucom reported that the training department "believed that the Spad-trained pilot is the best all-round pilot that can be developed on any one type of machine." A new Curtiss-built pursuit airplane (P–1) was to replace the MB–3A. "However, the maintenance of this ship [because of vibration of the

Interwar Years

S.E.5s from the 147th Aero Squadron (*above*) and an MB–3A bearing the 43d Aero Squadron insignia (*below*).

300-horsepower Wright engine] is of such proportion that its continued use is by no means a definite assurance."[68]

The department was satisfied with the equipment on hand, but it looked with concern to the future for single-seater transition planes. "Our service type Pursuit planes, like the MB–3 and the new Curtiss, are getting faster and more sensitive, while the training plane, JN–6H, is remaining stationary." As a result, Baucom continued, "the chasm between the training type and the service type is getting wider as time goes on. It seems, therefore, appropriate to make recommendations for providing single-seaters of suitable transition qualities. This can be done either by providing additional S.E.5s for replacements or by designing a machine of similar characteristics." He preferred the former.[69] That year, 1924, a new pursuit airplane, the PW–8 entered the inventory, the prototype of a pursuit category called Hawk, which represented an advance over previous aircraft performance.

Synchronizing students' progress through the fighter course was difficult because pilots entered the program at different levels of proficiency. Another factor that affected pursuit in particular lay in the requirement that students participate in demonstrations, races, and exhibition flying — where everybody wanted to see fighters. For instance, it turned out that Captain Baucom's mysteriously assigned eight students, about whom he had written Spaatz, were supposed to be trained in five weeks for a flying demonstration at Leavenworth.

These students, Baucom later crowed, "had never flown a single-seater. It meant taking them through S.E.5 transition and training, and Spad transition, and then qualifying them on MB–3's not only for Pursuit tactics but also for extended cross-country flying. Despite the great magnitude of this task, it was without doubt the most pleasant period of the entire Pursuit course. This was due to the fact that we had an absolutely free hand, with no outside interference, in the carrying out of our mission."[70]

While Baucom had reason to be proud of the successful outcome of his herculean task, the experience pointed up the fact that exhibition flying was a major activity of the Training and Operations Group. Moreover, the advanced students came in staggered numbers, some ready for transition work and others not, and some already proficient on S.E.5s, Spads, or MB–3s. So although classes began at an official time, instructors found it impossible to follow a definite schedule. In 1924, a group drafted a plan for three months of pursuit training and one month of combined exercises for all specialized branches of aviation. They thought that three months of specialized flying in a six-month advanced course was the minimum time required, assuming all students began together. The first month would be spent in basic transition flying on the DH–4B before the three-month period of specialized flying commenced, with one month additional transition to various pursuit types, and the final month devoted to combined tactics. The pursuit department believed this general scheme could be applied as well to the other specialties.[71]

Attack

No American air forces had been designated "attack" during the World War. As members of the Air Service drew together their ideas regarding doctrine, training, and combat functions afterward, they created the specialty and organized the 3d Attack Group (which had been the Army Surveillance Group) at Kelly Field. Fighter pilots dueled in the air, harassed enemy ground troops, and also protected against hostile aircraft, whereas attack squadrons were to fly close to the ground, equipped for ground strafing with bombs and machine guns. According to widely disseminated doctrine, "attack never seeks combat in the air and is heavily armored as a defense against ground fire attack." Pursuit aircraft hovered protectively over the battlefield "for the purpose of keeping the enemy offensive aviation from engaging in combat with friendly Attack aviation."[72]

Through the early 1920s the Attack Group, the Pursuit Group, and the Advanced Flying School — all at Kelly Field — competed for space, facilities, and manpower. Attack was a bit of a stepchild in this squeeze, the 3d Attack Group remaining at Kelly Field after the 1st Pursuit and 2d Bombardment Groups departed for stations without school squadrons. Also, in general, attack ranked low in esteem among pilots. When he was in training in the late twenties, Truman Landon elected to go into attack partly because, he later

Interwar Years

Truman Landon, shown here as a second lieutenant, made attack his choice for specialized training. The Douglas O-2, an H-model shown here, was adapted to this purpose.

recalled, he loved the roar of the attack plane's "sweetest sounding" engines and the idea of flying low. But even then, when attack was not so new, most of his fellows preferred to fly fighters or bombers.[73]

Like the pursuit, observation, and bombardment programs, in 1922 the attack course required approximately 432 hours. It consisted of attack of ground troops (use and methods of employment of machine guns and bombs), cooperation with Infantry and Cavalry units, attack raids against enemy concentration centers and troops in the rear of the front lines, antiaircraft and searchlights, and 250 hours of flying offensive missions against ground personnel, aerial gunnery, and communications with ground troops.[74] In one sense the attack training program benefited, if the tactical unit did not, by having an operational entity near at hand from which expertise could be drawn and airplanes could be borrowed.

Attack airplanes were intended to be fast, highly maneuverable aircraft equipped with considerable firepower, but for this purpose the school used observation planes modified for ground strafing.[75] They began with the DH–4, followed by trial flights on the bulky GA–1, a disappointing craft that caused the service to fortify the firepower of the familiar DH–4Bs. An adapted O–2, an airplane chosen in 1924 for observation purposes, became the next replacement.[76] Once again the Attack Group was shortchanged, as it received only one of the eight sent to Kelly Field in 1926.[77]

Bombardment

U.S. forces had not participated in bombing raids against Germany during the war, although members of the Bolling Mission had expressed interest in long-range bombardment. Late in 1917, Maj. Edgar S. Gorrell had submitted a strategic bombing plan taken directly from the RAF to General Foulois. Neither then nor in the early postwar years did American airmen enthusiastically, or at least publicly, embrace the concept of strategic bombardment. Though defined by the Air Service as part of the offensive "air forces," bombardment was largely restricted, along with the rest of aviation, to an auxiliary role.

Aircraft design reflected this conservative posture in not demonstrating any notable technical innovations for several years. Postwar bombardment training began with the ubiquitous DH–4. Between 1921 and 1927 the Martin MB–2 became the standard bomber. In 1927 the 2d Bombardment Group introduced soon-to-become-familiar Keystone twin-engined bombers.[78]

In 1921 approximately 25 percent of the Primary Flying School graduates specialized in bombardment.[79] At this time the 1st Day Bombardment Group trained cadets; a year later the bombardment course was part of the Advanced Flying School at Kelly Field. Besides piloting, students studied radio, aerial photography, signaling, bombing, bombsights, infantry contact and message dropping, artillery adjustment, camera obscura, gunnery, reconnaissance, and miniature range, among other concepts. The subject matter indicated bombardment's supporting role. In 1922 the curriculum included 132 hours of theory of bombing, 30 hours of bombing raids, 10 hours of antiaircraft and searchlight study, and 260 hours of flying and its related activities. Training regulations

Martin MB–2 bomber

dictated the types of planes and engines, instruments, navigation, gunnery and photography, and pilotage relevant to bombardment. Instructors found the need, and under peacetime conditions had the time, to devote greater attention to transition and solo flying for new pilots. Three-airplane formations flew cross-country bombing raids, practicing compass navigation, flights above clouds, instrument use, and reading meteorological charts. The Advanced Flying School experimented with night flying, a challenging task that became listed as a regular part of the course in late 1924.[80]

Tactical Unit Training

Congressional funding for the military in the years following the Great War dropped so dramatically that the Air Service's tiny cadre of officers was nearly incapable of manning both the training system and the tactical units at home and overseas. In March 1924, for example, Maj. H. S. Martin, VI Corps Air Officer, explained to Major Spaatz, then commanding Selfridge Field, why headquarters had turned down a request for more personnel:

> With reference to the question of assigning additional officers to Selfridge Field, the Chief of Air Service has advised that conditions existing at Selfridge and Chanute are not exceptional and that they appreciate the serious shortage of officers. This is due in part to the fact that the Air Service has been required to keep organizations in our foreign possessions up to full strength no matter what the effect on organizations in this country.[81]

As Martin reminded Spaatz, War Department policy dictated that garrisons in the foreign possessions would be maintained as close as possible to authorized strength. At the same time, the Air Service could not sustain itself, much less grow, without a functional training program. Therefore it was imperative to staff the schools at a level sufficient to cycle enough students through a system from which very few graduated. So, from the 1924 graduating class at the Advanced Flying School, for example, the Air Service assigned sixteen of the thirty-four officers as instructors.[82] This allocation of manpower resources left the tactical units coming in a poor third. Thus the dilemma: not only were these squadrons vital because they were the fighting units in wartime, they were also potentially in the forefront of training because they used combat equipment and experimented with combat tactics. As the war just ended demonstrated, an effective fighting force called for pilots familiar with combat organizations, practices, and equipment. Part of the postwar training program therefore included a period of apprenticeship in a tactical unit, even though the shortage of officers in these squadrons remained acute.

By 1923, a pursuit group, a bombardment group, and an attack group (each comprising four squadrons); one wing; eleven observation squadrons; and a

Postwar Retrenchment

group headquarters constituted the air forces in the continental United States. The overseas departments had composite groups that included observation, pursuit, and bombardment elements.[83] Besides providing the final phase of flight training for Regular Army officers assigned to the Air Service, tactical units also instructed National Guard and Reserve organizations.

Organized in the United States on August 22, 1919, the 1st Pursuit Group, which initially provided both advanced and unit training, moved from Selfridge to Kelly to Ellington Field in Houston, Texas, before returning in the summer of 1922 to its permanent home at Selfridge in Michigan. Pursuit Group instructors flew their twenty airplanes to the new station, then began training cadets and officers assigned to pursuit.

The 2d Bombardment Group, consolidated from units that had seen wartime service and been demobilized and reactivated afterward, also conducted advanced training until the establishment of the Advanced Flying School in 1922. The Bombardment Group then moved from Kelly to Langley Field where it remained through the interwar years. It began unit training with DH–4s and the Martin bomber.

Among the units designated as both training and combat organizations, the 3d Attack Group had the newest pedigree. After its birth in 1919 as the Army Surveillance Group, it patrolled along the Texas border, flying wartime DH–4Bs.[84] With its redesignation in 1921 as the most recent combat specialty, the Attack Group began to develop training methods and tactics experimentally and without direct reference to the Great War. It alone among the tactical units stayed at Kelly alongside the advanced school.

In the case of the last but most time-honored specialty within the Army — observation — one squadron was assigned to each corps area. In 1922, they were still flying the old warhorse DH–4Bs, but two years later the new O–2s started to appear on flying fields.

Although the redesignation of tactical units was still in progress, by 1921 the units were routinely providing one month of training to cadets and student officers who had just completed the advanced class. Once again, pilots studied the use and assembly of engines, airplanes, machine guns, bombs, bombsights, bomb release mechanisms, and pyrotechnic signaling devices. They also reviewed navigation and meteorology, oxygen apparatus and parachutes, cameras and photography, and radio and visual communication. Trainees flew the squadron's airplanes, learning the tactics employed in their particular specialty. The Air Service mandated, whenever possible, maneuvers or less formal liaison with other combat arms and other air squadrons. Very early in its planning, the Air Service decided that the final portion of tactical unit training should be spent in field exercises.[85]

Inspections of training units often laid out deficiencies in skills, usually caused by shortages or outright lack of equipment or facilities. Gunnery practice, for example, was inevitably hampered or nonexistent where no range

Interwar Years

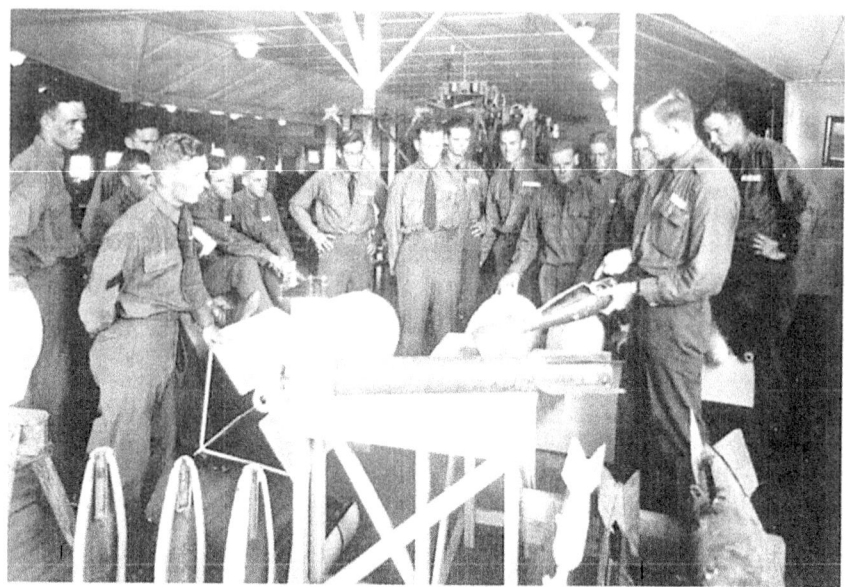

Future pilots received unit training in the use of aerial bombs.

was proximate. Pursuit training advanced in that regard with the establishment in 1925 of a gunnery camp at Oscoda, Michigan. The next year the 3d Attack Group gained access to a water gunnery range on the Gulf of Mexico, a great boon to the training program even though it lay some distance from Kelly Field. As late as 1929, however, Maxwell Field reported that it offered no instruction in aerial gunnery because it had no range. Similarly, many units could not conduct night flying because their airfields did not have the necessary equipment. In 1924, for example, the commanding officer at Selfridge Field defended his failure to increase the flying time of Reserve officers by pointing out that only officers already qualified on MB–3As would have adequate flying hours because this was the only airplane on hand in sufficient numbers.[86] Sophisticated equipment was in equally short supply. Although the bombsight had been under development for over a decade, one officer commented in 1925 that a bomber "can hit a town from ten thousand feet — if the town is big enough."[87] Even had they enough or the right kind of equipment, many squadrons had too few people to use it. The 3d Attack Group, for instance, was authorized 161 officers, but in what may have been an all-time low, on September 1, 1923, only 10 officers were on duty.[88]

Tactical unit training came under the supervision of station and unit commanders who conducted it differently depending on their mission and facilities. Therefore, a rigid training schedule formulated by the Training Division in the Office of the Chief of Air Service was not realistically possible.

Shortly after the war ended, General Menoher tried in 1920 to impress upon the Director of the War Plans Division of the General Staff the difficulties of mounting a complex training operation coordinated with other air functions and other branches of the Army. "Six years ago combat was a duel of individuals. Inevitably... combats between flights became the usual thing." Larger units necessitated more complicated training. "Unquestionably," Menoher predicted, "the future will see the combination tend to become larger and larger." Moreover, the Air Service did not fight alone. "We shall have to learn eventually to operate with brigades and divisions; but the detailed working out of such problems lies in the future, and can be handled only in a very general way in our training schemes."[89]

Nonetheless, the service worked conscientiously to publish manuals, pamphlets, and regulations that provided concrete guidance, and by 1925 training officials had the satisfaction of having drawn up a well-articulated program, including combined exercises and maneuvers during the tactical unit phase. In conjunction with and partly under the control of others besides the Training Division, this final segment had by 1923 been organized into three periods. By 1925 four phases constituted a training year: individual training occupied the first three months; the second four months focused on training the organization as a unit and on aerial gunnery; the third period (the addition to the earlier schedule) was given over to training the Organized Reserve, ROTC, and National Guard officers; and the last three months included field exercises with other branches of the Army and combined maneuvers with other aviation units.[90] Occasional events such as gunnery and bombing matches, group cross-country trips, and annual maneuvers supplemented regular training.

Early Recovery

The immense task of creating a training program for a still-new combat arm, tried in the ashes and unresolved hopes of the war, was not as long in the borning as General Menoher had feared in 1920. During the World War, perhaps the Air Service's most splendid accomplishment had been the construction of its training establishment. In this arena, wartime experience was intensely relevant to the future. Scarcely two years after Menoher wrote, the Air Service had coped with demobilization, seemingly unworkable budgets, perplexity and disagreement over its purpose, and the dissolution of combat units. At the same time, its officers managed to create a scaled-down but viable training program that incorporated a level of specificity beyond what Menoher thought possible and that would, in fact, be the model for American military aviation training into and beyond the next great war.

SEVEN

Boom and Bust: The Air Corps Years

"Gen. Foulois, in one of the shortest graduation speeches on record, congratulated the boys. He warned the girl friends sitting behind them that with the economic situation as it is, the boys faced the possibility of being out in the cold in 12 to 14 months."
— *The Flying Kadet*[1]

By congressional enactment the Air Service became the Air Corps on July 2, 1926. A five-year expansion program, the most fruitful part of the legislation, permitted the Air Corps to grow from fewer than 900 commissioned officers and 8,800 enlisted men (including cadets) to 1,518 officers, 2,500 aviation cadets, 16,000 enlisted men, and 1,800 airplanes. Expansion meant more than increased manpower, so the cost of procuring more advanced aircraft made Air Corps modernization an expensive proposition. Although it would feel the pinch of cost cutting by Congress, the Bureau of the Budget, and presidential economies, during the five-year program the Air Corps prospered compared to the rest of the Army. But because the money appropriated in yearly increments proved insufficient to purchase what the Air Corps wanted and felt it deserved, and because it did not reach authorized strength, airmen perceived themselves to be on a niggardly allowance.

The expansion program ran one year behind schedule. It was expected to extend from July 2, 1926 — the date of the Air Corps Act — to June 30, 1931. Since Congress appropriated no funds for the expansion when it passed the act, the five-year program actually began on July 1, 1927. In spite of the Chief of the Air Corps's futile attempt to complete the program in four years, it ended after all in five years, on July 1, 1932.

Unlike many U.S. banks, the Air Corps Training Center did not shut its doors during the grim days of the early thirties. To that point, the Air Corps showed steady progress in reaching manpower and aircraft inventory goals.

Interwar Years

Between mid-1928 and mid-1932, the aircraft inventory rose from 903 to 1,646 and officer strength increased from 1,014 to about 1,300.[2] Construction of the new Primary Flying School at Randolph Field and at other airfields expanded the training system. The Air Corps was able to siphon off enlisted men from the rest of the Army to meet the numbers mandated by the expansion program, although officer strength and serviceable airplanes did not reach the levels authorized by the 1926 act.

As the country moved deeply into economic depression, frugality translated directly into a reduced training program. An attempted follow-on to the five-year program did not materialize. When he looked at the possibilities for reaching the existing manpower authorizations in March of 1932, Assistant Secretary of War for Air F. Trubee Davison stated that the "question of producing the 400 pilots needed to bring the Air Corps to its authorized strength is largely academic in view of the attitude of Congress towards the size of the Army."[3] Faced with obduracy by a budget-slashing legislature, some Air Corps officers debated cutting back the training system in order to strengthen tactical units, but that approach was not formally adopted. In fact, approximately 5 percent more of the entering pilots at the Air Corps Training Center graduated in 1933 than in previous years. Otherwise, class size shrank so that only the approximately 150 men per year that the Air Corps could support on active duty graduated, and the number of trainees admitted into the flying program continued to drop in succeeding years.[4] The Air Corps further lowered the number of commissioned officers, while maintaining what it considered to be a barely adequate force, by giving graduate cadets their wings but not Regular commissions. Instead, beginning in February 1934, graduates remained cadets during their first year in a tactical unit; they then were commissioned during the second year as Reserve officers.[5]

Planning and Organization

When the Air Corps was established by law in 1926, President Calvin Coolidge appointed F. Trubee Davison, a wealthy attorney who had been a naval aviator and later a director of the Guggenheim Fund for the Promotion of Aeronautics, to the new position of Assistant Secretary of War for Air. Additionally, each division of the General Staff added an air section, and the legislation authorized two more general officers for the Air Corps. The 1926 act also directed that 90 percent of its officers below flag rank be "flying officers," and it specified that airmen command flying units.[6]

One of the two new brigadier generals, Frank P. Lahm (one of the first Army airmen and among those officers to be taught by the Wright brothers), took command of the Air Corps Training Center as an Assistant Chief of the Air Corps. When General Patrick reached the mandatory retirement age in December 1927, Maj. Gen. James E. Fechet succeeded him as Chief of the Air

Air Corps Years

Maj. Gen. James E. Fechet   Brig. Gen. Frank P. Lahm

Corps, Fechet remained in the job until his own retirement December 31, 1931. His successor was Brig. Gen. Benjamin D. Foulois, another pioneer airman and the first rated pilot to take charge of Army aviation.

An element of the headquarters staff, known at the time of the act as the Training and War Plans Division, held responsibility for Air Corps schools; gave supplemental training to ROTC, Reserve, and National Guard units; provided assistance to commercial and civilian aviation via an Airways Section; supervised the tactical and photographic units; and drew up plans for the air mission in peace and war. As part of a peacetime army, training suffused most of the organization's activities. By the end of the decade the division was again called Training and Operations, as it had been earlier. It then comprised an Operations Section, a Schools Section, and a War Plans Section.

The Army and Navy Departments contributed to the exhaustive studies, reports, boards, and hearings during that studious period surrounding passage of the 1926 act. General Patrick testified before a congressional select committee, and his office busily worked to revise the Air Service's organization chart and training pamphlets.[7] But once the legislation passed, many airmen considered it to be only half a loaf. Aeronautics had not achieved autonomy in controlling its operational or even its administrative affairs. The latter changes were neither far-reaching nor permanent. The position of Assistant Secretary of War for Air went unfilled in 1932 and 1933, for example, and the President

Interwar Years

abolished it altogether in 1934. The air sections on the General Staff were guaranteed for only three years. It seemed clear that the fundamental status of aviation within the Army had not altered. The postwar creation of the Air Service had declared the air arm to be a combat branch. Yet under the 1926 act, the General Staff still controlled the Air Corps, and much more significantly, division and corps ground officers commanded air squadrons in the field. Doctrinal disputes with the Navy regarding responsibility for coastal defense continued unresolved and acrimonious.

At the same time, the redesignation as a corps conferred a symbolic victory insofar as the Army no longer defined its air arm solely as a "service." Regulations explicitly acknowledged the concept of an offensive air force. As such, it could be argued that bombardment, pursuit, and attack enjoyed equal currency with observation, and that the Air Corps now had a legitimate strategic mission. TR 440–15 stated that "indirect support" might take place "in the area of the ground battlefield or at a distance therefrom." Army Regulation 95–10 assigned bombardment, pursuit, and airship units to a GHQ air force, and observation and balloon units to a GHQ air service, all part of a combat force, the GHQ Air Corps. Field armies and divisions also had assigned air force and air service units. This still-fanciful organization described a GHQ air force "primarily employed to execute offensive aerial missions for the purpose of defeating the enemy's aerial forces and maintaining control of the air.... Aerial missions may involve distant operations beyond the reach of the friendly ground forces, and closer operations within the normal battle area."[8]

In March 1926, as those in the Office of the Chief of Air Service examined their budget and the upcoming reorganization, Executive Officer Maj. Walter G. Kilner suggested that with little money and swiftly dwindling stocks of equipment, training should be curtailed. What he found particularly irksome, the Air Service had trained "more than enough pilots to take care of the demands of commercial aeronautics for sometime to come."[9] Everybody acknowledged the possibility that reductions might have to be made unless unforeseen resources came through, and aviation could not function adequately without proper and sufficient equipment. But as it was, so few people graduated from flight training that curtailing the program hardly seemed the answer.

Kilner's memorandum and request for a study may have been a stalking horse, since the idea of cutting back training met immediate and vocal opposition. Major Dargue, then Chief of the War Plans Section, pointed an accusing finger at the waste in time and money spent on nonmilitary activities (forest fire patrols; geological surveys; rivers and harbors surveys; and federal, state, and municipal photographic projects), and he added to the list of expendable activities "Special Projects" such as the various showpiece record-breaking flights and races, minor services and demonstrations provided to other branches and schools, support of commercial aviation, and unnecessary research and development. ("The Engineering Division now has so many irons in the fire that few get hot.")[10]

Aviation enthusiasts from the civilian community, who had routinely supported military aeronautics, again lobbied for Army aviation with some of the same zeal as in prewar days. In its review of General Patrick's 1926 Annual Report, *Aviation* magazine summarized the dilemma facing the Air Corps: "General Patrick complains that there is a serious shortage of good pilots among the officers of the Air Corps. Good pilots, he asserts are born rather than made," and, the article continued, statistics on graduation from flying training bore out his view. Only a fraction of the applicants became rated officers. Their numbers would be even further reduced were the training program to be cut. "The cost of flying training is exceedingly high," the journalist noted, and "while conditions in the service itself are such as to discourage officers from remaining in the Air Corps, there is little likelihood of any economies being possible in training activities, which swallow up a large part of the Air Corps appropriations."[11]

The size of the Air Corps training program continued to be debated after the five-year program actually began and the requested funds failed to materialize. In 1928 the new Chief, General Fechet, explained to a member of Congress that in attempting to meet the manpower levels authorized, the War Department encouraged officers from other branches to volunteer for the Air Corps. But the "results from this method were quite disappointing." Many of those who requested a detail could not pass the required physical examination, and of those who did, not many graduated. Furthermore, the Army had few vacancies because of a low attrition rate and a strength ceiling on Regular officers, a combination that kept potential candidates out of the Air Corps. The Reserves, a final source of manpower, also lived under a quota, and even those available for duty sometimes numbered more than the tactical units could employ because of a shortage of airplanes.[12] This avenue for procuring pilots ran into another roadblock when the Army Judge Advocate ruled that Reserve officers could not be redetailed in excess of 10 percent. One officer pointed out the bind this opinion created for the Air Corps: "Actually, then, we are training pilots, commissioning them in the Air Corps Reserve and feeding them to the commercial world as excellent pilots at a time when every effort is being made to increase the commissioned personnel of the Regular Army Air Corps."[13]

Another harsh indictment of the situation in which the Air Corps found itself came in late summer 1928 from Assistant Chief of Staff Brig. Gen. Campbell King who claimed that the pitiful number of graduates from flight training indicated that the "present 5 Year Program is utterly incapable of execution." It required much more extensive training facilities and, equally, an increase in training aircraft.[14] The Army continued its juggling act of apportioning airmen commissioned in the Air Corps, attracting pilots for detail to the Air Corps, and keeping enough qualified Reserve officers on extended duty. It debated the merits and briefly tried expanding the numbers by training enlisted

Interwar Years

pilots. The short tour of active duty for Reserve officers and the detail of officers who must return to the branch in which they had originally been commissioned meant potentially that, as Chief of Training and Operations Lt. Col. Frank M. Andrews commented in 1930, the "vast amount of money spent on training a pilot" could be "lost by separation of the pilot from the Air Corps."[15]

The Air Corps Training Center

The Air Corps Training Center at San Antonio, Texas, was at the center of the post-1926 expansion program. In 1922 it had been created in all but name in the consolidation of flight training at Brooks and Kelly Fields, only seven miles apart in San Antonio. Now, because Brooks Field could not meet the demands of the expansion program, in June 1927 March Field in California reopened as a primary flying school. Again the initial phase of flight training was split geographically, making coordination difficult.

Brig. Gen. Frank P. Lahm, newly appointed as an Assistant Chief of the Air Corps, took command of the Training Center. He established his headquarters at Duncan Field, near Brooks and Kelly. The Air Corps entered its new era, as of March 31, 1926, with 23 JNs, 66 PT-1s, 14 DHs, 7 VE-9s, 1 MB-3, and 2 SE-5s at Brooks Field. At Kelly there were 76 DHs, 14 NBS-1s, 1 JN, 7 O-2s, 1 VE-9, and 17 MB-3s.[16] On September 1, 1926, the Air Corps Training Center issued its first morning report.

Given the external constraints of strength ceilings and short reserve commitments, the Training Center experimented with expanding the intake of men accepted for flight training as one means of increasing the output.[17] The two primary schools had been set up to train a hundred students every four months. Experience showed, however, that flight training endured a very high rate of attrition owing to accidents, resignations, and large numbers of washouts. Although neither of the primary schools had enough instructors or equipment to provide thorough training, the Chief of the Air Corps decided to admit more than one hundred students per class. At the same time, the service gambled that the likelihood of graduating more pilots would be heightened by raising the educational requirements from a high school diploma to two years of college education.

As part of the intake process, besides the physical examination, training and medical officers gave applicants a personality study to determine aptitude, and submitted them to a well-tried physiological test in which someone twirled the candidate around until he was dizzy (in what was essentially a chair) in order to observe his eye movements and his reaction time and coordination. Additionally, the would-be pilot, in an early simulator called the Ruggles Orientator, sat in part of an airplane fuselage suspended in steel rings. Movement could be controlled in three directions by the trainee in the

Air Corps Years

Duncan Field, Texas, the Air Corps Training Center headquarters

"airplane" or by the instructor on the "ground."[18] While the data collected from these tests predicted success crudely, they indicated nonetheless that some men were unsuited to be pilots. The schools dismissed them from the program early. Thus the training program adopted the practice of admitting an overage into primary training to compensate for the elimination of those who immediately or within a few months were found to be unfit to continue. The ones remaining were most likely to complete the advanced course, and the numbers graduating from primary training might approach the hundred per class who, somewhat realistically, might be accommodated. The Air Corps thus established a mechanism for producing high output and little waste in an expensive education.

No sooner did the schools decide upon this seemingly sensible approach than in March 1928 the Office of the Chief notified the Training Center of its intention to alter the policy by requiring *all* the larger number of candidates except those physically disqualified be put through the entire course of flight instruction. Recommendations for dismissal would go through the Office of the Chief of the Air Corps. Senior officers at the Training Center balked. Instructors were already taxed to the limit. According to the Air Corps Training Center Commanding Officer Maj. J. E. Chaney, "by putting under instruction a total of 119 students at Brooks Field, for instance, with an acute shortage of instructors and equipment for even 100 students" two instructors had to be transferred from Kelly Field where they were also needed. If this policy were to be continued, the two primary schools could not accept more than the

Interwar Years

Among the criteria for admission to the Air Corps Training Center was performance in the Ruggles Orientator (*above*). The originator of this device, designed to test a potential airman's aptitude for flying, is shown in the overcoat in the photo at the *left*.

originally scheduled hundred per class. Moreover, Chaney snapped, the orientators requested for the school were no longer needed, and the motor skills and personality tests might just as well be discontinued.[19]

The effects of the upgraded educational requirements could not be felt immediately, but the percentage of graduates from the Advanced Flying School increased as a result of the larger incoming classes and an overall approach dubbed "Plan B." Inaugurated in July 1927, Plan B reconfirmed the concept of a staged system of flight training, although it reconfigured the time and training missions devoted to each phase. It was instituted after debating more dramatic alternatives. One possibility suggested by the commandant at the primary school at March Field was to cancel the advanced school altogether and either give the complete year of flight instruction at the two primary schools, or offer nine months of flying at the primary schools before sending graduates to tactical units. Without doubt, eliminating a specialized school would decrease

Air Corps Years

Brig. Gen. James E. Chaney, as a major, was the Air Corps Training Center's Commanding Officer.

overhead and streamline the program. But flight training only at the primary schools would not necessarily result in administrative efficiency if the schools had to manage all specialized training with their variety of equipment, and the idea of a single nine-month course would endanger proficiency standards. Although professionalization was not the issue at hand, such an approach would be a step backward in the twenty-year effort to systematize flight training. Another plan surfaced to give identical twelve-month courses at all three schools, each facility providing primary and all specialties, or primary and certain assigned specialties. This idea, which would occasion additional overlap and expense, also was jettisoned, although for a time the Air Corps used the two primary schools to test the "all-through" system at Brooks against the "stage" system at March Field.

The final scheme elected, Plan B, only nibbled around the edges of reorganization. It kept the structure of American flight training that had evolved during the war. Separate schools offered primary and advanced training. Potential pilots were required to meet higher educational standards for admittance into the primary program. Candidates found to be hopeless in the orientator or who scored low on the "neuro-psychic" tests, those inept at flying military aircraft, were rejected during primary training. Planners decided to shift transition training from the advanced to the primary flying schools. Because all intermediate flying (formally termed "basic") now occurred at the

Interwar Years

primary level, fewer men who entered the advanced school failed to complete the course, and a higher percentage of entering pilots graduated.

The year's course settled in to start with four months on a primary airplane — the JN or PT (primary trainer)[20] — instead of the earlier five or six months. The PTs slowly replaced Jennies, which were finally phased out in flight training by the end of the decade but remained for a while in ground school for use in practicing rigging and balancing the controls. The PT was a safe and serviceable plane, considering the still primitive condition of Army flying fields. As late as 1931 March Field was mostly an expanse of grass swept by the south Texas winds, and landing the PT-1 on the bumpy, grassy runway was not too hard because, as General Parrish commented in an interview years later, the plane's tail skid "slowed the plane down. The brakes weren't too good, and [the tail skid] kept it straight . . . that thing dragging in the sod."[21]

After concluding four months of primary, trainees continued in basic for the next four months, usually flying the DH-4 or its replacement O-2H. Now students trained as observation pilots and observers, although the service eventually eliminated the separate observer's course. If they successfully completed the eight-month session at the primary schools, student pilots spent four more months at the advanced school, specializing in pursuit, attack, bombardment, or observation.[22]

Throughout the 1920s, the Air Corps made a concerted effort to standardize the training program. A desire for high quality and uniform methods as well as the difficulty of accomplishing the five-year program with a shortage of instructors made it impossible to reinstitute flight training in universities using Army instructors. Flying instruction, Assistant Secretary of War for Air Davison told Senator Hiram Bingham (who had been instrumental in the World War I training program), "has [in 1929] reached a certain standard in the Air Corps after many years of development. Among the factors established in reaching the present stage is the importance of standardized methods of instruction. To maintain these standard methods requires a directing organization which exercises very close supervision over the flying instructors."[23] By this time, training fields submitted training plans to the Office of the Chief of the Air Corps where, after approval by the Adjutant General, they returned through the chain of command via the Commanding General of the corps area.

During the late 1920s and early 1930s, training plans suggested, among other things, the minimum number of hours required to learn various skills. Flight training became complicated at the end of the decade by the tightened budgetary restrictions occasioned by a worsening national economy. During the war, the pressure of time worked against the maintenance of high standards in pilot training. In peacetime, severely straitened finances took a toll on competence and safety. By 1930 the Chief of the Air Corps was cautioning his senior officers of the "urgent need for action . . . to assure that all airplane pilots continue adequate flying practice. Analysis of aircraft accidents indicates that a number of pilots exhibit a serious lapse of piloting ability, and many pilots have, over periods of years, failed to perform the minimum amount of flying necessary to maintain their ability to handle airplanes with reasonable safety under all conditions.... The cumulative effects of such neglect are becoming increasingly serious." He urged his commanders to ensure that all pilots log at least fifty flying hours annually, as the "infrequent flier is the most dangerous pilot."[24] Yet this much-needed admonishment conflicted squarely with an enforced limitation on the number of flying hours allotted to each pilot per year. In fiscal year 1931, for example, the Air Corps was authorized a total of 359,833 flying hours. By virtue of gasoline burned and the replacement of outworn engines and airplanes, flying hours translated directly into dollars.

Ceilings on the number of hours flown caused some perturbation over who should decide mission priority. Observation squadrons that performed various functions faced the problem particularly often. The Chief's office made it known, for instance, that for "suitable" periods of time, corps commanders should not call upon squadrons to engage in cooperative training with other branches. As Major Kilner expressed the Air Corps's viewpoint, "the diversion of an excessive number of observation airplanes primarily designed for observation missions and not for tow target missions, has resulted in reducing the efficiency of these airplanes for use as observation airplanes" and diverted a large number of Air Corps men from "their own tactical and technical training and from other more important cooperation missions with the Infantry, Cavalry, Field Artillery and Coast Artillery."[25]

Hours flown had to be completely accountable. Each field allocated a certain amount of time to different missions. At Maxwell Field, Alabama, for example, the month of June 1931 was scheduled for 250 hours of cross-country training, 100 hours of training for aircraft crews, 75 hours of practice flights, 50 hours of engineering flights, and 400 hours of annual gunnery and bombing practice.[26] This combat training competed with requirements such as the planned national air races in Cleveland, annual Air Corps demonstrations and maneuvers, Fort Benning exercises, Panama flights, annual machine-gun and bombing matches, and antiaircraft–Air Corps exercises. Also, hours devoted to special projects were levied against the allotments for each of the specialties.[27]

Some officers discovered that one of the few bonuses in becoming an instructor came from the fact that instructors were less carefully restricted in flight time. When Glenn O. Barcus joined the training staff at Randolph Field at its opening in 1931, he and his fellows averaged 70–75 hours of flying time every month, considerably more than would have been allowed had they been trainees.[28] Since the allotted flying hours were insufficient to cover the time required in all subjects, field commanders and instructors had to make judgments about the strengths and weaknesses of their trainees and apportion hours accordingly. Sometimes part of the published curriculum could be dispensed with as, for example, occurred at the Cavalry School at Fort Riley, Kansas. No cross-country training time had to be counted against their allocation because pilots logged many hours ferrying airplanes to and from depots and factories.

In order to evaluate the training program in June 1929, information about the proficiency of graduates of the Advanced Flying School was solicited from tactical units. Tabulated results of a questionnaire indicated that slightly more than half of those responding thought that most graduates needed further training before they were prepared to join a service squadron. The respondents also considered the standards to which pilots were held to be, on balance, too low. They summarized the specific deficiencies as no training in aerial gunnery, insufficient cross-country experience, not enough "big-ship" time, poor judgment, insufficient experience, no training in aerial bombing, little military knowledge, poor navigation skills, and lack of appreciation of responsibilities.[29] Problems associated with equipment and facility shortages were, of course, not amenable to solution by an alteration of training methods. Crucial training in bomb dropping and machine-gun fire, for example, could not be conducted without a suitable range near the Training Center, which explained the decision to incorporate this requirement into tactical unit training. One commanding officer suggested lengthening the training period, but this fairly straightforward solution was infeasible because funds were not available for an expanded training program.

Evaluation of the training program from subjective data proved inconclusive, so training officers also looked for answers in crash and fatality records. This frustrating exercise only succeeded in clarifying the limitations of quantification. Theoretically, success in battle might be accounted for by the simple means of accruing wins and losses. Yet the conflict now ten years past had failed to yield notable evidence of what constituted efficient or appropriate training during wartime. An even vaguer measurement of success applied to peacetime training. Proficiency in pilot training could never be fully described and quantified when it was so dependent on the human factor.

Primary Flying School (Including Basic Training)

The decision in 1927 to continue the system of a hierarchy of training schools led to the search for another primary flying field close to the hub of activity and the good weather in Texas. The investigation resulted in selection of an area seventeen miles northeast of San Antonio. Opened in 1931, Randolph Field was trumpeted as the West Point of the Air. This spot had an aviation history nearly as long as the air arm itself. Benjamin Foulois, Chief of the Air Corps during the first half of the 1930s, had been the solitary military pilot stationed at Fort Sam Houston more than twenty years earlier. In those days, Foulois had to repair and sometimes purchase his own equipment, and he had to learn to fly by begging for help in letters to Orville and Wilbur Wright. By 1932 the feisty

Air Corps Years

airman, now a major general, presided over a considerably expanded staff that included other general officers and a magnificent facility that housed the Air Corps Primary Flying School.

The school at Randolph Field was actually near the town of Shertz outside San Antonio, encompassing 2,368 acres of flat farmland. The first class at the new station, already in training at the older pilot schools, was conscripted to help complete construction of the facility. As the cadet newspaper reported, "Training started with a capital 'T' upon the upper classmen's arrival at Randolph. Dodo Days came back with a vengeance and the boys scrubbed floors, washed windows, sand-papered and painted beds and other such minor 'flying training.'"[30]

Despite this depiction of last-minute and make-do, which had become nearly customary with airfield construction projects, Randolph was built with care and some degree of elegance. It was the first facility intended specifically as a permanent air station, its physical design a departure from the normal configuration of Army installations. Randolph was laid out on a wheel rather than a grid pattern; streets and broad boulevards lined with buildings radiated outward from the officers' club at the hub. Officers' quarters rimmed a series of concentric circles. Housing and social facilities were commodious, all electrified and crisscrossed with underground telephone wires. The entire base was landscaped with trees, cacti, yucca, and other indigenous plants. Initially the field had only two flight lines (primary on the West and basic on the East), but by 1936 there were four spacious landing fields, straightforwardly named North, East, South, and West.[31]

The size of its aircraft inventory and the number of airmen who trained at Randolph Field failed to equal the compound's magnificence. By 1934 the school could graduate 150 cadets a year, although it had increased the number of flying hours by thirty-five and expanded the syllabus. Between the time in October 1931, when the first school troops reported to Randolph from the old primary schools at Brooks and March Fields, and March 1, 1935, when GHQ Air Force took form, slightly more than 2,000 would-be pilots reported to the Primary Flying School. Cadets constituted approximately 75 percent of the students reporting, and nearly 47 percent graduated.[32]

In addition to the small size of the aircraft inventory, a result of financial constraints, primary planes had to be long-lived because the school worked them hard and used them constantly. As of May 1, 1932, the Air Corps owned 251 primary training airplanes.[33] In 1936 Randolph had 35 PT-3s still in use. These early-model primary trainers were low-powered, open-cockpit biplanes, with the few instruments mounted on the outside of the fuselage. The PT-11D, which in 1933 the Chief of the Materiel Division pronounced "satisfactory," came to be considered the standard primary aircraft at the time, even though the Air Corps could afford to purchase only 29 of them. Aircraft design steadily evolved toward externally braced, metal construction, closed cockpit, low-wing

Interwar Years

A field of PT–11s stands ready for men engaged in primary flying training.

monoplanes with retractable landing gear. The PT–11D's design aimed in this direction: it featured a welded steel tubing fuselage covered with fabric, and fabric and wooden wings. It could cruise for three hours at a top speed of nearly 118 miles per hour. The Stearman-built PT–13 series was a follow-on from the early PT–3 two-seat trainers, with increased cruising time and higher speeds. The PT–13s were powered with 220-horsepower engines, constructed like the PT–11D.[34] These later-model primary trainers were considered to be so successful that they remained in use through World War II.[35]

The four-month basic phase had employed BTs (basic trainers) since 1929. These planes were intended to introduce the beginning pilot to combat aircraft. Initially BTs were modified observation or primary airplanes, but this solution proved unsatisfactory since, by the time the student had completed primary training, he had mastered the fundamentals of flying simple aircraft. Now, during the basic phase he needed to learn instrument flying and radio communication and to practice on faster aircraft with the new controllable pitch propellers and retractable landing gear.

In 1936 Randolph Field responded to the increased emphasis on instrument flying, requesting new basic trainers with a blind flying hood and instruments to include a bank-and-turn indicator, airspeed indicator, compass, rate-of-climb indicator, and directional gyro. Until then, the shortage in procurement funds had prohibited purchasing more sophisticated aircraft. In 1936 the Seversky Aircraft Company delivered the first true BT, the BT–8, the result of a design competition. It was a two-seat, low-wing monoplane powered with a 450-horsepower Pratt and Whitney Wasp Junior engine. The fuselage was

Air Corps Years

An early model BT-2 equipped with a blind flying hood appears above.

monocoque; the wings were multispar, reinforced inside with corrugated sheets of metal. The BT-8 evidenced noteworthy advances in design and construction, but it quickly proved to be too fast and difficult for novices to fly. A series of accidents convinced the Air Corps to discontinue it in favor of the BT-9. This airplane too, contributed to the high fatality rate in air training, but it remained in use nonetheless for several more years. An improved version, the XBT-12 was not tested and approved for purchase until 1941.[36]

Even though the old and new basic trainers had flaws, the basic phase at the primary school maintained an advantage over other elements of the training program in receiving new equipment and in having somewhat more of it. In 1935 George Brett, then stationed at Leavenworth, pleaded with the Training and Operations Division Chief Carl Spaatz for more training aircraft. He was told that he stood little chance of getting primary trainers of the type he wanted, whereas BTs were equipped with "all available instruments," according to Spaatz.[37]

Every four months the primary school admitted a new class for primary and basic training. Throughout the eight-month course, cadets attended ground school and took flying instruction. Primary and basic training each operated under the supervision of a Stage Commander and his assistant. Beginning students were assigned to flights, six in primary and four in basic. Under the Flight Commander (who oversaw each flight) were the instructors and their students. Instructors recorded the students' status on boards posted in the Stage House on the flight line. Each student also kept a log book of his own detailed training records. Others, in addition to the instructors, could evaluate the

Interwar Years

students. In particular, the Stage Commander subjected anyone whom an instructor recommended be released from the program to an elimination check.[38] Having the oversight of such an experienced officer was important because many instructors were themselves recent graduates. Beyond that, instructor training amounted to an informal course given at the field by more senior instructors. As a result, new instructors were only allowed to teach a few students in the primary phase.

What was called air work included navigation (52 hours), individual combat (5 hours), elementary formation flying (5 hours), instrument flying (12 hours), night flying (12 hours), performance (10 hours), and radio communication (9 hours). Those specializing in observation also took navigation (60 hours), combat (5 hours), night reconnaissance (15 hours), performance flights to altitude (5 hours), photography (5 hours), radio communication (6 hours), visual communication (4 hours), and reconnaissance (20 hours).[39] As always, the hours varied somewhat, depending on good flying weather, available aircraft, and individual student proficiency.

By the beginning of 1936 ground school at Randolph was, according to one of its graduates, "a bit of specialization on the West Point academics."[40] (A fair number of pilot trainees had graduated from West Point, so, unlike the flying cadets also in training, they held Regular Army commissions.) But, in fact, the classroom had become almost entirely technically oriented, and pilots-in-the-making had left considerations of military history and doctrine far behind. Students studied engines, aerodynamics, navigation, gunnery, maps, meteorology, wireless telegraphy, and pilot equipment. In the afternoons the future pilots marched in drill, shouldering rifles.

As it became the beneficiary of some of the newest navigational equipment coming into the Air Corps, Randolph experimented with blind flying. For the novice, this usually meant flying into cloud banks where neither land nor horizon could be seen, and where one could not visually ascertain the stability of the aircraft. Flying a BT–2 with a canopy drawn over the student's head caused greater anxiety. The accompanying instructor gave directions through the Gosport tube telling the student to make a left or right 90-degree turn, or warning him if he was losing altitude or going too slow or too fast. In early training the student normally sat in the front seat, but with blind flying the instructor took the front so he could see more clearly and, when necessary, snatch the controls to avoid disaster. Those who completed the training were awarded a diploma from the "Institute of the Blind."[41] Other challenging and sometimes frightening activities such as night flying, formation, and strange field landings also characterized the basic phase.

By and large, however, flight training did not engage the new young pilot at a high level of sophistication in either equipment or methods. The fundamentals of teaching men to fly had been established before the war, and postwar training was not much more complicated since the products of technical innovation were slow in arriving. Often, they were not used when they were to be had. As a young lieutenant in primary training at Randolph Field from October 1936 through the following spring, Albert P. Clark was asked later if he and his fellows had speed indicators and the like. "No," he recalled. "You flew by the wires, the sound of the wires."[42] Another pilot made the point that even when planes were equipped with instruments, as often as not airmen were disinclined to use them. "We just didn't look at the airspeed instrument," said Richard Montgomery about his student days. "A real pilot didn't do it.... We listened to the sound of the air going through the rigging, through the crisscrossed wires. It was very reliable.... When it reached the right tune — each type of airplane was a little different — the glide was just right. Not too slow, not too fast."[43]

An instructor at Randolph at about the same time, Noel Parrish remembered that few of the fledgling pilots had even climbed into an airplane before beginning flight training. Some "were all but panicky once they got to flying." It was tedious and repetitive to take them up over and over to practice basic turns, banks, and climbs, and "most of them did not learn too rapidly," Parrish admitted. As he summed up flying training at the time:

> Flying training was the most pragmatic type of training I have ever seen; you just did what worked. We had people coming up with all sorts of theories about training and notions. The test was whether you could turn out students any better or any faster than anybody else. Methods gradually improved, but only through tests, not through theories. Theories were interesting, but they were subject to immediate testing and proof, you see. Words, you had to speak — we had no electronic communications. You had to yell through a little tube, called a gosport-type helmet with a speaking tube that went into the ears. You had to throttle back on an open cockpit airplane; you couldn't hear with the engine going loudly so you throttled back. Those planes would immediately start settling, and you had to climb laboriously back up, so you learned to be very concise and use a minimum of speech. You used signals, tapping the stick, and motions. That is why old time aviators very often make motions all the time; they did that while flying to avoid losing altitude while talking.[44]

Advanced Flying School

Still gesticulating, no doubt, graduates of the Primary Flying School moved on to the Advanced Flying School nearby at Kelly Field. Kelly dated to World War I, built immediately after the United States declared war in April 1917. Its name honored the second U.S. Army airman killed in an airplane crash. Kelly Field had been the seat of the Advanced Flying School since 1922, but after the reorganization of the training program in 1927, the advanced school only gave specialized training. Under the new system, proportionally more men graduated, principally because more were eliminated earlier. Higher entrance requirements further improved statistics.

The four months of advanced flying brought the student pilot closer to combat in terms of (relatively) harsh physical conditions as well as familiarity with tactical aircraft. Compared to luxurious Randolph Field, Kelly was venerable but austere. The old operations shacks housed administrative offices. The World War I wooden hangars still remained, sitting beside the open grassy fields used as runways. Dirt and Bermuda grass covered everything, causing formation flights to kick up huge clouds of dust and giving new meaning to the term "blind flying."

As mentioned above, various committees and commissions convened to analyze Army aviation during the 1920s and early 1930s, and in response, the Air Corps reconsidered the structure and types of equipment appropriate to the training program. Regarding advanced training, perhaps the most revolutionary suggestion contemplated de-emphasizing specialization. Chief of the Training and Operations Division Carl Spaatz explained: "Thought is now being given to the idea of having the flying curriculum based on uniform instruction for all graduates . . . with less emphasis on particular types of aviation."[45] Spaatz had recently heard from a friend at the Command and General Staff School who expressed the view that the Army as a whole was enfeebled by the ingrained identification and loyalty of its officers to their individual corps rather than to the larger institution. Specialization within the Air Corps might produce the same splintering effect:

> We in the Air Corps are following that same questionable method, in having officers consider themselves pursuiters, bombardiers, attackers etc. The thing is initiated at the training center and carried thru and it starts the individuals off with prejudices that are bound to have an adverse effect thruout their service. . . .
>
> The training at Kelly should be so as to equip each officer to fly any type of plane and know at least the minor tactics and technique of each branch.[46]

Delaying specialization would contribute indeed to group cohesion, produce a unified concept of mission, and simplify the assignment to specialties. Employing a single type of training airplane into the advanced phase also boosted the effort toward aircraft standardization. Not having pilots trained on common ground was reminiscent of the unsatisfactory World War I experience wherein all pilots in the U.S.-based segment of the training program flew Jennies but all pilots in the European schools flew whatever castoffs were made available. That system, if such it could be called, enforced by wartime necessity, produced very unsatisfactory results when airmen finally entered combat.

In its own way, that unwieldy combination of teaching pilots to fly a generic training airplane plus a variety of specialized aircraft lived into the postwar years. By the early thirties, the attempt to avoid the hodgepodge in the advanced phase was addressed by adding the AT (advanced trainer) to the inventory, and by the middle of the decade each specialty also had dedicated aircraft. Ideally, the pilot in advanced training learned to fly both types of machines. In fact, the training program lived with the old compromise. Pilots specializing in attack, pursuit, bombardment, or observation used outdated versions of tactical aircraft and an amalgam of modified training aircraft left over from the middle to late 1920s. The program phased out the AT fairly quickly, discontinuing the designation and the idea for it fading until a replacement airplane called the BC (basic combat) appeared in 1937.[47] The idea of creating an all-purpose pilot flew in the face of the clear trend toward specialization demanded of a modernizing military. When standardization resurfaced, albeit attenuated in 1934, the Air Corps dismissed it in favor of maintaining distinct specialties and of introducing specialist training earlier rather than later.

As late as 1936, the advanced school's tactical aircraft were old-fashioned biplanes, including the 1930 Curtiss-built A–3Bs for attack; 1930 Keystone B–3As and B–5As for bombardment; the Thomas-Morse O–19s and Douglas O–25s, both from 1928–1929, for observation; old P–1s; and the Boeing P–12Bs and P–12Ds for pursuit. The school used BTs for instrument flying.[48] The big, open-cockpit Keystone bomber, as described by young pilot Richard Montgomery, "had so many criss-cross wires holding the wing together that we used to say that in the morning the crew chief would turn a canary loose inside there between those wings; and, if the canary could find its way out, one of the wires was broken somewhere."[49]

That elite group — culled from the many civilian applicants who entered flying training and who made it through the primary, then the basic course at Randolph Field, who went on to the advanced school at Kelly, and who passed

The instrument board of the Keystone B–3A airplane is shown with closed panels in the smaller view. The later model B–6A is shown in the larger image.

Interwar Years

The Advanced School's tactical aircraft included the Boeing P-12B (*top*), the Thomas-Morse O-19 (as shown in the flight group *center*), and the Douglas O-25 (*bottom*).

all tests — finally earned the right to wear wings. Some, mostly West Pointers, already held Regular Army commissions. By 1934 cadets still had to spend a year in a tactical unit before they were commissioned in the Reserves. These graduates of the Air Corps Training Center's heavier-than-air flying course now carried the rating Military Pilot.[50]

The first cadet class to enter at Randolph Field graduated from Kelly in the summer of 1932. The young pilots were exuberant and proud of their hard-won status, yet the graduation exercises provoked sobering thoughts. The student newspaper, *The Flying Kadet*, described the event:

> Gen. Foulois [Chief of the Air Corps], in one of the shortest graduation speeches on record, congratulated the boys. He warned the girl friends sitting behind them that with the economic situation as it is, the boys faced

the possibility of being out in the cold in 12 to 14 months. (At this point several of the girls looked pale and some of the boys had a sickly grin on their faces.) — He warned the boys to keep out of debt, and stressed the fact that the most important phase of their makeup would be a good reputation. He urged them to start their active duty period with the resolve to do what is wanted cheerfully and as effectively as possible and not to be afraid of asking questions.... The year as a flying cadet was over.[51]

Tactical Unit Training

After receiving his wings, the new pilot reported to a tactical unit. Presumably, this period of apprenticeship offered the greatest opportunity for cooperative activities with the line of the Army, when debates over missions might subside in favor of constructive partnerships between air and ground officers. Lt. Gen. Orval Cook recalled the collegiality between young Air Corps and antiaircraft officers who trained together. Other bonds formed informally through friendships or a shared sense of status. Pilots and armored officers stationed at Fort Knox, Kentucky, for example, discovered a kinship based on mutual involvement in newly developing military technologies whose employment and tactics were not easily understood or accepted by many in the old Army. As Cook remarked, "There were a lot of people who thought the day of the horse was not over."[52]

Cooperative training was far from perfect, but a number of senior ground and air officers worked uncompetitively to improve the process. Maj. W. R. Weaver, commanding at Maxwell Field, reported to the IV Corps Area Commanding General in the summer of 1929 that much of the Air Corps's participation in joint missions had little training value "in as much as they did not involve a knowledge of the organization, movement and disposition of troops or their tactical employment on the part of the Air Corps." Airmen should be integrated into Army training as a whole if the benefits of aviation were to be realized.[53] At the same time that one airman was requesting more joint training, a ground officer, the Commanding General of II Corps, argued for the singularity of air training. While it was efficient and reinforced the mission to have ground and air training occur concurrently, nonetheless "undisturbed unit training of these Air Corps units is essential," Maj. Gen. H. E. Ely maintained. "Unless they are allowed a suitable period during which they will not be called upon for cooperation missions with other branches, they cannot properly train themselves to give efficient cooperation with these other branches."[54]

The Adjutant General ruled that all airmen — whether in tactical squadrons or photo sections or group, wing, and post headquarters — would receive not less than six months of training in a tactical unit. This period would not be interrupted by miscellaneous activities or assignments unconnected with air

Interwar Years

training. Commanding officers of all bureaus and stations were notified that "time should be allotted for training of Air Corps Units with other arms, in which the primary purpose shall be the training of the Air Corps. Such training should be in contradistinction to cooperation with other arms, and should be arranged as part of the Air Corps Unit Training period."[55] Tactical unit training thus came to be a requirement for all heavier-than-air officers and the singular part of the program that incorporated a period of joint training, part of which included air exercises.

By 1931 structure had been applied to unit training. That year's training directive outlined the total number of flying hours allotted and the air and ground subjects for each specialty, with minimum hours specified for each pilot per year. During the 1932 training year when flight time was heavily curtailed because of strict budgets, headquarters authorized 200 flying hours annually for each attack and bombardment pilot, 180 for each observation pilot, and 220 for each pursuit pilot.[56]

In 1927 the 9th Observation Group, located at Mitchel Field, Long Island, New York, comprised the 1st and 5th Observation Squadrons and the 61st Service Squadron. The remaining observation squadrons spread throughout the corps areas. Air instruction for these units included engineering flights, preliminary training, communication, photography, artillery spotting, aerial navigation, night and formation flying, individual combat maneuvers with and without camera guns and flown with pursuit planes when available, liaison exercises with ground troops, cross-country flights, aerial gunnery and bombing, and the mandated period of field exercises. Ground courses in communications, photography, adjustment of artillery fire, and liaison were tasks unique to observation.[57] At this time, pilots trained as both pilot and observer. This scheme foundered, as so often occurred, over equipment shortages. During the first four months of 1930, for instance, the 9th Observation Group had sixty-eight officers in training and seven airplanes.[58]

All pursuit squadrons — the 17th, 27th, 94th, and 95th Pursuit Squadrons and the 57th Service Squadron — were assigned to the 1st Pursuit Group at Selfridge Field, about fifteen miles from Detroit. Air instruction included engineering or test flights to obtain data, preliminary training, formation flying, aerial navigation, night flying, aerial employment and tactics, ground attack, aerial gunnery and bombing, cross-country flights, and field exercises. Ground instruction encompassed aerial gunnery, theory and practice of bombing, aerial navigation, meteorology, oxygen equipment, organization, supply and maintenance, night flying, air tactics, parachutes, engineering, Field Service Regulations, combat orders, lectures on missions and roles of other branches, and uses of federal troops in civil disturbances.[59]

The 2d Bombardment Group at Langley Field, Hampton, Virginia, consisted of the 11th, 20th, and 96th Bombardment Squadrons and the 59th Service Squadron. The 49th Bombardment Squadron was also part of the group

Air Corps Years

Langley Field was an important link in the airfields positioned for coastal defense. In this 1924 photo, an aircraft is seen taking off for the capital, Washington, D.C., 135 miles to the north.

but it was stationed at Phillips Field at the Aberdeen Proving Ground in Maryland where it assisted in Ordnance Department experiments. Langley Field was said in 1927 to be "the most strategical in the United States, in view of the fact that it permits of easy access to all points along our eastern coast line."[60] Two squadrons there were equipped with Keystone LBs and one squadron had Martin MB–2s. Hugh Knerr, who commanded the Bombardment Group at the time, later described how this "most strategical" unit carried out its training:

> We had airplanes of sorts, but no means for making them effective, due to the lack of bombsights. We devised our own do-it-yourself methods while the Engineering Division at Dayton conducted the long-range studies on the Norden, Seversky, and Ingles proposals. Meanwhile, I was able to get fair results with strings rigged in the bombay, over which I could draw a bead on the target. I guided the pilot above with strings tied to his arms.[61]

Working more with creativity than with state-of-the-art aeronautics limited airmen's abilities to practice skills in the published curriculum — the engineering flights, formation flying, camera obscura, bombing and machine gunnery, aerial navigation, bombing raids on simulated targets, night and cross-country flying, and field exercises. Even so, emphasis lay on flying rather than

263

Interwar Years

ground instruction, and here, as elsewhere, the commanding officer was permitted to judge the proficiency of his pilots and waive classroom attendance. All bombardment pilots trained in the functions of pilot, bombardier, and machine gunner.[62]

Finally, by 1927 the 3d Attack Group, composed of the 8th and 90th Attack Squadrons and the 60th Service Squadron, had moved from Kelly Field to Fort Crockett at Galveston, Texas. The open space along the Gulf of Mexico literally brought a breath of fresh air after the earlier cramped quarters that had made strafing and fragmentation bombardment both dangerous and difficult. Though better situated, the new station nonetheless had few amenities, having recently been a cow pasture. For some time there were no runways. New equipment inevitably improved training. As Maj. Gen. Truman Landon, then a young pilot with the Attack Group, recalled:

> We started out working with ground panels because we had no radio in the airplanes. While I was at Crockett, we started getting radios and, of course, initially they were a novelty. We would fly around and try to make them work. First, we could receive but we could not transmit.... So really, our first communication was visual from the ground. The only thing we could do from the air was make some maneuver with the airplane to indicate that we understood or did not understand, until we got the radios. Once we got the radios, it advanced pretty well.[63]

In joint field exercises the attack squadrons frequently trained with the Cavalry at Fort Brown. Although the curriculum included night flying, in fact the group had no lighting equipment, making flights perilous and thus rarely undertaken. As always happened, nasty winter weather, even on the Gulf, curtailed flight training, and wind patterns made gunnery firing in any but a southerly direction highly inaccurate. Most commanding officers complained about all the miscellaneous activities that interrupted training. "All of these are valuable and should not be disregarded," Maj. John Jouett, commanding the 3d Attack Group in 1929, stated politely, but they "make it more or less impossible to carry out a scheduled program of ground instruction for a tactical organization, consideration being given to the fact that flying training comes first in the Air Corps tactical duties."[64]

Instrument Flying

A new emphasis upon navigation and instrument flying invigorated the training program of the 1930s. Aerial navigation borrowed its instruments and techniques from marine navigation, although in an attempt to distinguish itself from maritime endeavors, for a time the Air Corps adopted the term "avigation." In the earliest days of flight, as long as the pilot had visibility and visual landmarks by which to steer, he only had to develop sufficient flying acuity to correct for the drift from his course caused by the winds, learn map reading, and navigate by iron compass (railroad tracks). But on long cross-country flights, or in low visibility, in cloud banks, or over water, the pilot faced considerable difficulty in figuring out where he was, where he was going, or how to get home. In the Great War, American airmen had discovered the limits of their ability to find their way over unfamiliar territory when they flew over the villages and battlefields of France.

By the end of that war, airmen had use of a compass, ideally positioned directly in front of and below the pilot to avoid parallax. Rudimentary airspeed indicators, altimeters, and driftmeters had been developed, but they were unreliable and seldom used. Given the technology available, it is understandable that training instructions to the Army pilot of early 1918 cautioned: "Do not trust any altitude instrument. Learn to judge altitude, especially in landings. Barometric conditions may change in a cross-country flight so that even a barometer that is functioning properly may read an incorrect altitude. Also the altitude of the landing place may be different from that of the starting place."[65]

During the early 1930s, as the Navy and the Air Corps engaged in verbal hostilities regarding the coastal defense mission, the Air Corps decided it needed to learn to navigate over water. Away from land, no visual landmarks cued the flier as to his location. At sea level, a navigator could calculate distance and direction by use of a sextant, but once aloft, one could not sight from the horizon. For this purpose, the bubble sextant, developed immediately after the war, created an artificial horizon.

For the most part, the Army pilot concerned himself with overland navigation. His most valuable new tools were directional radio and a range station system and transmitter carried in the aircraft. In time, stations along the airways beamed both aural and visual signals that could be picked up by indicators in the cockpit. The radio compass measured the bearing of a radio signal. Map reading still remained an indispensable navigational skill, and by 1937 the entire United States had been mapped, thanks in part to cooperative efforts between the Air Corps and other government agencies. The Commerce Department published a series of air maps that were of inestimable value to both civilian and military pilots. Thus at the end of the decade, the basic means of aerial navigation (map reading, dead reckoning, and celestial navigation), accomplished by magnetic compass and sextant, were augmented by new instruments — airspeed indicators, gyro compass, directional radio, landing lights and beacons, charts, altimeters, and driftmeters.[66]

Navigation training employed a new aid called the Link trainer. It was not the first flight simulator the Army used, but at the time of its introduction it was the most sophisticated. Unlike the earlier Ruggles Orientator, the Link became integral to the training process itself, rather than merely a test device to determine potential fitness to fly. Its manufacturer first advertised the Link Aviation Trainer in 1929. The Navy purchased one in 1931, but the Air Corps

Interwar Years

The subject in the top photo is receiving instruction in instrument flying with use of the Ocker Box. The navigator trainee at the left is learning to use a sextant.

was slower to see its advantages. After the disastrous experience of flying the air mail in 1934 and the concomitant push to improve instrument training, the Army, too, ordered the trainer.

The first six Model A Link trainers arrived on June 23, 1934.[67] One each went to March, Mitchel, and Selfridge Fields. Another was shipped to Duncan Field in San Antonio, the original headquarters of the Training Center. Wright and Langley Fields, where classes in navigation had begun, received the remaining two. The Model A was a mock fuselage with wings mounted on a turntable. The cockpit controls included a compass and airspeed, rate-of-climb, and bank-and-turn indicators. A student sat in the "airplane" operated by an instructor. The two communicated via radio signals and light beacons. The instructor could also create the sensation of air turbulence, but the Model A did not have the capacity to simulate instrument-landing conditions.

By the end of 1936 the Air Corps owned twenty-one Link trainers. In that year the Model C was introduced with many new instruments — magnetic compass, airspeed indicator, bank-and-turn indicator, rate-of-climb indicator, directional gyro, artificial horizon, altimeter, radio compass indicator, marker beacon indicator, and tachometer. For radio communication, the student could speak with men on the "ground" through earphones and a microphone. He could see his instruments by cockpit lights when the hood was pulled over. The

Air Corps Years

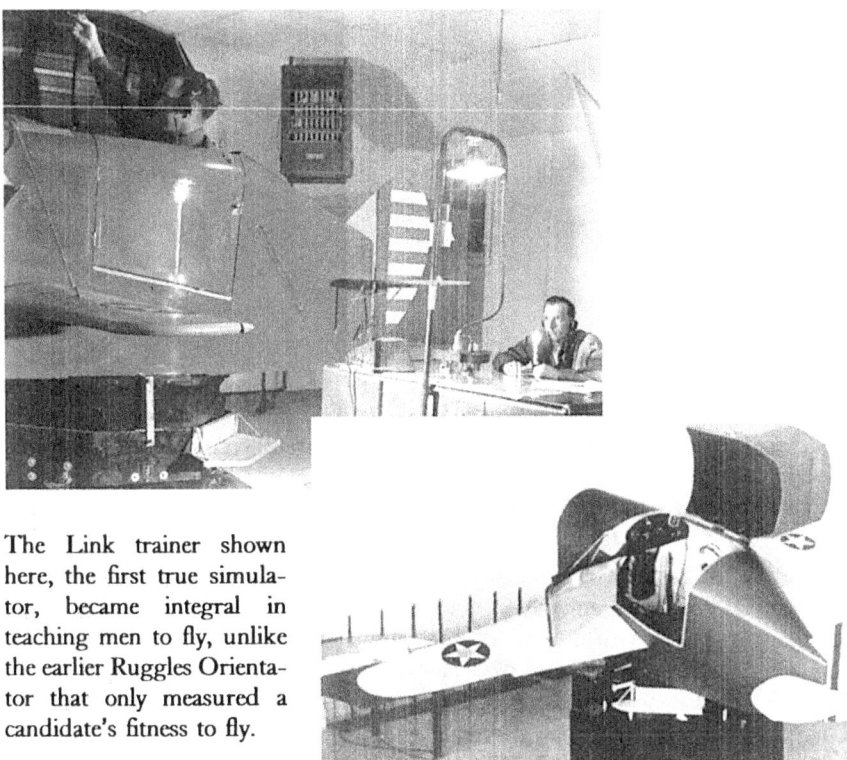

The Link trainer shown here, the first true simulator, became integral in teaching men to fly, unlike the earlier Ruggles Orientator that only measured a candidate's fitness to fly.

instructor worked from a table holding six radio range beacons and two aural marker beacons. He created problems in which the student had to negotiate around and above obstructions, flying by instrumentation alone. The student's responses could be charted by an automatic course recorder that plotted the course flown.[68] Subsequent models replicated changes in aircraft angle and weight when climbing or descending and included controls for simulated instrument landing system approaches and automatic direction finder indicators.[69]

Unfortunately, it had taken catastrophe to integrate new navigational tools into training. In February 1934 President Roosevelt charged the Air Corps with the responsibility for delivering the U.S. mail, a service previously provided by commercial carriers. During the 78-day Air Corps mail operation, military pilots flew on badly equipped aircraft over unfamiliar routes during one of the worst winters on record. The loss of life was staggering. In 66 crashes, 12 pilots died.[70] The Air Corps had been unable to equip its planes, most of which were seriously out of date, with the latest navigational aids. Most pilots had not trained in blind flying, were unfamiliar with the expensive new radio equipment, and had flown only during the daytime and in decent weather.

Interwar Years

Bombers such as this Keystone B-6A were used to carry the mail when the Air Corps took over this service previously provided by commercial carriers. However, modern navigational aids had not been installed in the available aircraft, and frequent disasters resulted.

When the ill-fated air mail experiment ended in June 1934, the Air Corps was forced to defend itself against charges of incompetency. Ultimately, except for the grievous loss of life, the positive effects outweighed the negative. The Congress, public, press, and War Department all recognized that military aviation failed to accomplish its mission because it was ill-prepared. The Air Corps needed new, up-to-date airplanes, equipped with the full range of modern instruments, and pilots should be trained to use them.

The Air Corps had not been oblivious to its failure to achieve all-weather proficiency even before the ruinous air mail experience. Although a formal school could not be opened without specific authorization, navigation training units had been set up at Rockwell and Langley Fields during late October 1933. Ironically, training in these units was interrupted when members of the second class to be enrolled were pulled out to contribute to air mail operations.[71] In another attempt to expand the use of instruments, Chief of the Air Corps Foulois had tried, against the pressure of time, to institute a crash program at Wright Field in Dayton, Ohio. There, a new course in instrument landing began in March 1934. Pilots qualified on training airplanes and the new Martin bomber. Graduates of this course were to be dispersed throughout tactical units to assist in setting up instrument courses. Using Public Works Administration funds, the Air Corps purchased forty-eight trucks to be used as instrument landing and guiding stations.[72]

In the aftermath of the air mail contract, the Air Corps made a full-scale attempt to incorporate instrument flying into the training program, even without an increase in the military budget. It hoped that the nucleus of officers trained at Rockwell, Langley, and Wright Fields would establish courses of instruction at their permanent stations. In March 1935 the Office of the Chief of the Air

Air Corps Years

This Douglas O-46A was among the first aircraft completely equipped for instrument flying with the advent of newer and more numerous devices.

Corps issued Circular 50-1 prescribing standards for instrument flying that all Regular Army and Reserve pilots on extended active duty should be qualified to perform.[73] Instrument flying, "cloud flying," and day and night navigation became part of the Air Corps Training Center curriculum, taught during the additional thirty-five flying hours allotted per student in 1934.[74] Yet the assumption on the part of the commandant of the Air Corps Tactical School that, as a result of these efforts, "routine training at the Air Corps Training Center is of such scope as to qualify students in 'blind' cross country flying with and without radio aids" was far too sanguine.[75]

Training in instrument flying did, however, become more widespread. In 1935 General Foulois ordered each pilot in a tactical unit to log 5-10 more hours of instrument flying and 15-20 more hours of night flying per year.[76] New airplanes coming into the inventory were equipped with radios and better instruments. The commandant at Wright Field notified the Air Corps chief in the spring of 1937 that "radio compasses, Type E-4A, are available at Air Corps Depots and can be installed in airplanes wired for this instrument as rapidly as these airplanes can be brought to the Depots. Airplanes Types O-46A, BT-9, and A-17, will be completely equipped for instrument landing training with the installation of the Type E-4A radio compass."[77] Besides the Air Training Center, the Air Corps Tactical School and the Army schools at Leavenworth also started navigation courses. Instrument training spread through tactical units, many of which initiated their own navigation schools.

If instrument flying insinuated itself into the training program, supply problems kept it superficial. Demand for blind flying equipment and radios far outdistanced the amount available and affordable. Manpower and fiscal resources were still spread thinly, and the diffuse, decentralized system of navigation training prevented standardized, in-depth instruction. The hope for a special navigation school, after the demise of the schools in the bombardment

Interwar Years

groups at Rockwell and Langley Fields, did not come to fruition.[78] However, when the Air Corps expansion program loosened the purse strings in 1939, additional instruments became standard on new aircraft, and airmen at least had some notion of their usefulness.

The Beginning and End of "Normalcy"

Largely in reaction to the futile bloodletting of World War I, military budgets remained slim during a decade of general national prosperity. The catastrophic stock market crash of 1929 brought further retrenchment in military spending as the country reeled from the shock on Wall Street to face the precipitous decline into the Great Depression. As the system of finance capital and insecure international investments crumbled, America turned further inward, ignoring worldwide economic depression and political aggression such as the 1931 Japanese invasion of Manchuria.

The Air Corps made many gains in the first postwar decade that would remain intact, but the much-anticipated grand expansion stemming from the Air Corps Act would not be realized. As the service moved into the years of national economic scarcity, its Reserve commitment offered one of the few early depression programs providing needed government employment. Among the nonmilitary activities of the Air Corps — about which airmen frequently grumbled — relief work became a stark necessity. In 1929 the Air Corps staged a demonstration on behalf of the American Red Cross at Bolling Field. It parachuted medics, food, clothing, and medical supplies. That year other squadrons delivered supplies to flood-stricken areas in Alabama and northern Florida.[79]

While contributing to civil affairs, the Air Corps attempted to keep up its prescribed training duties. But in his annual report for 1929, the Chief offered up a sad litany of lacks. Less than two years later the Air Corps could not commission *any* cadets graduating from its Training Center into the Regular Army. The War Department placed a ceiling on the number of officers, and the Chief of the Air Corps admitted that "very few officers are offering to transfer from other Arms."[80] Of the West Point graduates, for example, only a small number elected flying training, and those who did were subject to the usual high attrition rates. Airplanes that remained in the inventory often sat outside, weathering without hangars.

Nevertheless, given the factors under its control, the Air Corps could take some pride, having thoroughly established, if not perfected, its training methods. It could advance no further without technical progress and dissemination of the fruits of that development. In the fall of 1928 one of General Foulois's inquiries met with the stunning reply: "No definite plans have been developed for inaugurating training in aerial navigation at the present time."[81] Years later Lt. Gen. William Tunner's description of his early flying days underscores that remarkable statement: "If there was any one rule in flying, it was to know, every second, where you were going to land if you had to. A flight was not so much from San Diego to Sacramento as it was from this pasture here to that cornfield there to whatever that flat place was up ahead. You felt as though you were trying to steal second base when you lost sight of a field."[82] While this eyes-only approach to navigation did not materially change for some time, Lt. James H. Doolittle made a promising start in 1929 when he flew the first blind cross-country flight from Mitchel Field, relying solely on instruments. The Air Corps was then on the cusp of developments in navigation and aircraft design that would deepen the proficiency of its pilots and expand their roles.

Training, tactics, and doctrine had and would always depend on performance capabilities of airplanes and equipment. In 1923, for example, the 3d Attack Group changed its training because of aircraft limitations. No plane had been developed with the speed and maneuverability required for the attack mission. So, forced to employ slow, heavy observation aircraft, the group shifted to a smaller, differently spaced and configured formation to accommodate their large, bulky airplanes. The Air Corps awaited significant technological advances and another war before altering some of the patterns established during the previous conflict, not yet a generation old.

Instrument flying received a boost when Lt. James H. Doolittle, shown here in the cockpit of a blind flying aircraft, made the first cross-country instrument flight in 1929 that included a blind takeoff and landing.

Interwar Years

If refinement in combat training did not demonstrate any significant advances during the 1920s and the early 1930s, the sober, steeped-in-death occupation of military aeronautics nonetheless took on a heady circus atmosphere. Aviation enjoyed tremendous popular appeal, and the Air Corps catered to the public's enthusiasm with races and exhibition flights. Before Glenn Barcus became an instructor at Randolph Field, he was a member of the 1st Pursuit Group stationed at Selfridge Field. "I will be honest with you," he later confessed. "The mission was to fly at cities and fly at airshows mostly.... We would indulge in maneuvers occasionally. That was still pretty much of an airshow."[83]

Officers who elected to join the Air Corps and had the luck, talent, and temperament to graduate from flight training exhibited characteristics that set them apart from many other Regular Army lieutenants. Air Corps officers still tended to be a daredevil lot who, much as in aviation's earliest days, faced down death in flying primitive aircraft. They relied more upon their own skill and confidence than upon knowledge of aerodynamics or instruments. General Parrish's reminiscences about his basic training at newly opened Randolph Field recall the Air Corps of the early 1930s. His graduating class numbered ninety-six, less than half of the men who had entered with him.

> The figure that rather appalled me was that out of that 96, within a year, fifteen were dead. Crashes — mostly pilot error, and most of it was from high-spirited behavior. We, of course, had no radios in the planes, and people would take chances on weather.... Doing stunts, flying under things, flying low, especially, and pulling up. Most of us had no strong desire to get up where it was terribly lonesome and fly around, other than to do a little acrobatics, but to get down low where people could see us, because we attracted an awful lot of attention. Everybody came out to watch.... This was too much temptation and led to a few crashes of people doing stunts, pulling out, flying low, and pulling steep climbs, and things of that sort — doing acrobatics at low altitudes. Some of it was engine failure, things of that sort would cause it, but it was a very risky life.[84]

In their recruitment drives, senior officers spoke of the need for athletic and aggressive men in the Air Corps. They openly acknowledged the danger and high fatality rates of military aeronautics. Yet official pronouncements usually left out the fact that many times a pilot brought disaster upon himself. Recruits tended to be cocky, individualistic, high-spirited young men exhilarated by the drama and show. The Air Corps worked to standardize and professionalize flight training, but even the most elegantly planned and executed training program had to accommodate this temperamental bias.

EIGHT

Training an Air Force: The GHQ Era

It must be remembered that the airplane is more than just another supporting weapon. It is a fact which has apparently been recognized by most of the great world powers, that the airplane is an engine of war which has brought into being a new and entirely different mode of warfare—the application of Air Power.
— Maj. Gen. Frank M. Andrews[1]

What has waggishly been called the "fog of peace" renders war planning, doctrinal abstractions, and training methods theoretical. Strategists and tacticians, procurement and acquisition specialists, and war fighters themselves must make assumptions about future conflicts, and convince military and civilian bureaucracies to support and fund their concept of an appropriate force structure and inventory. During the interwar years, Army airmen dreamed about and planned for an air organization able to field mobile tactical squadrons under its own command structure. But because the Air Corps lacked viable proof of its utility during wartime, it failed to persuade the ground-based Army to give it complete autonomy over its own units during peacetime.

Thus, the creation of the GHQ Air Force made 1935 a banner year, as the air arm moved a step closer to that longed-for reality. Air strike elements were, for the first time, concentrated under an air commander. As part of the legislation, Congress also authorized greater numbers of aircraft and men for the Air Corps. GHQ Air Force commanded the 1st, 2d, and 3d Wings and the 21st Airship Group. The three heavier-than-air wings included nine groups with thirty squadrons—twelve bombardment, six attack, ten pursuit, and two reconnaissance.[2] Once the GHQ Air Force was established, oversight of tactical unit training shifted to that entity.

One drawback to the reorganization was the reintroduction of an old split, with two air organizations answering separately to the Army Chief of Staff,

Interwar Years

rather than the single hierarchical structure that many senior airmen had urged so strenuously. The General Staff failed to place GHQ Air Force in the chain of command under the Chief of the Air Corps, thereby causing a measure of internal ambiguity and competition within the air arm. George Brett gave GHQ Air Force Commander Frank Andrews a succinct assessment of the divisive situation in October 1937: Chaney at the Training Center, Brett remarked, "is fighting for the existence of what he considers to be the most important thing. You are fighting with Chaney because you feel as if your phase of it is the most important thing.... Robins [head of Materiel Division] and Chaney present their ideas to Westover [Chief of the Air Corps]. You have to present your ideas to General Craig [Chief of Staff]."[3] Furthermore, Brett went on, the placement of a commanding general at the Training Center in Texas and another chief of training at headquarters in Washington resulted in potentially competing authority. The Air Corps was forced nonetheless to operate under this balky system until the two organizations reunified under the prewar expansion program.

The new structure gave the Chief of the Air Corps responsibility for overseeing individual training at the flying schools. The Training Section reviewed the programs of instruction at the Primary and Advanced Flying Schools, the Air Corps Tactical School, and the Air Corps Technical School; reviewed training programs submitted by the War Department; supervised preparation and revision of pertinent training materials including manuals, regulations, circulars, and films; maintained various types of training records and statistics; and reviewed and recommended matters concerning the training of the National Guard and Air Reserve.[4]

Besides issuing administrative, policy, and training directives, the Office of the Chief of the Air Corps took a significant step in the professionalization of the air arm by mapping out a career pattern for its officers. It followed more than a decade of discussion and planning for professional eduction, which frequently stalled over the merits of the Air Service's preference for college-trained men.[5] Immediately after the war, the Air Service began to construct an educational system. An Air Service pamphlet stated that there was "no way of training officers, even in part, for the Air Service, except in the Air Service itself." Airmen floated the idea of an Air Academy, then, with greater success, an Army Air Service School of Application at Langley, Virginia. In 1920 the War Department authorized establishment of eleven Air Service schools. Among them was the Field Officers' School at Langley, which would be renamed the Air Service/Air Corps Tactical School, and eventually received the originally suggested name, the School of Application.[6]

In late 1937 Chief of the Air Corps Maj. Gen. Oscar Westover submitted to the General Staff a statement of Air Corps objectives that ratified pilot specialization and pinpointed desired stops in the now-established professional educational system along an upward path toward promotion and leadership.

The GHQ Era

Maj. Gen. Oscar Westover, Chief of the
Air Corps, 1935–1938

Following graduation from the Training Center, all officers would join a tactical unit for at least two years. Thereafter individuals might compete for additional education and training. Between two and four years after graduation from the Training Center, for example, some officers would specialize in communications, engineering, armament, or photography at the Air Corps Technical School. Four to eight years from graduation, officers might enroll in civilian universities for advanced instruction in meteorology, engineering, or business administration. After five to ten years of commissioned service, officers were selected for the Air Corps Engineering School, to specialize in logistics. Finally, after ten years of service, officers might attend the Air Corps Tactical School.[7]

Westover gave explicit voice to the fact that the technical nature of the Air Corps meant that an airman's career path deviated from that of officers in the rest of the Army. "The Air Corps is confronted with a problem peculiar only to this branch," he stated, "in that it is necessary to have officers trained in many of the technical specialties." Ideally, he believed, all Air Corps officers should attend the service's technical school. However, such a requirement was never codified. Nominally, all Air Corps officers were pilots, and during a period of severe personnel shortages in both the training schools and tactical units, too few could be spared for duty that took them away from flying. And, it must be admitted, very few pilots expressed much enthusiasm for attending a school that mostly provided technical training for the enlisted force. On the other hand, attendance at the Tactical School assumed considerable importance in the Air Corps because many airmen thought the school offered the only professional military education in the Army that formulated tactics and doctrine for aeronautics. By the mid-1930s the Tactical School was taking the long-

Interwar Years

The Air Corps Tactical School graduating class of 1928 included two future general officers in the GHQ Air Force: Frank Andrews, seen here as a major seated second from the left, front row, and George Brett, also a major, here seated second from the right, front row.

range bombardment mission very seriously, whereas nonsupport roles for air tended to be underplayed at the Army's Command and General Staff School.

The GHQ Air Force Perspective

After a decade of contemplation and a few years of intensive planning, the GHQ Air Force came into being on March 1, 1935. Two years earlier Brig. Gen. Oscar Westover, then Assistant Chief of the Air Corps, had assumed command of a GHQ Air Force (Provisional) as a result of the influential Drum Board[8] findings that the Army's intended responsibilities for coastal defense necessitated the creation of a unified strike force. Although air squadrons were not in fact reorganized for this purpose at the time, GHQ Air Force inched closer to consummation. The Air Corps's interests were additionally furthered as General Douglas MacArthur moved to reorganize Army ground forces into a more consolidated structure of four field armies that could be quickly mobilized in the event of war.[9]

The GHQ Era

The Army might have been propelled faster toward creation of an air combat force but for the negative publicity wrought by the air mail experience. Questions arose as to the viability of military aviation in the event of any emergency. In the wake of this seeming humiliation, another board, headed by former Secretary of War Newton D. Baker, met to reconsider the status of the Air Corps. Although blue-ribbon panels largely laid to rest charges of Air Corps incompetency, they waffled over the perplexing issues of the employment of army and naval forces in coastal defense and the extent of Air Corps autonomy. Nonetheless, in July 1934, the Baker Board endorsed the Drum Board proposal for a GHQ Air Force.

On December 31, 1934, the Secretary of War directed that the GHQ Air Force begin operations on March 1 of the new year. It would be headquartered at Langley Field; three regionally based wings would be located at Langley in Virginia, March Field in California, and Barksdale Field in Louisiana. GHQ Air Force assumed control of all tactical units of bombardment, pursuit, and attack, heretofore under the auspices of the nine corps areas. The Army promoted Lt. Col. Frank M. Andrews to the temporary rank of brigadier general and gave him command of GHQ Air Force. Andrews took with him to Langley the previously assembled staff from Bolling Field. He answered directly to the Chief of Staff in peacetime and to the theater commander in time of war. As mentioned, he was not part of the chain of command flowing through the Office of the Chief of the Air Corps.

Andrews had served as head of Training and Operations in the Office of the Chief of the Air Corps, but he brought to his new job more the goals and attitudes of an operational commander than a staff officer. His contribution would be less as a strategic thinker — although he was an articulate spokesman on behalf of long-range bombardment as it was being defined at the Tactical School — than as a spokesman for tactical operations. As a young pilot, Andrews had spent four years as a flight instructor at Kelly Field. By the time he commanded at Selfridge Field several years later, he recognized that his pilots had no instrument flying capability and very little navigational equipment with which to learn. As a member of the mass flight to France Field in the Canal Zone in 1932, he saw how the capriciousness of weather could paralyze flying missions, so he took a three-week blind-flying course at the Advanced Flying School.[10] Now, as chief of GHQ Air Force, Andrews tried to ensure that more intensive work in navigation, instrument flying, and gunnery became a formalized part of unit training. GHQ Air Force increased the number of required flying hours overall and instrument training in particular. GHQ Air Force pilots also took an operational readiness test on a quarterly basis.

It was unfortunate that training improved less as a result of wartime experience than from the painful and highly public scrutiny it received as a result of the Air Corps's failure in the arena of civil aviation. Yet the air mail debacle that contributed to the establishment of GHQ Air Force, coupled with

Interwar Years

Brig. Gen. Frank Andrews is shown here with his staff on March 6, 1935, being honored as he assumes command of the newly created GHQ Air Force headquartered at Langley Field in Virginia.

the personality of Frank Andrews, who won the affection and respect of both air and ground officers, did much to push training toward greater combat readiness. GHQ Air Force was designed as a mobile strike force equipped to repel attackers from offshore or those encroaching on American borders. Training a combat force required more than instilling individual piloting skills. Thus, Andrews emphasized strenuous crew training, combined training of Air Corps branches, and cooperative exercises with ground forces. Collective training permitted squadrons to fight alongside other air units and ground armies and, when necessary, to act alone. To perform their duties successfully, Andrews believed airmen needed the most sophisticated reconnaissance and bombardment equipment.

Earlier, at Selfridge Field, Andrews had remarked pointedly to his friend "Tooey" Spaatz, "Air Force units have no more place in a Corps Area than a battleship."[11] "Hap" Arnold, then in command of the 1st Bombardment Wing at March Field, likewise commented, "All training should come under the head of a commander of the Air Forces instead of Corps Area Commanders. Even with a Corps Area Air Officer on the Corps Area Staff, they do not seem able to recognize that there is a big difference in the training of observation units and air force units."[12] Although the intoxication with the concept of strategic air power that undergirded these and similar remarks considerably delayed

The GHQ Era

development of tactics and a workable infrastructure for air-ground cooperation, the centralization of air combat units under GHQ Air Force leadership, rather than under ground commanders, met many airmen's objections and generally improved tactical unit training.

The dual organizational structure, however, compromised the efficacy of the entire training program. The division of authority between the Office of the Chief of the Air Corps and GHQ Air Force resembled the gap between the Division of Military Aeronautics and the AEF in World War I, only without the geographical separation. During World War I, individual training was given mostly in the United States; now the Office of the Chief of the Air Corps held that responsibility. Previously, squadrons in Europe provided tactical training; it now took place in GHQ Air Force units. Inevitably, conflict erupted between the two air organizations, one of which determined training policy, assigned personnel, and apportioned equipment to the other.

Both the Chief of the Air Corps and the Commanding General, GHQ Air Force expressed frustration with the awkward arrangement, but for four years they could not convince the War Department to change it. Andrews became particularly impatient with the system of personnel assignment. An airman might be on duty in a squadron for only a few months before being sent to the War College or some other training site. Tactical units needed experienced officers for staff work, he protested, and young Reserve officers (who, it had earlier been argued, might perform ground duties if kept longer on extended active duty) were too inexperienced for command assignments in the squadrons. Until newer officers could be assimilated and trained, the bleeding off of more senior pilots to schools worked an avoidable hardship.[13]

Besides personnel matters, at times the two organizations found themselves at odds over budgets and acquisition. In late 1937 the Secretary of War directed both the Office of the Chief of the Air Corps and GHQ Air Force to draw up a five-year airplane replacement plan. Not surprisingly, their perspectives differed. The Office of the Chief of the Air Corps focused on procurement of advanced training planes and reclassification of other aircraft to meet training needs. Its program called for 495 more advanced training types than the GHQ Air Force projected. The latter agreed to reclassify some first-line aircraft for training purposes, and it added them to the aircraft already on hand. But at the heart of the GHQ Air Force position was a decrease in the numbers of new noncombat planes and medium bombers to be purchased, in favor of attack, pursuit, and heavy bombers.[14] This scheme echoed the Drum Board's earlier recommendation that airplane distribution reflect an increase in combat and long-range reconnaissance over observation and training aircraft.[15]

Other disagreements ruffled the harmony within the Army, such as the recurring antagonism between pilots and engineers. In late November 1933, for example, Maj. Hugh Knerr, then at Wright Field where much of the research and development work was proceeding, complained to Frank Andrews, then

Interwar Years

commanding at Selfridge Field,[16] that "it is very hard to make [the Engineering Section] understand that experimental equipment has no place in the equipment of Tactical Troops."[17] A month before, in a lengthy letter to Spaatz, Hap Arnold, commanding the 1st Bomb Wing at March Field, compared the types of aircraft he wanted with those on order. "I may sound pessimistic without reason," he wrote, "but I do believe that the Air Corps is confronted right now with a problem which requires thought by the best brains we have, and unless our policies are changed, it is highly possible that the Materiel Division as usual will steal the show and force the tactical units to take the planes that they themselves want us to have."[18]

The establishment of GHQ Air Force brought little respite for the personnel shortage in tactical units. In June 1936 the new command had only 409 of its authorized 1,350 Regular officers. There had been reason for hope since, by the fall of 1933, the Air Corps had activated all the squadrons authorized under the five-year expansion program: 4 attack, 12 bombardment, 13 observation, and 21 pursuit squadrons.[19] Although the creation of GHQ Air Force effectively reorganized these squadrons, there was no accompanying mechanism to produce trained pilots to man them. Because Air Corps training now emphasized combat skills to a greater extent, employing more Reserve pilots (a frequent proposal made a decade earlier as a way to increase the flying force) appeared less attractive. In December 1938 Senator Henry Cabot Lodge passed along the suggestion that many of the commercial airline pilots holding reserve commissions might be kept for longer periods in tactical units. Brig. Gen. W. G. Kilner, then Acting Chief of the Air Corps, explained that in spite of the shortage of trained pilots in the Air Corps, the advantage of using Reserve officers would be offset by the ongoing need to train them. Even though most commercial pilots held reserve commissions in the Army or Navy, Kilner observed, and had been trained in military flying schools, "when we consider all commercial pilots as a class, there are necessarily included individuals who vary greatly in flying training, experience and skill. The sending of these pilots to our tactical organizations for periods of active training would necessarily interfere greatly with the operation of those units."[20]

By the mid-1930s it became abundantly clear that those in the military, Congress, the Roosevelt administration, and various investigative boards lacked any shared concept of what constituted an air force. The Drum Board, for instance, had called for an inventory of 980 airplanes. In mid-1935 there were 450, of which considerably fewer than half could be considered up-to-date. On the first of January 1936, GHQ Air Force had 174 relatively modern tactical aircraft, including those in depot overhaul. That June, Congress authorized the Air Corps to bring its aircraft strength to 2,320, but that item did not appear in the President's fiscal 1937 budget request.[21]

Throughout the 1930s, a fruitful period for aeronautical research and development, pilots aired their frustrations to one another as they tried to

The GHQ Era

conduct routine training duties and at the same time increase their proficiency in newly advancing technical fields.[22] Knowing that more sophisticated equipment had been developed, but being unable to get it, seemed intolerable at times. Most units had too few of the old outdated aircraft. The usually imperturbable Spaatz sent a furious letter to a colleague in which he railed against a change of schedule for the 1933 exercises. Units at March Field had been "conducting very intensive flying training of the units concerned, necessitating flying day and night.... During this period equipment has received considerable punishment," he wrote. It took time after training to overhaul the equipment. Advancing the date for the next scheduled exercises would not allow for a careful inspection and repair of equipment. "If all the airplanes in question were fairly new there would be no difficulty," Spaatz exploded. However, "the B-2 airplanes have had a hell of a lot of flying over a period of four or five or six years and require much more careful maintenance than airplanes of more recent construction.... It seems to me that any exercise that requires as much intensive training by *previously intensively trained* combat crews as has been called for in this case must be either highly improbable or something else must be wrong with it.... I am very reluctant to attempt to train combat units for someone else to destroy."[23]

Once he moved to the Office of the Chief of Air Corps as head of Training and Operations, Spaatz became privy to the headquarters' perspective, and found he often could do little to change conditions that caused so much irritation in the field. Spaatz then wrote to George Brett, who fumed over the lack of training aircraft at the Command and General Staff School, that most Air Corps installations were also short:

> There doesn't seem to be any possible help for your airplane situation at present. All Pursuit Squadrons are down to about ⅔ of their authorized strength in airplanes, and out of this number have to come the airplanes in the depot for overhaul and out of commission locally. As a result, the squadrons are normally operating with less than 50 percent of their authorized strength. The Training Center is also short of the number of pursuit airplanes required for student training.[24]

Brett, undeterred, continued to press, and very slowly he and other commanders began to receive the trickle of new aircraft coming off the production line.

Nevertheless, a year and a half later the Adjutant General spoke of "the present shortage of airplanes, which is constantly growing and from which no appreciable relief can be expected for many months."[25] The complaints and justifications and explanations continued. Spaatz wrote in pique to the air detachment at Fort Leavenworth: "You state that your airplanes are flying an average of 36 hours per airplane per month. This is not unusual for airplanes in the Air Corps at the present time. At some posts planes are flying as much as 60 hours per plane per month. It will probably be another year or more

Interwar Years

As a flight instructor at Fort Leavenworth, Lt. Col. George Brett found himself constantly requesting more training aircraft.

before the situation on equipment improves."[26] It would take longer than another year.

But the formation of three GHQ Air Force wings, understrength and underequipped though they were, heralded a clear victory for the Air Corps's ability to work out new training strategies and employment doctrines. Each wing was a regionally based tactical organization that could incorporate pursuit, bombardment, attack, and reconnaissance squadrons.[27] Heretofore the specialties had operated fairly discretely, although they met (all too rarely) in combined training exercises during the tactical unit phase and in joint Army maneuvers. During the World War the Air Service had worked feverishly to form the system for individual training, but most airmen went into combat with little experience working together. After the war, the Air Corps adopted a functional approach through its specialties, and the training program made strides in articulating the piloting skills and aircraft for each task. The last link in that now-bureaucratized system of specialization came with the creation of tactical unit training in the GHQ Air Force. Its tables of organization commingled aviation specialties, so that the Air Corps began to resemble an air force rather than an assembled collection of weapons.

Andrews spoke frequently of the training advantages in placing different types of aircraft in the same wing. Maneuvers provided further opportunities

The GHQ Era

for combined training, in spite of the fact that groups sometimes had to be assembled from units spread throughout the three wings. From his experience trying to train fighter-bomber groups before they came under GHQ Air Force, Spaatz had written General Foulois:

> I am convinced, from watching the progress of the 7th Bombardment Group and the 17th Pursuit Group in their training during the past months, that they have developed much more rapidly with both Groups operated from the same field than could have been possible with the Groups operating at different stations. A friendly spirit of rivalry between Pursuit and Bombardment places both Groups on their metal and results in a higher standard of training throughout. The personnel of each Group become familiar with the procedure of the other Group.... Of course, joint training can be accomplished by having the units operate together for a certain period of each year but I do not believe the results in this method of training will compare with the results obtained from side by side operations.[28]

The advantages of "side-by-side operations" could also be seen, as Spaatz noted elsewhere, "when new equipment with marked changes in performance is about to be provided, which may necessitate some drastic changes in tactics."[29] Arnold made the same point, since the 1st Wing that he commanded was "spread all over California."[30] Although the GHQ Air Force reorganization the next year brought a redesignation of some air units and alleviated some of the problems of split command, the component elements of GHQ Air Force wings still remained geographically dispersed. As of March 1, 1935, for example, elements of the 1st Wing were stationed at Hamilton, March, Rockwell, and even Brooks Fields. By the following year all so-called observation squadrons had been removed from GHQ Air Force, and in the case of the 1st Wing, reconnaissance squadrons became integrated into the bombardment groups.[31]

Training a Wing

The venerable 1st Bombardment Wing now headquartered at March Field, California, whose lineage dated to World War I, was as well placed and well trained as any during the GHQ Air Force years. Its experience illustrates the evolution of a depression-era force transitioning to a war-fighting capability.

At the time it converted from a primary flying school to an operational base, March Field's aircraft inventory included a few Curtiss Condor B–2s, Keystone B–4s, B–5s, and B–6s, and, as Arnold later said, "a miscellaneous collection of planes for our Fighter Group; mostly P–12s."[32] Nevertheless, units of the 1st Wing had been in the forefront of operational training before conversion to the wing structure, frequently using ingenuity in lieu of the latest

Interwar Years

Replacing the early model Curtiss Condor B-2 bomber (*bottom*) was the newer model Martin B-12 (*top*).

equipment. Units had, for instance, originated new command and control practices by equipping a Fokker transport airplane with radio equipment for use either as a forward command post on the ground or in communication with the lead pilot in formation when airborne.[33] Not surprisingly, therefore, the wing's 7th Bombardment Group received the first of the newer model B-12s and B-12As; those planes in turn were relinquished for the B-18s in mid-1937. In 1939 the original production of B-17s (thirteen YB-17s) was assigned to the 38th Reconnaissance Squadron. The 17th Attack Group flew A-12s until 1936 when they were able to fly the newer A-17, which was similarly armed but could carry twice the number of thirty-pound bombs.

Besides conducting peacetime training, March Field assumed the greatest burden of Civilian Conservation Corps supervision in the Air Corps. During the depression years of the early 1930s, nonmilitary matters, which curtailed tactical training, consumed valuable summer months. As a result, some squadrons and commanders with considerable Conservation Corps duties logged only a few hours compared to other pilots.

Otherwise, March Field was well situated for military purposes. Removed

The GHQ Era

from urban development, it enjoyed space and topographical diversity. Although the 1st Wing could not rely on the extensive network of airports and lighted, radio-equipped landing fields of the populous East, its bomber and fighter units spent considerable hours in night flying, thereby, as Arnold noted in his memoirs, coming closer to combat conditions than their eastern cousins:

> Flights starting from the baked plains of March Field could soon be over hot deserts, the high mountains, the great salt flats, and the Pacific Ocean. Thus, we were able to take advantage of rugged training conditions impossible in the East. . . . I put crews, and whole squadrons, on airdromes away from their home stations for weeks at a time, under field conditions which no other American airmen were to know until Brereton's units joined Tedder and Coningham in time for Alamein [World War II].[34]

More than any other factor, its superb bombing and gunnery range made March Field the envy of other air bases. Across the San Bernardino Mountains, on the baked clay of the Mojave Desert, lay a smooth, flat, dry lake bed that presented an ideal landing field and bombing range. The Air Corps purchased Muroc Dry Lake, and bombing practice and air-to-ground and air-to-air gunnery became a regular element of training. The very perfection of the range, however, according to Arnold, afforded peculiar training experiences:

> The clay-covered lake is so smooth that the early pilot had to receive

The expansive March Field, California

> special instructions before going in to land. A speed of 100 miles an hour on it seemed slower than 50 miles an hour on a normal field. It is so vast that there was no way the pilot could tell whether he was rolling straight ahead on landing, or turning, unless he was careful to pick out checkpoints on a mountain from five to seven miles away. This caused one ground loop after another until finally our pilots learned.[35]

Learning to land on this eerie landscape reminded pilots that one repeatedly relearned basic techniques in accordance with the eccentricities of topography, geography, climate, winds, and machinery.

The importance of appropriate training facilities cannot be overestimated. Because they lacked firing ranges close at hand, many installations had no adequate means to train pilots, to say nothing of bombardiers or gunners, once the need for separate training courses for those specialties developed. However enamored they might have been with the image of the lone warrior (the fighter pilot), airmen recognized that an effective fighting force relied not only on individual skills but also on group proficiency. As Chief of the Air Corps during the early 1930s, Benjamin Foulois advocated placing greater emphasis on combat crew training. He was echoed during the second half of the decade by Frank Andrews, who as Chief of GHQ Air Force spoke on numerous occasions about the necessity for crew training in the tactical units. In a January 1939 address before the National Aeronautic Association in St. Louis, he stated:

> The combat crews to fully man each airplane must be trained and available, and they must have sufficient experience to prepare them thoroughly in their particular specialty. A superior pilot is of little value if his bomber cannot place the bomb on the target, and both will fail unless a competent navigator succeeds in directing the airplane to its objective.... It is a rarely recognized fact that an airplane of a type in production can be built much faster than a crew can be trained to man and maintain it.[36]

The point was simple: gunnery and bombardment training required firing ranges.

Training the Specialties

Despite the recognition that combat training would become more realistic if relevant units trained together and the fact that the wing structure allowed for it, except for maneuvers and exercises, most training remained mission-specific and geographically distinct. A pursuit squadron, for example, did not routinely sharpen its skills alongside a bomber formation. However, tactics for each of the four specialties — attack, pursuit, bombardment, and reconnaissance — became more refined as the definition of their roles and the priority accorded

Fighters (Pursuit)

The legacy of the World War had given pursuit an envied glamour and preeminence, but by the mid-1930s advocates of strategic bombardment had begun to dominate, and pursuit became more narrowly defined as defensive. Doctrine and aeronautical design reinforced one another, tactics and doctrine prescribed by the limitations of aircraft design. Since fighter aircraft were subject to enormous stress from the torturous maneuvering required in aerial combat, for some time they continued to be constructed with heavily reinforced double wings. The drag from the wings made them slower than the monoplane bombers. Many airmen concluded therefore that the fast, heavily armed big bombers could hold their own against enemy pursuit interception, and bombardment missions could be flown without shorter-range fighter escort.

Although pursuit had its defenders (such as Capt. Claire Chennault who wrote treatises on fighter tactics at the Air Corps Tactical School), there was no unanimity of opinion regarding the pursuit mission. Its uncertain status could be sensed from the vague policies and tactics found in training manuals, the relative disinterest in pursuit at the Tactical School in the mid-1930s, and the debates regarding aircraft design. A board of officers — convened at Wright Field in January 1933 to make recommendations for new pursuit aircraft — supported the development of a high-speed, low-weight, single-seat pursuit machine, but also recommended working toward the "growing demand for a long-range, multi-place airplane of high fire power to accompany bombardment airplanes on distant missions."[37]

One of the last biplane fighters, the Curtiss P–6E, first flew in 1931. The Boeing-built P–26, or the Peashooter, tried out the following year and went into production in 1934. It was an all-metal monoplane with landing flaps plus the old features of an open cockpit and fixed landing gear.[38] In many ways its odd admixture of old and new symbolized the limbo in which pursuit languished at the time. Nonetheless, the 1933 board concluded that the P–26 was the best contemporary airplane to meet the requirements for single-seat pursuit aviation. For a two-place fighter, it gave a favorable nod to the P–25, but that airplane did not proceed beyond the experimental. A modern fighter was not developed until the Seversky P–35 appeared in 1937–1938. A low-wing, cantilevered monoplane, it had retractable landing gear, controllable-pitch propeller, trailing-edge landing flaps, and a stressed-skin construction with a closed cockpit.[39]

Airmen were fervent in their enthusiasm for the newest aeronautical developments. Yet advanced aircraft and equipment required special training, not only to learn performance characteristics and refine tactics but, both before and after an airplane came into the inventory, to determine needed modifica-

Interwar Years

tions. Checking out new aircraft was not necessarily turned over to a cadre of test pilots. For instance, two fighter pilots, Capts. Hugh M. Elmendorf and Frank O'D. "Monk" Hunter, both of whom had been on the pursuit board in 1933, made a test flight shortly thereafter in one of the new models with a closed cockpit. The plane crashed after going into a long spin from a high altitude, killing Elmendorf, who was piloting. Spaatz, also a fighter pilot who was close to both men, speculated that "the closed cockpit, with gasoline fumes, or carbon monoxide gas, together with combating at altitudes from fifteen to seventeen thousand feet, may have caused Elmie to pass out. . . . With the cockpit closed, the customary blast of fresh air is lacking."[40]

Another fighter pilot, Capt. Ira Eaker, also experimented with instruments and tactics in single-seaters, but with happier results in his case. "I devised and built a baby-buggy top for the cockpit of the P–26," he later recalled. "I then discovered if I covered the cockpit shortly after take-off and began a slow, climbing turn to the left, I was able to climb through several thousand feet of overcast without difficulty. The next step was to have additional

While test-flying a new pursuit aircraft, fighter pilots Capts. Frank O'D. Hunter (*above*) and Hugh M. Elmendorf (*right*) crashed, and Elmendorf lost his life. In 1940, Elmendorf Field in Alaska was named in honor of this early test pilot.

The GHQ Era

planes flying formations on me." After careful practice, his squadron was soon flying six-plane formations. When Frank Andrews arrived for an inspection, he asked to participate in the exercise, both as a covered leader and member of the formation. Thereafter blind flying in formation became standard practice in pursuit training in GHQ Air Force.[41]

By and large, however, fighter pilot training in GHQ Air Force built upon well-established tactics. The pilot was required to demonstrate superior maneuvering ability and aerial marksmanship. Although he was permitted some initiative, doctrine at this time held that he was to "avoid rather than seek combat with other classes."[42] During its first year as a GHQ Air Force unit, the training curriculum of the 20th Pursuit Group at Barksdale, Louisiana, listed acrobatics, aerial gunnery and bombing, air navigation, individual combat, formation flying, instrument flying, night flying, performance flights, and radio communication. Navigation absorbed the largest number of hours (forty-five in both basic and advanced phases). Of the aforementioned subjects, the unit phase only included navigation, formation flying, night flying, radio communication, and a couple of hours of performance flights. The other flying hours in unit training were taken up with combat exercises, techniques of tactics and employment, and field exercises.[43]

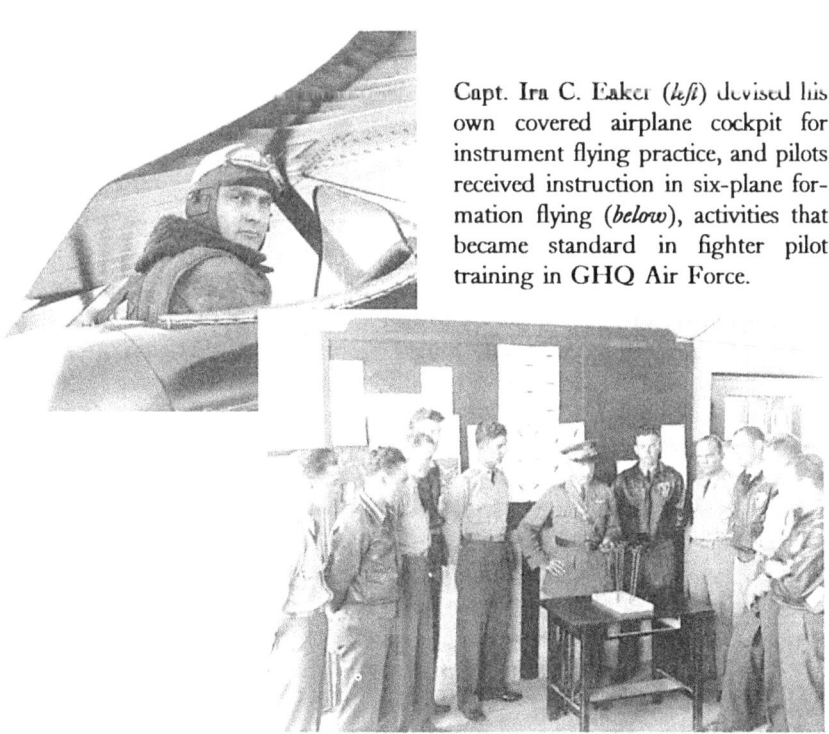

Capt. Ira C. Eaker (*left*) devised his own covered airplane cockpit for instrument flying practice, and pilots received instruction in six-plane formation flying (*below*), activities that became standard in fighter pilot training in GHQ Air Force.

Interwar Years

Attack

Attack and bombardment were the primary offensive forces; bombardment ranging beyond the immediate battle area, and attack strafing and bombing occurring at low altitudes. The 1935 issue of TR 440–15 described attack training as follows:

> In attack aviation, the pilot possesses great courage and stamina in flying, as the missions make him the target for ground weapons of all kinds as well as of hostile pursuit aviation. The attack pilot combines his skill as a pilot with his accuracy as a gunner and bomber in a single physical operation. The attack gunner excels in skill with his machine guns, both as offensive weapons against ground targets and hostile pursuit planes attacking from the air. All members of the team should have a working knowledge of the tactics and technique of the ground arms in order to obtain the maximum destructive effect on the enemy and the maximum support for friendly units. The technical use and tactical effect of chemicals must be thoroughly understood.[44]

From the middle to late 1930s, attack aviation in GHQ Air Force was localized in the 17th Attack Group of the 1st Wing at March Field (34th, 73d, and 95th Squadrons) and the 3d Attack Group of the 3d Wing at Barksdale Field (8th, 13th, and 90th Squadrons). The 37th Attack Squadron was attached to the 2d Wing as part of the 8th Pursuit Group.[45] As elsewhere in the Army during the depression years, attack aviation operated on a shoestring. General Landon recalled that when he was posted to Fort Crockett, where the 3d Attack Group was headquartered before it moved to Barksdale Field in 1935, pilots were limited to four hours of flying time per month, the minimum required to receive flight pay.[46] Conditions had begun to improve by 1935 with the arrival of new aircraft.

The year before, forty-six Curtiss-built A–12 Shrikes came into the inventory. This airplane carried a two-man crew, was powered with a Wright Cyclone nine-cylinder, radial air-cooled engine, could fly at a maximum speed of 175 miles per hour for 3½ hours, and was armed with four machine guns and 400 pounds of bombs.[47] Neither these airplanes nor the A–11s were fitted with instruments, so they could not be used for navigational training at the time. Nonetheless, squadrons in the 3d Wing were richer than those of the 17th Attack Group at March Field, which had to fly old P–12 series planes for some time. The 3d Attack Group was also the first in 1935 to receive the new Northrop A–17s, which, including the improved A series with retractable landing gear and more powerful engine, became the standard aircraft for attack aviation.

Like pursuit squadrons, an attack squadron (assuming available personnel) was to train at least twenty combat crews, each consisting of one pilot and one gunner. Officers were desired; if none were available, the gunner could be a

The GHQ Era

A-12 Shrike

cadet or an enlisted man. Training directives warned against devoting too much unit time to individual training. Rather, unit training stressed crew coordination in instrument flying, radio, air and night navigation, and combat firing.[48] The published curriculum for attack units had become standardized by the middle 1930s, but practices varied, depending upon facilities, available aircraft, other adjacent air units, and the makeup of ground units stationed nearby. Gunnery and bombing training relied upon ranges located proximately. Squadrons of the 3d Wing in Louisiana had to go to Galveston, Texas, or to Florida for aerial gunnery. Units stationed at March Field, California, on the other hand, had the nearby Muroc Dry Lake. The single attack squadron at Langley Field engaged in exercises with the larger tank and mechanized units, including the 7th Cavalry Brigade and battalions of the 66th Infantry (light tanks).[49] Although Langley lacked extensive facilities for gunnery training, the 37th Attack Squadron could join units of the pursuit and bombardment groups in practice firing over the Atlantic Ocean and Chesapeake Bay, while keeping an eye out for any ships straying into range.

Bombardment

From the middle to late 1930s, the Air Corps pinned its hopes on long-range strategic bombardment. Innovation in aeronautical technologies coincided with developing strategic doctrine, especially as espoused by vocal proponents at the Air Corps Tactical School. Europeans had generally acceded to the doctrine of massed bombing raids, contrary to American theorists who argued for the use of heavily armed unescorted bombers flying in high-altitude daylight formations to bomb specific targets. American airmen's yearning for an autonomous role coalesced around the bombardment mission, yet their views collided with announced American security policies and General Staff biases.

Through 1938 the U.S. diplomatic posture remained one of political

Interwar Years

neutrality and a military armed solely for defense. Secretary of War Harry Woodring finally agreed to an increase in the air arm to 2,300 planes by the projected date of June 1940, but he remained unconvinced that developing a heavy bomber was preferable to fielding a larger number of smaller planes. Naturally, the General Staff still favored close air support, a viewpoint reinforced by recent reports of the successes of aerial combat in Spain, Ethiopia, and China. Through the mid-1930s the Army also elected to spend procurement funds on existing rather than experimental weapons. In contrast, GHQ Air Force leadership pressed for an advanced long-range bomber, specifically the Boeing B–17. This airplane could fly fast and far and, equipped with the new bombsights developed by the Norden and Sperry companies, could bomb specific bottleneck targets. Airmen's arguments recalled pre–World War I debates over the morality of bombing civilians behind the lines. Precision bombing was not only possible and more effective against military or industrial targets than mass bombing, they claimed, but it was vastly more humane than the German terror bombing of London had been in the earlier war.

The compromises meted out by these contrasting views took the Army Air Corps into World War II. Not everybody believed that long-range bombardment could succeed without pursuit escort. Composite groups put together for air maneuvers and exercises, for example, such as the first concentration of air forces of the 1st Wing in March 1935, solved a communication problem of bombardment with accompanying pursuit. Within the bombardment groups, squadrons trained with their old bomber fleet and on the few B–17 Flying Fortresses, and they cherished Air Corps concepts of strategic bombardment while publicly stating that the large bomber was "useful for coastal patrol." In the spring of 1936 Andrews underscored the importance of celestial navigation training missions for B–17 crews, reasoning that the airplanes would be employed in long-range reconnaissance over land and water.[50] In 1938 the Army Chief of Staff and the Chief of Naval Operations agreed to restrict Air Corps operations to no more than one hundred miles offshore, which limited long-range air missions and supported the reconnaissance over the strategic function for heavy bombers.

In 1933 Hap Arnold had grumbled to Tooey Spaatz about "the same old point" — not having aircraft able "to carry a full load of bombs and enough gas to attack a point 200 miles distant."[51] That year the Boeing B–10 went into service. It was an all-metal, two-engine monoplane incorporating all the newest design features, capable of higher altitudes and greater range and speed than any previous airplane. It was so promising that the next year's specifications requested a multiengine bomber capable of a 1,000-mile range and with a 2,000-pound bombload capacity: ultimately, the Boeing B–17. Unfortunately, its experimental model crashed on final testing, so the Air Corps could only wrest agreement to purchase thirteen planes. They arrived at Langley Field on March 1, 1937.

The GHQ Era

On this small fleet, the Air Corps could not hope to train all its bombardment squadrons, but it could begin to reevaluate its training concepts within the confines of official military policy. Encouraged by the heady existence of GHQ Air Force, the Air Corps underscored the importance of crew training. A large airplane carrying a several-man crew necessitated separate functions and teamwork, as opposed to the generalist pilot who could do everything. Crew training was in fact a hollow concept because the Air Corps conducted very little. Essentially, as before, pilots performed the roles of bombardiers, gunners, navigators, or observers. The Air Corps created no specialist career tracks for these jobs. During peacetime, and until large numbers of heavy bombers came into the inventory, the new shift in perspective remained mostly conceptual.

The TR 440-15 of 1935 outlined the characteristics necessary for members of a bombardment crew:

> In bombardment, the pilot must be characterized by great determination and endurance, by the accuracy of his individual and formation flying, by his willingness and ability to fly and navigate long distances on unknown routes over land or water at high altitudes regardless of weather, darkness or light. The bomber must concentrate on his task to the obliteration of every other thought and outside disturbance, for the hostile attack by pursuit aviation and antiaircraft artillery will be greatest just prior to and during the bombing operation. The gunner must be a team player first and an individualist last, for the safety of himself, as well as the team, will depend on the direction of his fire on the proper enemy machine, which is ordinarily not the one directly attacking the gunner's plane.[52]

At this time, according to regulations, each bombardment squadron should train, if possible, not less than nine combat crews, each three-man crew consisting of a pilot, a bombardier-navigator-gunner, and a radio operator. Ideally, all should be officers, but if they were unavailable, officers could be replaced by flying cadets or enlisted men. Bombardiers and radio operators should be qualified as gunners; in the event an airplane was larger than a three-seater, an additional gunner should be trained. Earlier training plans of the 1st Bombardment Wing specified a crew consisting of pilot, bomber, radio operator, and gunner. The subsequent reduction in aircrew size undoubtedly resulted from a lack of manpower.

Crew training in GHQ Air Force bomb units came to mean that certain trained pilots regularly performed nonpiloting tasks (excluding radio operation) so that a crew remained a stable unit. Increasingly, men specialized in specific jobs, such as airplane commanders or navigator-bombardiers, even though the latter were also rated pilots. By 1938 the Air Corps had selectively trained a few enlisted men as bombardiers for three-man squadrons, but it was not ready to endorse the practice wholesale, as the service could not easily relinquish its fundamental belief that all fliers should be officers, and all flying officers

Interwar Years

should be pilots.

GHQ Air Force more successfully trained bombardiers than navigators. Teaching sophisticated navigational skills proved daunting because formal instruction had taken place only briefly in the six- to eight-week courses at Langley, March, and Rockwell Fields between 1933 and 1935, and because many aerial navigation instruments remained experimental through the end of the decade. Despite the erratic nature of their training, bomb crews made concerted efforts to fly practice missions over water and in fog, using dead reckoning and available instruments for celestial navigation.

The frontier defense training of the 19th Bombardment Group, headquartered at Rockwell Field, indicates the kinds of activities undertaken. More than 2,500 square miles of ocean, punctuated by five large islands, gave ample space for overwater aerial navigation. The weather varied considerably, providing challenges for training. From May through October a thick cloud, which lay all along the coastal plain and out to sea for several hundred miles, gathered at night then burned off somewhat during the day. In daylight and nighttime flights, pilots used celestial navigation and radio communication for takeoffs and landings in the thick fog. During the winter months the fog lifted, to be replaced by frequent rainstorms and ocean squalls. In this environment, pilots gained additional experience flying amphibious aircraft. The group also devised methods for monitoring the course of patrolling airplanes and bombers as they tracked mobile sea-going vessels marked as targets. Training was so successful, the group reported, that "on flights of 150 miles errors of more than one-half mile in lateral deviation at the objective and two minutes in the predicted time of arrival at that objective are considered excessive and seldom occur, except in the early stages of training."[53] If correctly reported, unit training in celestial navigation achieved remarkable accuracy.[54] The GHQ Air Force Training Directive for 1938–1939 required that navigators be qualified to establish position in the air by celestial means to within twenty-five miles. By dead reckoning they should be able to "navigate toward an objective within a limit of error of one and one-half minutes in estimated time of arrival for each hour flown, and within a lateral deviation of one degree [one mile in sixty]."[55]

The 2d Bombardment Group at Langley Field was also situated near water, but it did not contend with the fog-bound flying of units training over the California coast and Pacific Ocean. The Langley units concentrated on flying by instruments through overcast, developing a means of calculating positions of aircraft in formation descents so as to prevent collisions.[56]

The navigation school in the 1st Wing at March Field closed on July 1, 1935, thereby signaling the end of centralized instruction. Subsequently, those who had learned at the advanced navigation training units in the bombardment groups or at the Tactical School provided training in the tactical units. These brief courses were conducted intermittently, but while navigation training was not formalized by the establishment of an official school, it did become an

The GHQ Era

integral part of tactical unit training. What limited it, more than anything else, was the lack of aircraft suitable for long flights and overwater operations and of more sophisticated aerial instruments, some of which — such as improved bubble sextants, drift and ground speed indicators, compass gyro, and automatic pilots — were still in development in the middle to late 1930s. Not until August 1940 did the Air Corps open a separate navigation school; a year later it finally began to commission nonpilot flying officers.

All bombardment units flew training missions at night. The practice of using illuminating formations that dropped flares to guide the bombers superseded earlier tactics of sending out formations of aircraft carrying their own flares. These units developed a means of evading sound-locating antiaircraft fire by dispatching small formations of bombardment aircraft from different directions and altitudes.[57] Besides this nightly activity, exercises and maneuvers permitted bombers to train with other types of units. During the 1935–1936 training year, for example, the 7th Bombardment Group engaged in exercises in the San Joaquin Valley, joined GHQ Air Force maneuvers in Florida the first two weeks in December, and went to Muroc Dry Lake in March to practice individual and formation bombing. The 31st Bombardment Squadron participated in exercises in August at Medford, Oregon, in what were essentially logistics operations.[58]

Observation and Reconnaissance

Observation was the oldest aerial mission. After the establishment of GHQ Air Force, the Army divided observation units (including balloons) between those remaining with Army corps areas and a few renamed reconnaissance squadrons attached to GHQ Air Force. Because GHQ Air Force was designated as a strike force and observation aircraft were employed defensively, not many observation units shifted to the GHQ Air Force. The operation and therefore the training of these two elements diverged. Observation squadrons cooperated with ground forces, particularly in positioning artillery fire. A group, normally of three squadrons with forty-four airplanes, was assigned to each field army, and one group to each corps.[59] GHQ Air Force reconnaissance units trained for long-range surveillance flights in order to search out and determine the nature of the target, the best route of approach, the location of antiaircraft and enemy airfields, and the type of bombs to be carried. Reconnaissance crews photographed the terrain and maintained constant surveillance during operations.[60]

For some time the missions of reconnaissance and observation units were not distinguished. The 1935 training regulations described the generic purpose and characteristics of observation as follows:

> The great variety of the duties imposed upon the personnel of observation aviation requires men of high character, initiative, judgment and courage, and of wide knowledge of the tactics and technique of the other arms and services. There must be thorough indoctrination of all members of the

Interwar Years

> combat crew — pilot, observer and gunner — that the fundamental purpose of their existence is to get accurate information and report it to their commander. Whether on distant reconnaissance, artillery registration, or any other mission, the personnel of observation must have always in mind the thought that obtaining formation is only a means to the end. The end is to deliver information to the commander who needs it. . . . Though observation should, when possible, avoid detection and combat, its personnel must be ready to fight their way through when necessary. Observation airplanes normally operate singly, but the operations of hostile pursuit may require observation to fly in formation for mutual protection.[61]

The two-man observation crew was to consist of a pilot and an observer, both officers. In a larger, possibly GHQ Air Force reconnaissance airplane, a radio operator or gunner, ideally an officer but usually an enlisted man, joined the crew.

In 1936, an article in the *Air Corps News Letter* described the observer as the forgotten man of aviation. "Little glamour attaches to his role," the author opined, "yet the length of time required to produce a fully trained competent observer is as great as that to train a pilot. He must know organization and tactics of large ground units. He must fly with any pilot to whom he is assigned; he is at a great disadvantage if anything goes wrong with the pilot or airplane; he works in cramped space often in extreme discomfort; he must keep one eye in the air against hostile Pursuit, and the other on the ground. His is a responsible and unenviable lot and he deserves a big hand."[62]

To the dismay of some airplane pilots and instructors, a number of lighter-than-air fliers joined observation squadrons. Balloons had remained part of the Air Corps, but balloon and airplane pilots heretofore had not trained together and the technologies, techniques, and tactics bore little resemblance to one another. When a young officer during the 1930s, Maj. Gen. Samuel Anderson was an instructor at Kelly Fields. There he trained some of the pilots who switched from aeronautics to aviation. "Unfortunately," he commented, before catching himself, "not unfortunately — that's a bad, bad word to use in connection with what I was going to say. I had more than my share of ex-balloon pilots when I was in observation. In fact, I think I had them all."[63]

What would come to be a crippling lack of preparation for war, peacetime training of associated arms never attained a satisfactory level of mutual respect and problem solving. By and large, armor and infantry units stayed unconvinced, or perhaps unaware, of the value of aerial observation. Even members of the Air Corps were overheard to grudgingly acknowledge that the cheapest, and possibly best, observer was a man with glasses standing on a high promontory. More damning, many airmen themselves seriously undervalued observation (and other elements of tactical aviation) in favor of strategic bombardment. As a result, the U.S. Army was to fight World War II without

The GHQ Era

A K–6 camera is seen mounted atop the observation and reconnaissance airplane (*below*). Men trained in aerial photography also received instruction in darkroom technique (*right*).

ground-air coordination equal to that of the enemy's.[64]

One of the perennial areas of disagreement between air and ground forces during the interwar years concerned training time spent in cooperative missions, in particular, towing targets for artillery gunnery practice. The Army had decided years earlier that the final six months of aviation unit training would be devoted to joint field exercises, but the Air Corps resented dedicating resources to an activity that did nothing to advance its own combat proficiency. In the summer of 1934 the IX Corps Area Adjutant General complained to Washington that once again airmen were uncooperative.

> It is desired to emphasize the belief that actually carrying on ground missions with ground troops is the finest training to which the Air Corps personnel can be submitted; and, in fact, the efficient rendering of such missions is, of course, the ultimate goal of Observation Aviation.... It is, therefore, to be expected, and it is perfectly natural, that ground troops should ask for, and should have, cooperative missions more or less continuously thr[oughout] the training year, which will necessarily break into the unit training period of Observation Aviation.[65]

Interwar Years

The subject had already been broached by the IX Corps Area air officer, who requested an increase in allotted flying hours for cooperative missions as a way to accommodate ground commanders' demands. Major Spaatz, recently posted to the Office of the Chief of Air Corps, responded in time-honored bureaucratic fashion by citing paragraph and line of the training directive, and then grumpily agreed, noting that "why anything more should be necessary . . . I don't know. To my mind, it would be difficult to decide in cooperative missions whether the Air Corps officer was getting the training or whether it was the ground troops."[66] Officers in reconnaissance squadrons attached to bombardment groups had little reason to express similar resentment.

The training program for observation outlined at the Tactical School in May 1938 listed the skills for which the observer should be trained to include organization and tactics of associated arms, communications, photography, sketching, intelligence, knowledge of all types of observation missions, maps, navigation, gunnery, pilotage, and pilot-observer report. For his part, the pilot should have knowledge of the observer's duties and demonstrate competence in pilotage, navigation, instrument flying, and gunnery.[67]

The GHQ Air Force Training Directive for 1938–1939 stated that reconnaissance aircraft fly individually or in two-plane formations. Pilots and copilots were to be trained in instrument flying, bombing, and navigation. Depending upon the size of the aircraft, a reconnaissance crew might also include a bombardier, navigator, aerial gunner, radio operator, aerial photographer, and engineer.[68] Neither crews of the magnificence nor possessing the broad and deep training mandated for them by the training directives could be assembled or accomplished at that time. As with other specialties, much smaller crews flew with considerably overlapping duties, and they frequently lacked sufficient equipment to practice all the required skills.

In January 1936, the commanding officer of the 15th Observation Squadron at Scott Field, Illinois, described the aircraft and manpower of his organization. Of the squadron's eight combat airplanes (O–19s), four were in the depot for overhaul. Additionally, they had one cargo plane, one primary training plane, and one basic training plane used exclusively for meteorological flights. Fifteen rated pilots and two in training manned the squadron. Besides attending to their own training, Regular officers trained Reserve officers. In a line from his report, Maj. W. C. Goldsborough expressed a widespread sentiment in the Air Corps: "The present allotment of aircraft . . . is entirely insufficient for the training requirements of this organization."[69] A couple of years later, at the end of the period under scrutiny, the situation remained much the same.

The GHQ Era

The End of an Era

With the following words, Roosevelt notified the country in October 1937 that its President was no longer thinking in the strictly isolationist terms of the past:

> There is a solidarity and interdependence about the modern world, both technically and morally, which makes it impossible for any nation completely to isolate itself from economic and political upheavals in the rest of the world, especially when such upheavals appear to be spreading and not declining.[70]

It is historical commonplace to recall that war clouds had formed by late 1937. In fact, what Winston Churchill called a gathering storm had loomed from the early depression days of the 1930s. Japan seized Manchuria in 1931, quit the League of Nations in 1933, and renounced naval armament limitation treaties set to expire in 1936. After coming to power in 1933, Adolf Hitler rearmed Germany and sent Nazi troops into the demilitarized Rhineland. Benito Mussolini's Italian army overran Ethiopia in 1935. In 1936 a vicious civil war in Spain pitted the Loyalists against General Franco's insurgent fascists. The bombing of Guernica taught a contemporary lesson about the deadly uses of air power.

While these events were not greeted with total complacency in the United States, shudders abroad did not shake Americans' entrenched disengagement from foreign affairs. The President's "quarantine" speech therefore met with fury in many quarters. A severe recession in 1937 and the election in 1938 of a Congress ever more reluctant to support Roosevelt's programs postponed any thought of increases in industrial productivity for defense or in military budgets. Although the Army executed rather than determined U.S. military policy, it did contribute to planning for a possible national emergency with its protective mobilization plan of 1937. This plan posited a structure for individual and unit training as manpower levels rose to the several million men to be inducted from the civilian population after the mobilization of the National Guard, Regular Army, and Reserve forces.[71]

At this time, Maj. Gen. Frank M. Andrews, who in the eyes of many airmen had done much to solidify, strengthen, and modernize flight training and make it more responsive to wartime demands, was nearing the end of his tenure. During the summer of 1938, his last as head of the GHQ Air Force, his airmen came together for maneuvers in the Northeast — 300 airplanes and 3,000 officers and men headquartered at Mitchel Field, Long Island — to carry out war games directed toward defense of American industrial centers.

On the continent, British Prime Minister Neville Chamberlain, in what would be a futile hope to maintain "peace in our time," met with the German Fuehrer. Shortly after their ineffectual Munich agreement, riots erupted in Austria, and Jewish homes, businesses, and synagogues were looted in Germany. As the Reich's territorial greed appeared to be unappeased, Roose-

Interwar Years

velt sent a strongly worded warning to Hitler and began the subtle conversion of his own people and the U.S. Congress to the possibility of change in the American policy of neutrality and the necessity for greater military preparedness.

Though directed toward industrial production to aid European friends, Roosevelt proposed an extravagant expansion in airplane production and manpower. The new Chief of the Air Corps Brig. Gen. Henry H. "Hap" Arnold came away from a White House meeting in late 1938[72] in which he exulted, "An authorized expansion of our officer strength in April, to 2092! And in June, an enlisted growth to 21,500!" This to support a 10,000, and then a 20,000, per year airplane production schedule in addition to an "immediate" Air Corps inventory of 7,500 combat airplanes of which half would be in reserve, and another 2,500 training aircraft.[73]

The fantastic leap in manpower and aircraft procurement promised to the Air Corps would be acted upon in January 1939. The ensuing buildup toward twenty-four combat groups would itself fail to meet the perceived emergency after the fall of France in May 1940. Force levels would continue to be revised upward. Thus, the close of 1938 marked the end of an era for the Air Corps. Arnold's glee after meeting with Roosevelt can be understood, for on June 30, 1938, he commanded an organization of 1,434 regularly commissioned officers, fewer than half that number of Reserve officers on extended active duty, and a yearly trickle of graduates from the Air Corps Training Center in Texas. At the beginning of the year, the Air Corps had 1,226 "modern" combat planes. But only the B–17 was equal in performance to aircraft already in service in England and Germany, and the Air Corps had only the thirteen B–17s it had acquired the previous year.

From the end of World War I until the expansion authorized by the President, the Army had been in slow retreat. War materiel was used, repaired, and reused; the budget process discouraged and undercut research and development on new weapons. In this environment, aviation had managed to remain relatively healthy, and by 1935, experiment and testing had brought new airplanes and equipment into the inventory. Increases in military appropriations as the new decade approached would allow greater numbers of planes to be purchased and spread throughout the schools and tactical units. Frank Andrews spoke with a decided note of authority and confidence when, in early 1938, he directed a memo to the Secretary of War:

> It must be remembered that the airplane is more than just another supporting weapon. It is a fact which has apparently been recognized by most of the great world powers, that the airplane is an engine of war which has brought into being a new and entirely different mode of warfare — the application of Air Power. It is the only weapon which can engage, with equal facility, land, sea, and other air forces. It is another

The GHQ Era

means operating in another element for the same basic purpose as ground and sea power — the destruction of the enemy's will to fight.[74]

The push for the strategic bomber was not merely a craven political concern for defending the Air Corps's status within the institution. It reflected thinking about the American experience in World War I regarding the need for an industry capable of producing equipment with which the nation could go to war.

It must be said that the responsiveness to Air Corps claims on this point did not represent a tardy recognition on the part of the General Staff, Congress, or the administration of the autonomous role of air power. Mostly, it reflected the general modernization of the Army, which was becoming more motorized and mechanized. In 1936 for example, the 1903 Springfield rifle was finally retired in favor of the Garand automatic. Acquisition of aeronautical advances was swept along by this process. More B–17s were on order. To come were new attack (redesignated light bombardment) and pursuit aircraft, as well as heavy and medium bombers. A fleet of trainers would enter the inventory. The 1930s saw great improvements in air-to-ground and air-to-air communications, in navigational instruments, and in bombsights. The tactical units displayed greater proficiency in navigation and instrument flying. Along with the rest of the Army, the Air Corps began to equip and train its people for warfare with weapons of increased firepower, range, speed, and mobility.

Members of the Air Corps had for years been demanding greater autonomy within the Army, if not outright independence from it. Yet professionalization had to precede autonomy. The former was well on its way to accomplishment by the end of the 1920s, and by the time of the establishment of GHQ Air Force, the complexity and sophistication of the air arm had increased remarkably since its infancy a scant twenty years earlier. Autonomy, at the end of the 1930s, could not be defined as independence from the parent, but as the articulation and demonstration of special, not to say unique, functions and roles. This articulation came through doctrine espoused at the Tactical School and in a rash of manuals, regulations, and syllabi that spelled out training and tactics. The demonstration came in the successful creation of a system of flight training at the Training Center, and in the less well-established follow-on in tactical units.

At the end of its third decade of existence, the pace of air training remained relatively luxurious. The Training Center inducted small classes three times a year. Each student pilot worked through primary, basic, and advanced phases, somewhat at his own pace, and then went on to a tactical unit to continue full-time training in the peacetime military. This system allowed thoroughgoing immersion in high-quality individual training. But on balance, the training program had yet to prepare its people for the skills aerial warfare would require, to turn competent, motivated, and well trained individuals into a combat air force. GHQ Air Force helped remedy that inadequacy, but the

Interwar Years

familiar American disinterest in international affairs and a preoccupation with the domestic economy did little to reinforce any moves toward modernization and combat readiness within the military. Necessary equipment and manpower were late in arriving. Thus, in the Air Corps a set of dichotomies resulted: a combat air force without service equipment, an increasingly technical service chronically short of technically skilled people, and well-established concepts of specialization in an all-pilot institution. The Air Corps remained essentially a one-skill occupation.

These impediments were to change in the future, although pilot bias would last well past the first generation of the independent Air Force. What would be unrecoverable from the interwar years was the sense of connectedness among airmen. "Looking back on it," remembered Noel Parrish many years later, "I think of it as a great deal more personal and a great deal more satisfying...than any organization I can imagine today. It was a small group of people, and we felt a little isolated from society, but we liked each other so well.... Everybody... was part of the same team."[75] For the country and the air force, this epoch was over.

Part IV

Rearming 1939–1941

IV

> Many of the cadets had never been near a plane before, but all of them wanted to fly. They came to the Army as volunteers, from the farms, the schools, the factories, the offices of America.
> — Willard Wiener, *Two Hundred Thousand Flyers*[1]

The international disarmament efforts that had punctuated the twenties and thirties had failed. The seemingly inevitable war finally began. On September 1, 1939, Germany invaded Poland. A brief, false calm fell during the winter — the "phony war," it was called. President Roosevelt proclaimed U.S. neutrality and, under terms of the Neutrality Act of 1937, placed an immediate embargo on armaments to warring nations. In his annual budget request to Congress, in January 1940 the President called for $1.8 billion for national defense.

Then, that May, German mechanized ground and air forces moved through the Low Countries and into France, quickly obliterating the French air force. The blitzkrieg that brought France to her knees in June sent reverberations across the Continent, and those on the western side of the Atlantic no longer could stand apart from the hostilities among the European powers. President Roosevelt declared a limited national emergency, and two months later Congress lifted the Neutrality Act.

The Battle of Britain between July and mid-September 1940 dramatically deepened the sense of crisis stemming from the fall of France, and the United States responded with all-out military assistance to Great Britain and passed the Lend-Lease Act early in the new year. Allied nations would not be required to pay cash for war materiel purchased from the United States. To that end, Roosevelt believed that America's chief contribution would be in supplying aircraft.

In addition to providing planes and materiel, U.S. aviation schools were made available for British and Canadian pilot and navigator training. Members of the 19th Bombardment Group began training British pilots and crews on B–17s in early 1941. Thereafter, six civilian contract schools agreed to build additional schools to train RAF pilots, and by the summer five of the schools were in operation.[2] A program conducted by Pan American Airways at Coral Gables, Florida, also trained British navigators. At the time of Pearl Harbor, the Army Air Forces had started training Chinese airmen and a handful of others from foreign nations.

American military planners had not sat idly watching the crumbling of European states without realizing the implications of war for the United States. They were already at work, intensively revising existing war plans. But they labored under conflicting national security aims and enormous political opposition to active engagement. As late as October 1938, the esteemed General George C. Marshall delivered an address at the Air Corps Tactical School in which he stated:

> It is ... literally impossible to find definite answers for such questions as: who will our enemy be in the next war; in what theater of operations will the war be fought; and what will be our national objective at the time?[3]

As to the contribution of air power or other new technologies of warfare, Marshall proclaimed on another occasion (thereby departing from Roosevelt who had faith in their deterrent effect): "We expect *too much of machines*."[4]

By the time of the German invasion of Poland that set the conflict in motion, the United States did in fact have a national objective and U.S. war plans had coalesced. They assumed the possibility of fighting a two-ocean war in which the United States would be pulled into combat against aggressor nations in both Europe and Asia. Given the continuing isolationist mood of the country, however, defense of the United States and the Western Hemisphere still dominated war planning. GHQ Air Force had already demonstrated that the range of the B–17 allowed it to protect either coast in an emergency and to reach threatened parts of Latin America and Hawaii. In April 1939 Brig. Gen. Barton K. Yount, who would head the air training organization throughout the war, expounded on this capability and what it meant to American national security:

> Throughout the world today there are on the drawing boards numerous [aircraft] designs capable of flying over 9,000 miles non-stop, but so far as we know there is today no foreign Air Force able *effectively* to bomb any point in the United States from either Asia or Europe and return to its home.... [But from] commercial air bases in this Hemisphere mass attacks could readily be launched against vital elements of our military defenses even prior to a declaration of war. Such bases could be made ineffective only by the use *by us* of a powerful Air Force.[5]

Defense entailed more than increased efficiency and combat readiness of existing military units. It meant larger expenditures for new equipment, especially aircraft. Air Corps expansion had begun in earnest with the Pres-

Part IV

ident's order in November 1938 for a 10,000-airplane production schedule to be accomplished within two years, with greater increments to follow. Congress responded with what one historian called a "golden rain" of money for aircraft procurement, appropriating $57 million for new equipment in April 1939, a supplemental $89 million for immediate expenditure, and $44 million more in contract authorization in July.[6] At this early stage, aircraft production and air training were not knitted together. Because the government earmarked much of the American production line for British and French rather than for American air forces, initially neither the administration nor the War Department considered it neccessary to balance aircraft and pilot production.

The terrible summer of 1940 brought more active wartime contingency planning, spurred on by the President's call for production of 50,000 airplanes, of which 36,500 would be allocated to the Army and 13,500 to the Navy. American military observers, including Maj. Gen. Delos C. Emmons of the Air Corps, visited Britain, confirming the view that the United States should anticipate fighting a coalition war with the immediate goal of defeating Germany, a commitment well beyond hemispheric defense. Finally codified in the summer of 1941, the joint war plan named RAINBOW 5 led in turn to the War Department Victory Program of September 1941. The victory program estimated that 215 American Army divisions would be needed to defeat Japan and the Axis. As part of the outline of new military requirements, the first war plan for air, known as AWPD–1, looked toward an air strength of 26,000 combat and 37,000 training planes by 1944.[7] AWPD–1 assumed that, in the effort to defeat Germany first, an American air force would see action before ground troops were fully engaged, and that the ultimate goal would be to subdue the German economy by bombing her electric power system, transportation network, and petroleum industry.

As the United States moved reluctantly toward belligerency, it began retooling the American economy for war production and reorganizing the military. The Army Chief of Staff brought in veteran artilleryman Leslie J. McNair to organize and train the Army for war. McNair recognized that a modern combat force had to meld basic military principles of warfare with the new technologies of mobility and striking power. He therefore directed training plans be drawn up that emphasized realistic combat training in tactical units.[8] The Air Corps followed suit, drafting training plans for crew and operational training in its tactical units.

Initially however, because it had to build a force from the ground up, the Air Corps stressed individual and specialized flying training rather than crew and unit training, cooperative training with other arms and with the Navy, or exercises. "At this phase of the Air Corps expansion," announced General Andrews, "unit tactical training has had to give precedence to individual training of pilots and mechanics. As soon as our training resources permit, we will return the emphasis to tactical combat training."[9] Since, according to a

1939–1941

revised 1939–1940 Training Directive, "the requirements of the Expansion Program are of a nature so urgent that specialized individual training must receive first priority in all units," pilot hours devoted to navigational or cross-country flying were restricted "to that required to maintain individual proficiency."[10] It was necessary to scale down the number of hours spent in unit and combat-related training in order to maintain equipment availability for the increasing number of pilots entering the program.

Furthermore, in evaluating the existing units and training requirements in mid-1940, Arnold, then Chief of the Air Corps, notified his people that no GHQ Air Force or overseas units would be built up, wings would not be reinforced, and some base squadrons would be used in the training program. Several operational stations were turned over to the training center. While these decisions indicated maximum use of resources for training purposes, Arnold considered all new training units to be provisional rather than permanent, presumably to allow for future reconversion and strengthening of the tactical units.[11]

In a series of steps under the expansion program, the functions of air planning, training, supply, and operations came increasingly under a concentrated air authority. On March 1, 1939, the GHQ Air Force, which was to become the Air Force Combat Command, was placed under the administration of the Office of the Chief of the Air Corps. June 20, 1941, marked the creation of the U.S. Army Air Forces. Lt. Gen. Henry H. Arnold, who assumed command, oversaw the Air Force Combat Command and the Office of the Chief of the Air Corps, a move that reunified the two halves of the air forces at headquarters level. Arnold had constituted an Air Staff for the purposes of coordinating planning and operations with the senior body, the Army General Staff. The Training Section under the Training and Operations Division was subdivided into Flying Training, Technical Schools, and Training Literature. Further realignment took place under the 1941 reorganization.[12]

The training establishment inevitably expanded and decentralized as the Air Corps swelled. On July 8, 1940, three regional training centers were authorized. The Air Corps Training Center at Randolph Field was redesignated the Gulf Coast Training Center. Soon thereafter the Air Corps activated the West Coast and Southeast Training Centers. The old dual authority within the air arm for individual and tactical unit training remained intact after the creation of the Army Air Forces. The Office of the Chief of the Air Corps, largely through the leadership of General Yount,[13] retained responsibility for individual training at the three training centers and civilian contract schools. The Air Force Combat Command held jurisdiction over the four numbered continental air forces charged with regional defense and the training of combat units.

From early 1939 through late 1941, the several War Department, presidential, and congressional national defense plans, policies, and budgets affecting the Air Corps were formulated through a series of escalating requirements under the rubric of the expansion program. Successive goals were set in terms of the number of U.S. combat groups to be formed and the number of pilots to be trained. The Army did not have a sound, established basis upon which to calculate manpower and aircraft strength. Because doctrine remained unsettled, the proportionate numbers of fighters, bombers, and observation aircraft and of the specialist pilots and aircrew members to be trained to fly them could not be projected easily. The Army had never accurately determined the basic calculus between pilot and aircraft production. Until the formulation of AWPD–1 as part of the victory program in September 1941, no air war plan incorporated the tactical and strategic elements upon which to base realistic estimates of requirements for combat groups to be formed, the types or quantity of aircraft, or the number of men to be trained under the expansion program.[14] Effectively, with little comprehension of what their mandate entailed, Congress and the President directed the Air Corps to increase exponentially in size and training rate, as quickly as possible.

The Air Corps responded to the ever-escalating demands with considerable speed and no little chaos, yet ultimately came close to the astronomical goals set for it. After a slow start, owing to what has been called "confusion of objectives,"[15] the Air Corps began to win steady agreement for a balanced program to include trained pilots, crews, technicians, equipment, and facilities. From the time of the fall of France, events moved swiftly, bringing a near daily readjustment of requirements for the Air Corps's expansion program, with the result that new aircraft and personnel production schedules tumbled over one another almost faster than war planners could draft a detailed blueprint for them, and certainly faster than any one could be seen to completion.

The 24-group program of April 1939 required 1,200 graduate pilots a year. To achieve that number, given the high elimination rates in flight training, the Air Corps had to recruit 12,000 applicants annually, of which 2,200 would be found to be eligible to enter the program, and of whom half could be expected to graduate. Even though pilot production requirements were to skyrocket beyond this small beginning, the measure of its difficulty could be seen by the fact that the total number of officers in the pre-expansion Air Corps was well under 2,000, all trained out of one small training center that barely graduated 200 men annually. The 41-group, 7,000-pilot-a-year program was submitted to the War Plans Division almost a year later, in March 1940.[16] This program brought Arnold much of what he had been lobbying for: a $106 million training program to include new facilities and training aircraft. The Air Corps could anticipate purchase of 800 new primary trainers, 800 basic trainers, and 600 advanced trainers.[17]

Just as the details of the 41-group program were being worked out in May 1940, German Panzer divisions rolled through the Low Countries and into France. The resulting 54-group program, termed the First Aviation Objective, called for 12,000 pilot graduates a year. For one of the newly established training center headquarters, this amounted to producing 364 single-engine and 647 twin-engine pilots, 254 enlisted bombardiers, and 1,314 gunners every five weeks, and 133 cadet navigators every six weeks.[18] A sense of the magnitude of the program was indicated by plans for achieving the pilot output at 28 primary civilian flying schools, 7 basic Air Corps schools, and 11 advanced Air Corps schools.[19] New gunnery stations, unable to provide specialized instruction until December 1941, and cadet reception centers were also budgeted. The Materiel Division estimated that the aircraft production quota under the Objective would mean an operating air force of 7,378 tactical and 6,882 training aircraft. Under the approved plan, the Air Corps was authorized procurement of over 14,000 airplanes as part of the War Department munitions program.[20]

The Air Corps was still striving to implement the 54-group program in early 1941 when the War Department authorized an 84-group, 30,000-pilot-a-year program. These figures were derived from the aircraft industry's capacity — the size of force necessary for a 36,500 airplane-a-year production rate, manufactured on full-shift schedules — rather than from operational considerations. For training purposes, meeting the new quotas meant graduating from one of the three regional training centers 455 single-engine pilots, 808 twin-engine pilots, 358 cadet bombardiers, 100 observers, and 656 gunners every five weeks, and 133 cadet navigators every six weeks.[21] Congress voted funds for twenty new flying schools, one more gunnery station, and an additional cadet reception center.[22] By the time this Second Aviation Objective was announced, the AAF was already discussing raising the training rate to 50,000 pilots by a projected date of mid-1942. The 84-group program was official when Japan shattered the American air fleet in the Pacific on December 7. Seventy of the 84 groups then were activated, but most were understrength and underequipped.[23]

Between the time of Arnold's White House meeting with the President and his advisers in November 1938 and the Japanese attack on Pearl Harbor on December 7, 1941, the United States reeled under the devastating news from Europe and reacted with increased military appropriations and ordnance requirements. The U.S. Army Air Corps training program in particular experienced unprecedented expansion as a result of the opening of the fiscal purse. Accelerated airplane production became a pivotal element in the nation's economic conversion, and air training, first for continental defense and increasingly for a coalition war against the Axis, took precedence among the activities of the Army and its air arm.

As Chief of the Air Corps, Arnold gave unmistakable priority to the training mission. In his memoirs, he recalled meetings with members of the Roosevelt administration in 1940 in which he was "still having a hard time convincing the people in the upper brackets that our training program must expand evenly and be coordinated with our airplane strength." From Arnold's perspective, "it was just as essential to have a balanced production of trained combat and maintenance crews as it was to have planes."[24] Behind his arguments lay the certainty, based on experience, of the extraordinarily time- and manpower-consuming nature of flight training, and the conviction that preparedness should mean the building of American air forces as surely as it should entail support and supply of materiel to friendly nations.

At the beginning of the catastrophic summer of 1940, the training directive still gave first priority to individual training. A year and a half later the Army Air Forces had jumped from the 41-group to the 54-group program, was officially working on the 84-group, 30,000-pilot program, and was contemplating a 50,000-pilot production program. The training directive issued November 19, 1941, reflected the shift of emphasis and the grimmer mood on the eve of war: "The objective of all training is *combat*."[25]

NINE

Individual Pilot and Aircrew Training

> Training of combat and enlisted and commissioned specialists is the bottleneck in the production of air power.... Slow though the production of airplanes may be, the production of experienced pilots will be even slower.... For the next two years our attention will be focused sharply upon the details of training.
> — Maj. Gen. Delos C. Emmons, September 28, 1938[1]

GHQ Air Force Commanding Maj. Gen. Delos C. Emmons addressed these remarks to the Army War College class of 1939–1940. In a speech at nearly the same time, his predecessor Frank M. Andrews, then Assistant Chief of Staff for Training and Operations (G–3), underscored the importance of training in the creation of balanced forces.[2] Another senior airman, Col. Carl A. "Tooey" Spaatz, after going overseas to inspect training bases and evaluate British fighter and bomber tactics, drew the same conclusion and used almost the same words as Emmons, calling training the "neck of the bottle."[3] From the beginning of the expansion program, the senior Air Corps leadership hammered away at the need to match the President's airplane production goals with the requisite numbers of trained men.

Once given the green light to do so, the training establishment came together in hurried fits and starts, mirroring the confusion of a time when grim realities overtook every new plan. Through the war years, the continental-based training system was nominally centralized under the Flying Training Command, but this organization did not arrive whole overnight. The process began in earnest after a meeting of November 14, 1938, at the White House, when the President announced his resolve of turning out 10,000 American aircraft annually, an event that Arnold later called the Magna Carta of Army air power.[4] Not five months before, the entire Air Corps numbered 20,196, but the gathering crisis and events in Europe through the spring of 1940 prompted successive forecasts that by late 1941 would produce an excess of 2.1 million men in the Army Air Forces (AAF) alone. Quotas for pilots and other aircrew members were determined in turn by the continually revised estimates of the number of air combat units required for a two-front war and the production capability of the American aircraft industry.

Planning figures for combat groups grew with each German triumph. By July 1, 1939, when Hitler had bloodlessly absorbed the remains of Czechoslovakia and two months before the German attack on Poland, the Air Corps established a 24-group program with a training requirement of 1,200 pilots a year. The graduation rate of 100 a month was to be achieved by the end of June 1941. On March 20, 1940, only weeks before the Germans invaded Denmark and Norway, a revised aviation program overtook the 24-group plan to prescribe 41 groups and a rate of 7,000 newly trained pilots a year for realization by the last day of 1941. When the German army overran France in six weeks in May and June 1940, a 54-group scheme known as the First Aviation Objective replaced the 41-group plan; pilot production figures rose to 12,000 a year in this round of expansion, which was put in train by mid-July 1940. Ephemeral too, in the event, it gave way to an 84-group program with a projected 30,000 pilots a year in March 1941.[5]

Astronomical increases in the aircraft and pilot production levels forced vast changes in the administration of the training program. Air training in the United States centered around San Antonio, Texas, with three main installations at Brooks, Randolph, and Kelly Fields. In this aggregate, known as the Air Corps Training Center, the expansion immediately changed the shape of things as well, beginning at the top with the center's commander, General Yount.

Yount, who acquired a second and a third star during the war, was hardly a recognized figure outside his command. He and the Air Corps chief graduated together from the U.S. Military Academy in 1907, the same year the Army bought its first heavier-than-air flying machine. A veteran of China service with the 15th Infantry, Yount transferred to the Signal Corps's Aviation Section in August 1917 and the next month assumed command of the School of Military Aeronautics at Austin, Texas, although he was not a qualified pilot. He learned to fly only after World War I, remaining in various command positions in the Air Corps throughout the interwar period.[6] Soon after Roosevelt stepped up the demand for airplane production, Arnold summoned Yount from the Training Command headquarters at Randolph to head the Training and Operations Division in the Office of the Chief of the Air Corps. Yount arrived in Washington in February 1939 as the Air Corps was calculating the requirements for the 24-group plan. Shortly, Yount would return to Texas to head the expanded and decentralized Training Command for the duration of the war.

Yount's headquarters division subdivided responsibility for Air Corps training among three major regional commands established to supervise activities at individual sites. Facilities lying east of the 92d degree of west

Pilot and Aircrew Training

Lines of training aircraft awaiting students stretch nearly to the horizon at the Air Corps Training Center located at Kelly Field.

longitude, a line roughly following the Mississippi River valley, came under the control of the Southeast Air Corps Training Center at Maxwell Field, Alabama. The central section of the country, lying between the 92d meridian and the 108th, the line bisecting Montana in the north and running along the Arizona–New Mexico border in the south, was under the Gulf Coast Air Corps Training Center at Randolph Field; everything west of that line fell to the West Coast Air Corps Training Center headquartered at Moffett Field in California. The Army activated the three commands on July 8, 1940.[7]

The aviation recruit's first experience with military life came in preflight. In this stint (whose length varied), he learned basic military courtesies, personal hygiene, rifle drill, marksmanship, the rudiments of code communications, and a respect for Army ways. As an aviation cadet, he was lower than the lowest enlisted rank in the Army, and as yet nowhere near an airplane. By 1943 this elementary phase took place in eleven regional centers, each of which eventually generated multiple satellites. These installations processed more than two million men for the AAF during the war.[8] From preflight, the pilot recruit moved through the now well-established pattern of flight training. Because the Air Corps had much less experience to draw upon, training programs for aircrew members were more erratic.

1939–1941

The scene at Randolph Field as several flights of student pilots approach their planes and the training day begins.

Individual Pilot Training

Already in place was a progressive system of training from cadet to combat-trained pilot: primary, basic, and advanced, followed by operational training in a tactical unit. In the "all through" method, the same instructor taught a student throughout a training stage. The expansion program accelerated the entire process and eliminated the personalized nature of pilot training. By 1940 there were too many men to train and too few to give instruction, so change in the training system came in magnitude rather than technique, comparable, as the official history of the AAF in World War II so aptly states, to a shift "in production methods from piecework to production line."[9] Otherwise, the training establishment added, as mentioned above, preflight indoctrination in cadet reception centers for some of its pool of candidates; conducted the basic phase at dedicated facilities; adjusted elements of curricula and length of courses; and in the most significant departure from previous practice, delegated the primary phase of flight training to civilian flying schools.

During the first days of the expansion program the Air Corps shortened pilot training from twelve to nine months.[10] In May 1940, all three stages reduced from twelve to ten weeks each, with classes entering every five weeks. Generally the program cut hours in military and ground school training, these

Pilot and Aircrew Training

As the demand for flying personnel increased dramatically in early 1940, increased activity became the norm at Randolph Field.

being offered in preflight at the reception centers. When the primary course shrank from twelve to ten weeks, it reduced flying hours from sixty-five to sixty and eliminated well over a third of the time spent in ground school.[11] The omitted flying time in basic and advanced training came primarily from cross-country navigation.

In October 1940, training plans to accomplish the First Aviation Objective estimated that more than 9,300 students would be in flight training at any one time.[12] By the end of the year enough pilots were in the pipeline that it appeared some primary flying schools possessing the requisite facilities could be converted to basic schools. Likewise, where basic schools' capacities exceeded requirements, they would become advanced schools. All existing and projected civil and military schools would be pushed to their maximum capacity.[13] In considering how to reach the new quotas, training officials reevaluated curricula to determine what subjects should be taught in each phase of pilot training, what should be shifted to another segment, and what should be eliminated altogether.

The redefinition of advanced training from the eve to the end of the expansion program illustrated the condensation of pilot training. The first classes graduating from the Advanced Flying School at Kelly Field did not specialize in pursuit, attack, bombardment, or observation until they joined a

317

GHQ Air Force unit.[14] At the time of the planning for the First Aviation Objective, officials considered combining advanced and specialized training, and changing the system from one of classes to one consisting of less structured stages. The approved training program announced on October 12, 1940, kept the formalized class system but moved specialized training into the advanced phase.[15] By 1942 advanced students specialized in pursuit at single-engine schools and bombers and transports at twin-engine schools.

Expansion necessitated a many-fold increase in the training pool. During the twenties and through the depression, the air arm could have coaxed more civilians into volunteering were it not for Army personnel ceilings and the Air Corps' limited resources. But the huge demands for personnel of the expansion years put an enormous strain on the recruitment process. Thus, procurement efforts redoubled. The service also discussed lowering the physical and educational requirements. While that practice was not approved officially, of necessity it occurred in fact.[16]

An external mechanism for recruitment, the Civilian Pilot Training Program (CPTP), never enjoyed much favor with the military, nor successfully made the transition from a New Deal economy-boosting endeavor to war preparedness. In early 1939 Congress passed legislation for a program of civilian pilot training, using the classrooms of American colleges and universities and facilities of flying schools certified by the Civil Aeronautics Administration. The program was of a piece with the National Youth Administration and the Civilian Conservation Corps, with which the Air Corps was already unhappily familiar. But the Air Corps was not much better pleased with the new government-funded aviation program, for reasons oddly enough best explained by President Franklin Roosevelt in his January 7, 1941, press conference:

> The Army and the Navy don't think we are getting enough out of these schools in the way of military and naval pilots . . . [I]t may mean . . . that all these people who go into these schools, largely at the expense of the Government, with thereby some obligations on their part to serve in the Army or Navy — which there never has been up to the present time.[17]

Not only did graduates of the CPTP lack military piloting skills — a major impediment from the Air Corps' point of view — but an obligation to enlist, as Roosevelt noted, was not a requirement. Therefore, the CPTP never became a very active recruitment device, even after the program shifted more purposefully toward the defense mission.

Another attempt to increase the number of applicants for pilot training, the Aviation Student Act, signed into law June 4, 1941, was short-lived legislation that allowed enlisted men to be trained as pilots. Enlisted pilots did not have to meet the educational qualification of two years of college or its equivalent, as did commissioned pilots. After completing flight training, the enlisted pilot received a warrant as staff sergeant pilot.[18] The Chief of the Plans Division wanted to begin flight training for enlisted pilots at the earliest opportunity, but he warned that "it should not be conducted so as to delay or work a detriment to the aviation cadet training program."

The first 200 students began training at the Gulf Coast Training Center on August 23, 1941. Designated classes were composed entirely of enlisted trainees, and their facilities were separate from those of the cadets who would be commissioned.[19] Segregated classes and the attitude of senior staff who had supported it reluctantly soon demonstrated that this solution to pilot procurement was fundamentally unworkable. It was contrary to the existing structure based on rank and status, and in particular the historical fact that in the prewar Air Corps most pilots were commissioned officers. To serve as a flying officer but *not* be a pilot, as happened with observers in World War I, was itself considered a more lowly occupation. Not long after the enactment of the Aviation Student Act, one AAF officer stationed overseas pointed out that the British were long accustomed to class distinctions, which made the noncommissioned pilot a useful member of an air force, but American society operated differently.

> There have been serious questions of morale among the enlisted pilots in our own service who held Reserve commissions and Reserve pilots who were on so-called "extended active duty." It is known that many of these pilots have chafed under the realization that, while they performed the tactical flying missions, they did so with status so inferior to that of the regular commissioned officer that they felt they were neither fish nor fowl. The fact that they had voluntarily assumed the status of enlisted pilot or that of Reserve pilot on extended active duty did not alter the situation.[20]

A highly structured hierarchy in which leadership and status were based on rank rather than function, although atypical in the more fluid American civilian society, was central to military efficiency and at the heart of the commissioned–enlisted pilot issue. The matter was finessed in July 1942 when the category of "aviation student" was eliminated in favor of the peculiar new grade of Flight Officer, who did not have command responsibilities. At the same time, the educational qualification for commissioned pilots was reduced. Much of the discussion about the Flight Officer Act centered on the notion that although commissioned combat pilots might not be selected for their high academic accomplishment, they were distinguishable nonetheless from other airmen, including flight officers, who lacked the ephemeral qualities of command.[21]

Even when the Air Corps found it easier to recruit pilot candidates, high attrition rates in flight training always kept the proportion of graduates to applicants extremely low, and during the 1920s and 1930s authorized quotas were seldom reached. To reduce the number of accidents and fatalities, the prewar air arm concentrated on slow, standardized training, and pushed eliminations toward the front end of the program. But training for an emergency meant many more students, too few instructors, and shortened courses. Inevitably, new pilots received less individualized instruction. Coupled with larger aircraft and bigger aircrews, the potential for substantial losses in single accidents was greater. In releasing a statement by one of its pioneer aviators, now Brig. Gen. Herbert A. Dargue, the War Department publicly acknowledged what airmen had always known. "Let us face this situation," Dargue stated, "with a calm realization that preparation for war takes its toll as well as war itself and that there is no more hazardous profession at arms than that which the combat flier has elected to follow."[22]

Nevertheless, as it developed, the rate of fatal accidents per 100,000 flying hours in the continental United States was lower between 1939 and the end of 1941 than in any preceding years or during the war.[23] The negative results of foreshortened pilot training would be seen less in high fatality rates than in poor preparation for combat. Howell Estes, assigned training duties through much of the war, admitted the fact of "people just not having enough training." But, given the pressure on the Training Command, "there you are, you've got a requirement to provide a certain level of force, and you've got to do it within a certain period of time."[24]

Primary

> Earlier... we flew with each student in accordance with his personality and tried to teach him to be the best goddamned pilot that ever came down the road. But once the expansion started, we had to do it all on a production line basis.[25]

Primary, or what was sometimes termed elementary pilot training, was not only the first step in a cadet's career, it occasioned the first major administrative adjustment in the training program. In the early hours of the expansion program, Arnold called together a group of officers to consider ways to raise pilot production. Two clear possibilities emerged. Ideally, some felt, the Air Corps should duplicate what had already worked well: build more schools on the model of Randolph Field, known as the West Point of the Air, and extend the length of classes.[26] Another group recommended that the Air Corps contract with nine private flying schools. Arnold favored the latter proposal, since the time and cost of constructing many new flying fields seemed prohibitive.[27]

The Air Corps chose to contract initially with nine of the twenty-one flying schools approved by the Civil Aeronautics Authority.[28] The schools agreed to provide facilities and training personnel; the Air Corps supervised training, determined textbooks and a standardized curriculum, and eventually furnished the aircraft. The instructional program copied the one developed at Randolph Field, with some small change in the number of flying hours spent in dual and solo. The first group of cadets to begin flight training under this arrangement,

Pilot and Aircrew Training

Class 40–A, entered on July 1, 1939, and by the following spring the first nine contract schools were in full operation. At the time of Pearl Harbor, the formula for primary training had not altered, except that the course length had declined from twelve to ten weeks, with each class numbering approximately 178 students as opposed to the originally specified 50, and the three training centers oversaw 41 civilian primary schools.

During these 2½ years, the civilian contract program expanded in the number of sites as a response to the series of revised pilot production quotas. Training officials made every effort to secure locations in the southern United States where mild weather permitted a longer training year. All of the Sun Belt did not, however, lend itself to flying. Even though the weather was warm and the land flat, flying fields in Oklahoma and Kansas, for example, all but shut down during months of high winds and vicious dust storms that tore up propellers and engines. The California schools with their year-round moderate climate and gentle topography turned out to be most congenial to the demands of nearly round-the-clock primary training.

The 7,000-pilot program was to be realized by July 1, 1941. To achieve this goal, class size increased, nine additional schools opened, and two northern schools closed. The geographic clustering of schools under three regional centers (the Southeast, Gulf Coast, and West Coast Centers) was in progress. Once established, each training center supervised an approximately equal number of civilian schools. Graduates of the primary schools then reported to basic and advanced schools in the same region.[29]

Increases in numbers of schools and graduates continued under the 12,000- and 30,000-pilot programs. The first class under the 12,000-pilot program in the Southeast Air Corps Training Center had to be postponed because of a shortage of instructors.[30] Not only were there too few instructors, those already employed at the civilian flying fields were not usually versed in military practices, as described by Willard Wiener in an early history of the program:

> Many instructors were old-timers who had been flying since the days of the Jennys (JN–4Ds). They could take a ship up and put it down anywhere in the world. But now, for the first time in their lives, they were being required to fly with precision and to teach precision flying. They were finding out that precision flying was something quite different from "just teaching a guy to fly." Precision flying involved problems some old-timers had never heard of, thought about, or ever wanted to hear about. Men with years of experience, who could fly anything with wings, had a tough time getting the knack of it.[31]

Army men supervised the civilian schools and taught the "old-timers" about military flying. Since the primary program was modeled on the methods developed at Randolph Field, the first group of military instructors trained at Randolph and then dispersed to the civilian schools to oversee those programs.

1939-1941

"Just teaching a guy to fly" was no longer good enough, as this crash of a JN-4 aircraft into a structure at Kelly Field, Texas, shows. More advanced military tactics would now be required.

Richard Montgomery, assigned to Randolph during the expansion years, described the practice and its problems:

> As we expanded, of course, we began to take in graduates of the flying school, and we brought them right back as instructors. They were students one day, and the next day they were instructors. We gave them a "souped up" instructors' course at Randolph. We would supervise the instructor school between classes, fly with them and teach them as students, what our standards were and what we required, how we did chandelles, and what our procedures were, etc. We tried to standardize their instruction.... Then people began to go out to the other two flying training commands, and it diluted what we had. It was a constant problem of training. For the most part, the civilian schools trained their own instructors at their own expense, albeit under Air Corps supervision.[32]

The situation, from the civilian perspective, looked slightly different, again to quote Wiener:

> The idea of Army men checking up on civilian veterans was a source of some of the early friction. The majority of Army supervisors were youngsters in aviation, with only a few hundred flying hours to their credit, and yet they were telling men who had flown thousands of hours how to handle a ship. There was plenty of rough going.[33]

The military–civilian partnership, though successful overall, remained uneasy.

Training officials frequently discussed the merits of three equally divided training phases. During late 1940 and into mid-1941, officers at the West Coast Center recommended shortening primary training, which some considered a "partial waste of time and effort," and lengthening advanced training. Assistant Chief of the Air Corps Brig. Gen. George E. Stratemeyer responded that the shortage of planes in all phases of training would not permit any radical changes in the schedule. Officers in the West Coast schools observed that

A military instructor checks the adjustment of the parachute leg straps on a flying cadet as he and his fellow students stand before a Stearman PT-17 while training at Randolph Field.

students became bored during primary ground school: "Too many students entered primary school with an interest only in flying an airplane and with a view that theoretical training was a necessary evil."[34] This sentiment was hardly new. Typically, cadets and officers wanted to fly more than to study.

The Army Air Corps appreciated the contribution of the civilian flying school owners, who had been quick to agree to the experiment before Congress budgeted funds to pay them. Obviously, some schools rose to the pressures of time and numbers with more ease than others. Some schools had better facilities and climate, some had more workable equipment. For example, primary schools in the Gulf Coast used Stearman biplanes that generally performed well. Although the early Stearman PT–3s hardly met the standards of the more advanced PT–13s, both were used simultaneously for a time. At the Spartan School of Aeronautics in Tulsa, Oklahoma, for instance, one class learned on the PT–3s and the next on the PT–13, and the classes continued to alternate. Lieutenant John Carpenter learned to fly at the Tulsa school after graduating from West Point in 1939. His was one of the classes assigned PT–3s with the old J–5 engine. "And boy, if you don't think it was cold in November and December in those doggone open cockpits," he recalled. "Fortunately we had face masks and a few other things. We didn't have too many instruments to worry about."[35] The PT–3s were eventually retired as obsolete, but the PT–13 biplane remained the standard primary trainer during the war years.

The Ryan School in San Diego, California, employed yet another primary trainer, the PT–16, one of the few training monoplanes purchased by the Air Corps. With their underpowered Kinner engines they were, according to training officials, a "constant source of trouble, both in maintenance and operation." The PT–16s were withdrawn from use after only two years, but during this interim the Ryan School had trouble keeping enough planes in commission at any one time to graduate its students on time.[36]

In their subsequent assessment, the Chief of the Air Corps and his senior

1939-1941

Joy shows on the faces of graduates of primary training who are moving to new quarters to begin the basic course.

staff pronounced themselves pleased with the huge contract program for primary training. Likewise, the dubious prospect of turning over the training of military pilots to men who had no background in combat flying or military indoctrination worked remarkably well, according to many young officers engaged in the process at the operational level. It seemed to Lt. Jacob E. Smart, who was in charge of the civilian program for the Gulf Coast Training Center during the expansion years, that the contract program enabled the Air Corps "to employ men whose physical condition and whose age would not admit them to the military service as pilots, but who could nonetheless serve as instructors, and very able instructors, of primary students. It was a wise course to follow, I thought."[37]

Basic

> Primary can make mistakes — Basic will correct them; Basic can not make mistakes because Advanced has no time to fiddle with corrections of technique.[38]

Lyon's comment goes to the heart of the three-phase program. During primary, potential pilots were screened for ability and a large number were eliminated. Those remaining received some military indoctrination and were introduced to the fundamental techniques and maneuvers of flying a low-powered airplane. When they moved on to basic, cadets left "kite flying," as some called primary, and started more intensive training as military pilots. That experience, according to future bomber pilot Philip Ardery, was "like a plunge into a pool of ice water."[39] Assuming he was not eliminated, the performance of each man henceforth would determine whether he went on to a single- or twin-engine advanced school, or whether he would be assigned a noncombatant role. As the Expansion Program progressed, the press of time became so great and the training program in its totality so compressed that the pilot entering advanced training had to be assigned already to a particular specialty. Thus, the basic phase became the determinative point in a pilot's career, and the air leadership felt keenly that this training must be under the firm control of military officers.

No matter which direction his career took after this point, each man entering basic faced the sine qua non of flight training: transitioning to more challenging aircraft. To some young pilots now comfortable with the slow primary trainers, the larger, faster BT–13 appeared to be, according to one airman, "a very tricky airplane — a cadet killer."[40] That reputation came more from the fear of flying higher-performance aircraft than from shortcomings of the trainers, since the BT–13s and BT–15s, which became the standard basic trainers, fulfilled their intended role satisfactorily. In fact, basic training aircraft caused the least difficulty in the training program. Basic faced fewer equipment shortages than the massive primary phase and required less specialized aircraft than did advanced. The BT–13 and BT–15 replaced the earlier BT–9 and the similar B–14, most of which were transferred to Randolph Field in 1940. The original Vultee Valiant, the BT–13, was a fixed-wing, welded steel construction, cantilevered monoplane powered with a Pratt and Whitney 450-horsepower Wasp engine. It carried a crew of two and had a 516-mile range. A Wright R–975 450-horsepower radial engine powered the BT–15.[41]

The first proposed basic schedule under the expansion program prescribed a twelve-week course with approximately 400 students entering every six weeks. Flying instruction began with an initial period of dual and solo work in fundamentals of landings, stalls, spins, forced landings and maneuvers. It was followed by a "diversified" phase to include acrobatics, accuracy approaches, strange field landings, instrument flying under the hood and with the Link trainer,[42] and night flying. Cadets took ground school classes and military training when they were not on the flight line.[43] Over time, this course became shorter, the number of students increased, and additional basic schools opened, yet the outline of the basic curriculum remained fundamentally the same.

When war broke out in Europe, the Air Corps ran one basic school, at Randolph Field in the Gulf Coast Training Center. The needs of the 7,000-pilot-a-year program forced the Training Division to reevaluate the school's capacity. The ten-week course went into effect, and it appeared Randolph could

1939–1941

Basic training included instruction in blind flying, in which the pilot's cockpit was covered with a canvas hood while a safety observer flew in the rear to watch for other approaching aircraft.

handle a maximum load of 900 students. Not until the fall of 1940 would the other two training centers (in Montgomery, Alabama, and at Moffett Field, California) share the burden. The year 1941 saw further additions. Goodfellow Field near San Angelo, Texas, began with Class 41–E on February 15; Cochran Field in Macon, Georgia, opened May 15; and the segregated school for blacks at Tuskegee Institute began November 8. Each training center also operated an "experimental" basic school that was civilian-run and therefore met some opposition from military training authorities.[44] The 30,000-pilot program anticipated further expansion, but those schools were not functioning at the time the United States entered the war.

The curriculum at the basic schools duplicated, where facilities and equipment permitted, that of the largest school at Randolph Field. It was modified somewhat as the Air Corps exploded in size and as training schedules compacted. The schools were divided into Departments of Flying, Ground School, and Military Training. Hours in the latter two reduced over time, particularly when preflight training was inaugurated in the reception centers. At the beginning of the expansion program, the number of hours in the flying phase shrank dramatically from 109 to 75. As a result of the 7,000-pilot program, air time went to 70 hours and the transition and diversified phases reduced, necessitated by the ten-week course. Navigation and formation flights disappeared from the curriculum, plus, at the end of June 1940, the Office of the Chief of the Air Corps decided that lesser amounts of equipment and time as well as too few instructors forced the discontinuance of Link trainer sessions. Flare landing practice was eliminated and the amount of time spent in night flying curtailed. Because they were made of magnesium that was used in the wartime industries and therefore in short supply, flares were expensive. Moreover, they were being discontinued in favor of wing-tip lights. Officers involved in basic and advanced training debated where night flying should fit into the program. Officials at the advanced schools claimed they were too taxed

Pilot and Aircrew Training

to supervise night flying. Besides, such a critical skill, they argued, should not come at the end of pilot training. Night flying remained part of basic.[45]

Another revision of the curriculum, again entailing some readjustment of hours spent in various subjects, came in December 1940. For the first three weeks students practiced previously learned maneuvers in the higher-powered basic training airplanes, and then started on new techniques such as power-on and power-off spins and forced landings. In the seven-week diversified period, cadets took three sessions of night flying, one hour each night split between dual and solo. They "flew" on the "Jeep," as they called the Link trainer. (This was again listed in the curriculum to be provided whenever the trainers were available.) Instrument training was central to the basic program, but students seldom received thorough instruction. During basic, one cadet in Class 40-A flew only by the needle, ball, and airspeed indicator because his instructor expressed little confidence in the early versions of the gyro horizon and directional gyros that required continual maintenance. The instructor counseled the students not to "pay any attention to these new fangled instruments, they're no good."[46]

Near the end of the basic phase, cadets began navigation and formation flights in BTs or ATs. Three-plane formations assembled, the instructor in the lead plane communicating by airplane movements and arm signals. Into the war years, pilots were still learning in this primitive way. In his basic course, John Frisbee went up in a three-ship formation only a few hours after he soloed. His instructor explained the hand signals and gave directions: "We'll fly around for awhile until you get the hang of it, then go over to an auxiliary field and shoot touch-and-go formation landings."[47] Apparently the instructor did not feel the necessity, or did not think he had the time, for a more relaxed introduction to formation flying.

Until the revised curriculum of June 1941, cadets were allowed ten hours of transition onto advanced-type aircraft. A shortage of advanced airplanes caused this provision to be dropped, and it appeared also that the basic curriculum was already too full to provide transition into advanced work.[48]

In ground school cadets studied engine and airplane operations, weather and navigation, and took "buzzer classes" (wireless telegraphy). One instructor described a typical buzzer class, which was reminiscent of kindergarten, in the Randolph cadet newspaper, *Form One*:

> I turn on the sending machines . . . and things run smoothly for about five minutes. Then, about half of them don't like their headsets, the tone is different or something, so they start milling around looking for a new position. It usually takes about five minutes to get them settled down again but after that things go fairly peacefully until the test. . . . The seven and eight word tests come off fine. In the middle of the nine word the inevitable happens. . . . Someone slams down his head set (he's missed several letters) creating such a disturbance that the rest of the class is in

an uproar. As a result we start all over again from the beginning.

Near the beginning of the expansion program, from middle to late 1939, almost 500 students registered in the basic course, and about half that number graduated. The training program would not reach its peak for another two years, but at the time of Pearl Harbor, 11,269 cadets had graduated from the basic schools and 3,183 were still in training.[49] The entire program, including the basic phase, lived with an endemic shortage of instructors, and ground classes in particular suffered. Supposedly the basic phase introduced in-depth instrument and Link trainer instruction, but some schools or classes had no equipment. Training authorities bemoaned the absence of a separate instrument flying course. On the other hand, the basic phase of the training program relied upon a reasonable supply of well-designed training aircraft, had a well-proven curriculum, and was run by experienced airmen headquartered at the most well-appointed air field in the service.

Advanced

> The acute shortage of planes suitable for advanced training seriously jeopardizes our ability to carry out our pilot training objectives.[50]

At the outset, the expansion program compressed advanced training, like primary and basic, into a three-month course; the 111 flying hours reduced to 75; and the number of hours of ground instruction dropped from 77 to 68. Specialization in attack, bombardment, pursuit, or observation shifted from the advanced phase to tactical units. Under the 7,000-pilot program, the Air Corps planned a separate five-week course of specialized training, but this scheme did not materialize before the advanced course once again absorbed specialized training.

To take care of the initial expansion, the Corps reactivated Brooks Field as a substation of Kelly for advanced training. New aircraft procurement had barely begun, so cadets trained on whatever could be scrounged. John Carpenter was among the many who trained at civilian flying schools for primary, Randolph for basic, and Kelly and Brooks Fields for advanced training. In the last stage, he and others checked out on "various and sundry things" such as BC–1s, BT–8s, and the old P–12, which, according to Carpenter, "was the greatest airplane I ever flew in my life, just a great old biplane."[51] He was also among the first group to try out the new AT–6 that became one of the most reliable training airplanes of World War II.

After completing the advanced course, the graduate of 1939 or mid-1940 joined a tactical unit that determined the mission he would fly. The reasons for eliminations and for assigning specialties had always seemed shadowy to the individuals going through the program. Now, during the buildup, assignment to fighters, observation, attack, or bombers related directly to the current but changeable production quotas for tactical squadrons. To take one man's experience as an example, Charlie Bond had mostly flown fighters during his training and been checked out in the P–12. The day after graduation when assignments were posted, he found to his astonishment that he was to report to the 2d Bombardment Group at Langley Field, Virginia. "I charged over to my instructor," he recalled, "and demanded to know why I had been condemned to bombers. He explained that there were insufficient graduates to meet the expanding needs of the bombardment units, and my name was skimmed off the top of the roster because it was in the first half of the alphabet."[52] Bond would eventually make his way back to fighters as a member of Chennault's Flying Tigers in China. But at the time he graduated, the Air Corps was under particular pressure to increase the number of bomber pilots.[53]

Bond and Carpenter were among those whose fate was decided during or at the end of advanced, rather than at the end of the basic phase. A significant change, not only in the timing but in the way the training system handled specialization came with the 12,000-pilot program. In the summer of 1940, General Arnold notified the Adjutant General that one of the planned sites for advanced training would "have a capacity of two hundred cadets in training. A class of one hundred will enter every five weeks for a ten weeks' course."[54] The course would reincorporate specialization, but it would be accomplished by streamlining the functions. Rather than training at separate facilities in the separate missions of pursuit, bombardment, attack, or observation, at the end of the basic course students would be assigned specialties more generically in an advanced course of single-engine or twin-engine training.

When the Air Corps drew up the plan that placed specialization earlier in the training sequence, it also expected students to transition into advanced aircraft during the basic phase. But by the time the new curricula were distributed in December 1940, it was apparent that what was intended to be a simpler system encompassed such vast irregularities that a number of alternate schedules had to be arranged, depending upon whether a student had received some or no transition flying or, in advanced twin-engine, whether he had any background in navigation or bombardment. Headquarters soon abandoned these complicated multiple schedules. Transition dropped out of the basic and into the advanced phase. The system never in fact became simple. Twin-engine training in particular, which included programs for pilots, navigators, and bombardiers, never completely unsnarled because twin-engine aircraft remained in such short supply that virtually none of this training occurred before the United States went to war. As a result, most pilots in advanced training flew single-engine aircraft, although there were a couple of bastardized programs, such as those at two of the California schools that mixed portions of the single- and twin-engine curricula because they possessed neither the facilities nor the equipment to do either.

At the time of Pearl Harbor, eight advanced and one flexible gunnery school were ready to open. Of the facilities already providing instruction, few

1939–1941

were operating according to the plans drawn up under the expansion program. As announced on October 12, 1940, under the 12,000-pilot program, Ellington Field was to serve for single-engine training in the Gulf Coast Training Center, and Kelly was to be converted along with a new school at San Angelo, Texas, for twin-engine training. In December 1941, although Kelly was designated for twin-engine and navigation, it provided single-engine; Brooks was designated for twin-engine and observation, but provided single-engine; Ellington had just begun twin-engine training; and the school at Victoria, Texas, was giving single-engine training.[55]

In the Southeast Air Corps Training Center, Craig Field was to be single-engine, and Barksdale and Maxwell Fields were to be advanced twin-engine schools. Only Maxwell, always called simply the advanced school, began instruction in 1940. Designated a twin-engine school in all plans, it operated as single-engine and graduated only pursuit pilots until October 1941 when it began training RAF students. The Office of the Chief of the Air Corps gave Barksdale Field, Louisiana, priority for twin-engine planes, but because of aircraft shortages it operated a dual program. Even when instructing initially in single-engine, it was seriously handicapped by having no fighter planes.[56]

This dreary litany of unrealized plans may have been even worse in the West. The Western Training Center received no twin-engine planes at all until December 1941. The school at Stockton, California, failed to meet the terms of *any* published program of instruction because it had no gunnery range for the pursuit course and no facilities for the required twenty-four hours in bombing. This school and the one at Mather Field therefore taught a twin-engine course

The Advanced Flying School at Maxwell Field, Alabama, graduated this group of RAF pilots in January 1942.

in ground school and a single-engine course in flying. Despite this peculiar compromise, they managed some twin-engine air time for instructors who flew as copilots on depot transports.[57]

Twin-engine pilot training took unmistakable precedence on paper even though it barely existed in fact through the end of 1941. Only Barksdale Field in Shreveport, Louisiana, actually turned out twin-engine pilots. Other schools possessing the rare twin-engine aircraft used their time and equipment to train instructors. Even the Barksdale graduates formed an instructor nucleus for teaching at other schools. However, the planning and methods for twin-engine training that developed at Barksdale determined the approach adopted subsequently in bombardment training after the United States entered the war.

Looking back, Lt. Gen. Earle Partridge described the unstructured conditions he encountered when setting up the advanced twin-engine school at Barksdale Field:

> We just got going with a small class, kept them for instructors and just pulled ourselves up by our bootstraps in all the schools there. This was true elsewhere in the country. I wasn't given any [timetable or deadline] at all nor told what to do, no curriculum, nothing. I wrote the gunnery manual personally. We had to teach them to shoot, and the T-6 had a gun on it. I stayed there until the next spring [1941] when I had put together a flying school complete, running, ground school, everything.[58]

This school was unusual in that it qualified pilots in navigation and bombardment.

The December 1940 curriculum called for 24 hours of transition flying, 18 hours of formation flying, 13 hours of navigation, 5 hours of instrument work, and 5 hours in night flying. Every man first had to master the unfamiliar task of controlling two engines instead of one. Cadet John Frisbee found that "going from single-engine to twin-engine aircraft with retractable gear, constant speed props, and the more complicated systems was the most difficult transition of the entire flying training program."[59] Back on the ground, in the 10 hours on the Link trainer, the student went beyond the basics to study radio range orientation, beam flying, and letdowns.[60] Thereafter he furrowed his brow over the technicalities of dead reckoning and celestial navigation. Finally, according to the curriculum at least, he qualified in aerial gunnery and checked out as a bombardier. The program of instruction of June 1941 that was sent out to the phantom twin-engine schools also required five hours of bombing practice.[61] Obviously, most of this never happened.

The curriculum for advanced single-engine training came much closer to accomplishment. The 1941 program of the Gulf Coast schools included 10 hours day and night transition in ATs and 9 hours in pursuit aircraft, 6 hours formation in ATs and 12 in fighters, 6 hours of navigation flying in ATs and 3 in pursuit airplanes, 2 hours in gunnery in ATs and 21 in fighters (10 hours

1939–1941

The curriculum for advanced training during the expansion program included ground classes in flexible gunnery (*above*) and engines (*right*).

with ground targets and 13 hours with aerial targets), and 10 hours of instrument flying and 10 additional hours with the Link trainer when available. Ground school consisted of 123 hours of which 40 were spent in military training. Although the training centers often discussed reallocation of hours, they changed little in this curriculum until after the outbreak of war.[62]

All the advanced schools experienced an acute and chronic shortage of appropriate aircraft, but twin-engine training was especially ill-favored. Pilots trained on a wide variety of airplanes, including B–10s, B–12s, B–18s, AT–7s, AT–8s, and AT–10s. The desirable B–18s were relatively rare. Lend-lease, other elements of the training program, and tactical units also demanded the scarce twin-engine aircraft. Navigation trainees, for example, competed with advanced school pilots for airplanes and instructors. At one point, the GHQ Air Force was directed to transfer twin-engine combat aircraft to the schools for transition training but the shortage of operational aircraft through 1941 made compliance impossible. That year the United States agreed to ferry aircraft destined for Britain between factories and stations in the northeast. Advanced twin-engine graduates from Barksdale Field had first call on ferrying duty and thereby gained additional flying time.

Through 1940, single-engine advanced schools used the BC–1 and also some basic trainers. In 1940 the safe and reliable North American AT–6, known as the Texan, made its appearance. This metal-frame, cantilevered low-wing monoplane with retractable landing gear became the backbone of advanced single-engine training.[63] No aircraft was failsafe, however, and the AT–6 was difficult to pull out of a spin. Nor was it well built for instrument training, which caused problems in implementing a 1941 policy that all except

Pilot and Aircrew Training

Advanced schools used the single-engine AT-6 to train men for pursuit and the twin-engine AT-11 to prepare them for bombing.

primary and single-seat pursuit aircraft should be usable for instrument training in flight. Every fighter squadron was supposed to have at least five airplanes that could be modified when needed for instrument training.[64] But the AT-6 had such a cramped cockpit that an auxiliary panel had to be mounted when it was used for instrument training.[65]

Official directives of the basic and advanced programs repeatedly listed instrument training, but in practice — in operations as well as training — instrument flying often fell by the wayside. For example, a senior Air Corps observer studying RAF tactics in 1941 reported not "a single instance of instrument take-off in fighter aircraft."[66] Training airplanes were not equipped routinely with sophisticated instruments, but this lack did not set instrument practice apart from many other vital aspects of advanced military training. Often, airmen had no live ammunition for gunnery practice nor, for that matter, ranges on which to bomb and shoot, had they had the ammunition. The most fundamental equipment was often missing since, as mentioned, very few twin-engine aircraft were assigned to advanced flight training.

The disconnect between expectation and performance owed to facility and equipment shortfalls persisted through the interwar years and into the war to come. Military men, during peace and war, often relied on ingenuity rather than standardized techniques that time and ample equipment might have permitted. But since training officers could neither afford to defer action until aircraft arrived nor wait for a body of well-proven tactics before drafting training directives, they crafted elaborate programs of instruction for a technically complex enterprise but, in fact, seldom trained realistically for combat. In sum, in the advanced pilot program, single-engine training experienced predictable problems; twin-engine training barely existed.

1939–1941

Aircrew Training

In coping with pilot production quotas, sheer numbers confounded the Air Corps. For navigator, bombardier, and flexible gunnery training, the situation was more dire. The Air Corps had only scattered experience, no administrative structure in place, and no coherent body of training literature upon which to build. During the expansion years, aircrew training programs remained discouragingly ineffectual, at best a preamble to what developed after the United States went to war. The First Aviation Objective set a goal of 4,888 navigators per year. That number stood in depressing contrast to the total of 339 navigators who had been trained by midsummer 1941. Bombardier and flexible gunnery training fared even worse. Three separate bombardier schools opened between July 1940 and December 1941, producing a total of 122 instructors and 204 cadet bombardiers. No flexible gunnery schools existed until December 1941.[67]

At the outset, the Air Corps did not step up to the line in procuring or training aircrew members, in part because it was disinclined to train many nonpilots and therefore did not expect to recruit many. However, once the need became apparent, the service found a relatively simple means of procuring nonpilot flying officers. Under the expansion schedules, the Air Corps had to bring in enough candidates to fill its mushrooming pilot training program and compensate for the high washout and fatality rates. It soon discovered a valuable use for the large pool of pilot eliminees—as navigators and bombardiers. On June 3, 1941, Congress abetted the process by approving a bill to replace the old grade of flying cadet with the new one of aviation cadet. Air Corps cadets now had parity with Navy and Marine trainees and were permitted to occupy nonpilot and nonflying billets.[68] An aviation cadet who was a pilot eliminee could be moved to an aircrew position, theoretically without a loss of status because he would remain an officer. After graduation, an aviation cadet in one of the nonpilot occupations (including nonflying tasks such as engineering, communications, and photography) received a commission as did the graduate pilot.

The Air Corps had already created pilot replacement centers to give preflight training and hold its unassigned pool of pilot candidates. Under the 12,000-pilot program, it formed a similar pool for navigators and bombardiers at Maxwell Field, consisting mostly of pilot eliminees who had already gone through the five-week Pilots' Replacement Center course and some "selected" civilians. While awaiting assignment, these men received brief training in combat crew duties and flexible gunnery. Chief of the Training and Operations Division Brig. Gen. Davenport Johnson declared it to be "obvious that this instruction will not by any means prepare this personnel to the extent that would enable them to undertake the responsibilities that would normally be assigned to them in combat units, and that a considerable amount of individual training would necessarily be required in the tactical organizations." Johnson suggested that the center be expanded, and he furnished a list of subjects for a new curriculum.[69] As a result, by June 1941 a reconnaissance school within the center offered an eight-week course, after which trainees were to proceed for five weeks of flexible gunnery. By September plans were laid for three redesignated Air Corps Replacement Centers (Aircrew).[70]

Perhaps the fact that, initially at least, cadets in training as navigators and bombardiers were mostly pilot eliminees, men who had already been accepted and received some flight training and in whom the service therefore had an investment, allowed the Air Corps to tolerate the notion of nonpilot flying officers. At the same time, the strong pilot bias, rooted in the earliest days of aviation, may have contributed to the Air Corps's dilatory implementation of centralized training programs for navigators and bombardiers. The Air Corps had always considered navigation and bombing to be among the several skills that pilots should master. The crew concept developed in GHQ Air Force units in the mid-1930s usually meant that one qualified pilot routinely performed the navigation function, for example. This individual had been trained, as had other pilots, in pilotage, navigation, bombardment, and aerial gunnery. Usually he kept up his flying hours to maintain his pilot rating and receive flight pay. At the time, the Air Corps leadership considered (but withheld approval) for training navigator-bombardiers who had not been accepted for pilot training — a change of immense cultural import finally brought to pass by wartime pressures. The Air Corps took a step in that direction during the expansion years.

In 1939 new policy dictated that a number of enlisted men would be trained as bombardiers.[71] However, by the time a bombardier school opened to train them, a pool of pilot eliminees also awaited. Moreover, the tactical units still conducted much of the bombardier training, just as before the advent of the bombardier school. The GHQ Air Force felt such a pressing need for bombardiers that it capitulated to the plan for the training centers to train and forward to the units both enlisted men and aviation cadets. But it was unwilling to relinquish its own men to the training centers as instructors, and it insisted that both enlisted and officer bombardiers should conform to the same standards of bombing accuracy.[72] The Air Corps thus created a fledgling specialty that was open to both officers and enlisted men. Well into 1941 the Training and Operations Division wrestled with the perplexing problems of selecting enlisted men, determining what educational standards should be required, whether they should be selected before or after they demonstrated any aptitude for the job, whether the standards for graduation should be the same for officer and enlisted bombardiers, and whether enough enlisted volunteers could be found in GHQ Air Force units to make the exercise worthwhile.[73]

Not long before Pearl Harbor, the pool of pilot eliminees for navigators and bombardiers began to dry up, and planners had to find new ways of filling aircrew positions. Beginning in November 1941, the Army allowed navigators to take in-grade training.[74] It also promised rapid promotions to enlisted bombardiers, the War Department being cognizant, as one officer stated, "of the importance of providing proper awards and careers for bombardiers."[75] At an October meeting, senior staff deliberated setting up a permanent career field for bombardiers. That appearing unlikely, they recommended opening up navigator and bombardier training to Reserve officers, "Branch Immaterial," and to National Guard officers.[76]

The Air Corps never capitulated entirely to the idea of training a flying officer in a single, narrowly defined specialty. It seldom had the excess manpower to do so. It assumed, for example, that all aircrew members except pilots would possess the combat skills of flexible gunnery, even though no specialized schools provided this training until December 1941. Early on, planners considered training bombardiers and navigators in a dual capacity. By combining navigator and bombardier training into a single course, aircrew members with overlapping skills could be produced by a program with finite resources. To this end, a plan introduced in the fall of 1941 envisioned gradually integrating the bombardier and navigator schools without interrupting the interim specialized training of bombardiers and navigators that was then underway to meet the First Aviation Objective. Yet, training men in two fairly dissimilar fields presented numerous conceptual and practical problems. Since the expeditious flow of individually trained specialists was not occurring either, working out the intricacies of such a scheme became moot. The Air Corps abandoned the idea of combined training in December, although it would resurface later.[77]

The appeal of multifunctional training lay in its efficiency. But when the Air Corps squeezed most all types of bombardment training into the schools at Barksdale Field, too many programs competed for scarce resources. Similarly, the reassignment of pilot eliminees into navigation and bombardment assured a ready supply of men into those specialties almost to the time of Pearl Harbor. But this streamlining, too, produced negative consequences by overlooking lowered morale and incentive brought about by the reduced status of the nonpilot. Only a very small percentage of those applying for aircrew training opted to become bombardiers. When a cadet washed out of pilot training, he could volunteer to be retrained as a bombardier, but he seldom made the choice with enthusiasm. The Air Staff — being rated pilots themselves — recognized that navigators and especially bombardiers were held in relatively lower esteem. Thus, when it launched a publicity campaign late in 1941 to attract more volunteers into nonpilot aircrew training, it made every effort "to eulogize bombardier navigators."[78]

Cross-training also failed to address the differences in aptitude required for the three specialties. New cadets selected for navigation generally demonstrated a mastery of more demanding intellectual skills than either pilots or bombardiers. Moreover, many believed high school graduates specifically tested for aptitude would make better bombardiers than would pilot eliminees. Capt. William Garland of the Department of Bombardier Instruction at Ellington Field called for an even higher standard when he asserted that "students assigned to bombardier training should be selected from the upper brackets of college students with technical background in training instead of the lowest bracket of wash-outs which has been true in the past."[79] And Capt. "Skippy" Harbold, Director of Navigation Training at Barksdale Field, took the same tack with respect to navigators: "We are more interested in getting youngsters with more educational background, who have not necessarily attempted the flight training course. We believe we can get a better quality in this manner, and also they will not have the mental attitude of washout to overcome." Maj. Dick Nugent argued the brief for the other side: "You people are getting the cream of both the eliminated pilot trainees and civilian applicants for Navigation training." As opposed to the "college boys of today who specialize in Music Appreciation, Botany, Bible Study and kindred subjects which require no Mathematics," the pilot washout went before two separate boards who judged his suitability for future training. "Secondly," Nugent counseled, "he has at least a fundamental start on his military training. He knows how to right face and left face. He knows what a salute is. These two advantages far outweigh . . . the mere fact that he lacks military piloting ability."[80] Until shortly before Pearl Harbor, the viewpoint Nugent expressed held sway.

Navigator

During the 1930s, technological advances in aircraft instrumentation and radio direction-finding, plus the humiliating but enlightening experience of flying the air mail, gave airmen a growing awareness of the importance of navigation training. Nonetheless, as late as 1940 one estimate put the total number of well-qualified pilot navigators who were actually performing this function in tactical units at eighty-five. Training had always been erratic; textbooks on dead reckoning and celestial navigation were then seriously outdated.[81]

Since it was apparent that isolationism and neutrality were fast waning in the United States, the bombardment mission lay at the heart of an air force turning towards combat. Consequently, the Air Corps needed navigators for the medium and heavy bombardment and attached reconnaissance squadrons that were scheduled for activation. To train them, an administrative structure, where none existed, had to be developed. The preparation given heretofore in the training centers and tactical units had proceeded haphazardly and failed to produce large numbers of people in a new and relatively complicated technical specialty.

Before it came to a final decision, the Plans Division submitted a report on the advisability of opening a centralized aerial navigation school — a post-graduate course for already-certified pilots. The idea harked back to earlier proposals made by the Chief of the Air Corps and Commanding General of the

1939-1941

GHQ Air Force that had not then been acted upon for lack of funds. The 1940 study laid out several additional options for navigator training: navigators could be trained in special training units at duty stations, or at a centralized navigation school, or by detailing personnel to commercially operated schools. The study implied that the Air Corps would not realize the benefit of a large expenditure of time, personnel, and money on an Army navigator school, and the Plans Division recommended that training be continued in tactical units.[82] This suggestion, wisely perhaps, was nearly the only road *not* taken. In a desperate and ultimately futile effort to reach navigator quotas, the Air Corps first contracted with civilian outfits, then established its own central navigation school, then put navigation alongside advanced pilot schools in the three training centers. More important than administrative trial-and-error, the Air Corps quickly abandoned the idea of wasting its hard-won stock of graduate pilots on another specialty; navigators would be pilot eliminees or civilians who had not entered pilot training.[83]

As with primary pilot training, the Air Corps felt it lacked the resources — in this case, expertise — to create a new program, so it turned first to civil aviation to train navigators. On August 10, 1940, the first class of aviation cadets (pilot eliminees) entered the Pan American Airways (Pan Am) navigation school at Coral Gables, Florida, under the supervision of the Southeast Air Corps Training Center. Pan Am's experience in long overwater flights in the Caribbean and Pacific seemed especially applicable to an Air Corps that had recently argued for a strategic role in coastal defense. Pan Am provided facilities (although matters such as housing remained a bone of contention between the company and the Air Corps), instruction, and aircraft. The government furnished equipment such as aircraft octants, air-speed indicators, drift meters, aperiodic compasses, altimeters, navigation watches, textbooks, aeronautical charts, air navigation forms and tables, individual navigation equipment, and parachutes. At the outset, the course lasted 12 weeks and included 240 hours of ground instruction in navigation, 60 hours of ground instruction in meteorology, and 50 hours of flight instruction.[84]

It quickly became apparent that the first class of cadets, who met the educational qualifications for pilot training but had been eliminated in the primary phase, had insufficient grounding in mathematics. Instructors asserted that the educational standards for admittance into pilot training had become somewhat lax under the pressure of the expansion program but that they should be strictly adhered to for individuals sent forward as navigator trainees. Boards were set up at Maxwell Field to examine the apptitude of the newly created pool of cadets. Students in the pool took a special course in spherical trigonometry, and in a further effort to educate cadets in this relatively abstruse field, in early 1941 the navigation course extended to fifteen weeks.

Every branch of the training program faced a shortage of instructors and aircraft. The fact that the Air Corps had so few trained navigators to provide instruction at the beginning of the expansion program compelled the service to turn elsewhere. Unfortunately, the civil program too lacked sufficient training aircraft and people. Pan Am used one Sikorsky four-engine and four Commodore twin-engine flying boats, which amounted to a ratio of ten student navigators for every training mission. There was an element of self-training, since typically one student acted as master navigator and the others critiqued his work during flight.[85]

At the inception of the expansion program, finding facilities, staffing, and developing a curriculum for a military school took longer than contracting with a civilian-run operation. But a revised 12,000-pilot program required greater output, so the Air Corps decided it would wait no longer to open its own navigation school. It seemed sensible to train navigators where there was already a supply of twin-engine aircraft (an idea easier in conception than actuality). Barksdale Field stood first in line to receive twin-engine equipment, making it the logical place for the navigator school. Planning for the military school had begun by September 1940 and training commenced that November.[86]

Initially the Army's program of instruction was an amalgamation of the Pan Am model, the experience of the 19th Bombardment Group that had provided navigation training between 1933 and 1936, and navigation training units in other bombardment groups thereafter. The tentative 10-week program of July 1940 was elongated to 12 weeks during that first year and further extended to 15 weeks in February 1941. The several revisions of the curriculum mostly tinkered with the number of hours and instructional materials in ground classes, which averaged well over 80 percent of the total hours in the course.[87] As of September 1941, the hours required to complete the 15-week navigation course totaled 469½ in ground school (202 in dead reckoning, 201½ in celestial navigation, 66 in meteorology) and 100½ in flight (56½ in dead reckoning, 12 in day celestial navigation, and 32 in night celestial navigation).[88]

As it became more sure of itself, the AAF began to move away from the Pan Am approach, which better suited commercial than military aviation, in favor of a system taught by the RAF in its advanced navigation course in Canada. A trainer developed by the post engineer at Barksdale Field and tested and constructed by the Materiel Division enhanced ground school instruction. In this simple device, the student sat on a seat attached to a rolling base. An operator simulated conditions for which the student worked out navigation problems.[89]

The Barksdale school faced some special variants on the usual training program difficulties. For one thing, after the Air Corps opened its own navigation school it had so few students in training at any time that it needed relatively few instructors. On the other hand, those it had were inexperienced since most often they themselves were recent graduates. Also, navigation training planes were intended to have three sets of controls to accommodate the

1939–1941

Navigation ground school included instruction in radio transmission.

three-student teaching method. Until delivery of the AT–7, navigation training relied on whatever aircraft could be found, all of which required modification. As with instructors, the supply of training aircraft was adequate at times only because so few students were enrolled.

Finding sites for airfields where good weather prevailed — always a concern in flying — turned out to be a considerable problem for navigation training. The constant low overcast skies in swampy Louisiana obscured visibility, so navigation by dead reckoning and use of the drift meter were problematical. It seemed prudent to move the school to a more suitable location, so the Air Corps chose the new twin-engine airfield under construction at Albany, Georgia. Since schools in the Gulf Coast and West Coast Training Centers were also designated for twin-engine training, they too were slated to train navigators. The Barksdale school closed in July 1941, having graduated fifty-two navigators. Graduates of the Barksdale and Pan Am schools fanned out to the three new locales. Course materials developed at Barksdale and its successor at Turner Field in Georgia were disseminated.[90]

Even so, navigator training was not yet off to a good start. The other two training centers graduated very few classes by the end of 1941. The intention to combine navigation and twin-engine pilot training augured for the selection of Ellington Field in the Gulf Coast Center. Here, too, weather conditions suggested a change, and the navigation school relocated slightly to the north to Kelly Field. Presumably, the drier central Texas air would be an improvement. Instruction began August 2, 1941.[91] The first class under supervision of the West Coast Center entered at Mather Field on the same date as at Kelly Field. Upon graduation, all were assigned to the Ferry Command four-engine school or the Air Force Combat Command, both located at Albuquerque, New Mexico. The second class, which graduated December 6, 1941, likewise was assigned except for ten men who remained in California as instructors at Mather Field.[92]

In late 1940, Maj. N. B. Harbold, who had helped set up the navigation programs at Barksdale and Turner Fields, prepared a detailed study of navigation training within the Air Corps. Harbold was one of only a couple of officers whose experience dated to the inception of navigation training in the 19th Bombardment Group in 1933. Regarding the newest efforts, he cited administrative problems, particularly the lack of coordination in disseminating information, and the shortage of equipment at the training centers and in GHQ Air Force units. As a result, Harbold stated, the service had not standardized training requirements for the rating of aerial navigator. He suggested that a more centralized authority within the Training Division should oversee navigator training, and that the navigation schools should be separated from twin-engine pilot training.[93] At that time, and for some while thereafter, the Air Corps lacked physical plants, instructional resources, or requisite equipment to develop navigator training along these lines.

In the summer of 1941 the Training and Operations Division queried the field regarding the efficacy of navigation training. Few respondents requested fundamental departures from techniques listed in navigation school curricula. Air Force Combat Command, for example, required graduates to be able to fix their position celestially within twenty miles of actual position, so the command had little quarrel with the school standard that called for an even higher level of accuracy. But units suggested that graduate navigators should become more familiar with charts and catalogues and accumulate more experience in taking and plotting radio bearings and establishing findings from them.[94] Expressed needs varied from unit to unit. Medium bombardment squadrons mostly employed dead reckoning, and so stressed the importance of this skill; heavy bombardment demanded more celestial navigation. In other words, they cited the need for greater proficiency in techniques already outlined in training directives. Most of all, they requested more navigators.

Throughout the tenure of their collaboration, the Air Corps evaluated procedures and revised the curriculum of the Pan Am school. Chief among its deficiencies, Pan Am provided an insufficient amount of air training. At one point aviation cadets at Coral Gables logged 50 hours of air time compared with 100 hours in Air Corps schools.[95] The military was emphatic about the importance of flying experience, but it recognized that Pan Am did not have the training aircraft to comply. Despite its shortcomings, Pan Am provided an invaluable service. During the expansion years, more than 80 percent of Air Corps navigators came from the school, even though the number was halved in the spring of 1941 when Pan Am also opened its doors to the RAF. Shortly afterward, the Air Corps extended its own navigation program into the three training centers.

1939-1941

All in all, the military navigation schools established to the time of Pearl Harbor were poorly sited, understaffed, and almost nonequipped. Navigators had little opportunity for adjunct crew or gunnery training. The Air Corps had begun to address some difficult personnel issues and draw up comprehensible and well-justified programs of instruction, but as one training officer sensibly stated, "the effect of inadequate flight experience cannot be counterbalanced by improved Programs of Instruction."[96]

Bombardier

Supposedly, the Office of the Chief of the Air Corps was responsible for individual training of pilots and aircrew members. Tactical squadrons, on the other hand, were only to conduct combat or unit training and proficiency checks. At the beginning of the expansion program, most training resources were devoted to individual training because the tactical units were neither equipped nor staffed adequately. But even in individual pilot training, which was the primary focus of early expansion efforts, the Air Corps could not build facilities, train instructors, and procure aircraft fast enough to meet the escalating requirements. Navigator training lagged much farther behind, as just described, and in the case of bombardiers, despite official plans, most individual training took place in tactical units.

The formal effort by the Office of the Chief of the Air Corps got under way at Lowry Field near Denver, Colorado, which trained three classes of instructors beginning in July 1940 and graduated its first class of bombardiers in April 1941. The bombardiers reported to B–17 squadrons, and the instructors joined the staff of the new bombardier school at Barksdale Field, Louisiana, that opened in May.[97] At the same time, tactical units were sufficiently eager to have bombardiers for crews being activated that they were willing to continue individual training of enlisted and officer bombardiers even while the Office of the Chief of the Air Corps set up a school. Thus, a handful of bombardiers began training at Lowry Field, and GHQ Air Force "coincidentally" trained the rest. Despite the establishment of school programs at Lowry and Barksdale, fairly standardized bombardier instruction operating throughout the training system did not commence until after the United States went to war. The domestic air forces continued to train many of the bombardiers, and in fact this arrangement continued to a lesser degree until the end of 1943.[98]

In March 1940 the GHQ Air Force published a curriculum to guide its units in training enlisted bombardiers. But without detailed training manuals and dedicated facilities, little standardization was possible. Reports from the field spilled over with frustration. For example, during the 1939–1940 training year, officers from the 2d and 25th Bombardment Groups at Langley Field pointed to inadequate bombing ranges and the fact that they possessed only one aerial camera.[99] Faced with these complaints, one officer from GHQ Air Force headquarters figuratively threw up his hands and stated the obvious: "The

Pilot and Aircrew Training

acquisition of additional bombing ranges and the establishment of central schools for training bombardiers and bombsight maintenance personnel should facilitate individual and unit training in all bombardment units."[100]

The experience of the 17th Bombardment Group (Medium), which converted from attack to bombardment in 1939, illustrates both the accomplishments and failures in one GHQ Air Force unit. When based at March Field, the group enjoyed an advantage shared by few other units, a superb range at Muroc Dry Lake, about eighty miles away. It borrowed three officers from the 19th Bombardment Group and two qualified enlisted bombardiers from the 38th Reconnaissance Squadron as instructors. The first class of enlisted bombardiers began ground training on the M–1 bombsight in October 1939 and dropped its first practice bombs a month later. The initial success was not a predictor of others to follow. Another squadron was equipped with the O–1 bombsight, but in this case none of the instructors knew how it worked. Night bombing that commenced during the spring had to be curtailed owing to the lack of spotting equipment. In mid-June 1940 the group discontinued bombing training altogether when it moved to McChord Field in Tacoma, Washington.[101]

It was about this time, as mentioned, that the Office of the Chief of the Air Corps started training bombardiers in a ten-week instructors' class at Lowry Field in Denver, Colorado. Beginning with a group of pilot washouts, three classes graduated 122 men between July 16, 1940 and March 15, 1941. Thereafter a "test" class of cadet bombardiers graduated 34 men.[102] This effort generated a curriculum that, although revised in some details and not routinely implemented, served as the basic bombardier course:[103]

Bombardier Ground Training	Course Requirement
Laws of physics	4 hrs
Theory of bombing	8 hrs
Theory of bombsights	4 hrs
Bombing technique	4 hrs
Military instruction	60 hrs
Electricity	4 hrs
Bombing trainers	12 hrs
Instruments and their calibration	6 hrs
Use of computers and conduct of missions	6 hrs
Forms	8 hrs
The "M" series bombsights	40 hrs
Gyroscope	8 hrs
Elementary navigation	8 hrs
[Automatic flight control equipment]	44 hrs
Scoring method	8 hrs
Bomb rack control, bombs, and fuses	16 hrs
Causes of errors and analysis of results	8 hrs

1939-1941

Train, formation, through-cloud, and overcast bombing; bombing with assumed defective bombsight	8 hrs
Theory of probabilities and bombing accuracy	12 hrs
Bombing tactics	12 hrs
Pedagogy	8 hrs
Examination on listed subjects	60 hrs
Total ground hours	348 hrs
Bombing Trainer Practice: Air Instruction	
Dry runs	40 per student
Bomb releases	200 per student

The Air Corps had always been uneasy about combat personnel who were not officers. The prewar conversion to a combined enlisted and officer aircrew was bumpy, especially so in bombardier training. Field commanders argued that officer and enlisted bombardiers should be held to identical proficiency standards. Yet the Office of the Chief of the Air Corps attempted to distinguish some elements of the curriculum that were applicable to each group. In September 1940 it notified the Southeast Training Center, which had been made responsible for programs of instruction and texts for bombardier training, that officer bombardiers should be qualified as Air Force Reconnaissance Observers. To this end, ground classes should include such subjects as antiaircraft defense, employment and organization of ground forces, employment of naval forces, aerial photography, maps, and codes. It was not practical to set aside an additional block of hours for air training, but when possible, provision should be made for "reconnaissance training incident to normal bombardier training missions." To meet the requirement, the training center proposed to use the time scheduled for gunnery, training that would be transferred to a flexible gunnery school prior or subsequent to bombardier training.[104]

A technique used in ground school for bombardiers was practice firing on electrically activated targets from a high chair that simulated the speed and drift of a plane in flight.

Pilot and Aircrew Training

Having established some guidelines that would govern bombardier training, the Training Division decided to transfer future classes to Barksdale Field where it presumed weather conditions would be satisfactory. The first class began training May 3, 1941, but the bombardier program replayed the experience of navigator training. The school found the climate unsuitable, and it transferred again. As one training center historian stated frankly: "The school at Barksdale began feebly, soon bogged down in a dismal fashion and was finally removed from Southeast Air Corps Training Center's control in November, 1941."[105] The painful recital of the Barksdale school experience and that of its brief successor at Ellington Field summed up bombardier training to the time of the attack on Pearl Harbor.

The original plans suggested locating schools at both Ellington and Barksdale Fields where bombardier training could employ existing inventories of B–18s. In the event, Ellington only opened as a bombardier school when Barksdale closed, and it never had enough B–18s. For a time, whether classes would be composed of officers or enlisted men remained unresolved. First it appeared that the school would train enlisted men; then as part of the curriculum revision of July 1940 the Office of the Chief of the Air Corps decided it should also train officers. Therefore, part of the course had to include hours in reconnaissance observation, which in turn meant relocating the flexible gunnery course. The bombardier school was "designed to function as a branch of the advanced twin-engine school," yet Barksdale never received enough twin-engine trainers to serve the multitudinous activities taking place on the base. What aircraft did arrive had to be shared among all the programs. Also, the field had difficulty obtaining bombing ranges. The flow of students from the aircrew replacement center got "badly out of joint" with the schedule of classes at the bombardier school even though there appeared to be an adequate supply of candidates from civilian applicants and pilot eliminees.[106]

As occurred in all aspects of the program, bombardier training was shorthanded on instructors, and those employed often were poorly trained. Some twin-engine pilots still in training were assigned to fly student bombardiers, but this solution appeared unsatisfactory from the perspective of both pilot and bombardier. Headquarters considered loaning additional twin-engine pilots from the Combat Command. But the Combat Command claimed it could not spare pilots: "There is, at present, a decided shortage of two-engine pilots.... From this meager number, it is necessary to develop and train 13 bi-motored combat groups, furnish ferry pilots ... and other extraneous activities [including] field maneuvers with ground forces.... It can readily be seen that it is impossible to meet [the] request."[107]

Despite numerous obstacles, training officials made progress in their simultaneous efforts to establish a bombardier training program and evaluate and revise it for a standard curriculum. In the process, the Training and Operations Division received detailed reports and data, along with "comments and recommendations as to corrective measures to be taken for any unusual trends indicated," which allowed the division to publish "corrective instructions to the schools."[108] It based the program of instruction on the original instructor's course at Lowry, but it lengthened the course from ten to twelve weeks and allocated a greater number of hours to air training. The last classes spent 85 hours in the air, 373 in ground instruction, and made 145 bomb releases in the qualification phase and 55 in the combat or tactical bombing phase.[109] Some useful new techniques evolved such as a method of low-altitude bombing using an E–6B computer that required double drift solutions. It reduced the amount of time spent on a bombing run.[110]

Barksdale Field was heavily oversubscribed throughout the expansion years because, at one time or another, it hosted the bombardier school and the bombing approach section (pilots who flew bombardiers) as well as the twin-engine pilot and navigator schools, all the while providing unit training. Everybody vied for the precious multiplace aircraft. The commanding officer of the Southeast Training Center concluded that "demands made on the Bombardier School are increasing in scope although the means to accomplish the requirements are being decreased."[111] As of the week ending October 18, 1941, for example, 80 bombardiers were under instruction, but of the school's 9 assigned airplanes equipped for bombing, only 5 were in commission. Effectively, the ratio of students to aircraft in commission was 16:1. According to one report, the "demand for airplanes for bombing missions has been increased to the point where the ships are used continuously throughout the twenty-four hours, until mechanical maintenance or prescribed inspections force them to be removed from schedule." This was particularly true for the B–18A, which was equipped for greater bombing accuracy than the B–18.[112]

Bombardier training used the B–18 series almost exclusively. But there were only 197 of them throughout the continental United States, and only 25 were officially assigned to "training activities." Apparently GHQ Air Force loaned an additional 25 to Lowry Field for instructor training. In March 1941 the AT–11 went into production; when available, it would be used thereafter for bombardier training.[113]

Bombsights were in as short supply as aircraft. Students were introduced to the Sperry and Norden precision sights as well as to nonprecision and some foreign-made bombsights. The shortage of equipment, and the fact that bombardiers might practice with bombsights that were not available in the tactical units to which they were assigned, caused an obvious disjointedness between training and operations. Training units generally employed the Sperry bombsight at a time when the B–18 was the principal bombardment training airplane. But most airmen found the Sperry to be unreliable. Moreover, the soon-to-become standard trainer, the AT–11, was not engineered to use it.[114]

The Barksdale school began dismally, as its historian confessed, and it ended dismally. The heavy overcast conditions that caused the Air Corps to

Pilot and Aircrew Training

Barksdale Field attempted to host both a bombardier and a navigator school while providing twin-engine pilot and unit training as well. Competition for resources was extreme, and the achievement of training goals proved elusive.

close the navigation school also occasioned the move of the bombardier school. The Training Division asked each of the training centers to survey its existing and planned facilities for one that might be favorable to bombardier training.[115] The Air Corps selected Ellington Field, and the first class of twenty-six graduated the end of December 1941, two months after the school opened. This was the only class to graduate from Ellington before the school moved once again in search of better flying weather, this time to Albuquerque, New Mexico. At the time of the attack on Pearl Harbor, the three successive schools had harvested a crop of 122 instructors and about 200 bombardiers.

Gunnery

Little of note can be said about aerial gunnery training prior to Pearl Harbor except that the Air Corps made a start on it and accumulated information that would be put to productive use in the future. Fixed gunnery was taught in the pursuit course in advanced single-engine schools. The flexible gunnery program had to be much bigger because heavy bombers carried a crew of eight,

1939–1941

and light bombers, a three-man crew, all of whom except the pilot were supposed to be competent gunners.[116] Yet navigator and bombardier schools provided almost no flexible gunnery training, and specialized gunnery schools did not open until December 1941. Until that time, the Office of the Chief of the Air Corps evaluated potential training equipment, surveyed its people for ideas and current practices in gunnery training in tactical units, and worked closely with the RAF to establish a gunnery program that drew upon European combat experience.

In connection with the revision of TR 440–40, in July 1940 Col. Walter R. Weaver, then President of the Air Corps Board, posted a frank and negative assessment of the competence of aerial gunners. Reports from those with experience in tactical units caused Weaver to conclude that the "standard of proficiency of aerial flexible gunners in the Air Corps is extremely low and that, unless some drastic remedial action is taken immediately, there is but little prospect of improvement in the near future." Discussion with a number of officers, Weaver went on, "indicates clearly that the gunnery training system as a whole is at fault, and that corrective action within the framework of the present system is a practical impossibility." Training under the existing regulation did not qualify a man as a combat gunner, and the situation was not likely to change as long as gunnery training remained the responsibility of the units. All recommended the establishment of a flexible gunnery school.[117]

By the fall it appeared that the Air Corps would have the funds to establish two gunnery schools. The Chief asked the Southeast Air Corps Training Center to assess possible sites, equipment, and personnel needs, and to put together a tentative program of instruction.[118] According to the training center historian, "it became clear almost immediately that there was an inadequate store of knowledge in this headquarters on the subject of flexible gunnery." Help came from Lt. Col. Thomas M. Jervey, the ordnance officer at March Field, who drew up a proposed course. Through his effort, the Office of the Chief of the Air Corps approved a program of instruction for gunnery training in January 1941.[119]

In response to the 30,000-pilot program, organization of the proposed schools had begun by the spring. Three fixed gunnery schools would be lodged with pursuit schools, and three

Col. Walter R. Weaver

Pilot and Aircrew Training

flexible gunnery schools would be equipped with twin-engine aircraft for towing and firing and single-engine aircraft for camera gun targets. It appeared that aircraft for the gunnery schools would be dedicated to advanced single- and twin-engine, bombardier, navigator, and flexible gunnery training. Only reception centers were not included in this scheme.[120]

A proposed facility at Panama City, Florida, where the Gulf of Mexico and nearby forested land promised excellent air-to-air and air-to-ground firing ranges, was approved in April and completed in December 1941. Another nearly uninhabited site near Las Vegas, Nevada, was chosen, and troops arrived in June 1941. The school graduated three classes of instructors but had not begun training gunners before December 7, 1941. The third school, also near the Gulf at Harlingen, Texas, afforded excellent conditions for overwater aerial firing, a warm climate, and ready accessibility to rail transportation. School personnel arrived in September, but again training did not commence before Pearl Harbor.[121]

While the facilities were under construction, the Air Corps staged meetings and conferences to discuss training requirements. Two Air Corps officers went to England to study RAF schools and the employment of aerial gunners in combat. They carefully reviewed the British syllabus and incorporated elements into the tentative American curriculum, and they compared organization and proficiency standards of the two air services. Maj. William L. Kennedy prepared a detailed description and analysis of thirteen RAF training aids. Training officials also conferred with the Materiel Division regarding training devices available from manufacturers in the United States, inspected a moving target track installation being built at Fort Eustis, and suggested improvements on the Waller trainer being built in New York City.[122] From these investigations the Training Division formulated an equipment list and program of instruction for flexible gunnery.

The first course outlined in 1940 was to be four weeks long. The 1941 course that went into effect after a year's intensive planning added a fifth week. Students were to spend the first two weeks in classroom instruction and turret operation; the third week in sighting and lectures on matters such as range estimates, repairing guns, aircraft recognition and camera gun drill; and the last two weeks in ground and aerial firing at stationary and moving targets.[123]

The final stage, operational training, brought pilots and aircrew together.

TEN

Operational Training

It is desired that every means be utilized to bring the units... to a high state of combat proficiency as soon as possible. To insure the early combat training of all units it is considered essential that any duties and training, including specialized ground training, not contributing directly to the attainment of combat efficiency be held to an absolute minimum.

—Maj. C. E. Duncan, April 25, 1940[1]

Thus, in a directive of April 1940, the GHQ Air Force Commander was notified of an evolving shift in the training goal. The successive pilot programs always authorized larger numbers of units for activation than the service could fill with people or aircraft. At the time Arnold sent his unmistakable directive, his staff was explaining the Air Corps's personnel predicament to Senator Henry Cabot Lodge:

> The present program has expanded the number of combat units in the Air Corps probably threefold, and to meet the requirements for increased activities of the Materiel Division, the Training Center, and provide the necessary experienced personnel for supervising the build-up of the new combat units, has resulted in spreading the experienced personnel of the Air Corps so thin as to make the present program the maximum effort which can be undertaken without greatly increasing the casualty rate in the Air Corps and without the danger of considerable wastage in the expenditure of funds.[2]

Expertise was at a premium and the units would have preferred fully trained aircrew members when they arrived at their squadrons, but the training system could not yet meet the demand. As a result, operational groups were hard put to meet Arnold's challenge when they were forced to spend time on individual as well as unit training.

The situation pointed to the dilemma of an organization on neither a peacetime nor wartime footing. Airmen recognized the ambiguity. In mid-1941, for example, a few Army air officers went to Britain to observe training and combat practices. They left a country girding to fight and arrived in one that had recently held out against the harrowing night bombing of London and other cities and the withering losses of aircraft and men during the Battle of Britain. The different organizational requirements of an air force at peace and at war became clear instantly. "For training purposes in peace-time conditions under our present [American] system," one report stated, "it is believed not so necessary to allow too great a degree of latitude to Squadron commanders.... In wartime use, it is believed as great latitude as possible should be given Squadron commanders."[3]

The American air training establishment was clumsily but rapidly restructuring itself in preparation for a war in which it was not yet engaged. Theoretically the training function split neatly: one component of the air arm was responsible for individual training of pilot and nonpilot airmen, and another, for combat training of fully qualified aircrews. But in fact, for some time both the tactical units and the Training Center schools provided individual training for pilots and aircrew members, and each approached the task differently. The Training Division aimed for standardized, predictable methods and curricula that could most efficiently and safely produce the greatest number of graduates. It relied on structure, having neither the time nor the personnel for a highly individualized program of instruction. The tactical units, on the other hand, also strapped for experienced people, expected graduates of the schools to be proficient in their specialties and ready to coordinate with others as members of crews and squadrons. Speed rather than orderliness in qualifying aircrew members for the new units being activated was of the essence.

The contrasting standards of judgment of the two training agencies appeared especially glaring in the case of navigators. The issue of comparable training in Air Corps schools and tactical units came to a head in late summer of 1941 when it was revealed that the 16th Reconnaissance Squadron required 30 hours of training before a navigator was qualified in dead reckoning and celestial navigation, while at the same time the schools (theoretically) provided approximately 500 hours. It must be said that this enormous discrepancy was atypical, for the Air Corps made every effort to coordinate training methods among the various units and the schools.[4] Nonetheless, when the issue of comparability surfaced on this occasion, the Chief of the Air Staff took the position that a navigator should not be rated unless his training was substantially equivalent to that offered in the schools. The Commanding General of the Combat Command, on the other hand, argued that proficiency should be the determinant, not completion of any arbitrary number of hours. Ultimately they agreed that specified objectives should be set within tactical units and that in addition navigators should be certified as expert aerial gunners or aerial sharpshooters.[5]

Operational Training

Similarly, standards for rating pilots in tactical units were reevaluated as time went on. Eligibility requirements for qualification as first pilot (aircraft commander) in multiengine aircraft moved away from the concepts of an academic curriculum. In late 1941, passing prescribed transition courses on the aircraft employed in tactical units substituted for number of flying hours and time in service.[6]

When specialized training moved back into the advanced phase, pilots ostensibly became familiar with the job and the aircraft they would fly in tactical units before they graduated from the schools. Because the advanced schools had so few tactical airplanes, however, pilots usually were unfamiliar with existing tactical aircraft or the newest types coming into the inventory. As the product of the massive aircraft procurement program started to flow from the factories by late 1941, the operational squadrons could foresee, for nearly the first time, having airplanes built for specific missions. The tactical units, rather than the schools, were left to give transition training on these aircraft as well as the earlier models.

As in the advanced schools, tactical units found it easier to provide individual transition training to single-engine pilots than to multiengine pilots. For some time single-engine squadrons employed aircraft that predated the expansion program and were similar to those used in advanced training. Single-engine pilots transitioned onto P–26s, P–35s, and P–36s until newer pursuit models began to arrive in 1941. Three pursuit groups were partially equipped with the P–40 at that time; later in the year they received P–39s and P–43s. A transitional model between the earlier aircraft and the later high-performance planes was discussed but never ordered.[7]

Multiengine pilots, on the other hand, came to the tactical units with almost no experience flying twin-engine aircraft. Until the eligibility requirements changed, it took this group longer to accumulate the number of flying hours required to qualify a pilot for one of the several bombers then available. In some instances there were no dedicated aircraft, as was the case, for example, with newly redesignated light bombardment units that had previously been called attack groups. Many of

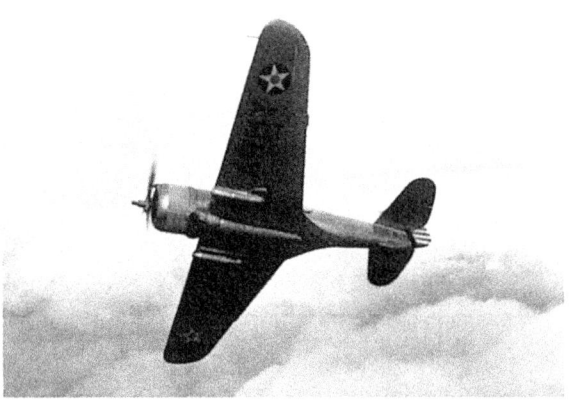

Some single-engine pilots transitioned onto P–36s during tactical unit training.

353

1939-1941

The B-18 was the most successful transition aircraft used in twin-engine tactical unit training.

the Training Center graduates began flying basic training aircraft or the BC-1 and flew as copilots on twin-engine types (usually the B-18) to meet the eligibility requirements for first pilot. The B-18 was the most successful and widely used transition airplane. In fact, it was earmarked for multiengine specialized training of combat crews in heavy bombardment, attack-bombers, strategic reconnaissance, two-engine fighters and pursuit, and cargo and personnel transport. Pilot transition training also employed the multiengine B-10, C-33, C-39, C-40, OA-8, OA-9, and A-18.[8]

To a lesser extent, navigators shared with pilots the problem of transition to new aircraft. In the summer of 1941 pilot John Carpenter of the 19th Bombardment Group began working with new navigators coming from the Pan Am school in Coral Gables. He described their adjustment to navigating in B-17s:

> They trained in a bunch of old flying boats. They took off at 65, cruised at 65, stalled at 65, and landed at 67. We got these newly graduated navigators, and they couldn't move their lines of position forward fast enough to navigate in a B-17. They had just been flying along at 65 miles an hour, sort of like driving your car, you know. Even though the B-17 speed wasn't high, it was about twice as fast as they had been accustomed to flying. It took them a while to adjust, but pretty soon they were navigating with the best of us.[9]

Through the expansion period, the newly graduated navigators were assigned to tactical units for approximately six months of training before commissioning. Instruction in the squadrons tended to be informal. Trainees were seldom supervised; they more or less rode along on piloting, bombardment, or gunnery missions. As often as not, the pilot acted as navigator on the training flights, and the navigator and bombardier tutored each other in their respective functions.[10]

Bombardiers received even less attention than navigators, and unrelenting criticism rained down upon them for a reputed lack of proficiency. A February 1941 report summarized the statistical analysis of average bombing error during fiscal 1940 and of trends in bombing accuracy as revealed in an Air Corps Board study. It concluded that "present bombing accuracy is not commensurate with either the capabilities of the equipment or the training effort involved." The cost of the training program was shockingly high considering that it resulted in the "*ineffectiveness* of a bombing force using poorly trained bombardiers." Maximum allowable error for the rating of expert bombardier had been changed early in 1940. A bombardier was expected to qualify after a short "relatively easy" initial course followed by frequent combat bombing under diversified conditions. "Unfortunately," the study continued, "target practice or qualification bombing has continued to predominate," and bombing accuracy had continued to drop. This retrogression was "further aggravated by the training of large numbers of enlisted bombardiers." The covering memorandum to the study was unforgiving in its assessment of the low quality of student bombardiers in tactical units. "It is obvious that continued training of this type of personnel is a waste of time and money, and their assignment as bombardiers to combat crews misleading, as they are incapable of hitting even large targets with any degree of certainty."[11]

In part, the woeful state of affairs reflected the lopsided emphasis on pilot training in tactical units after which, according to one early report, "training of the remaining members of the combat teams should be concurrent when practicable. When not practicable, it should be incidental until the pilot load is fully met."[12] Training multiengine pilots meant using some of the training resources of the other specialties. When John Carpenter, as noted earlier, said that navigator cadets were "navigating with the best of us," for example, he was referring to the fact that pilots too learned dead reckoning and celestial navigation and were also cross-trained as bombardiers, gunners, and radio operators. In recalling some of the methods pilots devised in navigation training in the 19th Bombardment Group, Carpenter's comments indicated the often creative, improvisational quality of tactical unit training.

> One idea was to get two or three of us down in the nose of a B-17 blindfolded. Somebody else would take off and fly for three hours. Everybody takes off his blindfold, no maps, no nothing. Where are we? You might be surprised that by the end of that summer if we could see the terrain I could fly for 10 minutes anywhere in the United States, and tell you where we were. We were flying then about 8,000 or 9,000 feet, mean

1939–1941

sea level, so we were fairly close to the surface.... Of course, most of our practice was on celestial navigation since we were faced with some long, overwater flying.[13]

Although training in the tactical units tended to be uneven, it was not without goals and objectives, and for pilot training at least, the GHQ Air Force gave it structure. The stated objective of the training directive for 1940–1941 was training and equipping units for field operations by the spring of 1941. Specialized and individual training of newly graduated pilots during a twelve-week training period received first priority. Those in light bombardment were to receive 174 hours of ground and 60 hours of air instruction; in medium and heavy bombardment, 180 hours ground and 64 hours air; in pursuit, 172 hours ground and 60 hours air; and in reconnaissance aviation, 184 hours ground and 68 hours air. Some air and ground training was common to all; hours spent on some skills varied according to specialty. Each subject listed a minimum number of hours, yet "the flying hours to be devoted to any phase or form of training is discretionary with the Wing Commander." After the twelve-week period of individual training, pilots began unit training. Here they had no set number of hours or limitations on time. It was a constant and ongoing process of flying missions aimed at combat proficiency of the entire crew. Nonetheless, given the emphasis on individual and specialized training, followed secondly by unit training, cooperative work among crew members and elements of a squadron remained nearly a postscript to be determined by Wing Commanders "within the limits of available funds."[14]

When he was preparing his 1940 training directive to the GHQ Air Force, Arnold sent Maj. Gen. Daniel Van Voorhis, commander in the Canal Zone, some general guidance for operational training. Procedures for air base defense should be included in training, Arnold urged, particularly for overseas stations that were especially vulnerable. "Airplanes should be camouflaged," he expounded. "They should be dispersed on the airdromes or on outlying airdromes when they are parked." Arnold's concern about dispersal of aircraft and the use of satellite airfields probably came from one of the earliest lessons of the European air war and also, if followed, might have reduced the damage to U.S. air forces in the first Japanese attacks in the Pacific. In its initial raid on Polish airfields in September 1939, the Luftwaffe had been unable to obliterate the Polish air fleet because it had been secreted and camouflaged on a number of airfields. But in the events that brought an American declaration of war in December 1941, the AAF concentrated its aircraft at Clark Field as a protection against sabotage.

Arnold also discussed night flying under blackout conditions: "Our present plan whereby we use full illumination of the field for landing and take-off cannot possibly be followed out in time of war... and we must step by step learn to take off with little or no light and land as nearly as we can under the same conditions." Arnold recognized that "quite naturally, however, we cannot

Operational Training

Pilot ground school included instruction in aerial manervers.

do this at once."[15] As the AAF was to learn from the British, night flying under near-blackout conditions was extraordinarily dangerous and required special equipment and training in instrument flying. "Hundreds" of British pilots were killed, reported one memorandum, having "crashed soon after takeoff due to what one experienced pilot expressed as lack of courage to go on instruments as soon as the airplane left the runway; the pilots would seek to keep a dim flare path in sight and would slip off on a wing while looking back." Installing more elaborate lighting systems became one partial means to offset the lack of systematic training.[16]

The GHQ Air Force Training Directive for the 1940–1941 training year highlighted the development of combat skills under simulated wartime conditions. It emphasized air intelligence, instrument flying, night flying, and high-altitude and cooperative missions. Training reports from the field also indicated a shift from individual training toward unit combat readiness. For instance, the 18th Bombardment Wing at Hickam Field in the Hawaiian Department placed "maximum emphasis" on the "training of stabilized combat crews to perform missions involving night bombing, night reconnaissance and instrument approaches to targets." Its headquarters staff assured the Chief of the Air Corps that "every effort will be made to exercise that ingenuity necessary to accomplish as much as possible within equipment limitations." Unit commanders exercised their own judgment because they operated without authoritative training guides and each combat group prescribed its own techniques.[17]

Specialties

Under the expansion program, new groups were spinoffs of already existing units. A cadre from the older 20th Pursuit Group, for example, started up the new 35th Pursuit Group in 1939 and, later, the 14th and 51st Pursuit Groups. Recent graduates from the schools brought each of these units to full strength.[18] In January 1939 the GHQ Air Force had two heavy bombardment groups, two attack groups, two pursuit groups, two medium bombardment groups, and four reconnaissance squadrons. The projected tables of organization called for five heavy bombardment groups plus one squadron, two attack bombing groups plus one squadron, six pursuit groups plus one squadron, and two medium bombardment groups.[19] Between April 1939 and August 1941 the number of authorized groups rose from twenty-five to eighty-four.[20]

Dive Bombing

During the expansion years the subject of dive bombing came up frequently in communiqués between the General Staff, the Air Staff, and officers of the Materiel and Training Divisions. The attention paid to dive bombing is peculiar, considering that during the war to come the AAF neither trained for nor employed dive bombers operationally to any great extent since dive bombing remained principally a Navy mission. Yet before and into the early days of war, the Army drafted curricula for dive bombing and procured dive bombers from the Navy. What amounted to a flirtation with the practice indicates that the Air Corps did not fashion its training program, initially at least, exclusively on its own doctrine and experience. It listened carefully to reports of air operations in Europe and responded also to the concepts and requirements of ground commanders.

Based upon exercises conducted during the 1930s, the Air Corps had reason to fear that bombing from high altitudes at a steep angle was accomplished at an unacceptable cost, and once into the war, its experience would confirm that judgment. The developing doctrine that took the AAF into the war stressed daylight, high-altitude *level* bombardment. Americans were convinced that they could attain sufficient precision, especially with the Norden sight, to bomb a moving target.

Nonetheless, believing that it promised considerable bombing accuracy in support of ground forces, Army Chief of Staff General Marshall insisted that the Air Corps try dive bombing. Apparently, Air Corps Chief Arnold showed little enthusiasm, although in May 1940 he cited the success of the Ju 88 German dive bomber (essentially designed as a medium bomber) that had "revived again the serious discussions of the relative merits of dive bombing as against horizontal aerial bombing."[21] Arnold requested that tests be undertaken to investigate each method. In September he prepared a lengthy memorandum for the Chief of the Training and Operations Division detailing recent British experience in flying bomber formations over German-controlled territory. He considered it doubtful that "the high altitude (above 10,000 feet) horizontal type bombing can be performed with the same certainty as we now find possible in peace time operations." He went on to recommend that all light bombers be equipped so they could be employed as dive bombers.[22] The following month, however, Arnold cited reports from British pilots who participated in light bombing missions against German troops in Holland. In these reports the AAF observer concluded that "dive bombing tactics against troops as well defended as the Germans would meet with excessive losses." The zone between 50 and 500 feet was "somewhat of a suicide area."[23] There was likewise every indication that the German Ju 88 was highly susceptible unless heavily escorted by fighters.

For all the collective War Department indecisiveness regarding dive bombing, the Air Corps continued to observe naval exercises, to set up a demonstration group to test glide bombing and low-altitude bombing against high-altitude level bombing, to experiment with attachments to the Norden sight, and to conduct dive bombing tests jointly with the armored force. The latter tests were "not considered exhaustive enough to arrive at any definite conclusions."[24] The discussion was still ongoing after the United States entered the war, but in a less intensive form because the AAF then had its own operational experiences to draw upon. The experimentation with dive bombing not only indicates the seriousness, if reluctance, with which the air arm contemplated altering its own dogma and training tactics at the time, but also confirms that it was flexible in its willingness to redirect its training practices in light of operational reports.

Light Bombardment (Attack)

Air Corps light bombardment units (formerly called attack) trained for low-level bombing missions in support of ground forces. Airmen were less opposed to low-level bombardment than dive bombing, but they doubted that low-altitude horizontal bombardment could be successfully accomplished with available technology. In mid-1940, according to one GHQ Air Force staff officer, the "basic and associated items" required for the attack mission "have not yet been supplied or apparently developed."[25] Another report laid out the technical problem as it affected training:

> Low altitude bombing is all bombing for which the bombsight cannot be synchronized.... The difference between low altitude bombing and normal bombing is that in the latter the dropping angle is calculated continuously by the bombsight and the bomb is released automatically,

whereas in the former the dropping angle must be computed and then set on the sight and the bomb is dropped manually.... The technique of low altitude bombing is so different from normal bombing that special training is necessary to master it. It is very necessary that as much of this training as possible be conducted with moving targets.[26]

The Air Corps compared the fire power of various aircraft and guns to compile data on bombing accuracy. It hoped to derive an index of changes related to different materiel and training techniques. The War Department directed that a series of combined exercises involving air and ground forces be carried out between February 11 and June 17, 1941. It planned further tests afterward to determine what types of bombs were best used against mechanized vehicles.[27]

During the 1930s the Air Corps had seen useful advances in aircraft with low-level attack capability — the A–3 and A–12. For a time, the Third Attack Group favored the B–18, but that weighty bomber was less successful than the light bombers to follow. Nonetheless, the Air Corps was not able to come to a conclusive decision. According to one assessment, the A–18 was "extremely unsatisfactory from the standpoint of maintenance and operations." Col. George Brett, then GHQ Air Force Chief of Staff, told the Chief of the Air Corps that even old parts for the airplane were hard to obtain. More to the point, it was necessary "to perform maintenance on an airplane for several days following the participation by that airplane in a single mission. As a result of these conditions, it has only been possible to average approximately forty-five minutes tactical operation a day on six airplanes."[28] Until at least mid-1940 test

A–20As on the line in 1941 await their light bombardment trainees.

flights at the Aberdeen Proving Ground were carried out with the A–17A; thereafter the flight characteristics of the A–20A were evaluated.[29]

Decisions regarding aircraft procurement and optimal fire power for attack missions could not be resolved without greater clarification of tactical doctrine, and the Air Corps had begun to recognize this gray area in its thinking. One evaluation from overseas stated a principle that would become widely recognized, namely that "slow-flying aircraft that we once visualized as being necessary for close cooperation with ground forces is no longer of any value unless superiority of the air is maintained."[30]

Heavy Bombardment

During the war, the backbone of the Army Air Force's heavy bombardment fleet would be the B–17 Flying Fortress.[31] It could navigate with the then-standard bubble sextants and with directional radio transmitters and low-frequency radio compasses that had some utility across the U.S. airways but very little over water. By 1941 the pilot and one or two copilots were training with cruise control. Eventually, lead aircraft on combat missions were equipped with the Norden bombsight that was accurate under ideal conditions for level bombing up to 30,000 feet. The M–1 (Norden) had a pilot direction indicator that told the pilot which way to turn to stay on the course set by the bombardier. Equipped with an electrically operated release mechanism, the sight reduced the time lag between the moment of decision of when to drop the bombs and their release.[32] According to pilot John Carpenter, who flew B–17s, "the defensive armament left a great deal to be desired," and the AAF would continue to have trouble with defensive gunnery. However, Carpenter thought the basic .50-caliber, air-cooled machine gun "was great.... Before the war, most of our firing was a ring and bead sight." Until hydraulic and electric turrets were installed after war began, the gunner manually directed the guns from a bathtub-shaped space in the belly.[33]

In January 1941 the Training Centers received information on the constitution of combat crews for existing service-type bombardment airplanes and of the crews being established for experimental aircraft. B–17 and B–24 crews were to consist of eight members, all of whom, with the exception of the pilot and copilot, also were expected to man the guns.[34] The GHQ Air Force had some confidence that it would soon have on hand a large number of bombers and necessary spare parts. Filling the crews with trained personnel presented the greater difficulty. At the end of May 1941, Brig. Gen. Carl A. Spaatz, Chief of the Plans Division, sent a detailed memo to Acting Chief of the Air Corps Maj. Gen. George Brett calling attention to the problem of providing experienced bombardment pilots for combat units. Seasoned airmen had to be spread throughout a rapidly expanding tactical organization. Officers who would otherwise be available for four-engine first-pilot training, Spaatz remarked, had also to be assigned to Headquarters GHQ Air Force, numbered

1939-1941

The operational training of eight-man crews required to staff the service-type B–17s of of the GHQ Air Force at Langley was carried out at Hendricks Field in Florida.

air forces, Bomb and Interceptor Commands, wings, group headquarters, and pursuit organizations. "This necessity," Spaatz pointed out, "obtains in comparable organizations throughout the Air Corps."[35]

"Due to the acute shortage of commissioned personnel in the GHQ Air Force," read the 1940 training report of the 25th Bombardment Group, "Squadron Staffs during this training year will not exceed *three* key staff officers," assuming even that number were available. Assigning full combat crews on B–17 type airplanes held first priority "at all times." The training year divided into four three-month periods to accommodate training sequences of new personnel. During the first year (or longer) new pilots took transition training on single-engine aircraft, receiving a certificate of proficiency for each type of airplane flown. Pilots going on to qualify on B–18 type airplanes at this time had to fly 100 of the required 500 hours on multiengine equipment, and those rated as airplane commanders had to become proficient in supervised landings and local and cross-country flights.

The 25th Bombardment Group established one set of requirements for pilots reporting from organizations equipped with the same type of airplanes as its own, and another for those who had not previously flown the Group's aircraft. "Normally," stated the training report, "all pilots when first reporting to this Group for duty will be assigned to a squadron, and in turn, to a Flight

Operational Training

for progressive training as Engineer-Pilots, Fire Control Officers, Navigators, First Pilots and Airplane Commanders, while receiving concurrent training as Squadron Staff Officers." Each member of the combat crew was to become proficient in his particular assignment yet, in the event of casualties, be capable of replacing others to enable the crew to complete its mission.

Men eligible for transition training as an airplane commander for B-17-type aircraft had to have been rated officer pilots for at least four years and to have flown not less than 1,250 hours, of which 300 were on multiengine airplanes and 100 were as pilot. The first week of transition training consisted of ground study, a local flight to demonstrate controls and instruments, and a conference on airplane maintenance. The second week the pilot left the academic curriculum behind and "considers the airplane as a class room." He navigated using the radio compass and drift meter and practiced targeting with the bombsight; he trimmed the airplane for flight on various combinations of engines; he received instruction in landings, flaps, and slow-speed flying; he flew bombing runs at medium and low altitudes; and he coordinated with the bombardier. The third week included an instrument training flight and instrument flight check, a performance flight to 25,000 feet and a bombardment mission from 20,000 feet, gunnery practice to illustrate the duties of the fire control officer, and formation flights. During the fourth week the pilot made a full load performance flight, night landings, and a night landing check, and he took a navigation flight that included at least two landings outside the home airdrome. After receiving a final check by the group, squadron, or flight commander and successfully completing a written questionnaire, the pilot was certified as having completed B-17 transition training.

All rated pilots were supposed to be qualified to navigate, but, stated the training directive, "the requirements of the expansion program are so urgent that *specialized individual* training must receive first priority in all units," and individual navigation flying was restricted. Pilots did, however, have to log hours in instrument and night flying and, within available range facilities and ammunition allowances, practice combat firing. Copilots were trained to the performance level of airplane commanders, but they functioned as the engineer officer on the crew. Navigators had to be qualified in dead reckoning and celestial means and were to function as alternate airplane commander. Bombardiers had to meet the requirements under existing training regulations; they operated on the crew as alternate navigators. The fire control officer was to recognize airplane and surface craft silhouettes of all nations, use and be familiar with trajectories of flexible guns and fuse bombs, supervise and train aerial gunners, and be certified as an engineer-pilot of B-17s. Each member and alternate member of the combat crew was to qualify as an aerial gunner in accordance with training regulations. A combat crew might also have a radio operator, an aerial engineer, and an armorer.[36]

Despite regulations, combat crews were assembled on the basis of available men and machines. Because pilots were cross-trained, and specially trained navigators and bombardiers were so few, pilots on bomber crews often functioned in these latter roles rather than as pilots or copilots. This practice was at times dispiriting to men who had earned their wings but were performing functions authorized to individuals eliminated from pilot training. Some Air Corps officers charged that morale was lowest in heavy bombardment units. Not only were bomber pilots frequently doing jobs there other than piloting, they were part of large teams and consequently lacked the independence and singleness of action of the fighter pilot. Furthermore, the rewards of working as a combat team were partially negated by the fact that combat crews constantly changed owing to substitutions during practice. One officer explained the drawbacks of breaking up combat crews: "This means a resulting lack of efficiency due to the fact that men are not used to working together and must expend much effort and time in acquainting themselves with each other's peculiarities before each of these conglomerate crews can take off on missions."[37] Efficiency as well as morale suffered.

The gap between policy and practice in assignment of duties could also be seen between espoused doctrine and the training that reflected it. The American fixation on daylight precision bombing was tempered by copious reports from overseas that advocated night bombing. The devastating British losses in December 1939 in the daylight encounter between RAF bombers and German fighters and the subsequent British reliance on night operations, all of which were carefully analyzed by American observers, kept alive the notion that techniques of bombing and principles of employment should not be rigidly doctrinaire, but must vary according to ever-changing situations.[38] Nevertheless, during the expansion years Americans made "no radical change in bomb sights to facilitate night bombing," according to the Materiel Division. It informed the Training and Operations Division that "optics in the sperry sight have been greatly improved to improve the vision. Both bomb sight manufacturers are studying ways of reducing internal reflections inside the sight due to cross hair illuminating lights." Meanwhile, experimentation with parachute flares showed them to be too unreliable to introduce into training.[39]

Heavy bombardment units were less able (and less inclined) to train for night bombing than for other practices such as all-weather bombing. Flying in clouds or fog, or in the western states in the haze from forest fires, had long since alerted airmen to the necessity of navigating with instruments. The 2d Bombardment Group developed a "100 percent instrument approach" for bombing large targets through overcast. They began the approach toward a visible target, dropping into the overcast and releasing the bombs at a time and place determined by dead reckoning. Thereafter they calculated the probable error based on distance from the target.[40] In May 1939 the GHQ Air Force asked its units to conduct further bombing missions under conditions of poor visibility and low ceiling. The 7th Bombardment Group undertook twenty-eight missions over the Hamilton Field water target, bombing from 4,000 feet to the lowest altitude through overcast. The group reported difficulty in accurately determining the initial point over water, and frequently pilots were unable to see a target ahead with the M–2 bombsight. They suggested improvements in the bombsight, longer-delayed fuses to clear the formation from the danger area, and additional training for the pilot and bombardier in using overcast for protection, "dropping below the overcast only for an initial course and for subsequent correction."

The recommended changes in bombing tactics were not given the attention they warranted because at that time individual training held first priority in the units. Therefore, GHQ Air Force headquarters believed that a technique in dead reckoning approach should be investigated by the Air Corps Board and 23d Composite Group. In connection with equipment improvements, the Air Corps and Signal Corps joined in an experimental project at Stanford University for using an ultrahigh frequency obstacle detector to aid in bombing through overcast. Thereafter they proposed a cooperative project with the Materiel Division for a research-and-development project for aids in bombing through overcast.[41]

The 24th and 54th Bombardment Squadrons and the venerable 1st Pursuit Squadron comprised the 23d Composite Group, which GHQ Air Force headquarters recommended for testing navigational methods. The group functioned both as a service test and demonstration organization, and as an individual and crew training unit. Like other bombardment units, it conducted squadron navigation schools and trained formations in defensive fire against pursuit and antiaircraft protection and it trained light and medium bombardment aircraft in area or precision bombing. Bomber pilots were to become versed in the operation and maintenance of fighters, familiar with the performance and tactics of other types of aircraft, qualified in aerial firing, and trained to navigate by dead reckoning and with commercial aids to navigation.[42]

Pursuit (Fighters)

The training of pursuit groups was less bifurcated than that of bombardment.[43] As with other GHQ Air Force units, pursuit groups in the early expansion program offered a twelve-week course of specialized training to pilots newly graduated from the Training Center. "Insofar as practicable and consistent with requirements," read the 1940 training program of the 31st Pursuit Group, "unit training will be continued and a state of readiness for immediate field service will be maintained." Ground training for junior flying officers consisted of 100 hours the first year and 90 hours the second. The group authorized all pilots to spend no more than 260 hours in the air; combat pilots, not less than 240 hours; and staff pilots, no fewer than 160. All pilots had to be able to navigate with maps and instruments but without radio, and with radio but without maps, and successfully to fly a course of at least 250 miles. Each man was required to fly

1939–1941

at least one hour per month on instruments and make maximum use of the Link trainer. Pilots flew night training missions and practiced acrobatics and qualified in aerial gunnery through proficiency tests rather than by meeting published training directives. They assembled for formation in the single-plane element, the two- and three-plane element, and the three-plane V; flew interception missions at least once each month; participated in patrol missions; engaged in combat exercises against pursuit, light and heavy bombardment, and observation aviation; aided in convoy or special support operations; and trained to become familiar with attacks on ground objectives.[44]

The published curriculum just described, as invariably happened, was a model for action that could not be implemented easily. For instance, one of the pursuit groups in the III Interceptor Command reported in December 1941 that conditions on the runways prevented the scheduling of any night flying that month, that the ground gunnery range was only available four days during the month, and that no tow target airplanes could be had for aerial gunnery.[45] The Commanding General of the 18th Bombardment Wing acknowledged the impracticality of detailed training directives that were virtually impossible to accomplish: "Standardization of training, through the medium of authoritative training publications, should be limited to those fundamentals generally applicable to any combat unit of a particular branch of aviation — regardless of the particular situation confronting that unit."[46]

The AAF went to war with two new fighters, the P–39 and P–40.[47] The prototype P–39 was of unique design, with a completely retractable tricycle landing gear, an engine mounted behind the pilot, and a 37-mm cannon that fired through the propeller hub. Despite these features, the P–39 never performed well and was frequently modified, although it made a name for itself in close air support. The Curtiss P–40 became the most successful American fighter in the first two years of combat.[48] Units began receiving P–40s by the fall of 1940 and used them immediately in high-altitude firing tests of aircraft armament.

The tests were given considerable weight because flying pursuit at altitudes above 30,000 feet became a high priority of the training program. In mid-1941 the Secretary of War gave notice that training at altitudes of 35,000 to 37,000 feet above sea level could no longer await the development of pressurized cockpits. Since flying at this ceiling put enormous physical demands on pilots, the GHQ Air Force ordered that only young airmen in excellent physical condition be given high-altitude training and that they be restricted to no more than ten hours of training per month. Pilots would be allowed only thirty minutes of flying time at 25,000 to 30,000 feet for the first three hours and not more than one hour per flight at these altitudes for the remainder of the month. To fly above 30,000 feet, the pilot had to have been checked for high-altitude tolerance in a low-pressure chamber. For all high-altitude training flights, he would start taking oxygen at 10,000 feet or lower. Only a relatively small

Operational Training

The P-40 fighter aircraft

number of officers were permitted high-altitude training; they would become the nucleus of instructors.[49]

Observation and Reconnaissance

Historically, aerial observation was the specialty for which the Air Corps and other arms shared training responsibilities. As a result, it occasioned constant debate over the issues of command and control. The diffusion of authority continued through the expansion years. During World War I the Coast and Field Artillery supplied nonpilot aerial observers to Air Service observation squadrons. Many of these men had been trained in fire control in artillery schools. After the war, the Army did not train aerial observers as a separate specialty; only the Air Corps Advanced Flying School conducted observation pilot training. Pilots thereafter went for unit training to tactical squadrons nominally under control of Army corps and divisions.

On March 1, 1935, Air Corps tactical units, including two reconnaissance squadrons, were brought under command of the newly created GHQ Air Force. Excluded from this reorganization were most of the observation units which were allotted to continental ground forces and overseas departments. They remained in this status under the expansion program although their number was cut from fourteen to ten. Immediately after Pearl Harbor, training responsibility for observation squadrons was removed from the ground forces. Until then, observation units performed different functions under two types of organizations, and thus the requirements for training and the aircraft employed by Army corps and GHQ Air Force units varied.

1939-1941

Representatives of the Air Corps, GHQ Air Force, Infantry, Field and Coast Artillery, Cavalry, and Signal Corps met in 1940 to study the overall needs of observation aviation. The committee recommended development of both short-range, relatively light, slow aircraft for observation, liaison, and courier missions, and long-range, twin-engine aircraft for general reconnaissance and aerial photography. The Air Corps proposed to use A–20 light bombers for the last-mentioned application.[50] A crew training school opened at the same time at Brooks Field for pilots and observers. There, the curriculum moved from an early emphasis on artillery missions toward more general, long-range reconnaissance and photography. Pilots learned photography because they might be called upon to serve as a pilot or as a photo observer. Sometimes the pilot flew alone with an automatic camera and sometimes, with a cameraman shooting with a manual camera.

Graduates of the Air Corps school reported to observation squadrons for unit training, where pilots were to be matched with nonpilot observers. Reminiscent of the World War I experience, Army ground corps drew up plans for training selected officers as air observers.[51] Field artillery commanders assigned "willing" officers to the job of observing and adjusting artillery fire from airplanes. Now however, the Air Corps was not abjectly grateful for any men that ground commanders chose to send. General Arnold disliked detailing nonrated observers to air crews because "there exists a definite line of demarcation between combat crews of Army aircraft, properly trained and rated, and passengers not possessing technical qualifications.... Detail to duty involving frequent and regular flights is essential in the case of combat crews to perform tactical missions."[52]

Officers assigned to observation were trained in programs along the lines outlined in 1940 by the 97th Observation Squadron at Mitchel Field on Long Island. The squadron's crews consisted of officer pilots and observers, and enlisted gunners and radio operators. The plane they flew in training was the O–47B. In an interview many years later, observation pilot David A. Burchinal called the O–47 a "first rate airplane." He went on to describe its employment:

> You could combine the visual with the photographic and do a pretty good job. Its main purpose was not artillery spotting, but going in, finding enemy positions and troop concentrations, and airdromes, and bringing that intelligence back in photo-form, as well as debriefing-form, and make this information available to the commander.[53]

Newly graduated officers from the Air Corps Training Center who joined the 97th Observation Squadron began a 6-week, 65-hour course that included transition to observation-type aircraft; day, night, and photographic reconnaissance; artillery adjustment; infantry and cavalry missions; day and night navigation; and instrument flying. Thereafter they trained with the rest of the squadron in artillery spotting, aerial gunnery, air navigation, formation and

Operational Training

O–47 reconnaissance aircraft in formation

instrument flying, liaison exercises, night flying, performance flights to altitude, photography, radio and visual communications, reconnaissance, and artillery adjustment. Observation squadrons were to perform cooperative missions with other branches and to take part in passive airdrome defense using local cover, camouflage, and dispersion. They spent a minimum of two weeks in field training.[54] This reasonable-sounding program of training for close air support belied the fact that few sustained operational exercises actually took place. Before he transferred to the Air Corps, Howell Estes was commissioned as a cavalry officer whose only contact with the Air Corps occurred during maneuvers when "they used to fly over and drop flour bags on us to mark where a particular element in bivouac or whatever had been attacked."[55]

The training program for a GHQ Air Force long-range reconnaissance squadron was comparable in some respects, but its focus was stronger on bombardment than on surveillance and ground support. Reconnaissance crews expected to fly larger multiengine aircraft. Pilots of the 41st Reconnaissance Squadron at Langley Field were to be "thoroughly familiar with the operation of the bomb sight and bomb release mechanism." They were to be "trained in close cooperation with a navigator." Copilots served as engineer officers but were also to be capable of taking over as airplane commanders. Other crew members were qualified as navigator-observers, trained in dead reckoning and celestial navigation, and able to function as an alternate airplane commander. The bombardier-observer acted as the fire control officer. Combat crews carried

1939-1941

a radio operator, an aerial engineer, and an aerial photographer. Besides practice in photography and low-visibility flying, crews flew overwater missions and bombed waterborne objectives. All crew members were to be qualified expert gunners. Airdrome defense was part of reconnaissance as well as observation unit training. It entailed communications, using radio as little as possible in favor of visual means, and preflight planning.[56]

In April 1941 GHQ Air Force qualification standards for reconnaissance observers included study of maps and charts, ability to transmit and receive radio communications, visual communication by lamp signals, capacity to read naval signal flags of both international and U.S. types, and proficiency in reading panels used for air-to-ground communication and other types of signals. To do so, the observer had to develop a working knowledge of naval, ground, and air force organizations, functions, and techniques.[57]

Missions, types and availability of aircraft, preferences and demands by air and ground commanders, training materials, and theories of application of fire by the Air Corps and artillery all varied, and sometimes competed, depending on whether a squadron was earmarked for observation or reconnaissance.[58] The Air Corps's relative lack of interest in observation aviation kept it from thinking creatively about tactics, and as a result, unit training remained a hodgepodge that steadfastly refused to become standardized. A number of senior officers expressed frustration over the discontinuities. One of the authors of a training circular on observation tactics admitted that it was "drawn out of a clear sky and was not complete." About the time the circular was released in 1941, after observing summer maneuvers, GHQ Commander Lt. Gen. Leslie J. McNair, who as a ground commander arguably might have understood little air power doctrine, observed presciently nonetheless, "Training and employment of observation aviation today is progressing along lines almost identical to those of 1918, and is predicated on the assumption that we will have superiority of the air, and that observation aircraft will be able to operate over and behind hostile lines without interference from either ground or air."[59]

On the Cusp

Coincidentally, on the day Germany invaded Poland — September 1, 1939 — George Marshall was named Chief of Staff of the United States Army. He was by temperament and experience well equipped to plan and reorganize the military for impending war, being mindful of the lessons from World War I when he served on Pershing's staff in France. According to historian Russell Weigley, "Marshall transformed the office of the chief of staff into [a] command post.... He mastered both the grand strategy of the war and, to an extraordinary degree, the details of staff planning, including those of industrial mobilization and logistical support." Although he was not an air power enthusiast, more than many of the Army leadership, General Marshall

Operational Training

recognized the potential of air warfare and pushed forward the modernization of the Air Corps. Some of the limitations Marshall faced were those historic to the U.S. Army that, as Weigley writes, "had grown up defending a continent it could dominate completely, but which was still new to the military perplexities of the world at large."[60] Within a very short period of time, the Army had to shake off its leisurely practices and update the obsolescent equipment that had been its hallmark over the past twenty years. The first major Army maneuvers since 1918 did not take place until April 1940.

In several respects the Air Corps experience reflected that of its parent body. Air Corps training had always occurred in isolation from ground troops, with few combined exercises that might have exposed weaknesses and sharpened tactics and doctrine. From the postwar into the early expansion period, the Air Corps officially maintained its role in military affairs to be one of coastal and then hemispheric defense. As national policy moved away from neutrality toward participation in war, the Air Corps more vocally expressed the doctrine and began to train for long-range offensive operations. Like that of the rest of the Army, the larger strategic mission was untried.

The fervor of isolationism before World War I delayed American mobilization but did not keep the nation out of war. The stinging memory of that failure of preparedness was not lost on the planners of 1939 and 1940. There is every indication that military officers well below the level of the War Plans Division knew their country inevitably would be drawn into the conflict even though the general mood of the public remained opposed. The Air Corps therefore made considerable effort to remain abreast of strategy and tactics of the air war in Europe. Although airmen experimented with practices such as dive bombing, it is not clear that information from overseas substantially tempered American air doctrine or recast training practices. The AAF went to war convinced that advance planning and its own doctrine held the keys to aerial, and possibly battlefield, victory.

Planning and preparation of the training program that took the AAF into war stood as signal achievements of the expansion years. The organizational pieces had been put in place at the time of Pearl Harbor. The strategic plans and air requirements of AWPD-1, the first war plan for air that coalesced shortly before Japan attacked Hawaii and the Philippines, brought no substantive change to the training program. Moreover, it is doubtful that the Air Corps could have mobilized faster than it did. Between July and December 1939 the Air Corps graduated 982 men from flight training. The next year the total rose to 8,125. Graduates numbered 27,531 in 1941, a total that began to approximate the expansion program projection. However, nearly all of those graduating were pilots. No navigators or bombardiers graduated in 1939. Eighteen bombardiers (all of whom were listed as instructors), 44 navigators, and 20 nonpilot observers in flexible gunnery graduated in 1940. Compared to more than 27,000 pilots turned out in 1941, the AAF produced 206 bombardiers, 137

371

1939–1941

navigators, and 117 nonpilot observers trained in flexible gunnery.[61]

The AAF principally lacked modern aircraft on which to train. The lead time for industrial mobilization was such that the United States was at war before airplane deliveries began to satisfy the needs of the training units. Until that time, lower-performance training aircraft were ordered and arrived in greater numbers than the tactical aircraft used for unit training. The slow delivery rate of heavy bombers especially vexed the Air Corps because the bombardment mission had become the cornerstone of doctrinal thinking and held pride of place in the training program. By May 1941 the President stressed bomber production at the expense of other aircraft, and the offensive outlined by AWPD–1 was predicated on B–17s and B–24s based in England, with B–29s, B–32s, and B–36s to be deployed elsewhere. At the time of Pearl Harbor the tactical units did not yet possess a sizable fleet of B–17s, and in general, few tactical aircraft of any kind were delivered before 1942.[62] Furthermore, the agreement to supply and ferry airplanes to the British additionally strained twin-engine training. Pilot, navigator, and bombardier training depended on bombers, many of whose precious number were siphoned off for export.

Because of its technical nature, the Air Corps conducted staged training programs that increased incrementally in complexity and took more time than was required to train ground troops. All aircrew members underwent extensive training in Air Corps schools before being assigned to tactical units. Like the rest of the Army training system, the Air Corps shortened most of its courses over time and accelerated the pace of training. Abbreviating school courses to meet personnel production goals had the drawback of sending partially trained men into tactical units which then had to provide individual training for new members. The shortage of instructors and airplanes that continued to the time of Pearl Harbor meant that GHQ Air Force served as both a training and an operational command.

The test of it all began on December 7, 1941.

Part V

Training for War: Planning, Procuring, Organizing

V

> I knew of no secret Red Plan, or Orange Plan, or Rainbow Five Plan, but I knew fairly well what was printed in our daily newspaper, the Chicago *Tribune*. On my eighteenth birthday, only days before the attack on Pearl Harbor, a *Tribune* article described what it called the President's blueprint for total war involving ten million American servicemen on at least two oceans and three continents.
> —John Boeman, *Morotai: A Memoir of War*[1]

By late 1941 the likelihood that the United States would join the fighting on the side of the Allies had become increasingly evident to most Americans. But it took the burning wreckage of the American air and naval fleets in Hawaii and the Philippines to catapult the country into the conflict. The Japanese attack on Pearl Harbor on December 7, 1941, set in motion a vast and initially confusing mobilization of men, requirements for new equipment, and reorganization within the military, as all the economic, political, and military resources of the United States turned to war.

In his address to Congress a few weeks after the attack, Roosevelt called for the production of incredible numbers of military aircraft: 60,000 by year's end and 100,000 in 1943. Effective two months later, War Department Circular 59 established the AAF as one of three autonomous Army commands, along with the Army Ground Forces and the Services of Supply. In accordance with the newly recognized importance of the air arm to the war effort, the Troop Basis of January 1942 set the goal for expansion of the Air Forces from 350,000 to 998,000 within a year, thereafter to reach 2,000,000, and the ceiling was raised again in August. The AAF more than met the first schedule, numbering nearly 1,600,000 by the end of 1942 and ultimately swelling in size to become nearly one-third of the U.S. Army.[2]

Mobilization and reorganization were intended to implement war plans already in place. By the time of the Pearl Harbor attack, American military

World War II

policy had evolved from hemispheric defense to contingency plans in which the United States would fight a global two-front war against the Axis. Once conjoined in common cause against the enemy, the British and Americans pursued a course that sought to defeat Germany first.[3] Air power, as the President's call made plain, would constitute a critical element, for the Allies expected to launch bombing attacks aimed at Germany's industrial base before initiating a massive ground force invasion. In accordance with the general war plan known as RAINBOW 5 that was in effect at the time of Pearl Harbor, the AAF planned to carry out its strategic mission as codified in air war plans AWPD-1 and AWPD-42. Upon these documents, with their vision of offensive air operations, planners based their anticipated force structure, numbers and types of aircraft, and training requirements.

AWPD-1, drafted before the attack on Pearl Harbor, forecast that it would take one year to produce the aircraft and train the men and nine additional months for deployment and initial combat experience. The strategic bomber force would then be poised for a massive strike against Germany. To reach that level of readiness would require more than 37,000 training planes along with combat and transport aircraft, and more than 2 million men trained as pilots and aircrew, technicians, nonflying officers, and support personnel. AWPD-42, the revision of AWPD-1 that came shortly after Pearl Harbor, proposed an increase in men and equipment, particularly air transports and bombers, and 12,232 more trainers in 1943.[4]

AAF Commanding General "Hap" Arnold's rendition of the requirements derived from the war plans, submitted to the President the late summer of the first year of war, was that 60,670 combat and 32,647 training planes would be needed in 1943 in order to attain air supremacy over the enemy. American industrial might would ultimately stand as one of the greatest contributors to the prosecution of the war, such that by war's end the projected levels had come close to attainment.[5] Aviation historian Roger Bilstein attributes that success to the "efforts of individual companies to raise stock; multimillion-dollar loans from the depression-era Reconstruction Finance Corporation; innovative funding and support through federal bureaus such as the Defense Plant Corporation; cooperation between rival manufacturers to achieve maximum production for specific designs; and hard work by everyone involved, including aviation-related unions."[6]

Arnold and others would eventually regret a glaring oversight in the early planning for aircraft procurement. Up to the time of war, Air Corps boards considering specifications for pursuit aircraft did not think a fighter could be built with the maneuverability, range, armament, and altitude to accompany bomber formations over long distances. More important, American strategic bombardment doctrine held that heavy bombers were capable of reaching targets without the protection of friendly fighters — the old adage that "the bomber will always get through." Of the twenty-four groups in the Aviation

Expansion Program, only nine were to be fighters, five to be heavy bomber, six medium bomber, and two light or attack bomber groups. Painful experience ultimately forced AAF leaders to admit their error in planning for fighter escort, yet few ever relinquished the strategic doctrine itself. As a result, the AAF trained most of its pilots, navigators, bombardiers, and gunners for high-altitude precision bombardment. But when they entered combat, many of these men fought a different kind of war.[7]

As during the expansion years 1939–1941, the AAF framed its requirements in terms of training pilots and forming combat groups whose numbers derived from aircraft procurement levels. A proportionate number of bombardiers, navigators, and gunners were to be trained simultaneously. The program in effect at the time of the Pearl Harbor attack, for example, envisioned training 30,000 pilots for 84 groups, along with 4,888 navigators and 5,590 bombardiers.[8] Planners worked backward from the overall goals to arrive at the number of people that the regional training centers and their subordinate flying fields should train. Therefore, in the case of pilot training, each of the three training centers received a quota that for most of the war was based on a nine-week schedule for each of the three phases of pilot training. Anticipated elimination rates during primary, basic, and advanced and the relative number of graduates needed in fighter and bombardment units determined the total population in training and the distribution of students at the various flying fields. The same type of calculation applied to bombardier and navigator training. In October 1943, for example, the Training Command informed AAF headquarters that to meet training requirements for heavy bomber crew production, it anticipated elimination rates in pilot training would be 31 percent in primary, 13 percent in basic, and 2 percent in advanced. The elimination rate in navigation would be approximately 15 percent, and in bombardiers, 20 percent.[9]

Changes in technology and battlefront conditions repeatedly forced those in charge of training to revise their estimates of the numbers of people needed in the different types of combat groups, which in turn altered and diversified the training programs. Different kinds of aircraft and their assigned missions — very heavy, heavy, medium, and light bombers; single-engine and twin-engine fighters; photoreconnaissance and observation aircraft; and, for a time, dive bombers — required different types of training for the pilots. Moreover, heavy and very heavy bombers carried large crews that included several specialist gunners (who doubled as flight engineers, radio operators, mechanics, radar operators, armorers, and photographers) as well as nonspecialist gunners.[10] As the war progressed, planning schedules tipped increasingly toward greater employment of heavy bombers that, until the activation of very heavy units in 1943, required the largest and most diversified crew. Yet the gunner requirements were slow to reflect the preponderance of heavy and very heavy crews with their many gunners because the rate was tied to pilot requirements that in turn depended upon aircraft production.

World War II

The capacity and quality of existing training facilities, the availability of equipment and aircraft, and the number and competence of instructors and commanders further militated against a smooth and speedy response to the burdensome demands for personnel. In short order, the air forces had to build more training fields and gunnery and bombardment ranges. Besides chronic shortages of training aircraft, the extreme scarcity of high-octane fuel curtailed all types of specialized training into 1944. Obtaining instruments, bombsights, and gun turrets slowed the process further, since the training program clung to the bottom rung of the procurement ladder.

The agencies that shared administrative responsibility for training the new military giant were Headquarters AAF (which primarily determined requirements and policy), the Training Command (which was responsible for individual training), the First, Second, Third, and Fourth Air Forces (which conducted crew and unit training in the continental United States), and squadrons in the theaters of war (which prepared the men and aircraft for actual combat). Although wartime reorganizations consolidated training responsibilities, blurring and overlapping of responsibility among the administrative elements persisted, no matter how the organizational charts were drawn.

As Assistant Chief of Air Staff, Training, hard-working Brig. Gen. Robert W. Harper was one of the chief architects of the training program throughout the war. His headquarters staff coordinated individual training conducted by the Training Command with unit training conducted by the domestic air forces. Harper's organization issued training standards and directives based upon combat and intelligence reports, day-to-day experience in the training air forces, technical experiments carried out by the Proving Ground Command, and tactics developed by the AAF Board and the AAF School of Applied Tactics.[11]

The Training Command implemented the training directives and standards in its specialized schools.[12] The Command was the successor to the Air Corps Flying Training Command, established as a major command on January 23, 1942, and redesignated the AAF Flying Training Command a month later. The March 1942 reorganization, which placed the AAF on a par with the Ground Forces and Services of Supply, gave this command, in conjunction with the Directorate of Individual Training at Air Forces headquarters, oversight of all individual flight training. On July 7, 1943, the Flying Training Command merged with the Technical Training Command to become the AAF Training Command under Maj. Gen. Barton K. Yount.[13] The schools that provided individual instruction for pilots, navigators, bombardiers, and gunners were grouped under three regional flying training commands.

The First, Second, Third, and Fourth Air Forces provided the final phase of training in the United States. Special units, called Operational Training Units (OTUs), assembled cadres for new groups and supervised their training. Each unit went overseas intact. To keep these squadrons from being robbed of trained personnel, Replacement Training Units (RTUs) were activated to train individuals and crews as replacements.[14] By mid-1943, many fewer new groups were needed than individual and crew replacements for existing units, so RTU predominated.

The continental air forces thus became the last link in the stateside combat aircrew training chain that began with classification and sequentially went through preflight, Training Command specialized schools, and finally OTUs and RTUs. The challenge to the large and disparate system lay in standardizing instruction and maintaining coordination among all the administrative components responsible for some portion of the training function as the air forces simultaneously decentralized the command authority. In the report of his opening address to a training conference in September 1943, General Harper emphasized that the "smooth progress of training from one phase and one agency to another was one of the most important problems to be dealt with." The Training Command "should produce what the Air Forces want. The Air Forces should produce what the active theaters call for. Close coordination with Headquarters, Army Air Forces is highly desirable." The Air Staff, he assured his listeners, "is extremely desirous of being of assistance to see that real continuity of training is achieved."[15]

If there was one guiding principle for coordinating the training agencies, it was General Arnold's frequently repeated declaration that his headquarters would "tell the Command and Air Forces what to do but not *how* to do it."[16] AAF headquarters issued training standards that were carried out largely as the subordinate commands saw fit. While attempting to maintain the separation of policy and operational functions implied in Arnold's maxim, training officials made every effort to coordinate elements of the training network. Harper requested commanders of all air forces and commands to exchange visits with one another's units so that they might understand the specialized training required of the various air crew components.[17] Standardization schools existed in the Training Command and in the continental air forces. Central instructor's schools opened for nearly every major flying skill taught in the schools and air forces, e.g., single-engine, twin-engine, instrument, and gunnery. Major training conferences drew together representatives from all the U.S. training agencies and, frequently, from the combat air forces overseas.

In his classic *War and Politics*, Bernard Brodie warned that "one must not demand too much in the way of foresight" from those engaged in war. Senior leadership responsible for establishing a program should be asked, however, to take into account "the character and dimensions of uncertainty."[18] For training officials, significant contributors to the support apparatus underlying military operations, success depended not on the ability to see the future but upon structuring an efficient, flexible, and useful system that trained men to fight in the face of the inevitably shifting requirements of war. Their success or failure could not be evaluated cleanly in the traditional military terms of battlefield victory or defeat. Moreover, World War II training practices did not reflect the

World War II

thinking of a single commander, or even of a discrete handful of policymakers. The basic concepts took shape before the war and were now being adapted collectively to a much larger and more complex set of circumstances.

By any quantitative measure, the Training Command engineered a remarkable growth over a four-year period, at its peak achieving an annual training rate of more than 74,000 pilots.[19] In meeting the numerical goals imposed on it, the training program benefited from the American industrial engine, which quickly accelerated to produce the tools of war. As the volume of goods expanded, aviation technology also advanced; although the resulting new hardware meant new types of instruction, more and better equipment also provided the means to meet training objectives.

Miraculously, given the staggering personnel explosion and the move into higher performance aircraft, the number of training fatalities, always a nervously guarded secret, appears to have stabilized, although it remained at its customary high level. According to Arnold's report to the Secretary of War in February 1945:

> Twenty years' accumulation of experience, by a comparatively small and fixed group of men, brought the AAF accident rate down to 51 per 100,000 hours in 1940. Expansion instroduced a new and enormous block of inexperience, which would tend to reproduce the situation of the early 'Twenties. Vigorous preventive measures were taken against the expected rise. The degree of success can be measured by the fact that the accident rate has been held down and new all-time lows attained.[20]

Unlike quantity, a matter of counting planes or pilots or casualties, quality proved harder to assess. A torrent of reports from the battle zones told stateside training officers that many field commanders did not consider new crews properly prepared for combat. Many airmen, especially gunners, complained bitterly about the things they had *not* learned in schools or OTUs. Precombat training succeeded in doing what it had been doing for years: developing standardized methods for teaching individuals the fundamentals of flying, navigating, bombing, and shooting. On the other hand, there had been no peacetime means for evaluating how individuals would apply these lessons when war came — what a later generation called "combat readiness." A lecturer at the AAF School of Applied Tactics, speaking about AAF training problems during the war, acknowledged the previous lack of experience with modern weapons. He noted that training maneuvers on the eve of war had been conducted with "broomsticks and pipes."[21] Airmen flew planes as old as the P–26, the Army's first monoplane fighter, operational for almost a decade and hopelessly obsolete.

Training under wartime conditions was now held to the harshest standards. War quickened and intensified its pace. In all practical areas of individual and unit training the pressure of time caused the service constantly to balance the

Part V

competing demands of quantity versus quality, of theory contrasted with practical experience, of realism opposed to safety, and of standardization over improvisation. But even the crucible of war did not fundamentally alter the principles by which the AAF chose to train. Airmen were certain of the rightness and the efficacy of the strategic precision bombing mission. The war would be fought and won largely in those terms, they believed, and therefore the skills instrumental to that role should be at the heart of the training program. Nevertheless, as the firsthand experience of war came closer to fliers and planners, the applicability of theory, both strategic and tactical, became more tenuous. AWPD–1 presumed that an early strategic air offensive would soften Germany for a massive ground force, cross-Channel invasion. The AAF leadership expected to fight on those terms. It did not. Likewise, it hoped that the tactics employed in the theaters of operation would be the same as those the men had rehearsed in training. They were not.

In many ways, training was a valiant but impossible attempt to impose order, control, and predictability on an inherently fierce enterprise of chance and luck, experimentation, fury, uncertainty, and desperation. It would not easily or immediately produce either the level of competence needed of combat crews or the special requirements of the different theaters of operation.

ELEVEN

Picking the Men, Training the Pilots

"...got to be damned sure no boy's ghost will ever say, 'If your training program had only done its job'..."
— Inscription on cover of Ground Training Guide[1]

Air Corps officers should be pilots and pilots should be masters of all the necessary skills to make the airplane an effective weapon of war. That prewar reasoning — and the poetry and symbolism of flight — popularized the knight-errant of the air as the single romantic warrior riding his mechanical winged horse. But technology and war conspired to end the pilot's monopoly of the airplane. Advances in aviation technology brought more sophisticated instruments, navigational aids, bombsights, guns, and turrets. All required specialized training for their use. During the expansion years the air forces tried for as long as possible to ensure that the flying officers who used this new equipment had some pilot training by selecting its navigators and bombardiers from among pilot eliminees. By the time war was declared, however, the numbers required were too hopelessly high to continue such a policy. The AAF then developed extensive new tests for classifying aircrew volunteers into pilot and nonpilot specialties. Under duress, the service finally broadened its concept of an officer corps to include combat airmen who were not pilots.

Recruiting during the war marketed the team image — usually depicting a bomber crew rather than the solo fighter pilot. Even in pilot training, the singularity of the individual diminished as cadets flowed through the system in a process more closely resembling production-line than the old handcraft methods. Upon graduation, most pilots were assigned to bombers and thus flew with a crew. All aircraft, fighters and bombers alike, flew in formation. Moreover, the numbers of airmen now reached into the thousands. Membership in a small group of the elect who surveyed heaven and earth, usually alone, disappeared with the war. For all the change, pilots remained, nonetheless,

World War II

among the Army's elite. Stratification within the AAF was based on skill, with the pilot clearly at the top.

Manpower Procurement and Classification

The perceived ability of air power to strike at the enemy before ground and naval power could mass enough force to do so gave the AAF an early edge in wartime budgets and recruitment priorities. Yet even this advantage in money and manpower could not assure the quick creation of a proficient military flying force; training was too lengthy and attrition rates were too high among men and planes to permit rapid results. The AAF had begun to build its forces as a result of the Expansion Program, but a declaration of war further raised the quotas. The air forces had to acquire large numbers of fliers from a civilian population with little background in aviation, classify candidates with unknown skills into aircrew specialties, and begin elementary flight training for thousands of fledgling airmen.

Besieged with applicants from the beginning of the war, the AAF threw off earlier constraints and standards, drew in legions of human material for the war, then leveled off and gradually phased out the intake programs by war's end. At first, it increased the number of local procurement boards and gave them the authority to accept candidates. It eliminated the college educational requirement in favor of an on-the-spot qualification test, and it dropped the enlistment age from twenty-one to eighteen years. With its highly technical orientation, for a time the Air Corps enjoyed the pick of Army draftees who scored the highest on classification tests.

Thereafter, the primary means for procuring aircrew trainees came through the Air Corps Enlisted Reserve, which was established on April 1, 1942. The Reserve was essentially a pool of men already bound for duty in the air forces but whose training could be deferred for as long as six months. A nearly transparent subterfuge that hoarded a large reservoir of promising manpower for the Air Corps and denied it to other services, the system provoked considerable interservice jealousy and public criticism.[2]

The pendulum swung again when, from the end of 1942 through the fall of 1943, the lowered draft age and the elimination of voluntary enlistments, which brought more men into the ground forces, worked against the AAF. At the same time, pilot graduation rates at the end of 1943 began to equal the quotas, and a backlog began to form. In early 1944 manpower requirements were revised downward; the passing grade on the Aviation Cadet Qualifying Examination and the physical standards were raised. By the end of that year, all aircrew specialists were pouring out of the schools in the desired numbers, and the entire system slowly reversed itself, eliminating the need for a mechanism to procure aircrew.[3]

Throughout this period, the AAF did all it could to aid the procurement process through public relations campaigns. The service was thought to have such an unfair advantage in recruiting because of the glamorous image of flight that for a time it was not allowed radio advertising. A sense of its methods and success is indicated by a notice of March 14, 1944, when the manpower supply problem had abated, that "General Arnold directed yesterday that our practice of sending aircraft and crews returned from the combat theaters around over the country on tours is here and now stopped."[4]

All potential aircrew members accepted for classification and preflight schools, who might become officers after graduation, had to receive a passing score on the new Aviation Cadet Qualifying Examination. The test was based on data from World War I, subsequent research studies by the National Research Council's Committee on Selection and Training of Aircraft Pilots, the U.S. Navy, and the Technical Training Command, as well as on information gleaned from the Psychological Research Units in the Replacement Training Centers and suggestions of experienced flying officers.[5]

Most people who volunteered for flying training and passed the qualifying exam hoped to become pilots. Even though the AAF expected to train 50,000 pilots a year, then 70,000, and then more in what seemed to be ever-ascending numbers, not all enthusiastic applicants achieved their ambition. Some lacked the necessary skills while others fell victim to the need to train men for other aircrew specialties. During the expansion years from 1939 through 1941, bombardiers and navigators came mostly from the ranks of those eliminated from pilot training. This method of reclassification, however, resulted in very serious morale problems. "I lived to see in the combat zone," one pilot recalled, "the frustration of a would-be pilot relegated to a bombsight or a navigator's computer."[6] The system also wasted precious resources giving pilot training to men who were then siphoned off into other occupations. To increase efficiency under the pressure of time, the Training Command needed to find another way to match men with occupations.

In conjunction with physical examinations and personal interviews, a new implement of social science research was developed to classify men into the three flying officer specialties of pilot, navigator, and bombardier. Psychologists developed aptitude tests in which an individual's score, called "stanine,"[7] measured his likely performance in each of the three specialties. The air force found these tests to be extremely useful predictors of success. The rate of elimination during training of high-scoring men proved to be much lower than for men screened through other, more impressionistic, means. Those selected as pilots tested well on factors correlated with perception and rapid physical reaction, and they exhibited an ability to discriminate between visual objects and to visualize mechanical movements. They were well coordinated and were judged to be interested in and informed about aviation. Navigators demonstrated an aptitude for mathematics, exhibited accuracy in using tables and instruments, were interested and informed about science, showed strong reasoning facility, easily understood maps and charts, and quickly comprehended verbal material.[8]

Men waited anxiously for the results of the classification tests. According to one (perhaps biased) pilot who may, nonetheless, have represented a considerable cross section of the air forces: "Pilots used to say the classification tests sorted cadets into three bins: pilots were chosen from those who gave quick and correct answers; bombardiers were chosen from those who gave quick and erroneous answers; navigators were those who gave slow and correct answers."[9] Even if most airmen harbored the view that pilots were the best and brightest, in fact the average navigator had the highest stanine score of the three specialties.

Throughout the war the required passing score on the qualifying exam and the stanine levels changed, depending upon the need for trained men in each specialty. For example, in November 1943 a man could be classified as a pilot with a qualifying exam score of at least 180 and a pilot stanine rank of 5 or higher. A year later, when the AAF was not so short of pilots, the standard was raised to a score of 180 or more with a pilot stanine of 7 or more. These scores did not assure graduation from flight training; they merely improved the odds and lowered the elimination rates. The higher the entrance requirement, the greater the number of applicants needed but the lower the ultimate training cost.[10]

Replacement Training Centers for classification and preflight training of pilot, navigator, and bombardier candidates were already in operation by early 1941. They were renamed preflight schools and set up separately for each of the officer aircrew specialties after war began. Preflight instruction differed very little for each of the three specialties. Moreover, the schools adopted no standardized curriculum until April 1943, thereby keeping instruction unfocused. Matching in length each of the three stages of pilot training, the nine-week course, consisting of military, ground, and physical training, extended to ten weeks in May 1944 (as did the pilot course), when the three separate preflight schools again combined.[11]

Most cadets became restive in preflight. Future pilot Bert Stiles thought the "attempt at education was the saddest, poorest, most incomplete I ever ran into." Teaching potential pilots the principles of flight was a fine idea, he mused, but "the Army way of putting it across, giving the maximum of predigested information in the minimum time, with no time to think it over, or talk it over, is a pretty bad way of doing things, and only justified in an emergency."[12]

Individuals in preflight training (such as Stiles, who found ground classes stultifying), cadets who had been delayed and placed into "pools" at some point in their training, flight instructors, headquarters staff officers, and those in the Training Command all shared some of the frustration at being removed from the "action" in wartime. In this respect the training system suffered from a

Training Pilots

The time between arrival at the preflight schools and the pinning on of pilot's wings was one of rigorous selection processes and classification into the elite pilot program.

persistent morale problem throughout the war. Howell M. Estes later spoke of being "stuck" in the Training Command:

> ... The general tendency was to retain us there in spite of requests and efforts on the part of others to get us out and on the part of our own efforts to get out. The Commander of the Training Command felt that he had ... the biggest job and if that job wasn't done properly then obviously the combat job couldn't be done. Every effort that was made to get some of ... the more senior people out of the Training Command was fought off by him personally.... People used all kinds of dodges to try to find ways to get out.[13]

Instructors tended to be an especially discontented lot. They were, as one Air Force historian described their plight, "far from real war and its stimuli, sometimes malassigned and keenly aware of their unfitness for their task, and demoralized on occasion by lackluster leadership ... or by the sheer boredom generated by a standardized teaching system which allowed little room for individual initiative."[14] Civilians teaching ground classes or primary flight training found themselves to be outsiders in the military school system. Military instructors were frequently unsuited to be classroom teachers or infuriated because they were teaching cadets to fly rather than being in a combat assignment themselves. "The idea of sitting out the war as a damned flight instructor in Texas or somewhere tore me up," recounted fighter pilot Chuck Yeager. When he had completed more than his required share of combat missions near the end of the war, Yeager was in fact returned to an airfield in Texas to teach flying. There he encountered considerable jealousy from other officers "who had spent the war in Texas. They were assigned as pilot instructors, and they hated every minute of it."[15] Because all phases of the training program experienced such a continuous shortage of instructors, the service never satisfactorily juxtaposed superior teaching talent, personal preference, and morale.

Pilot Training

The leadership of the AAF were pilots; most volunteers for AAF aircrews *wanted* to be pilots. Even though the service now had to train other combat airmen, to say nothing of a complex of technical specialists and a host of support personnel, the pilot ranked highest in terms of prestige and responsibility. The AAF prized those who pinned on pilot's wings and had a healthy appreciation for the dangerous, slow, and intractable process required to turn a recruit into an airplane commander. The U.S. Army graduated 193,440 pilots between July 1, 1939, and August 31, 1945; another 40 percent of that elite group who had entered pilot training failed to make it through the course.[16]

After their initial selection for aircrew training, their classification as pilot cadets, and acceptance into preflight school, would-be pilots progressed through the three stages of training — primary, basic, and advanced — given in Training Command schools. From early 1942 through mid-1944, each of the three stages lasted nine weeks, but once the quotas were being met, pilot training was conducted at a less frantic pace, and each phase reverted to the previous ten-week length. After graduation from the advanced course, pilots learned to fly tactical aircraft either in special Training Command programs or in continental air force units. This postgraduate instruction, formally called transition, almost constituted a fourth stage of individual training that, like the others, usually lasted nine weeks. Because transition as a distinct phase sometimes occurred in OTUs of the training air forces, it served as a link between individual pilot training and operational training.

The fundamentals of the three-staged pilot course did not change greatly throughout the war. In general, pilots had to perform certain types of maneuvers satisfactorily and demonstrate familiarity with controls, engines, and instruments on specified types of aircraft before moving to the next stage. Eliminations rarely came for poor performance in ground classes. Training officials tinkered with the curriculum, adding hours here and compressing others there, to try to hurry the trainee to completion and get him into combat units with the skills that at any moment were considered most critical. At times they increased or slowed down the production of pilots in response to demand for specialists on the battlefield, to compensate for an oversupply or undersupply of students already in training, and to maintain an orderly flow of students.

Of the various programs of combat aircrew training, pilot training was both the most important and the easiest to manage because it did not require reconceptualization. It was built rather straightforwardly on the procedures and ideas from the interwar years, which had been modified during the expansion period primarily to allow for growth in the number of men to be trained. Once the United States actually went to war, it was the sheer demand for people that further strained the system.

Responsible for thousands of new recruits without previous flying experience, having an insufficient training staff that itself required instruction, and lacking time, the Training Command relied on standardization of procedures to produce an acceptably competent "product." Pilot training could no longer be a leisurely, individualistic process. Central instructor's schools did much to standardize methods of instruction and aid in evaluation of pilots for assignment. Even returned combat pilots recycled through four-engine instructors' schools that were allowed to retain their best as supervisors or instructors. Frequent inspections of the training units and standardization schools provided other means of ensuring continuity throughout the training program. Late though it was, by the fall of 1943 the service had taken steps to standardize instrument panels on aircraft as much as possible and to establish proficiency requirements by types of aircraft. With marginal success, the AAF even tried to develop an objective scale of flying skill, in an attempt to override the inevitable subjectivity in evaluating cadets in training.

The brute problems of expansion dominated pilot training but were accompanied by the need to accommodate the increasingly more sophisticated and technical aspects during advanced and transition training. In the advanced phase all cadets were assigned for the first time to single- or twin-engine training, which had different curricula. Twin-engine advanced training had been on the books during the expansion years, but in fact it had hardly existed despite its high priority. The advent of war brought financial resources to implement the avowed strategic mission and, with the additional funds, new higher-performance aircraft and intricate equipment as well as diverse requirements for pilots in light, medium, heavy, and very heavy bombardment.

Skills considered crucial under combat conditions received greater emphasis than before. Formation, high-altitude, and instrument flying became a part of nearly all areas of training, virtually irrespective of specialty. The first two especially dominated the pilot's time once he entered a training unit of the continental air forces, but he was introduced to them in the latter phases of his pilot course as early as equipment and his level of skill permitted. Instrument flying continued to be stressed throughout a pilot's tenure in stateside training.

Primary

> If a pupil is given too many rules of thumb instead of a real appreciation of what he is trying to do; if he is taught in a mechanical way; taught to fly by numbers; taught to follow rigid and precise flying patterns without variation, when something unusual happens he just does not know what to do and his reactions will be slow.[17]

At last the cadet climbed into an airplane. An experienced pilot would teach him to fly. Whether it was 1922 or 1942, the beginner still felt an anxious tightness in the chest when approaching the first solo flight, horror and denial on witnessing the first fatal accident, wonder and disorientation on the first night mission, and the temptation to fly by the seat of the pants rather than by instruments. Thoughts of combat brought a rush of exhilaration; fear came at the prospect of washing out, which was viewed as unequivocal failure. As one airman put it, "it sounded as though you turned colorless and just faded away, like a guilty spirit."[18] Another cadet captured what must have been a frequently felt kaleidoscope of emotions on the first flight with his instructor:

> As the plane climbed, he explained about co-ordinating rudder and aileron pressures in turns, about the "feel" of the airplane, about how too much rudder without aileron control made the airplane skid to the outside of the turn and too much aileron control without rudder made it slip to the inside. Demonstrating, he told me how to feel it in the seat of my pants when we slipped or skidded, and to note how the wind came through the side of the cockpit when a turn was unco-ordinated. Hanging on grimly, with an increasing queasiness, I tried to feel what he said I should.[19]

No matter that war brought a massive scale to the AAF training program, primary training was for each cadet an individual and personal initiation into the astounding and sometimes frightening experience of flight.

In the primary phase the student learned to fly a low-powered, stable airplane and to execute good landings and basic turns and maneuvers. On this most elementary level, flight training was the least military, allowing the AAF to make the greatest use of civilian resources. In his final report on the war, General Arnold commended the civilian–military partnership: "Training of personnel in time of war can be done on a large scale only by utilizing all the nation's facilities and experience . . . . The armed forces will never have all the facilities required to meet war programs. Civilian agencies must in some way be kept aware of their responsibilities especially during peace when planning and preparation for war are so distasteful to Americans."[20]

At the beginning of the Expansion Program the Air Corps decided to assign its primary flight training to civilian aviation schools. Forty-one of those

Training Pilots

The Ryan School, a civilian school that provided primary flight training, was located in San Diego, one of the birthplaces of flight and home to the Signal Corps Aviation School before World War I. The many primary flying schools were among the most successful military-civilian parterships of World War II.

schools were in operation on December 7, 1941. Thereafter, stepped-up pilot requirements meant opening more schools and exercising greater control. The government arranged to buy most of them from the contractors, who then rented them back at a reduced fee for instruction. At the height of primary flight training in May 1943, fifty-six schools were in operation. Almost immediately after reaching the overall pilot training peak in the fall, the service began to reduce its pilot production sharply. It took a step toward remilitarization in August 1944 with the opening of a service-run primary school at Goodfellow Field, San Angelo, Texas. At the end of the year, only ten civilian contract schools remained. On V–J Day there were two, and by the end of the war all were closed to military flight training.[21]

The AAF tried to synchronize the relatively laconic civilian-run primary training network with a military organization whose purpose was combat. Pilots had to learn to fly more aggressively, which meant, according to one cadet, "being positive in the use of the controls. Instead of easing into turns, we were to lay the airplane up in a steep bank and use lots of back pressure to get around the turn."[22] Many instructors who came through the Civilian Pilot Training

World War II

Program received some indoctrination in military methods in central instructor's schools in the Training Command before becoming instructors in the primary flying schools.[23]

The military presence was enhanced further by a detachment of officers and enlisted men at each school and a handful of AAF check pilots.[24] The latter flew twenty-, forty-, and sixty-hour evaluation rides with the cadets and proficiency check flights with anyone marked for elimination by the civilian instructors. The morale of these men, who were themselves as often as not newly commissioned Army pilots, was not as low as those teaching ground school classes, but as then-lieutenant John Frisbee moaned, "After mastering the AT–6 or the AT–9, and maybe getting a little time in operational aircraft, going back to a sparsely instrumented primary trainer that might do 120 mph nose-down and with the throttle firewalled was like repeating fifth grade at the age of sixteen."[25] The depressing fact that the military check pilots thought of themselves merely as "winged rubber stamps" (according to Frisbee) testified, on the other hand, to the general competence of the civilian flying instructors.

At the beginning of the war, aggregate classes of nearly 12,000 students entered every 4½ weeks to begin the 9-week primary course. Cadets belonged to a class that was numbered by the month and year of their expected graduation from advanced training. The Training Command attempted to complete this process seamlessly, by the rules, in the military way. One cadet, Eugene Fletcher, and his mates had been assigned alphabetically in preflight. Thereafter, in primary at the Mira Loma Flight Academy at Oxnard, California, men were lined up according to height, and then assigned accordingly, the tallest in Squadron 3 and on to Squadron 8 for those under 5 feet 9 inches.[26] From such an auspicious, or at least orderly, beginning, an individual's progress could be delayed for a number of reasons: personal circumstances requiring emergency leave; failings in technique that did not merit elimination, since additional training could correct them; or inclusion in a holding pool because no more students could be accommodated in the next stage of training.

During the 9-week primary course (becoming 10 weeks in May 1944), 28 or 29 hours of air work were dual (flying with an instructor), and 31 or 32 were solo (flying alone).[27] In his initial flights with an instructor, the student practiced stalls and spin recoveries, climbing and gliding turns, and forced and normal landings and takeoffs. In the intermediate phase he added figure eights and chandelles (abrupt climbing turns), cross-wind landings, maximum performance glides, stalls, and spins. What was called the accuracy phase stressed precision landings and approaches and power-on, power-off, and short field landings. Finally, according to TM 1–445, the diversified stage was subdivided into night flying, navigation, and instrument training with the Link trainer.[28]

The Director of Individual Training in Washington and the Flying Training Command constantly refined and revised the program, deciding how many hours should be devoted to learning different maneuvers. The latest edition of TM 1–210 was to be the manual for primary training. Training detachments kept detailed weekly reports, recording the time spent in dual and solo air work accurate to the minute for each class.[29] Despite the reams of paperwork that tracked students' progress and official directives laying out curricula, instructors in the field inevitably brought their own emphases and standards and operated with a certain amount of latitude. Training officials appeared to understand that it was not possible, and perhaps not always desirable, to conform strictly to the programs of instruction and mandated hours. The 1942 and 1943 curricula, for example, held that "a ratio between dual and solo hours will be determined for each individual student on a proficiency basis."[30] On January 30, 1943, when the program of instruction had just received another fine tuning, one memorandum stated: "It is not considered desirable to require that every dual flight include a minimum of one spin and one stall."[31]

Most of the techniques practiced in the very earliest period of flying a small primary training airplane were considered to be essential but hardly comparable in complexity to those that would be required of an airplane commander of a B–17 or a P–51 fighter pilot. Yet many graduate pilots later reflected upon the importance of the fundamentals perfected (or not) in primary training that would directly affect one's safety and performance in combat. Bomber pilot Philip Ardery, for example, had no doubts about the utility of what he had learned practicing turns over a road junction:

> ... all of which were to teach us the effect of wind drift on the path of flight. . . . Three years later I found the wind-drift principle applicable with extreme importance in forming up a unit of heavy bombers over an overcast. In England we would take our bombers off one at a time, climb individually through the overcast and, upon breaking out, form up circling over a radio beacon. Frequently the problem was made more difficult by strong winds. It was surprising how many pilots had come that far along and still appeared not to know the wind-drift principle.[32]

Most students learned on open cockpit primary trainers such as the Fairchild PT–19, a low-wing monoplane first flown in March 1939. After Pearl Harbor it and the similar PT–23 were manufactured by the ASHA group (Aeronca, St. Louis, and Howard Aircraft Corporations). The aircraft were troublesome because of excessive vibration, their partial-plywood wing construction, fragile landing gear, and the sliding cockpit cover that often failed to slide. The "B" model and the PT–26 were improved by lights and navigation equipment for night and blind flying. The Ryan PT–20, PT–21, and PT–22 (dubbed the "Maytag Messerschmitt") were commercial training aircraft modified for military purposes. These airplanes were also plagued by excessive vibration and rough engine performance. The other primary trainers were PT–13 biplane modifications, ending with the PT–17, PT–18, and PT–27.[33]

World War II

Representative of the commercial training aircraft modified for military purposes was the PT–22, the "Maytag Messerschmitt."

Design problems of several of the primary trainers were never satisfactorily resolved by the end of the war. However, airmen typically judged certain aircraft superior to others in what were essentially personal preferences. The British, for example, were sufficiently unhappy with the Stearman trainers they were given for their flying schools in the United States that at war's end they returned many of them unused.[34] On the other hand, the Assistant Chief of Air Staff, Training, General Harper, notified the Joint Aircraft Committee that the AAF Training Command especially valued the Stearman PT–17:

> The PT–17 has proven to be a most satisfactory type and maintenance difficulties negligible compared to the Fairchild wooden types. The wood aircraft will not stand up in the hot, dry climate where many of our schools are located, and much difficulty is being experienced with the PT–23 due to the vibration trouble.[35]

Flight instructors usually taught four students, two of whom would fly in the mornings while the others attended ground classes. At this early stage of their training, cadets spent more hours in the classroom than in the air. During the 9-week course, the curriculum called for approximately 85 hours in navigation, weather, aircraft and engines, and aircraft identification. The 10-week course of mid-1944 dropped weather and added code (signal communications), and because the Pacific war was then engaging the greatest attention, cadets studied naval vessel recognition as well as aircraft identification.[36]

As of V–J Day, the elimination rate during primary training averaged 27.5 percent, the highest of the three stages. The figures confirmed that getting through primary training posed the biggest hurdle to attaining pilot's wings. A cadet's future was most at risk during primary training, when the AAF had the lowest financial investment and training was the least dangerous. Brig. Gen. Cleo Bishop recalled his own experience in primary:

> There was an extremely high washout rate...I think, not so much that the people they washed out couldn't learn to fly, but they couldn't learn to fly fast enough. You were expected to solo within 8 or 10 hours, and if you

hadn't done it by 12 or so, they would wash you out. They just simply didn't have the time. They had more people than they did time.[37]

Most cadets believed that training standards were stiffened or relaxed arbitrarily, that the standards for qualification or cause for elimination were capricious. Cadets wondered uncertainly about technical or personal offenses that caused a man to wash out, or whether they themselves were about to be eliminated at any moment. One successful graduate speculated that two types of men washed out most frequently: "First, there were those who had previous civilian flight training, who evidently had trouble flying the 'Army way.'" The other group were transferred ground officers who, he thought, "did not seem as motivated as were aviation cadets."[38] Another graduate pilot concluded that there was one simple, unimpeachable criterion for elimination: "Whenever an instructor felt a student wasn't material to go on — for any reason whatever — the reason always given was that he 'lacked coordination.'"[39] A Navy pilot, who labored under the same type of system, summed up what most airmen came to believe, that there was an indefinable quality by which pilots were judged:

> These were tests, but not like any test that I had taken at school or university. You couldn't cram for it, and you couldn't fake it. You weren't even being tested on something that you had studied, really, but on what you were. If you were a flier, you passed; if you weren't, you washed out — fell out of the air, and became a lower order of being.
>
> It became clear that some people were natural fliers, and some weren't. The athletes usually were; they used their bodies easily and naturally, and they seemed to make the plane a part of themselves.[40]

Basic

Since it formed the intermediate stage between primary and advanced pilot training, the basic flying curriculum underwent frequent overhaul in response to the changing demands of the other two phases. Should trainees in basic perfect their proficiency in elementary maneuvers learned in primary? Or should they concentrate on more demanding formation flying, instrument and night work, and be introduced to combat aircraft, all of which foreshadowed advanced training? The Training Command eagerly replaced basic with advanced trainer aircraft when the opportunity arose. As for training tactics, however, several changes in the curriculum between 1942 and 1944 indicated that training officials were unable to agree on the ideal balance in basic flying, given the time allotted to the course. Until it returned to its prewar ten-week length in 1944, the emphasis shifted back and forth between proficiency flying and more advanced skills used in combat.

During the expansion period and war years, the military owned and operated all but four basic schools (including the segregated school at

World War II

Tuskegee, Alabama, for training Black pilots). Although the training costs at the Army and civilian schools were equivalent, the AAF found, as a contemporary pundit declared, that public and private management were "alike in all unimportant respects." The service considered it paramount to transform civilians into officers during basic flying. Military men considered civilian schools to be ill-equipped to undertake that critical indoctrination, so the experiment of contract basic training did not become widespread.[41]

As cadets progressed through the three stages of flight training, the enterprise grew more deadly. In primary, about 2 students were killed in accidents for every 100,000 hours flown, well below the Training Command average. By comparison, during basic the elimination rate dropped to less than half that of primary, but the mortality rate started to climb. Few airmen in basic escaped seeing or knowing about somebody who died in a training accident. Only two weeks into the first class, the basic school at Pecos Army Air Field, Texas, suffered its initial casualties. The "bloodiest period ever," according to a historian of West Texas aviation, began in June 1943, "when weekly crashes ... killed 25 young flyers."[42] The grisly statistics, here and elsewhere, demonstrated that flying, even in the relatively benign early stages, was a highly dangerous occupation. Possibly during primary, and certainly during basic, the cold-blooded numbers took on visceral meaning. As pilot Samuel Hynes reflected, "The reality of death comes to you in stages. First it is an idea — all men are mortal, as in the syllogism. Then it is something that happens to strangers, then to persons you know, but somewhere else, and at last it enters your presence, and you see death, on a runway or in a field, in a cloud of dust and a column of smoke.... At that moment the life of flying change[s]."[43]

In the first basic curriculum developed during the 9-week course, ground classes included a few hours each in aircraft identification and navigation, 20 hours in code, and 38 hours in weather. Near the end of 1942, some training officials favored a dramatic jump of from 10 to 24 hours in navigation and the addition of 6 hours in instruments. That total of 94 hours could not be accommodated until the course again expanded. In fact, hours spent on ground subjects changed little through the spring of 1944, except for increased time for instrument training.[44] As time went on, combat veterans came back to regale students in ground classes with lessons learned in air battles in the European and Pacific theaters.

As noted, between 1942 and 1944 the curriculum flip-flopped between an emphasis on the fundamentals of flying and an emphasis on combat maneuvers. The change of focus could be seen by the relative primacy accorded the phases termed transition (learning to fly the basic training aircraft) and diversified (learning more complicated maneuvers and practicing navigation and instruments). Under the March 1942 program, pilots flew 6 hours in night transition, 4 in radio-beam orientation, 8 in instrument without radio, 8 in instrument flying by student teams, 5 in day formation, 2 in night formation,

6 in day navigation, 2 in night navigation, and 6 in transition to advanced trainers. This shortened 9-week course of 1942 reduced the amount of time spent in the transition phase in order to stress the more demanding maneuvers and acrobatics, instrument, navigation, and night flying of the diversified phase.

The 1943 curriculum revision reversed that emphasis to promote proficiency in the fundamentals because training officers in the continental air forces complained that men coming out of the schools were not competent in essential piloting skills. Presumably, for pilots progressing to advanced training it had become automatic to take off and land smoothly and with precision; to make spirals, elementary figure eights, and steep and climbing turns with ease; and to recover from power-on and power-off stalls and from spins. But these comparatively straightforward maneuvers nonetheless took a great deal of practice that these men apparently were not getting. As a result, the January 1943 program leapt from 23 to 34 hours in day transition, and eliminated radio-beam orientation, night formation, and transition to advanced trainer.[45]

The pendulum continued to swing back and forth, back and forth. Emphasis on elementary flying skills led to criticism that pilots were unprepared in instrument and formation flying, which by 1943 experience had demonstrated to be essential in combat. Although the heavy, stable BT–13 was generally thought to be ideal for formation flying, one pilot described the difficulties in learning to fly close:

> Beginners in formation always overcontrol, fighting to hold the proper formation position with wild bursts of power, followed by sudden frantic yanking the throttles rearward when it appears that the wing of the lead plane is about to be chewed up by the propeller of the airplane flying the wing position. Beginners also try to hold lateral position using only the rudders. The airplane is likely to wallow through the air like a goose waddling to its pond.[46]

Nor did flying by instruments come naturally. Pilots still relied on their own eyes and ears and flew by the seat of their pants. Howell Estes, whose

The heavy, stable BT–13 was extensively employed for basic pilot training.

service in the Training Command spanned much of the war, recalled that instrument flying was the "one area where we were the weakest at the start of the war." Initially the service lacked "any standard program for instrument flying instruction, any standard procedures to use for various kinds of situations, any standard methodology for teaching an individual what the real capabilities of his aircraft and his instruments were under given weather conditions."[47] Most instructors were unfamiliar with instruments and often suspicious of their reliability. In his memoirs, C. V. Glines humorously confirmed the prevalence of these attitudes in recounting his own instrument flight check:

> "Any questions, Glines?" the check pilot inquired after he had relieved his boredom by demonstrating his aerobatic prowess.
> I had only one. "Sir, what are these two instruments that we're supposed to keep caged all the time?" One looked like a compass, and the other had a small airplane on it.
> "Don't mess with those things, Glines! Keep those gyros caged. They're for airline pilots."[48]

Glines's experience might be considered only an aberration or a joke; also, he might have entered basic during one of the times when it was deemphasizing instrument flying. But early in the war, training regulations themselves implicitly recognized that most pilots, even those with considerable experience, had acquired minimal proficiency in instrument flying. All pilots on active duty who were not recent graduates of the schools or who already held airline pilot or instrument flying certificates were directed to take a one-month course to qualify on instruments.[49] Although it is highly unlikely that the directive was carried out, by early 1943 instrument training was considered sufficiently critical that all students in basic had to take and pass an instrument flight check because the "importance of instrument instruction cannot be overemphasized."[50] Most significant, the AAF opened a central Instrument Instructor's School at Bryan, Texas, in March 1943 and instituted a new system of instrument flying.

Both Estes and Glines credited Col. Joseph B. Duckworth, who established the school, with achieving considerable savings in lives and greater exactitude in carrying out combat missions. Duckworth introduced the full-panel system of instrument training that the Navy already used. Most Army pilots had been taught the 1-2-3 System of aircraft control, using only the needle (turn indicator), ball (bank indicator), and airspeed indicator. They were ignorant of the uses of the directional gyro and artificial horizon, which were caged (locked up) in Glines's plane. Duckworth's full-panel system employed all these instruments plus the magnetic compass, the rate of climb indicator, and the clock. The full-panel system brought a revolutionary improvement in instrument flying throughout the service. Unfortunately, the short basic course

Training Pilots

that at the time emphasized proficiency also prohibited extensive practice. The longer ten-week basic course of May 1944 added more hours in instrument and formation work during the diversified phase.

The standard basic training airplanes were BT–13s and BT–15s, the Vultee Valiants. The BT was faster and more instrument-laden than the primary trainers were, giving it the look of a "gadget from Mars," according to football star and then-student pilot Tom Harmon. "We had all heard," he added, "what a killer ship the BT was supposed to be," and in fact it was in basic that his class lost its first member in a training accident.[51] The basic trainers were low-wing, cantilever monoplanes with a top speed of 180 mph and ceiling of 21,000 feet. A 450-horsepower Pratt and Whitney Wasp engine powered the BT–13, and a 450-horsepower Wright, the BT–15. Owing to the industrial demand for steel and aluminum, during the spring of 1942 plywood substituted for metal in flaps, outer wing panels, tail surfaces, ailerons, and parts of the fuselage.

No airplane was problem-free, and the BT–13 was subject to engine failures attributable to a faulty fuel system that required modification by the manufacturer.[52] Student pilot Charles Watry provided a personalized picture of the basic trainer's eccentricities: "Spin recoveries in the BT were slam-bang affairs. The BT–15 spun well but spins were what gave the BT its affectionate nickname, 'Vultee Vibrator.' Once in the spin the canopy shook and rattled as if it might come off, giving the impression that the whole airplane was vibrating."[53] Another cadet might have questioned whether the nickname was affectionate, since he had "never met anyone who enjoyed flying in them."[54]

Some students also flew advanced trainers during their basic course. By the latter part of the war a surplus of AT–6s permitted their use in basic.[55] Yet the AT–6, like the basic trainers, was a single-engine aircraft, and the preponderant call, by 1943 and 1944, was for multiengine planes and crews. The Training Command, at a time when the basic curriculum stressed combat maneuvers and flying more advanced aircraft, introduced twin-engine planes into the basic

As these three men, former football stars and now cadets, prepare to take off during basic pilot training, they stand before one of the Vultee Valiant trainers.

program on an experimental basis. Small test classes consisted of a combination single-engine– and twin-engine–familiarization course, and another group trained solely on twin-engines. Although the Training Command recognized the advantages of beginning specialization during basic (the combination course provided insufficient experience in either type of plane), it discontinued the program in 1945 because it could not procure enough twin-engine aircraft.[56]

Cadet Charles Watry had nothing good to say about the small twin-engine Cessna AT–17 flown in the experimental program. He called it "a worthless airplane, without a single redeeming feature." Its wings, he scoffed, were "made of wood, the fuselage was fabric covered, and it had wooden controllable pitch props, for crying out loud!" Not only did he despise the plane, he dismissed the considerable amount of training in emergency one-engine flying that he considered to be statistically unlikely in combat.[57] But he, and other twin-engine advanced students, would soon discover that this type of training was routine, and for good reason.

Advanced

As early as the summer of 1940, advanced training meant specialization in single- or twin-engine aircraft. Yet of the eight schools providing advanced training at the time of Pearl Harbor, virtually all were single-engine because the training program owned a paltry number of twin-engine crafts. None of the latter type were even delivered to the Western Flying Training Center, for example, until December 1941. Aircraft did not materialize overnight with a declaration of war. Not until the following spring did specialized training begin to be more than a prayer and a paper plan. By that time twin-engine training was under way at ten schools, in response to the predominate need for bombardment pilots. At its height in September 1943 the ratio of twin-engine to single-engine pilots was 75 percent to 25 percent. Over the course of the war, 60 percent of pilot graduates were twin-engine trained.[58]

Single-engine. Although heavy bombardment dominated the other specialties and came to be viewed by subsequent generations as lying at the heart of the AAF's wartime identity and mission, a disproportionate number of cadets hoped to become fighter pilots. Apparently that preference was widespread among airmen as, for example, Samuel Hynes, who became a Marine Corps dive bomber pilot in the Pacific, later remarked that "to choose any course except single-engine planes... would have seemed cautious, unromantic, almost middle-aged, like wearing your rubbers or voting Republican."[59]

The single-engine advanced course mostly relied on the North American AT–6 Texan monoplane that began rolling from the factory in 1940 and continued in production through mid-1945. It was modified to upgrade armament and electrical equipment so that gunners, for example, trained in the AT–6A and AT–6B. That training was possible by the fall of 1943 when a

Training Pilots

Some AT-6s became available for training during basic during the latter part of World War II. Here, a pilot trainee performs a preflight check before boarding his single-engine AT-6D. An intercom microphone cord hangs from his leather helmet.

sufficiency of AT-6s warranted their use in flexible gunnery, in advanced single-engine, and even in basic schools. Over 4,000 of the D model alone saw service.[60] Even the relatively few fighter pilots designated for twin-engines flew the single-engine AT-6 in advanced training.

Training officials believed that at the advanced stage training aircraft should mirror combat versions as closely as possible. More often than not, tactical aircraft were becoming obsolete by the time their numbers or age made them available to students. As one example, small numbers of P-36 fighters that had come into service in 1938 remained in use throughout the war. Bill Colgan, who would enter combat as a fighter-bomber pilot over southern Italy, was one of the few cadets in his advanced class at Eagle Pass, Texas, who started on the P-36. This elderly airplane was short of spare parts, so Colgan and his ground crew resorted to improvisational methods to get it airborne:

> The normal starting system on this P-36 could not be used. Instead, it had to be started with a bungee. That was a long elastic line with a "boot" on one end and a "Y" in the line on the other end. The boot was placed on a prop blade in an overcenter position as one man held the prop there. Several men on each side of the Y ran down the ramp stretching the bungee. Once fully stretched, the man on the prop flipped it beyond center and jumped back; the bungee spun the engine and the boot flew off the prop and went whistling down the ramp between the men on both sides of the Y. If the engine doesn't start, it all has to be done again.[61]

Serendipity, improvisation, and discussion were never discarded for a by-the-numbers approach. By the time a pilot had reached Advanced Flying School, Single Engine, it was assumed that he would not wash out over issues

World War II

of competency, and he and his fellows and his instructors worked together to learn and practice. In fact, the insistence on informality and exchange was written into training plans as, for example, appears in the Program of Instruction for January 1943: "Flight instructors will be urged and encouraged to spend every moment possible in the student rooms holding critiques and discussions with their students."[62]

In ground school the fighter pilot was not introduced to the extensive array of dials on the mock instrument panel that confronted his friends who were now in twin-engine advanced classes. Nonetheless, he was supposed to gain an understanding in the classroom of the sensations experienced during instrument flight. Because the fighter pilot flew neither in large formations nor with a crew, he had to rely heavily on his own navigational ability; navigation was therefore an important topic in ground classes. Strangely enough, the study of weather did not become a part of the single-engine curriculum until the expanded 1944 course.

Reports from the European theater stressed the importance of drills in aircraft identification, and as the naval war in the Pacific assumed greater strategic importance, basic and advanced ground classes added ship recognition. A returning bombardier from the European theater commented in 1943 that nearly everybody forgot what they had learned in school about the shapes and markings of ships and aircraft. As a result, "we shot at anything that wasn't a B-17."[63] The 1944 training standards reiterated the importance of the subject, requiring recognition of forty operationally important aircraft and "those types of combatant and noncombatant naval vessels of operational importance."[64]

Curricula divided hours spent in the air into transition, formation, navigation, instrument, and acrobatics. Practice on the Link trainer was considered part of air rather than ground work. No significant changes occurred during the period in which the advanced course lasted nine weeks except that night formation disappeared from the curriculum, to the perhaps momentary displeasure of none other than General Arnold.[65] All advanced programs included formation flying, although less in single- than in twin-engine aircraft. Fighter pilots in training flew the close three-ship V formation for the purpose of checking proficiency, and thereafter the two-ship combat element with two elements (four aircraft) to each flight. Tactical formations of twelve airplanes amassed for the purpose of demonstrating "air discipline, precise and accurate position holding." Fighter pilots practiced acrobatics much more than formation flying. They also flew at least one low-altitude mission and, if oxygen equipment was available, one at 15,000 feet or above.

Single-engine pilots navigated principally by dead reckoning[66] and made at least one blind flight under the hood as another student acted as safety pilot.[67] Navigational skill was crucial to the pilot who flew alone. As late as May 1944 airmen in the China-Burma-India Theater wrote home about the dangers to pilots who were inept at basic navigation. In a part of the world with treacher-

Training Pilots

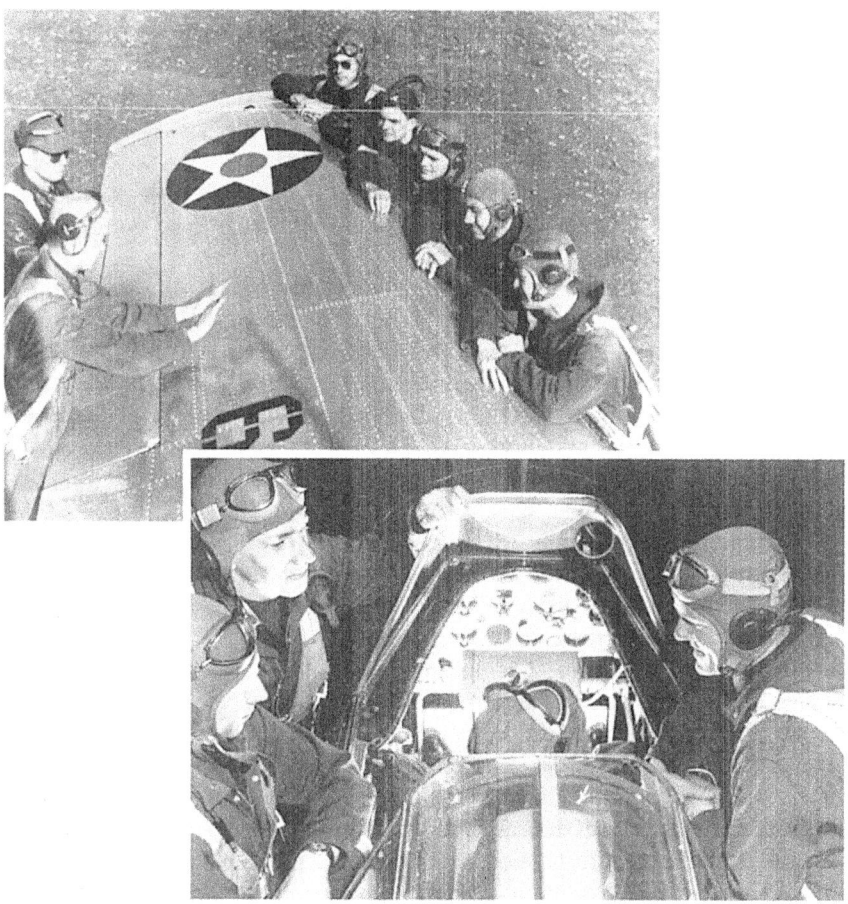

The curriculum during advanced training for single-engine pilots included coursework in formation and instrument flying.

ous terrain, unpredictable weather, limited radio facilities, and very poor maps, pilots who relied strongly on visual references often became "lost or 'confused' and since they have had limited training in dead reckoning appear not to trust the compass, which results in milling around in search of check points instead of striking out on a course in an effort to work a time distance problem."[68]

Grim statistics showed that fighter pilots died in training accidents more often than did pilots in other specialties; fatalities ran to about 70 for every 100,000 hours flown. The emphasis on acrobatics and the risk-taking personality of many of those who were assigned to fighters offer obvious explanations. Richard Montgomery, who was on the training side of the Air Staff, later commented about the assignment process: "We'd say, 'Well this

World War II

fellow is aggressive and he ought to go to fighters.'"[69] Another clue to the high accident rate in pursuit may be traced to the physical requirements. More men competed for fighters than there were open slots, so those chosen were presumed to be the best pilots. Early in the war combat commanders complained that fighter pilots failed to perform as well as expected, but many airmen believed that fighter pilots were picked because of their small physical size rather than their high aptitude, which would explain the lower competency and the higher accident rates.

The Training Command eventually changed the fighter course (if not the selection of small, aggressive men) to improve safety and competence. It stressed proficiency flying and de-emphasized combat maneuvers. In 1944 all phases of training expanded to ten-week sessions; additionally, fighter pilots continued for an extra five weeks in gunnery, transition on the P–40 or P–39, and some refresher training. Further opportunity to practice combat maneuvers in OTUs and RTUs after graduation helped give pilots more experience before going to war.[70]

Because most fighter pilots flew single-seat aircraft without gunners, their success, to say nothing of their continued survival in battle, depended largely on their shooting ability. Pilots selected for fighter, reconnaissance, and dive bombardment units were drawn from those demonstrating superior marksmanship in the first weeks of the gunnery course at single-engine advanced schools.[71] The three training centers also operated gunnery ranges for single-engine advanced students. Unlike the movable, i.e., flexible guns in bombardment airplanes, the pursuit ship fired a stationary gun mounted on the leading edge of the wing or nose of the airplane (or both). On a gunnery simulator the student could calculate deflection of a fixed gun when firing at a moving target. Gunnery practice went from skeet shooting to ground range work and to aerial firing. From the air, the pilot could aim at a ground target to simulate strafing or at a towed target for air-to-air combat. Students participated in "individual attack," an exercise in which they flew in a loose string slightly above the instructor and then individually peeled off to make simulated attacks against the instructor's plane. Thereby they learned to

An aviation cadet fires from a Link fixed gunnery trainer to simulate the ideal curve of pursuit when firing at a moving target.

judge range for firing, rate of closure, and use of gunsights.[72]

Twin-engine. With the exception of the relative few who flew twin-engine fighters, the advanced twin-engine course trained pilots for bombardment units. Students had not previously controlled two engines that demanded adjustments even in taxiing. Moreover, although the advanced course focused upon individual proficiency, in learning to fly a multiplace airplane that might carry a copilot, a navigator, a bombardier, and a number of gunners, the pilot was introduced to the crew concept. A man had to communicate with others at the same time that he performed his own tasks. He had also to rely unquestioningly on the ability of others. If a twin-engine pilot was to become an airplane commander, he had to demonstrate qualities of leadership and a sense of responsibility for his eventual crew's performance and coordination.

Although he was not yet in charge of a full crew, in training the pilot began working as part of a team. Usually two students flew together, one acting as first pilot, the other as copilot, and then switching positions on other flights. Formation flying also furthered group cohesiveness. One bomber pilot recalled that he and his cohorts became very close during advanced training, mostly by flying formation together: "We had to fly on each other's wings from time to time and that requires complete reliance upon each other.... When I hit combat it didn't take me long to learn that in those days a pilot's life absolutely depended upon his ability to fly good formation, because its purpose was mutual protection."[73]

Hours in the air began with the transition phase. Starting his advanced twin-engine course at Blackland Army Air Field outside Waco, Texas, John Boeman found that "the loops, rolls, and fighter maneuvers of previous training no longer applied." Instead, he was introduced to a plane with entirely different performance characteristics. After becoming comfortable with the twin-engine trainer, the cadets simulated engine failure by flying with one engine shut down, the practice that cadet Charles Watry had disdained in his two-engine experimental basic course. They then went on to formation cross-country navigation, and then to advanced instrument flying in a hooded cockpit.[74] In response to reports from the field, the Training Command emphasized high-altitude flying for all specialties and, at least in 1943, expected twin-engine pilots to take one flight up to 30,000 feet. The curriculum listed the standard ten hours in the ubiquitous Link trainer.[75]

During most of 1942 and 1943, student pilots practiced transition, formation, navigation, and instrument flying at night as well as during the day. They flew navigation flights off the airway at night and under the hood. The training formation was the basic three-ship element with approximately five feet of clearance between wing tips. Because of the closeness of the airplanes when visibility was low as well as the danger of disorientation in the dark, even experienced pilots, according to one historian, approached night formation "with dread."

World War II

Most single-engine students flew the well-regarded AT–6, but twin-engine pilots learned on a slew of aircraft, none of which combined the desired balance of speed, maneuverability, and docility. The AT–7 was the first designated twin-engine advanced trainer, initially procured for navigator training. The partially wooden AT–10 fared badly in the cracking, drying heat where many of the training fields were located, and it had poor control characteristics. The AT–11 was a C–45 modified to carry flexible guns and bomb racks, but it had a service ceiling of only 17,500 feet, possessed no oxygen equipment, had a limited bombload, and its flying time was only six hours. The Cessna AT–17 and AT–9 were converted commercial types with serious design defects: part of the AT–17 wing was severely understrength which contributed to a large number of fatal crashes before it was modified; the AT–9 suffered from leaky gas tanks and faulty landing gear. The AT–21 was an example of an advanced trainer whose delivery was delayed because the service could not decide whether there was a more pressing need for a bombing crew trainer, a gunnery trainer, a navigation trainer, or a transition trainer. A few other airplanes, such as the AT–18, were outfitted for specialized use and modified with either machine guns and tow target provisions or navigation equipment.[76] Eventually the training program received tactical aircraft, especially B–25s, but also some B–17s and P–39s, which succeeded in reducing or eliminating the transition phase in later replacement crew training.

Variations of the B–25 medium bomber poured from the factories at a great clip by the middle of the war. The surplus was a boon to the training program, allowing General Harper, the chief of training on the Air Staff, to redeploy training aircraft:

The AT–7 was the first designated twin-engine advanced trainer.

If 25 percent of the aircraft requirements for Advanced Twin-Engine training is furnished in B-25 type aircraft an overage in twin-engine trainers will exist. It is desired to place these excess airplanes in basic schools. No additional AT-17 or UC-78 aircraft are desired, and attrition aircraft should be furnished in AT-10 aircraft.[77]

The training version of the B-25 (also designated the AT-24) mimicked the combat plane, but it was lighter and faster because it was stripped of armament and armor. Cadet Watry found it to be "as honest an airplane as there was. It was a joy to fly." Initially, the instructor occupied the left seat, one student in the copilot's seat and a second cadet observing from the navigator's compartment. As always, different aircraft required different training procedures, as Watry learned:

> Landings did not pile up very fast in the B-25 training. We did not make touch-and-go landings, but taxied back to the takeoff end of the runway at the auxiliary fields following each landing. Once we were airborne we had to fly around for 30 minutes to allow the brakes to cool, which further reduced the number of landings that could be made in a training period.[78]

New aircraft also meant new controls, and adjusting to the change could be nearly fatal, as Watry likewise discovered. The first time he flew as a B-25 copilot, he was supposed to reduce the prop as the pilot cut back the throttles from takeoff power to climb. Accidentally pulling the mixture control lever backward, he realized his mistake barely in time to avert a crash. In the B-25, the mixture controls were located where the prop controls were in the AT-17, the airplane he had flown previously. "Never mind that the prop control knobs were colored differently," he admitted. "I had acted from habit, not logic."[79]

Observation. The Training Command opened schools for a third category of combat pilot training with which the AAF was least enamored — observation. Pilots in the advanced course flew single- or twin-engine observation type aircraft: they qualified as an observation pilot in a single-engine pursuit airplane or in a bi-engine, light-bombardment–type aircraft like those employed in observation squadrons, reconnaissance, and air support units. They also prepared for qualification as an Aircraft Observer, a rating obtained after an additional five-week course of instruction.[80]

Ground classes covered signal communications (the overwhelmingly largest subject that was allotted sixty-two hours of classtime), navigation, aerial photography and interpretation of aerial photographs, aircraft and naval vessels and ground forces identification, intelligence, observation aviation and missions, air forces and combined arms, antiaircraft artillery, and artillery tactics and missions. Pilots also became familiar with equipment maintenance and took physical and military training.[81]

Air work began with transition, both day and night, in single- or twin-engine advanced trainers and then, if possible, in combat-type aircraft. Cadets spent equal time in single- and twin-engine airplanes when both were available. Night flying stressed runway operations with and without wing lights and under simulated blackout conditions. Observation pilots flew a few hours in formation, but it was considered less important for them than it was for other specialists. Typically, the two-ship element was the basic formation for single-engine training, and the three-ship V for two-engine.

Using primarily dead reckoning and pilotage, the trainee flew navigation missions at low (500 feet), medium (between 2,000 and 5,000 feet), and high (above 15,000 feet) altitudes. This aspect of training was crucial for the observation pilot, as training standards stated: "It is of utmost importance to photographic missions that the pilot strike the various points on his line of flight exactly. Missions will, therefore, be graded on ability to cross each check point exactly." On navigation flights, pilots employed the E–6B computer to determine course, speed, and winds. Instrument flying was concentrated in the first four weeks of the course. To graduate, the pilot had to pass an instrument check flight on a radio range, if available, or on the Link trainer. Combat maneuvers received fleeting attention. Because the job of observation pilots primarily entailed photographic work, adjusting artillery fire, visual and night reconnaissance, and communications, they mostly practiced flying in clouds and evasion tactics.[82]

Transition Training

At what point should pilots learn to fly combat aircraft? The American system of training held that a fledgling pilot was incapable of handling high-performance planes during the early stages of his career, so he moved carefully and systematically from simpler to more demanding aircraft and maneuvers.[83] Yet the nagging pressure of time competed with training policy. Owing to the availability of large numbers of AT–6s, the Training Command transitioned some students onto advanced trainers during basic flight training. Even had the command wanted to introduce cadets to combat aircraft during the school courses, however, not enough tactical aircraft could be spared for use on a systematic basis. Combat units had priority for service airplanes; then came the training air forces. Thus, initially at least, the first units to receive tactical aircraft transitioned newly arrived men.

But the four domestic air forces were primarily responsible for assembling and training crews, not providing individual pilot training. The latter was the job of the Training Command. The AAF resolved this administrative division of duties not by adding transition to the existing course (although it did lengthen individual pilot training) but by adding another discrete stage to the end. A set of specialized schools and courses under the Training Command, and to a lesser extent, under the air forces, transitioned newly commissioned graduates of the schools to combat aircraft. All pilots completed this postgraduate instruction before taking up duties with a crew for operational training.

Training Pilots

Between August and December 1942, the Training Command established specialized schools for four-engine heavy bombardment transition and for B–25 and B–26 medium bombers. In early 1943 it initiated both single- and twin-engine fighter transition on P–38s (also called P–322s in training) and P–40s at the end of advanced training. Transition to light and dive bombardment and other fighter aircraft, on the other hand, remained lodged with the training air forces throughout the war. Headquarters AAF administered very heavy bombardment.

Transition training thus became a function spread throughout the system and thereby was less coherent administratively than were the earlier phases of individual instruction. In order to develop some consistency in methods and standards for first-pilot qualification, those responsible for individual proficiency during transition training collaborated through a series of conferences and maintained close liaison between the Training Command and training air forces.[84] In four-engine training, for example, no single standardization school was established for both the Training Command and Second Air Force, but Headquarters AAF called upon both training agencies to agree upon a standard checklist, emergency procedures, and instrument check. The September 1943 training conference at Fort Worth specifically addressed these matters and the curriculum and practices in four-engine training wherever it was offered.[85]

In the early stages of pilot training, gross numbers drove the system. At the transition stage, wartime demands for different types of aircraft set manpower goals. Yet time lagged between notification of operational requirements and

Specialized schools were required for four-engine heavy bombardment transition training, such as that for the B–17 bomber shown here in flight.

implementation of training programs. To cite one of the more intractable situations, quotas for night fighter pilots rose or fell according to changing circumstances in the combat theaters. Yet the Training Command never found a way to respond quickly at an end point in the training cycle of an especially small program, namely one that drew upon a tiny pool of twin-engine-trained fighter pilots who required transition to the scarce twin-engine P–38. Other specialties presented other impediments. Very heavy bombardment, for instance, as a latecomer to the war, incorporated some of the most advanced technologies, such as radar, that required developing skills for which there was little programmatic experience.

Reasons other than limited training resources kept transition training from becoming a true rehearsal for operations. During transition, pilots were to be introduced to controls, engines, instruments, flying characteristics, and the performance of combat aircraft they had not flown previously. Transition could not complete this familiarization process, however, because training versions did not duplicate tactical aircraft. The P–322, for example, lacked the superchargers and contrarotating propellers of the operational P–38. Bombers on training flights seldom flew fully loaded; they therefore handled differently than they would when armed for combat missions. In addition, fighter pilots were not allowed to engage in a training dogfight with live ammunition. The AAF made other compromises in the interests of safety over realism. To reduce the accident rate, for instance, one commanding officer sent his students down a brightly lighted airway and back instead of down the standard, unlighted triangular course for night cross-country flying.[86]

Bombers. Transition was both the last step of individual proficiency training and the first step in operational training. The bombardment pilot started his formal transformation into an airplane commander during transition. By the spring of 1944, transition included six hours on the duties and responsibilities of the airplane commander. Graduate pilots reviewed the responsibilities of an officer and the code of military law; the organization and policies of operational and replacement training; the duties, responsibilities, and obligations to crew members; the duties of ground and air crew; and the importance of air discipline. Returned combat pilots gave an account of their crew training and experiences in the theaters.[87] Despite the indoctrination, an inspector who visited the specialized four-engine school at Liberal Army Air Field, Kansas, observed that proficiency as an airplane commander was difficult to obtain during transition. "It is a new phase of training to students graduating from advanced flying schools, and who have had no command experience whatsoever." Pilots were new to the equipment and carried no crew during the transition phase, and because the instructor was in command of the airplane at all times, the inspector explained, "it is difficult to set up practical experiences for the students."[88]

Because of its high priority, heavy bombardment set the pattern for transition training. Units of the Second Air Force conducted it until August 1942 when it moved under the umbrella of the Training Command. The command established four-engine schools at Sebring, Florida, and Smyrna, Georgia. At the end of the year two more schools opened; in January 1943 a central instructor's school that also trained Second Air Force instructors began operations, and expansion continued. To meet the requirements of the Second Air Force, which in turn felt heavy pressure from Europe for heavy bomber crews, in September 1943 the Training Command began retransitioning medium bomber pilots and it raided instructor schools for students. Also, it compressed the course from time to time by reducing the number of flying hours prescribed in training directives.

For heavy bombardment, pilots transitioned onto B–17s or B–24s. Through the midpoint in the war the tactical training units asked for an equal number of B–17– and B–24–trained students, but the Training Command had a greater number of B–17 aircraft. It finally acquired more B–24s and converted schools from one type to the other as the need arose. The curriculum in each was similar, but because the B–24 was a more difficult airplane to fly, the B–24 course added hours during 1942 and 1943. The B–17 course lasted 9 weeks and had 105 flying hours, in comparison to the B–24 course which went to 125 flying hours before reverting to the B–17 course specification in February 1944.

Air work included day and night transition; instruments; formation, high-altitude, and navigation missions; and 15 hours of practice in the Link trainer. Technical instruction in both theoretical and practical engineering maintenance dominated ground classes. Navigation, radio, meteorology and weather, aircraft recognition and range estimation, naval forces and ship recognition, first aid, oxygen, code review and signal lamps, chemical warfare defense, and athletics occupied the rest of students' time. Transition and instrument training accounted for the increased hours in the B–24 course. Revisions in the course between April 1943 and February 1944 increased instruction in the use of the autopilot and allotted more hours to navigation and airplane commander duties. The 1944 program included bomb approach training.[89]

In heavy bombardment transition the troubling new issue of copilot assignment emerged. Suggestions for ways of providing copilots to the Second Air Force, which trained heavy bomber crews, came from all quarters, with no very satisfactory solution. When the Training Command established its four-engine schools, some single-engine pilots joined the operational training units of the air forces in the United States, retrained in bombardment, and became copilots, assisting graduates of the four-engine schools. These single-engine graduate pilots, relegated to copilot duties, had already been trained in now irrelevant fixed gunnery and pursuit tactics. Their morale plummeted and the accident rate soared.[90] As the four-engine schools met their quotas, some of these graduates also became copilots. Here too, assigning men who had made

World War II

their way successfully through a long and demanding training sequence, only to be foreclosed from becoming airplane commanders, produced bitterness and inefficiency. In January 1944 the AAF initiated a career copilot specialty. Partway into their pilot training, a number of second lieutenants were selected as copilots. They then took 4½ weeks of flexible gunnery training plus a ground course to qualify them in engineering aspects of four-engine aircraft.[91]

Pilots assigned to medium bombardment squadrons in the Third Air Force also transitioned onto one of two airplanes, in this case the B–25 or B–26. (A version of the B–26, the AT–23, had been modified for training as a high-altitude tow plane.) Because of its extremely high wing loading, early models of the B–26 landed at dangerously fast speeds. Mortalities were so high that the B–26 Marauder developed a fearsome reputation among pilots, who called it the flying coffin. The accident rate dropped when the Training Command was able to establish specialized transition schools.

Even so, B–26 transition schooling got off to a rocky beginning. For the first three months the thirty-seven airplanes allotted to the Training Command were all in overhaul. At the same time, the command worked hard to overcome negative publicity about the airplane. Initially it selected only the best twin-engine graduates for B–26 training, and for a time it assigned only volunteers. The B–26 was the subject of the first training film produced about combat aircraft (containing information such as flight characteristics, emergency procedures, and essential technical features). In 1942 experienced B–26 crews visited the advanced flying schools to entice upcoming graduates to volunteer as B–26 pilots and to provide some transition flying to the Director of Training or his representative. AAF headquarters forwarded glowing reports from the combat zones to the advanced and transition schools in order to "impress on future pilots the efficiency of this airplane under combat conditions."[92] The combined efforts helped to allay pilots' suspicions and fears. The course finally settled into place in 1943 and came to resemble four-engine transition, with the same stages of air work.

The B–25, on the other hand, proved to be one of the most reliable and widely used medium bombers of the war. It was considered to be such a forgiving airplane that the question was raised as to the necessity for creating a specialized transition school. In the category of famous last words, at the end of December 1942 the Director of Individual Training on the Air Staff stated categorically, "No repeat no B–25 postgraduate transition school is to be established."[93] Nonetheless, shortly thereafter the Training Command opened two such schools, although eventually both were converted into advanced twin-engine schools using the modified B–25 (AT–24).[94]

The smaller programs of light, dive, and very heavy bombardment also set up transition courses. None of them transferred to the Training Command. The Third Air Force mostly provided light bombardment transition during Phases I and II of operational training and Phase I of replacement training. Students

Training Pilots

The Martin B–26 was used for transition training on twin-engine bombers, although this aircraft was dangerous to land.

took a brief dual flight prior to solo on the Douglas A–20, variants of which were used with great success by Allied air forces. The H model was employed effectively in close air support and intruder missions in the Pacific, and a later version, designated the A–26 Invader, went into service as a ground attack and tactical bomber in 1944.[95]

Neither the A–24 nor the A–31, flown initially in dive bombardment transition, became operational. Dive bombardment squadrons were finally equipped with A–36s, a dive bomb or attack version of the P–51. Because this aircraft was not available for some time, most pilots transitioned onto single-engine P–39s; their experience resembled pursuit training. In August 1943 dive-bomb units were redesignated fighter-bomber and placed under the III Fighter Command, at which time these units were trained the same as fighter units.[96]

The B–29 Boeing Superfortress, which became the mainstay of the bombardment campaign in the Pacific at the end of the war, came into the training system in October 1943. One group of the XX Bomber Command, activated in November, remained stateside to form the nucleus for new units trained by the Second Air Force. As it appeared in February 1944, the transition course included six hours of landings and air work and three hours each of check flights in emergency procedures, day landings, night landings, and instrument flying during takeoff and low-visibility approaches.[97]

Fighters. Fighter transition included two groups, P–40 training for single-engine pilots and P–38 training for twin-engine pilots. P–40 transition had the advantage of drawing pilots who had been flying single-engine aircraft from their earliest training. When P–40s became available at advanced schools, many single-engine graduates had already partially transitioned onto the aircraft

World War II

Transition training on fighter aircraft began in early 1943 on the twin-engine P–38.

by the time they received their wings. Because these men had spent part of their advanced air work in transition, they were found to be deficient in some of the fundamental piloting skills. Training officials concluded that transition should not be integral to the advanced course, but rather should be tacked onto the end. Transition thus became a formalized phase although not taught in a specialized school, as was the case with heavy and medium bombardment. Near the end of 1943 the Training Command mandated an additional 4½, then 5 weeks of P–40 transition plus 30 hours of fixed gunnery training after the advanced course. Besides air work in transition and familiarization, formation, and navigation, gunnery practice included practice with camera guns and with fixed guns and sights, and aerial firing at towed targets. Tactics stressed simulated ground strafing.[98] Not only did proficiency improve, but the transition period functioned in the fighter program as it did in bombardment to screen those who would become airplane commanders from those less capable and who were reassigned as bombardment copilots or transport or utility pilots.

P–38 transition faced the significant hurdle of shifting single-engine pilots to twin-engine aircraft. The Training Command selected students from advanced single-engine schools for transition on twin-engine aircraft, the amount of time "to depend upon the availability of equipment and upon the time allowed for such transition training."[99] In January 1943 the command began a new advanced twin-engine fighter course. At the heart of the 9-week course were 23 hours of transition to P–38s. (The P–38 was interchangeably referred to as the P–322, but the two airplanes varied enough to cause some trouble, as one officer cautioned.) Unfortunately, P–38s were in such short supply that by the end of the year most pilots flew AT–6s with only a few hours in the twin-engine AT–9s and P–322s. Students who had already flown a minimum of ten hours day transition on the AT–6 in basic flying training received preference on the AT–9.[100]

Tactical units of the Fourth Air Force were also directed to conduct twin-engine fighter transition. The shortage of P–38s in both the Training Command and the Fourth Air Force occasioned a drastic accommodation. In February and March 1944 the Advanced Twin-Engine Fighter School and the IV Fighter Command began transitioning pilots on the P–39, a single-engine fighter. The Training Command sent its stock of P–38s to the Fourth Air Force. The best cadets in the advanced course who were then transitioning on P–39s were chosen for the rare openings in twin-engine fighter transition. The remaining single-engine graduates became bombardment copilots or took up other duties.[101] The lack of equipment had proved disastrous. Among the specialized pilot training programs, twin-engine fighter training was probably the least successful in terms of its capacity to train pilots and the numbers graduated.

Observation and Reconnaissance. Observation and photoreconnaissance transition training bore some similarities to the fighter course. For a time, pilots in both programs flew the Lockheed P–322, called the P–38 in fighter squadrons and the F–5 in reconnaissance units. The photoreconnaissance OTU of the Second Air Force gave transition training until the summer of 1943 when the new III Reconnaissance Command took over photoreconnaissance. Whereas the fighter program phased out P–322 transition, photoreconnaissance initiated it in RTUs in February 1944. At the same time, the Air Staff informed the Training Command that both programs employing P–322 aircraft might have to meet their requirements with other planes because the P–322s were aging and replacement parts were scarce.[102] Photoreconnaissance pilots took 60 to 80 hours of fighter training. Then they diverged from the fighter schedule, which included fixed gunnery and combat maneuvers, to concentrate on navigation by pilotage.[103]

Advance and Retreat

At the end of 1943 the training program reached its zenith, then began a sharp decline. On August 5, 1944, the Commanding General, AAF, personally approved the reduction of pilot training to 20,000 a year. Some 15 civilian-run primary schools were to be released from the program. General "Hap" Arnold and Assistant Secretary of War for Air Robert Lovett requested that no publicity be given to the news, as they correctly anticipated the great wave of adverse public opinion that would soon swamp them.[104] A month later, the Chief of the AAF Training Command, Lt. Gen. Barton K. Yount, announced: "All personnel in pilot training from preflight through advanced schools will be frozen on 1 October 44 in the phase of training they are undergoing at the time. No entrance will be made into pilot training until 1 January 45 and graduations from advanced schools will be made only as necessary to meet requirements." The AAF expected those requirements to mean an annual

World War II

Lt. Gen. Barton K. Yount, Chief of Training Command

graduation rate of 10,000 pilots. Yount projected a 10,000-pilot rate for the period between the defeat of Germany and the defeat of Japan. But in fact, after V–E Day the pilot quota dropped to a thousand a year.[105]

By necessity, throughout the war the AAF trained extremely large numbers of people in a notably short period of time. Yet it kept intact the fundamental system of pilot training with which it went to war and insisted, even under enormous wartime pressures, upon a course of training much longer than that used by the Germans or, late in the war, the Japanese. It turned to mass production methods and relied heavily upon standardized means of teaching prescribed subjects, with clearly delineated rules for elimination and qualification. It drew upon the expertise of aviation psychologists in a number of areas, for example, in devising standardized tests for classification and in drawing up objective scales for evaluating an individual's performance in relation to that of others. In the latter case, the service tried to identify specific elements by which to rate one's flying skill so that passing or failing did not depend exclusively on the instructors' opinions. Tests of measurable skills became yardsticks at all flying schools and, afterward, in qualifying men for combat aircraft. Check pilot manuals came into use throughout the program.

Despite the herculean efforts that brought uniformity to the massive educational system, complete standardization of testing procedures and elimination of subjectivity and bias did not and could not have occurred. Certain qualities could be measured; others could not. Training officers were not so unsophisticated as to believe that human behavior and future performance could be reduced to a set of quantifiable and predictable variables. No one could completely quantify flying skill, which was composed of judgment as much as physical coordination. In pilot training, standardization and objectivity had their useful limits.

No wartime structure as vast as that formed to train pilots could have eliminated failures and mistakes and some lack of vision. But it is not clear in retrospect that the AAF could have done a substantially better job. Defenders could point to the fact that training occurred on hastily constructed, poorly equipped airfields, with too few aircraft, mentored by inexperienced instructors (who were themselves in chronic short supply), and judged against a strategic mission that lacked tried tactics and techniques as the means to accomplish it. These conditions mirrored those in the theaters of war. Nonetheless, despite these deficiencies, inspections, combat reports, and training conferences indicated that pilot training was more successful than training in other parts of the system. It should not have been otherwise since the fundamental methods of pilot training were as old as American aviation itself.

The training establishment found it difficult to attain the proper balance during wartime between thoroughness and speed, as expressed in the tension between proficiency flying and combat tactics, and between realism and safety. These issues were constantly revisited. In a peculiar reversal of the general direction in training, for example, in 1942 one officer urged that primary and basic training aircraft be sent to single-engine advanced schools so that pilots could be taught how to better estimate relative closing speeds in aerial combat.[106] Although that technique, like many others in training, was safer on slower aircraft, its adoption would have postponed other aspects of operational training, notably transition onto combat aircraft. The American system of pilot training was based upon a slow and steady evolution toward complexity, but it was seldom accomplished in a strictly linear fashion. Training officers were sometimes thought to emphasize combat techniques at the expense of careful, accurate flying; at other times they were accused of failing to teach tactics that simulated the stress and uncertainty of combat. Just as training officers could not articulate with exactitude the characteristics that defined a good pilot, or measure all the skills by which a man would qualify during training, they were unable to eradicate all of the ambiguities about the definition of an effective training program in time of war.

At the height of pilot training in 1943, the AAF was training some fifteen times as many men as the Japanese were.[107] The service was highly successful in achieving its manpower goals by this time, and combat attrition rates were lower than had been feared. As a result, the Army Air Forces leadership reined in pilot training and began to consider its postwar force structure. It was not yet clear how many men would be needed, but little doubt existed that the pilot would continue to be the backbone of a strategic-minded air force. As long as war remained to be fought, the pilot was only one member — though perhaps the main one — of an aerial army.

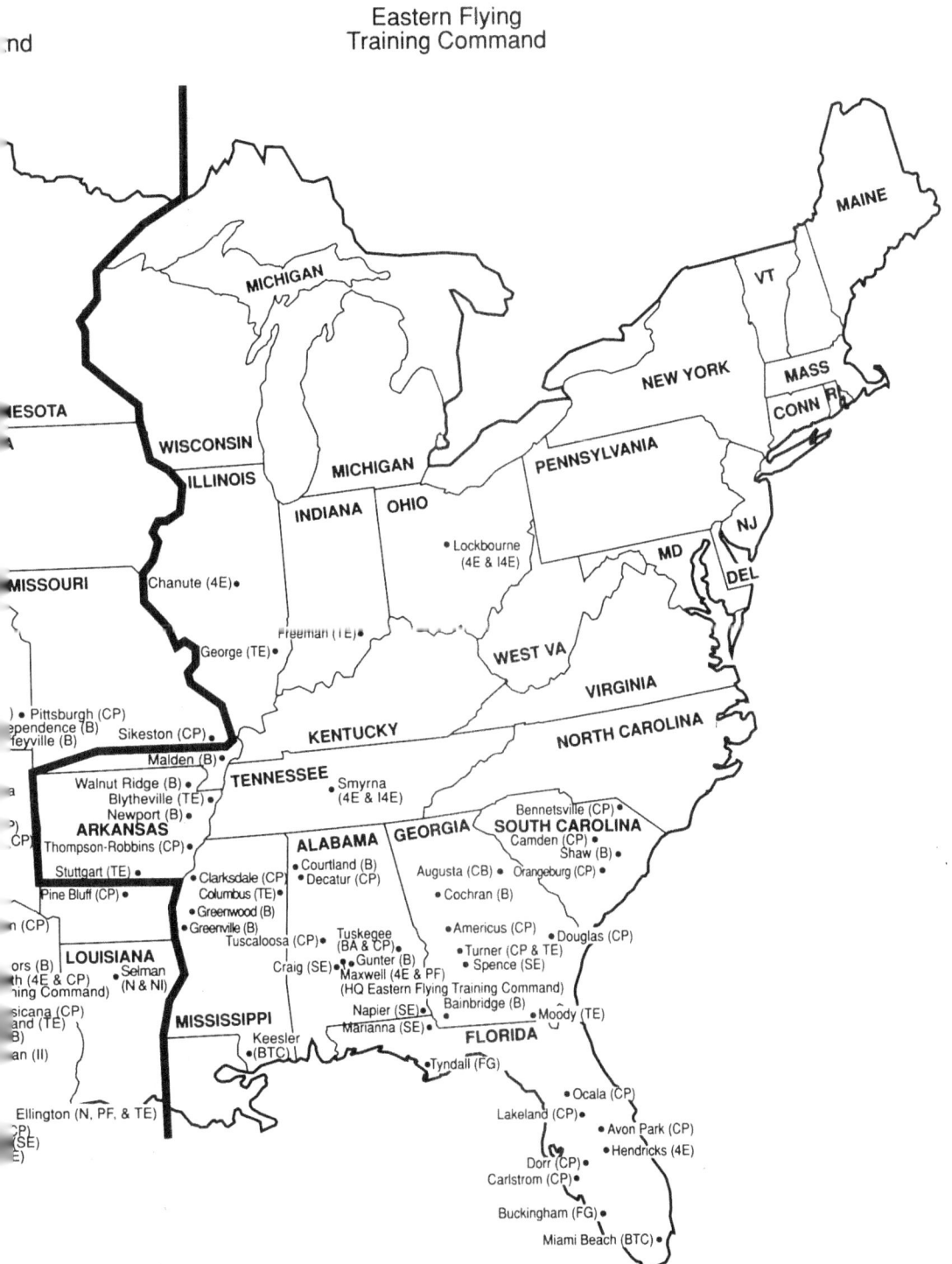

TWELVE

Not Just a Pilot's War: Individual Training of Navigators, Bombardiers, and Gunners

> This is an age of specialization. No rational man can hope to know everything about his profession.
> —Third Report of the Commanding General of AAF to Secretary of War, 12 November 1945[1]

Well over half of the American combat pilots would fly as airplane commanders of a crew composed of as many as nine other men — a copilot, navigator, bombardier, plus specialist and nonspecialist gunners. Just as for pilots, individual training of bombardiers, navigators, and gunners came in Training Command schools. Navigators and most bombardiers received their commissions at the end of specialized training; gunners were enlisted men. Navigators and bombardiers, like pilots, faced a battery of classification tests to determine their specialty before they began preflight training. Some men who had been chosen but then dropped from pilot training were reclassified and retrained as navigators and bombardiers.

While pilot training compressed in length early in the war, both the navigator and bombardier courses ultimately elongated when training officials found that men failed to achieve acceptable levels of proficiency in the time allotted. The change did not signify, however, that nonpilot flying-officer training reached the same length or depth (or importance) as that for pilots, which was more than twice as long as either of the others. The navigator course went from a total of 15 to 18 weeks in January 1943 and then to 20 weeks in December 1944 once quotas were being met. The bombardier course was 12 weeks at the time of Pearl Harbor, was shortened twice thereafter, then leapt upward again to eventually become 24 weeks. The expansion accompanied the implementation of a new curriculum by which all bombardiers also qualified in dead reckoning navigation. Gunners spent the least time in Training Command specialized schools; their course added one week, extending it from 5 weeks to 6 in April 1943.

Training plans stated production goals for navigators, bombardiers, and gunners relative to the total number of pilots to be trained. The ratios among the four specialties shifted in response to wartime requirements in the combat theaters. In September 1942 the authorized ratio of pilots to navigators was 9 to 1. At about the midpoint in the war, in June 1943 when all schools were working at full capacity, Maj. Gen. Barton K. Yount informed Arnold that the Training Command was aiming at a production rate based on the ratio of 1 navigator to 4.7 pilots and 1 bombardier to 6.3 pilots. The August 1942 proportion of 11:10 (gunners to pilots) tended to be representative since there were 6 gunners, for example, on a 10-man heavy bombardment crew.[2]

The skills required of navigators, bombardiers, and gunners were closely linked, resulting in considerable cross-training. In what was called dual training, many navigators learned elements of bomb-dropping, and by the midpoint in the war all bombardiers also qualified as dead reckoning navigators. Experience also demonstrated that all bomber crew members (excluding the pilot) should receive flexible gunnery training. That idea remained codified in training directives even though facilities were too overtaxed for most of the war to keep up with such ambitious requirements. Therefore gunnery training for officer aircrew members depended less on combat requirements than on when it could be fitted into the rest of the program. Sometimes bombardiers and navigators took gunnery before preflight, sometimes before their specialist course, sometimes afterward, and sometimes not at all.

Navigator Training

At the outbreak of war, the demand for navigators escalated disproportionately. Nearly every agency suddenly required navigators — the Air Transport, Troop Carrier, and Antisubmarine Commands; as instructors in the navigator schools of the Training Command and in the OTUs and RTUs of the Second and Third Air Forces; and, of course, in the combat units. Moreover, production schedules became harder to meet as training requirements for celestial and dead reckoning navigators differed among the medium, heavy, and very heavy bombardment units, and from one theater to another.

Navigator training had already begun at the time of Pearl Harbor, including the contract school run by Pan American Airways. But the Army-operated navigator program shared facilities and aircraft with the pitifully underequipped advanced twin-engine pilot schools, which themselves had priority in equipment and air time. Too few navigators graduated, and the quality of

Training Navigators, Bombardiers, and Gunners

training suffered. During the late summer and fall of 1942 the Training Command assumed virtually all individual training by establishing separate navigator schools. This was an essential step, although one still incapable of meeting production goals for some time. The schools remained strapped for instructors, aircraft, instruments, training aids, and facilities. Once again the system failed to turn out well-trained navigators. Nor did it graduate enough of them and had to resort to the expedient of dual navigator-bombardier training that, when formalized, extended the course from 15 to 18 weeks.

Desperately attempting to meet manpower quotas, training officers contemplated returning to the old, short schedule. On the other hand, all evidence indicated that even the 18-week course was too brief for thorough navigation training. So in the face of a serious shortfall in numbers of graduates, some officials also considered further lengthening the already elongated course. The compromises of dual training and an 18-week course remained in place, however, until late in the war. Into 1944 the schools ran at maximum capacity and then became overloaded, including double classes at some installations; more men entered preflight; and military cadets absorbed much of Pan American Airway's training capacity.[3] A better resolution of the tension between manpower quotas and proficiency occurred about at the end of 1944. By then the personnel requirements were being met and the course became 20 weeks long, putting the navigator program back in phase with pilot training.

The Training Command taught men to navigate over land and water using the four methods of dead reckoning, pilotage, and celestial and radio navigation. The men learned to compute the effects of various factors on a course, plot the projected course onto charts, and maintain the log and check position periodically in flight.[4] Most of this information was communicated in ground classes. Instructors explained theory, the use and calibration of instruments such as compasses, driftmeters, and altimeters, the E–6B dead reckoning computer, the sextant for celestial navigation, and maps and charts. Cadets worked out twenty-six practical classroom problems, followed by a similar number of ground missions that simulated flight situations.[5]

Navigation by pilotage meant following approximate compass headings and visible terrain features in consultation with maps. Students learned proper logbook procedure and took some training on a navigation simulator.[6] Dead reckoning *"is the firm foundation of all navigation"* trumpeted an RAF Bomber Command bulletin distributed to American airmen.[7] This cornerstone of navigation was a means of approximating position by the best information available, including previous position, time, speed, heading, and drift. Navigation computers, position graphs, and tables and charts gave accurate measurements and calculations to support rule-of-thumb judgments. In class projects involving pilotage and dead reckoning, the instructor might select a landmark with easily describable characteristics, such as a town with railroad

World War II

Future navigators received instruction in celestial navigation (*above*) and map and chart reading (*right*) in ground classes admininsterd by the Training Command.

tracks coming from certain directions or one with odd topographical features. Cadets were to locate the place by longitude and latitude. Then, given further information, they calculated their estimated time of arrival at certain points along a route by figuring ground speed by pilotage and by elementary dead reckoning.[8]

Depending on where he was located and the type of unit to which he was assigned, a navigator typically relied most heavily upon one method of navigation in preference to any other. Long overwater flights or night bombing missions, for example, required the more advanced techniques of celestial navigation. Here, using the principles of spherical geometry, one calculated the altitude above the horizon by observing planets, certain stars, and the moon and sun, and mathematically converted these values to determine the position of the aircraft. In a celestial navigation simulator, the pilot and student navigator sat in an airplane "cockpit" surrounded by a dome of stars and planets picked out in tiny lights. While the pilot held the heading, the navigator worked through his charts, almanacs, and tables, sighted through his sextant, computed and drew lines, and made entries in his log to determine his fix and give the pilot directions.

Owing to a lack of equipment through 1942, schools provided the least preparation in radio navigation. Officers serving in the European and the Pacific theaters sent word home that graduate navigators as often as not displayed no flicker of recognition when confronted with radio equipment on tactical aircraft. During 1943 a form of low-frequency radio called Loran (long-range navigation), which was fundamentally a ground-based rather than airborne system, appeared in the approved curriculum. By the summer of 1944 it was being taught during the last week of the course.[9] But once again training did not conform with practice because of equipment and facilities shortages. A Loran chain planned for installation along the southern coast of the United States for use by the Training Command and the Second Air Force had not materialized by the beginning of 1945.[10]

Everybody agreed that graduate navigators must be masters of pilotage and dead reckoning. "Every navigator" in the 97th Bombardment Group (Heavy) of the Northwest African Strategic Air Forces counseled trainees to hone those skills, yet cautioned them to be "thoroughly impressed with the limitation of Dead Reckoning as the sole means of navigation."[11] Their view coincided with an inspection report of Second Air Force units that criticized navigators for thinking "in terms of DR, celestial, radio and pilotage as *separate* and distinct *systems*." Navigators should "get the airplane from one point to another using a *combination* of *all* possible information which might assist them."[12]

Particularly in a highly complex, mathematically oriented specialty that had little corollary in the civilian world, students' competence depended in large part on the quality of instruction. All the aircrew specialties faced severe difficulties in putting together cadres of instructors. At least most instructors in pilot training had been pilots previously, even though many of them had only recently graduated from the advanced flying schools. Experienced instructors in navigation were much rarer. Henry Hatfield, for example, who began teaching aerial navigation to aviation cadets when the war started, had until that time taught college-level German. He had first to learn the new subject himself, he recalled. "I don't suppose I knew much more than the kids I was teaching." Given his experience with wartime training, he mused, "I've never understood how we won the war."[13]

One of the well-worn methods for addressing instructors' shortcomings was to revise the curriculum at the central instructor's school. At least one instructor came to the conclusion that the problem lay not in the educational system, but in students' natural sloppiness and, presumably, the laxity of instructors who did not warn against the dangers of accumulated errors in navigation problems. It was imperative that navigators constantly check and recheck their fixes and that they calculate and reevaluate throughout a flight. After all, "if you are on course, the time you reach your destination is of secondary importance; while you may have a perfect ETA, [you may] never see your destination." Navigation is "a form of logic," he went on, that "if blindly accepted and followed, as taught in the book, with no further effort on the part of the navigator, will result in poor and often disastrous missions."[14] Similarly, the Director of Unit Training at Headquarters AAF admonished his counterpart in Individual Training to communicate to students the necessity for careful maintenance and calibration of instruments; he was concerned about precision, not theory.[15]

World War II

In the limited time allotted to air training, three navigation trainees flew together with an instructor. They took turns directing the pilot, mostly by dead reckoning, then added other methods on other missions and follow-the-pilot procedures. During the war, navigator cadets flew training missions on AT–18s, AT–38As, B–18s, B–34s, and C–60s. Typically they trained in the AT–7, modified with a rotating dome for taking sextant readings and three sets of navigation instruments. Ideally, navigation trainers were long-range aircraft capable of training flights of 4, 8, and 12 hours' duration. Unfortunately, the cramped, relatively short-range AT–7 was not such an airplane, and only late in the war did the longer-range tactical aircraft become available.[16]

Navigators were supposed to be capable of firing the guns, but because many could not be squeezed into the gunnery schools, or accommodated at the most opportune point in their training, most had ineffectual preparation before going into the OTUs or to war. In the fall of 1943, General Harper, who was responsible for training on the Air Staff, estimated that approximately two-thirds of bombardiers and navigators were then receiving gunnery training.[17] That figure was probably higher than the average throughout the war years.

Navigator training varied considerably over time. Greater standardization came about with the opening of a central instructor's school in October 1943, but well after that time there was no one standard textbook or official syllabus. Failure to achieve consistency in training could not be blamed on a lack of effort by the training units. Even without an approved textbook, there were films, navigator compartment and instrument mock-ups, and dead reckoning and celestial navigation simulators. Training conferences convened to sort out the numerous requirements.

Besides instructor's schools, another frequently suggested means to improve training in all aircrew specialties was to employ men with combat experience as instructors. Such a move would help to avoid a canned approach to learning, according to one officer:

> So many of the navigators have stereo-typed procedures, first you do that, then you do this, with complete disregard to a changed situation. It is believed that this is caused by inexperienced instructors; that is instructors are often picked from graduating classes and have no practical experience to give them a good perspective. Consequently each school develops a set of pet theories that are now passed on to the students as rules.... In reality a certain procedure may be best under certain conditions and another procedure may be best under other condition[s].[18]

Ironically, the writer went on to make a recommendation that contradicted his final sentence. In schools, navigators spent a great deal of time on the B–3 driftmeter "which is practically extinct," he wrote, and "virtually none on the B–5 driftmeter which was used on all types of long range aircraft." But a report from the African theater advocated the totally contrary view that the "B–3 or

Training Navigators, Bombardiers, and Gunners

As Assistant Chief of Air Staff, Training, hard-working Brig. Gen. Robert W. Harper (photographed here as a major) was one of the chief architects of air training programs during the Second World War.

B–2 driftmeters are the only satisfactory driftmeters used by this group. All B–5 driftmeters in this group have been replaced by B–3's when possible."[19] In this case, there was no good or bad piece of equipment, or right or wrong approach to instruction, as measured by combat operations. The B–3 had a gyroscope that made it easy to use in rough air. The B–5 was not gyrostabilized and was simpler than the B–3 in other respects. It was, as a result, less likely to malfunction and was more commonly installed in combat aircraft. Both had their uses.

Variation and shortages of equipment, lag time between advances and availability of new technologies, and the need to shift focus to meet specific theater requirements contributed to a highly complex training pattern and shortcomings in the program. Air Adjutant General E. L. Eubank's inspection of tactical units in the fall of 1942 revealed that graduate navigators tended to be familiar only with the sextant on which they were trained in the schools rather than the variety being procured.[20] Sometimes navigators were well trained on one piece of equipment and then assigned to airplanes without it. The schools could not easily address systemic problems resulting from shortages and delayed equipment deliveries. But some critics charged that graduate navigators often lacked proficiency in fundamental navigation skills. One officer cited examples of navigators who did not know how to plot the line of the sun above a certain angle when, as in the South Pacific where he was stationed, the sun's altitude was above 80 degrees a significant part of the time. Errors in making calculations and inability to align instruments could be laid directly at the door of the schools.

Much criticism, of both small matters and broader issues of training, could not be rectified until the manpower quotas came closer to realization, attention could be given to theater specialization, and combat returnees were available

World War II

to fill positions in the training program. These changes were occurring by January 1944 when combat navigators filled the central instructor's school and were assigned at the navigator schools. Techniques improved, but not until August of that year did navigator training reach its peak.

Bombardier Training

Before the war, many bombardiers, like navigators, came from the ranks of pilot eliminees. Once the AAF developed stanine tests, it selected bombardiers on the basis of their aptitude. Unfortunately, many airmen nursed the conviction that the best of those admitted to flight training became pilots, and navigators were credited with above-average intellectual ability, leaving bombardiers with the somewhat unfair reputation of men who had flunked the more demanding and high-prestige positions.

The AAF had launched a zealous publicity campaign during the expansion period to enhance the image of bombardiers and thereby obtain better-qualified trainees. A cadet bombardier stared forthrightly from the cover of *Life* magazine in May 1942, and the six-page photo spread lauded the bombardier as the "key to victory." Hollywood featured actor Pat O'Brien in the (quickly forgotten) film *Bombardier!*[21] By 1943 the service felt it had achieved some success in promoting the roles of nonpilots. Moreover, bombardiers were no longer strictly pilot washouts. In fact, publicity aside, the bombardier became increasingly central to an air force that made daylight precision bombing its central tenet. Nonetheless, during the war the Training Command received stinging criticism for its failure to produce proficient and motivated bombardiers.

Like navigator training, bombardier training first began alongside advanced twin-engine pilot schools; it proceeded to independent specialized facilities. Bombardier schools at Barksdale Field, Louisiana, and Ellington Field, near Houston, opened in 1941. Along the humid Gulf Coast it was difficult to peer through the hazy overcast with the optical bombsights, so the Barksdale school moved to Albuquerque, New Mexico, and Ellington went to Midland, Texas. There, miles of flat, treeless brushland and cloudless skies offered ideal bombing ranges. Other schools and a central instructor's school, mostly located in the Southwest, followed. At the peak, about thirteen schools trained bombardiers; by January 1945 the AAF was able to cut back to four fields.[22]

At the time the United States went to war, the bombardier course was twelve weeks long. It immediately dropped to ten weeks in January 1942 and to nine weeks the following month. As graduation numbers improved, it reverted to twelve weeks in 1943, then to eighteen, and by the end of 1944 the AAF scheduled a twenty-four week course.[23]

The Training Command did not provide all individual training, even after it established specialized schools. By mid-1942 the Second Air Force was

Training Navigators, Bombardiers, and Gunners

training bombardiers after their preflight course. Supposedly the instruction paralleled that of the Training Command, in order to ensure that cadets qualified on the same basis as those commissioned after the school course. At about the same time, the Third Air Force was also authorized to train enlisted and officer bombardiers.[24] By and large, however, Training Command bore the heaviest responsibility.

Faced with the insistent demand for more people, the AAF revived the prewar notion of dual training as navigator/bombardiers. Men might be fed into tactical units more quickly if they could perform two functions in a consolidated training program. The Training Command began by testing bombardiers for navigational aptitude, and from the 50 percent who scored a 5 (of 9) or better on the stanine, it selected candidates for dual training. In drawing up a course for the fully dual-trained individual, the command decided that navigation instruction should precede bombardier training (although at least twenty hours of navigation should be reserved until the end as a refresher after elapsed time), that the training period for each phase could not be reduced below that currently in effect in the individual specialist courses, and that cadets should be commissioned at the end of their lengthy two-part training.[25]

Members of the Training Command and Air Staff expressed serious

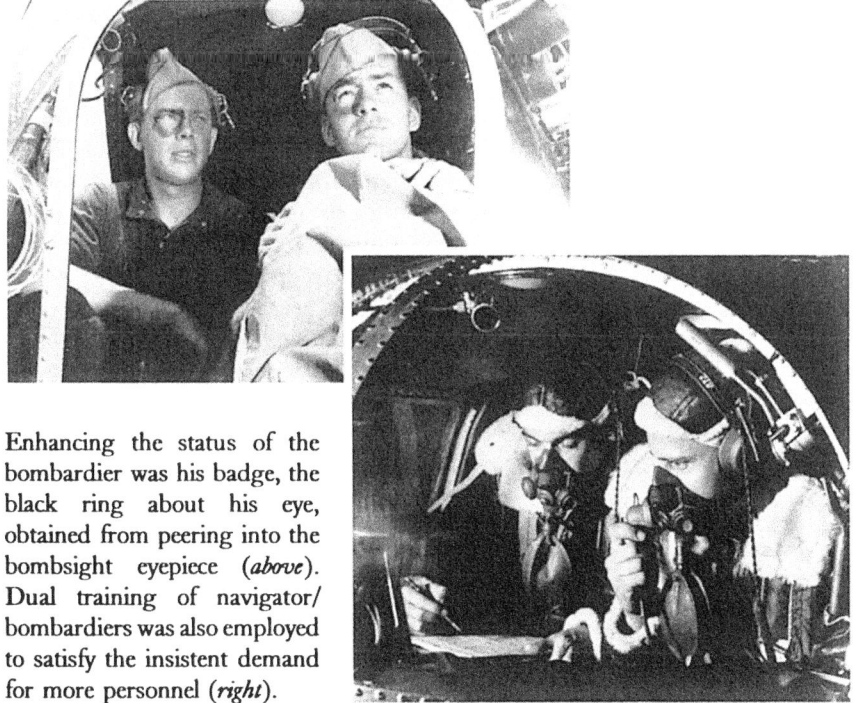

Enhancing the status of the bombardier was his badge, the black ring about his eye, obtained from peering into the bombsight eyepiece (*above*). Dual training of navigator/bombardiers was also employed to satisfy the insistent demand for more personnel (*right*).

World War II

reservations about implementing such a full-scale program, and all sorts of suggested variations poured forth. If the point was to speed up the time in getting graduates into operational units, sending people through both complete courses obviously was not the answer. Moreover, as one senior officer pointed out, "if a dual man is put in every combat crew position that now holds a single specialist, no saving in personnel is effected."[26] But replacing two individual specialists with one dual-trained individual could ease the manpower crisis in some types of units. Medium bombardment squadrons of the Third Air Force were equipped with the less-complex nonprecision D-8 bombsights that required less training, and the relatively shorter range of these aircraft meant that all men performing navigational duties need not be fully trained celestial navigators.

In other words, dual training worked when it could be done fairly quickly, and it could be done quickly when an individual was not trained in all aspects of both specialties. Thus, the bombardier school at Carlsbad, New Mexico, began a short course for navigators in D-8 nonprecision bombing in late 1942. Conversely, the Third Air Force instructed bombardiers in the rudiments of dead reckoning navigation. Because bombardiers comprised 75 percent of the nonpilot officers assigned to the Third Air Force, it immediately became clear that the Training Command should be providing the bulk of training, which in this instance was to bombardiers, and should leave the lesser amount, which was to navigators, to the Third Air Force. The instruction therefore reversed in February 1943, with Carlsbad becoming a bombardier–dead reckoning navigator school rather than a navigator–D-8 bombardier school. By the summer, the Training Command had converted the entire bombardier program to dual training: all bombardiers were trained as dead reckoning navigators. Since not all navigators were trained as bombardiers, however, dual training more profoundly affected the bombardier than the navigator training program.

The timing and specifics of the bombardier course reflected the aforementioned changes in policy, among others. Usually the bombardier began his training with preflight, and at some point before the end of his course he was expected to take flexible gunnery. In July 1942 the Training Command decided to send bombardiers for gunnery before preflight. At that time too many students had badly clogged the training program: some 4,300 potential bombardiers and navigators were awaiting assignment to preflight.[27] In this situation men would have forgotten most of what they had learned during the elapsed time between preflight gunnery training and their eventual graduation. The plan quickly shifted to post-preflight gunnery training, then to post-specialized training. In point of fact, accelerating requirements precluded gunnery training at *any* point when it would further delay delivery of graduate bombardiers and navigators to the air forces. Moreover, the demand for enlisted gunners was so great through the spring of 1944 that the gunnery schools could not absorb other specialties. As a result, in April 1944 the gunnery requirement

Training Navigators, Bombardiers, and Gunners

for navigators and bombardiers was dropped altogether.

Once they began specialized training, student bombardiers spent well over three-quarters of their time in ground classes. The number of hours devoted to any particular subject depended on whether the course had then been converted to dual training and the type of equipment available for instruction. Training aids included films, mock-ups of instruments, and navigational computers. The first simulator, the A–2, was an unwieldy, ten-foot-high, three-wheeled contraption through which the trainee aimed the bombsight. The more sophisticated Link A–6 bombing trainer projected the image of the ground at which the trainee would aim and fire at an indicated bomb impact point.[28]

Air training divided into Course 1, the qualification stage, and Course 2, the combat stage. Most cadets flew in the Beech AT–11 fitted with a Plexiglas nose and a bomb bay. During Course 1 the student learned to operate the bombsight and other equipment and to practice bombing. He and another trainee worked together, one operating the bombsight and the other filming the results through a hole in the floor. At this point a student either passed his qualifying tests or was eliminated. Course 2 aimed to present as many operational problems as possible; instructors and unit commanders, working within the constraints of their facilities and equipment, largely determined its content and scope. At Midland Army Air Field, for example, bombardiers had an ample 23 bombing ranges where they could drop their 100-pound M38A2 practice bombs. These steel containers held 3-pound black powder spotting charges and about 85 pounds of sand. As many as 1,000 bombs were dropped daily, six or seven days a week.[29]

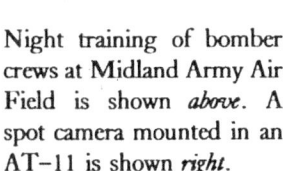

Night training of bomber crews at Midland Army Air Field is shown *above*. A spot camera mounted in an AT–11 is shown *right*.

World War II

Bomber crews trained at night and under blackout conditions. In early 1942 the Coast Artillery supplied antiaircraft searchlight batteries for bombardier training. As one officer stated simply, "the advantages to both services are obvious." As part of the training for air strikes in Europe, the Training Command also initiated a project whereby student navigators and bombardiers located, identified, and bombed realistically constructed targets under blackout conditions.[30]

Whatever order and standardization appeared in school curricula and training guidelines derived in large part from the consensus obtained at training conferences. Conferees at a meeting at Flying Training Command headquarters the end of 1942 established qualification standards for D-8 bombardiers and bombardier–celestial navigators, and they revised the curriculum for precision bombardier–dead reckoning navigators that became standard in bombardier schools. That 18-week course consisted of 51 hours in dead reckoning problems and procedures; 45 hours in the basic theory of bombing and bombsights; 17 hours on bombing accessories (bomb racks and controls, bombs and fuses, the intervalometer, and aerial cameras); 39 hours with trainers (including the automatic bombing computer) and training aids; 17½ hours of bombing analysis; 16 hours of bombardment aviation, 8 on the C-1 autopilot and 25 on bombsight calibration and problems; 16 hours on instruments and 17 on computers; and a number of hours on nontechnical areas including military training.[31]

The bombardier's most important pieces of equipment were the manual computing devices, usually the E-6B aerial computer, and the bombsight. The Sperry and Norden precision bombsights and the nonprecision Estoppey D-8 and British T-1 were all taught in bombardier courses and used in the training air forces. Tests by the Proving Ground Command found the T-1 to be highly inaccurate above 12,000 feet, and it required maintenance equipment not supplied to units in the combat zones.[32] The D-8 was manufactured by the National Cash Register Company; it was used until October 1943 when it was discontinued in favor of the Norden M-9. Its designer, George Estoppey, was a civilian engineer at McCook Field in the twenties. Simpler to use than the precision sights, the D-8 was designed primarily for lower altitude bombing.[33]

The precision optical sights, coupled with an automatic pilot, promised unprecedented accuracy. Perhaps unfairly, the Army preferred the Norden to the Sperry S-1 bombsight. There were reports that optical arrangement of the Sperry made its employment difficult at altitudes of 25,000 to 30,000 feet, although it was superior at very high altitude, that is, at 30,000 feet and above. Most critical, virtually all AAF bombers were installed with the C-1 autopilot, which did not function well with the S-1. Also, relatively few training airplanes were equipped for the bombsight, and the schools lacked A-2B bomb trainers with Sperry equipment.[34] By the end of 1943 the S-1 was no longer being manufactured, except for replacement parts.

Training Navigators, Bombardiers, and Gunners

The development of bombsights can be seen here from the nonprecision D-8, originally developed in the 1920s (*upper left*), to the precision Sperry S-1 (*above*), and finally to the the Norden bombsight (*lower left*), whose internal mechanism is being described by an instructor to a group of trainees.

    The M-series Norden bombsight, on the other hand, underlay the AAF's absolute faith in high-altitude precision bombing. It was to let the bombardier put a bomb in the proverbial pickle barrel. AT–11 training planes could also use the sight and the Honeywell C–1 autopilot. Once the Norden company could supply them in sufficient quantity, by 1944 the M–9 had become the principal precision bombsight employed throughout the AAF, both in combat and in the schools. By that time, too, the extreme secrecy surrounding the device had lifted.[35]

    Initially, however, only the schools and some units of the Second Air Force had precision sights. Because the bombardier schools mostly trained precision bombardiers, the training and assignment of men headed for heavy bombardment squadrons of the Second Air Force were coordinated. One bombardier school had Sperry equipment and could send its graduates to the Second Air Force units also equipped with Sperry sights. The others could go to units equipped with Norden sights. But because the schools gave very little nonprecision training, bombardiers going to light and medium units of the Third Air Force that used low-altitude nonprecision sights were ill-prepared. In

World War II

early 1943 the Training Command notified those responsible for drawing up manpower requirements that if the command knew who were to be assigned to medium bomber units at least one month in advance of graduation, it could concentrate D-8 training for these men. At the same time, it anticipated the wholesale changeover in the bombardier course to a dual-trained precision bombardier–dead reckoning navigator.[36]

If dual training became standard by the midpoint in the war, it happened more by necessity than by general accord. Dual training was an expedient, not necessarily the desired end, useful principally to supply medium bomber units near the beginning of the war. It worked because medium bombardment was less technologically demanding of its crews, and training in both specialties did not have to be as thorough. All types of units ultimately received some fully dual-trained men, but in varying proportions. After the reinstatement of dual training, one theater requested a bombardier–celestial navigator for each crew, some theaters required no celestial navigators, and some wanted one bombardier–celestial navigator for every three bombardier–dead reckoning navigators. Based on the number of lead crews considered necessary, by 1943 the agreed-upon standard for medium bombardment was a completely dual-trained specialist for 25 percent of the crews and a bombardier–dead reckoning navigator for the remaining 75 percent. Each B-17 and B-24 would have a celestial navigator and a bombardier–dead reckoning navigator. B-29 units reaped the greatest benefits — two precision bombardier–celestial navigators per crew. Training requirements were further layered when very heavy crews were to have two triple-trained specialists (precision bombardier–celestial navigator–radar officer).[37] Even these excruciatingly analyzed and reassessed paper plans did not work out neatly in practice. The Third Air Force found it difficult to maintain the desired 25 percent ratio of celestial navigators for medium bombardment crews because the flow of specialists from the schools did not allow full assembly of combat crews at the first stage of their training.[38]

Bombardiers were in greater supply than navigators, so the diversion of needed personnel to dual training was less onerous for the bombardier program. But the merits of the approach were debated on other grounds besides meeting manpower goals. Dual training resulted in lower proficiency for both bombardiers and navigators. Col. Edgar P. Sorensen, on the Air Staff in early February 1943 when the program had just begun, registered one such objection:

> Any diversion of functions or duties which detracts in any way from the one-hundred percent concentration of the bombardier's attention and best effort will be detrimental to the effectiveness of our striking force. Some of us have worked hard for several years to put the bombardier on a pedestal and in his proper place. Today our bombing effectiveness is far inferior to what it should be, largely because of incomplete training of the bombardier....
>
> In emphasizing the extreme importance of the bombardier and his

Training Navigators, Bombardiers, and Gunners

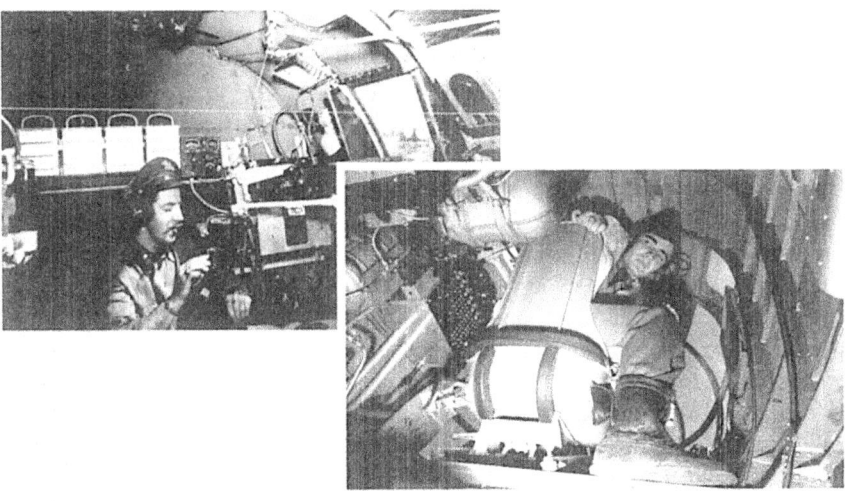

Original plans called for supplying one celestial navigator for each B-17 and B-24. Instead, most heavy bomber crews had a dual-trained bombardier–dead reckoning navigator. Shown here are a navigator and a bombardier in his "nest" in the nose.

function, no belittlement of any other crew member is necessary. The fact still remains that unless we can hit the target the bombardment mission fails. To place a mission's bombs on a target well within the enemy's territory is a costly adventure. Only through the hands of the bombardier will we get a return for the risk and cost of the mission. The maximum in training and capabilities of the bombardier is not too good under war conditions.[39]

As Colonel Sorensen indicated, the execution of the strategic mission rested upon the bombardier's success. During the final bomb run the bombardier controlled the airplane, which had to maintain a straight and level course no matter the opposition, until the target lay in the bombsight crosshairs and the bombs were released. To perform that nerve-wracking job, the bombardier needed to develop capability in train bombing (release of two or more bombs in succession from a single sighting), be proficient at computing bombing probabilities, and be competent at scoring and analyzing bombing results. He had to understand the effects on the plane and control surfaces by the C-1 autopilot linked to his bombsight. When he was also to be a dead reckoning navigator, he had to learn pilotage, map reading, and daylight dead reckoning. He was to have some working knowledge of radio and visual communication to the extent of sending and receiving radio telegraph code and blinker signals. When flexible gunnery school quotas allowed, he was to qualify as an aerial gunner.[40]

Without doubt, such an individual must be highly trained, and any time diverted to learning navigation lessened the amount spent perfecting his skill as a bombardier. The progress of the war, however, altered bombardment tactics and the role of the bombardier, helping the partially dual-trained man to become a useful and adequately prepared member of the aircrew. By the time of the Bomber Offensive in the European theater, American heavy bombers flew in huge, tight, defensive formations during daylight hours. The bombardier in the lead plane determined the course for the entire formation and, in what was called salvo bombing, the rest of the bombardiers released their bombs on cue from him. As a result, comparatively few bombardiers required training as precision bombardiers.

Training officials continually established, then revised qualification standards and evaluated vague concepts of professionalism. As applied to bombardiers, assessing bombing proficiency appeared fairly easy: did bombs hit the target? Bombing error, known as circular error probable, was expressed in terms of the radius of a circle centered on the target in which half of the bombs dropped were expected to fall. Chief of the Air Staff Maj. Gen. George E. Stratemeyer spoke with the unequivocal voice of the AAF when he stated in 1942 (when nothing had yet been proven) that the "results of operations of Air Forces in this war have demonstrated beyond question the absolute necessity of precision bombing from high altitudes."[41] But, he went on, "thousands of tons of bombs have been released with little or no effect by Air Forces employing non-precision bombing methods." Stratemeyer claimed that poor bombing accuracy came not from faulty bombing tactics or equipment but from sloppy practices.[42] Other training officials seconded his view, complaining that once men had qualified during Course 1, many were inclined to relax, merely to maintain an acceptable circular error score. They showed little incentive to work toward greater exactitude. Therefore, proficiency standards changed in 1943 from recording circular errors to counting only hits and misses, requiring a minimum of 22 percent hits on the target from the approximately 60 bombs dropped.[43]

The Training Command also addressed the issue of proficiency by the familiar means of trying to upgrade the quality of instructors. In the case of pilot and navigator training, long experience in the former could be drawn upon, and some (minimal) instruction in aerial navigation dated from the mid-1930s. The bombardier was largely a creature of war. The schools had relatively few experienced teachers or combat veterans. Training Command Chief of Staff Brig. Gen. Walter Kraus outlined his views on this matter in August 1942:

> It is considered undesirable to bring into the bombardier schools relatively inexperienced personnel of any age and particularly older men who would be placed in a position of command despite their low level of experience in the field of bombardier instructors. In addition, since the instructors carry their group of students completely through the course, they must be able to cope with the severe strain of long hours, cramped positions,

irregular living habits, high altitude and other flight phenomena. In the conduct of bombardier classes, students ask questions which cannot be answered by instructors who have not bombed under the conditions of the course and whose experience has been obtained mostly from text books.

Perhaps combat-experienced instructors were even more urgently needed in bombardier training than in pilot or navigator training. "By virtue of the fact that bombardiering is a comparatively new science and requires flexible minds which are alert and open," Kraus reasoned, "older, and therefore higher ranking, officers should [not] be assigned as bombardier instructors."[44]

Even improving teaching staffs and the relatively objective method of qualifying a bombardier according to his bombing score could not ensure that a man developed the qualities of professionalism that the AAF hoped to see in its officers. The service could try to devise tests by which the bombardier had to meet a higher standard — in this case, hitting the target rather than keeping an acceptable average score — but changing attitudes was a subtler problem, and like other issues of morale and professionalism, was one that a large organization mobilized for a short-term emergency could not easily solve. Speaking to this dilemma, an inspector at the bombardier schools noted the number of bombardiers who were "doing good bombing" but who were nonetheless uninvolved and uninterested in the work and the mission to be accomplished. He offered the idealistic recommendation that "an unwavering policy be followed at bombardier training schools which will immediately wash-out a man for indifference. Indifference at this time, when we are at war, cannot be tolerated."[45] In fact, *especially* in time of war, a military service could not afford to dismiss indispensable men for indifference.

In the final analysis, the greatest weakness in bombardier training came from materiel shortages rather than incompetency or failure of commitment of the men, even in the morale-plagued bombardier program. To borrow from historian Stephen McFarland's summation:

> Stateside training could do little to duplicate wartime conditions. Training flights were too short and too low. Shortages of aircraft, bombsights, equipment, and facilities were constant until the last year of the war. Shortages of oxygen and 91-octane fuel kept training flights at low altitudes. Trainees dropped bombs individually rather than in salvo. They did not fly formation bombing missions until the operational training unit stage. Training did little to prepare bombardiers for nonvisual or partially visual bombing. Some instructors had never seen combat. The AT–11 bombardier trainer could not reach the combat altitudes of heavy bombers and handled so differently through the [stabilized bombing approach equipment/automatic flight control equipment] systems that signifcant retraining had to accompany any transfer to a combat unit.[46]

World War II

Flexible Gunnery Training

Flexible guns, usually machine guns positioned in a turret or other type of swivel mount, turned in both the horizontal and vertical planes. Virtually every member of a bomber crew except the pilot was supposed to be trained to fire the guns. The programs of instruction for copilots, bombardiers, navigators, radio operators, radio mechanics, crew chiefs, armament specialists, armament personnel, turret and gunsight maintenance men, and airmen selected as nonspecialist gunners all (for a time) included flexible gunnery. The Training Command could never train all these people, and the schools finally had to be dedicated mostly to producing combat gunners.

The AAF found an ideal gunnery range on the open, nearly uninhabited miles of flatland around Las Vegas, Nevada, which became home to the first specialized flexible gunnery school. It and the second school at Harlingen, Texas, began graduating students soon after Pearl Harbor. Most of the gunnery schools that followed were in the hot, dusty Southwest — places like Kingman, Arizona, and Laredo, Texas, and the Yuma Army Air Field that converted from an advanced pilot school in November 1943. The southwestern schools possessed the advantages of mild winters and nearly deserted surroundings. Kingman was well served by transportation; Tyndall Field near Panama City, Florida, was near the Gulf of Mexico.[47]

In the midst of feverish construction activity during the summer of 1942, the Director of Military Requirements, Brig. Gen. Muir S. Fairchild, warned field commanders not to pester higher headquarters for more facilities. Gunnery schools were being built as fast as possible, and in the meantime training units would have to fall back upon their own resources: "Ingenuity, perseverance, and forceful action, that obtains results must be substituted for requisitions and letters reporting inability to accomplish objectives."[48] At the high point of training in 1944, seven specialized gunnery schools churned out graduates, but almost until that time the schools relied largely on their own resources.

Except for aircraft maintenance training, the gunnery schools graduated the largest number of AAF officers and enlisted men during the war. Their numbers might have been even higher had all the aircrew and ground crew specialists whose programs called for a course in flexible gunnery taken it. As it was, the schools were nearly filled with those in training as specialist or nonspecialist aerial gunners. The Director of Individual Training reported in late September of 1942 that there was a "tremendous" shortage of flexible gunners throughout the air forces, exacerbated by the fact that new gunnery schools were still under construction. As a result, existing facilities would operate with a 10 percent overload.[49] In December the manpower shortage was addressed further by shifting away from a volunteer basis for recruiting gunners. The following August, the chief of training on the Air Staff, General Harper, estimated that approximately 40,000 bombardiers and navigators were in gunnery training, 16,667 were being trained in each of the specialties of

Training Navigators, Bombardiers, and Gunners

Training in flexible gunnery, machine guns capable of both horizontal and vertical movement and positioned in a turret, typically occurred at locations in the Southwest, as at this site near Casper, Wyoming, in 1943.

radio operator mechanics, armorers, and airplane mechanics, and 50,000 were being trained as nonspecialist career gunners. At that time, General Arnold directed that specialist and nonspecialist career gunners should take priority over bombardiers and navigators.[50]

An initial reception center screened enlisted aircrew members, then posted them to a school for training in their assigned trade. The Training Command and Air Staff worked to secure specialized gunners who had first completed courses in factories or in the Technical Training Command. In early 1942 planners wanted those already trained in radio, aircraft mechanics, or armament in the approximate proportions of 30 percent, 30 percent, and 40 percent. The general requirement for bombardment crews was 17.5 percent engineers, 30 percent radio operators, and 53.5 percent gunners.[51]

The number and ratios of men trained as specialist or nonspecialist gunners shifted according to immediate wartime requirements. In the fall of 1943, for example, General Arnold informed one of the bomber group commanders that, as a temporary expedient, nonspecialist gunners would replace the assistant radio operator mechanic gunner, the assistant airplane and engine mechanic gunner, and, in ten-man crews, one of the two armorer gunners. In general, that instruction meant that gunners on heavy bombardment crews would include one radio operator mechanic, one armorer, one airplane mechanic, and three nonspecialist gunners.[52] As nonflying combat veterans became available by the turn of 1944, the training establishment heartily welcomed any volunteers who wished to retrain as gunners. According to General Harper, "Many of these men have been strafed, have had friends killed, and are generally 'combat-wise.' They would be of inestimable help in the gunnery schools from a morale standpoint and should become exceptionally good combat crew members."[53]

From the first tentative program of instruction in September 1940 through its subsequent iterations, the gunnery course included familiarization with equipment, sighting problems, ground range exercises and firing, and air range exercises and firing. In April 1943 the gunnery program increased from five to six weeks to accommodate the increased numerical requirements for graduates, the less-experienced and less-motivated student who was no longer a volunteer, and the anxious concern by training officials to upgrade the quality of gunnery instruction. The 312 course hours at that time were devoted to orientation, description and nomenclature of machine guns, sights and sighting (theory and trainers), ballistics and bore sighting, aircraft recognition, range estimation (theory and trainers), basic tactics, review, BB ranges, shotgun firing, .22-caliber ranges, moving target ranges (.30-caliber and .50-caliber), malfunctions, turret drill, turret maintenance, air-to-air firing, and physical and military training. To maintain proficiency of radio operators and mechanics during gunnery training, more hours were added in radio and visual code.[54]

The variety of the 1943 program suggested the multiple though still unstandardized methods of instruction that continued to be employed. The schools taught a number of different sights and systems of sighting. The training establishment tried to standardize instruction by drawing up a set of guidelines to be met for graduation from flexible gunnery schools, and it emphasized practicality over highly complicated theoretical instruction. Yet one year later, at the peak of training, both in terms of numbers of graduates and the scrutiny accorded to the gunnery program, the commanding officer of the instructors school, the central clearinghouse, expressed his frustration that "at present, this school does not know what training films are being used in training of flexible gunners. At no time in the past has a standard list of films to be used been distributed to individual gunnery schools."[55]

In addition to films, training devices included several types of cameras and simulators of varying sophistication. By 1944 when .30-caliber weapons could be phased out with the greater supply of .50-caliber guns, and when turrets were available on training aircraft, the M–6 Bell Adapter came into use. The Spotlight trainers, employed in turret practice, projected a moving spot of light on the wall at which students tracked and fired. An automatic photoelectric counter recorded hits and misses out of the total shots fired. The E–14 or Jam Handy simulator projected films of combat scenes, approaches, attacks from various angles, and breakaways. The student estimated the range and "fired" a beam of light from a model machine gun at 600 yards. A sound like that of gunfire notified him if he was in range, and a second projector could demonstrate the correct aim.[56]

Training Navigators, Bombardiers, and Gunners

The newest and most elaborate simulator, the Waller trainer, had been first ordered and then canceled by the Navy because of its considerable expense. The Waller projected films on a concave screen and indicated points of aim and presumed hits on photographs of attacking planes. The cost, its uncertain training value over the cheaper Jam Handy, the objections by the central instructor's school experts to the films provided, and high maintenance requirements brought its worth into question. It was, however, quite realistic with its combat noise and vibration. And it was extremely popular, if for no other reason than its carnival shooting-gallery appeal. When he tried it, Maj. James Gould Cozzens enthused that the "effect is great fun. You seem to be sitting in space in the tail turret of a bomber and the attacking planes appear three dimensional and scare the hell out of you." Entertainment, Cozzens suspected, as much if not more than its training value, was instrumental in "persuading the AAF officers involved that it must be good."[57]

Gunners had to know not only how to sight and track, but also when and when not to fire. Ground classes in all aircrew specialties — pilot, navigator, bombardier, and gunnery — included aircraft and naval vessel recognition. Because flexible gunners manned most of the armament, they most of all had to know what they were looking at. In July 1943 the Bombardment Branch at Headquarters AAF reminded the Training Command of the critical importance of this skill, and having concluded that gunners' proficiency in this area was quite poor, directed that the schools provide a minimum of thirty hours of instruction in aircraft recognition.[58] A British Air Ministry pamphlet distributed by the Training Command led instructors step-by-step through the subject, beginning with the admonition that it was too important to tackle in a "haphazard" way: "After all, what purpose would there be in teaching a man how to use a lethal weapon without teaching him what to shoot at? Letting an enemy pass or deliver an attack because one cannot be certain that he *is* an enemy, or shooting down a friendly aircraft because one thinks he is an enemy, are [*sic*] likely to be direct results of not taking aircraft recognition seriously."[59] The American schools used the Renshaw System of Identification of Aircraft during much of the war. Gunners studied slides, photographs, and models and practiced on simulators when they were available. In battle, aircraft frequently did not present themselves in silhouette and so had to be identifiable head-on and from side and angled views.

For moving target practice outside the classroom, airmen shot clay pigeons from moving trucks with shotguns on skeet and trap ranges, fired at stationary and moving targets with BB machine guns and .22-caliber rifles, gauged gunsight alignment on a harmonization range, fired at moving targets from mounted machine gun pedestals and from guns in turrets, and, at some schools, engaged in night firing.[60] They learned how to care for and maintain and, while blindfolded, strip and reassemble the .50-caliber Browning M-2 air-cooled machine gun. At one point the schools were directed to train gunners in the

World War II

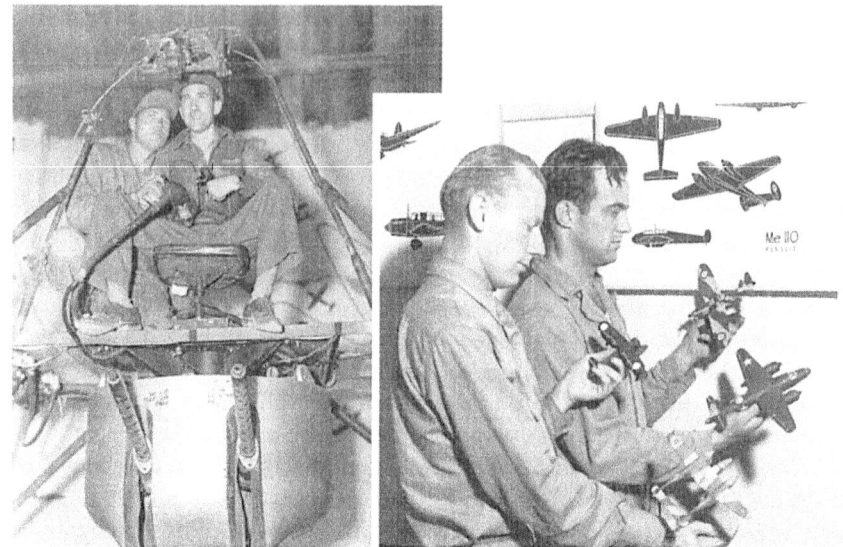

Ground school included simulators for gunnery training plus classes in friendly and enemy aircraft recognition.

operation of .45-caliber submachine guns.[61] Students spent a great deal of time on the malfunction range diagnosing equipment failure.

One of the controversies in the flexible gunnery program emerged as a result of the difference in techniques employed in ground gunnery practice and in aerial battles. Fairly early, combat experience demonstrated that gunners scored few hits in aerial exchanges with enemy fighters but came home with many holes in their own aircraft. Mathematicians who were set to work on the problem of sighting and firing between two moving objects in the air discovered that the bullet traveled ahead of where it was aimed, so for the gunner's bullets to travel along a line that would intersect with an attacking fighter's line of flight, the angle of deflection lay between the fighter and the tail of the bomber. In 1943 the new system of position firing, based upon this revelation and experiments in air-to-ground firing at the central instructor's school, went into effect. In ground practice the gunner was told not to fire until he was within striking range (600 yards) and at the angle he would be in an airplane when firing at a fighter, aiming behind rather than in front of the fighter.[62]

The new method of sighting had its critics, however. For one thing, on moving target or moving-base ranges, the gunner led ahead of his target as if he were shooting ducks. In position firing, the gunner fired behind the target to compensate for the pursuit curve (the flight path of the fighter). In essence, the gunner was practicing two contrary systems of sighting. To eliminate the

Training Navigators, Bombardiers, and Gunners

Moving target practice employed enemy markings on targets to encourage a gunner's spirited response.

anomaly, skeet ranges, renamed basic deflection ranges, were rearranged during 1944 so that students could fire directionally at incoming targets and replicate the pursuit curve.[63]

Some cited another drawback of position firing to be its departure from the relative speed system, a torturously complicated process in which the gunner estimated the difference in speed between his and the enemy's plane to arrive at the angle of deflection. Because automatically calculating computing sights (sights that automatically calculate wind, range, and other variables) used relative speed principles, some training officials insisted that gunners should understand and use the methods employed by the newest and most sophisticated equipment. If a gunner learned careful tracking and how to operate sights correctly, he would not be forced to rely on "guesstimation." Other officers disagreed, pointing out that the computing sights were often inaccurate under combat conditions or were liable to malfunction and that some guns were not equipped with automatic sights. Maj. Gen. Follett Bradley, who had come into Army aviation at the end of World War I and was a close associate of Arnold's at AAF headquarters while this debate simmered during 1944, took a practical point of view: "The position system is taught for the same reason that a man is taught to swim — not that swimming is man's normal method of traveling through water, but to save his life in case his boat is sunk. Training schedules must continue to include the position system," he concluded, "but always with emphasis that automatic sights are to be used if available and functioning."[64]

The inexactitude of position firing raised the specter of increased friendly fire, the phenomenon that had initiated the research into principles of sighting

World War II

in the first place. The doctrine of high-altitude precision bombing, without long-range escorts, relied on a massive barrage of defensive firepower from the tight bomber formation. Maj. Gen. Haywood S. "Possum" Hansell, Jr., one of the architects of AWPD–1 and a committed advocate of the strategic doctrine, later said that the idea that the bomber formation could prevail "was based on hope and not on existing fact. We had no power operated turrets.... We had no .50-caliber defensive guns. We had no gunners who could hit anything. And yet our entire doctrine hinged on the defensive fire power potential."[65] In operations in Europe, these issues remained unsettled until, as General Bradley foretold, new computing and compensating sights superseded position firing, until long-range escort fighters provided defensive cover to bomber formations, and, perhaps most important, until the AAF achieved air superiority over the Luftwaffe.[66]

In the meantime, air-to-ground gunnery training adopted position firing: using the deflection of position firing, gunners fired at four targets that simulated positions on the pursuit curve as the bomber flew a straight course at low level.[67] Air-to-air practice, in which gunners usually aimed at towed target sleeves, even less effectively approximated combat firing. A variety of planes, including the single-engine trainer AT–6, but very few first-line tactical aircraft, carried out this work. In an understatement, one officer commented, "certainly with an AT–6, air to air firing can hardly be called realistic."[68] The plane was too light to carry a .50-caliber gun, so the gunner had to fire a hand-held .30-caliber gun from the rear cockpit.

Improvement came with the equipping of the twin-engine Lockheed AT–18 with Martin turrets. In 1942 the schools started receiving more combat planes, mostly B–34s and B–26s and a very few B–17s. The AT–23 was a modified B–26 used successfully for high-altitude towing. Tests conducted at the Kingman, Arizona, Flexible Gunnery School demonstrated the plane's high-altitude maneuverability, permitting twenty-five runs to be made on the B–17 in forty-five minutes, with all B–17 gun positions firing. All of the training air forces also used these airplanes for gunnery exercises. During 1944 the schools received old P–40s and P–39s and a modified P–39, the P–63. TB–24s (modified B–24s with remote control turrets) sufficed for B–29 training until some of the actual bombers became available in May 1945.[69]

Firing at towed sleeves hardly replicated combat conditions. As when he fired on a skeet range, a gunner aiming at a relatively stable towed target did not direct his fire as he would at an attacking fighter flying the normal pursuit curve. One officer on duty with the Fifteenth Air Force recommended that air-to-air firing be eliminated in training altogether: "The only real benefit the gunner derives from this training is air orientation and handling of his gun in the aircraft, all of which can be had in air-to-ground firing with targets placed on a pursuit curve."[70]

The most realistic air-to-air practice came from simulated fighter attacks.

Training Navigators, Bombardiers, and Gunners

In air-to-air gunnery practice (*left*), a hand-held .30-caliber gun was fired from the rear cockpit of an AT–6 trainer.

In air-to-ground practice (*right*), gunners fired at flagged ground targets.

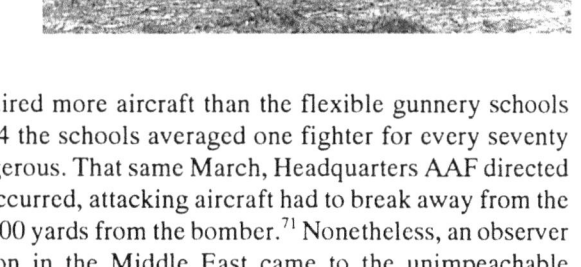

To do this routinely required more aircraft than the flexible gunnery schools possessed (in March 1944 the schools averaged one fighter for every seventy students), and it was dangerous. That same March, Headquarters AAF directed that when such training occurred, attacking aircraft had to break away from the attack at 250 rather than 100 yards from the bomber.[71] Nonetheless, an observer of the battlefield situation in the Middle East came to the unimpeachable conclusion that "unless a man has had a least four hours practice in tracking an actual pursuit ship from a bomber in flight, he will be about one third trained for combat. This is the most difficult kind of practice to obtain."[72]

It was difficult to provide that practice without more tactical aircraft in the training inventory, but it was made easier by new training devices for air-to-air firing that rendered tow target practice nearly obsolete. In more primitive form, camera guns were not new. They went back many years and were used experimentally in at least one gunnery school in 1942. During 1944 and 1945, they were installed in the top turrets of many bombardment training aircraft. Students could "fire" the camera which would record on film their hits and misses against attacking fighters. The frangible bullet introduced further realism. Made of lead and plastic, upon contacting an armored attacking ship, it splattered to leave the physical mark of a hit. A radio device in the attacking airplane also detected the audible score and communicated it to the gunner by a wing lamp signal. Good in theory, this technique proved flawed in practice: shards of frangible bullets found their way into cooling ducts, and the projectiles caused damage to control surfaces. Camera guns, but not frangible bullets, were widely used by the end of the war.[73]

Even with a nearly endless series of conferences and discussions addressing aerial gunnery, and with notable improvements in technology and methods, participants in a training conference in the spring of 1944 acknowledged that the flexible gunnery program remained the weakest link in the AAF training chain. In consultation with General Yount, General Harper took immediate steps to strengthen the control of the Training Command over the flexible gunnery program. He also authorized direct liaison among agencies whose functions were relevant to gunnery—the Assistant Chief of Air Staff for Training, the four continental air forces, all gunnery schools in the Training Command, the AAF Board, the Proving Ground Command, and the Materiel Command. Further coordination occurred when all flexible gunnery schools, including the central instructor's school, were placed under the 75th Flying Training Wing. Brig. Gen. E. B. Lyon was appointed to the new position of special assistant to General Yount for flexible gunnery.[74]

In their 1944 postconference reorganization of gunnery training, Harper and Yount confirmed the role of the instructor's school as the central clearinghouse for both technical and personnel information. It was intended to coordinate gunnery technique and approve equipment. But the school suffered from exceptionally poor morale that undercut its utility. Many instructors had, themselves, only recently graduated and now, as privates, were teaching either officers or enlisted men who had completed armament training in factories. Other elements of the training program shared the headache of low instructor morale and competence and redressed it similarly by raising the standards required for appointment as an instructor and assigning available combat returnees to the schools. As it developed, however, the gunnery instructor's school became more than a force for standardization of instructional methods and the publication and distribution of teaching materials — all of which it did. By 1944 it school served as technical adviser on flexible gunnery problems to all the training agencies, developing theory and "conducting complete and continuous experimental research on flexible gunnery, flexible gunnery training methods, and flexible gunnery training aids, such activity to include psychological research."[75]

The need to standardize gunner instruction led to administrative reform and an effort to procure turrets for the gunnery schools that eventually culminated in a specialization policy by which men and schools trained for specific guns. In a testy memorandum written six months into the war, General Yount protested that the lack of equipment rendered aerial gunnery training equivalent to that given in 1917 and 1918, in large part because manpower schedules called for 2,000 gunners a month at a time when not a single airplane in the gunnery schools mounted a turret.[76] Under the circumstances, statements coming back from overseas such as the "training that the Turret gunner is getting back in the States is getting a great deal of adverse criticism over here," was fairly mild.[77] By the end of the year the situation had improved: a number of AT-18s were equipped with Martin turrets, and some other advanced trainers also mounted turrets. Shortly, six types of turrets were in use throughout the training air forces.[78]

Because all the schools were not then in full operation, training officials had the opportunity to redirect the program to avoid the emerging chaos by specializing schools according to type of aircraft and armament: specific turrets were assigned to specific aircraft, which in turn were assigned to specific schools. The Texas schools at Laredo and Harlingen became B-24 schools using the Martin upper, Sperry ball, and Consolidated tail turrets; Las Vegas and Kingman took the Sperry upper and ball turrets for the B-17; Tyndall Field at Panama City trained half for B-25s with Bendix upper and lower turrets (the Bendix lower turret was almost immediately discontinued) and half for B-26s with the Martin upper; and Buckingham Field at Fort Myers was 60 percent B-26 (B-34) training with the Martin upper and 40 percent with the General Electric Central Fire Control turret for light and dive bombardment. The central flexible gunnery instructor's school, of course, trained on all types. Graduates of the heavy bombardment schools went into the Second Air Force; those from Panama City and Fort Myers, to the Third Air Force.[79]

Under the specialization policy, each crew member trained for a specific gun position. On a B-17, for example, the bombardier was assigned the chin turret; the navigator, the side nose guns; the engineer, radio operator mechanic, and armorer, the waist guns; and the three career gunners, the ball, upper, and tail guns. This approach, however, made each man dangerously inflexible in a combat situation because he knew only the operation of his own gun. As a result, policy changed in 1944 to require a gunner to be familiar with, if not exhaustively trained in, all the gun positions on the plane to which he would be assigned.[80]

Tighter administration, standardization of training methods, and refinements in the specialization policy did much to improve the gunnery program. Even after the reorganization and implementation of the specialization policy, however, gunners were criticized for their unfamiliarity with equipment they would eventually fire in combat. Sometimes ineffectiveness resulted from malassignment, as when one gunnery instructor opined that with "deep regret I had to leave Fort Myers and see both the men I had trained in Sperry, and many others, being converted to do maintenance and instruction work on Martin turrets."[81]

Although complaints about gunners' poor attitudes and low levels of competency continued throughout the war, by mid-1944 Maj. Gen. Robert W. Harper had far fewer occasions to lash back at critics with the (probably misguided) disbelief that he had displayed in October 1943 toward an air inspector's report: the "statement that turret gunners have never been in turrets can not be accepted by this office. It would be just as inconceivable as a statement that a graduate of a twin engine school had never flown a plane."[82] Even better than the support of senior leadership and the indication that the shortage of equipment was being redressed, gunners slowly came to see themselves and to be viewed with greater respect as their accuracy and professionalism grew under fire. PFC Don Moody graduated from the AAF aerial gunnery school at Harlingen, Texas, in May 1944 and went on to prove himself with the 307th Bombardment Group (Heavy) in the Pacific. He developed the pride he would take with him while in training, as he wrote his family:

> Before I came down here I'd always thought that the gunners, aerial mechanics, radio operators were about the lowest branch in the Air Corps, but I've found out different.... Those guys deserve just as much credit or *more* than the pilot, bombardier, or navigator. They not only protect the plane but keep it on its course, put it together if anything happens, and bring it back again. The pilot just maneuvers the plane where the navigator tells him to. The navigator has a *lot* of responsibility, but where would any of them be without the protection, radio directions, and maintenance of the other six guys?[83]

Summation

Individual training of aircrew specialists became a triumph of numbers. But the accomplishment was slow and arduous in coming. Pegged to pilot production, requirements in the other specialties were revised steadily upward, yet all lagged far behind the pilot program in attaining the goals enumerated in training plans. The pilot program reached its zenith in late 1943 and began to scale back, whereas the peak for navigators, bombardiers, and gunners was realized nearly a year later, in August and September 1944. At that time the Training Command graduated navigators at the rate of 25,600 per year (or more than 2,500 a month), 18,500 bombardiers a year, and gunners at a weekly rate of 3,500 (or approximately 180,000 per year). More than 45,000 bombardiers, 50,000 navigators, and approximately 300,000 gunners were the wartime product of specialized schools.[84]

Compared to the pilot program, the other combat specialties built upon shakier foundations. Effectively, the individual nonpilot aircrew training programs grew from nothing. Training practices were not formalized and in some respects, not even well formulated. (Certainly the latter applied to gunnery training, if much less so in navigator training.) Insufficient equipment and few facilities existed at the beginning of the war. When, in 1939, each medium and heavy bomber was directed to carry one navigator, the number of qualified navigators in the entire GHQ Air Force was only 166. Bombardiers were in equally short supply. In 1940 Air Corps schools had graduated 18 bombardier instructors. The additional 104 instructors and 206 graduate bombardiers the following year still numbered too few with which to go to war.[85] Many more gunners were needed than men trained in other specialties, and gunnery training remained the bottleneck in the system well into the war.

The numbers achieved in all the specialties represented a compromise between high manpower requirements and thorough training. The flow through the system was best accomplished if all the programs of specialized training could be synchronized. But it was impossible to coordinate the schedules, as each specialty had unique requirements. In the face of heavy demands for navigators, for instance, the Training Command nonetheless lengthened the navigator course from fifteen to eighteen weeks. The inevitable and anticipated effect was to forestall the time it took for badly needed navigators to reach the theaters of war. In this case, in a decision in favor of high training standards, the Training Command determined that the competence of navigators and the safety of aircrews were worth the price of slowed production. Once calls for manpower slackened, both the navigator and bombardier courses further extended; the navigator school became twenty weeks in December 1944 and the bombardier course, eighteen and finally twenty-four weeks in 1945.[86]

In other cases, the Training Command condensed training in order to meet quotas. For example, it developed the specialization policy for gunners and dual training for bombardiers. In the gunnery program a lack of flexibility among gunners resulted: a man came to know a great deal about one weapon but could not move easily between gun positions. And for every hour devoted to dead reckoning navigation, the dual-trained bombardier spent less time on the mechanics of precision bombardment. In the former case, the air crewman became more specialized; in the latter, less.

Other factors besides length of training jeopardized individual aircrew training. Until late in the war most schools had to be creative in developing their methods of instruction. The Training Command could not immediately and simultaneously build the facilities, train the instructors, supply the equipment and training aids, and standardize instruction in each of the programs. Also, specialized schools usually received new equipment last, causing obvious delays and ineffective training.

Of all the programs of aircrew training, gunnery training stabilized most slowly. The geographically diverse schools provided different opportunities for ground firing ranges and aerial firing over ground or water. Turrets remained in critically short supply, training materials were often outmoded, and the Army Air Forces could dedicate far too few modern planes to gunnery. Into 1944, computing sights were scarce, and the newer and more precise compensating sights were virtually nonexistent in the schools.

Many of these problems resembled those in other types of training. Most detrimental to the development of the flexible gunnery program, however, was the absence of clear-cut methods and principles for the schools to follow. Training experts disagreed not only over the proper methods for sighting flexible guns and the ranges at which they were effective, but also over the reliability of the computing sights in use. They even raised questions about the utility of the well-established practice of firing at towed targets.[87] The training air forces and schools did not provide uniform instruction in the tactics recommended by the AAF Board, which was responsible for its development. As late as 1944 one senior officer commented, "We have agencies to determine all necessary factors of the problem, but there is no coordination or control except of the very loosest kind. Each organization has its own ideas and gives them free rein. One unit teaches one tactic and another unit a different one."[88]

Some observers thought that the Training Command grappled too slowly with the admittedly difficult training demands. Training guidance, as found in manuals and directives, seemed hopelessly vague with respect to fundamentals such as the sequence and presentation of courses of instruction and descriptions of types and uses of equipment. Moreover, the command inadequately addressed personnel policy such as the training of enlisted and commissioned instructors and trainee elimination.[89] The program suffered badly because of the relative inexperience in flexible gunnery, the effective absence of specialized schools before the United States entered the war, the difficulties in procuring equipment, and the problems in developing theory and the training practices that would follow from it.

One must conclude, however, that in all the individual aircrew training programs success outran failure. Accomplishment could be measured in ways other than the dominant one of high manpower output. Much more specific training directives, based on considerable experimentation, made for improvement and greater standardization of training practices during the latter part of the war. Those responsible for the bombardier program, for example, instituted a more precise means of measuring bombing proficiency of school trainees. Certainly it was an achievement on a relative scale. Given the difference in cloud cover and wind velocity between the southwestern United States and the combat theaters, no system of measurement could be truly objective or replicate combat conditions. Much more significant, discovery of the pursuit curve and its translation in training into position firing, as well as the increased use of camera guns, resulted in enormous improvements in accuracy in gunnery training and, later, in combat. Finally, the capitulation to the insistence of numbers proved workable: not all bombardiers in large bomber formations had to be precision trained and equipped with Norden bombsights, making the utility of dual-trained bombardiers evident.

Despite all the expressed reservations about the thoroughness of their training, thousands of navigators and bombardiers completed their schooling and pinned on the wings of second lieutenants. Enlisted men finished their course and qualified as gunners. These individually trained aircrew members left the schools to join pilots in forming newly created crews and squadrons... and training continued.

THIRTEEN

Crew and Unit Training: Organization, Technology, and Doctrine

In our new planes, with our new crews, we bombed
The ranges by the desert or the shore,
Fired at towed targets, waited for our scores —
And turned into replacements and woke up
One morning, over England, operational.

— Randall Jarrell, "Losses"

To wake up "one morning, over England, operational," graduates of the Training Command's specialized schools — pilots, navigators, bombardiers, and gunners — went through crew and unit training in the four continental air forces before being shipped overseas. The final portion of stateside training was intended to pick up where specialist training left off, that is, to put a group of individuals together to work in crews and squadrons. Combat airmen were assigned to bombers (heavy, very heavy, medium, light, and dive), fighters (interceptor, escort, night, and fighter-bombers), or photographic and tactical reconnaissance planes. Each type of aircraft varied in performance characteristics, crew size and composition, mission duration and types of targets, and tactics utilized. The AAF set up separate programs for heavy, very heavy, and medium and light bombers, but did not distinguish functionally in the cases of fighters and reconnaissance planes.

Over the course of the war a few new courses were added, most notably for radar, and existing courses incorporated information regarding advances in aircraft and engine design that increased altitude, range, rate of climb, armament and bombload, speed, and maneuverability. Manufacturers, AAF design and testing divisions, and tactical units in the field modified aircraft in conformity with their applications in the theaters of war, and the training

World War II

programs gradually adapted as well as they could to the variations in practice.

In keeping with the espoused strategic mission, heavy bombardment dominated in the operational units of the AAF: at the height of combat group strength in December 1943, of the 80 operational bombardment groups, 27 were heavy, 25 were very heavy, 20 were medium, and 8 were light groups, compared to 71 fighter, 13 reconnaissance, 29 troop carrier, and 5 composite groups.[1] The priorities these numbers represented dictated personnel requirements in the training system in the United States. In one of the earliest wartime training schedules, in January 1942, multiengine pilots made up 54.5 percent of the total number.[2]

Most combat airmen, in other words, were members of a crew. But all of them, including the smaller number of single-seat fighter pilots, took their final, formalized program of stateside training in the new wartime system of Operational and Replacement Training Units. Operational Training Units (OTUs) trained pilots and crews in tactical units that deployed overseas as a group, whereas the Replacement Training Units (RTUs) turned out pilots and crews that went overseas for assignment as individual replacements or whole crews to tactical units already in theater.

Crew and unit training in the United States reflected as accurately as possible not only the personnel requirements but also the operational concepts and tactics of the combat theaters. The head of the Bombardment Branch at Headquarters AAF (one of several offices reporting to the Assistant Chief of Air Staff, Training) expressed this principle in December 1942 when he informed the Second Air Force, then tasked with the sole responsibility for training heavy bomber units, of the "exact conditions to expect in all theaters so the training and preparation of units and crews would be complete."[3]

The U.S. training establishment gathered vast amounts of information in its efforts to construct programs responsive to operations. Many sources delivered data upon which to base training. These included returned combat personnel; daily and weekly intelligence digests from the numbered air forces overseas and from offices in AAF headquarters, the continental air forces, and AAF domestic commands such as the Air Service and Training Commands, the Antiaircraft Command, and Army Ground Forces; exchanges of officers between U.S.-based commands and the combat air forces; reports of inspections, conferences, interviews, and letters; "marriages" of overseas units with those in training to correspond on issues of an instructional nature; direct suggestions and requests from operational commanders to senior training officials; training intelligence reports on a variety of subjects, many provided by Allied air forces; analysis reports prepared by the Operations, Commitments and Requirements Division at AAF headquarters; and subject manuals and tactical doctrine used in combat units that were distributed to home-based squadrons. As examples of the materials it found especially useful, in 1943 the Second Air Force cited the Special Combat Intelligence Memoranda which

Crew and Unit Training

Combat airmen received their final, formalized training in the United States, training as members of a crew. Most of them trainied on multiengine aircraft that typically might have as officers a pilot, bombardier, and navigator, shown here from left to right.

were devoted to topics such as evasive action on the bomb run, enemy antiaircraft equipment and its dispersion, and enemy fighter craft and tactics, and a briefly published series of "Digests from Combat Theatres on Training and Tactics."[4]

The amount of paper that tracked operations reached mountainous proportions. Training officers sifted through it in an attempt to coordinate their programs with the ever-shifting concerns and requirements reported by operational commanders and airmen overseas. Information was synthesized in official training standards issued from headquarters; directives from the training air forces for their fighter and bomber groups; training literature published as texts, field manuals, technical manuals, and orders; and training films and film strips.

Yet this process of defining operational concepts for training purposes proved daunting, for the body of information was huge but also conflicting. Tactics varied from theater to theater, and within a theater as well, dependent on mission and circumstance. Confronted with considerable subtlety and multiplicity in operational practices, AAF headquarters therefore directed that the continental air forces remain with the fundamental tactics for each type of aircraft that were considered applicable to air operations in *all* theaters. Medium bombardment, more than others, took into account the different tactics employed in Europe and in the Pacific. But in general, the training program for each type of aircraft did not so much change as stress, deemphasize, or refine one practice or another in response to combat reports. In the main, training officials worked to standardize techniques and increase proficiency of men and crews throughout the training air forces. Not until the AAF could foresee the end of the war in Europe did the picture change to permit a shift toward theater specialization.

Organizing and Administering Operational Training

The four continental air forces assembled and trained new combat units and replacement crews. Because they protected vital strategic borders, the First Air Force (located in the Northeast) and Fourth Air Force (in the Northwest) performed defense as well as training duties for some period into the war. As long as their defense mission dominated, they conducted fighter operational training. Ultimately all four continental air forces trained both fighter and bomber units. The Second Air Force was, however, charged primarily with heavy bombardment and the Third Air Force with medium, light, and dive bombardment as well as pursuit replacement crews. Under a March 1942 reorganization, the Second and Third Air Forces reported directly to Headquarters AAF. The First and Fourth Air Forces came under the Eastern and Western Defense Commands, but for training purposes at least, they looked informally to the Air Staff for guidance. A 1943 reorganization officially sanctioned that chain of authority when one of the six assistant chiefs of the Air Staff became the director of training and supervised both the Training Command and all four numbered air forces.[5]

One further consolidation late in the war drew together the administration of the training air forces. Discussions between Arnold, Marshall, and their deputies resulted in the establishment of the Continental Air Forces, activated in December 1944, that held authority over the four domestic air forces and the Troop Carrier Command. The Air Staff considered including the Training Command under the penumbra of the Continental Air Forces, but ultimately decided that the formation of operational units and replacement crews, the redeployment of units returning to the United States, and the conversion of some heavy bomb groups to very heavy units would be hindered by coordination with individual training.[6]

The AAF borrowed the OTU and RTU systems from the British. New combat units were formed from OTUs, and individual and crew replacements from RTUs. At the beginning of the war, no established, orderly method was

Crew and Unit Training

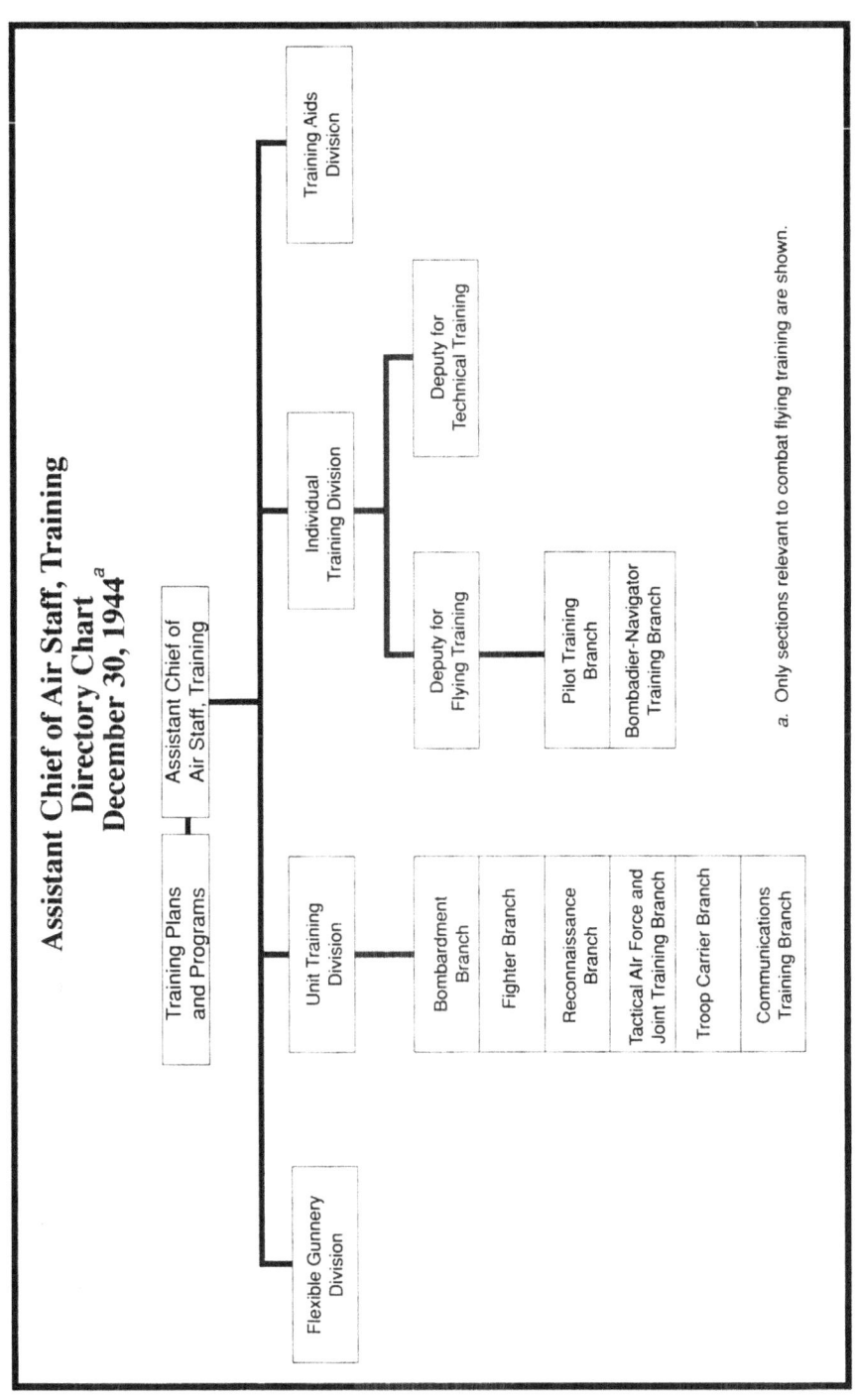

Assistant Chief of Air Staff, Training
Directory Chart
December 30, 1944[a]

a. Only sections relevant to combat flying training are shown.

in place to familiarize men with combat airplanes and equipment, assemble crews, and train them as units to be sent overseas. Training took place on an ad hoc basis in squadrons that, as often as not, were then raided for trained individuals or crews for new or replacement units. This practice was deleterious to the evolving training program and did little to ensure proficiency and cohesiveness of the units sent into combat.

The training system chose not repeat the time-consuming World War I experience of building facilities and shipping equipment for combat units to organize and train extensively near the front lines. (Each theater did, however, train newly arrived men in its own tactics and equipment.) In the present conflict, even America's European allies were unable to complete all air training on their home soil. Moreover, air warfare had changed drastically from the previous war. Larger, faster, more heavily loaded aircraft required expanses of open country for bombing and gunnery ranges; navigation training necessitated a variety of terrain and large bodies of water. To meet these requirements, the British found their answer in the British Commonwealth Air Training Plan by which RAF units would be trained in areas of less densely populated Canada. Instead of sending school graduates directly to combat squadrons, the RAF established OTUs in Canada where aircrew members flew operational aircraft, then trained as units before going to the front. Some American airmen who began working with the Royal Canadian Air Force returned to the States after Pearl Harbor to report on British–Canadian training practices. The AAF was sufficiently impressed with the newly created British OTU–RTU system that it quickly adopted this approach.[7]

OTUs were formed of overstrength parent groups that, by late 1942, included an initial cadre, most of whom were trained in a month's course at the AAF School of Applied Tactics, supplemented by new graduates of Training Command schools. After six weeks of intensive training, the parent split off 20 percent of its now experienced personnel (some of the School of Applied Tactics graduates and some newer people) to form a satellite group. Untrained men flowed into both the parent and satellite groups to fill the remaining slots. Every six weeks the parent group activated a satellite whose training continued for six months before it was ready for movement overseas.

Crew members took refresher training and were introduced to new equipment while OTUs organized and came up to their authorized strength. Because the purpose of crew training was to fuse individuals into a group, ideally all those in a new crew were to report at the same time. But manpower requirements, particularly those for navigators, were too burdensome for the Training Command to meet on schedule, so more often than not, a full bomber crew did not begin working together at the outset. After the initial phase of individual training and transition to operational aircraft, men proceeded to training missions that required them to function as a crew. A third phase linked crews together as elements of a flight, squadron, and group.[8] By late 1943, cooperative training with naval and ground units, and between fighters and bombers, became more commonplace.

Succumbing to early wartime demand for personnel, crews were often rushed precipitously from OTUs into combat zones. There, tactical squadrons were forced to provide training that should have been completed previously. Not only did extensive theater training delay entry into combat, but the accident rate of crews flying their aircraft to deployment points overseas reached unacceptably high levels because men were unprepared for the challenges of navigation and the exhaustion of human and fuel reserves on the long overwater journey. Although they wanted badly to reach their authorized combat strength, operational commanders came to realize the price paid by cutting into training time, and they enjoined AAF headquarters to reduce OTU preparation only under extreme emergency. Additionally, in furtherance of safety and crew cohesion, headquarters directed that whenever it became necessary to replace a key crew member, i.e., a pilot, navigator, or radio operator, the requirement would be filled by an entirely new crew.[9]

In RTU training (the term RTU will be used exclusively in this narrative although after August 1944 the name Replacement Training Unit changed to Combat Crew Training School), individuals and crews designated as replacements did not train as part of a larger group and therefore took a shorter course (usually two rather than three months) before they transferred to the theaters. Early in the development of the RTU system, planners compiled data on comparative attrition of individual crew members, by theater and type of aircraft, over time. With that information, as well as current production figures and requirements, a headquarters group determined the tables of organization, meeting each month to set the allocation of Training Command graduates and replacement crews by theater and aircraft type for the following three months. (Until well into the war the manpower allocation could not be settled sufficiently far in advance, and was consistently subject to change, so that crews could not be trained for the particular theater where they would be assigned.)

Supplying the mandated number of crews, which took into account the anticipated attrition rates, received first priority. Thereafter, fresh men and crews replaced the war-weary; the next available were assigned as fillers to bring units to full authorized strength. Finally, any additional overage became multiple crews per plane.[10] Each training air force conducted its RTU program differently, depending upon its mission at the time. For redeployed units, the training organizations determined the specific manning, again considering the number of individuals available from the continental air forces and AAF commands, while keeping in mind the time available for remanning, the mission of the unit, and its combat effectiveness upon arrival in the theaters.[11]

The RTU approach came to dominate in all the training air forces by mid- to-late 1943, because fewer new units had to be constituted than men and crews

World War II

to replace those killed or who had completed the required number of missions (which varied according to theater).[12] By the following spring the air forces came close to meeting commitment dates for shipment of replacement crews overseas, particularly for heavy bombardment units. The easing of the pressure for personnel, along with the policy of leaving replacement crews on one base throughout their training, and the allocation of substantial numbers of aircraft for training increased the speed and efficiency of the entire system.[13]

During the period when each of the four continental air forces provided the bulk of one or more types of OTU-RTU training, the Second Air Force conducted B-17 and B-24 training. The Third Air Force trained medium (B-25 and B-26), light (A-20 and A-26), and dive (A-24 and A-36) bombardment crews and single-engine fighter pilots. The First Air Force gave fighter operational training on the P-47 and some on the P-51 before the end of 1943, and it used O-46s, O-47s, and liaison aircraft to tow targets for antiaircraft practice. The Fourth Air Force, also undertaking the defense mission, therefore also concentrated on fighter training. It gave operational training on the P-38 and P-39; fighter replacaement training on the P-38 and, to a lesser extent, the P-39; and it towed targets with B-34s and various observation and liaison craft. The Third Air Force trained reconnaissance crews using the A-20, B-25, P-43, P-40, P-39, P-51, and liaison and rotary-wing types.[14]

About the time the RTU program predominated in late 1943, the discrete division of duties among the air forces no longer applied; all air forces then began training for both bombers and fighters. Increased diversification permitted mock combat exercises between bombardment and pursuit groups. Even though training under simulated combat conditions was still, as one officer admitted, "at best largely make-believe," it was an important step toward greater realism in training. When word of approval for pitting fighter against bombardment units arrived from AAF headquarters in June 1943, one wag penned on the memorandum, "Hooray."[15]

Heavy Bombardment

Through most of the war, "heavies" were the centerpiece of crew and unit bombardment training in the domestic air forces. Between December 1942 and August 1945, 12,217 B-17 and 14,708 B-24 crews were trained, compared during the same period with 5,887 medium, 1,602 light, and (for its shorter period of activation) 2,347 very heavy bombardment crews.[16] The Second Air Force remained principally responsible for the conduct of heavy bombardment training. Immediately after the United States declared war, the AAF reorganized its tactical units to give all Second Air Force theater operations to the Fourth Air Force on the West Coast and to make the Second an "operational training command." The Second Air Force organized its crew training on the OTU-RTU model, initiating both programs in the first year of the war.

Because bombing operations in combat units were normally done on a

Crew and Unit Training

More bombardment crews trained on the B-24 than any other heavy bomber.

group basis, in training, too, the group became the fundamental unit.[17] The Second Air Force began by creating two heavy bombardment groups. These and all subsequently activated bombardment groups and replacement crews were assigned to the 15th, 16th, or 17th Bombardment Wings, each of which supervised one of the three phases into which the Second Air Force divided OTU–RTU training. By the end of 1943 training officials replaced this vertical system with a horizontal structure in which all three phases of training took place in one wing, each wing specializing in one type of aircraft.

That change in the Second Air Force mirrored the larger reorganization occurring as the AAF began to spread bomber and fighter training among the continental air forces. The First Air Force, for example, shifted from the Eastern Defense Command to assume a training rather than tactical mission under jurisdiction of the AAF. It would add bombers to its ongoing fighter training. In keeping with the then existing OTU–RTU vertical structure that entailed moving from one station to another for each phase of training, planners initially thought that the Second Air Force would continue to provide first-phase training before units transferred, with their assigned aircraft, to the other air forces for second- and third-phase training. Such an undertaking threatened

World War II

to become so unwieldy and time-consuming, however, that the AAF decided that an entire OTU or RTU course would take place within a single air force: the First and Fourth specializing on B–24s, and the Third on B–17s. The Second added fighter unit training in exchange for relinquishing a portion of its heavy bombardment responsibility and began training very heavy bombardment crews (those flying B–29s and B–32s).

Training directives and policies issued by AAF headquarters to the Second Air Force guided the other air forces' heavy bombardment programs, and all were urged to consult with one another. Each air force, however, exercised considerable autonomy in carrying out its training. Because the other air forces' heavy bombardment programs were smaller, they did not, for example, copy the Second's administrative wing structure although they did emulate its three-phase system. Converting groups of the First and Fourth Air Forces, organized for combat, from OTUs to RTUs, occasioned additional delay. Nonetheless, the redistribution successfully deployed units where they could take advantage of existing facilities and good weather, lessened personnel requirements levied on the Second Air Force, gave all RTU training at one base (thereby eliminating troop movement that slowed training time), and allowed combined bomber-fighter exercises in all air forces.[18]

Very Heavy Bombardment

The Second Air Force capitalized upon its concepts and experience with heavy bombers in developing very heavy bombardment training. After the Normandy invasion and the shifting of battlefield requirements, the Second converted from heavy to very heavy training. It abandoned the relatively generalized OTU–RTU approach, at least for its initial training effort, because the first B–29 group, the 58th Bombardment Wing, specifically trained to become the combat nucleus of the XX Bomber Command that would operate in the China-Burma-India Theater. On June 15, 1943, the 58th was activated at a Second Air Force base at Marietta, Georgia. It was intended to serve as an experimental project whose training and subsequent combat experience would point the way toward future developments in the program. Once AAF headquarters turned over control of its training and that of the 73d Bombardment Wing (for replacement crews) to the Second Air Force in the late fall, the program became more than experimental.

The groups of the second very heavy bombardment wing, the 73d, awaited training until the high-priority groups of the 58th Bombardment Wing completed training and went overseas. These two wings of the XX Bomber Command trained on four Second Air Force stations, and in 1944, when more widespread training commenced, the number of OTUs expanded to eight in Kansas and Nebraska.

Even with the luxury of a high level of experience among its air crews in training (virtually unknown elsewhere during the war), a training air force well

Crew and Unit Training

Very heavy bombardment crews trained on B–29s.

versed in the methods of strategic bombardment, and a training program tailored directly for a particular theater, a substantial part of B–29 Superfortress training in the Pacific remained to be completed in the active theater of war.

Just as training for the other types of aircraft was distributed among the training air forces by the turn of 1944, by early 1945 very heavy bombardment training likewise spread beyond a single air force. The III Bomber Command, which had begun with medium and light bombardment and had then taken on some heavy bombardment training, now divested itself of medium and light training (which the First Air Force assumed) and added responsibility for training some very heavy units. It eventually converted four bases to B–29 training. The command anticipated further expansion, but the capitulation of Japan curtailed the need.[19]

Medium and Light Bombardment

Medium and light bombardment employed the OTU–RTU organizational pattern. The Third Air Force followed the Second's lead in dividing its medium and light unit training into distinct phases. It also initiated OTU before adding RTU, then phased out OTU altogether around the beginning of 1944. However, instead of a wing structure like the one Second Air Force used for heavy

World War II

Medium and light bombardment crews trained in B-26s (*upper left*) and B-25s (*lower right*).

bombardment crew and unit training, the Third Air Force's training units came under the jurisdiction of the III Bomber Command. Similarly, the I and IV Bomber Commands and the I, III, and IV Fighter Commands conducted operational training in the other air forces.

The III Bomber Command established its first medium bombardment OTU on March 2, 1942, at Jackson, Mississippi. The parent group in the medium program, the 21st Bombardment Group, did not begin training, however, until mid-June. With the parent organization itself barely functioning, the AAF directed it to spin off three satellites. Originally the Air Staff expected the new groups to be trained and ready for debarkation on September 1, 1942, but the schedule slipped to one group for each of the first three months of the new year. At the same time, headquarters ordered the Third Air Force to expand further to two light and one dive OTU.[20]

The difficulty the Third Air Force faced in activating new units arose from the serious manpower drain for immediate replacements overseas and from diversion to heavy bomb groups that enjoyed first priority in reaching full strength. The III Bomber Command, according to its historian, saw crews "snatched away from the O.T.U. groups at an alarming rate" and decided upon the investiture of RTUs to stem the tide from the OTUs. Using the same formula as the heavy bombardment program, medium OTUs had expanded from a ten- to twelve-week cycle, whereas crews in RTUs spent the shorter two months in training, thereby becoming available for combat units sooner and leaving OTUs intact. These plans were in the process of accomplishment when, in early 1943, requirements doubled. The two existing B–26 RTUs had the capacity to handle the increased load, but the B–25 group did not, so a B–25 OTU converted from an OTU to an RTU.[21]

Fighters

For the two air forces — the First and Fourth — that conducted most fighter training, administrative complications were more troublesome than usual because training responsibilities were superimposed upon their primary defense mission. Insofar as they performed a training role at the outbreak of war, pilots in the Fighter Commands of these air forces flew practice missions to defend U.S. coastlines against enemy attack. A significant change occurred on June 1, 1942, when the AAF gave the First and Fourth the additional job of activating and training fighter units for theaters outside the continental United States.

General Bradley, then commanding the First Air Force, presented a plan for carrying out this "difficult dual mission." He determined that the First Air Force would establish OTUs in which the parent and satellites divided responsibility for training and defense. Of the two air forces, the Fourth had the higher concentration of experienced fighter personnel. Probably for this reason, although its evolution from a defense to a training organization followed the same timetable as did that of the First (the First established only OTUs), the Fourth was instructed to take on OTU and RTU functions simultaneously.[22]

The awkward dichotomy of roles lessened as the likelihood of an attack on home shores diminished. In September 1943 the First Air Force was released from the jurisdiction of the Eastern Defense Command. About the same time, the I Fighter Command discarded the OTU system in favor of RTUs, a process occurring throughout the training air forces. After February 1944 the command, along with the Second Air Force, operated only P–47 RTU Fighter Groups. Similarly, only a lingering fear of a Japanese foothold in the Aleutians remained by the fall of 1943, and the Fourth Air Force also relinquished its defense mission. It then trained P–38 and P–39 groups and individual pilot replacements, trained a few P–40 pilots, and beginning in early 1944, conducted all night fighter training.[23]

Both the III and IV Fighter Commands practiced "continuous" training of pilots within a single squadron: each trainee spent his entire operational training time with a single instructor, rather than moving through the "stage" system of shifting to new instructors at specified points in training. The entire operational training program, in fact, went to this timesaving mechanism in the conversion from a vertical to a horizontal system, as the bombardment units called it.[24]

The Second Air Force applied its bombardment model to its fighter program, dismissing what it considered to be the sketchy directives from the First and Fourth Air Forces. It set up a highly standardized three-phase program, specifying what types of missions should be flown, if possible, during each period.[25]

World War II

Among the fighter units that trained on P-39s was this group of flying sergeants of the 312th Fighter Squadron.

Combat "Readiness"

Following on the heels of the Training Command's individual specialist programs, training in the continental air forces moved a step closer to the reality of aerial warfare. Men flew combat aircraft, bombed moving (usually towed) targets, did battle with "enemy" forces and supported their own in exercises, and participated in cooperative missions with fighters and bombers. Operational training squadrons had access sooner than the schools to equipment and modifications incorporating the latest technical developments. The Desert Training Center conceptually became an internal theater of operations where the Navy and Army air and ground forces worked out methods of operation and assignment of forces.

The AAF created a workable organizational structure for unit training and cooperated at all levels to improve and standardize procedures, but many problems remained unresolved until late in the war. The by-now familiar shortages of tactical aircraft and equipment, and of knowledgeable and experienced instructors, afflicted the domestic air forces just as they did the Training Command. Figuring out the completion dates for each of the specialties coming out of the schools and relating them to time of activation of OTUs and RTUs, and later for unit redeployment training of those going from the European to Pacific theaters, made for staggering scheduling problems. The

Crew and Unit Training

Training Command could not graduate navigators and bombardiers in a predictable sequence that facilitated smooth operational and replacement training. A navigator often did not come into OTU until the end of training. He found that in the interim the pilot of the crew had taken over much of the navigator's function during air work, so he and the bombardier mostly spent their time explaining to one another how to go about their partially shared tasks.

The differences in tactics from one theater to another had obvious ramifications for redeployment training because men who learned to cope with one set of circumstances now faced different ones, making some of their previous training irrelevant. Similarly, combat returnees reassigned to the training program were often committed to views based on their personal experiences, but they were not necessarily in touch with the broader perspective required to teach airmen whose eventual destination was unknown. The training program had to watch carefully for bias in its instructor indoctrination of these men.

During at least the first half of the war, tactical units devoted considerable time to individual rather than to crew training. After the manpower quotas were met and enough aircraft and equipment were on hand, the Training Command could offer courses of sufficient length and with sufficiently sophisticated equipment to send fairly well prepared graduates into the training air forces. Until then pilot transition onto combat aircraft often took place in OTUs and RTUs. David Burchinal was among the young officers in the 330th Bombardment Group in Alamogordo, New Mexico, responsible for training what they called the 90-day wonders (new graduates from advanced training on single-engine types) on B–24s. Training officers introduced new pilots to 4-engine aircraft in the 30-day primary phase, which was followed by second and third phases consisting of 30 days each. Then men deemed combat-ready embarked for the theaters. It was obvious at the time, Burchinal later commented, that "you can't make a pilot in a four engine airplane out of a kid like that. No experience, no anything, but we were doing it."[26]

Questionable proficiency, resulting from shortages of time, equipment, and personnel, was in evidence throughout unit training, as a report from the Second Air Force attested late in the summer of 1943:

> Not only is the Second Air Force required to take co-pilots and give them very short transition period to first pilots, but all deficiencies in all crew members between training and combat must be made up by the Second Air Force.... Likewise, the crew training problem is further aggravated by the fact that very often aerial gunners assigned to the Second Air Force have never fired guns in the air, radio operators have never operated a radio set in a B–17 or B–24, either on the ground or in the air, and aerial engineers have had no practical maintenance experience on a B–17 or B–24. At times, the situation has become so acute that individual crew

members must be trained by the Second Air Force from available basic personnel.[27]

That kind of accusation was leveled at every part of the training establishment. At the very time that the Second Air Force was castigating the Training Command in the report cited above, Brig. Gen. Frank A. Armstrong, Jr., who led some of the first strategic bombing missions in Europe, spoke for field commanders in his blunt plea that gunnery instruction must improve because "the lives of all our air crew members [are] absolutely dependent on gunnery."[28] Both the Training Command and the training air forces were fully cognizant of gunners' low level of proficiency. But awareness alone was not sufficient to bring about a satisfactory resolution, as evidenced by the fact that operational commanders were still declaring at the end of the war (perhaps unfairly) that AAF gunners seldom hit anything but each other's planes. In partial explanation, as they struggled with these and other deficiencies, each element of the training system considered its failures to be magnified by factors beyond its control.

Until fairly late in the war, the training air forces geared instruction toward fundamental and general skills, honed to maximum proficiency, given the time and equipment available. They could not impart tactics relevant only to a particular theater of war. At a training conference in the late summer of 1944, General Yount of the Training Command told his audience that in the field "they find it necessary to have a theatre indoctrination training course. This is perfectly natural, and I think that it will exist no matter how well our training at home may be. Every theatre differs, and I am glad to hear General Harper [Assistant Chief of Air Staff, Training] say that we are getting to the point where we can soon begin to train for the specific theatres."[29]

Theater-specific training inevitably affected programs run by the continental air forces far more than those of the Training Command. The first successful experiment using returnees to form an OTU nucleus aimed for a specific theater began in the summer and fall of 1943 when a group back from the South Pacific became the core of a new unit to be trained for service there.[30] By the next fall, planners could anticipate commitments following the defeat of Germany and therefore schedule units farther in advance of movement overseas. Training still had to include basic skills common to all theaters, but, initially at least, it could begin providing familiarization with subjects particularly relevant to the theater of operations where a unit was committed. The Training Aids Division disseminated pocket guides to the Pacific area, Arctic climates, and for locales in the China-Burma-India Theater. They published handouts such as "Living Off the Southwest Pacific" and "Aircraft Operations in the Desert." Meteorological conditions in the Pacific theaters figured prominently in training intelligence reports of 1944.

The end of that summer General Harper stated that "our chief difficulty heretofore in training specifically for a theater has been overcoming delays which are caused by matching a trained crew with specific airplanes and meeting demands for increased crews for other theaters." Henceforth, he informed Lt. Gen. George C. Kenney, commander of MacArthur's Allied Air Forces in the Southwest Pacific, one Second Air Force P–47 base, one Third Air Force B–25 base, and two B–24 and three P–38 bases operated by the Fourth Air Force would begin special training for Kenney's command. At the same time, Chief of Air Staff Lt. Gen. Barney Giles urged Harper to prepare a general plan for redeployment for the Pacific war. Navigators, for example, should concentrate on the skills needed to guide formations on long overwater missions against pinpoint targets. Such a plan was then in progress. In late September, the chief of the Unit Training Division on the Air Staff notified his colleagues that all the continental air forces were modifying their training directives to emphasize overwater navigation, instrument flying and night operations, and recognition only of surface vessels and Japanese aircraft.[31]

As the war advanced, the training system as a whole evolved toward higher quality and more dedicated training. By late 1944 enough aircraft and people had arrived in the theaters (the Eighth and Fifteenth Air Forces had two crews for each of their 2,000 and 1,200 heavy bombers, respectively), hard lessons had been learned about tactics applicable to training in the air over Europe and across the Pacific archipelago, technological improvements (such as the development and deployment of the P–51 Mustang as a long-range escort fighter and the fairly widespread use of radar) were aiding mission accomplishment, and the war was being successfully prosecuted on both fronts such that the vigor of the complaints about training had begun to diminish. Criticism focused less on gross deficiencies than on finer points of training. Although achieved only late in the war, the continental air forces could adapt crews and squadrons to the demands they would likely meet.

Training, Doctrine, and Tactics

For the first time, airmen went to war with an approved strategic mission articulated in war plans, even if, under official Army doctrine, their "sphere of action" remained under the commander of field forces.[32] The Army Air Forces expected to carry out that mission — the destruction of selected targets in the enemy's industrial infrastructure — through daylight, high-altitude, precision bombardment. Air tacticians put their faith in the heavily armed B–17 Flying Fortresses and B–24 Liberators, flying in large numbers in close defensive formations. The first B–17 daylight raids on Rouen in August 1942 confirmed airmen's expectations — Americans bombed targets with some accuracy at little cost to themselves. Analyzing the results of these first encounters, Eighth Air Force commander Maj. Gen. Carl A. Spaatz cabled General Marshall: "The Army Air Forces early recognized the fact that the effective use of air power *on a world wide basis* required the ability to hit small targets from high

World War II

altitudes." Spaatz found the initial sorties to be encouraging portents of future success, "especially considering that the crew training prior to operations was much less than desired."[33]

Members of the U.S. training establishment listened carefully to operational commanders in order to clarify their own thinking. Two months after Spaatz's report, Maj. Gen. Robert Olds, then Second Air Force Commanding General, concluded much like Spaatz that the "development of long-range heavy bombardment aircraft, capable of self-defense which permits its use for daylight operations, is based on sound tactical principles, and has been advocated for a number of years by the foremost exponents of United States heavy bombardment aviation." Having endorsed the American strategic doctrine, he went on to direct his wing and group commanders to train their heavy bomber groups toward this end, to build up "well-coordinated units properly armed and manned, that will be able to operate during daylight in the face of an aggressive enemy, regardless of fighter opposition that he attempts to place in your way, and hit accurately with individual or pattern attacks such stationary or moving targets as you may be instructed to destroy."[34]

Dedicated though it was at the outbreak of war to proving the efficacy of the strategic doctrine of daylight, high-altitude, precision bombing, the AAF had no intention of limiting its resources to heavy bombers. The concept of balanced forces could be traced at least as far back as the early days of the Air Corps Tactical School in the 1920s. The school maintained departments for each type of mission (bombardment, fighter, etc.) unified under a core course entitled "The Air Force" that emphasized coordination of all the forces working together. At the time the United States entered the conflict in December 1941, the AAF expressed its long-range strategic doctrine in AWPD-1, intended to be carried out in operations by heavy and very heavy bombers. But it also had an aircraft inventory capable of lower-altitude bombing, ground support, and fighter interception.

In fact the AAF's most recent and practical prewar experience applied to training the tactical rather than the strategic forces, coming as it did from the great 1941 Army GHQ maneuvers in Louisiana and the Carolinas. Even though these maneuvers had pointed the way and the AAF had laid down rules for tactical operations, the formal articulation of doctrine evolved through several iterations. AAF doctrine that acknowledged air power to be properly a part of the ground battle, albeit under control of *air* commanders — of vital importance from the AAF's point of view — and that codified, in Field Manual (FM) 100-20, the employment of tactical air forces, awaited the bitter lessons of the fighting in North Africa before this publication appeared in July 1943. The document held that the first priority to be attained was air superiority over the battlefield (a lesson learned in World War I and reiterated after the 1941 maneuvers), followed by interdiction of the enemy's lines of supply and communications, and then ground support. In a tactical air force, aircraft and

Crew and Unit Training

practices inapplicable in a strategic mission (fighter-bombers and dive-bombing, for example) were crucial to success.

Despite varying missions, tactics, and circumstances, and therefore some significant differences in training in the theaters of war, a consensus emerged that in the aggregate the AAF should train to fly during daylight, at high altitude, in formation, and while using instruments. From reports of airmen flying sorties under fire everywhere, the importance of formation flying surfaced repeatedly; those in heavy bomber crews pressed for its accomplishment at high altitude. Pilot Bill Carigan said of his experience in the 737th Bombardment Squadron of the Fifteenth Air Force, "We fought the war in formation, almost without exception."[35] From those men flying B–17s in North Africa: "The major fault found with training in the U.S. is the lack of practice in high altitude formation flying." From P–38 pilots: "High altitude formation flying and combat training are most important." And from those in P–40s: "Formation flying should be stressed to the utmost."[36] The Eighth Air Force's surgeon claimed that "most of the flying personnel newly arrived in this theater have ... never flown at the [high] altitudes at which operations are routinely carried out in this theater of operations."[37] Well into 1944, reports from the Mediterranean and European theaters and the China-Burma-India sector urged more practice in high-altitude close formation for heavy bombardment and in formation at somewhat lower altitudes for medium bombers. The Fifteenth Air Force specifically described the occasions in which high-altitude formations had and would save the lives and missions of pilots, bombardiers, and navigators who came into its squadrons.[38]

High-altitude formation flying was crucial during combat, so advanced training included instruction and practice in this technique.

World War II

From the beginning, stateside training did its best to implement the practical, and practicable, suggestions from combat personnel. Everybody understood the necessity for protective formations, so both fighter and bomber OTU and RTU devoted a considerable amount of time to taking off, assembly, and landing in formation. High-altitude flights became de rigueur for heavy bomber crews and for fighters escorting them. Of course many units would never fly high, and some airplanes *could* not. All the same, experience at high altitude was seen to be an important part of AAF training, oddly enough appearing even in Training Standards for light bombardment.

In May 1943 Chief of Air Staff General Stratemeyer directed that all rated flying personnel on duty with the War Department would undergo high-altitude tests in a low-pressure chamber.[39] Even so, ground-based indoctrination in high-altitude flight did not precede the desired number of hours in the air, owing, for one reason, to the shortage of high-octane fuel. In a priority list for 100-octane gasoline in May 1943, OTU and RTU training placed fourth among five potential users, coming in ahead only of "all other flying," which essentially amounted to the activities of the Training Command.[40] During the first five months of 1944, 91-octane gasoline comprised approximately two-thirds of the fuel delivered for all types of training. Flying above 20,000 feet with this lower octane fuel, aircraft often developed engine problems. In response, headquarters directed that high-altitude formation training be kept at a minimum altitude of 16,000 feet when 100-octane fuel was unavailable. By summer, the air forces had sufficient B–17s and B–24s to bring the total for all RTUs to the required level, and about 55 percent of the fuel allocation to the training air forces was the 100-octane type. These factors allowed minimum high-altitude requirements to be met by all heavy bombardment crews.[41] At that time, the chief of training on the Air Staff, General Harper, cautioned the training air forces that the

> minimum requirements established by this Headquarters are the absolute minimum which can be accepted. It is desired that as much of the training as possible of Heavy Bombardment and Fighter crews be conducted at altitudes where the use of oxygen is necessary. Except in exceptional circumstances, the shortage of 100 octane gasoline will no longer be accepted as an excuse for non-completion of high altitude requirements.[42]

For years training agencies had been attempting to redress airmen's unfamiliarity with and reluctance to use instruments, one of the deficiencies now targeted in combat reports. Acknowledging the problem early on, a board of officers investigating instrument training in the Training Command, OTUs, RTUs, and staging areas for overseas movement announced baldly in late 1942 that "with few exceptions, instructors, trainees and students were found to be incompetent."[43] Instrument training was on the books, however. The Second and Third Air Forces' curricula listed a certain number of flight hours that

Crew and Unit Training

medium and heavy bomber pilots had to acquire during adverse weather conditions. But dramatic improvement was not soon or easily realized. As late as January 1944, a report culled from experiences of the 21st Bombardment Wing recommended that "actual instrument flying *be required* of Combat Crews in third phase [RTU] training, plus additional instruction in meteorology by competent personnel. Eighty per cent of the [new] pilots...have never experienced actual instrument conditions." On their long overwater flights overseas, pilots talked incessantly about the weather, the report continued. Pilots "appeared very concerned about being on instruments for an hour or two even though the air was smooth and no hazardous conditions existed."[44] Airmen stationed in the Pacific theaters, too, where weather conditions were frequently poor and nearly always unpredictable, continually reminded the training establishment to concentrate on instruments and all-weather flying.

AAF headquarters increasingly tried to impress upon training units that instrument use was just as critical during the thick of battle as it was for navigation. In December 1942 it told the Second Air Force to intensify bomb approach training by manual control, using the pilot's direction indicator and secondarily to instruct pilots in the operation of the automatic flight control equipment (AFCE). The equipment soon assumed much greater importance; the Unit Training Division noted improved bombing accuracy in an Eighth Air Force raid on Vegesack, Germany, which was attributed to the first use of AFCE. The division began to push for greater reliance on instruments. Bomb approaches demanded "the most precise kind of instrument flying skill and...under the stress of combat our pilots can not be expected to measure up to the skill required." Use of automatic equipment could help overcome erratic flying during the bomb run, staff officers counseled. Without saying so directly, they accused airmen of seat-of-the-pants flying, and they mused over how to "sell" instrument flying to theater commanders.[45]

The Army Air Forces Commanding General himself implied that the

Instrument training used devices such as this cockpit mockup of a B-25.

World War II

Second Air Force dragged its heels on using the C–1 automatic pilot for no particularly good reason. Brig. Gen. E. L. Eubank, Director of Bombardment, spoke for General Arnold, for example, when he informed the Second Air Force that "in the OTU system, the C–1 pilot is not being maintained due to the fact that it has not been used." Second Air Force Commanding General Olds testily replied that the C–1 modifications were difficult to obtain and slow in coming, and that the instrument was temperamental at high altitudes even with modifications. Moreover, heavy bombardment aircraft were flown by "kids still trying to learn how to fly an airplane and bombing from a B–17F with a tail that wanders all over the sky and from a B–24 with controls so heavy and stiff . . . you can barely fly the thing on instruments."[46] The exchange demonstrated a disagreement less from principle than from vantage. In actuality, the senior leadership all agreed on the importance of instrument flying, but they demurred over the reasons for low levels of proficiency by aircrews in training.

No matter how hard the training establishment tried to change their habits, pilots tended to trust instruments much less than themselves or each other. As in other areas of training, no doubt, the problem could not necessarily be solved systemically. Fighter pilot Cleo Bishop commented many years after the war:

> I don't think I even learned how to fly instruments until after World War II. You would go to great lengths to stay out of the clouds, like flying on the deck or finding a hole or something. We used to get in clouds only as a last resort.[47]

Fighter pilots were not alone in their resistance to the technological solution. When a man commanded a crew and flew in formation over a prescribed route, he could look to others in the formation. Within the ship, a bomber pilot could call upon his navigator; a navigator in most crews could follow the lead aircraft. A night fighter pilot knew he could not navigate visually and counted on his radar observer. Many a day-fighter pilot out watching for the enemy believed that all he needed were his instincts, fast reflexes, sharp eyesight, and good luck. The World War II pilot never gave over his primary faith in himself and other men to the technical gods.

The need for improvement in another area of unit training, recognition of enemy and friendly forces and equipment, came to light glaringly early in the war, although the dangerous eventuality had been predicted in the 1941 Louisiana and Carolina maneuvers. British Air Chief Marshal Harry Broadhurst later told the story of the first RAF fighter escort of an AAF bombing raid over France: "I went on that mission and anybody that went near the American bombers was asking for trouble because they didn't know a Spitfire from a Me–109."[48] A good year later, observing the American troops assembled for the invasion of North Africa, General Dwight D. Eisenhower, Allied Commander in Chief in North Africa, worried that the air and ground forces' inexperience training together, and their potential failure to distinguish between German and

Allied forces, might have severe repercussions in combat.

The fighting in the Kasserine Valley in Tunisia bore out General Eisenhower's fears. American B–17s bombed noncombatant settlements far from the battlefield. Moreover, General Arnold later attributed much of the air loss to antiaircraft fire from American ground troops who did not recognize their own airplanes. In his reply to Arnold, Commanding General of the Army Ground Forces Lt. Gen. Lesley J. McNair countered that AAF airmen had likewise attacked U.S. ground troops.[49] The AAF redoubled its efforts to stress air-ground identification in training, which already included enemy aircraft, and would add naval vessel recognition. At the end of February 1944 the Twelfth Air Force sent reassurances that cases of mistaken identity had lessened, although they added the postscript, "Don't let up in recog[nition] training however."

Regrettably, the AAF never eliminated misfortunes of this kind. (Nor has any combat force since.) In the Mediterranean theater, Allied aircraft bombed an American-held town in Italy, and P–40s strafed their own 3d Infantry Division, killing and wounding more than a hundred men. In the summer months following the invasion of France, the AAF mistakenly attacked friendly ground forces on numerous occasions. While it fought the air war against Japan differently than the one on the Continent, in the campaigns across the Southwest Pacific the AAF also bombed very close to friendly troops, with resulting casualties.[50] Those terrible errors could not, of course, be laid solely at the door of the training program. Under fire, men found it hard to suppress the self-protective urge to shoot first, ask questions later. Col. V. L. Zoller of the Fifteenth Air Force acknowledged that human reaction in his unusual suggestion at a conference held in August 1944 that training should completely give up aircraft recognition drills because "if any airplane, regardless of who it is, comes within range of our formation and looks as though it is going to make an attack it is going to be fired [upon]."[51]

The call for widespread improvement in such areas of training as formation and instrument flying and enemy recognition went out to all types of units. Some tactics, particularly high-altitude flying, directly reflected AAF doctrine. Yet the American strategic concept itself, supposedly the principle guiding training, slowly began to crumble. One aspect gave way completely — the idea that heavy bomber formations could successfully fight to the target and return without friendly fighter escort. The terrible attrition during the bombing raids of 1943 badly shook the optimism of the previous year. The raids over Schweinfurt and Regensburg during Black Week in October, huge bomber formations flying deep into Germany without protection, witnessed the horrifying loss of 148 Fortresses and their ten-man crews. The AAF was by then working furiously to develop an escort fighter, and by 1944 P–51B Mustangs equipped with auxiliary drop-tanks were accompanying bombers over Germany.

World War II

Recognition of friendly and enemy aircraft was difficult under combat conditions, so aircraft identifiction was emphasized in training.

By this time, too, training in units headed for both strategic and tactical air forces took account of another deviation from doctrine: the possibility of night operations, particularly for the onslaught on Japan. Earlier, stemming from Winston Churchill's reluctant approval at Casablanca in January 1943, the AAF and RAF Bomber Command agreed to fight from their separate air doctrines, the AAF bombing during the day and the RAF at night. Yet day bombing remained at issue, even with Americans. Only a year into the war, AAF headquarters directed the Second Air Force to maintain a ratio of one night bombing mission for every two day missions. Until the Third Air Force received enough qualified pilots from the transition schools to follow the same ratio, it was to run one night mission for every three day missions. Certainly one impetus for these instructions relatively early in the war sprang from the need to teach instrument techniques. But another, especially as time went on, appears to have been preparation for night bombing. In October 1944 General Arnold directed that the means the AAF Board investigate to use in prosecuting the war against Japan include diversion of a specified proportion of the AAF to night operations.[52]

The most telling abandonment of AAF doctrine came from the failure of precision bombing against pinpoint military targets. The postwar U.S. Strategic Bombing Survey found that only 25 percent of the bombs dropped on visually sighted targets landed within 1,000 feet of their aiming point. Yet the AAF had not awaited that assessment before changing its practices. After March 1945, B–29s operating in the Pacific conducted area bombing. By then, those within the senior AAF leadership, the War Department, and the Roosevelt administration had debated using the heavy bomber against nonprecision targets in Germany. Their discussions were backed up by specific proposals for area bombing of urban centers, and also for admitted terror bombing of nonmilitary targets in order to destroy the enemy's morale. The RAF, exhausted and infuriated by German air raids on their own population, felt little squeamishness about civilian casualties and argued doggedly for that approach. Relatively early, in November 1943, General Arnold agreed in part, countenancing hitting area targets when weather or other deterrents foreclosed precision strikes. Thereafter the Eighth Air Force flew blind bombing missions using radar that essentially amounted to area bombing. Pragmatic, military, and political concerns, as well as moral sentiment, kept the AAF from more actively pursuing that type of warfare until late in the European war and during the final push against Japan.[53]

Training did not follow the turnabout in operational practices. The AAF never abandoned the belief that its people flying strategic missions should be trained in high-level, daylight, precision bombing. Without doubt, proficiency in formation flying, aircraft and naval recognition, and instrument flying during training ensured greater safety and promoted success in the air war. Yet all the training in pressure chambers and high-altitude practice missions did little to improve the poor record of accurate hits on targets from high altitudes. In fairness, the reluctance to relinquish doctrinal preconceptions was reinforced at the time by continuing improvements in the vaunted, highly secret Norden bombsight and by experiments with radar. Modifications to the Norden bombsight met increased speed and altitude requirements, but the visual sight could not eliminate the problems of visibility created by clouds and haze, typical even during the summer months in Central Europe. Knowing in theory and confirming in practice that bombing missions had to abort when the target was obscured, the AAF embraced radar as the answer to the problem of bombing through overcast. Radar did serve as a navigational aid, directing bomber formations to the target area when visual identification was impossible, but radar technology did not progress rapidly enough during the war to ensure precision bombing.

If the relatively straightforward and uniform initial advice to training people from the strategic forces ultimately proved inconsistent and was abandoned in the field, training for the tactical forces was, from the outset, even more complex. (Here, too, operations turned its back on theory.) No set of "standard" practices applied everywhere. Even among the mixed forces of bomber and fighter wings, requirements varied. The campaigns in Sicily and Italy were accomplished with considerable air-ground cooperation, for example, whereas in the Pacific theaters fighter-bombers launched bombing and strafing

World War II

Crews trained with the highly secret Norden bombsight had mixed success because they still could not bomb with precision from high altitude through overcast.

attacks against Japanese shipping. The mission and the nature of enemy opposition influenced tactics and armament. Medium bombers often flew at relatively high altitudes in Europe to escape the intense antiaircraft fire. But on island-hopping missions in the Pacific, where the bombers less frequently encountered flak, they could fly at lower altitudes. Here, modified B–25s replaced the transparent bombardier's nose with one carrying as many as eight guns and sometimes a 75-mm cannon.

Similarly, fighter tactics differed for tactical and strategic missions. British Air Vice Marshal Arthur Coningham argued that the hard-won "fighter-bomber mentality" of the Tactical Air Forces in the Mediterranean would be destroyed if fighters were shifted to the strategic bombing offensive gathering forces for 1944:

> The fighter-bomber force in the T.A.F. [Tactical Air Forces] has long experience and is highly specialized. The present standard is due to continuity and the inculcation of fighter-bomber mentality born of more than two years offensive trial and error with armies. It is a difficult task which has to grow on a unit... to take these units away from their specialized role and convert them into defensive escorts to long range bombers is unthinkable.
>
> It is appreciated that the heavy bomber forces must have adequate

fighter protection, but this can surely be assured without breaking up this unique force.⁵⁴

An immediate postwar analysis of Ninth Air Force operations found that the "combat tactics and techniques originally taught at training centers in the Zone of the Interior were considerably and continuously altered in combat."⁵⁵ Even had training officers known about all of the variations in tactics, operational training was not lengthy enough to incorporate all the practices developed in combat. Moreover, until fairly late in the war the AAF lacked sufficient manpower to tailor training toward a particular theater. A fighter pilot and those who trained him did not know whether the aspiring flier would be protecting Eighth Air Force heavy bombers against Messerschmitts, for instance, or be assigned to the Tactical Air Forces in the Pacific, which was virtually devoid of strategic targets, or be part of the Twelfth Air Force's Desert Air Force in the Mediterranean. That uncertainty lessened as the European war wound down after the invasion of France and theater specialization became possible.

On rare occasions when the AAF officially recast published doctrine, as in one of the most outstanding cases, the issuance in July 1943 of FM 100–20 (known as the AAF's Declaration of Independence), the results did not, according to historian Richard Hallion, "substitute a body of tactical recommendations or procedures" for the previous manual.⁵⁶ Thus, even a seminal document failed to provide a clear blueprint for field commanders or training officials. Of necessity, then, throughout much of the war, training standards adopted an across-the-board approach, determining the fundamental skills required for each specialty, putting together crews who could practice together, and leaving the ultimate application to frontline units.

In practice, the vision espoused by air leaders at home and abroad translated much more straightforwardly into training procedures than into operations. Training could be — and had to be — organized and standardized, whereas men in combat left doctrine behind when necessity called, often cobbling up battle tactics on the spot, then changing them as conditions dictated. Advice from the commanding officer of the 303d Bombardment Group to training units at home in May 1943 illustrated the point: "I am not enclosing our present plan of zones of search and fire, nor details about enemy fighter tactics, because I am convinced that they will have changed by the time this letter reaches you."⁵⁷

Distinguished airmen speaking at a conference forty-five years after the war ended recalled the untried nature of air power tactics and doctrine at the outbreak of war. Brig. Gen. Robert M. Lee was "not sure we had any well-established tactics and techniques.... There certainly wasn't any manual that was definitive." (In fact, *no* existing War Department manual directed air support operations.) Maj. Gen. Elwood R. Quesada considered the AAF experience in North Africa to have been "organized chaos. It was unbelievably

World War II

immature.... Most of the techniques that attracted attention were invented in the field."[58] In another interview, also many years later, Lt. Gen. Arthur C. Agan, who had been Chief of Tactical Operations for the Eighth Air Force, admitted that "as far as having a staff and having people actually concern themselves with tactics, we couldn't spare the people so there was a real ignorance concerning air tactics." Moreover, he contrasted training at home, founded upon doctrine, with the reality of combat experience. No matter what theory may tell you, he commented, "you do whatever has to be done to get the bombs down," so in the Mediterranean theater, "we ended up bombing from 8,000 feet."[59] An intelligence report of late 1943 made the point that those who planned a particular mission faced, in the event, "circumstances which they had never before experienced. Action experience turned out very different to the standard exercises."[60]

What one operational commander advocated often failed to apply elsewhere, leaving training officials reluctant to deviate from standard methods. When, for example, Brig. Gen. Haywood S. Hansell forwarded to AAF headquarters a copy of his rules governing B–17 units in the 1st Bombardment Wing of the Eighth Air Force, Brig. Gen. Robert W. Harper explained to the Deputy Chief of the Air Staff that while his office made constant revisions to AAF training standards in response to new ideas and practices in the theater,

> it would be unwise to submit the attached doctrine to the Second Air Force in the form of a directive as indicated by General Hansell. Such a directive would cause a great upheaval in the present training program without accompanying parallel benefits. The Second Air Force is now training crews under standards which are a combination of the best ideas from all theaters, with particular emphasis upon those received from the U.K., and their aim is to turn out a product which is basically sound in its training and capable of assimilating ideas and practices used in any of the many theaters to which it might be sent.[61]

It is significant to note that in its November 1943 tactical doctrine, the Second Air Force in fact adopted Hansell's practice of formation rather than individual bombing to protect against German fighter opposition. Much of the text derived from Eighth Air Force experience. Nonetheless, the generalized approach to training as stated by General Harper still applied, since by the time the Second Air Force published its guide, the tactics had demonstrated their application beyond the confines of the Eighth Air Force's type of warfare. Formation bombing had proved itself also in attacks against maneuvering ships.[62]

Not infrequently, operational commanders disagreed among themselves, once again leaving training officers in the United States to rely upon existing doctrine and precedent. As an example, in April 1943 the commanding officer of the 390th Bombardment Group (Heavy), a recently activated RTU, solicited advice from various officers in theater as to the skills that should be emphasized and the tactics that should be practiced in the upcoming three-month training period. He confessed to the commanding officer of the 306th Bombardment Group that two other officers from the same theater whom he had consulted "could not agree on any tactics and technique, except the necessity for stress on high altitude formation." In a lengthy response to the query, in which he touched upon size and assembly of formations, equipment, some navigational problems that could only be dealt with effectively in theater, bombing techniques that were "probably a little different from anything taught in the States," evasive action by the formation, range of fire, and the importance of aircraft identification, Lt. Col. Claude E. Putnam alluded to the lack of agreement among fellow officers by saying that the "past three months have seen considerable change in both our tactics and enemy tactics. It is highly probable that this change will continue but it is not believed that the broad principles of daylight, high altitude, precision bombing will be materially altered."[63]

Ultimately, war brought an inevitable, unbridgeable gulf between training and combat. When men left training, they went to fight, not merely to fly. Normal peacetime air training levied an exceptionally high casualty toll compared to other Army combat arms, and although death under these or any circumstances came neither fearlessly nor clean, now airmen did not go down only because of wind shears, or mechanical failures, or one's own mistakes. Now the enemy had a human face and intention, and he meant to kill. In April 1943, General Arnold returned home from a tour of combat units around the world with the "outstanding impression" that American airmen had a "burning desire to fight the enemy relentlessly." So at least he wrote General Kenney. Yet that view did not square entirely with the "lack of enthusiasm for combat in some individuals in our training units" that he attributed to the "discussion of the horrors of battle and of gruesome incidents on the war fronts described by wounded and war weary personnel returning to this country." Apparently, some of those patriotic Americans also brought tales of the horrific reality of human violence. Arnold therefore exhorted those fighting the war to help motivate young airmen, to tell "our trainees some of the glamours of war and the great personal satisfaction to be derived from actually hitting the enemy between the eyes."[64] Here, Arnold voiced an imperative that underlay all combat training, that men must be sent into combat knowing the truth, but not the whole truth. They must be indoctrinated to embrace the battle and hate the enemy. They must be taught how to fight, but not at what cost.

War brought another psychological element to unit training that was as old and unchanging as the military order itself — the drawing together of men in combat. Remarque's narrator in the World War I classic *All Quiet on the Western Front* spoke of it with simple eloquence: "It is as though formerly we were coins of different provinces; and now we are melted down, and all bear the same stamp."[65] Airmen on the battlefront during the Second World War reminded those at home that "crews must be trained as a team, not as a group of individuals."[66] The need to cement cohesiveness among members of the crew and squadron appeared in AAF OTU guidelines under subjects such as "duties of the airplane commander" and "learning to function as a unit." Esprit d'corps, group morale, and, preeminently, mission accomplishment depended upon mutual trust in the reliability and proficiency of each person.

Reminders of the bonds that grew between members of that irreducible air force unit, the crew, and, more commonly, among specialists such as the pilots in a squadron can be found in the memoirs, the poetry, and the gallows humor of the men who fought in the air. They started to pull together in their final training on home shores. In a novelistic treatment of his experience flying B–17s, Edward Giering spoke of the crew in training as being "meshed together like so many cogs of machinery to run as smoothly as the machine they flew."[67] In Philip Ardery's analogy, the heavy bombardment crew he joined in El Paso, Texas, took on a living identity: "When we arrived at Biggs Field we were only the skeleton of a heavy bomber group. Gradually sinew and muscle grew on the skeleton. It became a unit with as much individuality as a person has, and was as different from other organizations of like size and designated purpose as one human being is different from another."[68]

The fraternity among aircrew members had its limitations, at least according to some in the AAF hierarchy who contended that too close camaraderie lessened discipline and diminished the leadership role that the airplane commander must exercise. Moreover, particularly in retrospect, airmen tended to idealize the phenomenon. In actuality men still remained individuals; they rarely transformed into the organic whole or well-oiled machine of reminiscence. Also, a caste system among airmen persisted. An RAF gunner in a night fighter squadron later admitted, "I was not altogether happy about my own place in the crew. I always had the uncomfortable feeling that it did not really make much difference whether I was aboard the aircraft or not; and I had to stick pretty close to it to avoid being left behind."[69] Dive-bomb pilot Samuel Hynes confessed that flying with the radio and radar men of his crew in the Pacific was not "the bond that flying beside other pilots was. The pilots were my friends but the crew was just a responsibility, like relatives or debts." Nevertheless, the squadron was "more like a family than anything else," Hynes reflected. "Everyone moved in his own sphere, was free to dislike and fight with the other members . . . yet acknowledged that there were bonds that linked him to this group of human beings, and excluded all others. . . . It wasn't a hostile exclusiveness, just a recognition that ties existed."[70] Once into combat, if you were to die in the smoke and fear and noise of battle above the earth, it would not be with strangers.

FOURTEEN

Training for Strategic Bombardment

> With the beginning of crew training... the approach of war began to take on form and meaning. Now we would be in the air almost daily for long periods of time, operational accidents would claim their tolls, and we would have one day off every three weeks. Now all knowledge gained about the aircraft and its equipment might some day make the critical difference, and though we retained our high spirits and self-confidence, we also began to recognize the wild blue yonder for what it was. It would be our battlefield, and up there we would either win or lose, live or die.
> —Kevin Herbert, *Maximum Effort: The B-29s Against Japan*[1]

Men assigned to heavy and very heavy bombardment squadrons during the final, sobering phase of training that Kevin Herbert described knew they were vital to the American strategic air war: "Germany first," according to Anglo-American agreements, then Japan.[2] In Europe, the AAF initially pinned its hopes on the Eighth Air Force, the "mighty Eighth," as it has been called. In 1942 the Eighth Air Force, based in England, began building its bomber fleet in the expectation of spearheading the air offensive by bombing strategic military targets and wiping the Luftwaffe from the skies. To the great disappointment of some AAF strategists, however, demands for aircraft in other theaters forced the Eighth to postpone most long-range bombing missions for another year. But the delay gave field commanders an opportunity to practice some of the techniques for which they had little time or equipment at home — formation flying, turret firing, and navigation over landscapes laid out differently from those in the United States and over stretches of water such as the English Channel.

Finally, in late June of 1943, Eighth Air Force heavy bombers punctured

World War II

the Ruhr with their daylight raids on chemical and synthetic rubber plants and German submarine pens. During the remainder of the year, although the Mediterranean campaign siphoned off some Eighth Air Force units, those remaining in England bombed German shipyards, U-boat bases, ports, and industrial sites. The scale of these missions grew ever larger. In mid-December the number of bombers massing to attack the port areas of Bremen and Hamburg and the U-boat yard at Kiel reached 649. P–38 and P–47 fighters accompanied the bomber formations for as long as their limited range allowed, which sometimes meant to a not-too-distant target. Most frequently the fighters turned back when the bombers headed deep into Germany.

Late in the summer the first American Pathfinder group became operational. Pathfinder was a British method of navigation in which certain aircraft in a formation were fitted with British Oboe (a ground-to-air transmission that accurately fixed the location of the high-flying aircraft) and H2S (an airborne radar set with a rotating antenna that could scan the ground) or the U.S. version H2X radar (for a while called blind-bombing equipment). On November 3, eleven American Pathfinders with H2X led AAF heavy bombers to the port of Wilhelmshaven, where they successfully struck their intended target.[3]

As 1944 began, the strategic forces still had a traditional role to play before turning to tactical operations for the invasion of occupied Europe and Germany planned for the spring. To achieve air superiority over the Luftwaffe, Eighth Air Force commanders believed they needed to bomb German fighter plants and other strategic targets that lay beyond the range of American fighter protection. One report opined in February that "in recent months the range of our fighter cover has been greatly extended. In the future, we shall undoubtedly be able to go to central Germany with much lower losses." A now-unknown cynical reader scribbled in the margin "I don't believe this." In fact, a turning point in the war had been reached when the new long-range P–51 Mustangs began to provide relay escort to the target and back, allowing bombers to concentrate on accurate bombing instead of defensive firepower as they closed on the target and executed the bomb run.

If early 1944 signaled a change in the strategic air war in Europe, it also did so in the training that supported it. As the AAF prepared for the invasion of the Continent and reorganized their forces, the U.S. training establishment responded to the increased pressure by abandoning the time-consuming system of moving units and crews from base to base for each phase of training; reorganizing the training air forces to include both bombers and fighters in each; reducing the training period; and training mostly in RTUs (for crew training) rather than in OTUs (for unit training). B–17s and B–24s still comprised the operational force, although the Second Air Force had begun training the first very heavy bombardment wing the previous June. Because B–29s were to overtake the heavies in furtherance of the strategic mission, B–29 training built directly upon the model of heavy bombardment.

Training for Strategic Bombardment

Training for strategic bombardment with the B-29 very heavy bomber followed directly on methods used in heavy bombardment training with B-17s and B-24s.

In the Combined Bomber Offensive, the AAF elected to destroy small, pinpoint targets susceptible, they believed, to the methods of daylight high-altitude bombing. American bombers flew almost exclusively during daylight at an average altitude of 23,000 feet. They usually formed up in highly compact combat boxes of eighteen to twenty-one aircraft (based on a three-aircraft element and groups of three squadrons) when using visual identification methods and, because of the shortage of equipment when flying Pathfinder missions, in combat wings of two or three combat boxes. Patterns and bombing practices varied, of course, over time.

By early 1944, any number of studies described those tactics and the equipment employed on long-range missions that had been found to be more or less successful. Now, experience put doctrine and educated guesses to the test and added to the growing body of data informing training practices. Training directives emphasized daylight flights, at high altitude whenever possible. Aircraft flew in formations configured like those of the Eighth Air Force. Formations now concentrated all bombers as a single force rather than employing individual aircraft mainly as independent units. Although they never attempted to gather training formations in such gigantic numbers as in combat, training officers, especially those returned from Europe, insisted upon the importance of near-wingtip formations. Training for blind bombing, or what was renamed bombing through overcast, also began, although equipment shortages all but restricted radar to the very heavy bombardment training

World War II

program. Availability of long-range fighter escort, while not greatly changing training practices, seasoned the advice offered about bomber formation tactics, specifically the decreased need for evasive action on the bomb run. More important, all the training air forces now had both bombers and fighters, permitting mock battles between them.

Heavy and very heavy bombardment units operated at the center of the storm. The AAF expected their performance would validate American strategic doctrine. Although men trained at home for this mission, they frequently went on to fight a different war. Yet the AAF wavered little in its determination that its system for training men to fly and fight was correct. Improvements in aviation technology, the greater amount and higher quality of training aircraft and equipment, and the hunger for an independent airpower mission encouraged the AAF to begin and end the war convinced that its commitment to daylight, high-altitude, precision bombardment, and the training that supported it, had proved itself.[4]

Heavy Bombardment

The Second Air Force's heavy bombardment unit training most closely conformed to the ideal OTU–RTU pattern. To organize and train new units, overstrength "parent" groups sired OTUs. At the time the offspring began training, 20 percent of the officer and enlisted personnel transferred from the parent as the new unit's experienced nucleus; the remaining slots filled from the Training Command's schools. At regularly scheduled intervals thereafter, each parent would drop 20 percent of its replenished strength to form the nucleus of a new unit. The parent would supervise training during the first phase, devoted primarily to individual training of aircrew members. The satellite units, in what was called a vertical system, went to other stations in other wings for second- and third-phase training. The second phase emphasized teamwork, and the third phase included simulated combat missions and joint exercises. After completion of approximately six months of training, each OTU moved to a staging area for overseas movement. AAF Training Standard 20–2–1 summarized the state of readiness for heavy bombardment OTUs at this point:

> Upon completion of the prescribed period of operational training, heavy bombardment groups will be prepared to conduct offensive missions against the enemy.... To be capable of such action, a unit must represent a closely knit, well organized *team* of highly trained specialists of both the air and the ground echelons. Emphasis will be placed on perfecting offensive tactics, including coordinated attacks against all types of defended objectives. Proper attention will be given to the development of a strong defensive unit, proficient in flying all types of tactical formations, with a carefully organized system of air observation. The defending force

can surprise the enemy by being in such a position when interception is made that no exposed flanks are presented, and by opening fire on command with maximum fire power when targets come within range. Throughout operational training, the presence of an enemy capable of taking quick advantage of any weakness must be emphasized.[5]

Without a formalized RTU program, individuals or crews sent overseas separately as replacements robbed the parent groups and OTUs, which in the former case denuded groups of more experienced airmen, and in the latter, disrupted the integrity of units intended to be trained together and deployed intact. Initially in the Second Air Force, parent groups training OTUs also gave first-phase training to replacement crews. As with OTU, for a time each bomber wing conducted some portion of the three-month RTU course. Near the end of 1942 three of the 15th Bombardment Wing's six groups engaged in first-phase replacement crew training. Thereafter, crews spent a month at one of the stations of the 16th Bombardment Wing, whose principal function was to train crews as such, before they were assigned to a third-phase station in the 17th Bombardment Wing to train as part of a unit and prepare to take their places in heavy bombardment groups already in combat.[6] At this time replacement crews formed into provisional groups for the second and third phases. A provisional group commander shepherded the group overseas, stayed a short while to familiarize himself with problems of established units in the theater, then returned to the United States to be placed in command of an OTU in the process of formation. In time, the provisional group arrangement was abandoned along with the vertical training system.[7]

The general framework for heavy bombardment OTU–RTU fails to reveal the considerable organizational gross- as well as fine-tuning that finally made it work. For at least the first year of war training officials reconsidered and revised the administrative formula, with complications continually retarding attempts at consistency. Principally, the number of men and materiel flowing through the training system remained below anticipated levels. Initially, available qualified first pilots fell far short of the authorized manpower schedules. Combat units needed personnel so badly that many of these men bypassed OTU altogether to report directly to units on their way overseas. Moreover, throughout the war many OTUs trained for some period without navigators. For the first half of 1943, for instance, more than 75 percent of first-phase training in the Second Air Force proceeded without the navigator.[8] As of mid-March 1942, the entire Second Air Force employed only two qualified bombardiers, both of whom were assigned to the 39th Bombardment Group (Heavy), which had a total of five navigators. The 34th Bombardment Group (Heavy) possessed the greatest number of navigators — nine. The shortage of trained people matched the lack of facilities, aircraft, and equipment. Acquisition of more than the one bombing-gunnery range at Wendover, Utah, according to the command historian, caused "no end of grief." Despite the deficits, the training fields worked at a maddened pace, operating on a 24-hour-a-day, 7-day-a-week basis.⁹

OTU and RTU practices became more predictable during 1943, but problems continued to bedevil training. Units lacked aircraft, especially B–24s, and high-octane fuel. An officer from the AAF bombardier school noted in January 1943 that "without exception" all the OTUs in the Second Air Force that he contacted "were completely lacking in bombing equipment such as E–6B Computers, C–2 Computers, Bombing Tables, Bombsight Tools, Bombsight Spare Parts Kits, and C–1 Auto Pilot tools and spare parts." He saw no prospect of obtaining the equipment. "It is obvious," he concluded, "that even if the personnel were highly trained and perfect in their abilities, no accurate results could be expected without the above mentioned equipment."¹⁰ Men were not to be "perfect in their abilities," not only because they lacked equipment, but because they themselves arrived neither in the requisite numbers nor on schedule. The air forces blamed the Training Command for failing to graduate well-qualified specialists in a timely fashion and considered the flow charts emanating from AAF headquarters to be unrealistic. One Second Air Force officer noted caustically, "I don't think they can add in Washington."¹¹

In fact, shortages at home and heavy demands from the combat theaters created the scheduling nightmare.¹² In September 1943, Arnold was informing his training base commanders that the "drive for multiple combat crews is on," that all heavy bombardment groups would be authorized soon to carry seventy crews.¹³ By this time, the number of people coming into the training units approached levels required for the 273 Group Program that Arnold had wanted early in the year, yet the limited availability of aircraft limited the number of crews the Second Air Force could turn out. Although AAF headquarters did not allocate more aircraft to help the Second Air Force meet General Arnold's directives for the last quarter of 1943, it was able to hasten delivery of those on order. Representatives from headquarters also made a "thorough study of the situation" in the Second Air Force, resulting in "reshuffling" of crews, aircraft, and facilities. During negotiation over requirements and schedules that hectic fall, Arnold queried General Harper: "If I give you the untrained personnel by 17 November 1943, can you train four additional B–24 Groups over and above the present program?" Harper cautioned that any plan to increase production by using more untrained personnel had to be evaluated in terms of the "probable increased accident rate, lowering of training standards and increased maintenance difficulties" — factors always weighed in the training balance.¹⁴

During the fall of 1943 the AAF significantly changed the conduct of heavy bombardment training by spreading it among all four continental air forces. The restructuring of U.S. unit training went into effect concurrently with the training blitz to produce combat crews for the planned Allied invasion of Normandy. Units from the Mediterranean theater redeployed to the European theater in early 1944, the nuclei of these combat groups supplemented primarily by crews from the United States. The strategic forces in England and Italy approached their anticipated strength by the beginning of the year, although Arnold's desire to have two crews for each bomber was not attained by the Eighth Air Force until July, nor by the Fifteenth until December.[15]

To meet the requirements, during the spring and into the fall the continental air forces shortened the training period from the normal twelve weeks to ten. The abbreviated course absorbed the cut largely by eliminating hours in recreation and drill and canceling leaves and furloughs. The I Bomber Command, now training B-24 crews, was told to assign combat crew members for maintenance duties if necessary, not to wait for navigators' arrival before beginning crew training, to maintain a consistent station load of 162 combat crews, and to substitute one type of gunner for another when necessary.[16] The III Bomber Command, training for B-17s, stabilized its training load at 168 crews for each RTU station and managed to graduate crews with flying time in excess of AAF requirements, despite a shortage of first-line aircraft.[17] The accelerated program of 1944 would be the last major push in heavy bombardment training as such, because after the great offensive and its aftermath (when heavies were used in tactical operations), further redeployment of heavy bombardment units through the United States was largely for remanning and retraining for B-29s headed for the Pacific.

Training Curriculum

The heavily armed Flying Fortress, as its name implied, was built and armed as an offensive weapon capable of self-defense. Therefore it carried, during most of the war, a ten-man crew of whom six were specialist and "career" (nonspecialist) gunners. Except for the pilot, the other aircrew members, all officers — copilot, navigator, and bombardier — also trained in gunnery so that they could fire one of the arsenal of flexible .50-caliber machine guns or a gun turret.

During the last quarter of 1942 the heavy bomber crew added its tenth member, the armorer gunner. This composition remained standard until September 1944 when the AAF inaugurated a new crew system. At that time "A" and "B" nine-man crews replaced the ten-man crews. Because every airplane in a bomber formation did not, by then, require a precision bombardier, "A" crews went overseas without a bombardier, the armorer-gunner performing the bombardier's duties, except for sighting. The "B" crew included a bombardier, and the armorer gunner replaced the second career gunner.[18]

Aside from this change in the composition of aircrews, the AAF settled upon the essential nature of heavy bombardment crew training early in the war, based on the fundamental concepts of the American strategic doctrine. In developing the OTU program for the Second Air Force, AAF headquarters decided that crew training would begin with individual qualification and the creation of an organization that allowed members to "mutually trust and cooperate with each other." Five of the six requirements in unit training centered on formation flying under various conditions; the sixth entailed unit gunnery fire control.[19] By 1943 when OTU was well under way, Training Standard 20-2-1 (covering heavy bombardment) described the bomber as a "precision instrument" whose operation required successful piloting, accurate navigation, effective defense, efficient communication, and precision bombing. Those functions, it stated, "are performed by a group of highly trained specialists working together as a combat team. The success of each crew member in carrying out his assigned function depends to a large extent upon the cooperation and performance of other crew members." The Training Standard then spelled out the specific tasks and qualification requirements for individuals and the combat crew as a whole.[20]

Responsibility for the overall performance of his crew, in the air and on the ground, lay with the airplane commander.[21] "You are now flying a 10-man weapon. It is your airplane, and your crew," the Fortress Training Manual exhorted the pilot. "Your crew is made up of specialists," it went on. "Each man...is an expert in his line. But how well he does his job, and how efficiently he plays his part as a member of your combat team, will depend to a great extent on how well you play your own part as the airplane commander."[22] This injunction rallied the pilot to a greater conception of his duty than just guiding the airplane. The pilot's qualification requirements included the ability to take off and land in the shortest practicable distance; competence as a pilot-bombardier instructor in making both automatic (using the pilot direction indicator) and verbally communicated bombing approaches; capacity to make satisfactory dry runs and individual releases as a bombardier; ability to take off from a flare path under blackout conditions and to fly extended night missions; and qualification in day and night pilotage and dead reckoning without radio aids or lighted airways. Translated into Training Standard 20-2-1, the list expanded into specific skills for operation of assigned aircraft (including preflight checklist, general operations, formation, night operation, and instrument flying), navigation (the same as in the earlier statement, plus proficiency in radio navigation), piloting during the bomb run, communication using radio telegraph and visual blinker signals, and basic knowledge of weather analysis and weather conditions in each theater of operations.[23]

The copilot was a rated pilot. Usually he did not occupy the left seat either because he had been pulled out of the single-engine program to become a bombardment copilot or because he did not have the number of hours to qualify as a first pilot in four-engine aircraft. The B–17 Pilot Training Manual delicately reminded the airplane commander:

> Always remember that the copilot is a fully trained rated pilot just like yourself. He is subordinate to you only by virtue of your position as the airplane commander. The B–17 is a lot of airplane; more airplane than any

Training for Strategic Bombardment

one pilot can handle alone over a long period of time. Therefore, you have been provided with a second pilot who will share the duties of flight operation. Treat your copilot as a brother pilot.[24]

Because he might be asked in emergencies to fly the airplane, the copilot's qualification requirements fairly closely paralleled those of the pilot. Under normal circumstances, however, he attended to preflight inspections and the technical aspects of a flight and served as the pilot's executive officer.

Like the pilot and copilot, the other members of the aircrew came into OTU or RTU qualified in the skills necessary for graduation from Training Command courses. Because all might man gun positions, everyone's operational training included gunnery practice. They all took high-altitude indoctrination, were to be knowledgeable about defensive formations and evasive tactics, have some proficiency in radio telephone procedure and the pyrotechnic signaling equipment, and be able to recognize and understand the capacities of enemy craft.[25] To the extent any specialist substituted for another, he had to obtain some experience in the second skill, and during most of the war each man had a secondary duty.

The navigator's specialized requirements centered on his facility with the four basic means of navigation learned in Training Command navigation schools: pilotage, dead reckoning, celestial, and radio. According to one description of his combat role, the navigator

must keep the position of the aircraft established at all times by all methods of navigation, and log the time and position of any action or occurrence during the flight. Without error, the navigator must find the "initial point," direct the airplane

Every member of the aircrew received gunnery training. In-theater instruction continued to hone the men's skills, as shown here by ground schools for gunnery located in England.

to the target and identify the target upon reaching it.[26]

Changes in navigators' standards reflected developing technologies such as, in the last half of 1944, the ability to use the bombsight as a driftmeter and proficiency in making Loran fixes.[27] By this time, navigation training materials brimmed with information on such matters as wind velocity and direction at various altitudes and places and on the most-used types of navigation and the pitfalls to be encountered by units assigned to different parts of the world.

Reports from the combat theaters highlighted the important role of the underappreciated bombardier well past the time when many gunners had become togglers, and operations had come to deemphasize accuracy. For example, one training intelligence report of December 1944 fulsomely depicted bombardment as a pyramid that

> tapers to one point which represents the bombardier who is the single final factor in determining whether or not all the previous efforts of the organization shall be fruitful. He can neutralize the schemes of generals, the planning of colonels, the best pilotage and navigation available by an error of judgment, by a momentary carelessness during the crucial few seconds before he releases his bombs.[28]

The 1943 Training Standard stated that the bombardier should be able to bomb "under all conditions encountered in combat operations, including daylight and darkness, dusk and dawn, high and low altitudes, favorable and unfavorable weather and long and short approaches." In the classroom, bombardiers reviewed bombing theory and bombsights; pilot-bombardier coordination included five hours of bombsight operation on the trainer and two hours of dry runs. Pilot John Boeman described the simulator where he practiced with a bombardier who was learning to use the precision Norden sight:

> I sat driving a ten-foot-high wheeled platform as Joe [the bombardier], kneeling in front of me, aimed a Norden bombsight at a paper target atop an electrically driven "bug" crawling across a hangar floor. Our platform, moving slowly but several times faster than the bug, simulated a B–24 approaching its target. The bug's movement simulated wind effect on the airplane — unless its wheels were positioned so that it turned in a circle to simulate a boat on the water. As Joe sighted through the eyepiece, he turned azimuth and range knobs on the side of the sight to place the cross hairs inside it on the target image, thereby signaling electrically my pilot direction indicator. My job was to turn my bombing platform to keep the [pilot's direction indicator] centered and thus line up on the target, which the platform blocked from my view just as the nose of the B–24 would in the air. If all went well and Joe had the wind "killed" in his sight, his cross hairs remained on the target as our "airplane" approached its release point. Then the mechanism in the sight automatically signaled a plunger,

Training for Strategic Bombardment

simulating the B–24's electrical release system, at the base of the trainer, to score our accuracy with a mark on the paper target.[29]

In both theory and practice, bombardiers mastered train and formation bombing and flew at least fifteen bombing approaches, operating the aircraft manually from the pilot's compartment and using the C–1 automatic pilot when it was "in working order." In training, during the period when bomber formations practiced evasive action on the bomb run, the pilot controlled at altitude, and the bombardier took over during approach to the target and for at least twenty seconds before the bomb release.[30]

Although the bomber was designed to destroy its target with high explosives and incendiaries, it bristled with defensive armament. On models after the B–17D, the nose had a turret loaded with two .50-caliber machine guns, as did the top, lower ball, and tail turrets. The side waist guns were flexible .50-caliber guns. The AAF heaped additional armament on each modification. The final B–17G model carried thirteen and the B–24J ten of these guns.[31]

The B–17G bristled with armament, carrying thirteen guns and five gunners. Gunners were positioned to shoot from the nose (*upper right*), the waist (*lower left*), and the tail (*lower right*).

World War II

An aerial gunner's training varied, depending on whether he was a specialist gunner and what gun position in the aircraft he manned. The radio operator–gunner and the aerial engineer–gunner performed the most highly specialized jobs. The planning document outlining OTU qualifications listed the following requirements (besides flexible gunnery) for the radio operator: proficient in sending and receiving code at the rate of fifteen or more words per minute; proficient in operation and air adjustment of all radio equipment of assigned aircraft; proficient in maintaining equipment; able to locate position by radio fixes; proficient in sending and receiving visual code signals by a signal lamp; and knowledgeable about RAF Bomber Command signal organization for the United Kingdom and Middle East and with U.S. and British cipher devices.

Radio operators practiced those skills in the air only after they arrived in OTUs or RTUs. During a conference held at Second Air Force headquarters in late September 1943, a training officer reported that the technical schools were graduating radio operators who had yet to "see" a B–17 or B–24, and "haven't tuned a liaison set in the air."[32] As a result, not only did OTUs bear the entire load of practical training, they provided it in compressed form — because of the personnel deficit, as of September 1943 not one radio operator in training that year had finished all three complete training phases.[33]

The aerial engineer–gunner, according to Training Standard 20–2–1, had to be trained to "handle with coolness and intelligence all troubles encountered during flight. He must be thoroughly familiar with every detail of his aircraft and able to diagnose and cure, when possible, all flight troubles under possible conditions of extreme discomfort and danger." Because of his technical expertise, he shared considerable responsibility during preflight inspection with the pilot and copilot, and in the air he was capable of acting as copilot for all duties except piloting and navigating.[34] The armorer gunner manned a third gun position, and if three career gunners were aboard, one doubled as an assistant aerial engineer and another as assistant radio operator.

The training of aerial gunners caused the largest headache for the air forces, as it did for the Training Command, because, unlike other the specialties, gunnery relied on theories of sights and sighting that were still being developed and debated relatively late in the war. The position firing system came to be generally accepted for use by gunners without compensating sights, but its delay in implementation at every level of training gave Second Air Force's Commanding General Brig. Gen. Uzal G. Ent reason to despair in February 1944 that "I've been fighting this gunnery problem now for a year and we haven't licked it yet.... The original teachings that we gave [gunners] were faulty.... It has caused us a lot of dead people and a lot of lost aircraft."[35] Nonetheless, practical training did not wait for the resolution of theoretical problems. During 1943 the Second Air Force began turret training for its gunners on a 24-hour-a-day basis, and by 1944 the air forces had benefited from the Training Command's specialized turret program and were able to eliminate nearly all tow target missions by more widespread use of the gun camera. Moreover, the sighting controversy quieted once the training air forces could expect that their B–17s and B–24s would be equipped with compensating sights.

Unit training for heavy bomber crews divided into three phases. All members of a new crew devoted the first phase to perfecting their (supposedly) previously learned individual tasks. As in their earlier training, they spent time in "ever nebulous" ground school, as the Second Air Force history described it. Because the training units found it so difficult to anticipate when all members of a crew would come into OTU from Training Command programs and therefore be ready to begin flying together, ground school became an important mechanism for manipulating the training schedule, with the result that most airmen (except for pilots) spent more hours on the ground in first phase than in the second or third, and relatively few hours in the air. Depending on one's crew position, ground school varied from 83 to 125 hours in first phase.

An exchange of views among the senior staff during June 1943 illustrated the problems in bringing coherence and uniformity to first-phase training and the role played by ground school. General Arnold expressly asked the Second Air Force to find a way during the first four weeks of training for crew members, namely gunners and bombardiers who had had no opportunity to drop bombs, to be given some means to practice their team duties. At that time, the Second Air Force and the Training Command shared responsibility for transitioning four-engine pilots. Therefore, the bulk of air time was given to first-phase pilot training, and flying hours for other crew members were curtailed. The Second Air Force and the Bombardment Branch under the Assistant Chief of Air Staff, Training tried to respond to Arnold's request by expanding the ground program for nonpilots, particularly in gunnery, by using synthetic training equipment to the maximum degree.[36] The Second Air Force developed a ground training directive for combat crews in first-phase training that included instruction in such subjects as turrets, .50-caliber machine guns, and firing on gunnery ranges. Depending on crew members' duties, individuals also received specialized training in navigation, the theory of bombing, celestial navigation trainers, bomb trainers, communications, engineering, maintenance, intelligence, and medical matters. However, the Second Air Force noted that "to further complicate the training picture, there is a chronic shortage in personnel, such as the present one in navigators and radio operators." Because the shortage was so acute, many men did not receive any of the extensive first-phase ground training because they did not report for operational training until late in the second phase.[37] Even though operational commanders in the combat theaters decried the serious deficiencies among bombardiers and navigators in the operation of gunnery equipment, the Second Air Force could not train men it did not have.

Although he blamed administrative headaches on factors beyond the control of the Second Air Force, in 1943 its historian subscribed to the view that ground school also failed in part because "Group Commanders were preoccupied with flying training." Doubtless, most airmen preferred flying to studying. Based on 1942 OTU experience, in February 1943 AAF issued an exhaustive study, "Model Mission Flight Training," that laid out a specific set of missions to be flown during each of the three phases of unit training. The Second Air Force Flight Training Directive for Combat Crews dated August 30, 1943, listed twelve first-phase missions that focused principally on individual proficiency checks.

The emphasis on individual training in first-phase missions — amounting mostly to pilot transition — continued throughout the war, even after more and better equipment and sufficient fuel made it easier to conduct crew training missions earlier in the cycle. Because the training air forces were only part of the training chain, the Second Air Force, for example, still had to transition pilots onto heavy bombers. Philip Ardery was plucked from a Training Command basic flying school where he was an instructor pilot and reassigned to a B–17 OTU at Sebring, Florida. He later described the challenge he confronted in making that change:

> For a fellow with 1100 flying hours in a basic trainer it was a long hop to get checked out as the first pilot of a Flying Fortress and at the same time learn the things a tactical flier in heavy bombardment has to know. Learning to work with a crew of ten men, learning to fly a proper bombing run, learning a new type of formation flying, getting used to high altitude using oxygen, all were new to me.
>
> The chief purpose of the school at Sebring was not to teach tactics . . . . It was to teach pilots how to handle their individual airplanes.[38]

Although the few pilots already familiar with four-engine aircraft had the advantage of Ardery, even they had previously flown only stripped-down planes. Until they became accustomed to the heavier aircraft flown in unit training, the Second Air Force considered them "definitely dangerous."[39]

The Second Air Force began to take into account the tardy entrance of some specialists by using available aircraft to transition bombardment pilots. So, for example, in 1942 it set the radius of first-phase flights at fifty miles, and at one hundred miles in second phase, making the missions useful primarily for pilot transition rather than for navigation.[40] But the emphasis on pilot training to the detriment of other specialties produced an undercurrent of dissatisfaction. As one officer in the Eighth Air Force put it, the "trend so far appears to have been to stress pilot proficiency and flying hours during training at OTU's rather than coordinated missions involving training for every crew member."[41] When navigators arrived early in OTUs or RTUs only to find that missions made little use of them, they became bored, rusty at their skills, and developed what one officer termed a "fifth wheel complex." Many thought gunners' training ranked lowest among priorities. One combat report called gunners "orphans" of the system; another stated that "everyone is so busy, there is just no time for the gunners. They have ground school and plenty of it — of which 90% is of no value to them."[42]

By the second phase, combat crews were to begin working together on team tasks. The Second Air Force hoped to repeat uncompleted or unsatisfactory first-phase missions in addition to the thirteen to fifteen missions planned for the second phase. Ideally these exercises included high-altitude bombing and gunnery formation (usually three- and then six- and nine-ship formations), dead reckoning navigation, formation descent through overcast, and air-to-ground and air-to-air gunnery.[43] For instrument approaches, pilots in lead ships used the C–1 autopilot; an airborne element of the SCS–5 instrument blind-landing unit was installed in all operational training unit bombardment aircraft, except B–26s, for pilots to use in night landings.[44] New equipment brought greater technical sophistication to training, yet some of the basics were still missing. Training units did not have enough M38A2 practice bombs, for example. As of May 1943 the AAF authorized each crew a total of 375 bombs for the entire operational training period.[45] Pilot John Boeman described one of the bombing missions, probably in second phase, at a B–24 RTU:

> After a few ground trainer sessions, with the whole crew aboard we took loads of "blue pickles" (a term to describe practice bombs, possibly derisively derived from the myth that a bombardier could put it in a pickle barrel from twenty thousand feet with the Norden bombsight) over the range south of the base. One by one we dropped the 100-pound sand and smoke-filled, blue-painted, finned containers at big white crosses marked out on the desert floor. Men on the ground scored our accuracy by triangulation on the white smoke. Many constants on the bomb trainer became variables in the air — such as the need to jockey throttles attempting to hold a sometimes wallowing airplane at the bomb-run airspeed for which [the bombardier] had computed his ballistics of the bombs. Also, besides computing ballistics accurately, the bombardier had to set them into his equipment and position a series of switches correctly for the release system to function. When [the bombardier] called for "second station" with his cross hairs aimed at the target, and I flipped a switch transferring azimuth control of the airplane and release of the bomb to his bombsight, sometimes our scores were within a hundred feet or so — often not. Sometimes the bomb failed to release.[46]

Most missions required an entire crew to be present, and in general, third-phase missions presented more complex tactical problems on flights of much longer duration that required a greater variety of coordinated activity. Night as

World War II

well as day missions posed problems in navigation, formation, bombing, and gunnery — each flown above 20,000 feet. Camera bombing against industrial installations, for the purpose of target recognition, simulated combat strikes. Added realism came from practice against pursuit interception, when fighters were available. Again, cameras recorded action from the upper and lower B–17 turrets. Aircraft flew virtually all third-phase missions in formation, with crews taking turns as lead ships. Although it requested an additional month's training for selected combat crews, the Second Air Force in fact never gave specific training to lead crews. In the Fourth Air Force, at least by mid-1944, an instructor pilot flew in the lead of each element on formation flights.[47] Some third-phase model missions replicated those from second phase, such as the Second Air Force's instrument check that every pilot had to complete by the end of unit training, and instrument calibration, although the navigator assumed a greater role when it was performed in the third phase. Until the third phase, the bombardier could score the same circular error (average distance from the center of the target) as he had for graduation from Training Command's bombardier schools. On third-phase missions he was required to make 10 percent less error; any amount in excess of that caused his mission to be declared a failure and he had to repeat it. Headquarters set the standard of proficiency at 32.5 percent hits (bombs landing within the prescribed circular error distance) for bombing against standardized targets.[48]

As a cadet operates the bombsight, another would check his accuracy by filming the bomb's hit using a camera pointed through a hole in the floor of the airplane. As shown *above*, simulated combat targets were designed from actual enemy strongholds to give the bombardier-in-training a feel for the real thing.

Training for Strategic Bombardment

Once again the successful completion of the missions often hinged on factors outside the control of training units. For example, early in the program third-phase navigation flights suffered from lack of equipment and uncalibrated instruments, turning the flights into pilotage or inaccurate dead reckoning with very few celestial missions.[49] Prior to 1944, gunners still fired at towed targets, and because towing aircraft flew parallel to the bombers rather than approaching on a pursuit curve, tail gunners could not fire from their assigned position. Camera missions of various sorts appeared in the curriculum, but in actuality cameras were in extremely short supply. B–24s did not have the equipment through most of 1943; remote control switches for cameras at the bombardier's station were being installed on B–24Js in 1944. Mission No. 8, a 1,000-mile overwater flight, often could not be carried out because of too little gasoline and too few aircraft. Unless crews already had the airplane in which they would be flying overseas fairly early in third-phase training, three crews usually shared one aircraft, which did not allow time for all to crews make the 1,000-mile flight. Moreover, at this time only some B–24 crews, and virtually no B–17 crews, received 100-octane fuel for third-phase training, including the overwater navigation flight.

Gunners still practiced their skill against towed targets.

To the good, however, several joint exercises between air and ground forces occurred in 1943; those between heavy bombardment squadrons and Third and Fourth Air Force fighter units continued into the next year. P–38s, B–24s, fighter control squadrons, and antiaircraft units, for example, participated in a two-day exercise at Muroc Field in April.[50] Furthermore, operational training during this period began to be informed by changes in tactics employed by the Eighth Air Force, which sent information regarding its methods in following prescribed courses, rates of climb, and altitudes when forming a group after take off; for assembling a wing; and for optimal formations both for defense and for opposition.[51] Those procedures for heavy bombardment formations soon became the model employed throughout the training air forces.

World War II

Once heavy bombardment unit training was fully under way, it swung gently between the poles of standardization and flexibility. By 1943 the Second Air Force had pegged the number of flying hours per crew in each phase at sixty-five. Even as it welcomed the newly possible uniformity, the service came to realize that it paid an unacceptable price when standards were too rigidly applied. It became apparent, for example, that the accomplishment of a set number of flying hours had come to dominate at the expense of proficiency. Word spread therefore that the completion of a specific amount of flying time each day should not substitute for judgment about the quality of performance.[52] During 1944 the Second Air Force maintained minimum flying requirements in which quality rather than quantity, according to its historian, "found forceful expression." It tried not to abandon standardization in the process. On the contrary, operational procedures for each member of the B–17 and B–24 crew came out in booklet form in September, and the air force initiated other methods to check the proficiency and improve the techniques of instructors. At the same time, "horizontal" training replaced the "vertical" system, relieving the need to coordinate an individual crew's training with three organizations. The Second Air Force also took steps to centralize authority over its training bases.[53]

Greater flexibility became inevitable when other air forces took up heavy bombardment training. The administrative reorganization, however, signaled no reversal of established qualification requirements. By the time the I Bomber Command was training B–24 units in the spring of 1944, the RTU course lasted only ten weeks and its instructions stated that "training units are allowed complete freedom to schedule training, in order to best cope with local conditions and requirements." Nonetheless, individuals still had to attain the same level of competency established by the Second Air Force, and flight training followed a progression similar to that specified in the model mission program. Normally the first two weeks included individual proficiency checks, the beginning or completion of basic technical and special ground training, crew organization, and individual bombing and gunnery. Crews spent the third to seventh weeks on individual bombing, gunnery, and navigation; instrument practice; and flight and squadron formation. The last two weeks took up formation gunnery, navigation, long-range flights (including the 1,000-mile overwater navigation flight), final instrument checks, squadron and group formation, fighter missions, planned tactical missions, and completion of all requirements.[54]

When, during the same period, the III Bomber Command began training B–17 crews, it followed a set of twenty-two model missions that could be flown in any sequence chosen by the RTU at each base. The command developed a tailor-made program specifically for the European theater that included study of German and British aircraft, British control systems, and tactics applicable to the bombardment of Europe. The command also instituted

Training for Strategic Bombardment

the A and B crew systems. Availability of camera guns made possible great improvement in gunnery. Although it did not yet have all the requisite equipment, in May 1944 the Third Air Force directed all its bombardment training stations to ensure that all crew gunners, including the bombardier and navigator, fire a minimum of four gun-camera missions. Also, by that time the Third Air Force assigned P–40s to the III Bomber Command and promised additional fighter aircraft as they became available.[55]

By 1944 the air forces with some portion of heavy bombardment RTU responsibility generally taught tactics applicable to the European theater. "The only reason for the existence of a bombardment unit is its ability to drop bombs," the I Bomber Command unambiguously informed its crews. It defined the mission as high-altitude, daylight, precision bombardment, flown in dense defensive formations. In a heavy bombardment group in the European theater, the basic three-ship element flew a V formation; a squadron or flight was composed of two three-airplane Vs, and upon this arrangement larger groups assembled. A group of eighteen, for example, formed three squadrons in a combat box; one squadron flew high and behind the lead while another flew below and behind. Training formations never came close to the huge numbers of aircraft that darkened the skies over Germany or Romania, but they conformed to the style of those combat formations nonetheless; the I Bomber Command declared "complicated and locally-devised 'tactical' formations of odd nature" to be unacceptable.[56]

Very little controversy erupted over preferred formations, but questions arose regarding the wisdom of "packing in" formations as tightly as combat units. In 1943 the Second Air Force commanding general notified his wing commanders that they should "take the heat off formation flying" and allow pilots to fly as much as a mile or a half-mile apart until they felt "comfortable."[57] The Flying Fortress Pilot Training Manual stated:

> Close flying becomes an added hazard which accomplishes no purpose and is not even an indication of a good formation. Bear in mind that it is much more difficult to maintain position when flying with proper spacing between airplanes than with wings overlapping. Safety first is a prerequisite of a good formation.[58]

Yet instruction from AAF headquarters or the Training Command expressing concern for safety could not completely abate the zeal of group commanders to prepare their men for battle. Pilot John Boeman recalled that his instructors, who were combat returnees, commandeered good-sized training formations of twelve or fifteen or more airplanes, and they pushed the pilots relentlessly toward greater precision:

> "Get it in close! Get it in close!" demanded the instructors. "Don't leave room for fighters to fly through your formation. Keep it tight so your gunners give each other mutual fire support. The enemy always picks on

World War II

the sloppiest bomber formations first."[59]

Crews prepared in other ways for the planned invasion of the continent. They spent more time on ditching and bailout procedures, for example, although this training did not pertain exclusively to the European Theater of Operations. Faced with high-altitude fighter opposition, gunners with noncomputing sights applied the position sighting system, and those with computing sights had to "range" accurately. B–24 replacement crews in the European theater were flying Norden-equipped airplanes, and by early 1944 the AAF was in the process of converting the B–24 training program from Sperry to Norden sights. On the crucial bomb run, by this time, tactics advocated direct approaches without evasive action. The formation flew over the heavily defended target area for a relatively short time, and the short, straight approach permitted more accurate bombing. In training the pilot flew at high altitude, holding a straight course for forty-five seconds or longer until the lead bombardier signaled to drop.

By 1944 ground training functioned less as an expedient for coping with sloppy manpower schedules than as a well-focused reinforcement of what crews practiced in the air. Given the reduced time for unit training, the Second Air Force compressed the amount of technical training learned in the classroom, but it kept as much of the third-phase curriculum as possible.[60] The *Ground Training Guide* advised training groups of the I Bomber Command, "*Don't use lectures and theory more than necessary....* Keep training practical." It recited the many available training aids — synthetic devices (prominent among them, the bomb trainer still installed with the Sperry sight, the Celestial Navigator Trainer [type D–1] and the Navitrainer [type G–1], the Link trainer, and various gunnery trainers), small arms, and bombing and gunnery ranges — as well as the more passive classroom materials (charts, books, manuals, and films). The pilot should join a combat group thoroughly familiar with the full-panel system of instruments; the navigator should be well versed in radio aids and range orientation. Bombardiers should understand the tactics for pattern bombing; precision single-release bombing; bombing in train; formation; use of flight, squadron, or group circular error to determine tactics for each mission; and selection of bombs for individual targets. They should also know how to modify bomb release techniques in combat and make use of the automatic pilot on the bomb run instead of using the pilot and the pilot's direction indicator. The I Bomber Command reminded its crews that they enjoyed the "services of the entire AAF Fighter Command, equipped with the latest fighter aircraft" (an exaggeration), so presumably they should know tactics for fighter-bomber cooperation and defense against fighter interception. Despite the pronounced emphasis on teaching practical skills, ground school should also develop "initiative, resourcefulness, [and] fighting spirit."[61] Men must learn to work together to fight to and from the target, and to draw strength from one another in order to face another mission the next day.

Training for Strategic Bombardment

The most profound variation in the practice of heavy bombardment unit training stemmed from the different flight characteristics of the two heavy bombers. One pilot disclaimed any difference, finding the B–24 only slightly "hotter" than the B–17. According to another pilot, cadets were often told in their early training that "once you are a military rated pilot you can fly any airplane in the Army inventory." Fortunately, he thought, that simplistic view was "out of vogue with base commanders," and far from his own experience when he transitioned from B–17s to B–24s.[62] The experimental B–24 prototype differed dramatically from the B–17, and some of its hydraulics took time to add or were never added at all to the modified B–17s. The B–24s' greater speed and range offered such great advantages that more of them were manufactured and flown by American forces than B–17s or any other American combat aircraft.

Despite its appeal to the AAF leadership and operational commanders, training officials wrestled with the considerable reluctance of crewmen to fly the B–24, based on their statistically reinforced fear that it was accident-prone. The Liberator was bigger; it was faster at altitude and, perilously, also at landing; and it was more difficult to handle in an emergency than the Fortress. It was also harder to maintain, leading to a greater likelihood of airborne problems owing to poor maintenance. In 1943 the Second Air Force proposed to increase B–24 first-phase training to two months because the normal cycle sent Liberator pilots into combat situations "beyond their capacity." Washington rejected the suggestion, and the air force turned to the Training Command's method of first putting four-engine pilots into B–17s before transitioning them to B–24s. Thus the unit training time stayed the same for both bombers, even though AAF headquarters still acknowledged in December 1944 that a B–24 pilot needed approximately 25 percent more training to achieve the same proficiency as a B–17 pilot.[63]

The training establishment made a concerted effort to encourage pilots in the advanced course to elect B–24s and to reassure B–24 crews that they were flying a reliable, manageable airplane. One small example aimed at building heavy bomber crew morale, although it did not specifically name the B–24, was a poster tacked up in briefing rooms. It was captioned "The Odds Are With You," and it showed positive, improved statistics on attrition rates in the Eighth Air Force. One airman scoffed that as far as anyone could tell who looked hard at the overall losses in combat, coupled with the high training accident rates for B–24s, the poster *should* have read: "The Odds Are With You... You Are Statistically a Goner."[64]

Despite continuing anxiety about the aircraft, B–24 training improved between 1942 and 1944. Many Liberator pilots were convinced that the close attention by the training establishment paid off. The first new training practices were devised mainly by training officers at the base level. As a B–24 instructor at Alamogordo, New Mexico, in 1942, David Burchinal found that transitioning

World War II

A B-24 Liberator, the most numerous of the American aircraft built during World War II

pilots "didn't have the confidence in themselves, they would panic, they didn't have set procedures." Thus, the 330th Bombardment Group started a night school and kept an instructor pilot in the tower twenty-four hours a day. Night or day, somebody experienced could talk a pilot through new procedures and offer reinforcement while he was in the air. The approach proved itself by a lowered accident rate and the noticeable increase in confidence and proficiency of the crews.[65]

The most effective way to calm fear of the B-24 came by providing concrete information about engineering and performance characteristics — knowledge was therapeutic as well as life-saving. One young tail gunner in a B-24 RTU at March Field wrote his father, "I believe that I've gotten out of the fear [of] being scared of the plane now. The engineer told me what to be on the lookout for, and that helps a lot."[66] By this time, in the summer of 1944, training depended less on the energy and creativity of base training officers and trainee self-help. The airplane's manufacturer, Consolidated Aircraft Company, sent representatives and films to training bases to explain the highly complex component systems. John Boeman began B-24 training with what he called a six-week engineering course that took pilots "from nose to tail, and wingtip to wingtip." The operation of every part of the aircraft was demonstrated by "those who built the bomber, amplified by those who flew it in combat." For Boeman, ground school made him better acquainted with the B-24 "than any previous airplane when we went to the flight line a few weeks later, to begin our training flights." Men had gained confidence born of familiarity with their airplane by the time they began firing at wrecked automobiles strung out along a dry lake bed and shooting air-to-air bursts at high altitude with live ammunition.[67] Another pilot came away less impressed by his ground schooling. He agreed with Boeman that technical orders, Consolidated Aircraft presentations, and training manuals were invaluable, but the very extensiveness of the materials resulted in some conflicting and ambiguous information: "Which checklist to believe? Which film? Some bad writing, some ignorance, I finally concluded, caused the conflicting instructions."[68]

New techniques that countered the dangers inherent in flying the B-24 evolved over the couple of years in which its training program developed. AAF and Consolidated pilots conducted extensive tests to determine how to avoid stalling the aircraft when it slowed to land. They discovered that the pilot should approach the field with the nose down while losing altitude; under low overcast the airplane should come in with more power, but again with the nose down. The experts visited training bases to explain the revised technique of landing. Training officers also discussed power-off approaches, as was rumored to be done in some B-24 combat units, but there is no evidence that the practice was incorporated to any degree into training procedures.[69]

One other variant of heavy bombardment aircraft and its training came with the modification of B-17s and B-24s (as well as B-29s for very heavy bombardment) for long-range reconnaissance. Before the attack on Pearl Harbor, each type of bomber group had an associated reconnaissance squadron equipped with the same kind of aircraft. A year into the war, parent groups absorbed those squadrons, and reconnaissance as a separate function shifted to units training for four types of missions: photographic, tactical, liaison, and weather. Within that training program, photoreconnaissance, designed to produce crews for long-range missions, was the largest. Heavy bombers modified as photographic reconnaissance aircraft were the F-9 (B-17 with varied camera gear) and the F-7 (a B-24 with cameras in either the bomb bay or the nose or both).[70] A single training standard covering long-range photo, weather, and radio countermeasures, published in December 1944, required that crews meet the standards established for heavy bombardment, except for bombing. They were to fly above 20,000 feet and practice photography and radar. The ship carried two navigators, a navigator-photographer and a navigator radar observer.[71]

Very Heavy Bombardment

The B-29 Superfortress evolved from the continuing search for very long-range bombers dating from the 1930s. Design competitions for aircraft capable of improved speed, rate of climb, higher ceiling, greater load capacity, and longer distances led to the Boeing prototype, flown in September 1942. It not only met the specifications for altitude, distance, and so on with its advanced structural features, it was the first to have pressurized cabins to protect crews in the high-altitude environment, central fire control and remote control turrets fitted with computer-aimed guns, and extensive new radio and radar equipment. The

World War II

YB–29 service test took place in June 1943, and, despite extraordinarily grave problems in building the revolutionary "three billion dollar gamble," less than one year later B–29 squadrons were flying missions in the Pacific.

The very short elapsed time between the testing of the experimental model and the aircraft's entry into combat caused great difficulties in theater and squeezed crew training into a brief and intense period.[72] The advanced avionics further complicated the training program. Because of the aircraft's technical complexity, the B–29 crew took on a third pilot-trained individual, the flight engineer, who controlled the mechanical systems while the pilot and copilot flew the airplane. The same problems that frustrated all AAF training — shortages of aircraft, modifications to new equipment, inexperienced instructors, and the erratic flow of specialists into OTUs — applied here to an even greater degree. However, because of its high priority (from the beginning, for example, B–29s were slated for double crews) and its special technical requirements, the B–29 program drew less than others from the pool of recent Training Command graduates. Crew members tended to be men with as much previous experience as was possible to find. The Training Command scoured its instructor ranks for pilots, and some of the first group of pilots and navigators, who had at least two years of active military duty, came from the Air Transport Command. Pilots brought into the B–29 program had to have flown a minimum of 400 hours on four-engine aircraft, and navigators had to have made at least five overseas round trips as the principal navigator. The requirements escalated after a year's experience, when AAF headquarters instructed the Training Command to locate B–17 and B–24 pilots with a minimum of 1,000 hours flying time. Prerequisites changed over the duration of the war, but at all times, considerable experience flying four-engine aircraft, wherever it had been gained, remained in effect.[73]

The AAF began mulling over B–29 crew composition while the airplane still lay on the drawing boards. In August 1942 the Second Air Force, much involved in the planning for the program, anticipated that crews would be drawn from men already flying four-engine aircraft; pilots should have at least six months of operational experience. It proposed a twelve-man crew, of whom as many as seven could be officers; the other five were to be enlisted gunners. Ultimately the AAF decided upon an eleven-man crew. The pilot, copilot, flight engineer, and two navigator-bombardiers were all commissioned officers. The Training Command eventually provided a transition course to the three-man team of pilot, copilot, and flight engineer. The rest of the crew was composed of an engine mechanic gunner, electrical specialist gunner, power-plant specialist gunner, central fire-control specialist gunner, and radio and radar operators.[74]

Although the B–29 carried at least one more man than the heavy bombers did, in part owing to the advanced technology of the defensive armament, it required fewer gunners, and more of the men, including the gunners, were experts in the technical components of the airplane. For example, the flight engineer (now a pilot-trained officer in place of the heavy bomber's enlisted flight engineer) was a technician. Initially he was trained for mechanical and maintenance duties, but later his primary responsibility became the operation of the cruise-control system. The radar operator's "sole function," according to AAF headquarters, was to operate the BTO (bombing through overcast) equipment and "furnish information gleaned from the radar scope to the navigator and the bombardier when required." Thus, besides making calculations and operating the radar equipment, he was also partially trained as a dead reckoning navigator and a bombardier capable of bombing industrial-type targets with a 1,200 foot circular error.[75] (In practice, B-29s seldom bombed using radar.) All but one gunner had some type of technical training for manning the complex electronic or mechanical equipment.

Even though it had extraordinary firepower, for its combat role in the Pacific the long-range bomber no longer had to function principally as a fortress, defensively armed to the teeth. In fact, to permit greater bombload and fuel capacity, very early the XXI Bomber Command stripped the aircraft of some of its armament when it was flown in combat, most notably in the great fire raid on Tokyo of March 9 and 10, 1945. Whereas every crew member on a B-17 or B-24 except the pilot was expected to be able to fire one of the guns, only four of the eleven on a B-29 specifically took gunnery training.

In practice, because Japanese interceptors were to prove fairly ineffectual, the full might of B-29 firepower usually was not engaged. But not knowing the future, the Second Air Force initially designed B-29 training along the lines of heavy bombardment directives based on the European experience. AAF training standards stated that "tactics will be special variations of present heavy bombardment tactics and techniques of air attack by formations of bombers or lone armed reconnaissance bombers."[76] The Second Air Force drew up a series of training missions to be flown by all crews. Although the missions were not always flown in strict order, their content was standardized. Standardization boards were established to provide information on modifications and installations in the tactical B-29 not found in most training airplanes.[77]

Like heavy bombardment training directives, those written for the very heavy program became more specific as experience deepened. Standards promulgated in July 1944 directed individual crew members to accomplish the following tasks: the aircraft commander was to complete a minimum of twenty hours' formation above 25,000 feet and perform the prescribed instrument check; the copilot was to make a minimum of five landings from his own position and fly at least four hours on instruments under the hood; and the bombardier was to drop a minimum of twenty individual bomb releases from above 25,000 feet. Navigation missions by the crew included one of approximately 3,000 miles using cruise control and a second triangular course of at least 900 miles using radar alone. The pilot, navigator, and bombardier using

World War II

camera bombing methods practiced together for attacks on industrial targets; gunners performed at least four gun camera missions against attacking aircraft, and they practiced firing at high altitude from their primary and secondary gun positions. In a revised version in September, the Second Air Force emphasized overwater, dead reckoning, and celestial navigation and gave courses only on identification of Japanese and friendly aircraft operating in the Central Pacific. The training standard that followed at the end of December increased long-distance flights and operation at higher altitude, specifying the use of cruise control on all missions, and compelled the crew to use radar 50 percent of the time.[78]

When it became involved in very heavy bombardment training in 1945, the Third Air Force modeled its procedures very closely on those of the Second Air Force. Until the end of June, the III Bomber Command trained B–29 crews over a twelve-week period. In June it shortened the training to ten weeks; then in September it reverted to twelve when the training load lessened. The III Bomber Command divided training into three phases during which a crew flew a set of model missions similar to those of the Second Air Force. Also, a shortage of aircraft forced the command, like the Second, to employ B–17s as a "companion trainer" for certain kinds of missions, including those in radar bombing, radar navigation, and Loran. Again, as in the Second Air Force, succeeding curriculum revisions realigned the missions by consolidating short flights in favor of longer distances during which a variety of tasks could be performed.[79]

Even though those involved in stateside training attempted to remain constantly in touch with operations, every indication is that training adapted to combat realities at an evolutionary pace, and in some cases, it moved not at all. During the later part of the war, for example, when Lt. Col. John W. Carpenter III was assigned to the training division in Washington, he "had to know what the boys were doing." "So," he later explained, "I went over there and flew with them a little bit, not to any great extent. I was on two or three of the big burns over Tokyo, trying to evaluate what we were lacking in our training when we sent them over there."[80] At the time Carpenter went to the Pacific, Brig. Gen. Curtis E. LeMay was directing nighttime, low-altitude incendiary raids of the type that Carpenter presumably joined. Yet stateside training did not convert to low-level carpet bombing. As another example, in early 1943 the First Air Force had begun training in low-altitude bombing, expecting that the precision blind-bombing equipment AN/APQ–5 would be used in the Pacific theaters and on B–24s in Alaska and the Caribbean. However, this technique was employed mostly in individual night search and strike against ships.[81]

In other respects the training establishment accommodated to combat practices more directly. The first very heavy bombardment training standard to address radar stated that "emphasis will ... be placed upon defensive training from the standpoint of avoiding detection by enemy radar installations."[82]

Training for Strategic Bombardment

Whereas in Europe an extensive ground warning network called up vicious antiaircraft fire and alerted swarms of German fighters, the puny Japanese antiaircraft shield permitted bombers in the Pacific to de-emphasize defense and carry heavier bombloads. Some bomb squadrons removed guns from their B–29s to achieve greater speed and altitude.[83] One training response to this phenomenon was the so-called stripped eagle crew that carried fewer gunners. Similarly, as the lessons from the Pacific became better known, the AAF thought of converting a portion of its very heavy bomber force specifically to night operations. A double payoff came from the presumption that bomber crews trained for night operations could fly the easier daylight missions as well. In fact, the AAF did not train an exclusive night fighting bomber force.

Sometimes the slow institutional pace at which training changed proved its value, as in those occasions when field commanders jettisoned new techniques in favor of older ones for which men had been trained originally. For example, B–29 pilots in the Twentieth Air Force began to deviate from the large formations necessary in the European Theater of Operations to provide maximum defensive firepower. To protect themselves against Japanese fighters that rammed AAF formations, the men started flying in a very thin V so that only one airplane, rather than large chunks of the formation, was susceptible to enemy attack. But when General LeMay saw this formation, he vetoed it immediately. According to Twentieth Air Force pilot David Burchinal, the men "went back to the old formation he taught in Europe which was to uncover the guns and stagger the airplanes and fly in a bombing box. I must say, we weren't bombing very well in that formation," but, he admitted, "we weren't getting shot down or rammed either."[84]

Whether the crew used it to any extent, the B–29 had an awesome arsenal of firepower. Its nonretractable two upper, two lower, and combination cannon-gun tail turrets mounted ten .50-caliber machine guns and one 20-mm cannon. A central fire control system with sighting stations for the bombardier in the nose and in the gunners' positions operated the defensive armament by remote control. Computers calculated necessary deflections for firing at targets within range, and the guns could be aimed almost instantly to concentrate the fire for greatest protection. Gunners operated the turrets from their pressurized compartments.[85] That sophisticated equipment, as well as the scrutiny accorded very heavy bombardment, might have turned B–29 flexible gunnery into a model training course. But, typical of the rest of AAF flight training, gunnery was the program's Achilles' heel. Initially only a few gunners familiar with the central fire control went to Training Command schools, and parts were in such short supply that men learned on conventional Martin turrets.

In early 1945 the Training Command began inducting classes at two schools given over solely to B–29 gunnery. Here, it essentially invented a new twelve-week course. Although gunners had little opportunity for air work (typical for everybody but pilots), the command instituted a useful system in

World War II

which gunners trained in teams of five rather than as individuals. The bombardier who operated the forward turret acted as the gun captain; other members comprising the gunnery crew were a remote control turret mechanic gunner, two waist gunners (one of whom was an armorer and the other an electrical mechanic), and a tail gunner (who was not a technical school graduate). During the spring of 1945 the AAF changed the make-up of the gunnery crew. The stripped eagle crew only carried two waist gunners and a tail gunner (the other gunners and turrets were removed to lighten the plane for greater altitude and distance). By the last summer of the war, the AAF had thousands of graduates from flexible gunnery schools for whom no assignment was available in medium and heavy bombers. After a six-week conversion course, these men joined the pool of gunners assigned to B–29s.[86]

The Second Air Force struggled with some of the same difficulties that the Training Command faced. The training air forces dealt with equipment for which they, too, had little expertise, and in the case of B–29s, operational experience could provide few hints. The Second Air Force and the Training Command argued over whether the preponderance of gunners' time should be spent in schools learning how the equipment worked or in OTUs flying with other crew members. AAF headquarters favored holding back gunners from operational units while they finished the Training Command course. Thereafter the crews of five joined second-phase OTU or RTU training.[87]

Not surprisingly, gunners found air-to-air firing to be of the greatest training value, but as late as August 1944 the Second Air Force still used towed sleeves as air-to-air targets. This practice had long been considered unsatisfactory for heavy bomber training, and the more technically sophisticated B–29 made it even more irrelevant. Also (when they were available during training), fighters could not keep up with B–29s. In 1945 the III Bomber Command complained about having to use P–63s rather than the faster but rarer P–51. The Second Air Force coped by putting B–29 sights in the slower B–24s and B–17s. Once gunners actually defended against pursuit interception in the Superfortress, however, they found B–29 performance in simulating evasive action to be different from that of a heavy bomber. Other tactics, too, hinged on the difference in speed. Slower closing rates of fighters attacking B–29s from the tail reduced the airplane's vulnerability. Analysts therefore urged training units to caution against excessive firing at attacking but essentially unmenacing enemy fighters because the ratio of friendly to enemy damage was higher than it was when slower bombers were flown.[88]

One other change in gunnery procedures affected those who flew F–13s, the B–29 modified for very long range photoreconnaissance. The training units expected their crews to fly missions above 30,000 feet, so they required gunnery practice at altitudes above 25,000 feet. The most significant change in the training of these crews was, however, that photography replaced bombing. In this instance the navigator-photographer used the bombsight for photogra-

Training for Strategic Bombardment

Training in a B-24, a B-29 tail gunner checks his sights and remotely controlled .50-caliber machine guns, *above right*. A remote gunner's compartment can be seen at the *lower left*.

phy, navigation, and flash bombs, and the crew included two photographer-gunners.[89]

Radar in Strategic Operations

Training on radar most differentiated B-29 from heavy crew training programs because over the course of the war, all Superfortresses but relatively few B-17s and B-24s were radar equipped.[90] However, although formal stateside training was not under way, heavy bombers of the Eighth Air Force used radar on combat missions before the B-29s were operational.

The Eighth Air Force deployed to England in anticipation of an early combined bomber offensive with the RAF. Although that campaign was delayed, the Anglo-American partnership shared intelligence and technologies.

World War II

The British had begun using radar for defensive purposes in their extensive ground-to-air warning system. The U.S. War Department followed suit in setting up its own defense network to protect American coastlines. Once the AAF went to war with its strategic mission, the desirability of airborne radar became evident. Aiming for a cloud- or fog-shrouded destination, bombers needed some nonvisual means of navigation as well as the capacity to bomb accurately once formations reached the target.

In retrospect, it was the British night bomber force, not the Americans, who most benefited from electronic advances. The RAF, having sustained grave losses when flying daylight raids, determined to bomb only at night under the relative protection of darkness. Moreover, the deep bitterness felt by the British over the German bombing of London and other English population centers erased any scruples about precision accuracy on military targets. Lack of nighttime visibility provided a substantial incentive to adapt radar technology to the purposes of navigation and area bombing.

The RAF began with Gee, a medium-range radio navigational system. By December 1942 and during the following month, the British Pathfinder Force became operational with two new electronic aids: Oboe, a ground-to-air transmission that accurately fixed the location of high-flying aircraft; and an airborne radar set with a rotating antenna that could scan the ground, H2S (often thought to stand for Home Sweet Home, since it permitted the bomber to home in on a target).[91] A combination, most often Gee and H2S, allowed bomber formations to fly en masse to a target area (with Gee) and then identify the bomb release point (using H2S).[92]

American engineers, too, had been experimenting with nonvisual devices, but at the time the United States entered the war the equipment remained technically limited. Moreover, field commanders expressed reluctance about the increased weight that penalized aircraft performance, the necessity for extensive modifications, and the resultant congestion in aircrew compartments. Radar appeared to be useful mostly for B–25s when navigating under especially hazardous conditions, such as in Alaska, or locating surface vessels in places like the South Pacific.[93]

A study prepared in February 1943 under the auspices of the Armament Laboratory and Engineering Division of the AAF Materiel Center that compared British and American systems of navigation and blind bombing concluded that Gee worked well as a navigational aid "in a fixed theater of operations but cannot be used as bombing equipment in areas where accuracy is desired." Furthermore, while it considered American-made equipment to be most promising for blind bombing in the European theater, that equipment was still in the experimental and modification stages.[94]

Thus, the Eighth Air Force turned to its host to supply American bomber groups. In early 1943 the RAF parted with enough H2S sets to equip a few AAF bombers. Those fitted with the American version, the higher frequency

Training for Strategic Bombardment

H2X, began flying missions by the end of the year. A year later, approximately 80 percent of Eighth Air Force missions employed those radars. Bombers usually navigated over England with Gee and "splasher" beacons (for use with the radio compass); then, by the time they came to the Dutch or German coast, they used dead reckoning and radar. Aircraft with H2X or H2S carried two navigators and a bombardier. In H2X-equipped planes the dead reckoning navigator operated Gee; the radar navigator operated H2S.[95]

Unfortunately, the impressive technical promise of the equipment failed to bring success to American daylight raids. Formations could reach major targets deep inside Germany, such as Berlin, but once there they endured fearsome losses from enemy flak. Moreover, even in less heavily defended areas, H2X could not ensure that the bombs landed where they were intended. Investigation indicated that some 50 or 60 Eighth Air Force targets measured approximately 950 feet by 1,700 feet, a size small enough to put the proverbial bomb-in-the-pickle-barrel to a severe challenge. Moreover, those targets, according to a February 1944 study, "are in many cases in or near relatively small towns.... These towns are difficult H2X targets in that the radar signal from them is probably too small to be clearly recognizable among the myriad of small signals which appear."[96] Some of those towns were so isolated that AAF bombers had difficulty finding them at all, making accurate bombing over the target moot.

Even though his B–17s were equipped with it, in early 1944 Eighth Air Force commander, Lt. Gen. Carl A. Spaatz, considered H2X only "interim equipment... adequate for this period of 'growing pains.'"[97] The equipment and the training to use it failed, however, to mature coincident with the most intense period of the strategic air war in Europe, the months before and after OVERLORD, the invasion of France. Although H2X had permitted the Eighth Air Force to launch numerous missions when the weather would otherwise have grounded them, little benefits were realized in bombing accuracy. The American heavy bomber forces in Europe owed their success much more to the development of long-range fighter escort than to radar bombing.

The story of strategic operations in the Pacific was different, but here too the AAF's achievements were not owed to pinpoint accuracy attributable to radar. Ultimately, the more sophisticated equipment on B–29s was not much more effective than what had been used earlier in Europe. As in Europe, successful bombing sorties benefited more from advances in aircraft technology and changes in tactics than from radar.

Yet, at the time the AAF turned its gaze toward Japan, it still held to its strategic doctrines, and radar was instrumental to those precepts. The advanced AN/APQ–13 radar system and its follow-on variants were installed on Pacific-bound B–29s. The radar observer, pilot, copilot, navigator, bombardier, and tail gunner were all expected to understand and use, to one degree or another, airborne radar. By 1944 the dual-training scheme had been put into effect in the

World War II

bombardment programs whereby a single individual functioned as navigator and bombardier. A select number were to be fully trained as precision bombardier–celestial navigators. Because radar could be used for both navigation and bombing, two fully trained men, given further instruction in radar, were to be assigned to each B–29. The impossibility of achieving that ambitious goal soon became apparent, since the addition of a third skill to a crew member's training further delayed deployment. The AAF therefore decided to abandon dual training for the time and instead gave radar instruction to individually trained navigators and bombardiers.

The AAF divided training responsibility between the Training Command school at Boca Raton, Florida, that taught both the AN/APQ–13 (used against Japan) and the AN/APS–15 (similar to equipment on Pathfinders used against Germany) and on-the-job training in an OTU or RTU in the Second Air Force.[98] In 1944 a B–29 usually carried an AN/APN–4 (Loran, operating like Gee on a principle of synchronized pulses), an AN/APQ–13 radar set for navigation and high-altitude bombing, various radar countermeasure sets, a radar altimeter, IFF (identification, friend or foe) equipment, and other communications devices. For training, the Second Air Force used B–17s loaded with eight radar scopes — flying classrooms, as the command thought of them. When the III Bomber Command began very heavy bombardment training, it gave instruction on four airborne radar sets: AN/APQ–13 (airborne bombardment and navigational radar), AN/APN–4 or –9 (Loran), SCR–718 (radar altimeter), and AN/APG–15 (radar gun sight).[99]

Needless to say, the newer equipment was not trouble-free nor simple to operate. In an attempt to better understand and teach the sophisticated systems, the military turned to the civilian laboratories that had been instrumental in developing microwave radar. The senior Second Air Force communications officer attended a Bell Laboratories course, and the command employed another civilian graduate of the course as an instructor. Even so, the high-priority program would produce far too few qualified instructional personnel, too little equipment, too few facilities, and, it turned out, a barely nodding acquaintance with the technologies. At a training conference held in January 1944, the officer briefing the conferees on radar stated that airborne radar "has very rarely been seen in the Air Forces, very few of you probably have seen a real live functioning radar set."[100]

That state of affairs could not have been more dispiriting, given the high-level support for the program. Early in the war, Secretary of War Stimson had become sufficiently enamored with the prospects for new technology applied to military purposes that he decided to employ a radar adviser. Suggested by the influential civilian scientist and counselor to military leadership Vannevar Bush, Stimson hired Edward L. Bowles. "Eddie" Bowles, who had headed the communications division of the electrical engineering department at Massachusetts Institute of Technology, became a leading proponent for the use of radar and other forms of advanced communications in the Army. Bowles offered one possible explanation for the failure to establish a succesful radar training program, as indicated by the astounding assertion by the briefing officer mentioned above: "I am convinced that much of our radar equipment is over-classified. . . . It seems to me that when a piece of gear is being used in combat, and when it has been dropped all over enemy territory, there is little excuse for having a related synthetic trainer classified SECRET or CONFIDENTIAL."[101]

Probably more detrimental to training was radar's technical complexity. In a report of April 1944, the Committee on Radar Aids to Bombing headed by Dr. Julius Stratton found that "in the present state of the radar art greater instrumental accuracy can be obtained only at the expense of complexity, more difficult maintenance and increased demands on operator skill." The committee thought that the latter factor weighed most heavily against improvements in bombing accuracy. Dr. William B. Shockley, another consultant in radar and communications to the Secretary of War, drew a similar conclusion, as he informed the Commanding General of the Second Air Force: the instructional facilities and training were inadequate because the "rapid development of radar . . . has technically far outstripped training methods. The development of training methods and the carrying out of training has, furthermore, been severely hampered by the lack of equipment beyond that installed in combat aircraft." His conviction had only strengthened by the end of the year, as he wrote Bowles: "You will also see that the value of the radar bombing to date has been negligible compared to the visual bombing." Aside from technical roadblocks that hampered advances, he also postulated that the difficulties in training derived from military decisions: "One of the factors which prevented training from being carried out earlier was a policy from Headquarters AAF urging complete emphasis on getting on with operations."[102]

The criticism of its scientific advisers concerning the military mindset did not fall on deaf ears at the Air Staff. But those in uniform grappled daily with the numbers of people the system could realistically train, the absence of equipment, and the priorities of theater commanders. It was true that they were driven primarily by operational considerations, and as the events of war unfolded, some were becoming less sanguine than the engineers and physicists about the utility of the new technology (and the doctrine it upheld). The Requirements Division, for example, stated that it was unable to promise a smooth flow of radar trainees "in view of the dynamic 'state of the art' in radar."[103] As nearly everybody realized, effective use of the equipment depended on the resolution of training, technical, and maintenance problems. Those difficulties remained substantial such that radar's promise had only begun to be attained by the end of the war. By then the success claimed for strategic bombardment was measured by a different yardstick.

In the meantime, as Shockley predicted, training methods had to run to catch up with the new and rapidly evolving technology. One year into the

World War II

conduct of very heavy bombardment training, Second Air Force Commanding General Ent ruminated to a friend:

> I feel that with the present B-29's using the APQ-13, if we are to do precision bombing this bombing must be done visually in daylight. We are concentrating our training with this in view. This does not mean that we are slighting radar training in any way. To successfully bomb in daylight, I feel that we must go to small formations — four to six aircraft. The combat box idea used in the UK is not economical from the standpoint of bombing accuracy. I believe these large formations also are increasing the damage from flak and I am sure that smaller formations will not have so much self-inflicted damage. With the central fire control on the B-29, we are afraid that self-inflicted damage in large formations will be prohibitive. Another thing that precludes the use of these large formations in the B-29 is the pressurized cabin and the effect upon this pressure by constantly changing throttle settings.[104]

From observations such as these the Second Air Force rendered training directives laying out missions to be followed, types of formations to be used, standards of flying in daylight and darkness, criteria for visual and radar sighting, turret firing requirements, and so on. By the end of 1944 changes in radar technology forced revisions of training guides to emphasize "teamwork" among the crew using radar in navigation and bombing exercises. The Very Heavy Bombing Training Standard issued by AAF headquarters required radar on 50 percent of all missions; the Second Air Force directives aimed for radar-related tasks on all missions and levied special requirements on the navigator and the radar gunner.[105]

The training bases had, of course, to adapt their practices to the specific equipment and training materials they possessed and the need for continuous updating and modifications. To cite Loran training as an example, the officer in charge at one training base reported that

> instruction in the 2nd AAF is being hampered due to lack of current information concerning Loran; such as, new station location, new equipment, and information concerning enemy jamming. Flight training, a very important phase of training, is ommitted [sic] in the 2nd AAF due to location of bases.[106]

Having no effective solution to the problem just outlined, an officer at AAF headquarters responded lamely that "operators who receive suitable basic training in this country will be able to increase their knowledge and experience in short order when they reach theaters of operations where signal will be available."[107]

The III Bomber Command's program, based closely on the Second Air Force's, also centered many of its requirements around radar. Its manual published in March 1945 included radar operation on a 3,000-mile navigation mission; on a celestial navigation, high-altitude mission; and on camera bombing missions. By this time the shortage of equipment was not so acute. Yet, like the Second Air Force, the command struggled to stay abreast of technological and operational developments in radar, organizing an Operational Radar Section in April. One of the section's tasks became the coordination of AAF and III Bomber Command training standards in order to devise a set of training missions. Ultimately the section relinquished its oversight role and mostly responded to locally generated proposals because too many variables existed to standardize training throughout the command.[108]

Assessments

> We had three months to get ready, three B-17s, and only a couple of people who knew how to fly them.... By the time we had to go overseas I had worked with the navigators for only about a week or two.... The bombardiers came into Muroc a couple of weeks before we were due to go. They had never dropped a live bomb in their lives — because we had no airplanes to allocate to bombing training.... The gunners were supposed to have gone through a gunnery school, but they had never shot a gun from an airplane.... We had never flown formation until we got to England simply because we didn't have enough airplanes.[109]

In words written many years later, LeMay described the 305th Bombardment Group's questionable combat readiness at the end of its training. At about the same time that LeMay's group left for England to join the Eighth Air Force, Maj. Gen. Carl A. Spaatz, then commanding the AAF in Britain, sent a letter of commendation to those conducting training in the United States in which he characterized the first bombardment units sent to England as "excellent." While stateside officers naturally made public Spaatz's plaudit, they appended their own critique:

> The first bombardment units dispatched to operating units in the British Isles were trained to a degree far below Army Air Force standards. Even so, they are acquitting themselves creditably, proving beyond any doubt the soundness of our training system, and the combat fitness of our aircraft.[110]

The remarks by AAF headquarters were as fair as they were self-serving. Young airmen and high-ranking officers alike recognized that Americans went to war unprepared. They learned to fly under fire. During that first year, the AAF lived under a dreadful truce. It had joined a war to fight, and fight it would. Yet it lacked widespread expertise and a system equipped to prepare men for their jobs, so it sent them into combat trained to a level of competence it knew was too low.

World War II

At home, training officers held their breath as they watched the ninety-day wonders flying high-performance aircraft. The twin pressures of time and requirements collided with proficiency in the interplay between safety and realism. Second Air Force accounts reflect the constant anxiety over the high number of training accidents. The disastrous spring of 1942 saw a series of airplane crashes attributed to ill-trained pilots flying in poor weather. Word from above directed training units to cease flying under dangerous conditions. Accidents decreased, but the training program, admitted the command historian, "stultified." It was clear to training officers that

> in actual combat, these crews would frequently find it necessary to fly missions when weather conditions were anything but favorable, and, if they had no particular practice during their operational training, the likelihood of mishaps in the theatres of operations would be greatly increased.[111]

Similarly, instances in which commanding officers ordered their subordinates to avoid tightly packed training formations or not to direct antiaircraft fire above and in front of the flights evidenced the same tension between realistic training and keeping people alive. Although they tried every means thinkable to curb accidents — improving and standardizing instruction, disallowing certain types of flying, increasing supervision, and encouraging greater discipline among pilots — the training air forces never successfully resolved the safety–realism dichotomy because it was integral to a training system under the stress of speed. Lowered accident rates owed less to changes in training methods than to factors such as a reduced training load or a greater amount of equipment or gasoline.

Not until 1944 did the heavy bombardment training program have the men, aircraft, and equipment that enabled it to function largely as planned. Describing use of the A–5 Sperry bombsight in a combat group in the Mediterranean theater, one report claimed the training they got "was good, but entirely inadequate."[112] What was true in first-line combat units applied all the more at home. Maj. Gen. Robert W. Harper, Assistant Chief of Air Staff, Training, explained to General Arnold in May that until that month the "RTU system has been short as much as two hundred B–17 and three hundred B–24 aircraft at one time." Moreover, the "aircraft on hand were so old and required so much maintenance that it was very difficult to meet minimum requirements for the training of crews."[113] By mid-1944 aircraft inventories and requirements matched, and enough high-octane gasoline could be spared for training to ensure minimum requirements in high-altitude flights and for more long-range navigation missions. From about this time, too, the total number of practice bomb releases per crew increased, as did the use of camera guns in air-to-air gunnery and simulated bomb runs.

Following directly on the heels of and overlapping the heavy bomber

Training for Strategic Bombardment

program, B–29 training profited from a seasoned training air force, the latest technology, and A1 priority under the direct and watchful eye of General Arnold himself (who also directed Twentieth Air Force operations from Washington). Nonetheless, it tumbled immediately into the same sea of troubles as the heavy bombardment program. The anticipated virtues of the new aircraft made the AAF leadership eager to deploy it as soon as possible. Yet its advanced systems pushed the training program beyond its limits. The high hopes for the B–29, which caused the program to race forward precipitously, cut into both the quantity and quality of training as, at the beginning, the AAF simultaneously tested the aircraft, trained the crews, and tried to unscramble conflicting opinions regarding the most useful types of equipment and the value of visual versus radar bombing.

The B–29, as General Arnold subsequently admitted, "posed more problems" than any other World War II aircraft; the Second Air Force's maintenance personnel considered the airplane a nightmare. Leaving aside the headaches for the ground echelon, the plane's engineering problems affected flight crews in any number of ways, not least of which was the tendency for engines to catch on fire while in flight. The Second Air Force estimated that during the month of October 1944, for example, such fires accounted for 25 percent of B–29 training accidents.[114] Even the B–29 program's high priority failed to alleviate the kinds of shortages familiar to all aspects of AAF training: too few radar-equipped training aircraft, instructors, crew members, B–29s themselves (for which B–17s substituted for many of the air missions), computers for the central fire control system, and training facilities.[115]

The need to coordinate the efforts of the several training agencies, and of the training establishment as a whole, with the theaters of operations consumed the thinking of many officers. Give-and-take among policymakers, the training air forces, and field commanders remained essential. General Yount of the Training Command and some of his senior staff toured overseas combat units, and the command convened numerous conferences in its efforts to achieve greater standardization of training practices. The Second and Third Air Forces also considered standardization to be a chief goal, but unit and crew training included battlefield tactics, not just the operation of airplanes and equipment. So even though instructors in the training air forces were spread extremely thin and could not easily be spared even for educational purposes, some visited combat squadrons nonetheless. For instance, personnel involved in radar training went overseas in 1944, according to General Harper, for "familiarization training in H2X operations and tactics so that the training agencies will be fully cognizant of combat requirements."[116] Although General Harper was reputed to be a tireless and concerned administrator, some accused the Air Staff of failing to stay adequately in touch with operational realities, in part because of real logistical difficulties but also because of some myopia at the top. Lt. Gen. Richard Montgomery, for example, who himself served at AAF

World War II

The command staff of the subordinate training commands flanks Maj. Gen. Barton K. Yount in early 1943, as the Commanding General, Training Command is seated at his desk. The Training Command headquarters occupied the top six floors of this railway building in Fort Worth, Texas, at the time.

headquarters as a young officer, later admitted:

> I just didn't feel that we were completely up to date with the combat units. We didn't get in a B-17 and fly over and spend some time with the combat units. My thought now is that we should have. But those airplanes were hard to come by. The ones going over were being delivered by flight crews.... I should have been over there visiting with these combat commanders to find out what the deficiencies were and to come back and try to correct them. The Training Command did send some people overseas, but I think we in the Air Staff were not as quick to anticipate this and to learn our lessons too. We knew training, but we didn't know the combat side of it.[117]

Assessments of the relative effectiveness of visual and blind sighting lay at the heart of many reports written during 1944 by experts in radar and electronic communications. Statistics revealed that the AAF sustained considerable loss and enjoyed relatively little success using either kind of sighting. During 1943 the weather over the target, especially in southern Germany and Austria, restricted visual bombing missions to about five days a month. Individual bombing achieved greater accuracy than pattern bombing, but blind-bombing missions had to be flown with so many aircraft that the maneuverability necessary for individual sighting became virtually impossible. Moreover, according to one study issued in February, when the huge formations released on the leader, the target area stretched over approximately a mile and a half — hardly pinpoint. High winds of fifty or sixty knots blew the smoke drifts from the first bombs so far off course that often a second combat wing could not identify the release point. When the AAF bombed Bremen, its most frequently attacked target, Pathfinders led twenty-one of the total of seventy-five combat boxes. Yet, according to the report, "No bombs fell within two miles of the aiming point and only five combat boxes succeeded in getting their bombs within five miles."[118]

The dismal record of bombing accuracy with the Norden sight and with radar did not, however, significantly alter AAF doctrine or training. The training system could not by itself redirect training policy and practices, because to do so would call into question American doctrine itself. Training was instrumental to, but not determinative of, policy. The AAF always found it hard to measure air losses and gains with any great specificity. Yet the February 1944 report, written in a measured and highly cautionary tone, laid bare the fact that the American bomber forces had been brutally battered on their missions into Germany and may have paid their deadly price in vain, for all evidence indicated that up to that time the incidence of bombs landing on target was extremely low. A few months later, the same story came pouring from the Pacific in after-action reports of visual and blind bombing by B-29s. Those documents confirm the notion that while many associated with the strategic campaigns knew that their efforts were not bringing the desired results, they hoped against hope that if new technologies came on line and training improved — as so many radar experts saw as the root of the problem — daylight precision bombing might remain viable doctrine and practice.

In fact, it was an ill-fated vision, soon to give way completely to a different reality. B-29s equipped with the most sophisticated electronic equipment became the strike force in the Pacific, not in Europe. There, in that war, the AAF ultimately gave up every pretense of abiding by theory, even though it did not relinquish training by it. Not surprisingly, since he was one of the architects of AWPD-1, which had served as the strategic blueprint, Maj. Gen. Haywood S. Hansell first led the XXI Bomber Command in the Marianas on high-altitude missions. Like so much of the bombing in Europe, relatively few bombs fell on their intended targets. Later a noted poet, John Ciardi, who flew as a B-29 gunner, called the first few months of the campaign "wasted effort" because

World War II

"we lost all those crews for nothing. We had been trained to do precision high-altitude bombing from thirty-two thousand feet," he recalled. "It was all beautifully planned, except we discovered the Siberian jet stream. The winds went off all computed bomb tables. We began to get winds at two hundred knots, and the bombs simply scattered all over Japan. We were hitting nothing and losing planes."[119]

Shortly after LeMay took over from Hansell, the prosecution of the air war took a stunningly different turn, the strategic doctrine effectively going up in the conflagration that destroyed miles upon miles of Japanese cities. The familiar story began in earnest in March 1945 when LeMay's bombers flew their massive incendiary raid over Tokyo. Low-altitude, nighttime saturation bombing, unleashing a firestorm of heat and flames seen from a distance of 150 miles, obliterated the AAF's public commitment to high-altitude daylight raids against precision targets.[120]

With some reluctance, the AAF previously had sent its bombers to flatten some strategically insignificant cities in eastern Germany. Now, against Japan, the Air Staff and operational commanders decided the time had come finally for a wholesale conversion to area bombing. On training bases, on the other hand, it was as if much of this were not happening. The inevitable gap between training and operations seldom yawned so wide.

FIFTEEN

Crew and Unit Training for the Tactical Air Forces

> Heading the parade at one thousand feet were six squadrons of B-25 strafers, with the eight .50-caliber guns in the nose and sixty frag bombs in each bomb bay; immediately behind and about five hundred feet above were six A-20s, flying in pairs—three pairs abreast—to lay smoke as the last frag bomb exploded. At about two thousand feet and directly behind the A-20s came ninety-six C-47s carrying paratroops, supplies, and some artillery.... On each side along the column of transports and about one thousand feet above them were the close-cover fighters. Another group of fighters sat at seven thousand feet and, up in the sun, staggered from fifteen to twenty thousand, was still another group. Following the transports came five B-17s, racks loaded with 300-pound packages with parachutes, to be dropped to the paratroopers on call by panel signals as they needed them.
> —Lt. Gen. George C. Kenney to General "Hap" Arnold, September 1943[1]

Although the AAF most publicly espoused its commitment to strategic bombardment, during much of the war it was unable to showcase its strategic capability. Moreover, many field commanders and pilots preferred other types of aircraft to the big bombers, and other roles and missions rather than high-altitude, long-range bombardment. General Kenney, for example, had long been an *attack* man. By 1943, directing the air war in the Pacific across hundreds of miles of ocean dotted with tiny islands, Kenney used all the aircraft at his disposal in a variety of tactical roles. In the description above, he names some, though not all, of the airplanes assembled on the 5th of September to mount an assault on Nadzab. The Fifth Air Force, he wrote, was "as the kids said" finally "cooking with gas." General MacArthur himself swore that the work of the air

World War II

More A-20 attack airplanes were produced during World War II than any other tactical aircraft in this weight category.

forces in this preliminary move to take out Lae "was the most perfect example of discipline and training he had ever seen."[2]

By that time also, in fall of 1943, theoretical ground rules for tactical air warfare in Europe had been hammered out of the painful experience of fighting in Tunisia. The Mediterranean campaign had delayed the start of the long-anticipated Combined Bomber Offensive, which then completed the incapacitation of the Luftwaffe before the Allies unleashed their full might in the invasion of France. Like the ground battle, the tactical air war on the Continent peaked in the actions following Normandy. Blessed with overwhelming numbers, the Ninth Air Force led British and American units through the textbook steps as they first claimed air superiority, isolated the battlefield through air interdiction, and then gave over their efforts to close air support.

The AAF used fighters and fighter-bombers, medium and light bombers, and armed reconnaissance aircraft and even drew in heavy bombers to support the invasion. In a postwar analysis of its operations, the Ninth Air Force cited the "increased employment of specialized weapons and equipment," especially for its modified fighters used as fighter-bombers and its reconnaissance groups. Rockets, large fragmentation clusters, Pathfinders, Oboe, radar and radar photography, and new bombsights became important in tactical air strikes.[3] Similarly, in the southwest Pacific and in China, P-38, P-40, P-47, and P-51 fighters flew most of the ground support sorties. B-25s and A-20s bombed and strafed at treetop level using delay-fused demolition bombs, parachute-rigged fragmentation bombs (parafrags), and napalm.[4] In what amounted to short-lived experiments in dive bombing, the AAF used Douglas A-24s in the Pacific; in

Italy and the Sicilian campaign it was the A–36, a Mustang configured with dive-brakes and bomb pylons.[5]

Crews for those missions trained in OTUs and RTUs in the continental United States. During the period when each air force had clearly demarcated training responsibilities, the Third Air Force trained medium (B–25 and B–26), light (A–20 and A–26), and dive (A–24 and A–36) bombardment crews, as well as some single-engine fighter pilot replacements. The First and Fourth Air Forces conducted most of the fighter training along with their defense mission. Slowly the latter function dissipated, and the First and Fourth converted exclusively to training. A balanced air force finally became a reality in training as well as in operations when, by 1944, all the continental air forces trained fighters and bombers that flew together in exercises. The Second Air Force, for example, then trained fighter pilots principally to provide interception for heavy bombers.

The training directives for medium bombardment closely resembled those for heavy bombardment, even though the lighter-weight bombers performed in tactical rather than strategic operations, carried a smaller crew, and bombed at lower altitudes and over shorter distances. (They therefore required less specialist training for many of their bombardiers and navigators.) Over time the published curricula related more closely to combat practices and increasingly incorporated techniques applicable to specialized roles in both Europe and the Pacific. The medium bombardment program not only shared an affinity with many of the concepts and techniques of heavy bombardment unit training, it also experienced some of the same problems of manpower and equipment shortages. For example, so few gunners were available for medium bombardment units in late 1943 that they faced the choice between delaying the shipment of replacement crews and sending them out with incomplete training by relaxing proficiency standards. Neither medium nor reconnaissance air crews assembled until one-third of their training was completed, equivalent to the situation in heavy bombardment OTUs of devoting first-phase training to individual proficiency checks and ground school.[6] One survey of Third Air Force OTUs and RTUs in January 1943, which could as easily have described the Second Air Force program, spoke of the "irregular schedule by which crew members enter training units, unequal balance of the various crew members, shortage of airplane equipment and shortage of navigation equipment; no certain training period allowed."[7]

Light bombardment, operating in ground support activities, diverged quite clearly in mission and tactics from the strategic model. Fairly early in the war, the AAF ceased dive bombing, and those aircraft and crews were then trained and deployed as fighters. Reconnaissance lived as a poor relation, drawing its personnel, aircraft, and curricula largely from the bombardment and fighter programs, and ranking at the bottom in the esteem of pilots. It grew nonetheless from a small, ill-equipped force, trained in observation squadrons and flying

World War II

obsolete O-type aircraft, to assume a well-recognized place in the balanced air force. Both fighters and bombers were modified for photographic and tactical reconnaissance. Most reconnaissance training occurred in the Third Air Force and in fighter RTUs of the First and Fourth Air Forces.

Fighter training encompassed a variety of roles that evolved over the course of the war: escort, interception, dive-bombing (briefly), fighter-bombing, night fighting, photoreconnaissance, and tactical reconnaissance. Those functions expanded from a narrowly defined defense mission that presumed because fighter aircraft were fast and highly maneuverable, but had no long-range capability, they could be used most effectively against any invader who reached U.S. borders. Unlikely and xenophobic as the scenario may appear in retrospect, the nature and severity of that perceived threat, even midway through the war, can be sensed from a statement read into the *Congressional Record* by Representatives from the state of Washington. They feared that Japanese incendiary bombs might ignite fires across miles of the heavily forested Northwest, creating

> a blanket so thick and wide as to blind our defending air force to attack attempts of the enemy.
>
> And we may see whole cities destroyed or abandoned because of this conflagration; we may see thousands made homeless, and many killed; we may see our great west coast war plants crippled and stilled; and, worst of all, we may hear as the direct result of such a coastwise fire, the tramp of yellow invaders on our home soil.[8]

As long as these anxieties remained, fighters in the continental air forces based on each coast were deployed alongside the ground-based warning system to defend the Zone of the Interior.

Just like all other units, fighter OTUs and RTUs suffered from equipment and personnel shortages. But different principles applied to fighter and bomber operational training. Most notably, day fighter pilots flew alone, so the crew concept was absent. Even the cornerstone of the American system of flight training — dual instruction — was inapplicable to fighters because no biplace tactical fighters were available for training. For the fighter missions, therefore, teamwork meant association with the other pilots in a squadron and supporting the bomber formation or providing close air support for the ground forces. Formation practice with a lead pilot and wingman became a regular and crucial aspect of unit training.

Coordination between fighters and bombers improved when the Second and Fourth Air Forces began joint exercises in 1943. In May, for instance, the II Bomber Command dispatched a large group of B–17s to Fourth Air Force bases for a mock attack on the West Coast by an enemy carrier raid.[9] The AAF training standard for combined air forces training stated the first purpose to be "proficiency in teamwork essential to combined operations and ability to function without friction or confusion under unified tactical control."[10] More regular fighter-bomber exchange took place within the training sequence when the Second Air Force picked up fighter units and the First and Fourth Air Forces added heavy bombardment OTUs.

Of all types of joint training, that between air and ground forces fared worst. It never became predictable, in part because theater commanders kept up steady pressure for crews who had completed operational training. Moreover, air-ground training added one more variable to an already overcomplicated schedule, as Brig. Gen. Robert W. Harper, chief of training on the Air Staff, explained: "Due to the difference in commitment schedules to active theaters, training air units cannot be coordinated with training of ground units so as to bring an air unit to the state of training for a test at the same time as a ground unit is ready."[11]

The training programs varied administratively. Light bombardment and fighter pilot transition, for example, remained a responsibility of OTUs, whereas the Training Command formally assumed OTU work for pilots assigned to heavy and medium bombardment. The training programs also shared similarities — in organization and approach and in shortfalls of equipment, for instance — but they again diverged according to their own circumstances, their priority in the training scheme, and the requirements laid upon them.

Medium Bombardment

> The final stages in readying us for the real thing were tough. We had long navigation and overwater navigation flights, gunnery, skip bombing, altitude bombing and night flying. We averaged about eight hours of flying every other day, and that is a long time to be up in the air.[12]

Several thousand American-manufactured B–25 and B–26 bombers went into service with the AAF and Allied air forces between 1941 and 1945. These twin-engine aircraft usually bombed from level flight or a shallow dive at lower altitude than the heavies and B–29s did. Operating successfully as tactical bombers, most mediums in units of the Twelfth and Ninth Air Forces in Europe flew interdiction missions over predetermined targets. But medium bombers garnered special acclaim in the Pacific. In the first months of the war, General Arnold, in consultation with his staff and two naval officers, debated how to attack the Japanese homeland. They decided to launch a raid against Tokyo from a U.S. Navy aircraft carrier. "The next thing," according to Lt. Col. James H. Doolittle, one of Arnold's advisers in the planning sessions, was "to select an airplane that would go 2000 miles carrying 2000 pounds of bombs and take off short." Doolittle got the job of choosing the aircraft and training the crews. He chose the B–25 and talked his way into leading the crews he trained.[13]

Twenty-two airplanes arrived at Eglin Field in Florida on March 26 and 27,

World War II

A gunner trains in the nose turret of the medium tactical bomber, the Martin B-26 (*upper right*), and a tail gunner practices for tail defense on a skeet range on a training field in Florida, using a salvaged model of the same type of aircraft.

1942. The crews spent the next three weeks training and checking out their aircraft. They painted pseudo–flight decks on the runways to practice taking off from a restricted space; they flew navigation flights over land and water, including one from Eglin to Ft. Myers in Florida, then on to Houston, Texas, and finally back to Eglin. They trained extensively in low-altitude bombing (dropping some 800 sand-loaded bombs; all but two of the bombardiers also dropped at least two live bombs) and in short take-offs with a full load (under supervision of a naval officer). They flew one mission with fighters simulating attacks on the bombers and practiced evasive action and turret operation (minus what Doolittle called the "unnatural-to-use" lower turrets, whose removal also allowed the airplane to carry a greater amount of fuel). The armament officer removed the high-altitude bombsights that had been supplied, substituting a simple plate and sighting bar. (The bombardier dropped when the bar, set on a calibrated scale at a predetermined dropping angle, fell in line with the target.)[14]

The flight of sixteen B–25s took off from the U.S. Navy aircraft carrier *Hornet* on April 18. Although the formation reached the target successfully, on return almost all the planes went down in bad weather over China. Three of Doolittle's men who reached China perished in parachuting or crash landing. The Japanese killed three of the eight men they captured, and another died of disease while in prison. Despite calamity on a personal scale, the mission was enthusiastically reported at the time and energized the American efforts even though it did little to damage Japan. The exploits of medium bomber crews stayed in the news as later, again in the Pacific, the world learned of General Kenney's creative and effective low-altitude strafing attacks against Japanese naval vessels.

To deliver "accurate fire by bombing against enemy installations," as stated in the 1943 Training Standard, medium bombardment aircraft required "highly trained specialists working together as a combat team."[15] The same words precisely echoed those used to describe a heavy bomber crew, although a medium bomber carried fewer gunners than a heavy did (by 1944 most medium crews did not have both a bombardier and a navigator). The aggregate crew of a medium bomber was smaller, but the individual responsibilities of each crew member were essentially the same as those of a heavy bomber crew.

A significant departure from heavy to medium bombardment training, slowly incorporated into official training standards, arose from the tactics that General Kenney perfected in the southwest Pacific. Kenney directed his B–25s to strafe ships using forward-firing machine guns (he was not pleased with the results of their use on B–24s), and he employed a low-altitude skip-bombing technique. In his memoirs, Kenney recalled working out the possibilities for low-altitude bombing as he traveled from San Francisco, where he had been on duty with the Fourth Air Force, to Australia, where he would join MacArthur and the Fifth. "It looked as though there might be something in dropping a bomb, with a five-second-delay fuse, from level flight at an altitude of about fifty feet and a few hundred feet away from a vessel, with the idea of having the bomb skip along the water until it bumped into the side of the ship. In the few seconds remaining," he hoped, "the bomb should sink just about far enough so that when it went off it would blow the bottom out of the ship. In the meantime, the airplane would have hurdled the enemy vessel and would get far enough away so that it would not be vulnerable to the explosion."[16]

Medium bombers were modified for the new tactics. First, the B–25G was equipped with a 75-mm cannon in the nose. But since the powerful weapon's recoil when it was fired caused the plane's airspeed to drop alarmingly, pilots were unenthusiastic about the gun. The H model that went into service in the Pacific in 1944 added eight machine guns to the nose in addition to the six in defensive positions.[17] Very early the Third Air Force began to teach medium bomber pilots to fire fixed guns and to bomb using Kenney's methods.[18] Although conditions differed for the Eighth Air Force flying out of Great Britain, in early 1943 its officers also advocated the use of fixed guns on mediums, and bombing at treetop level.[19]

The 1944 Training Standard governing pilot qualification called for an understanding of the operation of the airplane (for example, stalling characteristics and single-engine and other emergency operations), familiarity with all

World War II

A B-25 practices low-altitude skip-bombing, a tactic used to effect in the Pacific.

crew duties, proficiency in low visibility patterns and approaches, high-wind takeoffs and landings, night operation, instrument flying, formation in all positions of the basic three-ship element, and navigation (low- and medium-altitude dead reckoning, day and night pilotage at medium altitude, and use of radio aids).[20]

The copilot played a lesser role than he did on a heavy bomber, which caused some senior officers to consider eliminating him altogether from the medium.[21] The copilot retained his position as part of the crew on most medium bombers nonetheless, and he became especially useful in night operations. Because he might be called upon to man a gun in an emergency, he had to be familiar with all the gun positions on the aircraft.

For some time, training standards listed a navigator and a bombardier among the crew. The most significant change in the composition and duties of the medium bomber crew came with the conversion to a dual specialist performing the functions of both navigator and bombardier. Because mediums flew in formations at lower altitudes, until late in the war their bombardiers trained on nonprecision sights that were simpler and more abundant than the Norden and Sperry sights installed in the high-flying heavy bombers. The AAF determined that the bombardier on medium aircraft could therefore master a secondary skill. Moreover, lower-altitude formations did not require that every crew include a navigator. By 1944 the Training Command had shifted to dual training whereby all bombardiers received instruction in dead reckoning navigation; those men became navigator-bombardiers on 75 percent of the medium crews. The man in a lead crew was to be a completely dual-trained navigator-bombardier.

By early 1943 medium bombers had gained considerable experience in the Mediterranean theater, and by year's end they had obliterated much of the Italian lines of communication. In the Pacific the Fifth Air Force was showing notable success against the Japanese fleet. The training system, however, lagged behind in integrating operational practices into its official programs of instruction. The February 1943 medium bombardment Training Standard dealt with qualification requirements for each member of a combat crew, but it only described training missions in a generalized way. The fact that the crew composition was in flux contributed to the vagueness of unit standards.

Although the decision had been made in early 1943 to reduce crew size by having one person perform the duties of navigator and bombardier, dual-trained graduates had not yet reached the Third Air Force. Units still trained their own navigators as bombardiers, and bombardiers as navigators. The latter was a slower process, owing to the greater complexity of navigation training. Navigators effectively took on training responsibility for bombardiers, which limited their opportunity to practice their own specialty, as one report pointed out:

> Bombing can be learned much faster and with less ground instruction than navigation and the navigators in medium bombardment groups were kept busy instructing bombardiers in navigation, both on the ground and in the air, so that the commissioned navigator was flying navigation missions as an instructor for bombardiers and was not getting any practice on flights in which he did the navigating.[22]

The different bombsights and tactics employed among the RTUs further contributed to the lack of uniformity in published directives. In the III Bomber Command, for example, some crews bombed with the D–8 sight, flying mostly, but not entirely, during the day from altitudes between 50 and 8,000 feet. Those missions emphasized evasive action, short approaches, controlled time of arrival over target, coordinated attack by formation in daylight and at night with flares and incendiaries, daylight incendiary attack, and coordinated navigation and bombing missions with demolition bombs. Other crews bombed with the N3A pilot's gun-bombsight, practicing skip bombing. At low altitudes the latter proved more effective than the D–8 releases.[23]

While training practices were tailored in part to match available equipment, they also evidenced an attempt by the training bases to prepare for specific combat missions despite the generalized perspective of the training standards. But training practices learned at the local level were not necessarily up-to-date either, like, for example, the first lessons from North Africa that suggested bombing at 8,000 feet or below was too dangerous because of devastating antiaircraft fire. In theater the aircraft reverted to a higher level, and American crews borrowed bombsights from the British. Then, flying high, their bombs usually missed the target. In the Pacific, on the other hand, low-level strafing was proving successful. In any event, because units were assigned insufficiently far in advance to train extensively for a particular theater, sometimes equipment used and techniques learned in training would be marginally useful when men joined tactical units other than those anticipated.

At the end of November 1943 the I Bomber Command, newly involved in medium bombardment training, delivered to the First Air Force Commanding General a long list of deficiencies encountered in training two medium OTUs. The complaint, running to pages, indicated problems in basic training at the beginning of the OTU program, lack of satisfactory or incomplete first-phase training in the Second Air Force, loss of time in transferring from one air force to another, maintenance and supply difficulties, and slow delivery dates and ill-equipped and poorly maintained aircraft assigned for immediate transfer overseas. Many of the problems were amenable to partial cure by the shift from the OTU program to RTUs and the elimination of the vertical training structure, a "fix" the First Air Force was quick to point out to the AAF Commanding General and his chief of training.[24]

Members of the training establishment debated who or what accounted for other deficiencies, most notably the low level of bombing accuracy, since precision bombardment was still considered to be the primary means of delivering firepower. Senior officers frequently criticized their subordinates about crews' sloppy bombing techniques, as expressed by poor circular error scores. In the spring of 1943, for example, Maj. Gen. Barney Giles wrote to the Commanding General of the Third Air Force that "we must not lose sight of the fact that accurate horizontal bombing is the very foundation upon which our Air Forces are built. Our training methods must be directed unceasingly toward producing bombardment units capable of delivering bombs against assigned targets with an acceptable degree of accuracy." Medium bomb groups must "strive to improve," he admonished, because "a lack of understanding of the bombing problem and a failure to appreciate the importance of proper training methods are undoubtedly responsible for this condition, to a large extent." Giles probably conceded less than was warranted when he admitted that the "type of bombsight with which you are presently equipped is not conducive to precision bombing."[25] At that time, the Third Air Force made do with a variety of nonprecision sights. Most bombardiers had not learned on Norden sights in bombardier schools and never saw one during their tenure in OTUs. Complaints from the battlefield bemoaned the ill effects of the training when bombardiers came into units equipped with Norden sights. It was not until the spring of 1944 that Maj. Gen. Robert W. Harper could confirm the fact that the Third Air Force had enough Norden sights to use them in training medium bomber units.[26]

The greater thoroughness of the 1944 Training Standard indicated the salutary results brought by administrative changes, better equipment, a concrete tactical doctrine, and a settled crew composition. Both medium and light bombardment, by this time, took note specifically of parafrag and skip bombing. Over the training period, on crews without a bombardier, the pilot released forty bombs at minimum altitude and ten from the wing position in formation at medium altitude (7,000 feet or above). A bombardier released forty individually aimed bombs at medium altitude, and the pilot released ten

Training for the Tactical Air Forces

from minimum altitude. The entire crew flew navigation missions, including at least one day and one night flight, without radio aids. A crew with a celestial navigator made one night and two day celestial flights. On gunnery practice missions, the pilot fired at least 200 rounds from each fixed forward-firing gun at ground targets; if the ship had 75-mm cannon, the pilot fired a minimum of 25 rounds. The standard stated the minimum number of rounds of ammunition the turret and flexible gunners should fire in air-to-air and air-to-ground target practice. All but the pilot had to complete at least four camera missions. By 1945 towed targets were no longer used; with the aid of camera guns, a gunner fired at least 200 rounds from his position on each of five air-to-ground or air-to-water targets.[27]

A specified number of missions were to be flown at night. The AAF's enthusiasm for daylight bombing led General Arnold to warn nonetheless in October 1942 that "during all of this very favorable publicity with regard to day bombing, the tendency will be to forget about night bombing in our training. This must not occur." The Director of Bombardment responded that both heavy and medium OTUs were attempting to conduct a ratio of one night mission for every two day missions.[28] Although the Third Air Force lacked the essential manpower to meet that standard then and for some time to come, night bombing instruction became a regular part of medium bombardment unit training. Combat reports and senior officers, including General Arnold, continued to emphasize the importance of night tactical operations, particularly in the Pacific war.[29]

College-football-star-turned-pilot Tom Harmon began his unit training in one of the III Bomber Command airfields where B–25 training began in 1943. "After the ships we had flown in training school," he recalled, "the B–25 was a treat. There was about as much difference between it and a trainer as between an old model T and a new Cadillac."[30]

Most other pilots also found the bomber to be a reliable, smoothly performing airplane. Variants occasioned slight changes in training and in crew composition. In some planes a solid nose equipped with what was essentially a standard Army 75-mm field gun replaced the bombardier's nose compartment. The aircraft also carried fourteen 0.5-in. guns and up to 3,200 pounds of bombs or a 2,000-pound torpedo. Because an airplane with this armament functioned more as an attack ship than as a traditional bomber, its five-man crew, consisting of a pilot, a navigator, and three gunners, carried no bombardier. The Far East Air Forces in the Pacific, which employed medium bombers in low-altitude strafing, modified their J models with attack armament before receiving production-line versions. The most widely produced B–25J returned to the transparent bomber-type nose with the bombardier's compartment. In training, the Js assembled a six-man crew: the J1 had a pilot, copilot, bombardier, and three gunners; the J2, a bombardier-navigator instead of the

World War II

Football stars from rival Texas universities are featured in this promotional photo for the Lubbock Army Flying School where men prepared for twin-engine bombing missions.

bombardier; and the J3, two pilots, a bombardier, and three gunners.[31] The AAF remained critically short of navigators through much of the war, so it assigned many fewer of them to medium than to heavy bombers. Between the two medium bombers, the B–25 received most celestial navigators because the plane was intended to fly longer missions over poorly mapped territory or long stretches of water.

The AAF revised its medium bombardment training standard in 1944 largely in response to the loudly expressed ire of the III Bomber Command that the previous general, all-encompassing requirements could not be met during the training period. The new standard was less vague but was still insufficient, according to the command, and it did not differentiate between B–25 and B–26 procedures. The command went on to develop its own methods. It had already been promulgating its own directives. Now it set up a series of model missions akin to those for heavy bombardment, and over the next several months the directives increased in specificity. They also required more time in formation flying and bombing than did the AAF standard.[32]

Variations in armament and crew composition reflected alternative B–25 missions. Yet until October 1944, training was not directed toward a particular theater of operations, so a crew could be expected to go wherever medium bombers were needed at the time. With the change, most B–25s went to the Pacific, and crews began learning radio procedures in the Far East and eliminating those for the United Kingdom. Japanese aircraft and naval recognition and overwater navigation received emphasis, as did instruction in tropical weather conditions and jungle survival.[33]

Just as the two heavy bombers performed quite differently, so did the two mediums. If the B-17 and B-25 were dependable warhorses, the B-26 Marauder, like the B-24 Liberator, was a difficult, accident-prone ship needing special handling. When he turned his attention to the AAF's fastest medium bomber, the B-26, Jimmy Doolittle, who had chosen the B-25 for the Tokyo raid, recommended to General Arnold "a change in our training methods, because we were training people to fly a normal airplane, and the B-26 was not normal."[34] Pilots expected to take off and land at a normal speed that was considerably less than the B-26's extremely fast 130 mph. The airplane achieved high performance at the expense of considerable wing loading that, along with its rather ineffectual flaps, made for a dangerously high stalling speed. Those in training heard horror stories such as an airplane loaded so heavily in the tail that it would stall at 180 mph, forcing the pilot to land at 200 mph, or other instances in which ice or frost on the wings brought the stalling speed to at least 160 mph.[35] One of Doolittle's 1942 reports called the B-26B "barely acceptable aerodynamically." He continued to paint a portrait of the airplane that was certain to alarm those contemplating pilot training:

> For satisfactory operation, and particularly for training, the B-26 must operate off of long hard runways without obstructions at either end. The take-off is slow and the climb, immediately after take-off, poor. Due to the high win[g] loading and powerful engines, the fuel consumption is high and the range characteristics correspondingly poor. The airplane lands fast and drops suddenly.

Some at headquarters drew even harsher conclusions:

> It is the opinion of this office that the B-26 airplane is most "glitter" and little "gold". Its operational weaknesses are: inadequate fire power; inadequate performance in high speed, climb, take-off and landing; too high a percentage of "out of commission"; and it will probably have a very high crash rate in the theater of operations.[36]

Indeed, the crash rate, both in combat and training, reached monstrous heights. The airplane was so treacherous for beginning pilots to fly that some officers argued that the B-26 should not be flown at all in OTUs. A year into the war only the best graduates of the twin-engine schools transitioned onto B-26s, these assignments even taking precedence over top-priority four-engine training. Moreover, all B-26 pilots were volunteers and anyone who wished to discontinue the program could transfer to B-25s.[37] The Air Staff wrung its hands over the poor morale of B-26 crews. The Bombardment Branch recommended midway into the war that "two experienced crews be withdrawn from combat to tour the medium bombardment training establishments in this country, to demonstrate and 'sell' the B-26 airplane."[38]

World War II

Modifications to the B–26 failed to lower the landing speed substantially, and pilots continued to be carefully chosen and carefully trained. More than once the Third Air Force reminded the senior leadership that the B–26 "requires more flying ability than any other type aircraft."[39] A B–26 training conference in March 1944 urged that pilots be especially well instructed in safety techniques — gaining directional control upon engine failure, operation of auxiliary and emergency equipment, and careful assembly in formation from take-off.[40] That sounded much like the more chatty advice, both in warning and reassurance, given by the officer in charge of B–26 training at a bombardment training center of the South African Air Force:

> For the quick emergencies that might be met each pilot must have a predetermined course of action. He must be ready instantly to cut both throttles if a motor quits just after take-off. This plane makes an excellent belly landing. He should also know when to call for immediate crew bail-out if conditions should warrant it. . . .
>
> The ratio of landing (or take off) speed to normal cruising speeds exceeds that of any known airplane. In flight the high wing loading means there is a short interval of time between when something happens and when the corrective action must be taken. . . .
>
> Caution enters [in formation landing]: This is a steep turn close to the ground. The leader must remember that the plane flying on the inside of the turn is going 10 m.p.h. slower than he, so he must not be stalled out.[41]

In fact, training accident rates dropped, and by 1944 B–26s of the Ninth Air Force enjoyed the lowest loss rate in the European theater, no doubt owed in part to the caution with which pilots were trained and the effort made to build the confidence of crews. The use of the airplane as a night bomber, however, brought a continuing rain of criticism from those flying it in theater — remarks such as "the B–26 is not a night or instrument ship. Forget about night flying. I dread it. Let the RAF do it."[42] Moreover, there is no evidence that the official nickname, the Marauder, eradicated the less-flattering appellations such as widow-maker, flying prostitute, or Baltimore whore (its wingspan was so short that it had no visible means of support).

The training program too experienced inordinate difficulty in achieving the proper focus and predictability of output, given the volatility of the requirements for this type of bombardment. Whereas the B–25 became the chief medium bomber used in the Pacific, the AAF employed the B–26 most successfully in Europe. Not until the end of 1943, when B–26s joined the Ninth Air Force for the upcoming invasion, did airmen begin to discover the mission and tactics most suited to the airplane. And not until later was III Bomber Command authorized to train specifically for the B–26's tactical role on the Continent. To this end the command had to cull most of the B–26s in the training air forces and the Training Command for the training blitz required in

Training for the Tactical Air Forces

the months after the invasion of France. The set of model missions to be accomplished during the equally divided two phases of unit training had to be changed drastically when the training period was cut from twelve to eight weeks. The command eliminated the phase system, reduced flying hours, and revised ground school instruction. All this reached a positive crescendo, allowing the Third Air Force to meet its quotas during the last quarter of 1944, only to find in the new year that it had far more B–26s than the AAF had much use for.[43]

Light Bombardment

> The 3rd Light Bombardment (Dive) Group . . . which used to be the 3rd Attack Group back home . . . had trained for years in low-altitude, hedge-hopping attack, sweeping in to their targets under cover of a grass cutting hail of machine-gun fire and dropping their delay-fuzed bombs with deadly precision.[44]

The venerable 3d Attack Group dated from 1919 when it was organized as the Army Surveillance Group. It patrolled along the Texas-Arizona border then as well as after it was redesignated the 3d Attack Group in 1921. In 1939, despite its pilots' resentment about the name change, it became the 3d Bombardment Group (Light). The 3d finally saw combat for the first time when it joined the Fifth Air Force in Australia in 1942, serving from that time until V–J Day. In the Pacific its pilots flew A–20s, A–24s, and B–25s, bombing and strafing enemy airfields, shipping, and supply lines. Thereafter the group fought in the Battle of the Bismarck Sea and moved with the Fifth Air Force through ensuing campaigns to end the war flying missions to Japan.[45]

The A–20 Havoc, one of the aircraft flown by the 3d Bomb Group, was the most widely employed light bomber of the war, from the time it first began attacking German airfields on the Fourth of July 1942 until after production ended in September 1944. The largest number were G models with the unglazed nose that housed armament; those planes were especially useful in the Pacific in intruder and close support missions. The final variants returned to the traditional bomber nose. The A–26 followed from the A–20, going into service in mid-1944 and showing itself to be an excellent ground attack bomber. The A–26 also had both attack and conventional noses, the former being extremely heavily armed.[46] A–20Gs with gun noses carried a pilot, an air mechanic gunner, and an armorer gunner; a bombardier joined crews of the J model with the bomb nose. The gun-nose A–26 had a pilot and an air mechanic gunner; the bomb-nose version, a pilot, a bombardier-navigator, and an air mechanic gunner.[47]

Just as heavy bombardment training standards set the pattern for medium OTUs, light bombardment RTUs were configured very much like the medium RTUs. Training for the A–20 resembled that offered for B–25s, and A–26

535

World War II

Tactical training in A–20 light bombers included strafing practice.

training was modeled after that made available for the B–26s. However, as combat missions and tactics varied in practice, so, to an extent, did training. The earliest training standard of December 1, 1942, No. 30–3–1, described the first objective of light and dive bombardment to be "proficiency in teamwork between ground and air units [to accomplish] the mission ... in support of ground forces."[48] To this end, the lightest-weight bombers carried a smaller crew and were modified with both the bomb and gun noses of the medium bombers, but unlike the mediums, they always flew low. They served principally, as stated in the training standard's unit qualification requirements, for information gathering, communication, and defensive operations. Crews had to be familiar with ground force equipment and their operating characteristics.

The first training standard also stated it to be the "policy to return to this country, a prescribed number of pilots with combat experience to be used in operational training units for the instruction of combat crew personnel." That policy was mostly observed in the breach during at least the first year of war, and in the case of light bombardment, the lack of any trained men at all, much less those with combat experience, to staff new OTUs meant that training started very slowly. At a time when other bombardment programs had functioning OTUs, air support commands conducted minimal light bomber training for crews and units already committed to combat units, with essentially no personnel left over to build a training program. When the picture changed in late 1943, the whole training system was in the process of shifting from operational to replacement crew training and to diversification of training within all the air forces.

The experience of the III Bomber Command, principally responsible for

Training for the Tactical Air Forces

both the medium and light bombardment programs, provides the most useful snapshot of A–20 replacement crew training during 1944. The command conducted light bombardment very much like medium bomber training, considering AAF guidelines to be too vague, and therefore developing its own directives. It initiated a set of model missions to be undertaken in two phases. The first, lasting four weeks, aimed at pilot transition; during the remaining eight weeks the entire crew flew together. When the command revised its directives on October 1, it eliminated the phase system and prescribed twenty-one standard missions to be carried out at the discretion of local training units. Those missions included pilot transition (day), instrument transition, pilot transition (night), precision bombing above 7,000 feet, low-altitude bombing, parafrag bombing, dead reckoning and pilotage navigation, low-level navigation, radio aids navigation, formation, single-engine operation, air-to-ground gunnery, fixed forward firing gunnery, gun camera gunnery, chemical spray, instrument calibration, weight and balance computations, maximum load take-offs, long-range cruise control, combined fighter-bomber, and preflight inspection.[49]

Earlier, in mid-1943 when both OTUs and RTUs were more a plan than a reality, a conference on light bombardment tactics found that the experience in the North African theater demonstrated that low- and minimum-level bombing against German installations was both ineffective and highly costly in terms of men and airplanes. Participants urged the Third Air Force to train its crews in higher-altitude bombing.[50] (Despite the danger of flying low, a number of A–20 combat crews criticized using the plane as a medium bomber.[51]) In a reversal of the recommended policy regarding the European theater, in the fall of 1944 the III Bomber Command was redirected to substitute minimum- for medium-altitude training and to make maximum use of parafrags, since all A–20 combat crews scheduled to complete training after January 1, 1945, would be sent to the Far East.[52] In short, experience had then shown that German antiaircraft gunners, but not Japanese, were deadly against aircraft attacking at low altitude.

Some A–20 crews still in training at the end of 1944 shifted to A–26s as the whole A–20 program began to phase out when new aircraft came on line. By this time the III Bomber Command had also begun to train A–26 crews separately. The command initially had no bomb-nose A–26s, so it trained bombardier–dead reckoning navigators for light bombers on B–26s at those fields.[53]

The AAF's struggle to achieve an adequate gunnery program had predictably deleterious effects on light bomber training. By the spring of 1944 the situation was improving in that gun camera missions, in which photography recorded the direction and results of fire, began to supplant firing at slow, towed targets. At that time all heavy, medium, and light bombardment replacement crews in the Third Air Force were expected to fly a minimum of four gun-camera missions. In reality, a lack of the requisite assessing devices

World War II

for A–26s and a shortage of instructional materials and fighter pilots to fly the attacking aircraft kept the III Bomber Command from meeting the requirements. Instead, A–26 crews flew four aerial tracking and aiming missions against fighters flying the pursuit curve. Fortunately, the supply of fighter pilots increased when the Third Air Force assigned some directly to the bomber RTUs rather than loaning them from the III Fighter Command.[54]

The hoped-for involvement of combat personnel in light bombardment training came in an unanticipated fashion in the last year of war, creating unforeseen complications. When, in January 1945, the I Bomber Command took over redeployment training of the 319th Bombardment Group, converting it from medium to light, it had to coordinate the training of an existing combat unit with that of new Training Command graduates. The latter ultimately amounted to approximately 70 percent of the whole. The command quickly discovered that recent cadets and battle-tried crews not only differed in knowledge and experience, which affected training requirements for each group, but the men also brought widely divergent attitudes to the business of training itself. According to the command historian, when an operational training base "that has been handling not just novices but casuals" met a "veteran group, proud, individualized, forgetful of continental flying, and not a little contemptuous of it," an immediate collision occurred. Moreover, the older pilots were guilty of numerous breaches of "air discipline": failure to follow traffic instructions from the tower, low flying that "terrified the entire population" of the town, disregard for safety rules, and disdain for less experienced colleagues and for stateside flying altogether. At least everybody had in common a near total lack of interest in ground classes. Few made even a "slight effort to absorb the technical knowledge imparted in the classroom."[55] As in the rest of the training program, heroes and novices alike preferred to fly rather than study.

Fighters

> In the midst of a wild sky, I knew that dogfighting was what I was born to do. It's almost impossible to explain the feeling: it's as if you were one with that Mustang, an extension of that damned throttle. You flew that thing on a fine, feathered edge, knowing that the pilot who won had the better feel for his airplane and the skill to get the most out of it. You were so wired into that airplane that you flew it to the limit of its specs, where firing your guns could cause a stall.... Maximum power, lift, and maneuverability were achieved mostly by instinctive flying: you knew your horse.[56]

Here Chuck Yeager gives expression to the classic Homeric hero of Western tradition, translated into American terms — the fighter pilot as cowboy, fearless and cocky, pistols drawn and looking for a fight. Of course he had to know his

Training for the Tactical Air Forces

metaphorical horse; he was out there riding alone. Others described the fighter pilot's temperamental qualities of ego and drive more directly (and critically) than did Yeager, the man whose laconic style so many pilots would eventually emulate. Sprinkled through the normally staid training directives for fighters were such highly charged words as "viciousness," "vigilance," "aggressiveness," "violence," and "reckless." The fighter pilot must be a man of "belligerent spirit" filled with the "desire to kill."

Training had to rein in the young hotspurs, to teach them to discipline themselves, sharpen their skills, and turn their steeds into effective weapons of war. Over the course of the war, training was provided for all the tactical aircraft: from those advanced pursuit planes available at the beginning — the P-40 Warhawk (last in the series of the Curtiss Hawks), the Bell P-39 Airacobra, and the Lockheed P-38 Lightning — to those deployed later — the Republic P-47 Thunderbolt, the North American P-51 Mustang, and the Northrop P-61 Black Widow (designed specifically as a night fighter).

Each airplane presented a different face to the pilot. The P-40 had a reputation as "a pretty famous ground looper," as one airman put it. Not surprisingly, that tendency did not endear the plane to trainees, although some more experienced men in combat squadrons in the Pacific were happy to trade other planes for it because it performed well where landing fields were muddy.[57] The P-40 remained a mainstay among American fighters during the first two years of the war, coming to public acclaim and affection when Claire Chennault's famous Flying Tigers emblazoned its nose with shark's teeth. The P-39 flew in the Pacific and in Europe, although the nonsupercharged aircraft failed to measure up to competitive enemy fighters at both ground level and high altitudes and in vertical maneuvers.[58] The older pursuit aircraft came to be less in demand than the newest fighters, yet the training squadrons did not inherit all the castoffs. The United States exported a huge number of P-39s to the USSR, for example, so that in early 1943 the Fourth Air Force lost out to the Russians when a number of P-39s destined for training flew instead in the battle of Stalingrad.[59] As for the unusual twin-engine P-38, steep requirements in the active theaters led to its chronic shortage in training, becoming one of the most serious problems in the fighter program.

The famous P-47 Thunderbolt demonstrated its utility escorting bombers but came into its own as a fighter-bomber much used by the tactical air forces. The superb P-51 Mustang capped the search for a long-range escort, thereby requiring its pilots to have greater proficiency in navigation than was necessary when flying shorter-range aircraft in interdiction, defense, and aerial combat. In a modified form as the A-36, the Mustang appeared as a dive-bomber for close air support. Until the much desired P-51s came off the assembly line in sufficient quantity for combat units to have their fill, training units received, from their point of view, an unfairly tiny number. When he was in the First Air Force, for instance, Glenn Barcus lamented that "we just weren't able to get"

World War II

P–51s, "the best fighter in the war, the best.... Those were the days of frustration, too little and too late."[60] Some while later, in October 1944, the Third Air Force asked for 100 P–51s and 200 P–47s modified as two-place trainers. Having none to offer, the Requirements Division planned to substitute 200 P–40s.[61]

Until late 1943 the First and Fourth Air Forces were handicapped because they were responsible for both fighter training and coastline defense. When he joined the Philadelphia Air Defense Wing in December 1942, Barcus discovered that the air defense mission was "very cumbersome and absolutely impossible." Even though he knew that General Arnold gave first priority to training, carrying out both was "a terrible job. We knew we didn't have enough to do the job with, and we didn't have enough people."[62] Besides, the canals, lake ports, coastlines, and towns to be defended were often undesirable training locales. The heavy snowfalls, the many lakes that punctuated the topography and caused treacherous up- and down-drafts, and the restricted space for maneuvers along the Canadian border made winter training for fighter units in Michigan nearly impossible. In New York, some P–47 squadrons practicing to go to war had to fly in and out of LaGuardia Airport in the midst of commercial traffic.[63] Furthermore, the training techniques used for home defense differed from those required for combat operations in Europe or the Pacific. Interceptors working within an early-warning system of ground controllers and a radar net relied heavily on communications for countering invading aircraft, whereas a fighter pilot in battle likely would rely largely on his own navigational skills and take part in ground support or escort duties quite unlike those in effect at home. The split responsibilities of U.S.-based fighter organizations evaporated by the fall of 1943 when the First and Fourth Air Forces turned entirely to training.

In the familiar pattern, AAF headquarters articulated the broad goals, content, and length of fighter training through "standards of proficiency" expected of crews and units on completion of their OTU or RTU programs. The air forces modified the standards when forced by constraints or, in rarer instances, when sufficient time and equipment allowed them to surpass the objectives. In establishing fighter OTUs, on May 2, 1942, the AAF directed key personnel for the new units to attend what was then called the Fighter Command School at Orlando, Florida. The school provided significant intellectual underpinning to the training program, since it had been created to work out doctrine, tactics, and techniques, test new equipment, and develop and standardize air defense operational procedures.[64] One lecture at the Fighter Command School, redesignated the AAF School of Applied Tactics, summarized the offensive fighter methods and techniques to be used in tactical air forces as air combat, strafing attacks, dive bombing, glide bombing, skip bombing, diversionary sweeps, rockets, reconnaissance, and timely coordination. The syllabus differentiated fighter employment in the Mediterranean and

Training for the Tactical Air Forces

The School of Applied Tactics at Orlando, Florida, provided not only fighter aircraft training (the attack version of the P–51 Mustang is seen parked in the foreground) but also strafing practice using A–20 light bombardment aircraft.

European theaters, and for the invasion, and distinguished it also from what was described as "low level in the Southwest Pacific" and "hit and run in China." For defensive action in Europe, fighters provided escort in short-range medium bomber missions and long-range escort in relay. In the Mediterranean, patrol missions assumed primary importance. Fighters provided cover over the target and for ground force and naval operations.[65]

The cadre from the Fighter Command School served as an important link between operations and components of the training system. They learned from the distillation of raw data about tactics in the field, which the AAF also consulted for its training standards, and relayed their observations at the decentralized instructional level of the training units. There the general guidelines could be related to specific practices applicable to the aircraft assigned to a unit. But until theater specialization came into being late in the war, instructors could not, for instance, teach a P–51 pilot only to fly escort or only fighter-bomber tactics. Therefore, training still remained at a level of generalization that left much to the combat units. Even then, as P–40 pilot Bill Colgan learned upon joining the 87th Fighter Squadron in southern Italy, "there was no standard 'book' to be pulled out and studied here on air-to-ground missions — the procedures, tactics, and techniques — any more than there had been such a 'book' on air-ground operations back in training in the States."[66]

Fighter pilots going through OTUs learned the basic skills for all types of missions within a two-phase, three-month period. In the first two months they got to know their own airplane and practiced elementary unit flying. In the final

World War II

month they took up tactical flying. AAF Training Standard 10–1–1 of December 1942 directed that day fighter pilots attain proficiency in transition and familiarization with controls, instruments, and performance of the assigned aircraft; rapid takeoff, assembly, close and open formation, and landing; camera gunnery missions; firing against ground and aerial targets; aerial bombing; acrobatics at operational altitudes; individual combat missions; qualification under instrument conditions; navigation without radio aids to a point at the limit of the radius of the airplane; and the minimally required night flying. The December 1943 standard prescribed familiarity with tactics and equipment of all theaters, yet it noted that special attention should be given to a destination if one was known.[67] Soon thereafter theater specialization became much more the norm.

As typically happened, once OTUs were under way, schedules and training standards devised by the Air Staff ran afoul of the shortages of men and equipment in the field. In the Fourth Air Force, for example, the 354th Fighter Group, a newly activated single-engine P–39 group, did not begin operational training until January 1943. Twin-engine training on P–38s did not start until that April with the 360th Fighter Group. Effectively, OTUs were in session only the one year, whereupon facilities all turned to training replacement pilots.[68]

The fighter RTU program continued throughout the war, and because it was shorter and simpler, it eventually worked more smoothly than the OTU system did. Before the air forces instituted any formal system, however, replacement training too was rather chaotic as individual pilots were drawn haphazardly from existing units. Directives for fighter training had little obligatory impact on a system in pursuit of speed and manpower, one still lacking basic equipment or internal coherence. For example, the AAF's instruction in March 1942 to the III Fighter Command for its conduct of P–38 and P–47 replacement training could hardly be implemented since the P–47 did not arrive until October, and the P–38 not at all.[69] In some instances RTUs were established before OTUs, as happened in the Third Air Force; sometimes OTUs preceded RTUs, as they did in the First Air Force; and sometimes both were conducted simultaneously, as in the Fourth Air Force.

When both programs operated together, they competed for resources. In at least one case, a group was a combination OTU and RTU.[70] The administrative variations interfered with the goal of standardization and made it harder for the air forces to communicate easily with one another. The compulsive demand for people in the combat theaters curtailed the number of flying hours, so most of the pilots who went through replacement training that first year did not receive the full program laid out by directives. Nonetheless, the issuance of specific training guides helped somewhat to regularize replacement training and staunch the erratic flow of men out of existing units.

The III Fighter Command, one of the agencies responsible for replacement

Training for the Tactical Air Forces

training, put out its first memorandum in June 1942. It claimed its purpose to be

> to transform, in the shortest possible time, a graduate of the Army Air Forces Flying Schools from a Trainee into a fully qualified Fighter Pilot who will be capable of taking his place as a wingman in a tactical organization in any Theater of Operation.[71]

The minimum number of flying hours to accomplish this feat began with 40, went to 60, reached 81 in May 1944, then rose to 120 and more. Ultimately the air forces categorized pilots by number of hours flown in training, e.g., 60-hour pilots, 80-hour pilots, and so on. In March 1943, flying time for the 40-hour pilots included transition and familiarization; elementary formation, formation at altitude above 20,000 feet, supervised acrobatics, navigation at altitude, and instrument and night flying; ground gunnery and dive bombing; and aerial gunnery and dive bombing. The additional 20 hours for the 60-hour pilots included a few hours each in acrobatics, combat, navigation, fighter tactics, night and instrument flying, ground strafing, aerial gunnery, and dive bombing. Pilots received additional instrument and transition time in bi-place aircraft before going to fighters. They combined aerial and ground gunnery with dive bombing by dropping bombs en route to gunnery ranges.[72] By the fall of 1944, the number of aircraft and the amount of high-octane fuel had risen such that the Fourth Air Force could be given an allotment for P–38 pilots who had flown 120 hours. The First Air Force could then meet AAF requirements by either training 223 pilots at the 100-hour level, or 200 a month at the 120-hour level.[73]

But until mid-1943 many pilots went into combat units having finished only half the training. Almost none flew the minimum number of hours. The training stations made progress during this period nonetheless. The III Fighter Command successfully reduced the accident rate, increased flying hours, readjusted the curriculum to conform more closely to combat practices, and improved the instruction and training facilities. It also reached its goal of obtaining 600 fighter aircraft for replacement units in January 1944. By then, each squadron could train on a single type of fighter plane (for instance, three squadrons of the 338th flew P–47s, two squadrons of the 53d flew P–39s, and so on).

Although the administration of the fighter program varied from place to place and over time, the tactics taught in training evolved very gradually — more deepened than changed — as information came to light about aircraft performance and shifting circumstances in the theaters of war. Experience brought greater refinement to the study of the best angles for attacking various types of enemy aircraft. Fighter pilots learned general rules of thumb, such as the basic types of approaches and the importance of keeping the sun at your back. *Always* "check your tail." According to Chuck Yeager, pilots had this imperative drummed into their heads from the first days of operational training. As their instructors warned, "The German who gets you is the one you'll never see," and "To be surprised is to be lost."[74]

The basic lessons were not new; fighter pilots had learned them in World War I. But no set of rules, even the most fundamental, always applied. Although a fighter pilot knew in his bones about the advantage that higher altitude gave him when he met his opponent, when a man flying a P–40 faced the much more maneuverable Japanese Zero with its higher ceiling, the enemy inevitably had the benefit of altitude, making an overhead dive at him an unlikely possibility. And even the best and newest American fighter, the P–51 Mustang, flown by a superbly trained pilot, could not close in fast enough to beat the revolutionary jet-powered German Me 262. Thus, besides offering general instructions regarding fighter maneuvers, numerous reports compared American and enemy aircraft, outlining the best methods when any two confronted each other. For instance, in late 1942, as part of the ring of security around Australia, the AAF relied heavily upon its P–39s against Japanese bomber formations and fighters. One report of that time explained how the P–39 pilot should deal with the Zero's maneuverability and greater rate of climb and how to build upon the P–39's strengths.[75] Sometimes all a pilot could do was dodge, and sometimes he had to run. Combat reports described the best techniques for strafing and dive-bombing missions in island warfare, as differentiated from those tactics that worked against German fighters on the Continent.

Fighter as well as bomber training stressed the importance of formation flying. Early in the war tactical squadrons flew a three-element formation. But they soon learned that the three-ship V left one fighter without rear protection. The AAF quickly adopted the RAF's staggered four-ship formation consisting of two mutually supporting two-plane elements. That arrangement, or at least one of the pairs, became the standard in training.[76] Most of the time pilots took off in twos, owing to the size of training base runways, but they attempted rapid landings with four-ship flights. Pilots in training were not expected to fly in formation under instrument conditions, however, for fear that excessive fatalities would occur.[77]

During much of the war, escort fighters flew defensively. Fighter and bomber pilots received the same kind of instruction regarding formation flying: stay close to one another and close to the bombers. As P–40 pilot George Preddy recorded in his diary from northern Australia in May of 1942, "[A]ll flights will stay together as much as possible in the air; also we will try to stay organized as a squadron instead of being scattered all over the sky — a good idea!" Nearly two years later, now flying across the English Channel, Preddy still described the four-man flight as part of a larger whole in which "the whole outfit is in very close."[78] One training intelligence report claimed that the "best life insurance a fighter pilot has is a close, tight formation."[79]

Training for the Tactical Air Forces

By the time that report was aired in May 1944, however, instructions were changing. Fighter pilots themselves had argued they should be allowed to move from the bomber formation to destroy the enemy as well as, in fact, *protect* the bombers. AAF leaders now directed long-range escorts to confront the Luftwaffe. Eighth Air Force General Spaatz reasoned that if the bombers attacked oil targets, the Germans would engage and fight — and thereby be defeated. Thus, American fighters switched to the offensive, allowing individual pilots to exercise greater freedom of action in hunting down enemy fighters. For escort fighters, the World War I dogfight had returned.

The lessening of German opposition in the air brought significant change to the air war as the air forces provided more support to ground-based objectives and directed more of their energies toward the Pacific campaigns. The fighter employed as a fighter-bomber came to the fore. Between November 1943 and May 1945, the Ninth Air Force flew approximately 70 percent of its fighter sorties as fighter-bomber missions.[80] Training too took greater account of the tactics used in attacking ground and naval targets. Offensive maneuvers of strafing, dive bombing, and skip bombing at low altitudes joined high-altitude training for defensive action. Fighter pilots heard the predilections of different combat squadrons regarding angle of attack, bombing techniques, and evasive action. "It's a very different war at fifty feet off the ground," Chuck Yeager later commented. Dogfighting "on the deck" was perilous at best. In fact, George Preddy, whose career mapped much of the aerial war, began flying in the southwest Pacific almost immediately after Pearl Harbor, then flew P–51 escort for Eighth Air Force bombers, and finally turned to strafing trucks and bridges in ground support. He was killed on Christmas Day, 1944, when his own troops mistakenly fired on him as he appeared unexpectedly over their heads, barely above the trees, in pursuit of a FW 190.

Fighters gradually assumed some of the ground support activities of light and dive bombers. In operations at Guadalcanal, P–39s dive-bombed Japanese task forces and provided cover for the Navy's SBD dive-bombers (similar to the Army's A–24). The Army never developed any great enthusiasm for dive bombing and withdrew its A–24s from action, considering them too slow and short-ranged to be useful. Mostly, it left dive-bombing operations to the Navy. Since September 1942, General Kenney's Fifth Air Force had been using A–20s to conduct low-altitude strikes with parafrags armed with instantaneous fuses. Low-level skip bombing practices with light bombers continued in the Pacific, but in Europe the large attack bombers proved extremely susceptible to ground fire. Although the AAF did not phase out its light bombers, it found over time that many air support activities could realize greater success if two-seat bombers were replaced with high-performance, heavily armed single-seat fighters.

At home, training attempted to incorporate the changing combat practices. The III Bomber Command had an OTU memorandum for dive bombardment

World War II

Training in low-level bombing from A–20s prepared men to serve in the Pacific, where A–20s were used extensively.

in place by June 1942, but most pilots who would fly fighter-bomber missions trained in the fighter replacement programs. A handful of groups trained specifically for fighter-bombers — an "advanced" RTU, as it was sometimes called — that was essentially an add-on to the basic 60- to 80-hour fighter curriculum of the time.[81] The III Fighter Command assumed this training, appropriately since the III Bomber Command had been conducting light and dive bombardment, and the fighter-bomber groups had all been activated from those two types. In August 1943 the AAF redesignated some of its attack units as fighter-bomber groups and equipped them with fighter planes. During the fall the III Fighter Command transferred three groups: one went to California (Fourth Air Force) and thence overseas to the Eighth Air Force; two went to the Second Air Force, where one became a dive-bomber force deployed to Alaska. The three groups retained by the III Fighter Command that saw overseas service went to the Ninth Air Force in the British Isles early in 1944.[82]

Fighter-bomber training expanded upon the fighter replacement curriculum by adding a third phase of 60 additional flying hours and 50 extra hours in ground classes and a fourth phase of maneuvers. The III Fighter Command's training memorandum prescribed the additional flying hours to be spent in formation and combat exercises at 1,500 to 20,000 feet and at high altitude; acrobatics and combat above 8,000 feet with one hour to be held above 20,000 feet; and navigation at various altitudes. Gunnery practice at aerial and ground targets increased; pilots logged two hours in ground strafing. The curriculum required hours in night flying, formation, instruments, and navigation. Pilots practiced dive bombing, skip bombing, and low-level missions. The fourth

Training for the Tactical Air Forces

phase encompassed maneuvers with other Army units, accomplished by groups during the fall and winter months of 1943. During their training, the units anticipated flying the aircraft they would be assigned in combat. As with so much training, that hope failed to materialize. The five groups who began the fighter-bomber program in August operated seven types of planes, but some of the last groups to depart the III Fighter Command flew mostly P-39s and P-47s.[83]

At the end of July 1944, AAF headquarters reminded commanding generals in the field that graduates of operational training "might be sent to any theater upon completion of their training" and that there was "a limiting factor in the amount of time available for the training."[84] But as battlefront conditions changed, and as stocks of equipment and manpower levels steadily grew, by early fall General Harper was sending a different message regarding theater specialization. No longer would the air forces train fighter pilots for all eventualities. The First Air Force should "exert every effort" to prepare for the air war in Europe; returned pilots from that theater would be assigned to the First and traded from other air forces. Specialization for the European theater entailed extensive practice in long-range formation flights with auxiliary tanks and ground study of German tactics, equipment, and geography. The Second Air Force, on the other hand, mostly trained P-47 pilots for the south and southwest Pacific, China-Burma-India, and Hawaii. Pilots strafed and bombed at low rather than high altitudes and learned to recognize relevant enemy targets and friendly naval forces. They completed between 120 and 150 flying hours before debarkation.[85]

As elsewhere, problems with gunnery retarded progress in the fighter program. Adequate range facilities, sufficient ammunition and equipment, useful training aids, and consistent doctrine all developed slowly. The training memorandum of January 1943 specified 20 hours of gunnery in each of two phases. Practice included ground firing at offshore oil slicks and targets, dive bombing, and bomb dropping; aerial gunnery included firing a minimum of 500 rounds of ammunition, using five bombs, and firing another 500 rounds at high altitude. This curriculum stayed intact in the ensuing training directives, except for slight changes in the hours and scheduling and elimination of the phase system of training.[86]

Training directives were more likely to be put into practice as equipment shortages decreased and theory advanced. In the fall of 1943 conferees discussing fixed gunnery concluded that first and foremost pilots needed to understand a simple rule of thumb: targets were either "in range" (within 1,200 feet of the target on opening fire) or "out of range."[87] Gunsight-aiming cameras radically improved the evaluation of deflection firing; when they became available, the AAF added gun camera missions to the curriculum. Because the gunnery ranges were overbooked, in January 1945 headquarters eliminated the gunnery requirement below the 100-hour level so that pilots took gunnery only

World War II

once at an advanced point in their training.[88]

One of the chief difficulties in day fighter training, both in the Training Command and in the squadrons, lay in the acute shortage of twin-engine P–38s. The airplane had been designed for speed, and at the beginning of the war it held the distinction as the fastest American-built fighter. The British bought early models of the Lightning, but by the time the United States joined the conflict, the AAF had modified the aircraft to convert an interceptor to a first-line offensive fighter. The F model of February 1942 added drop-tanks for longer range and new Allison F–5 engines for greater speed. Increased horsepower and additional fuel tanks came with subsequent modifications.[89]

The airplane's three-pod configuration, with the pilot and armament in the middle and the liquid-cooled engines on the outside booms, was an innovative departure from the conventional single-engine fighter. But this redoubtable fighter of unusual design had no backup training version, and precious few operational aircraft could be spared. As of January 31, 1942, for example, nine P–38s were availabale on which to train a hundred pilots of the 55th Pursuit Group. That situation improved very little until well into 1944. For transition and instrument training, the air forces borrowed a handful of twin-engine AT–9s and also some AT–6s from the Training Command. Usually however, in a practice with obvious drawbacks, most fighter pilots heading for twin-engine combat groups had learned to fly single-engine P–39s. Common to the two airplanes were a tricycle landing gear and the Allison engine.[90]

Compounding the distressing shortage of aircraft, trainees were said to be "scared to death" of the P–38, and with good reason since an unusually high number of pilots went to their deaths flying it. Some of the pilots from the 39th Fighter Squadron (with the Fifth Air Force in New Guinea at the time) admitted that on their first P–38 flight in training "their knees were shaking so badly that they could not hold the brakes. In some cases, instructors have told groups of pilots that some of them would be killed before they had all checked off."[91] The training system could potentially rectify the problems caused by too few maintenance personnel, faulty training leading to pilot error, and insufficient training for crew chiefs and mechanics. It could and did mount a public relations campaign to reassure pilots about the virtues of the airplane. But it had to await modifications of structural weaknesses in the aircraft and engine that contributed heavily to the crashes. Investigations revealed that the stress of high speed caused loss of control, difficulty in recovering from dives, engine failures, and even a tendency for the airplane to come apart.[92] When those defects were corrected, accident rates dropped.

All types of training required some number of night missions since a pilot flying combat sorties in the northerly latitudes might have to take off or land in darkness. But the fighter program was the only one to train specifically for night fighting and for which a separate curriculum was developed. Northrop designed the P–61, alluringly named the Black Widow, specifically for the

Training for the Tactical Air Forces

The Northrop P-61 Black Widow, the first airplane specifically developed for night flying, was in scarce supply in the training program because of urgent combat needs.

task. It was the only American fighter with two- or three-man crews, all of whom were volunteers. But it took time to field the airplane and put together a training program that would, as it turned out, have little access to the aircraft.

The AAF had not waited to be bloodied before deciding that it needed a night fighter. Before U.S. engagement, it watched the RAF struggle to defend English cities and towns with an insufficient defense warning system and aircraft virtually unable to operate under blackout conditions. C. F. Rawnsley, who flew as a gunner in an RAF night fighter, later wrote of the harrowing period called the phony war:

> But although the Germans failed to show up, we did not lack an adversary. We had no homing beacons and there was no system of blind approach, no way in which we could be talked down to a safe landing. Our radio was feeble and short-range, and the blind flying instruments were astonishingly temperamental. Our pilots fought a war that was far from phoney against an enemy that was much deadlier than the Luftwaffe. Human frailty and inexperience, and inadequate and unreliable equipment joined forces with the relentless and ever-present law of gravity, and a foe so implacable just had to be given a name. We called him Sir Isaac Newton.[93]

Although convinced of its own ability to fight a successful strategic air war

World War II

against Germany during daylight hours, the AAF saw the daylight raids between England and Germany come to a halt after the Battle of Britain. It may have been that experience that in late 1940 sent Air Corps representatives to talk to Jack Northrop about designing a twin-engine night fighter with means to "see" enemy aircraft in the dark. Although work began immediately, the Eighth Air Force based in England did not wait for the outcome but asked the RAF to help train American night fighter squadrons. Once the AAF joined the air war in earnest, it too experienced strikes by night. In its postwar summary the Ninth Air Force discussed the impact of night bombing and intruder operations, which had been especially acute in the European Theater of Operations "where the enemy largely carried out his movements at night and generally went unhindered because of the small available night force."[94]

The United States had less reason than the British to fear attack upon home shores, but in June 1942, as part of the flurry to set up an American antiaircraft defense system, the Interceptor Command School (later known as the Fighter Command School and the Army Air Forces School of Applied Tactics) added Section X, the Night Fighter Department. A small group led by its director Maj. Donald B. Brummel and assistant director in charge of tactics and techniques, Capt. Leonard R. Hall (who had a background in electronics), went to England to glean what they could about the British night fighter organization and radar equipment. They returned to start an operational training program at the school.[95] This endeavor demanded a great deal of night fighter pilots. In late October, the Air Staff drafted a mission statement for the Training Command to use in selecting pilots (optimally, advanced twin-engine graduates with some background in fixed gunnery):

> The mission of night fighter pilots is the interception and destruction of hostile aircraft at night while operating in accordance with the instructions of a ground controller and the airborne radar operator in the night fighter. The night fighter pilot is therefore required to operate mainly at night under blackout conditions, flying entirely by instruments until within visual range of unlighted hostile aircraft.[96]

Clearly, instrument training dominated.

The first P–61, without radar, arrived in the training squadrons in September 1943, more than a year after the training program began, and the night fighter squadrons then in training went to the southwest Pacific without having flown it. The school had been unable to give dual transition on instruments or airborne intercept (AI) radar because it lacked equipment, nor could it substitute Link training because it lacked simulators. It made do with advanced training aircraft, and at that had to beg to get them; it hoped to get P–38s as a stopgap measure, but these aircraft too were in pitifully short supply.[97]

As 1944 began, the AAF reassigned the night fighter operation to the

Training for the Tactical Air Forces

Fourth Air Force because conducting an extensive training program interfered with the school's purpose of training only key personnel and developing tactics, techniques, and doctrine for air warfare. The ground-controlled interception (GCI) aspect stayed at the school, although trained ground controllers were attached to the Fourth Air Force as part of night fighter training.[98] In California the training continued, once again largely without the P–61. In May when Col. Ralph Snavely, commanding the 319th Wing that supervised the night fighter training fields, pleaded for P–61s because no RTU pilots were being transitioned onto the aircraft, the Fourth Air Force backed him up, adding that a total of only eight P–61–type aircraft were used by the two OTU squadrons in the command.[99]

The Fourth Air Force issued its first night fighter training regulation in June 1944. It elected a three-month, three-phase system to be given at three separate fields, followed if possible with unit training. The first, or primary, phase included familiarizing the pilot with A–20s and whatever twin-engine instrument ship was available and instructing the radar observer at the technical school in Boca Raton, Florida. In the second, or basic, phase, pilots and radio observers came together as a team for an introduction to day and night radar interception. Directives prescribed 57 flying and 43 ground hours. Crews spent daytime hours learning GCI, AI, air-ground gunnery, instruments, and the SCR–720 radar. They spent 25 hours at night working on AI and 10 on navigation. Mostly they flew an A–20 conversion, the P–70, half the time above 15,000 feet, half below 10,000 feet. The third phase comprised tactical flying, mostly at night, again with the P–70 equipped with SCR–720 radar. Here, when a pilot flew acrobatic maneuvers in the dark, he relied entirely on instruments. (However, acrobatics in the fairly heavy twin-engine P–70 was problematic even with very well trained pilots.)

Because night fighters drew crews from the Training Command, which possessed even less equipment and expertise than the OTUs had, the training units spent valuable time in transition and retraining. However, by November 1944 sufficient numbers of radar observers had joined the units to be included in the primary phase. By then, more P–61s had come to training units, RTU replaced OTU, and a training cycle took place on one field. By 1945 the quality of radar observer training improved, owing to the inclusion of combat returnees who contributed a perspective lacking in those who had learned only in theory. Some P–61s were equipped with gun turrets, bringing a new emphasis on flexible gunnery. In June the Training Command's flexible gunnery school began an eight-week course in P–61 gunnery, but it too suffered from lack of equipment.[100] When gunners joined crews in RTU, they were not scheduled for much flying time.

Despite the greater number of P–61s going to training units, into 1945 most trainees still flew A–20s or P–70s. High accident rates prevailed in these generally worn-out aircraft; they tended to go into flat spins that resulted from

World War II

high-speed stalls. Because the night fighter pilot was expected to recover from spins and other unusual attitudes by instrumentation, a matter of enormous difficulty for still-inexperienced pilots, these were life-and-death weaknesses. The ballyhooed P-61 also evidenced design and performance problems, requiring much modification throughout the history of the night fighter program.

The most satisfactory solution to the absence of P-61s or any other two-place fighter aircraft for training was the development of the piggyback P-38. Modifications enabled it to carry two men by attaching the radar scope and control panel to the back of the pilot's seat. Although training squadrons of the IV Fighter Command could not anticipate a wholesale conversion of P-38s to night fighters, they were delighted by the performance characteristics of the P-38, which they considered to be superior to the P-61.[101]

The AAF also modified P-38s and other fighters and bombers for reconnaissance. The III Reconnaissance Command provided photo, tactical, weather, and liaison reconnaissance training, using the OTU-RTU system. The ideal aircraft for tactical reconnaissance was highly maneuverable, had good visibility, and was capable of flying at medium range (1,500-2,000 miles) and altitude. Pilots assigned to tactical reconnaissance were supposed to train on the P-51, the aircraft they expected to fly in combat, but the shortage of P-51s left them with P-39s and P-40s instead. Training emphasized gunnery, instruments, and proficiency in directing the adjustment of artillery fire. The P-38 was the airplane of choice for photoreconnaissance, but because it was virtually unavailable, the RP-322 (a training variant) often substituted. For night photoreconnaissance, the absence of the P-61 meant trainees flew A-20s, in which pilots took a number of hours of instruction in aerial photography and increased instrument training for night flying.[102]

Assessments

> Let's don't just consider me as a prejudiced fighter guy. I will readily admit to that, but I did start out in this whole strategic concept and believed in it. But it is just a fact of life that fighters shoot down bombers. You know, I can go write my name with machinegun fire on the wings of any bomber right now.[103]

Fighter pilots' tendencies toward independence and bravura pointed up a dilemma for the training program at large that went beyond personality — safety. Selecting men with what was considered the right attitude and preparing them to risk their lives in aerial combat competed with the need to hold the number of training accidents to a tolerable level. That tension blew as a hot wind through all types of training, but it was more acute in the fighter program because of the requirement for acrobatic flying. A communiqué from the Seventh Air Force to the Commanding General, AAF in October 1944 clearly

laid out the uncomfortable choice:
> Replacement pilots received here display considerable weakness in acrobatics, many of them admitting that their training in the subject consisted of barrel rolls, chandelles and lazy 8's. Loops, immelmans, split S's, or formation acrobatics of any kind were prohibited. However, the training standard requires (and combat will likewise require) every pilot to be able to properly execute at operational altitude *all* acrobatics permitted in the airplane with which equipped. This condition, although conducive to good safety records at home, increases the accident rate in the theater where the results are most acutely felt.[104]

At the time of this statement, the experience level for replacement pilots going overseas had reached of 120 hours, and theater specialization was under way. Finally the training program was not hostage to the exquisite mercy of time. Even so, welcome sufficiencies did little to offset the dangers of acrobatic maneuvers and low-level flying called for by the turn toward offensive tactical warfare. Would the service choose for the inevitable fatalities owed to these tactics to occur in theater or at home?

The AAF had struggled with that issue from the beginning. Brig. Gen. Barton K. Yount, as head of the Training Command, wrote to General Arnold early in the war that they were in the process of "overhauling" the training given to instructors regarding acrobatics practice. The command had to prepare pilots for the more demanding next phase, OTUs, because "we are all in the same school, and any public school system which did not coordinate the grammar school and high school would be pretty 'punk.'"[105] Once they received students from Yount's grammar schools, OTUs tried to curb accidents by levying severe punishments on any pilot engaged in unauthorized dogfights, or anyone caught buzzing civilians, and by outlawing all acrobatics at low level.

Conferees at a meeting in the spring of 1944, concerned with simulated attacks on combat aircraft, suggested that all training breakaways in fighter-bombers be started *before* reaching a minimum range of 200 yards. The Fourth Air Force had, in fact, been keeping beyond a 1,200-foot approach with attacks broken off at 300 yards, and the Second Air Force had a 500-foot ruling. The air forces admitted that little use could be made of camera guns at those ranges. Yet, after the conference General Harper, chief of training on the Air Staff, seconded the training restrictions, since the "factor of safety must be paramount."[106] Fighter training units also disallowed operation at maximum power settings because training aircraft tended to be old and worn and usually operated on 91-octane gasoline, a deadly combination.[107] Furthermore, in order to fly at altitudes above 20,000 feet on 91-octane fuel, airplanes were stripped of armor plate and other impediments. Such an airplane handled differently than it would fully loaded in combat, further diminishing realistic training.[108]

The safety issue induced particular nervousness because the AAF felt the sharp sting of public and congressional criticism whenever the press ran stories about the high number of injuries and deaths incurred during air training. But the AAF leadership knew they would pay a price in lives, one way or the other. As expressed by the Seventh Air Force, the service's safety record had been achieved "at great expense not always readily apparent and should not be permitted at the expense of adequate training for combat."[109] In one of his memoranda, novelist James Gould Cozzens, then a major assigned to the Office of Information Services at AAF headquarters, editorialized that "valuable training would result from removing the present penalties on buzzing air-fields, control towers, cows, and goddam civilians driving cars down lonely roads."[110]

Training officials arrived at no neat solution. Nonetheless, given its relative affluence in time and equipment, the AAF decided in early 1945 to authorize more low-to-the-ground, on-the-deck fighter training, expecting that an increase in fatalities would occur. It did. The Air Staff anxiously braced for a reaction but, by summer, had heard relatively little. By then the service was scaling back its pilot training, and the files bulged instead with letters of complaint from parents, sisters, friends, and congressmen about young men accepted into pilot training who waited uselessly in a personnel pool or who found themselves reassigned to other Army duties.

No aspect of flying training escaped criticism by somebody: aircrews had not flown enough high-altitude missions; aircrews had not flown enough low-altitude missions; pilots could not use instruments; fighter pilots were leery of acrobatics and afraid of low-level dogfights; bomber pilots knew little about enemy fighter tactics; flexible gunners shot holes in their own aircraft; navigators got lost; bombardiers missed the target by miles; aircrews had too little aggregate flying time; pilots were undertrained in the techniques of strafing and bombing; crews were inept at emergency procedures; aircrews were not familiar with the equipment of the tactical planes they would fly; fighter-versus-fighter and fighter-versus-bomber operations were almost unknown; fighter pilots were weak in deflection firing, and in general, facility in gunnery by aircrews was woefully bad; target recognition was poor; large-scale formation flying was nonexistent; the seasoned advice of combat-experienced returnees was often ignored; the judgment of combat-experienced returnees was often immature.

Those frequently correct if sometimes conflicting accusations did not come as news to people at home. The training establishment left few stones unturned in its attempts to learn from the experience of field commanders and young airmen flying combat missions. Failures in training derived not from lack of information, under which training officers were buried, but from starvation of other kinds. A corresponding list of grievances from the training side of the house might have included lack of time, expertise, aircraft, instruments, armament, ammunition, gasoline, training aids, and facilities. Speed was paramount; according to a training directive in May 1942, "the objective of *all* training is to develop *as quickly as possible*, units which can successfully engage in combat."[111] Training officials acknowledged that problems, some of their own making, owed to poor maintenance, indifferent instructors, sloppy procedures, bad weather, constant modifications, and shifting requirements. They predicted the repercussions for training resulting from such handicaps, but they found solutions to be less obvious and often imperfect in outcome. How, for instance, could a program training huge numbers of men in several skills under wartime conditions systematically redress lapses in readiness and morale such as the one attributed to medium bomber crews: "[T]here is evidently a lack of finality of purpose. In other words, men are not given a definite goal to shoot at.... It is as important to build up mental assurance of a man's ability as it is to do the flying, the bombing, the gunnery, and the navigation."[112]

The principles and practices of training seemed to be fundamentally sound, but training officers would have made mistakes under any circumstances, in part because they responded to rules crafted by others and to situations outside their immediate control. Tactical doctrine, the basis for training, remained to be articulated after experience in the field and from historical precedent, not from the preconceptions of air power theorists. In summarizing its wartime experience, the III Fighter Command added to its list of factors contributing to the high accident rate the "lack of uniform and progressive training doctrine under accelerated training requirements."[113] Not until well into the war did the AAF fight with a well-balanced force and a flexible doctrine that underscored the foremost need to obtain air superiority before any further aerial warfare could be successfully sustained. In May 1944 one officer wrote to another on the Air Staff that they had finally learned that survival was owed "solely" to "superior position irrespective of how it is gained."[114]

But for the AAF leadership, those in training at home, and those fighting overseas, more was at issue than the development of battlefield tactics and doctrine. They had politics and emotions at stake. They fervently believed that advanced aeronautics and electronics technologies would allow them to transcend the limitations of the earthbound soldier. They thought they could shorten, even win, the war. They longed for professional recognition. They hoped to see their independence realized. They were human.

An End and a Beginning

> The [23 November 1944] memorandum rehearses the history of trainee procurement, with emphasis on the necessity of very large procurement to start with. No more than 24% of the candidates could meet the high mental and physical requirements. We had no data on replacements, and so used British figures.... "The survival factor was to be implemented by sending two replacement crews to the theatre for each crew lost in combat." The cut-back in the Training Program was initiated because "as early as December 1943 it could be seen that the Training pipeline was too large." This August, "due to the satisfactory progress of the European war", the reduction to the 20,000 rate was made and it was further decided that the November entrance into primary training would be at a 10,000 rate....
> —James Gould Cozzens, *A Time of War*[1]

The huge manpower buildup required at the beginning of the war was so successfully managed that midway through the conflict the AAF began a retreat from the numbers. Maj. J. G. Cozzens, in the statement above, summarized the mushrooming but unsteady progress of procurement, followed shortly by reductions in air training up to December 1944, a process that had been under way for some time.

In January 1944, AAF Chief of Staff Brig. Gen. Walter Krauss informed his commanding general that so many men were still enrolled in training programs that lowering the quotas as Arnold wished would have to be done gradually. The war had not ended, so downsizing occurred erratically, not only because of the many men already in the pipeline, but also in response to the shifting course of events in the theaters of war that kept the service from cutting back uniformly. For example, Arnold directed that effective March 30, 1944, college training for aircrew would be eliminated, but the AAF increased rather than decreased the B–29 program. Also, crew training for heavy bombers

Conclusion

remained active.[2]

At the same time as they juggled numbers and attempted to pacify irate congressmen and their constituents who complained about the unfairness of removing boys from air training, those in Washington and in the Training Command responded to the ongoing requirements from the theaters of war. The flood of reports citing deficiencies in training and the urgent pleas to upgrade proficiency standards poured in, even though most of the air leadership sympathized with the problems of matching training and operations during the complicated last phases of a global war. The Eighth Air Force Commander, Maj. Gen. Ira Eaker, for instance, wrote to the Director of Bombardment at AAF headquarters, Brig. Gen. E. L. Eubank, that "it gives me high hopes for the future to know that you are working hand-in-glove with us back there. To date we have had one hundred percent support from you and your people."[3] The process of reducing quantity and increasing quality of training continued to the last day of the war.

That day came with the capitulation of Japan in the aftermath of two atomic bombs dropped by the U.S. Army Air Forces. Airmen had, they believed, finally and triumphantly demonstrated the capability of their strategic forces. There appeared to be no further reason for restraint in proclaiming the preeminence of air power, as the Commanding General of the AAF forecast in his final report: "In any future war the Air Force, being unique among armed services in its ability to reach any possible enemy without long delay, will undoubtedly be the first to engage the enemy and, if this is done early enough, it may remove the necessity for extended surface conflict.[4]

Reinforcing Arnold's and his predecessors' notion of a new military equation, in 1947 the Army Air Forces became the independent United States Air Force. Yet the military men who flew airplanes from the early days of the century through World War II lived with a different reality from those who followed afterward. The mission in the nuclear age would demand an unwavering state of readiness; deterrence would obviate doctrine based on old-fashioned concepts of aerial combat; the strategic role would overpoweringly dominate the tactical.

When it came to training, however, the Air Force learned from and retained much from the first generation of its existence. Before the Great War, Signal Corps officers had formulated a hierarchical system of dual instruction for pilot training, beginning with primary and going through advanced stages. During the last phase, pilots were supposed to fly tactical aircraft and train with troops of the line. When the number of people and the size of budgets permitted, airmen implemented that system, and though subject to experimentation and administrative alterations, and although joint training remained fragmentary, it endured fundamentally intact into the era of the independent Air Force. The pace, but not the process, of pilot training changed over time.

As early as the air arm was able, it sent out specifications for military aircraft that differentiated between training and tactical airplanes and between types of tactical aircraft. Yet the U.S. military, compared to European air forces, was slow to institute specialization, in large part because American Army doctrine held that aircraft were useful solely for reconnaissance. Even so, by the outbreak of World War I the U.S. Aeronautical Division had managed to procure variations of the Scout reconnaissance airplane. Once into the war, American pilots flew innumerable different aircraft, and afterward the Air Service itself institutionalized specialization in air training. That approach, whereby pilots specialized in one of several missions with the aircraft supposedly appropriate to it, became permanently embedded in the American training system. Based upon its World War II experience and organization, the independent Air Force created a functional command structure that continued into the post–Cold War era.

Both increased specialization and evolving doctrine waited upon technological advances and shifting national security policies. Beginning with an emphasis on reconnaissance, dating from the use of balloons in the Civil War, the Army gave lip service and most of its people to observation aviation during World War I, while reveling in the role of the gladiator of the air, the fighter pilot. The Air Corps created a culture built around that Homeric hero even as it publicly espoused cooperative ventures with the line Army and trained for what it considered the more prosaic tactical roles. The development of aircraft capable of going fast and high and carrying a heavy load, and successful experimentation with new navigational and bombing equipment during the 1930s, allowed the Air Corps to redefine itself in terms of high-level, precision bombardment. During World War II the AAF banked its reputation on the efficacy of strategic bombardment, although it trained much of its force for tactical roles that they performed successfully throughout the war.

In an important sense, specialization and the World War II experience put the linkage between air power doctrine (upon which training was based) and operations to the test. Despite the avowed primacy of the strategic mission by the late 1930s, faculty at the Air Corps Tactical School had long recognized that victory in the air often might depend upon a variety of aircraft employed in any given mission. The school therefore placed the Air Force course, admittedly taught at a rather high level of generalization, but one which emphasized cooperative forces, at the heart of the academic program.

Once into the war, American operational practices bore out that approach. Fighters ranged alone and also escorted bombers; tactical air forces comprised mixed groups of aircraft working alongside one another. Training, however, was not organized that way. Composite forces too seldom trained together. Rather, training tended to remain discrete for each type of aircraft. And because directives repeatedly cited high-altitude, daylight, precision bombardment as the cornerstone of the air forces' mission, training followed accordingly. As a result, air commanders in the field were working out concepts of balanced

Conclusion

forces, whereas training practices had changed little from World War I when tactical squadrons were trained and deployed according to weapon type. Moreover, a doctrinal rigidity set in that outlasted the war.

At all times during military aviation's first half century, the number of people and the progression through the training sequence depended directly on the amount and type of equipment on hand. The disparity between training and tactical aircraft during wartime introduced an additional and very serious complication. Even at a time when theory was discounted in favor of practical flying experience in order to train aircrews in as short a time as possible, men frequently went into combat without a technical mastery of the equipment, or experience with the handling characteristics of combat aircraft, or an understanding of the tactics used on missions they would fly. No internal variable in the training system ever overcame the problems caused by equipment shortages. What had become abundantly clear to airmen by the end of the Second World War was that efficiency in training and the proof of its success during wartime would depend largely upon type and availability of equipment.

Airmen also learned very early that military air training consumed an inordinate amount of time and considerable manpower wastage, and therefore in emergencies it was not possible to field a trained force quickly. From 1909 through the first months of World War II, the air force lived with chronic personnel shortage. The Army, under the dictates of Congress, imposed manpower ceilings during aviation's first twenty years. Once the fiscal purse strings loosened, it still took time to feed the voracious hunger for trained men.

Nonetheless, even when it was subject to enormous pressure to produce trained airmen quickly, and even when it had money, the air arm *never* accepted all comers. It firmly believed that a certain type of man was best suited to aviation, and it could not afford to take the others. Not only did flying call for special physical acuity, but the inordinately high fatality rate in air training proved that the service had to select its people with particular care. The air force devised various means to determine aptitude, always to find that the measure of a pilot could never be taken scientifically. Airmen *knew* who were good pilots, just as the Wright brothers knew and taught their pupils to *feel* the wind and the camber of the wings and the sound of the wires. That reliance on self, instincts, and personal experience to achieve technical mastery imbued air force culture from its infancy. Despite the constant drive for standardization in air training, airmen believed in their hearts that they were engaged in an individualistic, improvisational enterprise for which only an exceptional few possessed the temperament, talent, and luck.

That attitude helps to explain the service's near-total domination by the pilot, even as the air force expanded in size, professionalizing and developing a broader corporate identity. The changing emphasis in pilot training reflected a growing maturity, beginning with its earliest days when a handful of self-taught young officers learned to fix and fly their own airplanes. They went on thereafter to establish specialization in types of aircraft and standardized training methods in each. Then, during the GHQ Air Force years and through World War II, the program emphasized crew training and coordination of tasks. That meant, of course, that most pilots flew with others. Medium bombers held a crew of three or four, heavies usually carried eight, and the B–29 had a large cast. At the most, one or two wore pilot's wings. But it took an excruciating demand for personnel on the eve of World War II for the Army Air Forces to jettison its preference for an all-pilot flying force. For the first time, most combat airmen were not pilots. Yet after every war the service rushed to eliminate rated nonpilot specialties.

Faith in the war-winning promise of aviation technology clearly differentiated the air force from its ground-based parent. Since they could not train or develop tactics for equipment they did not have, airmen's love of the machine and a belief in its military capabilities sometimes pushed them to make pronouncements without a proven basis in tactical experience. One could theorize without equipment even if one could not practice without it. As a result, fliers were sometimes subjected to doubt or harsh criticism from their practical-minded superiors who were schooled to marshal claims for victory around observable battlefield wins and losses. Airmen, on the other hand, retreated from those traditional definitions of success, and air training essentially became technical training in service to a vision, as airmen recast their concepts of professionalism. In the pilot, the quintessential twentieth-century American figure, the impersonal mask of the technocrat overlay the face of the maverick, the man who wrote his own rule book. His was a very different character, composing a different portrait in leadership from his military parent.

Military training is, of course, fundamentally not an intellectual exercise, and insofar as the air force was a technically oriented organization, the task of the training establishment was to teach people how to perform complex tasks using complex equipment. Also, military organizations customarily distinguish training from professional military education. Yet the air force, at least during the period under study, almost disdained education for its flying officers, partly for the good and substantial reason that the service was inventing a new field of highly dangerous military operations that engaged all its resources of time, personnel, and money. And it felt an airman (read pilot), like the professional athlete, must stay in training at all times. For a man to divert his energies elsewhere in ground pursuits such as the classroom was potentially to lose his life.

Beyond the intellectual boundaries of the individual airman however, training practices are connected to and justified by doctrine. An evaluation of the application of the classical principles of war such as objective, mass, offensive, surprise, mobility, and economy of force guides the tactician, who in turn informs the training officer about the skills his men should practice in

Conclusion

order to successfully wage war. Furthermore, the doctrine embraced by military leadership determines how resources are allocated.

Ultimately, airmen had to justify their assertions about mission and doctrine, purpose and effectiveness, to themselves, since they were the ones who lived or died, dependent in part upon the thoroughness of their training. They knew how dangerous flying could be.[5] Everybody who went through the program knew of somebody who was killed in a training accident. But the severest test of their training came during a conflict, and there is little doubt that when that trial came in World War II, airmen suffered the painful realization common to all military men that they could neither foresee nor train adequately for all exigencies in combat. One man noted how this applied to crew training: "The permanent crew system worked fine during the training and the overseas movement . . . and was a blessing to the storyteller in Hollywood, but . . . it fell apart under operational pressures."[6] The senior leadership of the Army Air Forces also confirmed the inevitable gulf between training and operations, and the limitations on their ability to narrow it. In January 1944, Lt. Gen. Barton K. Yount, who headed the Training Command throughout the war, spoke movingly to his colleagues who had come in from the battlefield to offer counsel and voice complaints:

> There is not a thing that you have said that is not true. All we need is about two years to train each one of these pilots to do just what you would like. I wish we had more time. . . . Gen. Arnold is enthused about giving us more time if we can work it out; but, to date, the problem has been to get more men to the front — "get them to the front — if they haven't had this it doesn't matter — we have to get them to the front." Every criticism you have made we are thoroughly cognizant of, and we have done our best to correct it. I am not saying that by way of alibi, because we know the shortcomings that our pilots have. You have sized the situation up very well.[7]

Modesty, perhaps, kept General Yount from self-congratulatory statements about the training establishment's accomplishments in both magnitude and quality. He acknowledged the validity of criticisms and made no excuses. At the same time, he implied that a system building itself while fighting a war enforced restrictions on its architects. And he seems to have deeply understood the limitations of human power, as he concluded his remarks, "I begin to understand why God travels along with you as your co-pilot."

Appendices

Accident Statistics of the Interwar Period

Accidents and Fatalities	1921	1922	1923	1924	1925	1926	1927	1928	1929	1930	1931	1932	1933	1934	1935	1936	1937	1938
Aircraft accidents																		
Aircraft hours, in thousands	75	65	66	98	150	158	141	183	263	325	396	371	433	374	449	501	513	591
No. of accidents	361	330	283	275	311	334	227	249	390	471	456	423	442	412	453	430	358	375
No. of fatal accidents	45	24	33	23	29	27	28	25	42	37	21	33	28	35	33	42	27	38
No. of fatalities	73	44	58	34	38	42	43	27	61	52	26	50	46	54	47	59	48	62
No. of individuals injured	98	97	89	55	119	79	60	52	72	82	76	89	82	83	75	69	53	63
No. of fatalities, by group																		
Regular Army, officers	35	21	28	15	17	14	18	12	9	19	8	13	9	15	12	21	15	15
Regular Army, enlisted men	17	15	18	8	11	10	5	5	26	9	3	9	13	14	12	17	16	13
Regular Army, flying cadets	13	7	1	3	1	0	9	4	10	9	3	6	3	8	3	3	3	9
Reserve Corps, officers	0	1	7	5	8	14	9	4	13	11	9	20	18	16	7	9	12	20
Miscellaneous	8	0	4	3	1	4	2	2	3	4	3	2	3	1	13	9	2	5
Accident Rates																		
Fatalities per 1,000 flying hours	.94	.86	.88	.35	.25	.27	.31	.15	.23	.16	.07	.13	.11	.14	.10	.12	.09	.11
Accidents per 1,000 flying hours	4.67	5.06	4.30	2.81	2.07	1.96	1.61	1.36	1.48	1.45	1.15	1.14	1.02	1.10	1.01	.86	.70	.63

SOURCE: Richard G. Davis, *Carl A. Spaatz and the Air War in Europe* (Washington, D.C.: Center for Air Force History, 1993), 15.

Appendices

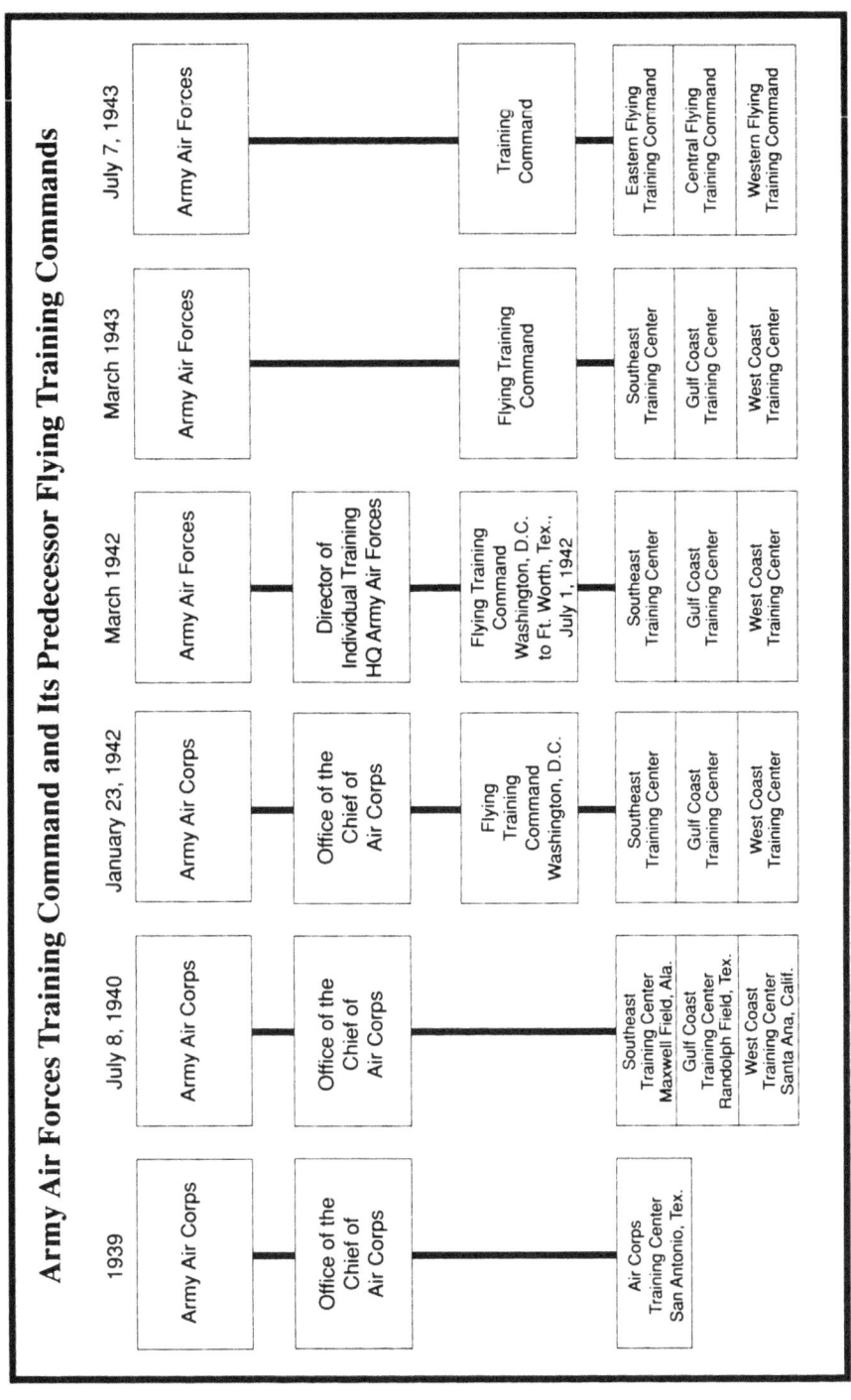

Appendices

Major Changes in Undergraduate Pilot Training, July 1939–January 1943

Training Type	Course Length (in weeks)	Flying Hours	Aircraft Used
July 1, 1939 (total time cut from 12 to 9 mos.)			
Primary	12 (cut from 16)	65	PT–3, PT–13, PT–21
Basic	12	75 (cut from 103)	BT–9
Advanced	12	75	BC–1
May 24, 1940 (total time cut from 9 to 7 mos.)			
Primary	10	60	PT–13, PT–17, PT–19
Basic	10	70	BT–13, BT–14, BT–15
Advanced	10	70 single engine[a] 86 twin engine[b]	AT–6, BC–1 AT–6, some BTs (BT–13s)
March 15, 1942			
Primary	9	60	PT–17, PT–19, PT–22
Basic	9	70	BT–13, BT–14, prob. some BT–15s
Advanced	9	70 single engine[c] 70 twin engine[c]	AT–6, P–40 AT–9, AT–10
January 1943			
Primary	9	60	PT–13, PT–17, PT–19
Basic	9	70	BT–13, some AT–6s in 1943
Advanced	9	Prob. 70 single engine 70 twin engine	AT–6, P–40 AT–6, AT–9, AT–10, AT–17, P–322 (P–38), UC–78

a. As many as 79 hours were allowed for some single engine courses. After December 1940, fighter transition combined with advanced (10 hours were given toward end of course).

b. December 13, 1940, marked the date when the first curriculum for twin engine training became distinguished from the curriculum for all advanced training.

c. Some sources say single engine and twin engine training was cut from 79 to 75 hours on March 15, 1942; however the Program of Instruction for twin engine states it is 70 hours.

SOURCE: "Major Changes in Undergraduate Pilot Training, 1939–1987," Hist and Rsch Ofc, Chief of Staff, HQ Air Tng Comd, Oct 1, 1987.

Appendices

Flying Training Graduates, July 1939–August 1945

Graduate Classification	No. of Graduates
Pilot	
Primary	233,198
Basic	202,986
Advanced	193,440
Single engine	102,907
Twin engine	90,533
Transition	108,337
Total[a]	768,991
Bombardier	
Precision	9,444
Instructor	14,571
Refresher	4,346
Total	28,361
Navigation	
Celestial	47,273
Dead reckoning	1,597
Instructor	2,815
Refresher	4,434
Total	56,119
Bombardier-Navigation	
Bombardier-Navigation	2,546
Bombardier-Dead Reckoning and Dead Reckoning Navigation	25,828
Instructor Bombardier and Dead Reckoning Navigation	106
Total	28,480
Flexible Gunnery	
Cadets and Enlisted Men	290,628
Gunnery Officer	1,175
Observer Nonpilot	866
Instructor	16,567
Total	309,236

a. Includes women, Americans in British schools, instructors, and other individuals not included in subsequent classifications.

SOURCE: *AAF Statistical Digest: World War II*, table 47.

Appendices

Location and Supervision of Pilot and Bombardier Training, July 1940

Location	Air Corps Training Center in Charge
Elementary (Primary) Pilot Training (10 wks. duration; 40% est. elimination rate)	
Albany	Southeast
Chicago	Southeast
Jackson	Southeast
Lakeland	Southeast
Tuscaloosa	Southeast
Dallas	Gulf Coast
Hicks	Gulf Coast
Lincoln	Gulf Coast
Muskogee	Gulf Coast
Sikeston	Gulf Coast
St. Louis	Gulf Coast
Tulsa	Gulf Coast
Glendale[a]	West Coast
Hemet[a]	West Coast
Ontario	West Coast
Oxnard[a]	West Coast
San Diego[a]	West Coast
Santa Maria	West Coast
Basic Pilot Training (10 wks. duration; 6% est. elimination rate)	
Montgomery	Southeast
Randolph	Gulf Coast
Moffett	West Coast
Advanced Pilot Training (10 wks. duration; 1% est. elimination rate)	
Maxwell	Southeast
Brooks	Gulf Coast
Kelly	Gulf Coast
San Angelo	Gulf Coast
Stockton	West Coast
Pursuit and Bombardment Pilot Training (5 wks. duration)	
Pursuit: Selma	Southeast
Bombardment: Barksdale	Southeast
Bombardment: Ellington	Gulf Coast
Bombardier Training (10 wks. duration)	
Barksdale	Southeast
Ellington	Gulf Coast

a. Graduates of these schools were to be sent to the Basic Schools of the Gulf Coast Air Corps Training Center.

SOURCE: Sked, "Pilot and Bombardier Training," Ofc Chief Air Corps, War Dept, box 1, entry 266, RG 18, NA.

Bombardier Requirements in Relation to the Group and Pilot Programs

Group Program	Pilot Program	Bombardier Requirement
25	1,200	1,093 (bombardier-navigator)
41	7,000	1,800 (aproximate)
54	12,000	2,500 (unofficial)
84	30,000	5,590
115	50,000	11,016
224	70,000	14,000
273	102,000	19,400 (by January 1944)

SOURCE: Baldwin, *Individual Training of Bombardiers*, chart 1.

Notes

Introduction

1. Scholars attempting to articulate a theory of air power lament the difficulties posed by the wide gap between conceptualizers and practitioners in the Air Force up to the present time. See, for example, Harold R. Winton, "A Black Hole in the Wild Blue Yonder: The Need for a Comprehensive Theory of Air Power," *Air Power History* 39, 4 (Winter 1992): 32–42.

2. James Gould Cozzens, *Guard of Honor* (New York: Harcourt, Brace, 1948), 53.

Part I

1. Col F. P. Lahm, "The Wright Brothers As I Knew Them," *Sperryscope* 8 (Apr 1939): 4.

2. Ibid., 24.

Chapter 1

1. Foulois to Cowan, Sep 27, 1910, 20247, RG 111, NA.

2. Ofc Memo 6, Brig Gen J. Allen, Chief Sig Off Army, Aug 1, 1907, Series 12: Establishment and Organization Letters and Memoranda, Aug 1907–Jun 1918, Record Group (RG) 18: Records of the Army Air Forces, National Archives (NA).

3. The Aero Club of America represented the FAI, which established standards for flight performance. Numbered certificates were issued for proven expertise in ballooning and, later, in heavier-than-air piloting. The Army formally adopted the FAI regulations for Army fliers in 1911. The FAI certificate remained the sole qualifier for pilots of airplanes, dirigibles, and spherical balloons until 1912, when the Aeronautical Division formulated a set of requirements for the rating of Military Aviator.

4. Ofc Memo 6, Allen, Aug 1, 1907.

5. *Dictionary of American Military Biography*, s.v. "Squier, George Owen."

6. Glassford's articles can be found in aeronautical and scientific publications of the period. Some correspondence relating to his hot air balloon experiments is in the Signal Corps correspondence files of the National Archives, and a few pieces can be found in the Glassford folder in the collection of the American Institute of Aeronautics and Astronautics (AIAA) held by the Library of Congress Manuscript Division (LCMD). Charles DeF. Chandler and Frank P. Lahm, in *How Our Army Grew Wings* (New York: Ronald, 1943), 43, credit Glassford's influence for assigning the balloon work to the Department of the Missouri.

7. Juliette A. Hennessy, *The United States Army Air Arm: April 1861 to April 1917* (hereafter *U.S. Army Air Arm*) (Washington, D.C.: Office of Air Force History, 1985), 13–14.

Notes

8. War Dept GO 145.

9. An Rprt, Chief Sig Off U.S. Army, FY Jun 30, 1907, entry 58, RG 111: Records of the U.S. Army Signal Corps, NA.

10. Timothy K. Nenninger, *The Leavenworth Schools and the Old Army: Education, Professionalism, and the Officer Corps of the United States Army, 1887–1918* (Westport, Conn.: Greenwood Press, 1978), 55–79; George Walton, *Sentinel of the Plains: Fort Leavenworth and the American West* (Englewood Cliffs, N.J.: Prentice-Hall, 1973), 170; Russell F. Weigley, *History of the United States Army* (Bloomington: Indiana University Press, 1967), 325–326; Paul Wilson Clark, "Major General George Owen Squier: Military Scientist" (Ph.D. diss., Case Western Reserve University, 1974), 110–116; "The Education of Army Officers," *Outlook* 91 (Jan 30, 1909): 229–230.

11. Proceedings, Academic Bd, 15110, RG 111, NA. "Aeronaut" was the term frequently used at the time for lighter-than-air operators.

12. Intvw, Frank Lahm, Thomas Milling, and Benjamin Foulois, by Gen Carl A. Spaatz, Jun 29, 1954, 9, K239.0512-767, U.S. Air Force Historical Research Agency (AFHRA).

13. Brig Gen Frank P. Lahm, "Early Flying Experiences," *Air Power Historian* 2 (Jan 1955): 2ff.

14. 16th Ind, J. Allen, Mar 30, 1907, 1193935, RG 94: Records of the Adjutant General's Office, NA.

15. Wright Bros to Bd Ord Ftns, Jun 15, 1907, frame 224, mcflm roll A1528, AFHRA.

16. Brig Gen James Allen, Chief Sig Off Army, to Recorder, Bd Ord Ftns, Oct 10, 1907, frame 242, mcflm roll A1528, AFHRA.

17. Chandler and Lahm, *How Our Army Grew Wings*, 82–83.

18. "President's Address," *American Magazine of Aeronautics* 1 (Nov 1907): 20–23.

19. Herbert A. Johnson, "The Aero Club of America and Army Aviation, 1907–1916," *New York History*, Oct 1985, 388; see generally, Bill Robie, *For the Greatest Achievement: A History of the Aero Club of America and the National Aeronautic Association* (Washington: Smithsonian Institution Press, 1993).

20. "President's Address."

21. "Summary of Meeting," *American Magazine of Aeronautics* 1 (Nov 1907): 16.

22. "Our Army and Aerial Warfare," *American Magazine of Aeronautics* 1 (Nov 1907): 57.

23. "Army News," *Aeronautics* 5 (Oct 1909): 133.

24. Foulois intvw, Jun 29, 1954, 13.

25. Lawson M. Fuller, Recorder, Bd Ord Ftns, to Wrights, Jan 3, 1908, frame 228, mcflm roll A1528, AFHRA.

26. Chandler and Lahm, *How Our Army Grew Wings*, 153.

27. "Army News," *Aeronautics* 5 (Aug 1909): 52.

28. Lahm, "The Wright Brothers As I Knew Them," 3.

29. Benjamin D. Foulois and Carroll V. Glines, *From the Wright Brothers to the Astronauts: The Memoirs of Benjamin D. Foulois* (New York: McGraw, 1968) 61–62.

30. Proceedings, Board of Officers Convened by Office Memorandum No. 18, Office of the Chief Signal Officer of the Army, Dated June 21, 1909, for the Purpose of Observing Trials of Aeronautical Devices, etc., Aug 2, 1909, frame 263, mcflm roll A1528, AFHRA.

31. Chief Sig Off to "My Dear General Edwards," May 8, 1909, 21347, RG 111, NA.

32. Lahm to Chief Sig Off, Feb 3, 1909, 21739, RG 111, NA.

33. Intvw, Lt Gen James H. Doolittle by Lt Col Robert M. Burch, Maj Ronald R. Fogelman, and Capt James P. Tate, Sep 26, 1970, 1–2, K239.0512-793, AFHRA.

34. For a detailed description based on local newspaper accounts of the Wright pilot training in 1910, see Air University History, Jan–Jun 1957, 239–277, mcflm roll K2509, AFHRA, reprinted in George N. Dubina, "Flying with Orville Wright, Montgomery, Alabama, 1910: A Chronology," Hist Study No. 62, Air University,

1974, AFHRA.

35. Lahm to Chief Sig Off, Jul 23, 1909, 21892, RG 111, NA.

36. Telg, Allen to Signals, Sep 29, 1909, 21892, RG 111, NA.

37. Col F. P. Lahm, "Training the Airplane Pilot," Twenty-First Wright Memorial Lecture, unrevised proof, 6, Frank P. Lahm File, Archives of the AIAA, LCMD; Milling intvw, Jun 29, 1954, 46.

38. Lahm, "Training the Airplane Pilot," 5–6. A complete record of Lahm and Humphrey's flights at College Park is given in the letter, Allen to E[rnest] L. Jones, Nov 9, 1909, 19331, RG 111, NA.

39. See, for example, H. H. Arnold and Ira C. Eaker, *This Flying Game* (New York: Funk & Wagnalls, 1936), 15.

40. Lahm, "The Wright Brothers As I Knew Them," 5.

41. Newspaper clipping, "Picturesque Sail at College Park," Oct 26, 1909, Lahm File, Archives of the AIAA, LCMD.

42. Lahm, "Training the Airplane Pilot," 6.

43. Foulois intvw, Jun 29, 1954, 29.

44. One of the reforms growing out of the Army reorganization of 1903 was the detail system in which officers of the line were placed with either the General Staff or special staff offices for as much as four years, when the officer returned to duty with his branch. The detail system aimed at broadening the perspective of the officer; his subsequent return to duty with the troops was to keep in touch with the field level and/or the activities of his arm. The "Manchu Law," passed by Congress in 1912, strengthened the restrictions placed on officers of the line below the rank of major, a provision obviously affecting the aviation lieutenants brought into the Signal Corps from other branches of the Army. This rule remained controversial and, over time, some branches were exempted. See Otto L. Nelson, Jr., *National Security and the General Staff* (Washington: Infantry Journal Press, 1946), 92, 107; Weigley, *History of the United States Army*, 332.

45. Allen to E. L. Jones, Jan 15, 1910, 19331, RG 111, NA.

46. Foulois, *From the Wright Brothers*, 69. The civilian mechanic O. G. Simmons, who remained with the Aeronautical Division until July 1911 when he left to take flying instructions from the Wrights, obtained FAI certificate 145, and went to work as a pilot for Robert Collier. Box 5, Foulois Papers, LCMD.

47. Foulois, *From the Wright Brothers*, 2.

48. Allen to E. L. Jones, Feb 13, 1909, 19331, RG 111, NA.

49. An Rprt, Chief Sig Off Army to Secy War, 1909, entry 58, RG 111, NA; Brig Gen James Allen, "Military Aeronautics," *Aeronautics* 6 (May 1910): 155–156.

50. Allen to Adj Gen, Dec 16, 1909, 22982, RG 111, NA.

51. Clipping, box 25, Foulois Papers, LCMD.

52. Foulois to Cowan, Sep 27, 1910, 20247, RG 111, NA.

53. Foulois, *From the Wright Brothers*, 76.

54. Foulois to Cowan, Sep 27, 1910.

55. Typescript, H. H. Arnold, "Pioneers of the Aerial Trails," c. 1925–1926, 13, box 227, Henry H. Arnold Papers, LCMD.

56. Foulois to Chief Sig Off, Apr 16, 1910, 23903, RG 111, NA.

57. Fld Svc Regs, Ofc Chief of Staff U.S. Army, War Dept, 1910, frame 2011, mcflm roll A1744, AFHRA.

58. An Rprt, Chief Sig Off Army to Secy War, 1910, 27, entry 58, RG 111, NA.

59. Chandler and Lahm, *How Our Army Grew Wings*, 193–194.

60. See corr in 26464, RG 111, NA.

61. Foulois to Chief Sig Off, Mar 8, 1911, 26464, RG 111, NA.

62. Ibid.

63. Foulois to Jones, Jun 17, 1925, box 5, Foulois Papers, LCMD.

64. F. H. Russell to Allen, Apr 3, 1911; Allen to Wright Co, Apr 8, 1911, 27098, RG 111, NA.

65. Foulois to Chief Sig Off, May 8, 1911, 27409, RG 111, NA.

66. Frank T. Coffyn to Chief Sig Off Maneuver Div, San Antonio, Tex., May

573

Notes

24, 1911, 27098, RG 111, NA.

67. Allen to Adj Gen, Apr 8, 1911, 27098, RG 111, NA.

68. Telg, Cowan to Milling, Apr 19, 1911, 27204, RG 111, NA.

69. Henry H. Arnold, *Global Mission* (New York: Harper & Row, 1949), 15.

70. Brig Gen T. DeWitt Milling, "Early Flying Experiences," *Air Power Historian* 3 (Jan 1956): 94–96.

71. Foulois to Cowan, Sep 27, 1910.

72. Arnold, *Global Mission*, 19.

73. Wright Co. Flying Rprt, Summary of Lt. Hen. H. Arnold's Training, 27098, RG 111, NA.

74. Wkly Rprts, Milling and Arnold to Chief Sig Off, 27098, RG 111, NA.

75. Ibid.

76. Arnold, *Global Mission*, 25–26.

77. Milling, "Early Flying Experiences," 96.

78. Russell to Allen, May 15, 1911, 27098, RG 111, NA.

79. Grover C. Loening, *Our Wings Grow Faster* (Garden City, N.Y.: Doubleday, Doran & Co., 1935), 19–20.

80. Russell to Maj C. McK. Saltzman, Aug 21, 1911, 28259, RG 111, NA.

81. Orville Wright to Saltzman, Sep 22, 1911, 28259, RG 111, NA.

82. Loening, *Our Wings Grow Faster*, 10.

83. Grover C. Loening, *Military Aeroplanes, Simplified, Enlarged; An Explanatory Consideration...* (Boston: Best, 1918), 137.

84. Long and bitter litigation between the two companies ensued over this point.

85. Jerome S. Fanciulli to Allan [*sic*], Dec 10, 1910, 25857, RG 111, NA.

86. Paul W. Beck to Allen, Mar 5, 1911, 25857, RG 111, NA.

87. Arnold, *Global Mission*, 24.

88. 2d Lt John C. Walker, Jr., to Squier, Feb 17, 1911, 25857, RG 111, NA.

89. Ibid.; Walker to Adj Gen, Dept of Calif., Mar 1, 1911, 25857, RG 111, NA.

90. Beck to Adj Gen, Dept of Calif., Mar 1, 1911, 25857, RG 111, NA.

91. Eugene P. Lyle III and Charles W. Diffin, *The City of Wings: A Narrative of Aviation and Its Swift Development as Seen in San Diego from the Earliest Days to the Present Time* (San Diego: The City Schools of San Diego, Calif., 1938), 20; Beck to AG, Dept of Calif, Mar 1, 1911.

92. Walker to Adj Gen, Dept of Calif., Mar 1, 1911.

93. Chandler and Lahm, *How Our Army Grew Wings*, 261; Arnold, *Global Mission*, 17–18.

94. Beck to Allen, Mar 7, 1911, 25857, RG 111, NA.

95. Robert B. Casari, *Encyclopedia of U.S. Military Aircraft: 1908 to April 6, 1917*, (Glendale, Calif.: Aviation, 1970–1972), 2: 2.

96. Foulois, *From the Wright Brothers*, 87.

97. Ibid., 88–89.

98. Arnold, *Global Mission*, 20.

99. Coffyn to Chief Sig Off Maneuver Div, San Antonio, Tex., May 24, 1911.

100. 2d Ind, Maj Gen Carter, Comdg, HQ Maneuver Div, San Antonio, Tex., May 27, 1911, 27098, RG 111, NA.

Chapter 2

1. Anthony Jannus, "College Park Flying Grounds," *Aeronautics* 8 (Jan 1911): 21.

2. Ibid.

3. Ofc Memo 13, Jul 3, 1911, 27836, RG 111, NA.

4. Allen to Chief Sig Off Maneuv Div, May 31, 1911, 27554, RG 111, NA.

5. Memo, Chief Clk Disbg Div, Jul 10, 1911, 27646, RG 111, NA.

6. Allen to Adj Gen, Nov 14, 1911, 28832, RG 111, NA.

7. An Rprt, Chief Sig Off Army to Secy War, 1911, 24, entry 58, RG 111, NA.

8. Memo, Chandler to Chief Sig Off, Jun 20, 1911, 27848, RG 111, NA.

9. Ibid.

10. Chief Sig Off to Chief Sig Off Maneuv Div, May 31, 1911, 27554, RG 111, NA.

11. Arnold, *Global Mission*, 33–34.
12. Chandler to Chief Sig Off, Nov 11, 1911, 26513, RG 111, NA.
13. Ibid., Dec 5, 1911, 28977, RG 111, NA.
14. Sig Corps Avn Sch Order 7, Dec 9, 1911.
15. News Bul, Sig Corps Avn Sch, Augusta, Ga., Dec 11, 1911, 29214, RG 111, NA.
16. When Chandler is cited hereafter, the reference is to Charles DeF. Chandler, not Lt C. P. Chandler.
17. Memo, Comdg Off Sig Corps Avn Sch to Adj Gen, Sep 5, 1912, Subj: Redetail of Officers of Organized Militia to Aviation School; 2d Ind, Div Militia Affairs, Ofc of the Chief of Staff, to Chief Sig Off, Sep 13, 1912, 31050, RG 111, NA.
18. Memo, Comdg Off Sig Corps Avn Sch, to Chief Sig Off, Dec 11, 1912, Subj: Training of National Guard Officers by the Signal Corps, 31823, RG 111, NA.
19. Records of Events, Sig Corps Avn Sch, Augusta, Ga., Mar 3, 1912, 29449, RG 111, NA; Mar 10, 18, 1912, 29214, RG 111, NA.
20. Ibid., Feb 4, 24, 1912, 29214, RG 111, NA.
21. Loening, *Our Wings Grow Faster*, 7.
22. Memo, Beck to Chandler, Feb 18, 1912, 29427, RG 111, NA.
23. "Requirements for Weight Carrying Military Aeroplane," Feb 8, 1912, and "Proceedings of a Board of Officers," convened by Order 9, Sig Corps Avn Sch, Augusta, Ga., Mar 14, 1912, both reports appear (accompanied by correspondence relating to specifications and order of military airplanes) in 29427, RG 111, NA.
24. Memo, Comdg Off Sig Corps Avn Sch, to Chief Sig Off, Sep 4, 1912, Subj: Designation of Military Aeroplanes, 31182, RG 111, NA.
25. *Aeronautics* 10 (Feb 1912): 67.
26. Casari, *Encyclopedia of U.S. Military Aircraft*, 2: 18.
27. "Army Aero News," *Aeronautics* 11 (Oct 1912): 123.
28. Chandler and Lahm, *How Our Army Grew Wings*, 233.
29. Chandler to Chief Sig Off, Feb 8, 1912, 29427, 111, NA.
30. Chandler and Lahm, *How Our Army Grew Wings*, 269. Inventories of airplanes in house and on order as of Fall 1912 can be found in 29278 and 31285, RG 111, NA.
31. An Rprt, Chief Sig Off Army to Secy War, 1912, entry 58, RG 111, NA.
32. Cir 5, War Dept Sig Ofc, Aug 15, 1911, 28227, RG 111, NA.
33. Memo, Chandler to Chief Sig Off, Jul 30, 1912, 31128, RG 111, NA.
34. Allen to Curtiss, Mar 5, 1912, 29427, RG 111, NA. Two pieces of legislation addressing the personnel needs of the Aeronautical Division were initiated in 1912, but no action was taken. An increase in numbers did not come until the War Department appropriation bill for 1913, which allotted the Signal Corps thirty officers. See Chapter 3 for a discussion of legislation predating H.R. 5304 that led, on July 18, 1914, to congressional authorization of the Aviation Section.
35. Reports and photographs relating to Hazelhurst's death are found in 30424, RG 111, NA.
36. Memo, Chandler to Chief Sig Off, Jul 30, 1912.
37. Cir 11, War Dept Ofc Chief Sig Off, Oct 26, 1912, 34358, RG 111, NA.
38. Sig Corps Avn Sch Order 61, Dec 20, 1912.
39. Milling, "Early Flying Experiences," 103.
40. Memo, Comdg Off Sig Corps Avn Sch to Chief Sig Off, Dec 9, 1912, Subj: Report of the Aviation School at Augusta from Nov. 18th to Dec. 8th, 31804, RG 111, NA; Wkly flight rprts [from Augusta], 28309, RG 111, NA.
41. Telg, Chandler to Signals, Feb 13, 1913, 32195, RG 111, NA.
42. Memo, Scriven to Chief of Staff, Feb 21, 1913, 32244, RG 111, NA.
43. Allen to Adj Gen, Jun 3, 1912, 30407, RG 111, NA.
44. Rprts of inst, 31256 and 31433, RG 111, NA.
45. Memo, 1st Lt Loren H. Call to Comdg Off Sig Corps Avn Sch, Nov 17,

Notes

1912, Subj: Weekly Reports of Instruction, 31636, RG 111, NA.

46. Memo, Comdg Off Sig Corps Avn Sch to Chief Sig Off, Mar 31, 1913, Subj: Weekly Report, 28309, 111, NA.

47. Curtiss to Allen, Nov 10, 1912, 31342, RG 111, NA.

48. [Edgar] Russel to A. C. Genung, Mar 22, 1913, 32425, RG 111, NA.

49. Arnold to Comdg Off, College Park, Md., and Chief QM Corps, Jun 14, 1913, 21892, RG 111, NA.

50. *Jackrabbits to Jets* (San Diego: San Diego Publishing Co., 1992), 10.

51. Curtiss to Secy War, Feb 20, 1911, with atchs, 26607, RG 111, NA.

52. Telg, Adj Gen, by order of Secy War, to Comdg Gen Eastern Div, Nov 13, 1912, 31342, RG 111, NA; telg, Adj Gen to Comd Gen Eastern Div, Nov 13, 1912, 1973981, RG 94, NA.

53. War Dept GO 75 of 1913 authorized the Aero Squadron; War Dept GO 77 of the same year prescribed record-keeping. See Memo, Col D. C. Shanks, Insp Gen, to Comdg Gen Western Dept, Feb 10, 1915, Subj: Report of Annual Inspection of First Aero Squadron and of the Signal Corps Aviation School, San Diego, Calif., 38014, RG 111, NA.

54. GO 79, Dec 13, 1913; Memo, Adj Gen to Col Hodges, 2096472, RG 94, NA.

55. Memo, Comdg Off Sig Corps Avn Sch, to Chief Sig Off, Mar 17, 1913, Subj: Weekly Rprts, 28309, RG 111, NA.

56. 2d Ind, Allen to Chief Sig Off, Western Div, Dec 27, 1912, 31342, RG 111, NA.

57. Wkly Rprts, Comdg Off Sig Corps Avn Sch to Chief Sig Off, Jan 1–Jul 14, 1913, 28307, RG 111, NA.

58. Chandler and Lahm, *How Our Army Grew Wings*, 266. Captain Cowan forwarded a general report on the Aviation School to the Chief Signal Officer on July 14, 1913, in which he discussed the strengths and weaknesses of each of the officers on duty at the Aviation School. Memo, Comdg Off Sig Corps Avn Sch, to Chief Sig Off, Subj: General Rprt, 28309, RG 111, NA.

59. *Jackrabbits*, 27.

60. Memo, to Allen, Apr 19, 1912, 29278, RG 111, NA.

61. Rprt of the Chief Sig Off, Brig Gen George B. Scriven, Oct 10, 1913, 809–810.

62. Memo, Comdg Off Sig Corps Avn Sch to Chief Sig Off, Jul 14, 1913, Subj: General Report, 28309, RG 111, NA.

63. Memo, Comdg Off Sig Corps Avn Sch to Chief Sig Off, Mar 31, 1913, Subj: Weekly Report, 28309, RG 111, NA.

64. Comdg Off Sig Corps Avn Sch memo to Chief Sig Off, Jul 14, 1913.

65. Reber to Dr. Hutton, Feb 18, 1914, 34587, RG 111, NA.

66. Memo, Scriven to Chief of Staff, Nov 11, 1913, 2096472, RG 94, NA.

67. Roy Kirtland to H. H. Arnold, Jul 16, 1938, box 222, Arnold Papers, LCMD.

68. "At the Army Aviation School," *Aeronautics* 13 (Dec 1913): 216.

69. Samuel Reber to Ira L. Reeves, Feb 26, 1914, 34610, RG 111, NA.

70. Sig Corps Avn Sch GO 1, Jan 2, 1915.

71. Arnold, "History of Rockwell Field," 38.

72. An Rprt, Chief Sig Off Army to Secy War, 1915, entry 58, RG 111, NA.

73. Memo, Brig Gen George P. Scriven, Chief Sig Off, to Chief War College Div, Gen Staff, May 16, 1916, Subj: Organization and Methods of Administration of the Aviation Section of the Signal Corps, frame 1503, mcflm roll A1528, AFHRA.

74. See reports pertaining to the work of the Technical Instruction School in 37653, 37912, 38347, 38348, 38705, and 38706, all in RG 111, NA.

75. For a thorough history, with reference to pertinent documents, of the NACA, see Alex Roland, *Model Research* 2 vols (Washington, D.C.: National Aeronautics and Space Administration, 1985).

76. Intvw, Maj Gen Howard C. Davidson by Thomas A. Sturm and Hugh N. Ahmann, Dec 5–8, 1974, 127–128, K239.0512-817, AFHRA.

77. Ibid., 152.

78. Lt. Henry H. Arnold to Scriven, Mar 9, 1915, and Scriven to Arnold, Apr 20, 1915, 38930, RG 111, NA.

79. Memo, Comdg Off Sig Corps Avn Sch, to Chief Sig Off, Sep 26, 1914, Subj: Report on Progress Made and Present Condition at Signal Corps Aviation School, 35718, RG 111, NA.

80. Davidson intvw, Dec 5–8, 1974, 123–124.

81. Memo, Ofc Chief Sig Off to Comdg Off Sig Corps Avn Sch, Feb 17, 1914, Subj: Use of Certain Aeroplanes, 34548, RG 111, NA.

82. Rprt, Capt A. S. Cowan, Comdg Off, to Chief Sig Off, "Annual Reports, Signal Corps Aviation School, 1914–1915," Aug 9, 1915, 39792, RG 111, NA.

83. Casari, *Encyclopedia of U.S. Military Aircraft*, 1: 1, 14–26.

84. Loening, *Our Wings Grow Faster*, 53.

85. Casari, *Encyclopedia of U.S. Military Aircraft*, 1: 2, 42–45, 55–62, and 3: 1–12, 23–60.

86. Memo, Cowan to Chief Sig Off, Mar 14, 1914, Subj: Tentative Specifications for Training Machine, 34791, RG 111, NA.

87. Shanks memo to Comdg Gen Western Dept, Feb 10, 1915.

88. "Army Aviation," *Aeronautics* 17 (Jul 15, 1915): 4.

89. F. G. Swanborough, *United States Military Aircraft Since 1909* (London: Putnam, 1963), 475.

90. Cowan rprt to Chief Sig Off, Aug 9, 1915.

91. On June 5, 1915, for example, the school had one Wright, five Martins of two different types, three different Curtisses, one Burgess-Dunne, and one true original pieced together by Loening. These were powered by thirty-one assorted engines. "Status of Aeroplanes on June 5, 1915," in "Material re: Charges Against Signal Corps, 1913–1915," 168.7119-7, Herbert A. Dargue Papers, AFHRA.

92. The so-called "dep" control, a means to guide or direct an airplane in flight, consisted of a manually operated wheel that controlled lateral and vertical motion; the foot worked the rudder. Its name comes from its originator, the French airplane designer Armand Deperdussin.

93. Of the sixteen aviators listed among the original personnel of the Aviation Section, seven were Infantry, six were Coast Artillery, and three were Cavalry officers. Office Memo 12, War Dept, Ofc Chief Sig Off, Jul 29, 1914, 35911, RG 111, NA.

94. H. H. Arnold, "Pioneers of the Aerial Trails," 1925–26, box 227, Arnold Papers, LCMD.

95. See "Wake Up, Congress," *Aeronautics* 12 (Jan 1913): 24.

96. Memos, E. Russel to Comdg Off Avn Det, Texas City, Tex., and to Comdg Off Sig Corps Avn Sch, both Sep 16, 1913, 33154, RG 111, NA.

97. "Army Aeronautics for 1913," *Aeronautics* 13 (Dec 1913): 20.

98. Frank M. Kennedy to Jones, Aug 10, 1945, 29214, RG 111, NA.

99. Maj H. H. Arnold, "The History of Rockwell Field," 1923, box 294, Arnold Papers, LCMD.

100. Henry Breckinridge to J.V.R. Butler, Nov 21, 1913, 33675, RG 111, NA.

101. Memos, Edgar Russel, May 13, 1913, 32658, RG 111, NA.

102. Rprt, Chamberlain, Insp Gen, to Comdg Gen Western Dept, "Report of Annual Inspection ...," Mar 19, 1914, 35013, RG 111, NA.

103. 3d Ind [handwritten draft], Chief Sig Off to Adj Gen, Feb 1916, 41804, RG 111, NA; Memo, Brig Gen W. W. Macomb, Chief War College Div, to Chief Sig Off, Apr 26, 1916, Subj: Organization and Methods of Administration of the S.C.A.S., frame 1516, mcflm roll A1528, AFHRA; Alan R. Hawley, "The Inner Trouble in the U.S. Army Air Service," *Flying* 5 (May 1916): 163–164.

104. Foulois, *From the Wright Brothers*, 125.

105. Efficiency Rprt, Lt Herbert A. Dargue, 1170424, RG 94, NA; Statement, Dargue Papers, Arnold Papers, box 2, LCMD.

Notes

Chapter 3

1. Mark Sullivan, *Our Times*, vol 4 (New York: Charles Scribner's Sons, 1932), 139.

2. Richard Hofstadter's *The Age of Reform* (New York: Random House, 1955) is the classic statement on the Progressive Era. For a discussion of its manifestation in the War Department, see C. Joseph Bernardo and Eugene H. Bacon, *American Military Policy: Its Development Since 1775* (Harrisburg: Stackpole, 1955), and Elting E. Morison, *Turmoil and Tradition, A Study of the Life and Times of Henry L. Stimson* (Boston: Houghton Mifflin Co., 1960).

3. 62d Cong, 2d sess, HR 448, Mar 9, 1912.

4. "Military Aviation," 62d Cong, 2d sess, HR 718, Apr 26, 1912.

5. Ltr, Secy War, Subj: Transmitting Pursuant to House Resolution 448 Information Relative to Military Aviators for the Army and for Other Purposes, in "Military Aviation."

6. Stimson, Secy War to Mil Comte of the House of Representatives, A Bill To Provide Military Aviators for the Army, and for Other Purposes, 62d Cong, 2d sess, HR 448, 80.

7. Open Ltr, Asst Secy, Aero Club of America, May 17, 1912, 29278, RG 111, NA.

8. *An Act to Increase the Efficiency of the Aviation Service of the Army, and for Other Purposes*, 63d Cong, 2d sess, HR 5304, Jul 18, 1914.

9. Rprt, War Dept, Ofc Chief Sig Off, "Report on Progress Made in Aeronautics in the Army Since about March 1, 1913," Jul 7, 1913, in House Committee on Military Affairs, *Aeronautics in the Army: Hearings on HR 5304, An Act to Increase the Efficiency of the Aviation Service of the Army, and for Other Purposes*, 63d Cong, 1st sess, 119.

10. House Committee on Military Affairs, *Aeronautics in the Army*, 58–59.

11. Davidson intvw, Dec 5–8, 1974, 152–153.

12. One of the few officers to disagree with this aspect of the Signal Corps proposal was Captain William Mitchell, later to achieve notoriety for his advocacy of aviation. He did not "think it very essential that the school itself should be with the troops." There was some irony to Mitchell's view in that he was not initially a flying officer but was brought into aviation specifically to introduce General Staff experience into aviation administration. One supposed perspective of General Staff officers was their opportunity to see the interworking of various branches of the line and support staffs.

13. 1st Ind, Maj Gen Leonard Wood, Chief of Staff, to Chief Sig Off, May 9, 1913, in House Committee on Military Affairs, *Aeronautics in the Army*, 133.

14. Memos, Brig Gen M. M. Macomb, Chief War College Div, to Adj Gen Army, May 6, 1916, Subj: HR 14867 and SR 5688, 64th Cong, 1st sess: Purchase of Aviation Field, College Park, Maryland; Brig Gen George P. Scriven, Chief Sig Off, to Chief of Staff, Jul 20, 1916, 2394360, RG 94, NA. See also the Army report relating to bills before the Senate and House concerning land in San Diego. "Aviation School and Training Grounds for the Signal Corps of the United States Army," in ltr, Act Secy War, transmitting Report of Commission . . ., 64th Cong, 1st sess, HR 687, Feb 14, 1916.

15. Reber to Scriven, Jun 3, 1912; 1st Ind, Reber, Jul 5, 1912, 29269, RG 111, NA.

16. Maneuv Rprt, Foulois, "Remarks," Aug 12, 1912, [3554], RG 111, NA.

17. Milling to Comdg Off, Avn Sq, Maneuv Div, College Park, Aug 22, 1912, 29269, RG 111, NA.

18. Rprts, Avn Sq Offs, 29269, RG 111, NA.

19. Reber to Chief Sig Off, Aug 20, 1912, 29269, RG 111 NA.

20. An Rprt, Chief Sig Off, 1912, entry 58, RG 111, NA.

21. Reber to Adj Gen, Aug 31, 1912, 31012, RG 111, NA.

22. 1st Ind, Scriven, Aug 30, 1912, 31012, RG 111, NA.

23. Foulois to Orville Wright, Nov 16,

1912, box 25, Foulois Papers, LCMD.

24. Milling, "Early Flying Experiences," 103.

25. Rprts, First Aero Sq, 28309, RG 111, NA. For a brief discussion of the activities along the border through the pre-war years, see Wayne C. Pittman, Jr., "Wings Over Texas: The First Aero Squadron At Texas City," *Journal of America's Military Past*, 20 (Summer 1993): 34–47.

26. Richard D. Challener, *Admirals, Generals and American Foreign Policy, 1898–1914* (Princeton, N.J.: Princeton University Press, 1973), 265–272; Cecil V. Crabb, Jr., *The Doctrines of American Foreign Policy* (Baton Rouge: Louisiana State University Press, 1982), 9–106; Walter Millis, *Arms and Men: A Study in American Military History* (New York: Putnam, 1956), 168; George Kennan, *American Diplomacy, 1900–1950* (Chicago: University of Chicago Press, 1951), 5–21; Fred M. Greene, "The Military View of American National Policy, 1904–1940," *American Historical Review* 66 (Jan 1961): 362.

27. Scriven to Chief of Staff, Aug 31, 1911, 28378, RG 111, NA.

28. Memo, Lahm to Chief Sig Off, Jun 3, 1915, Subj: Record of Aviation Service, 38980, RG 111, NA.

29. Capt. [R. J.] Burt to Chief Sig Off, Philippines Div, Jun 21, 1912, 30446, RG 111, NA. Burge later became an officer and retired from the service as a lieutenant colonel.

30. Beginning with Burge, who attained his FAI certificate in 1912, a total of twenty-two enlisted men of the Regular Army became pilots at the time the United States entered World War I. For a list of soldier pilots and information about them, see Hennessy, *U.S. Army Air Arm*, app 15, 249.

31. Memos, Comdg Off Sig Corps Avn Sch to Chief Sig Off, Jan 19, 1914, Subj: Applications M.S.E. McRae and Sgt. Boland, S.C. for Instruction in Flying; Ofc Chief Sig Off to Comdg Off Sig Corps Avn Sch, Jan 26, 1914, 34307, RG 111, NA.

32. Memo, Allen to Chief Sig Off, Philippines Div, Nov 25, 1912, Subj: Relief of Detailed Officers in Signal Corps, 31654, RG 111, NA.

33. Glassford to Chief Sig Off, May 29, 1912, 30711, RG 111, NA.

34. Memo, Arnold to Russel, Apr 16, 1913, 30711, RG 111, NA.

35. Memo, Dept Sig Off to Chief Sig Off, Jul 9, 1914, Subj: Annual Report of Signal Corps in the Philippines Department for the Year Ended June 30, 1914, 35718, RG 111, NA.

36. Memo, Scriven to Chief of Staff, Jul 17, 1914, 35766, RG 111, NA.

37. Memo, Maj Gen [illeg.], Comdg, Philippines Dept, to Adj Gen, Jul 24, 1914, Subj: Annual Report, 35718, RG 111, NA.

38. Memo, Dept Sig Off [Hawaiian Dept] to Chief Sig Off, Jul 1, 1914, Subj: Annual Report, 35718, RG 111, NA.

39. Robert B. Casari, "The First 59," *Journal of the American Aviation Historical Society* 16 (Spring 1971): 11.

40. Monthly Rprts, HQ Hawaiian Dept, 35075, RG 111, NA.

41. 1st Ind, Capt Chas. Wallace, by dir Chief Sig Off, Sep 22, 1914, 36296, RG 111, NA.

42. Memo, Scriven to Adj Gen, Jul 17, 1914, 35820, RG 111, NA.

43. Arnold, *Global Mission*, 3.

44. Davidson intvw, Dec 5–8, 1974, 141.

45. Chief Sig Off Allen to Hawthorne Hill, Jan 14, 1909, 19331, RG 111, NA.

46. Memo, Squier, Burt, and Beck to Chief Sig Off, Jan 17, 1912, 29427, RG 111, NA; "Military Aviation in France," *Aeronautics* 11 (Nov 1912): 147.

47. Cir 8, Brig Gen George P. Scriven, Ofc Chief Sig Off, "The Service of Information: United States Army," 1915, 46.

48. An Rprt, Chief Sig Off, 1915; Memo, Scriven to Chief War College Div, Gen Staff, May 16, 1916, Subj: Organization and Methods of Administration . . ., frame 1503, mcflm roll A1528, AFHRA.

49. Memo, Scriven to Adj Gen, Nov 24, 1914, Subj: Military Observer; 1st Ind, Adj Gen to Chief Sig Off, Nov 28, 1914, 36684, RG 111, NA.

50. Memo, Reber to Chief Sig Off, Mar

Notes

20, 1915, 38079, RG 111, NA.

51. Rprt, Lt Col George O. Squier, "Air Service with the British Army in the Field," Feb 26, 1915, 38395, RG 111, NA; rprt, C. DeF. Chandler to Bln Div, "Aerostation: Observation Balloons," Aug 4, 1915, Series 16: Balloon Bulletins, RG 18: Division of Military Aeronautics.

52. "Schools and Their Methods," *Aeronautics* (London) 9 (Jul 7, 1915): 2.

53. "The Supply of Aviators," *Aeronautics* (London) 8 (May 12, 1915): 320.

54. Several Portuguese officers received instruction at the Signal Corps Aviation School, for example, until Portugal became a belligerent in the war and training privileges were withdrawn in March 1916.

55. Walt Brown, Jr., ed, *An American for Lafayette: The Diaries of E.C.C. Genet, Lafayette Escadrille* (Charlottesville: University Press of Virginia, 1981), 74–102. For the first definitive history of the *Lafayette Escadrille*, see James Norman Hall and Charles Bernard Nordhoff, eds, *The Lafayette Flying Corps*, vol 1 (Boston: Houghton Mifflin Co., 1920).

56. "How France Trains Pilot Aviators," *Aviation and Aeronautical Engineering* 1 (Nov 1, 1916): 212–214.

57. Ibid.

58. House Committee on Military Affairs, *Aeronautics in the Army*, 119.

59. Chandler and Lahm, *How Our Army Grew Wings*, 209.

60. Milling, "Early Flying Experiences," 101.

61. Ibid.

62. The most thoroughgoing account of the "preparedness" movement is in John Patrick Finnegan, *Against the Specter of the Dragon: The Campaign for American Military Preparedness, 1914–1917* (Westport, Conn.: Greenwood Press, 1974).

63. An Rprt, Chief Sig Off, 1912.

64. Memo, Ofc Chief Sig Off to Secy War, Jan 5, 1915, Subj: Appropriation for Aviation Section, 37348, RG 111, NA.

65. As of December 31, 1915, the Army had ordered a total of 59 airplanes. Not until the 1916 appropriations did the situation change dramatically. See Casari, "The First 59," 1–17.

66. "Aero Squadron Leave San Diego," *Aeronautics* 16 (Jun 15, 1915): 116.

67. "Imperative Need of Training Airmen," *Boston Transcript*, reprinted in *Aerial Age Weekly* 1 (Jun 28, 1915): 342.

68. Address, Alexander Graham Bell, "Preparedness for Aerial Defense," Apr 1916, 5–6, frame 383, mcflm roll A1528, AFHRA.

69. An Rprt, Chief Sig Off, 1915.

70. 5th Ind, Scriven to Adj Gen, Sep 2, 1916, Series 12: Div Mil Aero, RG 18, NA.

71. The Second Aero Company also trained at Mineola, possibly with the Army, but it did not enter federal service on the border, as did the First.

72. Some dissension arose between the Aero Club and the Aviation Section regarding the allocation of the $13 million appropriation. The former maintained that approximately $9.6 million should be earmarked for the National Guard, whereas the Signal Corps claimed that it was to use the money as it saw fit. The Signal Corps position was upheld by the Judge Advocate General. The training of National Guardsmen was less satisfactorily accomplished than the training of men in the reserve program: the Aviation Section was reluctant to assume the training of as many National Guardsmen at San Diego as had been planned because the facilities and instructional staff were insufficient to accomplish the task. See *Aviation and Aeronautical Engineering* 1 (Oct 15, 1916): 192; Memo, W. A. Bethel, Act Judge Advocate Gen, to Secy War, Sep 29, 1916, Subj: Aviation Training of the National Guard—Appropriations Available, 353.9, entry 166, RG 18, NA; "Army Officer's [sic] Opposition to Training National Guard in Aviation Depriving Nation of Hundreds of Aviators," *Flying* 5 (Nov 1916): 422–423; Memos, A.S. Cowan to Chief Sig Off, Nov 17, 1914; Reber to Chief, Div Militia Affairs, Nov 30, 1914, 36447, RG 111, NA; "Plans for Training Aerial Reserve Corps and National Guard Fliers," *Aviation and Aeronautical Engineering* 1 (Oct 1, 1916): 157.

73. John Lovejoy Elliott, *University*

Presidents and the Spirit of Militarism in the United States (New York: American Association for International Conciliation, Jun 1915), 3–4; Philip A. Carroll, Military Training Association, New York, to Scriven, Jun 24, 1916, 211, entry 166, RG 18, NA; *Aviation and Aeronautical Engineering* 1 (Dec 1, 1916): 296; 1914–1917, folder 178, UA1.5.169, Lawrence A. Lowell Papers, Harvard University Library (HUL).

74. The physical location shifted from North Island to Coronado Heights, also on the littoral of San Diego Bay, when the owners of the North Island property requested the Signal Corps to vacate the tract as soon as possible after March 31, 1916. "Aviation School and Training Grounds," Report of Commission . . . , 64th Cong, 1st sess, HR 687, Feb 14, 1916. See also documents in 41803, RG 111, NA.

75. An Rprt, Chief Sig Off, 1916.

76. Wkly Rprts, Sig Corps Avn Sch and Rockwell Field, 319.1, entry 166, RG 18, NA.

77. Memo, Foulois to Chief Sig Off, Jun 25, 1916, Subj: Training of Aviators, 353.9, entry 166, RG 18, NA.

78. The Second Aero Squadron was based in the Philippines.

79. Memo, Squier to Chief of Staff, Sep 9, 1916, 2458193, entry 25, RG 94, NA.

80. Casari, *Encyclopedia of U.S. Military Aircraft*, 4: 19.

81. An Rprt, Chief Sig Off, 1914.

82. I. B. Holley, Jr., *Ideas and Weapons* (New Haven: Yale University Press, 1953; Washington, D.C.: Office of Air Force History, 1983), 81.

83. Bell Address, April 1916, 1.

Part II

1. John J. Pershing, *My Experiences in the World War*, 2 vols (New York: Frederick A. Stokes Co., 1931), 1: 27.

2. Ronald Schaffer, *America in the Great War: The Rise of the War Welfare State* (New York: Oxford University Press, 1991), 32.

3. For a thorough discussion of War Department organization, see Otto L. Nelson, Jr., *National Security and the General Staff* (Washington: Infantry Journal Press, 1946), 219–259.

4. Roger E. Bilstein, *The American Aerospace Industry* (New York: Twayne, 1996), 18.

5. Pershing, *My Experiences*, 1: 319–320.

6. Ibid., 2: 238.

Chapter 4

1. Lucien H. Thayer, *America's First Eagles: The Official History of the U.S. Air Service, A.E.F. (1917–1918)* (San Jose, Calif.: R. James Bender, and Mesa, Ariz.: Champlin Fighter Museum Press, 1983), Foreword.

2. Ibid.

3. Ltr, Chief of Air Svc AEF, to Dir Mil Aero, Aug 25, 1918, cited in memo, Maj W. W. Hoffman to Dir Mil Aero, Oct 28, 1918, Subj: Training, box 480, 321.9, entry 166, RG 18, NA.

4. The names of the three fields were changed as follows: Hicks became Taliaferro Field; Everman, Barron Field; and Benbrook, Carruthers Field.

5. Memos, Chief Sig Off to Comdg Off, RFC, Toronto, Canada, Aug 29, 1917, Subj: Agreement; Comdg Off, RFC Canada, to Chief Sig Off, Jan 7, 1918; Col D. L. Roscoe, Comdg Off, Taliaferro Fields, to Chief Sig Off, May 6, 1918, Subj: Reciprocal Agreement Between Chief Signal Officer and General Officer Commanding, RFC of Canada, box 216, C53.2 89, entry 206, RG 18, NA; S. F. Wise,

Notes

The Official History of the Royal Canadian Air Force, vol 1, *Canadian Airmen and the First World War* (Toronto: University of Toronto Press, 1980), 91–96.

6. Memo, Lt Col W. C. Sherman, Chief Air Svc Tng, to Chief T&O Gp (attn: Capt Miller), Aug 25, 1919, box 480, 321.9, entry 166, RG 18, NA.

7. The War Department appointed Maj. Gen. William L. Kenly head of the Division of Military Aeronautics on April 24, 1918; this action was confirmed by order of the President of the United States on May 20, 1918. Aviation was removed from the Signal Corps at this time, and General Kenly and the head of his counterpart agency, the Bureau of Aircraft Production, reported directly to the Chief of Staff. These branches were combined on August 27, 1918, and a Director of Air Service was appointed.

8. Memo, Air Div, Tng Sec, to Eqp Div, "Appendix E", Dec 22, 1917, atch to "Status of Military Aeronautics in America," in Royal D. Frey, "Evolution of Maintenance Engineering, 1907–1920," Hist Study No. 327, Air Force Materiel Command, 1960, vol 2, doc 31, K201.327, AFHRA.

9. Memo, Lt Col J. E. Carberry, Chief, Heavier-than-Air, to Chief Tng, Jun 30, 1918, Subj: Report of Heavier-than-Air Branch, Training Section, for period June 20, to June 30, 1918, box 464, 319.1, entry 166, RG 18, NA.

10. Canadian airmen joined the RFC and the Royal Naval Air Service until the fall of 1918 when both land- and naval-based Canadian air forces were formed.

11. Richard Maclaurin to Lawrence A. Lowell, May 14, 1917, Lowell Papers, HUL.

12. Hiram Bingham, *An Explorer in the Air Service* (New Haven, Conn.: Yale University Press, 1920), 21.

13. An Rprt, Chief Sig Off U.S. Army, to Secy War, 1917, entry 58, RG 111, NA. The capacity of each school by mid-August 1917 follows: MIT and Georgia School of Technology, 200; Cornell, Ohio State, and University of California, 300; Princeton, 600; and the Universities of Illinois and of Texas, 400. The schools were then receiving an average of 350 students a week. Memo, Maj Hiram Bingham, Ofc Chief Sig Off, to Adj Gen Army, Aug 14, 1917, Subj: Designation of Schools of Military Aeronautics as Training Camps for Candidates for Commission, folder: Memorandums, All Schools of Military Aeronautics, 1917, box 20, entry 18, RG 18, NA.

14. Memo, Brig Gen George O. Squier, Chief Sig Off, to Adj Gen, May 10, 1917, Subj: Aviation Schools, box 620, 353.9, entry 166, RG 18, NA; Memos, Admin Div, May 12, 1917, and Chief of Staff, May 19, 1917, box 620, 353.9A, entry 166, RG 18, NA.

15. Regs, School of Military Aeronautics, Massachusetts Institute of Technology, Cambridge, Mass., 1917, entry 206, RG 18, NA.

16. At Oxford, England, the model for the American and Canadian ground school program, the president of the Academic Board was an army officer who was also titled Chief Instructor.

17. Memo 119 for All Schools, Maj Hiram Bingham, Aug 30, 1917, folder: Memorandums, box 20, entry 18, RG 18, NA.

18. The titles of aviation offices changed with administrative reorganizations. Ground Schools Branch, which did not answer directly to the Chief of Training until the reorganization of March 27, 1918, is used here to refer to the office that administered the Schools of Military Aeronautics.

19. For examples of curricula and syllabi used in the schools during the tenure of the program, see folder: United States Schools of Military Aeronautics (Pilot Schools), in box 19, entry 18; folder: Methods of Instruction, Curriculum, Schools, in box 21, entry 19; and folder: New Curriculum, Oct 1918, in box 19, entry 18, all in RG 18, NA.

20. Memo 163 for All Schools, Sep 14, 1917, folder: Memorandums, box 20, entry 18, RG 18, NA.

21. Bingham, *Explorer*, 32.

22. Ibid., 18.

23. Telg, Squier to Pres, Harvard

Notes

University, Jun 5, 1917, Lowell Papers, HUL.

24. Memo, Ofc Chief Sig Off to Pres, Harvard College, Dec 21, 1917, Lowell Papers, HUL.

25. Brereton Greenhous, ed., *A Rattle of Pebbles: The First World War Diaries of Two Canadian Airmen* (Ottawa: Minister of Supply and Services Canada, 1987), xii–xiii.

26. Harold E. Hartney, *Up and At 'Em* (Harrisburg, Pa.: Stackpole, 1940; New York: Arno Press, 1980), 143.

27. Davidson intvw, Dec 5–8, 1974, 188–189.

28. Clarence D. Chamberlin, *Record Flights* (New York: Dorrance & Co., 1928), 193–94.

29. Robert M. Todd, *Sopwith Camel Fighter Ace* (Falls Church, Va.: AJAY, 1978), 20.

30. Bingham, *Explorer*, 20.

31. Memo, 1st Lt John C. Farrar to Col Carberry, Dec 24, 1917, Subj: Report on Ground Training in French and English Schools and Suggestions for Similar Training in U.S., folder: American E.F. Aviation Instruction Center, box 20, entry 18, RG 18, NA.

32. Rprt, Lt Col Dargue, on his tour of France and England, Nov 9, 1918, app to Frey, "Evolution of Maintenance Engineering," vol 3, doc 48.

33. Memo 54–A for All Schools, Aug 13, 1917, folder: Memorandums, box 20, entry 18, RG 18, NA.

34. Rprt, 1st Lt S. M. Clement to Capt Geo. A. Washington, n.d., Subj: Results of Observations While on Temporary Duty in France and England, February to April, 1918, folder: Report of Lt. S. M. Clement, box 19, entry 18, RG 18, NA.

35. Bingham, *Explorer*, 12–13.

36. Davidson intvw, Dec 5–8, 1974, 204.

37. Lt Col C. F. Lee to Lt Col L. Rees, Sep 6, 1917, folder: Information Section, Air Division, box 23, entry 19, RG 18, NA.

38. Lt Col Arthur Woods to "Sir," Jun 24, 1918, box 526, 333.1, entry 166, RG 18, NA.

39. Memo, Lt Col W. E. Gil[rune] to Chief Sig Off Admin Sec (attn: Col Arnold), Apr 22, 1918, box 621, 353.9, entry 166, RG 18, NA.

40. Woods to "Sir," Jun 24, 1918.

41. Bingham, *Explorer*, 30.

42. Memo 152 for All Schools, Sep 12, 1917, folder: Memorandums, box 20, entry 18, RG 18, NA.

43. Memo, 1st Lt S. T. Williams to Overseas Tng Mission, Subj: Information for Department of Military Aeronautics, Ground Schools Section (3d A.I.C. Issoudun, France), box 177, C53 45, entry 206, RG 18, NA.

44. Memo 118 for All Schools, Aug 30, 1917, folder: Memorandums, box 20, entry 18, RG 18, NA.

45. Todd, *Sopwith Camel Fighter Ace*, 21.

46. Ibid.

47. Davidson intvw, Dec 5–8, 1974, 191.

48. John M. Grider, *War Birds: Diary of an Unknown Aviator* (New York: George H. Doran Co., 1926), 25.

49. Memo, Maj Gen J. L. Chamberlain, Insp Gen, to Adj Gen, Nov 13, 1917, Subj: Memorandum for the Chief of Staff, Aviation Schools, box 526, 333.1, entry 166, RG 18, NA.

50. Arthur Sweetser, *The American Air Service* (New York: D. Appleton & Co., 1919), 97–98.

51. Hugh Knerr, "The Vital Era, 1887–1950," unpubl. ms., Air Force History Support Office (AFHSO), 75.

52. Essington was named Chandler Field for 2d Lt. Rex Chandler, who died after his airplane crashed into San Diego Bay on April 8, 1913. Mineola was christened Hazelhurst Field for 2d Lt. L. W. Hazelhurst, who was killed while making a flight at College Park, Maryland, on June 11, 1912.

53. "Aircraft Production," Senate hearings, 65th Cong, 2d sess, May 29, 1918, vol 1, 1918, 4.

54. Memo, Lt Col B. Q. Jones to Admin Div, Dec 6, 1917, box 620, 353.9, entry 166, NA; "Flying Fields," May 28, 1918, folder: Air Service 1918 Organization, box 222, Arnold Papers, LCMD.

55. Casari, *Encyclopedia of U.S. Mili-*

583

Notes

tary Aircraft, 2: 3.

56. Memo 26, Tng Sec, Sep 14, 1918, box 677, vol: List of Stencils, entry 206, RG 18, NA.

57. Memo, Comdg Off Avn Sch, Mt. Clemens, Mich., to Chief Sig Off, Sep 15, 1917, Subj: Schedule of Instruction in Flying for Primary Schools, box 620, 353.9, entry 166, RG 18, NA; Stencil 482-C, Tng Series, Admin Pers of Flying Dept, Showing Duties, Nov 14, 1917, box 620, 353.9, entry 166, RG 18, NA; Stencil T-84, Tng Sec, Flying Br, Oct 25, 1918, box 677, vol: List of Stencils, entry 206, RG 18, NA.

58. Chamberlin, *Record Flights*, 198.

59. Hartney, *Up and At 'Em*, 114.

60. Dean Ivan Lamb, "Suggestions for Fliers of the J.N.4-B," *Air Service Journal* 2, no. 19 (May 9, 1918): 657.

61. Signal Corps Aviation School, Rockwell Field, frame 980, mcflm roll 1571, AFHRA.

62. Maj George B. A. Reinburg, Comdg, "Instructions for Cadets Flying Solo at This Station," box 423, 300.7, entry 166, RG 18, NA.

63. Intvw, James P. Hodges, K105.5-29.

64. Intvw, John Macready, AFHRC mcflm roll K1215, 9-10.

65. Rprt, Lt Col H. C. Pratt, Comdg Off, Brooks Fld, to Dir Mil Aero, Aug 20, 1918, Subj: Inspection of Flying Schools, Exhibit "A," box 526, 333.1, entry 166, RG 18, NA.

66. Casari, *Encyclopedia of U.S. Military Aircraft: The World War I Production Program* (Chillicothe, Ohio: Casari, 1975), pt 2, 3.

67. F. Tredrey, *Pioneer Pilot* (London: Peter Davies, 1976), 96. Tredrey's account primarily concerns the development of the Gosport system in England.

68. A confirmed advocate of the Gosport method who had gone to France and had then come back to teach in Texas at Hicks Field became horrified by instructors who, with no experience, attempted to copy the Gosport approach: "Those Texas fields got the reputation of being 'man killing outfits,' 'aviators' graveyards.' The deaths among students and instructors were appalling." Hartney, *Up and At 'Em*, 112.

69. Telg, Kenly to Col H. H. Arnold, Jul 31, 1918, box 464, 319.1, entry 166, RG 18, NA; Memo, Maj J. R. Moulthrop to Exec Section, Stat Branch, Aug 5, 1918, Subj: Weekly Letter for Chief of Air Service, AEF, box 464, 319.1, entry 166, RG 18, NA; Memo, Col R. Smith-Barry to Comdg Off Tng Sec, Dept Mil Aero, Jul 31, 1918, box 620, 353.9, entry 166, RG 18, NA; Memo, Col R. L. Montgomery to Chief of Staff, Sep 30, 1918, Subj: Investigation of Gosport System, box 620, 353.9, entry 166, RG 18, NA; Memo, Maj Gen W. L. Kenly, Dir Mil Aero, to Chief of Staff, Oct 19, 1918, box 620, 353.9, entry 166, RG 18, NA; Memo, H. M. Hickam, Comdg Off, Dorr & Carlstrom Flds, to Dir Mil Aero, Nov 8, 1918, Subj: Use of V. C. Equipment, box 620, 353.9, entry 166, RG 18, NA; Memo, Maj John B. Brooks, Comdg Off, to Dir Mil Aero, Tng Sec, Nov 29, 1918, Subj: Gosport System, box 620, 353.9, entry 166, RG 18, NA.

70. Org Chart, Aerial Gnr Sec, and Org of Admin Br, Aerial Gnr Sec, box 613, 353.9, entry 166, RG 18, NA; Memo, Ofc Chief Sig Off to Comdg Off Sig Corps Avn Sch, Mineola, Long Island, N.Y., Jul 14, 1917, Subj: General Course of Instruction at Signal Corps Aviation Schools, box 620, 353.9, entry 166, RG 18, NA.

71. Briggs K. Adams, *The American Spirit: Letters of Briggs Kilburn Adams, Lieutenant of the Royal Flying Corps* (Boston: Atlantic Monthly Press, 1918), 48.

72. Rprt, Trip to Thirteen Flying Fields, June 2, 1918-July 8, 1918, box 526, 333.1, entry 166, RG 18, NA; Memo, Maj R. S. Potter to Maj Robertson (thu: Chief of Training), Sep 10, 1918, Subj: Status of Gunnery Training, box 613, 353.9, entry 166, RG 18, NA.

73. Thayer, *America's First Eagles*, 142-144.

74. Memo, Dir Mil Aero to Comdg Off, Oct 9, 1918, Subj: Mid-week Report, Stencil T-78, box 677, vol: List of Stencils, entry 206, RG 18, NA.

75. Memo, Maj Gen Chas. T. Menoher, Dir Air Svc, to Chief of Staff, Apr 21, 1919, Subj: Errors in Report No. 4, Statistical Summary Series, entitled "Length of Time Required for Training Aviators," issued by Statistics Br of Gen Staff, box 620, 353.9, entry 166, RG 18, NA.

76. Memo, Col M. F. Davis, Chief of Tng, Ofc of Dir Mil Aero, to Comdg Offs, All Flying Flds, Aug 23, 1918, Subj: Classification of Pilots, Stencil T-28, box 677, vol: List of Stencils, entry 206, RG 18, NA.

77. Memo, Maj J. R. Moulthrop to Maj Gen Kenly, n.d., box 586, 352.9, entry 166, RG 18, NA.

78. "Flying Fields," May 28, 1918. See also Diagram Showing Stages of Flying Training, Aug 21, 1918.

79. Memo, Lt Col J. E. Carberry to Dir Mil Aero, Jul 19, 1918, Subj: Training of Pursuit Pilots, box 620, 353.9, entry 166, RG 18, NA.

80. Conference on January 19th to Determine upon a Pursuit School Curriculum, box 620, 353.9, entry 166, RG 18, NA.

81. Group Flying for Pursuit Squadrons, Jan 30, 1918, Stencil 905, frame 1112, mcflm roll 1571, AFHRA.

82. Swanborough, *United States Military Aircraft*, 506–507; Resume of Pursuit Training from April 1st 1918, box 620, 353.9, entry 166, RG 18, NA.

83. Lt Col J. E. Carberry to Comdg Offs, All Flying Flds, Jul 18, 1918, box 620, 353.9, entry 166, RG 18, NA.

84. Memo, L. T. [McMenemy] to Chief of Tng, May 24, 1918, Subj: Army Corps Pilots, box 621, 353.9, entry 166, RG 18, NA.

85. Memo, Maj H. H. Arnold to Col Bennet, Jul 3, 1917; 1st Ind, Edward T. Donnelly, Adj Gen, Jul 18, 1917; Memo, to Capt McConnell, Aug 30, 1917, Subj: Observers, box 586, 352.9, entry 166, RG 18, NA.

86. Memo, Col W. S. Graves to Adj Gen Army, Jan 24, 1918, Subj: Aerial Observers, box 620, 353.9, entry 166, RG 18, NA.

87. Memo, Comdg Off 19th Fld Arty to Comdg Gen Southern Dept, Ft. Sam Houston, Tex., Sep 24, 1917, Subject: Training of Aeroplane Observers; 3d Ind, Oct 19, 1917, and 6th Ind, Oct 29, 1917, box 620, 353.9, entry 166, RG 18, NA.

88. Memo, War Dept, Ofc Chief of Staff, to Adj Gen Army, Jan 12, 1918, Subj: Training of Aerial Observers, box 621, 353.9, entry 166, RG 18, NA.

89. Pdn Chart for Schs of Aerial Obsn, Feb 15, 1918, box 464, 319.1, entry 166, RG 18, NA.

90. Memo, D. R. Mayes to Capt W. W. Hoffman, Aug 17, 1918, box 620, 353.9, entry 166, RG 18, NA.

91. Memo, Lt Col F. R. Kenney to Tng Sec, Jul 19, 1918, box 620, 353.9, entry 166, RG 18, NA.

92. Memo, Col M. F. Davis to Dept Exec, Jul 29, 1918, Subj: Student Material to Be Trained as Aerial Observers; Memo, Chief of Staff to Comdg Gen, France, Aug 9, 1918, Subj: Aerial Observers for Artillery; Synopsis, Kenly, Sep 28, 1918, box 621, 353.9, entry 166, RG 18, NA.

93. Memo, Lt Col H. A. Dargue to Dir Mil Aero, Sep 30, 1918, box 621, 353.9, entry 166, RG 18, NA.

94. Memo, Col M. F. Davis, Chief of Tng, by dir, Dir Mil Aero, to Comdg Off, All Flying Schools, Aug 15, 1918, Subj: Non-piloting Flying Personnel, Stencil T-17, box 677, vol: List of Stencils, entry 206, RG 18, NA.

95. Memo, Dir Mil Aero, to Maj Gen M. M. Patrick, Oct 16, 1918, Subj: Summary of Weekly Progress Reports, box 222, Arnold Papers, LCMD.

96. Schools for Observers [recommendations of a conference]; Kenly synopsis, Sep 28, 1918; Training of Aerial Observers, box 621, 353.9, entry 166, RG 18, NA.

97. Memo, to Chief of Staff, Jul 25, 1918, Subj: Inspection of Aircraft Factories, Flying Fields and Schools, box 526, 333.1, entry 166, RG 18, NA; Harold E. Porter, *Aerial Observation: The Airplane Observer, the Balloon Observer, and the Army Corps Pilot* (New York: Harper & Bros., 1921), 60–62.

98. Memo, 1st Lt Paul K. Yost to Chief of Tng, Sep 11, 1919, Subj: Summary of Aerial Observer Training in the United

Notes

States, July 1, 1918 to June 30, 1919, box 620, 353.9, entry 166, RG 18, NA.

99. Major [ ] to Maj Guy L. Gearhart, Sep 21, 1917, box 586, 352.9, entry 166, RG 18, NA.

100. Memo 170 for All Schools, Jan 31, 1918, box 613, 353.9, entry 166, RG 18, NA.

101. Memo, Lt Col T. L. Crystal to Chief of Tng, Apr 22, 1918, Subj: Artillery Observation; Memo, Capt W. W. Hoffman to Col Carberry, Jun 21, 1918, box 621, 353.9, entry 166, RG 18, NA.

102. Appendix A, Aerial Gunnery School for Observers, Aerial Gunnery School for Pilots, Folder: Methods of Instruction: Curricula, box 21, entry 19, RG 18, NA.

103. Memo, to Schs Sec, Feb 28, 1918, Subj: Information Covering Duties of Bombers, Bombing Pilots, and Course at Bombing School, box 613, 353.9, entry 166, RG 18, NA.

104. Bull 205 for All Schools, Mar 19, 1918, box 613, 353.9, entry 166, RG 18, NA.

105. Memo, to Col Bane, Feb 9, 1918, Subj: Conducting Bombing at All Primary Training Schools, box 613, 353.9, entry 166, RG 18, NA.

106. Memo, Maj I. Longanecker, by dir, Chief Sig Off, to Comdg Off, Ellington Fld, Houston, Tex., Feb 22, 1918, Subj: Bombing Pilots and Students, box 613, 353.9, entry 166, RG 18, NA.

107. Memos, to Col Jones, Mar 19, 1918, Subj: Report on Conditions Obstructing Bombing Instruction; to Chief Sig Off Admin Sec, Apr 3, 1918, Subj: Information Requested from Bombing Branch, box 613, 353.9, entry 166, RG 18, NA.

108. Description of Bombing Course, Ellington Fld, Houston, Tex., folder: Methods of Instruction: Curricula, box 21, entry 19, RG 18, NA.

109. Bul 205 for All Schools, Mar 19, 1918, box 613, 353.9, entry 166, RG 18, NA.

110. Memo, 1st Lt Frederick Blakeman, Off in Charge Inst, to Lt J. C. Farrar, Dec 5, 1917, Subj: Suggested Course of Instruction for Students in American Schools, folder: American E.F. Aviation Instruction Center, box 20, entry 18, RG 18, NA.

111. Quoted in Memo, Col S. D. Waldon, Ofc Chief Sig Off, to Chief Sig Off, Apr 8, 1918, Subj: Training of Pilots, box 620, 353.9, entry 166, RG 18, NA.

Chapter 5

1. James J. Hudson, *Hostile Skies: A Combat History of the American Air Service* (Syracuse, N.Y.: Syracuse University Press, 1968), 15.

2. Americans subdivided the combat specialties into three functions — pursuit, observation, and bombardment — but this division of tactical roles varied in different air services. Until 1915, for example, the French air service included observation, reconnaissance, and bombardment, and that year a *chasse* (pursuit) section was added.

3. Memo, Avn Bd to Chief of Staff, Jun 25, 1917; Bd of Offs, "Recommendation," Jun–Jul 1917, convened in Paris; and Memo, for Avn Bd, Exhibit L, Jun 29, 1917, Subj: Aviation Training in Europe, folder: Aviation Program in Europe: Proceedings of Board of Officers, box 34, William Mitchell Papers, LCMD.

4. Rprt, Lt Col Dargue, on his tour of France and England, Nov 9, 1918, app to Frey, "Evolution of Maintenance Engineering," vol 3, doc 48.

5. Cable, Bolling to Signals, Jun 27, 1917, and Memo, for Chief of Staff, Jul 2, 1917, Subj: Pilots and Mechanics to Be Trained in France, box 569, 350.1, entry 166, RG 18, NA; History of the Training Section, Gorrell's History of the American Expeditionary Forces: Air Service, 1917–1919, mcflm M990 (hereafter Gorrell), Series J, vol 1; History of the Air Service in Italy and of American Pilots on the Italian Front, Gorrell, Series B, vol 1.

6. R. C. Bolling to Chief Sig Off, Aug 15, 1917, Subj: Report of Aeronautical Commission, Gorrell, Series A, vol 1.

7. See monthly figures in History of the Training Section, Gorrell.

8. Cable, Pershing, copy attached to Bd of Offs, "Recommendations," Jun–Jul 1917.

9. Memo, Dir Air Svc Inst to Chief of Tng Sec, Gen Staff (thru: Chief Air Svc), Nov 15, 1917, Subj: School Program for Aviation Training, folder: American E.F. Aviation Instruction Center, box 20, entry 18, RG 18, NA.

10. Memo, Maj W. A. Robertson, to Dir Mil Aero (thru: Chief of Tng), Sep 18, 1918, Subj: Report on My Activities Overseas, box 613, 353.9, entry 166, RG 18, NA.

11. Thayer, *America's First Eagles*, 46–48.

12. The RMA brevet entitled the cadet to a commission in the U.S. Army. Those passing their tests at French schools received the French brevet, and those graduating in Italy passed the second Italian brevet which was equivalent to the RMA. Students completing tests at the foreign schools were eventually certified with the RMA.

13. Davidson intvw, Dec 5–8, 1974.

14. Training: French Aviation Schools at Avord and Pau, Stencil 193, Jul 3, 1917, box 177, C53.74, entry 206, RG 18, NA.

15. Thayer, *America's First Eagles*.

16. John Francisco Richards II, *War Diary and Letters of John Francisco Richards II, 1917–1918* (Kansas City, Mo.: George B. Richards, 1925), entry for Aug 15, 1917.

17. Davidson intvw, Dec 5–8, 1974.

18. Benj. Stuart Walcott to "My Dear Mr. Waldron," Jul 22, 1917, folder: 1917, box 2, Mitchell Papers, LCMD.

19. Rprt, "Pursuit Training: Section III: Conclusions as to Preliminary and Pursuit Training," Gorrell, Series J, vol 6.

20. Hartney, *Up and At 'Em*, 132, 135, 125.

21. Lt Col W. G. Kilner to Chief Tng Sec, Air Svc, Mar 15, 1918, box 620, 353.9, entry 166, RG 18, NA.

22. Memo of Agreement, Chief Sig Off U.S. Army and Gen in Charge of Italian Mil Commission for the Italian Govt, box 620, 353.9, entry 166, RG 18, NA.

23. Josiah P. Rowe, Jr., *Letters from a World War 1 Aviator* (Boston: Sinclair Press, 1986), 21, 56–58.

24. 8th AIC, Foggia, Italy, Gorrell, Series J, vol 7; Memo, Asst Chief Air Svc AEF to Chief Sig Off, Washington, D.C., Oct 12, 1917, Subj: Aviation Personnel, box 620, 353.9, entry 166, RG 18, NA.

25. Hudson, *Hostile Skies*, 34–35.

26. Memo, Capt Charles M. Fleischmann, to CO, Air Service in Italy, Dec 19, 1918, Subj: Foggia School, History of the Air Service in Italy, Gorrell.

27. Mowatt M. Mitchell, Summary of Italian Aviation Training, 1918, Gorrell, Series B, vol 15.

28. 8th AIC, Foggia, Gorrell. The Italian S.I.A. aircraft manufactured by Societa Italiana Aviazione was a two-seat biplane intended for reconnaissance.

29. "Final Report of the Chief of Air Service, AEF," in Maurer Maurer, ed., *The U.S. Air Service in World War 1*, vol 1, *The Final Report and a Tactical History* (Washington, D.C.: Office of Air Force History, 1978).

30. Memo, Ofc Chief Sig Off to Adj Gen Army, Aug 29, 1917, Subj: Training of Flying Officers in England, box 620, 353.9, entry 166, RG 18, NA.

31. "Training: A Report on the Special American Course at the British Machine Gun School at Grantham, Stencil 915," Feb 6, 1918, Folder: Information Section, Air Division, box 23, entry 19, RG 18, NA.

32. Report on the Air Service Flying Training Department in England, Gorrell, Series B, vol 4; A History of the American Air Service in Great Britain, Gorrell, Series B, vol. 2.

33. Grider, *War Birds*, 51, 55.

34. See Chapter 4. For a critique of English methods, see Training: Report on the Methods of Instruction at the Special School of Flying, R.F.C., Gosport, England, Stencil 885, Jan 29, 1918, Folder: Information Section, Air Division, box 23, entry 19, RG 18, NA.

Notes

35. The British considered the Avro vital to their all-through Gosport training method, but the airplanes were in short supply. As late as May 1918 one American officer reported he anticipated "some considerable time" before there would be enough planes to implement the system. Lt Col [B.] Q. Jones to Lt Col T. H. Bane, May 11, 1918, box 526, 333.1, entry 166, RG 18, NA.

36. Grider, *War Birds*, 55, 59.

37. Thayer, *America's First Eagles*, 61.

38. "Final Report," in Maurer, ed., *U.S. Air Service in World War I*, vol 1.

39. Lessons Learned from British System of Training, in Report on the Air Service Flying Training Department in England, Gorrell.

40. "Final Report," in Maurer, ed., *U.S. Air Service in World War I*, vol 1.

41. Rowe, *Letters*, to his Mother, June 2, 1918.

42. History of the Training Section, Gorrell.

43. Memo, J. Tulasne for Gen Squier, May 8, 1917, box 620, 353.9, entry 166, RG 18, NA.

44. Memo, Maj W. A. Robertson to Dir Mil Aero (thru: Chief of Tng), Sep 18, 1918, Subj: Report on My Activities Overseas, box 613, 353.9, entry 166, RG 18, NA.

45. Richards, *War Diary*, Nov 6, 1917.

46. In 1938, Carl Spatz changed the spelling of his surname to Spaatz. All text references will be spelled Spaatz. Original spellings are retained in the notes.

47. Victor H. Straham to his parents, Mar 21, 1918, Manuscript Collection, Western Kentucky University Library and Museum, Bowling Green, Kentucky.

48. Memo, Comdg Off 3d AIC to Dir Air Svc Tng, Dec 22, 1917, Subj: Cross-country Flying, box 620, 353.9, entry 166, RG 18, NA.

49. Dargue rprt in Frey, "Evolution of Maintenance Engineering."

50. Cecil [C.S.] Lewis, *Sagittarius Rising* (New York: Harcourt, Brace & Co., 1936), 47–48.

51. "Final Report," in Maurer, ed., *U.S. Air Service in World War I*, vol 1.

52. Memo, for Chief of Svc, Dec 16, 1918, Subj: The Air Service, box 480, 321.9, entry 166, RG 18, NA.

53. Rprt on Pursuit Training: Section III, Gorrell. The meter designation of Nieuports referred to the square meters in the wing surface.

54. Ibid.

55. Rprt, Air Svc Tng Sec, 3d AIC, Jul 21, 1918, box 464, 319.1, entry 166, RG 18, NA.

56. Bingham, *Explorer*, 142.

57. For a discussion of the evolution of air combat and the role of pursuit, observation, and bombardment, see Malcolm Cooper, "The Development of Air Policy and Doctrine on the Western Front, 1914–1918," *Aerospace Historian* 28 (Spring/March 1981): 38–51.

58. The term "bombardier" will be used hereafter, although it was not normally used at the time. Typically, the Air Service referred to bombardiers as bombing observers.

59. Memo, Chief Tng Sec, Air Svc AEF, to Chief Tng Div, Ofc Dir Mil Aero, Washington, D.C., Aug 11, 1918, Subj: Training of Flying Personnel, box 620, 353.9, entry 166, RG 18, NA.

60. Memo, Chief Tng Sec, Air Svc AEF, to Dir Mil Aero, Tng Sec (attn: Gd Schs Br), Aug 12, 1918, Subj: Information Concerning Training for Ground Schools Branch, folder: American E.F. Aviation Instruction Center, box 20, entry 18, RG 18, NA.

61. Thayer, *America's First Eagles*, 144.

62. Mason Patrick, *The United States in the Air* (Garden City, N.Y.: Doubleday, Doran & Co., 1928), 17.

63. John H. Morrow, Jr., *The Great War in the Air: Military Aviation, 1909–1921* (Washington, D.C.: Smithsonian Institution Press, 1993), 272.

64. Rprt, Pursuit Training: Section IV, History of Pursuit Training, Gorrell, Series J, vol 6.

65. Ibid.

66. Patrick, *United States in the Air*, 31.

67. The British-designed Bristol Fighter was a two-seat pursuit airplane used in combat by the British and gracefully retired to training by 1917. For a discus-

sion of its successes in combat, see Richard P. Hallion, *Rise of the Fighter Aircraft, 1914–1918* (Annapolis, Md.: The Nautical & Aviation Publishing Co. of America, 1984), 145.

68. Rprt on Pursuit Training: Section III, Gorrell.

69. "Final Report," in Maurer, ed., *U.S. Air Service in World War I*, vol 1.

70. Lewis, *Sagittarius Rising*, 35.

71. Davidson intvw, Dec 5–8, 1974.

72. Sweetser, *American Air Service*, 123.

73. Bennett A. Molter, *Knights in the Air* (New York: D. Appleton & Co., 1918), 81.

74. Bingham, *Explorer*, 146.

75. Richard Hallion makes the point that Americans learned not to fly large formations of tightly massed aircraft. "Following combat experience in the Battle of Chateau Thierry, the American air service decided to utilize large formations, but spaced out in such a fashion that if German fighters attacked, they could attack only one small unit at a time, and the rest of the large formation would be in better position for counterattack." Hallion, *Rise of the Fighter Aircraft*, 144. Colonel Thomas DeWitt Milling, who ended the war as Chief of Air Service, First Army, maintained that the change in tactics during this battle so proved its efficacy that the method was "immediately adopted by the Germans." Maurer Maurer, ed., *U.S. Air Service in World War I*, vol 4, *Postwar Review* (Washington, D.C.: Office of Air Force History, 1979), 9.

76. Maurer, ed., *U.S. Air Service in World War I*, 4: 67.

77. Bingham, *Explorer*, 151.

78. Ibid., 163.

79. Memo, Capt L. T. McMenemy, to Capt H. H. Salmon, Jr., Sep 21, 1918, Subj: Outline and Recommendations on Pursuit Training: Notes on D.H.4 Transformation, box 620, 353.9, entry 166, RG 18, NA.

80. Ltr, Maj Spatz, Jun 1, 1918, box 620, 353.9, entry 166, RG 18, NA.

81. Bingham, *Explorer*, 178, 174.

82. Draft Rprt on School at Furbara, 1st Lt H. S. Tierney, May 26, 1918, History of the Air Service in Italy, Gorrell.

83. "Final Report," in Maurer, ed., *U.S. Air Service in World War I*, vol 1.

84. Thayer, *America's First Eagles*, 249.

85. Memo, Comdr-in-Chief AEF to Secy Air Min, London, England, Jan 12, 1918, Subj: Training of American Cadets, Gorrell, Series A, vol 3.

86. Grider, *War Birds*, 95.

87. Peter Mead, *The Eye in the Air: History of Air Observation and Reconnaissance for the Army, 1785–1945* (London: Her Majesty's Stationery Office, 1983), 62–63, 75.

88. History of Aerial Observation Training in the American Expeditionary Forces 1917–1918, Gorrell, Series J, vol 3.

89. Mitchell journals and diaries, folder: 1917, box 2, Mitchell Papers, LCMD.

90. History of the Training Section, Gorrell.

91. History of Aerial Observation Training, Gorrell; Rprt 6, 1st Lt S. M. Clement to Capt George A. Washington, OIC Schs Br, Tng Sec, Air Div, Subj: Elementary Flying and Observers' School at Tours, folder: Rprt of Lt S. M. Clement, box 19, RG 18, entry 18, NA.

92. History of the Observers School: Second Aviation Instruction Center, Gorrell, Series J, vol 7.

93. Ibid.

94. Tours: 2d AIC, Gorrell, Series J, vol 7.

95. Pershing, *My Experiences*, 2: 195.

96. Memo, Chief Tng Sec Air Svc AEF, to Dir Mil Aero Tng Sec (attn: Gnd Schs Br), Aug 12, 1918, Subj: Information Concerning Training for Ground Schools Branch, box 177, C53 45, entry 206, RG 18, NA.

97. Memo, Chief Tng Sec Air Svc AEF to Chief Tng Div Ofc Dir Mil Aero, Washington, D.C., Aug 11, 1918, Subj: Training of Flying Personnel, box 620, 353.9, entry 166, RG 18, NA.

98. History of the Training Section, Gorrell; H. A. Toulmin, Jr., *Air Service, American Expeditionary Forces, 1918*

Notes

(New York: D. Van Nostrand Co., 1927), 291.

99. Bingham, *Explorer*, 172.

100. History of the Training Section, Gorrell.

101. Tours: 2d AIC, Gorrell; History of the Aerial Photography School: 2d AIC, Gorrell, Series J, vol 7; Rprt of Training at Radio Schools, 2d AIC: AEF, Jul 16, 1918, box 621, 353.9, entry 166, RG 18, NA.

102. "A Tactical History of the Air Service, AEF," in Maurer, ed., *U.S. Air Service in World War I*, vol 1.

103. History of Aerial Observation Training, Gorrell.

104. James L. Stokesbury, *A Short History of Air Power* (New York: William Morrow & Co., 1986), 31–35. In August 1914 the Germans enjoyed a considerable aeronautical advantage, having 384 military airplanes and 30 airships, or Zeppelins, against France's 123 aircraft including 10 airships. Of the French 113 airplanes, only 63 accompanied the British Expeditionary Force. See John Terraine, *The First World War, 1914–18* (London: Macmillan, 1965), 19. For discussion of German day and night raids carried out against England with large bombers during 1917–1918, see Raymond H. Fredette, *The Sky on Fire: The First Battle of Britain, 1917–1918* (Washington, D.C.: Smithsonian Institution Press, 1991).

105. Hudson, *Hostile Skies*, 52.

106. After arriving in France on November 27, 1917, Brig. Gen. Benjamin D. Foulois assumed command of all Air Service activities in the AEF. By General Order 81, GHQ AEF, Brig. Gen. (later Maj. Gen.) Mason M. Patrick replaced Foulois as Chief on May 29, 1918.

107. Rowe, *Letters*, 72.

108. Lawrence to "Mother," Sunday morning [1918], Lawrence J. Bauer Papers, Michigan Historical Collections, University of Michigan, Ann Arbor.

109. Edwin C. Parsons, "Chuck Kerwood, The Wildman of Aviation," in Charles W. Kerwood collection, Transportation History Center, University of Wyoming, Laramie.

110. Briggs Adams to "Father and Mother," Jan 12, 1918, in Adams, *American Spirit*.

111. Clermont-Ferrand: 7th AIC, Gorrell, Series J, vol 7; Beginning of American School at Aulnat-Puy-de-Dome, Oct 31, 1918, box 586, 352.9, entry 166, RG 18, NA.

112. Thayer, *America's First Eagles*, 58–59.

113. Clermont-Ferrand: 7th AIC, Gorrell.

114. Memo, OIC Bom Tng to Chief Inst Dept, Nov 13, 1917, Subj: Proposed Schedule of Training for Bombardment Teams at School of Bombardment, Aulnat (Puy-de-Dome), folder: American E.F. Aviation Instruction Center, box 20, entry 18, RG 18, NA.

115. Intvw, Prosper E. Cholet by Maj Thomas A. Keaney and Capt Robert S. Bartanowicz, Aug 10–11, 1976, K239.0512–939, AFHRA.

116. Memo, 2d Lt W. W. Williams to OIC Gd Schs Br, Tng Sec, Subj: Report on Visit to France and General Deductions on Ground Training, folder: Rprts on Ground Schools, box 19, entry 18, RG 18, NA; 1st Lt John [Farrar], VI: Report on Visits to Aviation Schools at Tours, November 19, Issoudun, November 21, Avord, November 23, and Pau, November 24, 1917, box 613, 353.9, entry 166, RG 18, NA; History of Day Bombardment Training: American E.F. 1917–1918, Gorrell, Series J, vol. 4.

117. Memo, Chief Tng Sec AEF to Dir Mil Aero, Washington, D.C., Aug 10, 1918, box 620, 353.9, entry 166, RG 18, NA.

118. History of Day Bombardment Training, Gorrell.

119. History of the Training Section, Gorrell.

120. "Final Report," Maurer, ed., *U.S. Air Service in World War I*, vol 1.

121. Capt Charles M. Fleischmann, History of Foggia, in History of the Air Service in Italy, Gorrell.

122. 2d Lt Joseph M. Aimee, American Pilots at Malpensa, in History of the Air Service in Italy, Gorrell.

123. Memo, for Chief of Air Svc, Mar

18, 1918, Subj: Reorganization of the Training Section, Gorrell, Series J, vol. 1.

124. Fleischmann, History of Foggia, Gorrell; Aimee, American Pilots, Gorrell.

125. 8th AIC Foggia, Gorrell.

126. Rprt, Lt Col B. Q. Jones to Lt Col T. H. Bane, box 464, 319.1, entry 166, RG 18, NA.

127. Cable, Foulois to Adj Gen, Jan 27, 1918, box 216, C53.2 89, entry 206, RG 18, NA; John J. Pershing to Maj Gen Sir H. Trenchard, Feb 6, 1918, box 620, 353.9, entry 166, RG 18, NA.

128. Charles H. Gibbs-Smith, *Aviation* (London: Her Majesty's Stationery Office, 1985), 176.

129. "Final Report," in Maurer, ed., *U.S. Air Service in World War I*, vol 1.

130. History of the Night Bombardment Section in France, Gorrell, Series B, vol 6.

131. History of the American Air Service in Great Britain, Gorrell.

132. Memo, Maj R. S. Potter to Aero Info Br, Nov 14, 1918, Subj: Report of Major R. S. Potter with regard to Gunnery Schools in France and England, box 463, 319.1, entry 166, RG 18, NA.

133. French Training Centers of Aviation at Pau and Cazaux, Stencil 23, Oct 30, 1916, folder: Information Section: Air Division, box 23, entry 19, RG 18, NA.

134. Memo, Maj W. A. Robertson to Dir Mil Aero (thru: Chief Tng), Sep 18, 1918, Subj: Report on My Activities Overseas, box 613, 353.9, entry 166, RG 18, NA.

135. Rprt 4, 1st Lt S. M. Clement to Capt Geo. A. Washington, OIC Tng Sec Schs Br, Subj: Gunnery, folder: Rprt of Lt. S. M. Clement, box 19, entry 18, RG 18, NA.

136. General Statement of Policy, Conditions and Equipment, Gorrell, Series J, vol 6.

137. History of the Aerial Gunnery School, Second Aviation Instruction Center, Gorrell, Series J, vol 7.

138. Memo, Chief Tng Sec, HQ Air Svc, Svcs of Supply, to Dept Mil Aero Tng Sec (attn: Gnr), Aug 5, 1918, Subj: Gunnery, box 613, 353.9, entry 166, RG 18, NA.

139. Memo, 1st Lt Stanley T. Williams to Overseas Tng Mission, Subj: Information for Department of Military Aeronautics, Ground Schools Section (3d AIC Issoudun, France), folder: American E.F. Aviation Instruction Center, box 20, entry 18, RG 18, NA.

140. Memo, for Chief of Air Svc, Mar 18, 1918, Subj: Reorganization of the Training Section, Gorrell.

141. Memo, Ofc Dir Mil Aero to Comdg Offs All Flying Flds, Jul 29, 1918, Subj: Notes on Training Squadrons at the Front, box 620, 353.9, entry 166, RG 18, NA.

142. Memo, Chief Air Svc to Chief of Staff, Jan 21, 1918, Subj: Scheme of Organization and Service Training of the American Air Service, box 620, 353.9, entry 166, RG 18, NA.

143. History of the Training Section, Gorrell.

144. B. M. Atkinson to Col J. E. Carberry, Mar 16, 1918, box 620, 353.9, entry 166, RG 18, NA.

145. "Tactical History," in Maurer, ed., *U.S. Air Service in World War I*, vol 1.

146. General Conditions in France, Apr 5, 1918, box 464, 319.1, entry 166, RG 18, NA.

147. "Tactical History," in Maurer, ed., *U.S. Air Service in World War I*, vol 1.

148. General Conditions in France, Apr 5, 1918.

149. Draft Rprt, Dir Mil Aero, box 463, 319.1, entry 166, RG 18, NA.

150. Memo, Maj. Gen. Chas. T. Menoher, Dir Air Svc, to Chief of Staff, Apr 21, 1919, Subj: Errors in Report No. 4, Statistical Summary Series, entitled "Length of Time Required for Training Aviators," issued by Statistics Br of the Gen Staff, box 620, 353.9, entry 166, RG 18, NA.

151. Intvw, Gen George C. Kenney by Col Marvin M. Stanley, Secy, AF Office Info, Jan 25, 1967, K239.0512-747, AFHRA.

152. Grider, *War Birds*, 109.

153. Peter H. Liddle, *The Airman's War, 1914-18* (UK: Blandford Press, 1987), 9.

Notes

154. Lee Kennett, *The First Air War, 1914–1918* (New York: Free Press, 1991), 22.
155. Edgar S. Gorrell, *The Measure of America's World War Aeronautical Effort* (Burlington, Vt.: Lane Press, 1940) (hereafter *Aeronautical Effort*), 14.
156. "Final Report," in Maurer, ed., *U.S. Air Service in World War I*, 1: 17.
157. Morrow, *Great War in the Air*, 338.
158. Ernest Toller, quoted in *The Penguin Book of First World War Prose*, Jon Gover, Jon Silkin, eds. (New York: Penguin, 1990), 273.
159. Draft Rprt, Dir Mil Aero, box 463, 319.1, entry 166, RG 18, NA.
160. Sweetser, *American Air Service*, 262, 263, 268; "Final Report," in Maurer, ed., *U.S. Air Service in World War I*, 1: 110.
161. Gorrell, *Aeronautical Effort*, 15.
162. Maurer, ed., *U.S. Air Service in World War I*, 4: 18–19.

Part III

1. Memo, Brig Gen C. E. Kilbourne, Asst Chief of Staff War Plans Div, for Asst Chiefs of Staff G-1 through G-4, Chief of Air Corps, Comdt Army War College, Comdt Comd & Staff Sch, Army Comdrs, and Comdg Gen GHQ, box 485, 321.9, entry 166, RG 18, NA.
2. It is not possible to cite the considerable literature on the topic, but a number of cultural and intellectual historians have portrayed World War I as a watershed. See Henry F. May's influential *The End of American Innocence: A Study of the First Years of Our Own Time, 1912–1917* (New York: Quadrangle, 1964), Paul Fussell's eloquent *The Great War and Modern Memory* (New York: Oxford University Press, 1975), and Robert H. Wiebe's classic, *The Search for Order* (New York: Hill & Wang, 1967). In Rob Kroes, "COMMENTARY: World Wars and Watersheds: The Problem of Continuity in the Process of Americanization" [*Diplomatic History* (Winter 1999): 73], the author states: "America's intervention in the war, the presence of its armies in Europe, the massive advent of its mass culture in subsequent years, allow us to look at World War I as a watershed."
3. Paul Kennedy, *The Rise and Fall of the Great Powers* (New York: Random House, 1987), 277.
4. Neil A. Wynn, *From Progressivism to Prosperity* (New York & London: Holmes & Meier, 1986), 199–200.
5. Memo, Lt Col Jacob W. S. [Wuest] to Chief of Air Corps, Washington, D.C., May 17, 1930, Subj: Overseas Air Operations, frame 128, 248.12602, mcflm roll A2711, AFHRA.
6. Rprt, Comte of Offs appointed by Secy War, Mar 27, 1923, 145.93–102A, mcflm roll A1443, AFHRA.
7. Intvw, Gen Orval R. Cook by Maj Richard H. Emmons and Hugh N. Ahmann, Jun, Aug 1974, 40–41, K239.0512-740, AFHRA.
8. Barry D. Karl, *The Uneasy State* (Chicago, Ill.: University of Chicago Press, 1983), 98.
9. John F. Shiner, "Benjamin D. Foulois: In the Beginning," in *Makers of the United States Air Force*, John L. Frisbee, ed. (Washington, D.C.: Office of Air Force History, 1987), 25–26.
10. Ltr, Lt Col F. M. Andrews to Oscar [Brig Gen Westover], Sep 21, 1933, folder: Aug 3–Oct 31, 1933, box 6, Spaatz Papers, LCMD.
11. Mar 15, 1935, mcflm roll 1458, AFHRA.
12. Kilbourne memo for Asst Chief of Staff War Plans Div et al.
13. "Report of Board on Revision of Training Methods," 1934, frames 549–550, 167.66–12, mcflm roll A1577, AFHRA.
14. Col W. H. Frank, Air Div Chief of Staff, by comd of Maj Gen Andrews, to Comdg Gens of all GHQ Air Force Wgs, May 25, 1938, Subj: GHQ Air Force Training Directive, 1938–1939, box 611,

353.9, entry 166, RG 18, NA.

15. John F. Shiner, *Foulois and the U.S. Army Air Corps 1931–1935* (Washington, D.C.: Office of Air Force History, 1981), 241; Maurer Maurer, *Aviation in the U.S. Army, 1919–1939* (Washington, D.C.: Office of Air Force History, 1987), 346.

16. Memo, Maj Carl Spatz to Chief of Air Corps, Jul 14, 1934, Subj: War-time Employment of Air Corps, folder: Jan 2–Jul 31, 1934, box 6, Spaatz Papers, LCMD.

17. Shiner, *Foulois*, 236.

18. Lt Col A. W. Robins to Chief of Air Corps, Dec 1, 1932, Subj: Air Corps Training, Heavier-than-air, box 610, 353.9, entry 166, RG 18, NA.

19. Laurence K. Loftin, Jr., *Quest for Performance: The Evolution of Modern Aircraft* (Washington, D.C.: National Aeronautics & Space Administration, 1985).

20. Roger E. Bilstein, *Flight in America 1900–1983* (Baltimore & London: The Johns Hopkins University Press, 1984), 83–123.

21. Gerald T. Cantwell, *Citizen Airmen: A History of the Air Force Reserve, 1946–1994* (Washington, D.C.: Air Force History & Museums Program, 1997), 18.

22. Rprt, MacArthur, quoted in *Washington Herald*, Dec 24, 1933, folder: Nat. Defense 1933–42, box 11, Andrews Papers, LCMD.

Chapter 6

1. Memo, H.M. Hickam to Exec, Feb 9, 1921, file B, box 481, 321.9, entry 166, RG 18, NA.

2. Memo, Col Milton F. Davis to Div Exec, Nov 29, 1918, with atch, A Project for the Size of a Separate Air Service During Peace, Nov 27, 1918, box 483, 321.9, entry 166, RG 18, NA.

3. Rprt, Dir Air Svc to Secy War, 1920.

4. Proceedings, Bd Offs appointed by Para 6, Pers Order 187, Ofc Dir Air Svc, Oct 18, 1919, Jan 23, 1920, folder B, box 481, 321.9, entry 166, RG 18, NA.

5. The 1920 act changed the title from Director to Chief of Air Service.

6. Memo, Lt Col W. C. Sherman to Chief T&O Gp, Jul 12, 1919, box 480, 321.9, entry 166, RG 18, NA. Sherman would have drafted peacetime training regulations, mindful of experience and doctrine coming out of the war. For an analysis of the nature and antecedents of American concepts of strategic aviation in particular at the end of World War I, as articulated by Lt. Col. Edgar S. Gorrell, see George K. Williams, "'The Shank of the Drill': Americans and Stategical Aviation in the Great War," *Journal of Strategic Studies*, Sep 1996: 381–431.

7. Memo, Maj E. A. Lohman, Comdg, n.d., box 617, 353.9, entry 166, RG 18, NA.

8. Memo, Col C. DeF. Chandler to Dir Air Service (thru: Chief T&O Gp), Apr 18, 1919, box 480, 321.9, entry 166, RG 18, NA.

9. Memo, Lt Col O. Westover to Col Davis, Apr 26, 1919, box 480, 321.9, entry 166, RG 18, NA.

10. James E. Hewes, Jr., *From Root to McNamara* (Washington D.C.: Center of Military History, United States Army, 1975), 52.

11. An Rprt, Chief Air Svc, FY ending Jun 30, 1922.

12. Maj H. A. Dargue, "Training a Major-General to Fly," *U.S. Air Service*, Sep 1923, 22, copy in AIAA files, LCMD.

13. Synopsis, Ltr, A.J. Clayton to Chief Air Svc, Jan 22, 1920, Subj: Entries into Flying Schools, box 586, 352.9, entry 166, RG 18, NA.

14. For a discussion of the status of the Reserve at this time and during the ensuing interwar years, see Cantwell, *Citizen Airmen*, 7–22.

15. Memo, Brig Gen Wm. Mitchell, by Lt Col W.C. Sherman, to Mr. MacFarland, Hist Sec, Library Div, Info

Notes

Gp, Sep 25, 1919, Subj: History of Field Operations Section, Operations Division, box 480, 321.9, entry 166, RG 18, NA.

16. Memo, Maj Gen Henry Jervey to Asst Secy War, Mar 1, 1919, Subj: Disposition of Flying Fields, entry 206, RG 18, NA.

17. Memo, Adj Gen Army to Dir Air Svc, Feb 25, 1920, Subj: Air Service Schools, box 575, 352.01, entry 166, RG 18, NA.

18. Russell F. Weigley, *Eisenhower's Lieutenants: The Campaign of France and Germany, 1944–1945* (Bloomington: University of Indiana Press, 1981), 4.

19. The Air Service Field Officers School was first organized at Langley Field, Virginia, in 1920. In November 1922 it was designated the Air Service Tactical School and became the Air Corps Tactical School in 1926. Here ideas concerning tactics and doctrine were studied, debated, and taught during the interwar years.

20. Memo, 1st Lt Ed. H. Guilford to Chief Air Svc, Jul 19, 1923, Subj: Proposed Training Regulations 205-5 (Observation Air Service Co-operation with Coast Artillery), folder: Training Regulations, 1922, entry 139, RG 18, NA.

21. Ltr, Bill Sherman to Barton [K. Yount], Nov 17, 1921, folder: Training Schools, 1921–1923, box 1, entry 139, RG 18, NA.

22. Memo, Lt Col J. E. Fechet, Chief T&O Gp, Aug 9, 1921, to Admin Exec, Subj: Comments on General Order Governing Training, box 619, 353.9, entry 166, RG 18, NA.

23. Robert Frank Futrell, *Ideas, Concepts, Doctrine: A History of Basic Thinking in the United States Air Force, 1907–1964* (Maxwell Air Force Base, Ala.: Air University, 1971), 22–23; Thomas H. Greer, *The Development of Air Doctrine in the Army Air Arm, 1917–1941* (Washington, D.C.: reprint, Office of Air Force History, 1985) (hereafter *Air Doctrine*), 16.

24. Draft of speech, Gen Mason Patrick, Ft. Leavenworth, Mar 27, 1924, box 1, entry 229, RG 18, NA.

25. Greer, *Air Doctrine*, 40.

26. Ibid., 30.

27. An Rprt, Chief Air Svc, FY ending Jun 30, 1921.

28. Memo, Chief Air Svc to Adj Gen Army, Jan 6, 1921, Subj: Reorganization Policy for the Air Service, box 485, 321.9, entry 166, RG 18, NA.

29. Memo, Chief Air Svc to Adj Gen Army, Washington, D.C., Apr 2, 1921, Subj: Strength of the Air Service, box 485, 321.9, entry 166, RG 18, NA.

30. Maurer, *Aviation in the U.S. Army*, 45.

31. Wkly News Letter (Div Mil Aero), week ending Sat, Dec, 7, 1918; Memo, Col A. L. Fuller to Exec, Mar 9, 1920, box 481, 321.9, entry 166, RG 18, NA.

32. Intvw, Brig Gen Noel F. Parrish by James C. Hasdorff, Jun 10–14, 1974, 34, K237.0512–744, AFHRA.

33. Robert L. Thompson, "Initial Selection of Candidates for Pilot, Bombardier, and Navigator Training," USAF Hist Study No. 2, Nov 1943, AFHRA.

34. "The Air Corps" (for Radio Speech), Feb 3, 1931, mcflm roll 167, AFHRA.

35. Memo, W. C. Sherman to Chief T&O Gp, Dec 19, 1919, Subj: Report on Pilots Fields, box 586, 352.9, entry 166, RG 18, NA.

36. Activities of Pilot Schools (March Field, Riverside, California, Carlstrom Field, Arcadia, Florida) During Fiscal Year, 1920, Jul 9, 1920, from Rprt, Tng Div, FY 1920, box 480, 321.9, entry 166, RG 18, NA.

37. Rprt, Tng Ops Gp, Jul 1–Dec 31, 1920, box 480, 321.9, entry 166, RG 18, NA.

38. "Transition" was the term later used both for the formal course of training or the more informal process of familiarization with new aircraft, usually tactical. In the 1920s the Air Service called the phase "transformation." This narrative substitutes the more modern "transition."

39. Memo, 1st Lt C. C. Moseley to Chief Tng Div, May 16, 1921, folder: 353.9 Misc. Training Projects 1921, box 3, entry 129, RG 18, NA.

40. Memo, Lt Col H. E. Hartney to Capt Drayton, Feb 27, 1920, box 586,

352.9, entry 166, RG 18, NA.
41. Knerr, "The Vital Era," 89.
42. An Rprt, 1922.
43. Distribution of 8760 Air Svc, box 485, 321.9, entry 166, RG 18, NA.
44. 5th Ind, Gen Mason M. Patrick to Adj Gen Army, Nov 10, 1923, box 619, 353.9, entry 166, RG 18, NA.
45. Prog Inst, for the General and Special Schools, 1922, vol 1, 248.122-7, mcflm roll A2704, AFHRA.
46. Proceedings, Bd Offs convened at Brooks Fld, San Antonio, Tex., box 619, 353.9, entry 166, RG 18, NA.
47. Distribution 8760 Air Svc.
48. Memo, H.M. Hickam to Comdg Off, Kelly Fld, Tex., Feb 25, 1924, Subj: Effect of Change in Proposed Air Service Organizations, box 485, 321.9, entry 166, RG 18, NA.
49. Capt James B. Burwell, "History of Kelly Field," *Air Corps News Letter*, 19, no. 1 (Jan 1, 1936): 17.
50. An Rprt, Chief Air Svc, FY ending Jun 30, 1923.
51. Ltr, [Maj Carl Spatz] to Gen [William Mitchell], May 21, 1923, folder: May 1–Jun 30, 1923, box 3, Carl Spaatz Papers, LCMD.
52. Memo, Kilner to Gen Menoher, Feb 21, 1920, entry 166, RG 18, NA.
53. Memo, Chief Air Svc to Dept Air Off, Philippine Dept, Manila, P.I., Mar 15, 1921, Subj: Air Service Tactical Doctrines, Principles and Methods, box 619, 353.9, entry 166, RG 18, NA.
54. An Rprt, 1921.
55. Ltr, Sherman to [Yount], Nov 17, 1921.
56. Air Svc Pilots Schools, Mar 1920 to Mar 15, 1921, entry 206, RG 18, NA.
57. Rprt, T&O Gp, Jul 1–Dec 31, 1920.
58. Memo, Lt Col H. H. Hartney to Chief T&O Gp, Jun 23, 1920, box 586, 352.9, entry 166, RG 18, NA.
59. Rprt, Dept Air Svc Off, Southern Dept, to Chief Air Svc, Washington, D.C., Jul 31, 1920, Subj: Semi-monthly Report on Activities and Tactical Training, period ending July 31, 1920, box 2, entry 129, RG 18, NA.
60. Memo, Moseley to Chief Tng Div, May 16, 1921.

61. Tng Reg 440-75, Air Svc, *The Pursuit Pilot*, 1923, folder: Nov 1–Dec 29, 1923, box 3, Spaatz Papers, LCMD.
62. Ltr, B. V. Baucom to Maj [Spatz], Sep 15, 1923, folder: Sep 4–Oct 31, 1923, Spaatz Papers, LCMD.
63. Ltr, H. W. Cook to Maj Carl Spatz, May 1, 1923, folder: May 1–Jun 30, 1923, box 3, Spaatz Papers, LCMD.
64. Ltr, [Spatz] to Gen William Mitchell, May 15, 1923, folder: May 1–Jun 30, 1923, box 3, Spaatz Papers, LCMD.
65. Memo, B.V. Baucom to Asst Comdt, [Air Svc Advanced Flying Sch], Feb 15, 1924, folder: Jan 2–Feb 29, 1924, box 3, Spaatz Papers, LCMD.
66. Tng Reg 440-75, *Pursuit Pilot*, 1923.
67. Ltr, Baucom to [Spatz], Sep 15, 1923.
68. Memo, Baucom to Asst Comdt, Feb 15, 1924.
69. Ibid.
70. Ibid.
71. Ibid.
72. Memo, Chief Air Svc to Dept Air Off, Philippine Dept, Mar 15, 1921.
73. Intv, Gen Truman H. Landon by Hugh N. Ahmann, May 31–Jun 3, 1977, K239.0512-949, 48, AFHRA.
74. Prog Inst, 1922.
75. Greer, *Air Doctrine*, 39.
76. Maurer, *Aviation in the U.S. Army*, 80–81.
77. Jon A. Reynolds, "Education and Training for High Command: General Hoyt S. Vandenberg's Early Career" (Ph.D. diss., Duke University, 1980), 72.
78. Enzo Angelucci, *The Rand McNally Encyclopedia of Military Aircraft, 1914–1980* (New York: The Military Press, 1980), 153; Michael J. H. Taylor, *Warplanes of the World, 1918–1939* (London: Ian Allan Ltd., 1981), 83.
79. Memo, Moseley to Chief Tng Div, May 16, 1921.
80. *Air Service News Letter* 4, no. 38 (Oct 8, 1920); Maurer, *Aviation in the U.S. Army*, 157; Prog, Air Svc Tng, n.d., 248.173-2, mcflm roll A2729, AFHRA; Prog Inst, 1922.
81. Ltr, Maj H. S. Martin to Maj Carl

Notes

Spatz, Mar 24, 1924, folder: Mar 1–Apr 30, 1924, box 3, Spaatz Papers, LCMD.

82. Reynolds, "Education and Training for High Command," 33.

83. Maurer, *Aviation in the U.S. Army*, 81.

84. Maurer Maurer, ed. *Air Force Combat Units of World War II* (Washington, D.C.: Office of Air Force History, 1983), 29–30.

85. Schedule of Tng for Air Svc Tac Units (supersedes Prog Air Svc Tng, 1919), box 577, 352.11, entry 166, RG 18, NA.

86. Ltr, Maj Carl Spatz to Maj H. S. Martin, Jan 24, 1924, folder: Jan 2–Feb 29, 1924, box 3, Spaatz Papers, LCMD.

87. Quoted in Lee Kennet, *A History of Strategic Bombing* (New York: Charles Scribner's Sons, 1982), 49.

88. Reynolds, "Education and Training for High Command," 37.

89. Synopsis, Memo, Chas. T. Menoher to Dir War Plans Div, Gen Staff, Aug 13, 1920, box 586, 352.9, entry 166, RG 18, NA.

90. Tng Prog for Air Svc Units, Dec 12, 1922; Tng Prog for Air Svc Units, 1925, box 610, 353.9, entry 166, RG 18, NA; An Rprt, Chief Air Svc, 1925.

Chapter 7

1. *The Flying Kadet* 9, no. 8 (Jul 30, 1932).

2. Shiner, *Foulois*, 102.

3. Memo, F. Trubee Davison to Chief of Air Corps, Mar 11, 1932, box 619, 353.9, entry 166, RG 18, NA.

4. An Rprt, Chief of Air Corps, 1933.

5. Maurer, *Aviation in the U.S. Army*, 348.

6. Ibid., 191–221; Shiner, *Foulois*, 31.

7. See, for example, correspondence from Maj. W. G. Kilner in the Chief's Office to Commanding Officers of units throughout the Air Service, soliciting their thoughts and recommendations, and their responses. Box 485, 321.9, entry 166, RG 18, NA.

8. Army Reg 95–10, *Air Corps*, box 483, 321.9, entry 166, RG 18, NA.

9. Memo, Maj W. G. Kilner, Mar 9, 1926, to Chief Tng & War Plans Div, box 619, 353.9, entry 166, RG 18, NA.

10. Memo, Maj H. A. Dargue to Chief Tng & War Plans Div, May 28, 1926, Subj: Comments on Executive Memorandum Dated March 9, 1926, box 619, 353.9, entry 166, RG 18, NA.

11. *Aviation* 21, no. 23 (Dec 6, 1926): 949.

12. Ltr, Maj Gen J. E. Fechet to Hon W. Frank James, House of Representatives, May 9, 1928, box 483, 321.9, entry 166, RG 18, NA.

13. Untitled study, n.d., box 619, 353.9, entry 166, RG 18, NA.

14. Memo, Brig Gen Campbell King to Chief of Staff, Aug 20, 1928, Subj: Air Corps Five Year Program, box 587, 352.17B, entry 166, RG 18, NA.

15. Memo, Lt Col F. M. Andrews to Chief, War Plans, Jul 22, 1930, Subj: Policies Announced in Circular 50–10, box 414, 300.5, entry 166, RG 18, NA.

16. Memo, Maj H. C. Pratt to Exec, May 14, 1926, box 619, 353.9, entry 166, RG 18, NA.

17. In August 1929 aviation cadet obligations increased to a three-year commitment, including the one year of flying training.

18. "The Ruggles Orientator," *Air Service News Letter* 2 (Apr 19, 1919): 2–3.

19. Memo, Maj J. E. Chaney to Chief of Air Corps, Washington, D.C., Apr 24, 1928, Subj: Change of Policy on Training, box 619, 353.9, entry 166, RG 18, NA.

20. The Air Corps bought 221 Consolidated-built PT–1s in 1925. This airplane had tandem seats and a Wright 180-horsepower engine. Succeeding models, the PT–3 and PT–4, were purchased in 1928 and 1929, and this aircraft remained the backbone of the training program until 1939. The new airplane designations indicated the function of the aircraft; in this case PT means "primary

Notes

trainer."

21. Parrish intvw, Jun 10–14, 1974, 34.

22. Memo, Brig Gen J. E. Fechet to Asst Secy War (Mr. Davison), Jul 30, 1927, Subj: Plans to Meet the Personnel Requirements of the 5-Year Program, box 483, 321.9, entry 166, RG 18, NA; Rprt, Chief T&O Div, Ofc of Chief of Air Corps, FY 1928, box 480, 321.9, entry 166, RG 18, NA; Ltr, Chaney, Jul 6, 1928; Burwell, "History of Kelly Field."

23. Ltr, F. Trubee Davison to Hon Hiram Bingham, Dec 13, 1929, box 619, 353.9, entry 166, RG 18, NA.

24. Memo, Maj Gen J. E. Fechet to Comdg Offs all Air Corps activities, Oct 15, 1930, Subj: Routine Training of Airplane Pilots, box 619, 353.9, entry 166, RG 18, NA.

25. 6th Ind, Maj W. G. Kilner, Exec, for Chief of Air Corps, to Adj Gen, Dec 19, 1930, box 616, 353.9, entry 166, RG 18, NA.

26. 8th Ind, Maj W. R. Weaver to Comdg Gen, IV Corps Area, Ft. McPherson, Atlanta, Ga., Jan 6, 1931, box 616, 353.9, entry 166, RG 18, NA.

27. Memo, Capt A. B. McDaniel to Exec, Jul 15, 1931, box 480, 321.9, entry 166, RG 18, NA.

28. Intvw, Lt Gen Glenn O. Barcus by Lt Col John N. Dick, Jr., Aug 10–13, 1976, 37, K239.0512-908, AFHRA.

29. Memo, Brig Gen W. E. Gillmore to Asst Secy War (Mr. Davison), Jun 30, 1929, Subj: Comparative Degree of Efficiency of Graduates of the Air Corps Advanced Flying School, Kelly Field, Texas, box 619, 353.9, entry 166, RG 18, NA.

30. *The Flying Kadet* 8, no. 10 (Nov 1931).

31. *Air Corps News Letter* 19, no. 7 (Apr 1, 1936): 3–5.

32. Capt John M. Weikert, "The Flying Training at the Air Corps Primary Flying School," *Air Corps News Letter* 19, no. 1 (Jan 1, 1936), 24.

33. Ltr, W. G. Kilner to Col Benjamin F. Castle, n.d., box 619, 353.9, entry 166, RG 18, NA.

34. Edward O. Purtee, *The Modification and Development of Training Aircraft for AAF Use, 1918–1945* (Wright Field, Ohio: Air Materiel Command, Nov 1946), 9–30.

35. The Consolidated- and Stearman-built PTs were, of course, not the only primary training aircraft, but they were significant in having been designed specifically for the purpose and in representing the evolution toward monocoque construction of monoplanes.

36. Purtee, *Modification and Development of Training Aircraft*, 42–53.

37. Spatz to George [Lt Col George H. Brett], Feb 21, 1935, folder: Jan 2–Apr 27, 1935, box 7, Spaatz Papers, LCMD.

38. Reynolds, "Education and Training for High Command," 146–153.

39. Extract, Tng Prog, Air Corps Primary Flying School (Staff and Faculty), Jul 1, 1935 to Jun 30, 1936, Oprd 165, box 614, 353.9, entry 166, RG 18, NA.

40. "The West Pointer at the West Point of the Air," *Air Corps News Letter* 19, no. 1 (Jan 1, 1936), 24.

41. Reynolds, "Education and Training for High Command," 150–151.

42. Intvw, Lt Gen Albert P. Clark by Maj Scottie S. Thompson, June 1979, 16, K239.0512-1130, AFHRA.

43. Intvw, Lt Gen Richard M. Montgomery by Capt Mark C. Cleary, Jun 28–30, 1983, 31–32, K239.0512-1526, AFHRA.

44. Parrish intvw, Jun 10–14, 1974.

45. Spatz to [Capt Kenneth] Walker, Jul 2, 1934, folder: Jan 2–Jul 31, 1934, box 6, Spaatz Papers, LCMD.

46. [Ken] to Maj Spatz, Jun 15, 1934, folder: Jan 2–Jul 31, 1934, box 6, Spaatz Papers, LCMD.

47. James C. Fahey, ed., *U.S. Army Aircraft (Heavier-than-Air) 1908–1946* (New York: Ships & Aircraft, 1946), 21, 24.

48. Burwell, "History of Kelly Field"; Fahey, ed., *U.S. Army Aircraft*, passim.

49. Montgomery intvw, Jun 28–30, 1983.

50. Cir 50–10, War Dept, Ofc Chief of Air Corps, Aug 16, 1938, *Training: Ratings and Requirements for Attainment Thereof*, box 414, 300.5, entry 166, RG

Notes

18, NA. In 1937 the Air Corps specified that the rating "Military Airplane Pilot" be held only by those who had been pilot-rated for twelve years and had logged 2,000 flying hours. See Maurer, *Aviation in the U.S. Army*, 379.

51. *The Flying Kadet* 9, no. 8 (Jul 30, 1932).

52. Cook intvw, Jun, Aug 1974, 40–41.

53. Memo, Maj W. R. Weaver to Comdg Gen, IV Corps Area, Jul 13, 1929, Subj: Report, Air Corps Training, Heavier-than-Air, box 617, 353.9, entry 166, RG 18, NA.

54. Memo, Maj Gen H. E. Ely to Adj Gen, Jul 14, 1930, Subj: Air Corps Training, Heavier-than-Air, 1929–1930, box 616, 353.9, entry 166, RG 18, NA.

55. Memo, Maj Gen [C. W.] Bridges, Adj Gen, to Comdg Gens all Corps Areas and Depts, to Chiefs War Depts Arms, Svcs, and Bureaus, and to Comdg Offs all exempted Air Corps Stas, Jul 20, 1931, Subj: Directive for Air Corps Training, box 610, 353.9, entry 166, RG 18, NA.

56. Directive, Ofc Adj Gen to Comdg Gens all Corps Areas and Depts, to Chiefs War Dept Arms, Svcs, and Bureaus, and to Comdg Offs all exempted Air Corps Stas, Jul 20, 1931, Subj: Directive for Air Corps Training, 248.177, mcflm roll A2730, AFHRA.

57. Tng Prog for Air Corps Units, 1927–28, to all Corps Area and Dept Comdrs, and Comdts Gen & Special Svc Schs, box 610, 353.9, entry 166, RG 18, NA.

58. Memo, Maj Gen H. E. Ely to Adj Gen, Jul 14, 1930, Subj: Air Corps Training, Heavier-than-Air, 1929–1930, box 616, 353.9, entry 166, RG 18, NA.

59. Tng Prog for Air Corps Units, 1927–28.

60. Distribution of Air Svc Tac Orgs, box 483, 321.9, entry 166, RG 18, NA.

61. Knerr, "The Vital Era," 108.

62. Tng Prog for Air Corps Units, 1927–28.

63. Landon intvw, May 31–Jun 3, 1977.

64. 2d Ind Maj John H. Jouett to Comdg Gen, VIII Corps Area, Ft. Sam Houston, Tex., Apr 29, 1929, box 617, 353.9, entry 166, RG 18, NA.

65. Tng outline for Flying Sch curriculum, Flying Sec, Mar 26, 1918, Stencil 1073, frame 982, mcflm A1571, AFHRA.

66. Monte Duane Wright, *Most Probable Position: A History of Aerial Navigation to 1941* (Lawrence: University Press of Kansas, 1972).

67. Lloyd L. Kelly, *The Pilot Maker* (New York: Grosset & Dunlap, 1970), 51–53.

68. *Air Corps News Letter* 19, no. 19 (Oct 1, 1936).

69. Kelly, *Pilot Maker*, 58–59.

70. Shiner, "Foulois: In the Beginning," in *Makers*, 30-31. The pilot fatality trend skyrocketed in 1934, undoubtedly attributable to the air mail operation. In 1933 the death rate per 1,000 had been 5.9; in 1934 it was 10.8. See Executive Office decimal file 240, Pay and Allowances Data, Hist Data, Dep Chief Staff Pers, RG 341, NA.

71. An Rprt, Chief of Air Corps, 1934.

72. Memo, Maj Carl Spatz, Chief T&O Div, to Brig Gen J. E. Chaney, Jul 3, 1934, box 480, 321.9, entry 166, RG 18, NA.

73. An Rprt, T&O Div, Jun 29, 1935, frame 1916, 167.6-11, AFHRA; Memo, Capt H. A. Halverson, Chief Tng Sec, to Maj Spatz, Jul 6, 1935, folder: May 1–Nov 8, 1935, and undated, 1935, box 7, Spaatz Papers, LCMD.

74. An Rprt, Chief of Air Corps, 1934.

75. 1st Ind, Col A. G. Fisher, Comdt Air Corps Tac Sch, to Chief of Air Corps, Jul 27, 1936, box 618, 353.9, entry 166, RG 18, NA.

76. Shiner, *Foulois*, 149.

77. Memo, Col Frank M. Kennedy to Chief of Air Corps, Apr 22, 1937, Subj: Instrument Landing Training, box 618, 353.9, entry 166, RG 18, NA.

78. See, for example, the memorandum from General Foulois to the Assistant Chief of Staff, War Plans Division, in which he reported on the results of fleet problem XVI, stressing the importance of "training all Air Corps tactical units in time of peace," through courses in naval operations and in a special service school for "navigation instruction." Synopsis, Memo, Gen Foulois to Asst Chief of Staff

Notes

War Plans Div, Aug 30, 1935, box 585, 352.17, entry 166, RG 18, NA.
  79. An Rprt, Chief of Air Corps, 1929.
  80. An Rprt, Chief of Air Corps, 1931.
  81. Memo, Capt Harold M. McClelland to Gen Foulois, Oct 2, 1928, box 575, 352.01, entry 166, RG 18, NA.
  82. William H. Tunner, *Over the Hump* (Washington, D.C.: reprint, Office of Air Force History, 1985), 5.
  83. Barcus intvw, Aug 10-13, 1976.
  84. Parrish intvw, Jun 10-14, 1974.

Chapter 8

1. Memo, F. M. Andrews, HQ GHQ Air Force, to Secy War, Jan 24, 1938, Subj: Air Corps Procurement Program, folder: Sec of War 1938-42, box 11, Andrews Papers, LCMD.
2. Maurer, *Aviation in the U.S. Army*, 327-330.
3. Geo. H. Brett to Frank M. Andrews, Oct 4, 1937, folder: Wilcox Bill 1937, box 11, Andrews Papers, LCMD.
4. War Dept, Ofc Memo 10-10, Jan 2, 1937, Miscellaneous: Peace Time Organization of the Office of the Chief of the Air Corps, folder C, 321.9, entry 166, RG 18, NA.
5. See Martha E. Layman, "Legislation Relating to the Air Corps Personnel and Training Programs, 1907-1939," USAF Hist Study No. 39, 1945, 125-126, AFHRA.
6. Robert T. Finney, "Early Air Corps Training and Tactics," *Military Affairs* 20 (Fall 1956): 159.
7. Memo, Maj Gen O. Westover to Asst Chief of Staff G-1, Nov 29, 1937, frame 2054, mcflm roll A1415, AFHRA.
8. In August 1933, MacArthur directed the Drum Board to assess the Office of the Chief of the Air Corps's Air Plan. See Shiner, *Foulois*, 121-122.
9. Mark Skinner Watson, *Chief of Staff: Prewar Plans and Preparations* (Washington, D.C.: Historical Division, United States Army, 1950), 26-31.
10. DeWitt S. Copp, *A Few Great Captains* (Garden City, N.Y.: Doubleday & Co., 1980), 124-126; Dennis G. Hall, "Andrews, Frank Maxwell," in Roger J. Spiller, ed., *Dictionary of American Military Biography*, vol 1 (Westport, Conn.: Greenwood Press, 1984), 33-36.
11. Ltr, Andy to Tuey, Nov 22, 1933, folder: Nov 1-Dec 26, 1933, box 6, Spaatz Papers, LCMD.
12. Copp, *Few Great Captains*, 112-113.
13. Memo, Maj Gen F. M. Andrews to Adj Gen, Washington, D.C., May 12, 1938, Subj: Detail of Officers to Service Schools, frames 2040-2041, mcflm roll A1415, AFHRA.
14. Memo, Comdg Gen GHQ Air Force to Asst Chief of Staff G-1, Dec 27, 1937, Subj: Assignment of Peace-time Air Corps Personnel in GHQ Air Force; Memo, Asst Chief of Staff Geo. R. Spalding to Chief of Staff, Jan 22, 1938, Subj: Five-Year Program for the Air Corps; Exhibit "A," Data on GHP Program as Submitted Jan 25, 1938, to Secy War (for purposes of correcting the memo for the Chief of Staff), Jan 22, 1938, Subj: Five-Year Program for the Air Corps, folder: War Dept Gen Staff, 1935-1942, box 11, Andrews Papers, LCMD.
15. Shiner, *Foulois*, 122.
16. When Andrews took command of GHQ Air Force, he selected Knerr as his Chief of Staff. The two men worked closely for the next three years, Knerr perhaps influencing Andrews's views about the importance of strategic bombardment, and Andrews serving as a model and mentor to Knerr. In 1927 Knerr commanded the 2d Bombardment Group at Langley Field. He participated in the flight of ten new Martin B-10 bombers to Alaska in 1934, and was a fierce advocate of the B-17 at GHQ Air Force.
17. H. J. Knerr to [Lt Col F. M. Andrews], Nov 21, 1933, folder: Nov 1-Dec 26, 1933, box 6, Spaatz Papers,

599

Notes

LCMD.

18. Ltr, H. H. Arnold to Tooey, Oct 24, 1933, folder: Aug 3–Oct 31, 1933, box 6, Spaatz Papers, LCMD.

19. Shiner, *Foulois*, 111, 244.

20. Ltr, Brig Gen W. G. Kilner, Act Chief of Air Corps, to Sen Henry Cabot Lodge, Dec 2, 1938, box 618, 353.9, entry 166, RG 18, NA.

21. Shiner, *Foulois*, 239–241; *Air Corps News Letter* 19, no. 6 (Mar 15, 1936).

22. For a general discussion of engineering progress during the period, see Roger E. Bilstein, *The American Aerospace Industry* (New York: Twayne, 1996), 36–48, 64–66.

23. Ltr, [Spaatz] to [Lt Col John H.] Pirie, Mar 25, 1933, folder: Jan 2–Mar 27, 1933, box 6, Spaatz Papers, LCMD.

24. Spatz to George [Brett], Dec 12, 1934, folder: Aug 2–Dec. 27, 1934, box 6, Spaatz Papers, LCMD.

25. Memo, Maj Vincent B. Dixon, Aug 6, 1935, Subj: Air Corps Training, Heavier-than-Air, box 610, 353.9, entry 166, RG 18, NA.

26. Spatz to [Capt Younger A.] Pitts, Feb 7, 1935, folder: Jan 7–Apr 27, 1935, box 7, Spaatz Papers, LCMD.

27. The three wings did not comprise the same number or type of squadrons. On March 1, 1935, for instance, the 1st Wing, headquartered at March Field, included two bombardment groups, one attack group, and one active observation squadron. The 2d Wing had two bombardment groups, two pursuit groups, and one active attached observation squadron. The 3d Wing at Barksdale Field had one attack group, one pursuit group, and no observation squadrons. The composition of the wings changed thereafter. Moreover, squadrons termed "observation" were those remaining in corps areas in support of ground forces or those in overseas departments. Reconnaissance squadrons were those under GHQ Air Force. See Maurer, *Aviation in the U.S. Army*, apps 5, 6, 7.

28. Ltr, Maj Carl Spatz to Gen [Foulois], Apr 21, 1932, folder: Jan 4–Jun 27 1932, box 5, Spaatz Papers, LCMD.

29. Ltr, Maj Spatz to [Lt Col John H.] Pirie, Feb 13, 1933, folder: Jan 2–Mar 27, 1933, box 6, Spaatz Papers, LCMD.

30. Arnold to Krog [Lt Col Arnold N. Krogstad], Dec 18, 1934, box 5, Arnold Papers, LCMD.

31. Maurer, *Aviation in the U.S. Army*, apps 5, 6.

32. H. H. Arnold, *Global Mission* (New York: Harper & Bros., 1949), 133.

33. David R. Mets, biog of Carl Spaatz, unpbl. ms., 80.

34. Arnold, *Global Mission*, 134.

35. Ibid., 137.

36. Address, Maj Gen Frank M. Andrews before the National Aeronautic Assoc, St. Louis, Mo., Jan 16, 1939, "Modern Air Power," untitled folder, box 16, Andrews Papers, LCMD.

37. Bd Proceedings, War Dept Ofc of the Chief of the Air Corps, Wright Fld, Dayton, Ohio, Jan 12, 1933, folder: Jan 2–Mar 27, 1933, box 6, Spaatz Papers, LCMD.

38. Loftin, *Quest for Performance*, 86–88.

39. Ibid., 95–96.

40. Maj Carl Spatz to Lt Col G. I. Jones, Feb 8, 1933, folder: Jan 2–Mar 27, 1933, box 6, Spaatz Papers, LCMD.

41. James Parton, *"Air Force Spoken Here": General Ira Eaker and the Command of the Air* (Bethesda, Md.: Adler & Adler), 95.

42. Tng Reg 440–15, 1935, box 485, 321.9, entry 166, RG 18, NA.

43. Tng Prog, HQ 20th Pur Gp, GHQ Air Force, Ofc Ops Off, Barksdale Fld, La., Jun 1, 1935, box 614, 358.9, entry 166, RG 18, NA.

44. Tng Reg 440–15, 1935.

45. Maurer, *Aviation in the U.S. Army*, apps 5, 6, 7.

46. Landon intvw, May 31–Jun 3, 1972.

47. Angelucci, *Encyclopedia of Military Aircraft*, plate 67.

48. Tng Prog 1935–36, HQ 3d Atk Gp, Barksdale Fld, La., May 10, 1935, GO 1; Tng Prog Jul 1, 1935–Jun 30, 1936, 17th Atk Gp, Air Corps, March Fld, Calif., Jul 15, 1935, Opord 53, box 614, 353.9, entry 166, RG 18, NA.

49. Memo, D. H. Torrey, Adj Gen, to

Comdg Gen V Corps Area, Ft. Hayes, Columbus, Ohio, May 11, 1938, Subj: Training of Attack Aviation with Tank and Mechanized Units, frame 1545, 248.176, mcflm roll A2729, AFHRA.

50. Memo, Maj Gen F. M. Andrews to Adj Gen, Washington, D.C., [Apr 1936], Subj: Stations for Reconnaissance-Bombardment Units, frames 1990–1995, mcflm roll A1415, AFHRA.

51. Ltr, Lt Col H. H. Arnold to Toohey, Jul 11, 1933, folder: Apr 3–Jul 31, 1933, box 6, Spaatz Papers, LCMD.

52. Tng Reg 440–15, 1935.

53. Memo, Capt Harold M. McClelland to Maj Spaatz, Jan 19, 1934, Subj: Frontier Defense, folder: Jan 2–Jul 31, 1934, box 6, Spaatz Papers, LCMD.

54. Wright, *Most Probable Position*, 183.

55. Memo, HQ GHQ Air Force to Comdg Gens of all GHQ Air Force Wgs, May 25, 1938, Subj: GHQ Air Force Training Directive, 1938–1939, box 611, 353.9, entry 166, RG 18, NA.

56. Maurer Maurer, "Aviation in the U.S. Army," unpbl. ms., 709–710.

57. Ltr, to Maj Willis Hale, Apr 19, 1932, diary entry of Jun 27, 1932, box 5, Spaatz Papers, LCMD; *Air Corps News Letter* 19, no. 15 (Aug 1, 1936).

58. "Seventh Bombardment Group Completes Training for F.Y. 1935–1936," *Air Corps News Letter* 19, no. 17 (Sep 1, 1936).

59. "The GHQ Air Force," *Air Corps News Letter* 19, no. 17 (Sep 1, 1936).

60. "G.H.Q. Air Force Observation," *Air Corps News Letter* 19, no. 19 (Oct 1, 1936).

61. Tng Reg 440–15, 1935.

62. "GHQ Air Force," *Air Corps News Letter*.

63. Intvw, Gen Samuel E. Anderson by Hugn N. Ahmann, Jun 28–30, 1976, 105, K239.0512–905, AFHRA.

64. Russell F. Weigley, *History of the United States Army* (Bloomington: Indiana University Press, 1967), 414.

65. Memo, Lt Col H. L. Walthall, Asst Adj Gen's Ofc, HQ IX Corps Area, Presidio, to Adj Gen's Ofc, Washington, D.C., Jul 10, 1934, Subj: Air Corps Training, Heavier-than-Air, box 610, 353.9, entry 166, RG 18, NA.

66. Ltr, Spatz to Mac [Lt Col L. W. McIntosh], Aug 14, 1933, folder: Aug 3–Oct 31, 1933, box 6, Spaatz Papers, LCMD.

67. Study, Capt Otto P. Weyland, "Training Program for Observation Aviation, The Air Corps Tactical School, Maxwell Field, Ala.," May 14, 1938, frames 1137–1170, 248.262–29, mcflm roll 2811, AFHRA.

68. Memo, HQ GHQ Air Force to Comdg Gens, May 25, 1938.

69. Memo, Maj W. C. Goldsborough, Comdg, HQ 15th Obsn Sq, Scott Fld, Ill., to Comdg Gen VI Corps Area, Chicago, Ill. (thru: Comdg Off, Scott Fld, Ill.), Jan 8, 1936, Subj: Recommendations, Air Corps Training, Heavier than Air, box 610, 353.9, entry 166, RG 18, NA.

70. *The Public Papers and Addresses of Franklin D. Roosevelt, 1937*, vol 6, 406–411, cited in Daniel Aaron and Robert Bendiner, eds. *The Strenuous Decade* (Garden City, N.Y.: Doubleday & Co., 1970), 449.

71. See discussion of the Protective Mobilization Plan in the 1938 annual reports of Secretary of War William H. Woodring and Chief of Staff Malin Craig, in Walter Millis, ed., *American Military Thought* (Indianapolis: Bobbs-Merrill, 1966), 418–435.

72. In Arnold's memoirs he recalls the day as September 28, 1938, although November 14 is normally given as the date on which Roosevelt called a conference of military and civilian advisers to discuss the rearmament. See Wesley Frank Craven and James Lea Cate, eds., *The Army Air Forces in World War II*, vol 6, *Men and Planes* (Washington, D.C.: Office of Air Force History, 1983), 9.

73. Arnold, *Global Mission*, 175–179.

74. Memo, F. M. Andrews, HQ GHQ Air Force, to Secy War, Jan 24, 1938, Subj: Air Corps Procurement Program, folder: Sec of War 1938–42, box 11, Andrews Papers, LCMD.

75. Parrish intvw, Jun 10–14, 1974, 76–77.

Notes

Part IV

1. Willard Wiener, *Two Hundred Thousand Flyers* (Washington: The Infantry Journal, 1945), 10.

2. For a study of the operation of one of those schools, see James C. Mesco, "Prepared for Combat: The Training of Royal Air Force Pilot Cadets by the United States Army Air Force [sic] at Cochran Field, Macon, Georgia, 1941-1942" (master's thesis, Georgia College and State University, 1998).

3. Address by General George C. Marshall at the Air Corps Tactical School, Maxwell Field, Ala., on Oct 1, 1938, quoted by Herman Wolk, "Independence and Responsibility: The Air Force in the Postwar World," 5, in "A Time of Change: National Strategy in the Early Postwar Era," *Colloquium on Contemporary History, June 7, 1989, No. 1* (Washington, D.C.: Naval Historical Center, Dept of the Navy, 1989).

4. Cited in Michael S. Sherry, *In the Shadow of War: The United States Since the 1930's* (New Haven, Conn.: Yale University Press, 1995), 35.

5. Speech, Brig Gen Barton K. Yount, "The Expansion of the Army Air Corps," Apr 6, 1939, in *Air Corps Newsletter* 22: 8 (Apr 15, 1939), 2.

6. I. B. Holley, Jr., *Buying Aircraft: Materiel Procurement for the Army Air Forces* (Washington, D.C.: United States Army Center of Military History, 1989), 179-180.

7. Watson, *Chief of Staff*, 126-182; Kent Roberts Greenfield, *American Strategy in World War II: A Reconsideration* (Baltimore: Johns Hopkins University Press, 1963), 86; Louis Morton, "Germany First: The Basic Concept of Allied Strategy in World War II" in *Command Decisions*, Kent Roberts Greenfield, ed. (Washington, D.C.: U.S. Army Center of Military History, 1987); Arnold, *Global Mission*, 245.

8. Charles E. Kirkpatrick, "Lesley J. McNair: Training Philosophy for a New Army," *Army History*, Winter 1990/1991, 11-17.

9. Speech, F. M. Andrews, Nov 27, 1939, untitled folder, box 16, Andrews Papers, LCMD.

10. Memo, to Comdg Gens of the four armies, Comdg Gens all Corps Areas and Depts, Comdg Gen GHQ Air Force, Comdts Gen and Special Svc Schs, Supt U.S. Military Academy, Chiefs all Arms, Svcs, and Bureaus of War Dept, and Exec Res Affairs, Subj: Air Corps Training, 1939-1940, box 519, 353.9, Air Adjutant General Files (AAG), RG 18, NA.

11. R&R, Exec to Chiefs T&O, Plans, Pers Divs, in turn, May 22, 1940; Memo, Geo. E. Stratemeyer for Brig Gen B. K. Yount, Chief Plans Div, to Chief of Air Corps, May 25, 1940, Subj: Bases to Be Turned over to the Training Center and New Locations for Units Evacuating Those Bases, box 519, 353.9, AAG, RG 18, NA.

12. Martha E. Layman, "Organization of AAF Training Activities, 1939-1945," USAF Hist Study No. 53, 1946, 4-14, AFHRA.

13. Barton Yount had been commandant at the Training Center at Randolph. He was a West Point classmate of Arnold's and, according to Pete Copp, in Arnold's "inner circle." Arnold sent Yount to Europe when war broke out to administer a plan for ongoing intelligence gathering overseas. Yount was also at Arnold's side in March 1940, testifying before the House Military Affairs Committee about Allied aircraft procurement. That summer, Arnold informed Yount about the highly secret Norden bombsight project. See DeWitt S. Copp, *Forged in Fire: Strategy and Decisions in the Air War over Europe, 1940-45* (Garden City, N.Y.: Doubleday, 1982), 1-20, 36, 81.

14. Holley, *Buying Aircraft*, 44, 211-231.

15. Watson, *Chief of Staff*, 129.

16. Craven and Cate, eds., *Men and Planes*, 434. An analysis of pilot and airplane production the end of September 1940 laid out a requirement for a 25-group program consisting of 1,722 pilots for tactical units, 1,500 for training centers, and 695 for overhead, for a total

Notes

of 3,917. Under the 41-group program, 3,788 would be required for tactical units, 2,275 for training centers, and 695 for overhead, for a total of 6,758. "Analysis of Pilot and Airplane Situation," Sept 30, 1940, box 380, 353, uncld, AAG, RG 18, NA.

17. Memo, Maj Gen H. H. Arnold to Chief Plans Div, May 14, 1940, box 178, 353.9, AAG, RG 18, NA.

18. Memo, Brig Gen Davenport Johnson, Chief T&O Div, to Comdg Gen West Coast Air Corps Tng Cen, Moffett Fld, Calif., Dec 17, 1940, Subj: Planning for Increased Training Center Production, box 178, 353.9, AAG, RG 18, NA.

19. Memo, Maj Edward P. Curtis to Chief of Staff, Mar 26, 1941, box 178, 353.9, AAG, RG 18, NA.

20. Tech Insts, Air Corps Mat Div to Asst Chief Mat Div, Wright Fld, Chief, Engineering Sec, Chief Fld Svc Sec, Chief Contract Sec, Chief Bldg & Gnds Sec, Chief Mat Planning Sec, Jul 19, 1940, Subj: Summary of Current Air Corps Programs, box 380, 353, uncld, AAG, RG 18, NA; Craven and Cate, eds., *Men and Planes*, 266.

21. Hist Sec, A–2, "History of the Army Air Forces Central Flying Training Command, 1 Jan 1939–7 Dec 1941" (hereafter Hist, Central Flying Tng Comd), AAF Central Flying Tng Comd Hist Study No. 63, Apr 1, 1945, Randolph Fld, Tex.

22. Craven and Cate, eds., *Men and Planes*, 467.

23. Wesley Frank Craven and James Lea Cate, eds., *The Army Air Forces in World War II*, vol 1, *Plans and Early Operations, January 1939 to August 1942* (Washington, D.C.: Office of Air Force History, 1983), 249–250.

24. Arnold, *Global Mission*, 205.

25. Memo, to Comdg Gens and Comdg Offs all activities under Chief of Air Corps, Nov 19, 1941, Subj: Air Corps Training Directive 1941–42, frame 1445, 248.176, mcflm roll A2729, AFHRA.

Chapter 9

1. Lecture, "The GHQ Air Force," Andrews Papers, box 16, folder: Speech and Article file, Source Material, 1939.

2. Address, F. M. Andrews to mtg Society of Military and Naval Officers of the World War, Sep 13, 1939, untitled folder, box 16, Andrews Papers, LCMD.

3. DeWitt S. Copp, *Forged in Fire: Strategy and Decisons in the Air War over Europe, 1940–45* (Garden City, N.Y.: Doubleday & Co., 1982), 44.

4. Arnold, *Global Mission*, 179. Arnold mistakenly places the date of the meeting on September 28, 1938.

5. Craven and Cate, eds., *Men and Planes*, 427–435.

6. Biographical Data from National Military Establishment [WD] service summary [Jul 1949], copy in AFHSO Library, Bolling AFB.

7. Craven and Cate, eds., *Men and Planes*, 465.

8. For a detailed account of this phase of World War II aviation training, see Eugene Fletcher's *Mister: The Training of an Aviation Cadet in World War II* (Seattle: University of Washington Press, 1992), 19–60, passim. An official account of the expansion of the training base is Warrant Officer Howard D. Williams's "Basic Military Training in the AAF, 1939–1944," USAF Hist Study No. 49, Nov 1946, AFHRA. See figure 1, "Trainees at Basic Training Centers" and table 2, "Attached Unassigned Recruit Strength . . ." for statistics showing the growth of the aviation recruit base through December 1944.

9. Craven and Cate, eds., *Men and Planes*, xxv.

10. Directive 1, Air Corps Expansion Program, Apr 18, 1939, box 3, 353, entry 548, RG 18, NA.

11. Craven and Cate, eds., *Men and Planes*, 458; Hist, Central Flying Tng Comd.

Notes

12. Directive 1, Air Corps Expansion Program, Apr 18, 1939; Craven and Cate, *Men and Planes*, 458; Hist, Central Flying Tng Comd.

13. Memo, Brig Gen Davenport Johnson, Chief T&O Div, to Comd Gen West Coast Air Corps Tng Cen, Moffett Fld, Calif., Dec 17, 1940, Subj: Planning for Increased Training Center Production [Similar letter sent to Southeast Air Corps Training Center], box 178, 353.9, AAG, RG 18, NA.

14. Directive 1, Air Corps Expansion Program, Apr 18, 1939.

15. Hist, Central Flying Tng Comd.

16. Robert L. Thompson, "Initial Selection of Candidates for Pilot, Bombardier, and Navigator Training," USAF Hist Study No. 2, Nov 1943, 58, AFHRA.

17. Dominick A. Pisano, *To Fill the Skies with Pilots: The Civilian Pilot Training Program, 1939–46* (Urbana: University of Illinois Press, 1993), 78.

18. J. Merton England and Chauncey E. Sanders, "Legislation Relating to the AAF Training Program, 1939–1945," revised, USAF Hist No. Study 7, Apr 1946, 56–57, AFHRA.

19. R&R, Chief Plans Div to Chief Mil Pers Div, Feb 21, 1941, Subj: Date of Commencement of Enlisted Pilot Training, box 178, 353.9, AAG, RG 18, NA; Memo, Col Walter F. Kraus, Tng Ops Div, to Comd Gen Southeast Air Corps Tng Cen, Maxwell Fld, Montgomery, Ala., Jun 19, 1941, Subj: Training of Enlisted Pilots, frame 1459, 248.176, mcflm A2729, AFHRA.

20. Memo, G–2 to Lt Col Evans, Jul 24, 1941, folder: M.A. Reports, box 1, entry 247, RG 18, NA.

21. England and Sanders, "Legislation Relating to the AAF," 79–93.

22. *Time*, Mar 31, 1941, copy in Dargue file, AIAA collection, LCMD.

23. *Army Air Forces Statistical Digest: World War II*, table 212.

24. Intvw, Gen Howell M. Estes, Jr., by Lt Col Lyn R. Officer and Lt Col Robert G. Zimmerman, Aug 27–30, 1973, 33, K239.0512-686, AFHRA.

25. Gen William L. Kennedy, quoted in Kenneth Baxter Ragsdale, *Wings Over the Mexican Border: Pioneer Military Aviation in the Big Bend* (Austin: University of Texas Press, 1984), 204.

26. Memo, Col John B. Brooks, Comdg, Randolph Fld, Tex., to Chief of Air Corps, Washington, D.C. (thru: Comdg Gen Air Corps Tng Cen, Randolph Fld, Tex.), Jan 18, 1939, Subj: The Capacity of the Air Corps Primary Flying School, box 3, 353, entry 548, RG 18, NA.

27. Arnold, *Global Mission*, 180–181.

28. The first civilian flying schools were the Alabama Institute of Aeronautics, Inc. (located at Municipal Airport, Tuscaloosa, Ala.), the Chicago School of Aeronautics (Curtiss Airport, Glenview, Ill.), the Dallas Aviation School and Air College (Love Field, Dallas, Tex.), the Grand Central Flying School (Grand Central Air Terminal, Glendale, Calif.), the Lincoln Airplane and Flying School (Municipal Airport, Lincoln, Neb.), the Parks Air College, Inc. (Parks Airport, E. St. Louis, Ill.), the Ryan School of Aeronautics (Lindbergh Field, San Diego, Calif.), the Santa Maria School of Flying (Santa Maria Airport, Santa Maria, Calif.), and the Spartan School of Aeronautics (Municipal Airport, Tulsa, Okla.). An Rprt, Chief of Air Corps, 1939, 11.

29. Richard F. McMullen, "The Role of Civilian Contractors in the Training of Military Pilots," Hist Study No. 2, Air Training Command, Scott AFB, Ill., 1955.

30. Hist, AAF Eastern Flying Tng Comd, Jan 1, 1939–Dec 7, 1941.

31. Wiener, *Two Hundred Thousand Flyers*, 13.

32. Hist, Central Flying Tng Comd.

33. Wiener, *Two Hundred Thousand Flyers*, 13.

34. Hist, West Coast Air Corps Tng Center, Moffett Fld, Calif., vol 2, Jul 8, 1940–Dec 7, 1941; Memo, Brig Gen Henry W. Harms, Comdg West Coast Air Corps Tng Center, to Chief of Air Corps, Washington, D.C., Feb 27, 1941, Subj: Planning for Increased Training Center Production, box 178, 353.9, AAG, RG 18, NA.

35. Intvw, Lt Gen John W. Carpenter III by Lt Col Arthur W. McCants, Jr., and

Maj Scottie S. Thompson, Jan 9–12, 1979, 32–33, K239.0512–1110, AFHRA.

36. Hist, West Coast Air Corps Tng Center.

37. Intvw, Gen Jacob E. Smart by James C. Hasdorff and Lt Col Arthur W. McCants, Jr., Nov 27–30, 1978, 30, K239.0512–1108, AFHRA.

38. Col E. B. Lyon, Comdg Off Basic Flying Sch, Moffett Fld, Calif., to Brig Gen Davenport Johnson, Feb 26, 1941.

39. Philip Ardery, *Bomber Pilot: A Memoir of World War II* (Lexington: University Press of Kentucky, 1978), 19.

40. John L. Frisbee, "On the Way to a Miracle," *Air Force Magazine* 65: no. 2 (Feb 1982): 75.

41. Purtee, *Modification and Development of Training Aircraft*, 54–55; Angelucci, *Encyclopedia of Military Aircraft*, 340.

42. For a description of the Link flight simulator, see Chapter 7.

43. Minutes, Bd recommendations regarding basic course, Dec 1938, with Exhibit "C": Proposed Program of Instruction for... Air Corps Basic Flying School, Randolph Fld, Tex., box 3, 353, entry 548, RG 18, NA.

44. The experiment in the West Coast Center was the conversion of the Cal-Aero Academy at Ontario, California. It opened March 22, 1941, with class 41–F. Class 41–G at Curtis Field, Brady, Texas, began training on May 3, 1941. This school had originally been a civilian school providing primary training. Bush Field at Augusta, Georgia, in the Southeast Air Corps Training Center, which began classes on May 26, 1941, was the third contract basic school.

45. Hist, Central Flying Tng Comd.

46. Estes intvw, Aug 27–30, 1973, 29–30.

47. Frisbee, "On the Way to a Miracle," 76.

48. Hists, West Coast Air Corps Tng Comd and Southeast Air Corps Tng Comd; Workers of the Writers' Program of the Works Projects Administration in the State of Texas, comp., *Randolph Field: A History and Guide* (New York: Devin-Adair, 1942), 25–80.

49. Leslie F. Smith, "Development of AAF and USAF Training Concepts and Programs, 1941–1952," USAF Hist Study No. 93, 1953, 64–65, AFHRA.

50. Arnold to Marshall, quoted in Copp, *Forged in Fire*, 83.

51. Carpenter intvw, Jan 9–12, 1979, 37–38.

52. Charles R. Bond, Jr., and Terry Anderson, *A Flying Tiger's Diary* (College Station: Texas A&M University Press, 1984), 14–15.

53. See, for example, the proposed distribution of Training Center pilot graduates put forward in a plan for specialized training at the end of March 1938:

Type	Percentage
Bombardment	
heavy	17.5
medium	18.6
attack	10.4
Pursuit	
single-engine	20.5
interceptor	4.1
fighter	8.7
Reconnaissance	
strategic	5.1
medium	5.7
army	1.8
Observation	
Corps and Army	7.6

Plan for Specialized Training, box 3, entry 247, RG 18, NA.

54. Memo, Maj Gen H. H. Arnold to Adj Gen, Jun 14, 1940, Subj: Advanced Flying School — San Angelo, Texas, box 519, 353.9, AAG, RG 18, NA.

55. Hist, Central Flying Tng Comd.

56. Hist, Eastern Flying Tng Comd.

57. Hist, West Coast Flying Tng Cen.

58. Intvw, Gen Earle E. Partridge by Thomas A. Sturm and Hugh N. Ahmann, Apr 23–25, 1974, 288–289, K239.0512–729, AFHRA.

59. Frisbee, "On the Way to a Miracle," 77.

60. Smith, "Developmenmt of AAF and USAF Training," 87–88.

61. Index Sheet, Chief of Air Corps to Comdg Gen Gulf Coast Air Corps Tng Cen, Randolph Fld, Tex., Jun 16, 1941, box 517, 353.4, AAG, RG 18, NA.

62. Hist, Central Flying Tng Comd.

Notes

63. Angelucci, *Encyclopedia of Military Aircraft*, 340.
64. Air Corps Policy 206–A, Feb 7, 1941, Digest of Air Corps Policies, vol 2, box 1, entry 198, RG 18, NA.
65. Purtee, *Modification and Development of Training Aircraft*, 65.
66. Mil Att London Rprt 42494, dated Mar 5, 1941, Synopsis of Mil Att Rprts for G–3, folder: M.A. Reports, box 1, entry 247, RG 18, NA.
67. Craven and Cate, eds., *Men and Planes*, 468–471.
68. England and Sanders, "Legislation Relating to the AAF," 50–56.
69. Memo, Brig Gen Davenport Johnson to Comdg Gen GHQ Air Force, Langley Fld, Va., Mar 24, 1941, Subj: Training of Non-Pilot Commissioned Members of Combat Crews, box 178, 353.9, AAG, RG 18, NA.
70. Memo, Capt J. P. McConnel, Adj HQ Southeast Air Corps Tng Cen, to Comdg Off Air Corps Advanced Flying Sch, Maxwell Fld, Ala., Jun 12, 1941, Subj: Bombardier-Navigator Replacement Center: Personnel Study to Accompany Program of Instruction Bombardier-Navigator Replacement Center (Reconnaissance School), Southeast Air Corps Training Center, box 178, 353.9, AAG, RG 18, NA; Ben R. Baldwin, "Individual Training of Navigators in the AAF," USAF Hist Study No. 27, Jan 1945, 34–35, AFHRA; Ben R. Baldwin, "Individual Training of Bombardiers," USAF Hist Study No. 5, May 1944, 9–11, AFHRA.
71. Memo, Brig Gen B.K. Yount to Adj Gen, Oct 16, 1939, Subj: Training of Airplane Combat Crews, box 529, 353.9, AAG, RG 18, NA.
72. 1st Ind, HQ GHQ Air Force, Bolling Fld, D.C., to Chief of Air Corps, Washington, D.C., May 14, 1941, box 529, 353.9, 18, AAG, RG NA.
73. Memo, Col Walter F. Kraus, Exec T&O Div, to Comdg Gen GHQ Air Force, May 7, 1941, Subj: Training of Bombardiers and Aerial Gunners; 1st Ind, HQ GHQ Air Force, Bolling Fld, D.C., to Chief of Air Corps, Washington, D.C., May 14, 1941 box 529, 353.9, AAG, RG 18, NA.
74. Smith, "Development of AAF and USAF Training," 150; Baldwin, "Individual Training of Navigators," 28–29.
75. Memo, Capt L. G. Seeligson to Adj Gen Ofc, Jul 22, 1941, box 529, 353.9, AAG, RG 18, NA.
76. Memo, Lt Col H. L. George, Asst Chief of Air Staff, to Chief of Air Staff, Oct 9, 1941, Subj: Aerial Bombardiers, box 529, 353.9, AAG, RG 18, NA.
77. Baldwin, "Individual Training of Navigators," 111–112.
78. Memo, George to Chief of Air Staff, Oct 9, 1941.
79. Ltr, Capt William M. Garland to Fred [Maj F. L. Anderson, Jr.], Jun 4, 1941, box 529, 353.9, AAG, RG 18, NA.
80. Ltr, Maj R. E. [Nugent] to Skippy [Capt N. B. Harbold], Oct 24, 1940, box 529, 353.9, AAG, RG 18, NA.
81. Hist, Eastern Flying Tng Comd, 490. Figures citing the number of qualified navigators varied. Baldwin, "Individual Training of Navigators," puts the number in 1939 at 166.
82. Memo, Col Carl Spaatz, Chief, Plans Div, to Chief of Air Corps, Feb 6, 1940, Subj: Air Corps Centralized Aerial Navigation School, box 170, 352.9, AAG, RG 18, NA.
83. On October 14, 1940, the Personnel Division stated that "at the present time there is more than a sufficient number of cadets being currently eliminated from further pilot training who are qualified for training as *navigators*, and who are recommended therefor by the academic boards, to fill the classes, as presently scheduled." Box 529, 353.9, AAG, RG 18, NA.
84. Hist, Eastern Flying Tng Comd.
85. Ibid.
86. Ibid.
87. Baldwin, "Individual Training of Navigators."
88. 2d Ind, HQ Air Corps Advanced Flying Sch, Kelly Fld, Tex., to Comdg Off, Gulf Coast Air Corps Tng Cen, Randolph Fld, Tex., Sep 11, 1941, box 178, 353.9, AAG, RG 18, NA.
89. Hist, Eastern Flying Tng Comd.
90. Ibid.
91. Hist, Central Flying Tng Comd.

Notes

92. Hist, West Coast Flying Tng Cen.
93. Hist, Eastern Flying Tng Comd.
94. 1st Ind, HQ Air Force Combat Comd, Bolling Fld, D.C., to Chief of Air Corps, Washington, D.C., Aug 7, 1941, box 177, 353.9, AAG, RG 18, NA; 2d Ind, HQ 19th Bom Wg Caribbean Air Force, Albrook Fld, C.Z., to Comdg Gen Caribbean Air Force, Albrook Field, C.Z., Aug 2, 1941, box 178, 353.9, AAG, RG 18, NA.
95. Baldwin, "Individual Training of Navigators," 107.
96. 5th Ind, HQ Southeast Air Corps Tng Cen, Maxwell Fld, Montgomery, Ala., to Chief of Air Corps, Washington, D.C., Sep 26, 1941, box 178, 353.9, AAG, RG 18, NA.
97. Philip A. St. John, *Bombardier: A History* (Paducah, Ky.: Turner, 1993), 15.
98. Baldwin, "Individual Training of Bombardiers," 18–20.
99. Memo, Maj Wm. L. Ritchie, Adj HQ 25th Bom Gp, Aug 6, 1940, Narrative Report Covering Conduct of Aerial Bombing, Regular Practice Season, Training Year 1939–1940, box 517, 353.4, AAG, RG 18, NA.
100. 7th Ind, Maj P. M. Whitney, Asst Adj Gen HQ GHQ Air Force, Aug 21, 1940, to Adj Gen, Washington, D.C., box 517, 353.4, AAG, RG 18, NA.
101. Col William H. Crom, Comdg HQ 17th Bom Gp (M), to Comdg Gen 1st Wg GHQ Air Force, March Fld, Riverside, Calif., Jul 18, 1940, box 517, 353.4, AAG, RG 18, NA.
102. Baldwin, "Individual Training of Bombardiers," 36.
103. Ibid., 40–41. At this time no specific number of hours was listed in the curriculum for "bombing trainer practice." Presumably the ommission is owed to the uncertain availability of equipment.
104. Memo, Brig Gen Jacob E. Fickel to Comdg Off Southeast Air Corps Tng Cen, Maxwell Fld, Montgomery, Ala., Sep 26, 1940, Subj: Training Program for Officer Bombardiers; 1st Ind, Brig Gen W. R. Weaver, Comdg HQ Southeast Air Corps Tng Cen, Maxwell Fld, Montgomery, Ala., to Chief of Air Corps, Washington, D.C., Dec 12, 1940, box 529, 353.9, AAG, RG 18, NA.
105. Hist, Eastern Flying Tng Comd.
106. Ibid.
107. R&R, T&O to Chief of Air Corps, Jul 7, 1941; Memo, Maj Gen Geo. H. Brett, Chief of Air Corps, to Comdg Gen GHQ Air Force, Bolling Fld, D.C., Subj: Two-engine Pilots for Training Activities; 1st Ind, HQ Air Force Combat Comd, Bolling Fld, D.C., to Chief of Air Corps, Washington, D.C., Jun 28, 1941, box 529, 353.9, AAG, RG 18, NA.
108. R&R, [illeg] Chief T&O Div to Plans, Apr 25, 1941, box 517, 353.4, AAG, RG 18, NA.
109. Baldwin, "Individual Training of Bombardiers," 46–47.
110. Hist, Eastern Flying Tng Comd.
111. 1st Ind, Maj Gen W. R. Weaver, Comdg Southeast Air Corps Tng Cen, Maxwell Fld, Montgomery, Ala., to Chief of Air Corps, Washington, D.C., Oct 24, 1941, box 529, 353.9, AAG, RG 18, NA.
112. Academic Rprt, Bombardier Class SE 41–B, Bombardier Sch, Barksdale Fld, La., wk ending Oct 18, 1941, box 529, 353.9, AAG, RG 18, NA.
113. Baldwin, "Individual Training of Bombardiers," 115.
114. Ibid., 118–121.
115. Memo, Brig Gen Davenport Johnson to Comdg Gen West Coast Air Corps Tng Cen, Moffett Fld, Calif., Jul 1, 1941, Subj: Bombardier Training, box 529, 353.9, AAG, RG 18, NA.
116. Henry H. Simms, "Flexible Gunnery Training in the AAF," USAF Hist Study No. 31, Mar 1945, 13, AFHRA.
117. Memo, Col W. R. Weaver to Chief of Air Corps, Washington, D.C., Jul 20, 1940, Subj: Aerial Flexible Gunnery, box 529, 353.9, AAG, RG 18, NA.
118. Memo, Maj C. E. Duncan, Exec Ofc Chief of Air Corps, to Comdg Off Southeast Air Corps Tng Cen, Maxwell Fld, Montgomery, Ala., Sep 19, 1940, Subj: Training of Aerial Gunners, box 529, 353.9, AAG, RG 18, NA.
119. Hist, Eastern Flying Tng Comd.
120. Memo, Maj L. P. Whitten, Chief Fld Svc Sec, to Comd Off Prov Air Corps Maint Comd, Apr 14, 1941, Subj: Air Corps Training Schools (Proposed Set-Up

Notes

under the 30,000 Pilot Training Program), box 170, 352.9, AAG, RG 18, NA.

121. Simms, "Flexible Gunnery Training," 7–9.

122. Ibid., 28–29; Hist, Eastern Air Corps Tng Comd; Avn Bul 19, "Tentative Lessons," May 8, 1941, synopsis of Mil Att Rprts prepared for G–3; General comments on existing conditions in the Air Force Combat Command in comparison with the Royal Air Force, Reference: Comments of A.F.C.C. Bombardment Pilots Just Returned from England," Jul 28, 1941, folder: M.A. Reports, box 1, entry 247, RG 18, NA; Memo, Brig Gen W. R. Weaver to Chief of Air Corps, Washington, D.C., Oct 12, 1940, Subj: Training of Aerial Gunners, box 529, 353.9, AAG, RG 18, NA; R&R, Mat Div to Info Div, May 8, 1940, Subj: Demonstration of Firing on Leak-proof Fuel Tanks by Glenn L. Martin Co., box 516, 353.1, AAG, RG 18, NA; Memo, Maj E. M. Power, Chief Engineering Sec, to Chief Experimental Engineering Sec, Wright Fld (attn: Lt Col G. Gardner), Sep 18, 1940, Subj: Vitarama Machine Gunnery Trainer, box 160, 353.9, AAG, RG 18, NA; R&R, T&O Div to Mat, Plans, Exec, Pers, Sep 24, 1940, Subj: Appointment of Board to Investigate Proposed Devices for Gunnery Training; R&R, T&O Div to Exec, Oct 17, 1940, Subj: Report of Board on Gunnery Training Device; Memo, Maj Carroll, Chief Experimental Engineering Sec, to Chief Mat Div, Oct 25, 1940, Subj: Pre-Flight Reflex and Gunnery Trainer, box 160, 353.9, AAG, RG 18, NA; Memo, Col Edward F. Witsell, Adj Gen Southeast Air Corps Tng Cen, to Chief of Air Corps, Washington, D.C., Oct 8, 1941, Subj: Aerial Gunnery Training Cards, box 516, 353.1, AAG, RG 18, NA; R&R, Chief of Air Corps to Chief AAF, Oct 21, 1941, Subj: Aerial Gunnery Training, box 529, 353.9, AAG, RG 18, NA; R&R, O.P.E. Chief, Mat Div, to T&O Div (thru: Plans Div), Jan 13, 1941, Subj: Technical Instructions, Serial No. TI–415, Aerial Gunnery Training Equipment, box 516, 353.1, AAG, RG 18, NA; R&R, T&O to Mat, Mar 24, 1941, Subj: Equipment Required for Training Machine Gunners, box 529, 353.9, AAG, RG 18, NA.

123. Simms, "Flexible Gunnery Training," 6; Memo, Brig Gen Davenport Johnson, Asst Chief of Air Corps, to Chief of Air Corps, Apr 28, 1941, Subj: Conference Reference Equipment for Bombardier, Flexible Gunnery, and Pursuit Gunnery Training, box 529, 353.9, AAG, RG 18, NA.

Chapter 10

1. Memo, Maj C. E. Duncan, Exec Ofc Chief Corps, to Comdg Gen GHQ Air Force, Apr 25, 1940, box 178, 353.9, AAG, RG 18, NA.

2. Summary of Intvw, Sen Henry Cabot Lodge by Gen Barton K. Yount and Col Carl A. Spaatz on May 7, 1940, May 8, 1940, box 157, 353, AAG, RG 18, NA.

3. General comments on existing conditions..., Jul 28, 1941.

4. See the reports on navigator training in the Hawaiian Air Force, the Caribbean Air Force, the Alaska Defense Command, and the Air Force Combat Command forwarded from the West Coast Air Corps Training Center to the Chief of the Air Corps, Sep 24, 1941, box 178, 353.9, AAG, RG 18, NA.

5. Baldwin, "Individual Training of Navigators," 58–61.

6. Betty Jo McNarney, "Pilot Transition to Combat Aircraft," USAF Hist Study No. 18, Sep 1944, 36, AFHRA.

7. Ibid., 45–46.

8. Ibid., 39–43.

9. Carpenter intvw, Jan 9–12, 1979, 52–53.

10. Baldwin, "Individual Training of Navigators," 48, 141–142.

11. Memo, Lt Col Robert Kauch to Chief of Air Corps, Washington, D.C., Feb 15, 1941, Subj: Increase in Bombing

Notes

Error during Fiscal Year 1940, 3 Incls: Exhibits A, B, C, box 173, 353.4, AAG, RG 18, NA.

12. Tentative Principles Governing Specialized Training, c. Nov 30, 1938, box 3, entry 247, RG 18, NA.

13. Carpenter intvw, Jan 9–12, 1979, 53–54.

14. GHQ Air Force Tng Directive, 1940–1941, to Comdg Gens, all GHQ Air Force Wgs, Jun 1, 1940. This directive was parroted by the training programs published by the units. See, for example, Cir 6–2, Training Program, 1940–1941, of the HQ 1st Wing, box 529, 353.9, AAG, RG 18, NA.

15. Ltr, Maj Gen H. H. Arnold to Maj Gen Daniel Van Voorhis, Aug 12, 1940, box 178, 353.9, AAG, RG 18, NA.

16. Rprt 1003, Ofc Naval Intel, May 29, 1941; Memo, G–2 to G–3 Sec, Jul 22, 1941, folder: M.A. Reports, box 1, entry 247, RG 18, NA.

17. Rprt, Ofc Comdg Gen HQ 18th Wg, Hickam Fld, Territory of Hawaii, to Chief of Air Corps, Washington, D.C. (thru: Comdg Gen Hawaiian Dept, Ft. Shafter, Territory of Hawaii), Jul 12, 1940, Subj: Combat Readiness, box 178, 353.9, AAG, RG 18, NA.

18. Gerald T. White, "Combat Crew and Unit Training in the AAF, 1939–945," USAF Hist Study No. 61, Aug 1949, 2, AFHRA.

19. Precis of information in conference with Col Carl A. Spaatz, Jan 11, 1939, box 3, entry 247, RG 18, NA.

20. Craven and Cate, eds., *Men and Planes*, 600.

21. Memo, Maj Gen H. H. Arnold to Adj Gen, May 24, 1940, Subj: Horizontal vs. Dive Bombing, box 517, 353.4, AAG, RG 18, NA.

22. Memo, Maj Gen H. H. Arnold to Chief T&O Div, Sep 3, 1940, box 173, 353.4, AAG, RG 18, NA.

23. Memo, Maj Gen H. H. Arnold to Gen Marshall, Oct 9, 1940, box 517, 353.4, AAG, RG 18, NA.

24. R&R, H. H. Arnold, Chief of Air Corps, to Plans, Jul 8, 1940, Directive to demonstration group to initiate dive bombing tests, box 158, 353.4, AAG, RG 18, NA; Memo, Col Wm. Ord Ryan to Chief of Air Corps, Chief Mat Div, Chief Plans Div, and Chief Coast Arty Corps, Jun 19, 1940, Subj: Dive Bombing Demonstration at Norfolk Naval Air Station, box 517, 353.4, AAG, RG 18, NA; R&R, Mat Div to Gen Echols, Mar 4, 1941, Subj: Glide Bombing and Our Bomb Sights, box 173, 353.4, AAG, RG 18, NA; 2d Ind, War Dept Adj Gen Ofc to Chief Army Air Forces, Jul 18, 1941, Subj: Combined Tests to Develop Doctrines and Methods for Aviation Support of Ground Troops; 9th Ind, HQ 3d Air Force, Tampa, Fla., to Comdg Gen Air Force Combat Comd, Bolling Fld, D.C., Sep 4, 1941, Subj: Combined Tests..., box 515, AAG, RG 18, NA; Memo, Col Wm. R. Grove, Jr., Maj Charles R. Johnson, and Capt Benjamin F. Witsell to Dir Air Supp, Sep 9, 1942, Subj: Dive Bombardment versus Low-Level Bombardment Aircraft, box 172, 353.4, AAG, RG 18, NA.

25. Memo, Col C. W. Russell to Chief of Air Corps, Jul 27, 1940, Subj: Status of Low Altitude Bombing Equipment, box 158, 353.4, AAG, RG 18, NA.

26. [Incomplete] Memo, box 158, AAG, RG 18, NA.

27. See, for example, Rprt, 13th Bom Sq, Savannah Air Base, Aug 25, 1941, box 515, AAG, RG 18, NA.

28. Memo, Col Geo. H. Brett, Chief of Staff, HQ GHQ Air Force, to Chief of Air Corps, Feb 3, 1939, Subj: Materiel for Training, Proposed Air Corps Expansion, 353.9B, AAG, RG 18, NA.

29. Memo, Col J. B. Rose, Comdg Ord Dept, Aberdeen Proving Gnd, to Chief of Ord, War Dept, Jul 10, 1940, Subj: Assignment of Airplanes, Aberdeen Proving Ground, Md., box 158, 353.4, AAG, RG 18, NA.

30. Air Eval Rprt 9, Apr 15, 1941; Synopsis of Mil Att Rprts for G–3, May 21, 1941, folder: M.A. Reports, box 1, entry 247, RG 18, NA.

31. For a discussion of heavy bomber development during the 1930s that led to the B–17, see Frank Robert van der Linden, "The Struggle for the Long-Range Heavy Bomber: The United States Army

Notes

Air Corps, 1934–39" (M.A. thesis, George Washington University, 1981).

32. Maurer, *Aviation in the U.S. Army*, 390.

33. Carpenter intvw, Jan 9–12, 1979, 52–55.

34. Memo, Lt Col Harry A. Johnson, Chief Tng Div, to Comdg Gen West Coast Air Corps Tng Cen, Moffett Fld, Calif., Jan 9, 1941, box 178, 353.9, AAG, RG 18, NA.

35. Memo, Brig Gen Carl Spaatz, Chief Plans Div, to Gen Brett, May 29, 1941, Subj: Letter from Lt Gen Delos C. Emmons, Apr 25, 1941 to Mr. Robert A. Lovett, Assistant Secretary of War for Air, box 178, 353.9, AAG, RG 18, NA.

36. An Tng Prog, 25th Bom Gp, GHQ Air Force, Langley Fld, Va., FY 1940, box 530, 353.9, AAG, RG 18, NA.

37. General comments on Existing Conditions . . . , Jul 28, 1941.

38. For reports dealing with night bombing see, for example, Memo, Maj Gen H. H. Arnold to Chief of Staff, Sep 18, 1940, box 173, 353.4, AAG, RG 18, NA; Mil Att London Rprt 42601, Mar 19, 1941; Final Rprt of Maj R.F.C. Vance, Mar 29, 1941; Air Bul 20, May 20, 1941; Memo, for Col Evans, Sep 3, 1941, Subj: Bombing Operations and Tactics; Memo, Col R. L. Walsh, Asst Chief of Staff, G–2, to Asst Chief of Staff, G–3, Oct 15, 1941, folder: M.A. Reports, box 1, entry 247, RG 18, NA.

39. R&R, Mat Div to Maj Archer, T&O Div, Mar 11, 1941, Subj: Night Bombing and Bullet-proof Windshield, box 173, 353.4, AAG, RG 18, NA.

40. Maurer, *Aviation in the U.S. Army*, 395.

41. Rprt, Lt Col Ralph Royce, Comdg HQ 7th Bom Gp, to Comdg Gen 1st Wg GHQ Air Force, March Fld, Calif., Jan 10, 1940, Subj: Bombing Under Conditions of Poor Visibility and Low Ceiling, box 158, 353.4, AAG, RG 18, NA; Memo, Maj A. J. Lyon to Asst Chief Mat Div, Wright Fld, Dayton, Ohio, Jan 2, 1940, Subj: Bombing Through an Overcast, box 518, 353.4, AAG, RG 18, NA; Memo, Maj [C.] E. Duncan to Comdg Gen GHQ Air Force, Apr 16, 1940, Subj: Bombing Through Overcast — Request for B–18 Airplane, box 173, 353.4, AAG, RG 18, NA; Memo, Col C. W. Russell, Chief of Staff HQ GHQ Air Force, to Chief of Air Corps, Sep 6, 1940, Subj: Bombing Under Conditions of Poor Visibility and Low Ceiling, box 173, 353.4, AAG, RG 18, NA; Index Sheet, Act Chief of Air Corps to Chief Sig Off, May 9, 1940, box 516, 353.1, AAG, RG 18, NA.

42. HQ 23d Comp Gp, Maxwell Fld, Ala., to Comdg Offs HQ & HQ Sq, 1st Pur Sq, 24th Bom Sq, and 54th Bom Sq, Jul 1, 1940, Subj: Group Training for 1940–1941, box 530, 353.9, AAG, RG 18, NA.

43. The designation "Pursuit" was changed officially to "Fighter" in May 1941. Both terms were used during the expansion period.

44. Memo 3–1, HQ 31st Pur Gp, Selfridge Fld, Mich., Feb 1940, "Operations: Training Program, 1939–1940," box 530, 353.9, AAG, RG 18, NA.

45. White, "Combat Crew and Unit Training," 6–7.

46. Memo, HQ 18th Wg, Ofc of Comdg Gen, Hickam Fld, to Chief of Air Corps (thru: Comdg Gen, Hawaiian Dept, Ft. Shafter), Jul 12, 1940, Subj: Combat Readiness, box 178, 353.9, AAG, RG 18, NA.

47. Developments in fighter aircraft lagged behind those of bombers, which critically affected tactics and doctrine for pursuit aviation and underlay the disagreement regarding pursuit escort. For a history of the long-range fighter, see Bernard L. Boylan, "Development of the Long-Range Escort Fighter," USAF Hist Study No. 136, 1955, AFHRA. In Charles R. Hyer, "The Curtiss P–36 in Squadron Service," *Journal of the American Aviation Historical Society*, Winter 1988, 242–257, can be found a description of the P–36 in operational training. Along with the P–35, the P–36 was a winner of the Pursuit Competition of 1936. How the arguments over bombers and fighters intertwined can be seen in two fine biographies of leading proponents of pursuit (Claire Chennault) and bombers (Kenneth Walker), written by Martha Byrd:

Chennault: Giving Wings to the Tiger (Tuscaloosa: University of Alabama Press, 1987), and *Kenneth N. Walker: Airpower's Untempered Crusader* (Maxwell Air Force Base, Ala.: Air University Press, 1997).

48. Angelucci, *Encyclopedia of Military Aircraft*, 225-226.

49. Memo, Maj Edward M. Morris, Comdg 33d Pur Sq (Ftr) GHQ Air Force, to Mat Ofc 8th Pur Gp (Ftr) GHQ Air Force, Langley Fld, Va., Aug 22, 1940, Subj: High Altitude Firing Tests on P-40 Type Airplane, box 516, 353.1, AAG, RG 18, NA; Memo, Adj Gen to Comdg Gen GHQ Air Force, May 19, 1941, Subj: Combat Training at Extreme Altitudes; 1st Ind, HQ GHQ Air Force, Bolling Fld, D.C., to Adj Gen, Washington, D.C., May 28, 1941, box 178, 353.9, AAG, RG 18, NA.

50. Robert F. Futrell, "Command of Observation Aviation: A Study in Control of Tactical Air Power," USAF Hist Study No. 24, 1952, 8, AFHRA.

51. Tng Memo 1, Program and Schedule — Air Observers' School, HQ I Army Corps, Apr 5, 1941, box 515, 353, AAG, RG 18, NA.

52. Memo, Maj Gen H. H. Arnold, Chief of Air Corps, to Chief of Staff, Apr 8, 1940, Subj: Combat Observers, box 517, 353.16, AAG, RG 18, NA.

53. Intvw, Gen David A. Burchinal by Col John B. Schmidt and Lt Col Jack Straser, Apr 11, 1975, 39-40, K239.0512-837, AFHRA.

54. An Tng Prog (Rev Jul 1, 1940), 97th Obsn Sq (Corps & Army), Air Corps, Air Base Mitchel Fld, Long Island, N.Y., FY 1940-1941, box 530, 353.9, AAG, RG 18, NA.

55. Estes intvw, Aug 27-30, 1973, 39.

56. Memo, Capt Curtis E. LeMay, Ops Off 41st Recon Sq (LR) GHQ Air Force, to Flt Comdrs 41st Recon Sq (LR) GHQ Air Force, Langley Fld, Va., Mar 6, 1940, Subj: Training Directive, 1939-1940, box 530, 353.9, AAG, RG 18, NA.

57. 1st Ind, HQ GHQ Air Force, Bolling Fld, D.C., to Chief of Air Corps, Washington, D.C., Apr 25, 1941, box 178, 353.9, AAG, RG 18, NA.

58. "The Air Corps Theory of probable error is based on the axiom that a bombing airplane, unlike artillery, cannot duplicate the conditions of firing.... Firing conditions for successive rounds of artillery fire vary due to changes in muzzle velocity and atmospheric conditions. Such variations are so minute compared to the variations of successive bombing approaches that it is still accurate to consider that for artillery the conditions of firing during 'fire for adjustment' remain constant for successive rounds; whereas, for bombing, conditions change considerably between successive approaches." Memo, Maj F. O. Carroll, Chief Exper Engineering Sec, to Chief Mat Div, Feb 21, 1940, Subj: Comments on Attached Letter from Commanding General, GHQ Air Force, to the Chief of the Air Corps, Jan 24, 1940, box 173, 353.4, AAG, RG 18, NA.

59. Futrell, "Observation Aviation," 10-11.

60. Russell F. Weigley, *History of the United States Army* (Bloomington: Indiana University Press, 1967), 423.

61. *Army Air Forces Statistical Digest: World War II*, table 47.

62. Holley, *Buying Aircraft*, 237-245; Haywood S. Hansell, Jr., *The Air Plan That Defeated Hitler* (Atlanta, Ga.: Higgins-McArthur/Longino & Porter, Inc., 1972), 78; see charts.

Part V

1. John Boeman, *Morotai: A Memoir of War* (Manhattan, Kans.: Sunflower University Press, 1981), 1-2.

2. Robert R. Palmer et al, *The Procurement and Training of Ground Combat Troops* (Washington, D.C.: Office of the Chief of Military History, 1948), 21-22.

3. See Morton, "Germany First," in

Notes

Greenfield, ed., *Command Decisions*, 11–47.

4. Haywood S. Hansell, *The Air Plan That Defeated Hitler* (Atlanta, Ga.: Higgins-McArthur/Longino & Porter, Inc., 1972), 98, 88, 109.

5. Arnold, *Global Mission*, 335–336.

6. Roger E. Bilstein, *The American Aerospace Industry: From Workshop to Global Enterprise* (New York: Twayne, 1996), 71.

7. The history of air operations during the war is full of examples of campaigns and missions that did not follow the strategic dictum. Most notable is Kenney's conduct of the Pacific air war. Most familiar is LeMay's low-level incendiary raids against Japan, although it had been LeMay who effectively carried out the daylight strategic campaign over Germany by use of rigid, defensive box formations and salvo bombing. But in many instances, the Eighth Air Force in the European theater, intended as the showcase for high-altitude precision bombardment, often fought differently. The massive B-24 raid over Ploesti, for example, about which author Pete Copp said, "no mission before or after was put together with such care, forethought, and training," was planned by Col. Jacob E. "Jake" Smart as the "antithesis of a high-altitude precision raid: a massed coordinated strike at extremely low level." Copp, *Forged in Fire*, 414–415. Focus on the strategic debate also fails to take account of the effectiveness of air-ground warfare. An interesting perspective on tactical air power in the making of doctrine can be found in Thomas Alexander Hughes, *Over Lord: General Pete Quesada and the Triumph of Tactical Air Power in World War II* (New York: Free Press, 1995), 51–82.

8. Baldwin, "Individual Training of Navigators," 5; Baldwin, "Individual Training of Bombardiers," 2.

9. 1st Ind, HQ AAF Tng Comd to Comdg Gen AAF (attn: Asst Chief of Air Staff, Tng), Oct 5, 1943, box 478, uncld, AAG, 353, RG 18, NA.

10. As of August 30, 1943, the Training Command conducted courses for the following (not including courses for non-flyers or technical courses): Aviation Cadet Examining Boards; Basic Training Centers (for Aviation Cadets; for Enlisted Men); AAF College Training Program; Classification Centers; Pre-flight Pilot Training; Pre-flight Bombardier Training; Pre-flight Navigator Training; Primary Pilot Training; Basic Pilot Training; Advanced Pilot Training (Advanced Single-engine Fighter; Advanced Two-engine Fighter; Advanced Two-engine Standard; Advanced Two-engine with Fixed Gunnery); Medium Bomber Transition; Heavy Bomber Transition; Advanced Bombardier Training (Straight; Dead Reckoning); Advanced Navigator Training; Bombardier-Navigator Training; Flexible Gunnery Training (for Aviation Cadets; for Others); Observer (Non-pilot) Training; Central Instructor Schools; Glider Pilot Training; Women Pilot Training; Advanced Liaison Pilot Training; Colored Pilot Training; The Enlisted Reserve. R&R Sheet, Statistical Control Div to Asst Chief of Air Staff, Tng, Aug 30, 1943, Subj: Information Regarding Training Course and Process Flow Charts for AAF Personnel. Box 888, 353, uncld, AAG, RG 18, NA.

11. The AAF Board was composed of high-ranking officers charged with evaluating procurement and tactics. The School of Applied Tactics in Orlando, Florida, began as the Fighter Command School set up to study air defense tactics. Its mandate expanded to all aspects of air tactics, hence the name change, and its further redesignation in October 1943 as the AAF Tactical Center.

12. A Training Standard issued by HQ AAF for any type of aircraft was a general descriptive statement of missions and tasks and did not differentiate among the various categories of aircraft. Programs of Instruction, usually issued by the Training Command or one of its regional headquarters, specified hours apportioned to subjects or missions.

13. Layman, "Organization of AAF Training Activities," 20, 38.

14. See F. J. Hatch, *Aerodrome of Democracy: Canada and the British Com-*

monwealth Air Training Plan, 1939–1945 (Ottawa, Canada: Directorate of History, Department of National Defence, 1983).

15. Rprt, Trng Conf, HQ 2d Air Force, Colorado Springs, Colo., Sep 20–22, 1943, box 478, uncld, AAG, RG 18, NA.

16. A variety of training documents cited this axiom from time to time. Arnold articulated it in a memo to commanding generals of all air forces on April 10, 1943. See Layman, "Organization of AAF Training Activities," 40.

17. Memo, Brig Gen Robert W. Harper to Comdg Gen 2d Air Force, Oct 5, 1943, Subj: Cooperation and Coordination between the Various Air Forces and Commands, box 900, 353, uncld, AAG, RG 18, NA.

18. Bernard Brodie, *War and Politics* (New York: Macmillan, 1973), 35.

19. Craven and Cate, eds., *Men and Planes*, 577–578.

20. Second Report of the Commanding General of the Army Air Forces to the Secretary of War, Feb 27, 1945. The number of training fatalities during World War II, as differentiated from the total number of deaths in accidents or from all eliminations from training, is difficult to substantiate. To date, the U.S. Air Force is reluctant to make available statistics regarding training fatalities. The most reliable figures come from *Army Battle Casualties and Nonbattle Deaths in World War II — Final Report* (Washington, D.C.: Dept of the Army, GPO, June 1, 1953) that gives total AAF accident deaths as 25,844.

21. Lecture Outline, Lt Col Robert B. McCutcheon, HQ AAF Sch Appl Tac Dept Insp, "Training Problems in the AAF, Objective: To Emphasize the Vastness of the AAF Training Program So That Air Inspectors May Render More Effective Aid in Helping Solve the Many Problems Which Present Themselves," box 890, 353, uncld, AAG, RG 18, NA.

Chapter 11

1. Ground Training Guide: Heavy Bombardment, I Bom Comd, Mitchel Fld, N.Y., box 76, 353.01, AAG, RG 18, NA.

2. The subject of pilot and aircrew procurement is complex and lengthy and deserves extended treatment that cannot be provided in this book. For a discussion of the beginnings of the Air Force Reserve that had its beginnings before World War II, see Cantwell, *Citizen Airmen*, 1–22. The inharmonious relationship between the AAF and the Department of Commerce's Civil Aeronautics Administration in manpower procurement is described in Pisano, *To Fill the Skies with Pilots*.

3. Natalie Grow, "Procurement of Aircrew Trainees," USAF Hist Study No. 15, Aug 1944, 54–135, AFHRA.

4. R&R, Brig Gen William E. Hall, Dep Chief of Air Staff, to Ofc Tech Info, Mar 14, 1944, Subj: Tours of Returned Aircraft and Crews, box 922, 353.1, uncld, RG 18, NA.

5. Thompson, "Initial Selection of Candidates," 25.

6. Ardery, *Bomber Pilot*, 20.

7. Stanine was a contraction of *sta*ndard *nine*, based on a numerical scale of 1 to 9.

8. *Stanines: Selection and Classification for Air Crew Duty*, Reported by the Aviation Psychology Program, Office of the Air Surgeon, HQ AAF, box 381, 353, uncld, AAG, RG 18, NA.

9. William Carigan, *Ad Lib: Flying the B-24 Liberator in World War II* (Manhattan, Kans.: Sunflower University Press, 1988), 74.

10. *Stanines*.

11. Craven and Cate, eds., *Men and Planes*, 558–559.

12. Bert Stiles, *Serenade to the Big Bird* (New York: W. W. Norton & Co., 1947), 191.

13. Estes intvw, Aug 27–30, 1973, 686, 37.

14. Martin R. R. Goldman, "Morale in the AAF in World War II," USAF Hist Study No. 78, Jun 1953, 13, AFHRA.

15. Chuck Yeager and Leo Janos,

613

Notes

Yeager: An Autobiography (New York: Bantam Books, 1985), 44, 80.

16. Craven and Cate, eds., *Men and Planes*, 577–578.

17. Tng Intel Rprt 1977, British Aircrew Tng Bul, "Prevention of Flying Accidents," box 264, 319.1, entry 294, RG 18, NA.

18. Hynes, *Flights of Passage*, 16.

19. Boeman, *Morotai*, 12.

20. Third Report of the Commanding General of the Army Air Forces to the Secretary of War, Nov 12, 1945, 63–64.

21. McMullen, "Role of Civilian Contractors," 1–24.

22. Charles A. Watry, *Washout! The Aviation Cadet Story* (Carlsbad, Calif.: California Aero Press, 1983), 90, 93–94.

23. See Chapter 9. See also Smith, "Development of AAF and USAF Training," 46–47.

24. At various points in the training sequence, student pilots were given a flight test on some portion of their training by a "check pilot" who was not the student's instructor.

25. John L. Frisbee, "The Cloud with the Mild Blue Lining," *Air Force Magazine* 64: no. 7 (Jul 1981), 88.

26. Eugene Fletcher, *Mister* (Seattle: University of Washington Press, 1992), 63.

27. Smith, "Development of AAF and USAF Training," table 2, 41.

28. Prog Inst, Elementary Flying Training for Pilot Trainees, Feb 1, 1943, box 368, 337, entry 294, RG 18, NA.

29. See for example, Weekly Tng Rprts, 57th AAF Flying Tng Det, Ocala, Fla., Jul 1942, box 530, 353.9, AAG, RG 18, NA.

30. Prog Inst, AAF HQ Flying Tng Comd, Fort Worth, Tex., Elementary Flying Training for Military Students to Be Given in Army Air Forces Flying Schools, Tentative, Dec 15, 1942, box 380, 353, uncld, AAG, RG 18, NA; Flying Tng Comd Memo 50-8-1, Prog Inst, Elementary Flying Training for Pilot Trainees, Jan 15, 1943, box 910, 353, uncld, AAG, RG 18, NA.

31. 2d Ind, Maj Allen B. Black, HQ AAF Flying Tng Comd, to Comdg Gen AAF (attn: Dir Indiv Tng), Jan 30, 1943, box 902, 353, uncld, AAG, RG 18, NA.

32. Philip Ardery, *Bomber Pilot*, 15–16. The wind-drift principle was to steepen the turn downwind and shallow it out upwind.

33. Purtee, *Modification and Development of Training Aircraft*, 34–41.

34. F. J. Hatch, *Aerodrome of Democracy: Canada and the British Commonwealth Air Training Plan, 1939–1945* (Ottawa, Canada: Directorate of History, Department of National Defence, 1983), caption p. 150.

35. Memo, Brig Gen Robert W. Harper to Joint Acft Comte, Sep 2, 1943, Subj: Requirements for Training Aircraft, box 478, 353, uncld, AAG, RG 18, NA.

36. Smith, "Development of AAF and USAF Training," table 1, 40.

37. Intv, Brig Gen Cleo M. Bishop by Lt Col John N. Dick, Jr., Jul 6–7, 1976, 10, K239.0512-904, AFHRA.

38. Watry, *Washout*, 92–93.

39. Ardery, *Bomber Pilot*, 14.

40. Hynes, *Flights of Passage*, 25.

41. R&R, Col R. S. Macrum, Bud & Fisc Ofc, to Mgt Con, May 11, 1944, Subj: Comparison between Army-Navy Flying Training and Commercial Flying Training, box 902, 353, uncld, AAG, RG 18, NA; Smith, "Development of AAF and USAF Training," 52; McMullen, "Role of Civilian Contractors," 40–41.

42. James L. Colwell, "Wings Over West Texas: Pecos Army Air Field In World War II," *West Texas Historical Association Year Book*, 1987, 56.

43. Hynes, *Flights of Passage*, 75.

44. Smith, "Development of AAF and USAF Training," table 1, 56; Prog Inst, Clas Cen, Ground Training in Pilot Pre-Flight, Elementary, Basic, Advanced Schools, Dec 15, 1942, box 380, 353, uncld, AAG, RG 18, NA.

45. Prog Inst, Basic Flying Training for Military Students to Be Given in Air Forces Flying Schools, Tentative, Dec 15, 1942, box 380, 353, uncld, AAG, RG 18, NA; Flying Tng Comd Memo 50-9-1, Prog Inst, Training, Basic Pilot Training, Apr 21, 1943, box 910, 353, uncld, AAG, RG 18, NA; Smith, "Development of

AAF and USAF Training," chart 57–58; Craven and Cate, eds, *Men and Planes*, 570.

46. Watry, *Washout*, 118.

47. Estes intvw, Aug 27–30, 1973, 686, 51, 56.

48. C. V. Glines, "Duckworth's Legacy," *Air Force Magazine* 73, no. 5 (May 1990): 178–179.

49. AAF Reg 50-3, *Training*, Instrument Flying Training, box 531, 353.9, AAG, RG 18, NA.

50. Prog Inst, Basic Flying Training for Pilot Trainees, Jan 15, 1943, box 368, 337, entry 294, RG 18, NA.

51. Tom Harmon, *Pilots Also Pray* (New York: Thomas Y. Crowell Co., 1944), 41–42.

52. Purtee, *Modification and Development of Training Aircraft*, 54–57.

53. Watry, *Washout*, 107.

54. Hynes, *Flights of Passage*, 66.

55. Memo, Maj Gen Barney Giles for Gen Arnold, Oct 14, 1943, Subj: Subject of Actions of the AAF Aircraft Requirements Board, box 449, 337, entry 294, RG 18, NA.

56. Craven and Cate, eds., *Men and Planes*, 571.

57. Watry, *Washout*, 113–114.

58. Smith, "Development of AAF and USAF Training," 100.

59. Hynes, *Flights of Passage*, 76.

60. Purtee, *Modification and Development of Training Aircraft*, 64–65; Fahey, ed., *U.S. Army Aircraft*, 21; Memo, Brig Gen Robert W. Harper to Joint Acft Comte, Sep 2, 1943, Subj: Requirements for Training Aircraft, box 478, 353, uncld, AAG, RG 18, NA.

61. Bill Colgan, *World War II Fighter-Bomber Pilot* (Blue Ridge Summit, Pa.: Tab Books, 1985), 17.

62. Prog Inst, Advanced Flying Training — Fighter Single-engine — for Pilot Trainees, Jan 15, 1943, box 368, 337, entry 294, RG 18, NA.

63. Smith, "Development of AAF and USAF Training," 69, 71.

64. AAF Tng Std 90-3, HQ AAF, 2-engine Advanced Pilot Training, box 915, 353, uncld, AAG, RG 18, NA.

65. Smith, "Development of AAF and USAF Training," 72; Memo, Col A. S. Barnhart and Col L. S. Smith, Dir Indiv Tng, to Comdg Gen AAF Flying Tng Comd, Oct 27, 1942, Subj: Southeast Training Center Bulletin No. 39, Night Formation Flying, box 902, 353, uncld, AAG, RG 18, NA.

66. Dead reckoning was the basic means of navigation whereby the pilot estimated his position by visual means plus calculations involving earlier known position, elapsed time, speed, heading, and effect of winds.

67. Prog Inst, AAF HQ Flying Tng Comd, Advanced Flying Training, Fighter Single-engine, Pilot Trainees, Tentative, Dec 15, 1942, box 380, 353, uncld, AAG, RG 18, NA.

68. R&R, Asst Chief of Air Staff, OC&R (Ftr and Air Defense Br) to Asst Chief of Air Staff, Tng, May 20, 1944, Subj: Training and Dead Reckoning Navigation, box 907, 353, uncld, AAG, RG 18, NA.

69. Montgomery intvw, Jun 28–30, 1983, 72.

70. Smith, "Development of AAF and USAF Training," 66, 73–77.

71. Memo, Brig Gen Robert W. Harper, Asst Chief of Air Staff, Tng, to Comdg Gen, AAF Flying Tng Comd, Jun 29 1943, Subj: Selection of Pilot Trainees at Advanced Single-engine Schools Who Are to Receive the Course of Instruction in Fixed Gunnery, box 473, 353, uncld, AAG, RG 18, NA.

72. Smith, "Development of AAF and USAF Training," 78–79; "History of AAF Flying Training Command, 1 Jan 1939 to 7 Jul 1943," frame 1112, mcflm roll 2286, AFHRA.

73. Arderey, *Bomber Pilot*, 23–24.

74. Boeman, *Morotai*, 22.

75. Prog Inst, Advanced Flying Training, Twin-engine Pilot Trainees, Dec 15, 1942; Smith, "Development of AAF and USAF Training," 87–93.

76. Purtee, *Modification and Development of Training Aircraft*, 67–78.

77. Memo, Brig Gen Robert W. Harper, Asst Chief of Air Staff, Tng, to Joint Acft Comte, Sep 2, 1943, Subj: Requirements for Training Aircraft, box 478, 353,

Notes

uncld, AAG, RG 18, NA.

78. Watry, *Washout*, 128–130.

79. Ibid.

80. Directed Prog Inst, Advanced Flying Training (Observation), Single-engine or Twin-engine, Jan 15, 1943, box 368, 337, entry 294, RG 18, NA.

81. Prog Inst, For Ground Training in Observation [Dec 15, 1942], box 380, 353, uncld, AAG, RG 18, NA.

82. Prog Inst, Advanced Flying Training (Observation), Single-engine or Twin-engine, Pilot Trainees, Dec 15, 1942, box 380, 353, uncld, AAG, RG 18, NA.

83. The Royal Canadian Air Force experimented with a class of sixty student pilots who flew their entire course of training, from primary through advanced, exclusively on the advanced Harvard (ATC–6) aircraft. American training officials were politely receptive to the idea, which harked back to the earlier British Gosport system that the United States had also failed to adopt at the time.

84. McNarney, "Pilot Transition to Combat Aircraft," 54–67.

85. Memo, Lt Col William Fane to Comdg Gen, AAF (attn: Asst Chief of Air Staff, Tng), Aug 21, 1943, Subj: Specialized Four-engine Training for Returned Combat Personnel to Be Assigned Second Air Force, box 917, 353, uncld, AAG, RG 18, NA; 1st Ind, HQ 2d AF to Comdg Gen, AAF (attn: Asst Chief of Air Staff, Tng), Sep 4, 1943, box 478, 353, uncld, AAG, RG 18, NA.

86. R&R, AFTFS thru AFRIT (attn: Col Smith), Sep 17, 1942, Subj: Flying Safety Activity vs. Training Efficiency, box 53, 353.9, uncld, AAG, RG 18, NA.

87. Memo with enclosures to Comdg Gen, AAF Western Flying Tng Comd (same ltr to AAF Eastern Flying Tng Comd and AAF Central Flying Tng Comd), May 19, 1944, Subj: Course of Instruction on Duties and Responsibilities of Airplane Commander, box 917, 353, uncld, AAG, RG 18, NA.

88. Memo, Lt Col Irvine H. Shearer, Fld Air Insp (Tac), to The Fld Air Insp, Jun 2, 1944, Subj: Report of Evaluation of Air Crew Trainees and Pilots of the Army Air Forces Training Command, box 485, 353.01, uncld, AAG, RG 18, NA.

89. McNarney, "Pilot Transition to Combat Aircraft," 68–75, 122–169.

90. 1st Ind, HQ AAF Tng Comd, to Comdg Gen, AAF (attn: Asst Chief of Air Staff, Tng), Sep 11, 1943, box 475, 353, uncld, AAG, RG 18, NA.

91. Memo, Col L. S. Smith, Dir Indiv Tng, to Comdg Gen, AAF Flying Tng Comd, Aug 26, 1942, Subj: Training of Four-engine Pilots in Four-engine Schools, box 176, 353.9, AAG, RG 18, NA; Memo, Col Philip Doddridge, Adj Gen, AAF Tng Comd, to Comdg Gen, AAF (attn: Asst Chief of Air Staff, Tng), Dec. 6, 43, Subj: Training of Four-engine Co-pilots, box 909, 353, uncld, AAG, RG 18, NA; Memo, Lt Col H. P. Bonnewitz to Comdg Gen, AAF (attn: Asst Chief of Air Staff, Tng), [Feb 15] 1944, Subj: Co-pilot Training Program, box 909, 353, uncld, AAG, RG 18, NA; McNarney, "Pilot Transition to Combat Aircraft," 169.

92. Daily diaries, Tng Sec, Oct 16, 1942, and Unit Tng Div, Bom Br, Feb 13, 1943, box 127, 319.1, entry 294, RG 18, NA.

93. Daily diary, Dir Indiv Tng, Dec 22, 23, 1942, box 242, 319.1, entry 294, RG 18, NA.

94. McNarney, "Pilot Transition to Combat Aircraft," 78–91, 170–203; Angelucci, *Encyclopedia of Military Aircraft*, 288–289.

95. McNarney, "Pilot Transition to Combat Aircraft," 91–98; Angelucci, *Encyclopedia of Military Aircraft*, 289.

96. McNarney, "Pilot Transition to Combat Aircraft," 98–101.

97. Ibid., 75–77.

98. AAF Tng Std 90–2, Fighter Aircraft Transition and Fixed Gunnery Training of Day Fighter Pilots, box 915, 353, uncld, AAG, RG 18, NA; "History of the AAF Flying Training Command, 1 Jan 1939 to 7 Jul 1943," frame 1118, mcflm 2286, AFHRA.

99. Memo, Col L. S. Smith, Dir Indiv Tng, to Comdg Gen, Flying Tng Comd, Jul 20, 1942, Subj: Transition Training on Twin-engine Airplanes for Students in Advanced Single-engine Schools Scheduled for Assignment to Fighter Units

Equipped with P-38 Airplanes, box 176, 353.9, AAG, RG 18, NA.
100. Tng Comd Memo 50-10-5, Prog Inst, Advanced Pilot Training, Fighter Twin-engine (Pilot Trainees), Nov 11, 1943, box 910, 353, uncld, AAG, RG 18, NA.
101. McNarney, "Pilot Transition to Combat Aircraft," 204; Smith, "Development of AAF and USAF Training," 76, 103.
102. Memo, Maj Gen Robert W. Harper, Asst Chief of Air Staff, Tng, to Comdg Gen, AAF Tng Comd, Mar 29, 1944, Subj: Fighter Transition and Reconnaissance Requirements Class 44-D through 44-F, box 917, 353, uncld, AAG, RG 18, NA.
103. McNarney, "Pilot Transition to Combat Aircraft," 111, 224.
104. Memo, Brig Gen Donald Wilson to Chief of Staff (attn: Asst Chief of Staff, G-3), Aug 10, 1944, Subj: AAF Pilot Training Program, box 483, 353.01, uncld, AAG, RG 18, NA.
105. Memo, Lt Gen B. K. Yount to Comdg Gen AAF (attn: Asst Chief of Air Staff, Tng), Sep 7, 1944, Subj: Flying Training Program between the Defeat of Germany and the Defeat of Japan, box 484, 353.01, uncld, AAG, RG 18, NA; Craven and Cate, eds., *Men and Planes*, Foreword, xxxi.
106. Memo, HQ Flying Tng Comd to Comdg Gen West Coast Air Corps Tng Cen, Mar 27, 1942, box 177, 353.9, AAG, RG 18, NA.
107. R. J. Overy, *The Air War, 1939-1945* (New York: Stein & Day, 1980), 143.

Chapter 12

1. Third Report of the Commanding General of the Army Air Forces to the Secretary of War, Nov 12, 1945.
2. Craven and Cate, eds., *Men and Planes*, 579-584; Baldwin, "Individual Training of Bombardiers," 2, 16; Baldwin, "Individual Training of Navigators," 195; Memo, Lt Col H. P. Bonnewitz to Comdg Gen AAF (attn: Asst Chief of Air Staff, Tng), Mar 28, 1944, Subj: Bombardier/Navigator Production, box 895, 353, AAG, RG 18, NA; Memo, Maj Gen Barton K. Yount to Comdg Gen, AAF (attn: Asst Chief of Air Staff, Tng), Jun 25, 1943, Subj: Maximum Training Effort, box 477, 353, uncld, AAG, RG 18, NA; "History of AAF Flying Training Command, 1 Jan 1939 to 7 Jul 1943," frame 877.
3. Baldwin, "Individual Training of Navigators," 14-16, 24, 71-73.
4. AAF Tng Std 90-7, Navigator, [Mar 15, 1944], box 915, 353, uncld, AAG, RG 18, NA.
5. Baldwin, "Individual Training of Navigators," 82.
6. 1st Ind, HQ AAF Flying Tng Comd to Comdg Gen AAF (attn: Asst Chief of Air Staff, Tng), Apr 19, 1943, box 910, uncld, 353, AAG, RG 18, NA.
7. Tng Intel Rprt 1543, RAF Bomber Command Bulletin: Use of Navigation Instruments, Dec 23, 1943, box 137, 319.1, entry 294, RG 18, NA.
8. Prog Inst, Training of Aerial Navigators for Military Students to Be Given in Air Corps Flying Schools [July 15, 1941], reproduced in Baldwin, "Individual Training of Navigators," app 1.
9. Baldwin, "Individual Training of Navigators," 91; Smith, "Development of AAF and USAF Training," 158.
10. Consolidated diary, 2d AF, Nov 4, 1944, box 240, 319.1, entry 294, RG 18, NA; Daily diary, 4th AF, Nov 29-30, 1944, box 245, 319.1, entry 294, RG 18, NA; Consolidated diary, 2d AF, Jan 17, 1945, box 76, 319.1, entry 294, RG 18, NA.
11. Memo, Capt William W. Foster, Jr., Gp Nav, to Comdg Gen 5th Wg (U.S.), Jun 20, 1943, Subj: Navigation Training, box 479, 353, uncld, AAG, RG 18, NA.
12. Baldwin, "Individual Training of Navigators," 93-94.
13. Henry Hatfield in Studs Terkel,

Notes

"The Good War": An Oral History of World War Two (New York: Pantheon Books, 1984), 353.

14. Memo, Maj Gen Robert W. Harper, Asst Chief of Air Staff, Tng, to Comdg Gen, AAF Tng Comd, May 1, 1944, Subj: CNT Training; 2d Lt Lew M. Warden, Jr., "Supplements for Navigational Training," box 907, 353, uncld, AAG, RG 18, NA.

15. Daily diary, Unit Tng Div, Bom Br, Feb 16, 1943, box 127, 319.1, entry 294, RG 18, NA.

16. Baldwin, "Individual Training of Navigators," 84, 162–163.

17. Memo, Brig Gen Robert W. Harper, Asst Chief of Air Staff, Tng, to General Perrin, Sep 3, 1943, Subj: Bombardier-Navigator Training, box 903, 353, uncld, AAG, RG 18, NA.

18. Memo, Capt Alfred C. Ball to Capt T. E. Hoffman, Air Insp AAF, West Coast Sector, Air Tng Comd, Hamilton Fld, Calif., Sep 14, 1943, Subj: Navigators [sic] Shortcomings, box 473, 353, uncld, AAG, RG 18, NA.

19. Memo, Foster to Comdg Gen, Jun 20, 1943.

20. R&R, Brig Gen E. L. Eubank, AFRBS thru AFRIT, Sep 9, [1942], box 531, 353.9, AAG, RG 18, NA.

21. James L. Colwell, "Hell from Heaven! Midland Army Air Field in World War II," pt 2, *Permian Historical Annual* 26 (1986): 63–64.

22. During the war, bombardier schools operated at Barksdale Field, La.; Big Spring, Tex.; Carlsbad Field, N.M.; Childress, Tex.; Deming Field, N.M.; Davis-Monthan Field, Ariz.; Ellington Field, Tex.; Geiger Field, Wash.; Gowan Field, Idaho; Hobbs Field, N.M.; Kirtland Field, N.M.; Lowry Field, Colo.; Midland, Tex.; Roswell, N.M.; San Angelo, Tex.; Victorville, Calif.; and Williams Field, Ariz.. For a brief description of these schools, see *Bombardier: A History*, 14–17.

23. Baldwin, "Individual Training of Bombardiers."

24. Memo, Col L. S. Smith, Dir Indiv Tng, to Comdg Gen Air Force Flying Tng Comd, Jul 16, 1942, Subj: Bombardier Training; 1st Ind, HQ 3d AF to Comdg Gen AAF (attn: AFPMP–2), Jul 8, 1942; 2d Ind, War Dept, HQ AAF, to Comdg Gen 3d AF, Aug 15, 1942, box 529, 353.9, AAG, RG 18, NA.

25. Baldwin, "Individual Training of Bombardiers," 60–114; Baldwin, "Individual Training of Navigators," 112–125; 1st Ind, Maj Allen B. Black, Asst Adj Gen HQ AAF Flying Tng Comd, to Comdg Gen AAF (attn: Dir Indiv Tng), Nov 16, 1942; R&R, AFRDB to AFRIT–2A, Dec 26, 1942; Memo, Brig Gen L. S. Smith, Dir Indiv Tng, to Comdg Gen Air Forces Flying Tng Comd, Dec 26, 1942, Subj: Bombardier-Navigator Training; 1st Ind, HQ AAF Flying Tng Comd, to Comdg Gen AAF (attn: Dir Indiv Tng), Jan 20, 1943, box 473, 353, uncld, AAG, RG 18, NA.

26. R&R, Brig Gen L. S. Smith, Dir Indiv Tng, to Dir Bom, Jan 9, 1943, Subj: Combined Bombardier-Navigator Curriculum, box 473, 353, uncld, AAG, RG 18, NA.

27. Memo, Col L. S. Smith, Dir Indiv Tng, to Comdg Gen Air Forces Flying Tng Comd, Jul 27, 1942, Subj: Gunnery Training for Potential Bombardiers and Navigators, box 529, 353.9, AAG, RG 18, NA.

28. Stephen L. McFarland, *America's Pursuit of Precision Bombing, 1910–1945* (Washington, D.C.: Smithsonian Institution Press, 1995), 159.

29. Colwell, "Hell from Heaven!" 53–55.

30. R&R, Ofc Chief of Air Corps to Chief AAF, Dec 31, 1941, Subj: Bombardier Training; Memo, Lt Col P. M. Whitney, Asst Air Adj Gen, to Chief of Coast Arty, Jan 26, 1942, Subj: Anti-aircraft Searchlights for Bombardier Training; 1st Ind, War Dept, Ofc Chief of Coast Arty to Chief AAF, Jan 29, 1942; box 529, 353.9, AAG, RG 18, NA; Memo, Col L. S. Smith, Dir Indiv Tng, to Comdg Gen, Flying Tng Comd, Aug 22, 1942, Subj: Navigation and Bombing Under Blackout Conditions, box 517, 353.4, AAG, RG 18, NA.

31. Tng Comd Memo 50-11-1, Prog Inst, Bombardier Training, Sep 29, 1943, box 895, uncld, 353, AAG, RG 18, NA.

32. Memo, 2d Lt Robert L. Clinton, Jr., to Dir Dept Armament, Air Forces Tech Sch, Mar 23, 1943, Subj: Conference, box 447, 337, entry 294, RG 18, NA.

33. *Bombardier: A History*, 31.

34. Memo, Maj Gen Barney M. Giles for Gen Arnold, Oct 21, 1943, box 449, 337, entry 294, RG 18, NA; Memo, Col J. H. Mills to Comdg Gen AAF (attn: Asst Chief of Air Staff, Tng), Apr 12, 1943, Subj: Training of Bombardiers on Sperry Sight, box 895, 353, uncld, AAG, RG 18, NA.

35. See McFarland, *America's Pursuit*, 154–157.

36. Memo, Col J. H. Hills, Adj Gen HQ Flying Tng Comd, to Comdg Gen AAF, Dec 28, 1942, Subj: Bombardier Training; Comment 1, Dir Indiv Tng to Dir Mil Pers (thru: Dir Bom), Jan 2, 1943; Comment 2, Dir Bom to Dir Mil Pers, Jan 6, 1943; Comment 3, Col E. S. Wetzel, Mil Pers Div, to Dir Indiv Tng, Jan 29, 1943; Comment 4, Brig Gen L. S. Smith, Dir Indiv Tng, to Mil Pers Div (attn: Maj Wisler), Feb 3, 1943, box 473, 353, uncld, AAG, RG 18, NA.

37. Baldwin, "Individual Training of Navigators," 26; Memo, Brig Gen Robert W. Harper, Asst Chief of Air Staff, Tng, to Comdg Gen, AAF Flying Tng Comd, Apr 14, 1943, Subj: Bombardier Standard; 1st Ind, Brig Gen Robert Wrd. Harper to AAF Sch Applied Tactics, Sep 24, 1943, box 473, 353, uncld, AAG, RG 18, NA.

38. Memo, HQ 3d AF to Comdg Gen AAF (attn: Asst Chief of Air Staff, Tng), Feb 5, 1944, Subj: Training of Celestial Navigator-Bombardiers, box 473, 353, uncld, AAG, RG 18, NA.

39. Memo, Col Edgar P. Sorensen, Asst Chief of Air Staff, A–2, to Chief of Air Staff (thru: Dir Bom), Feb 13, 1943, Subj: Training of Bombardiers, box 895, uncld, AAG, RG 18, NA.

40. Air Tng Std 90–6, HQ AAF, Bombardier, box 915, 353, uncld, AAG, RG 18, NA.

41. In fact, bombing accuracy figures from high-altitude precision raids in Europe were to prove appallingly bad, and carpet bombing was not uncommon. In the Pacific, airmen turned to low-altitude missions and area bombing. A thorough analysis of the efficacy of the Combined Bomber Offensive can be found in Richard G. Davis, "The Combined Bomber Offensive: A Statistical History," unpubl. ms., AFHSO.

42. Memo, Maj Gen George E. Stratemeyer to Comdg Gen Flying Tng Comd, Aug 4, 1942, Subj: Precision Bombing, box 517, 353.4, AAG, RG 18, NA.

43. Baldwin, "Individual Training of Bombardiers," 53–55; Craven and Cate, eds., *Men and Planes*, 582.

44. 1st Ind, Chief of Staff Brig Gen Walter F. Kraus, HQ Air Forces Flying Tng Comd, to Comdg Gen AAF (attn: Dir Indiv Tng), Aug 5, 1942, box 529, 353.9, AAG, RG 18, NA.

45. Memo, n.a. to Comdg Gen Air Forces Flying Tng Comd, May 27, 1942, Subj: Attitude of Bombardier Students, box 529, 353.9, AAG, RG 18, NA.

46. McFarland, *America's Pursuit*, 163.

47. Simms, "Flexible Gunnery Training," 7–10.

48. Memo, Brig Gen Muir S. Fairchild, Dir Mil Rqrs, to Comdg Gen 2d AF, Jun 29, 1942, Subj: Continuation of Training of Graduates from Flexible Gunnery Schools, box 529, 353.9, AAG, RG 18, NA.

49. Memo, Col L. S. Smith, Dir Indiv Tng, to Comdg Gen Air Forces Flying Tng Comd, Sep 24, 1942, Subj: Request for Gunnery Trainees, box 528, 353.9, AAG, RG 18, NA.

50. Craven and Cate, eds., *Men and Planes*, 590–591; Rprt, Visit to United States Flexible Gunnery Schools, May 16, 1943, box 923, 353.02, uncld, AAG, RG 18, NA; Memos, Brig Gen Robert W. Harper, Asst Chief of Air Staff, Tng, to Comdg Gen AAF Tng Comd, Aug 3, 1943, Subj: Expansion and Improvement of the Flexible Gunnery Training Program; Gen H. H. Arnold to Asst Chief of Air Staff, Tng, Aug 3, 1943, Subj: Expansion and Improvement of the Flexible Gunnery Training Program, box 483, 353.01-E, uncld, AAG, RG 18, NA.

51. Memo, Col Luther S. Smith to Comdg Gen, Tech Tng Comd, Mar 17, 1942, Subj: Gunnery Training, box 529,

Notes

353.9, AAG, RG 18, NA.

52. Ltr, Gen H. H. Arnold to Gene [Col Eugene H. Beebe, Comdg, 308th Bom Gp], c. Sep 14, 1943, box 488, 353.41-G, uncld, AAG, RG 18, NA; R&R, Brig Gen Robert W. Harper, Asst Chief of Air Staff, Tng, to Asst Chief of Air Staff, Pers (attn: Col E. S. Wetzel), Aug 25, 1943, Subj: Personnel for Expansion of Flexible Gunnery Program, box 473, 353, uncld, AAG, RG 18, NA.

53. R&R, Brig Gen Robert W. Harper, Asst Chief of Air Staff, Tng, to Asst Chief of Air Staff, Pers, Mil Pers Div, Feb 28, 1944, Subj: Eligibility for Flexible Gunnery Training, box 903, 353, uncld, AAG, RG 18, NA.

54. Flying Tng Comd Memo 50-13-7, Flexible Gunnery Training, Increasing Length of Course to Six Weeks, Apr 5, 1943, box 380, 353, uncld, AAG, RG 18, NA; "History of the AAF Flying Training Command, 1 Jan 1939 to 7 Jul 1943," frame 885.

55. Memo, Lt Col Richard R. Waugh, Comdg AAF Instructors Sch (Flexible Gnr), to Comdg Gen AAF (attn: Asst Chief of Air Staff, Tng), Apr 1, 1944, Subj: Training Films, Aerial Gunnery, box 924, 353.41, uncld, AAG, RG 18, NA.

56. "History of AAF Flying Training Command, 1 Jan 1939 to 7 Jul 1943," frame 910; Tng Intel Rprt 2486, XI AAF Ops Analysis Sec, AAF Flexible Gunnery Training Course, Aug 29, 1944, box 137, 319.1, entry 294, RG 18, NA.

57. James Gould Cozzens, *A Time of War: Air Force Diaries and Pentagon Memos, 1943–45*, Matthew J. Bruccoli, ed. (Columbia, S.C., & Bloomfield Hills, Mich.: Bruccoli Clark, 1984), 74.

58. Daily diary, Bom Br, Jul 20, 1943, box 127, 319.1, entry 294, RG 18, NA.

59. Directorate of Educational Services of the British Air Min, in Air Min Pam No. 137, "Hints on Teaching Aircraft Recognition," reproduced and distributed by Tng Intel Svc, AAF Tng Comd, box 132, 319.1, entry 294, RG 18, NA.

60. "History of the AAF Flying Training Command, 1 Jan 1939 to 7 Jul 1943," frames 893–894.

61. Memo, Col Luther S. Smith, Dir Indiv Tng, to Asst Chief of Staff G-3 (thru Asst Chief of Air Staff A-3), May 27, 1942, Subj: Training in Use and Operation of Sub-machine Guns in Army Air Forces Schools, and Ammunition Therefore [sic], box 529, 353.9, AAG, RG 18, NA.

62. Tng Intel Rprt 2486, Aug 29, 1944.

63. "History of the AAF Flying Training Command, 1 Jan 1939 to 7 Jul 1943," frame 212.

64. Memo, Maj Gen Follett Bradley to Maj Gen Barney M. Giles, Chief of Air Staff, Mar 28, 1944, Subj: Flexible Aerial Gunnery, box 903, 353, uncld, AAG, RG 18, NA.

65. Lecture, Haywood S. Hansell, Jr., "The Development of the United States Concept of Bombardment Operations," Air War College, Air University, Sep 19, 1951.

66. Simms, "Flexible Gunnery Training," 51–52, 57–63; Ltr, Maj Gen Barney M. Giles to Caleb [Brig Gen C. V. Haynes, HQ I Bom Comd], Nov 19, 1943, box 924, 353.41, uncld, AAG, RG 18, NA.

67. Tng Intel Rprt 2486, Aug 29, 1944.

68. Ltr, Col K. P. McNaughton to Tommie [Col T. J. DuBose, Dir Indiv Tng], Jan 30, 1943, box 479, 353, uncld, AAG, RG 18, NA.

69. Smith, "Development of AAF and USAF Training," 43, 255; "History of the AAF Flying Training Command, 1 Jan 1939 to 7 Jul 1943," frames 917–940; Rprt, Training Conference, HQ 2d AF, Colorado Springs, Colo., Sep 20–22, 1943, box 478, 353, uncld, AAG, RG 18, NA.

70. Memo, Lt Col J. M. Ivins to Comdg Gen AAF (attn: Asst Chief of Air Staff, Tng), May 9, 1944, Subj: Suggested Changes in Basic Training of Flexible Gunners, box 903, 353, uncld, AAG, RG 18, NA.

71. Daily diaries, Tng Comd, Mar 14 and 29, 1944, box 243, 319.1, entry 294, RG 18, NA.

72. Ltr, Mr. Borden to Mr. Olyarnik, Dalhart, Tex., Apr 12, [1943], excerpts in B22 box 903, 353, uncld, AAG, RG 18,

Notes

NA.

73. Craven and Cate, eds., *Men and Planes*, 593–594; Memo, Wg Comdr E. M. Donaldson, RAF, to Comdg Gen Flying Tng Comd, Sep 17, 1942, Subj: Visit by Wing Commander E. M. Donaldson to Southeast Coast Training Center Gunnery School, Eglin Field in Accordance with Orders of the General Commanding Training Command, box 528, 353.9, AAG, RG 18, NA; Simms, "Flexible Gunnery Training," 64–69; Smith, "Development of AAF and USAF Training," 251; "History of the AAF Flying Training Command, 1 Jan 1939 to 7 Jul 1943," frames 215–219.

74. Memo, Maj Gen Robert W. Harper, Asst Chief of Air Staff, Tng, to Comdg Gen AAF Tng Comd, May 5, 1944, Subj: Flexible Gunnery Program, box 903, 353, uncld, AAG, RG 18, NA; Ltr, Brig Gen E. B. Lyon to Maj Gen R. W. Harper, May 11, 1944, box 924, 353.41, uncld, AAG, RG 18, NA; Memo, Lt Col Irvine H. Shearer to Fld Air Insp, Jun 2, 1944, Subj: Report of Evaluation of Air Crew Trainees and Pilots of the Army Air Forces Training Command, box 485, 353.01, uncld, AAG, RG 18, NA.

75. Memo, Brig Gen Robert W. Harper, Asst Chief of Air Staff, Tng, to Comdg Gen AAF Tng Comd, Nov 23, 1943, Subj: Redesignation of Purpose of AAF Instructors School (Flexible Gunnery), box 76, 353.01, uncld, AAG, RG 18, NA.

76. Simms, "Flexible Gunnery Training," 74; Memo, Maj Gen B. K. Yount to Gen Arnold, Jun 18, 1942, Subj: Training of Aerial Gunners, box 529, 353.9, AAG, RG 18, NA.

77. Ltr, Mortimer Lee, Sep 21, 1942, excerpt (from the British Isles) in box 903, 353, uncld, AAG, RG 18, NA.

78. Simms, "Flexible Gunnery Training," 75; Smith, "Development of AAF and USAF Training," 252.

79. Simms, "Flexible Gunnery Training," 77; Smith, "Development of AAF and USAF Training," 242–243; "History of the AAF Flying Training Command, 1 Jan 1939 to 7 Jul 1943," frames 881–882.

80. Memo, Maj R. H. Thom to Comdg Gen AAF Asst Chief of Air Staff, Tng (attn: Lt Col Elmer A. Dittmar), Apr 22, 1944, Subj: Training of Career Gunners, box 903, 353, uncld, AAG, RG 18, NA; Simms, "Flexible Gunnery Training," 83.

81. Rprt on Tyndall Field, Feb 5, 1943, excerpts in box 903, 353, uncld, AAG, RG 18, NA.

82. Memo, Brig Gen Robert W. Harper, Asst Chief of Air Staff, Tng, to Air Insp, Oct 15, 1943, Subj: Individual Training of Air Crewmen, box 478, 353, uncld, AAG, RG 18, NA.

83. Quoted in Watry and Hall, *Aerial Gunners*, 85–86.

84. Craven and Cate, eds., *Men and Planes*, 522, 584, 589; Simms, "Flexible Gunnery Training," 97; *Army Air Forces Statistical Digest: World War II*, table 47.

85. Baldwin, "Individual Training of Bombardiers," 131.

86. Craven and Cate, eds., *Men and Planes*, 579, 586.

87. Rprt, Flexible Gunnery Training Conference, HQ 3d AF, Tampa, Fla., Apr 12, 13, and 14, 1944, box 381, 353, uncld, AAG, RG 18, NA.

88. Memo, Maj Gen Follett Bradley to Maj Gen Barney M. Giles, Chief of Air Staff, Mar 28, 1944, Subj: Flexible Aerial Gunnery, box 903, 353, uncld, AAG, RG 18, NA.

89. Memo, Maj Elmer A. Dittmar to General R. W. Harper, May 1, 1943, Subj: Standardization of Flexible Gunnery Schools, box 380, 353, uncld, AAG, RG 18, NA.

Chapter 13

1. Craven and Cate, eds., *Men and Planes*, 424; Robert Frank Futrell, *Ideas, Concepts, Doctrine: Basic Thinking in the United States Air Force, 1907–1960*, reprint, 2 vols (Maxwell AFB, Ala.: Air University Press, Dec 1989), 1: 132.

Notes

2. Memos, Col W. W. Welsh to Chief of Air Corps, Jan 26, 1942, Subj: 50,000 Pilot Training Program; Lt Col E. P. Curtis to Chief of Air Corps, Feb 20, 1942, Subj: Pages from the Chief of Staff's Notebook, box 177, 353.9, AAG, RG 18, NA.

3. Daily diary, Tng Sec, Bom Br, Dec 7, 1942, box 127, 319.1, entry 294, RG 18, NA.

4. Hist, 2d AF, 1943, 432.01, mcflm roll A4093, AFHRA.

5. HQ Ofc Inst 50-2, Training, Supervision of Operational Training, Jun 18, 1942, box 157, 353, bulky files, 1939–1942, uncld, AAG, RG 18, NA; Layman, "Organization of AAF Training Activities," 46; Simms, "Flexible Gunnery Training," 46.

6. Herman S. Wolk, *Planning and Organizing the Postwar Air Force, 1943–1947* (Washington, D.C.: Office of Air Force History, 1984), 115–116.

7. F. J. Hatch, *Aerodrome of Democracy: Canada and the British Commonwealth Air Training Plan, 1939-1945* (Ottawa, Canada: Directorate of History, Department of National Defence, 1983), 10–103.

8. White, "Combat Crew and Unit Training," 12–15; Smith, "Development of AAF and USAF Training," 280–285.

9. Daily diary, Unit Tng Div, Bom Br, Oct 26, 1942 and Oct 19, 20, 1943, box 127, 319.1, entry 294, RG 18, NA.

10. Memo, Col F. Trubee Davison, Asst Chief of Air Staff A-1, to Asst Chiefs of Air Staff A-3, Op Plans, Prog Planning, Mgt Con, to Dirs Mil Rqrs, War Org and Movement, Air Defense, Air Support, Bombardment, and to Comdg Gen Anti-Submarine Comd, Feb 12, 1943, Subj: Allocation of Replacement Combat Crews and Flying Training Command Graduates, box 446, 337, entry 294, RG 18, NA.

11. Diary, Dir Mil Rqrs, Nov 22, 1942, box 242, 319.1, entry 294, RG 18, NA; Consolidated diaries, HQ 2d AF, Gen Staff secs, Dec 27, 1944, box 240, 319.1, entry 294, RG 18, NA.

12. White, "Combat Crew and Unit Training," 17; Smith, "Development of AAF and USAF Training," 282.

13. Memo (with enclosures), Col Raynor Garey to Comdg Gen AAF (attn: AFACT), Apr 1, 1944, Subj: Study of Attrition in RTU Bombardment Training; Justification for Staff and Tow Target Aircraft; Benefits to Be Expected from Allocation of New Aircraft for Training, box 895, 353, uncld, AAG, RG 18, NA.

14. White, "Combat Crew and Unit Training," 18–19.

15. Memos, Col William C. Bentley, Exec, Asst Chief of Air Staff, Intel, to Col DuBose, Jun 16, 1943; Col T. J. DuBose, Chief Air Crew Tng, Asst Chief of Air Staff, Tng, to Col William C. Bentley, Jun 26, 1943, box 900, uncld, AAG, RG 18, NA.

16. White, "Combat Crew and Unit Training," 29.

17. By the middle of the war, four flights of three crews and airplanes each comprised a heavy bombardment squadron, and four squadrons made a group. Usually two or three groups made a wing.

18. Memos, Brig Gen Robert W. Harper to Comdg Gen 2d AF, Sept 9, 1943, Subj: Transfer of OTU Groups to First and Fourth Air Forces; Brig Gen Robert W. Harper to Chief of Air Staff, Oct 15, 1943, Subj: Heavy Bombardment Training in the Third Air Force; Col L. B. Kelley to Brig Gen R. W. Harper, Oct 26, 1943, Subj: Items for General Arnold's Attention; Brig Gen Robert W. Harper to Asst Chief of Air Staff, OC&R, Ops Div, Acft Br, Oct 29, 1943, Subj: Transfer of B-17 Type Aircraft, 143.03, mcflm roll A1326, AFHRA; Memo, Brig Gen Robert W. Harper to Comdg Gen 1st AF, Oct 25, 1943, Subj: Heavy Bombardment Training Program, box 483, 350.01, uncld, AAG, RG 18, NA; Daily diary, Bom Br, Oct 28, 1943, box 127, 319.1, entry 294, RG 18, NA.

19. Hist, 2d AF, 1944, 432.01, mcflm roll A4096, AFHRA; Hist, HQ III Bom Comd, Jan 1–Sep 2, 1945, vol 1, 440.01, mcflm roll A4134, AFHRA.

20. Memo, Col H. M. Jones, Adj Gen HQ 3d AF, to Comdg Gen AAF, Jun 14, 1942, Subj: Establishment of Three Medium, Two Light and One Dive Bombardment Operational Training Units, box

157, 353, bulky files, 1939–1942, uncld, AAG, RG 18, NA.

21. Hist, III Bom Comd, 1943.

22. Memo, A. J. Rehe, Adj Gen, by order Secy War, to Comdg Gens 1st, 2d, and 4th AFs, Jun 1, 1942, Subj: Training Directive, box 157, 353, bulky files, 1939–1942, uncld, AAG, RG 18, NA.

23. Hist, I Ftr Comd, Dec 1941 to Jul 1944, vol 1, 421.01, mcflm roll A4046, AFHRA; Org docu, HQ (with Comd and Staff), Dec 7, 1941 to Jan 1, 1944, 1st AF docus, frame 683, mcflm roll A4037, AFHRA; "Fighter Training in the Fourth Air Force, 1942–1945," 4th AF hist study II-2, vol 1: Narrative, 450.01-4, mcflm roll A4148, AFHRA.

24. Hist, III Ftr Comd, Dec 7, 1941, to Apr 30, 1944, 442.01, mcflm roll A4135, AFHRA; "Fighter Training in the Fourth Air Force, 1942–1945."

25. 2d AF hist, 1944; Hist, 2d AF, 1945, vol 2, 432.01, mcflm roll A4100, AFHRA.

26. Burchinal intvw, Apr 11, 1975, 47–48.

27. Rprt, 2d AF, Four-engine Training, copy atch to memo, Aug 25, 1943, box 477, 353, uncld, AAG, RG 18, NA.

28. Rprt, HQ 2d AF, Tng Conference, Colorado Springs, Colo., Sep 20–22, 1943, box 478, 353, uncld, AAG, RG 18, NA.

29. Minutes, AAF Tng and Gnr Conf, HQ 4th AF, Jul 31 to Aug 1–2, 1944, box 368, 337, AAG, RG 18, NA.

30. White, "Combat Crew and Unit Training," 66–67.

31. Memo, Maj Gen Robert W. Harper to Comdg Gen Trp Carrier Comd, Aug 28, 1944, Subj: Theatre Specialization Training; Memo, Maj Gen Robert W. Harper to Chief of Air Staff, Sep 8, 1944, Subj: Training of Crews for the Far East Air Forces; R&R, Col R. R. Walker, Asst Chief of Air Staff, Tng, to Asst Chief of Air Staff, OC&R, Sep 24, 1944, Subj: Training in the European Theater after the Capitulation of Germany, box 481, 353, uncld, AAG, RG 18, NA; R&R, Comment No. 1, Lt Gen Barney M. Giles to Asst Chief of Air Staff, Tng, Aug [18], 1944, Subj: Training of Navigators, Comment No. 2, Maj Gen Robert W. Harper to Chief of Air Staff, Sep 5, 1944, box 473, 353, uncld, AAG, RG 18, NA.

32. Under official doctrine, air power had always held a subordinate role. *Air Service Information Circular* 1 (June 12, 1940) stated "It is the role of the Air Service, as well as that of the other arms, to aid the chief combatant: the infantry." By 1935, Training Regulation 440–15 recognized the validity of some independent air operations, although by 1940 the major employment of air assets was still considered to be in support of ground forces. Air doctrine, in April 1942, was contained in War Department Field Manual 31–35, *Aviation in Support of Ground Forces*. Not until the publication in July 1943 of War Department Field Manual 100–20, *Command and Employment of Air Power*, was it stated (in uppercase letters): "LAND POWER AND AIR POWER ARE CO-EQUAL AND INTERDEPENDENT FORCES, NEITHER IS AN AUXILIARY OF THE OTHER." The first and, to date, final analysis of air force doctrine during World War II is found in Futrell, *Ideas, Concepts, Doctrine*, 1: 127–189.

33. Memo, HQ AAF to General Marshall, Aug 22, 1942, Subj: Army Air Forces Bombardment Doctrine, box 517, 353.4, uncld, AAG, RG 18, NA.

34. Hist, 2d AF, Dec 1941–Dec 1942, 432.01, mcflm roll A4091, AFHRA.

35. Carigan, *Ad Lib*, 89.

36. Booklet, comp from Rprt, A–2 Sec, Northwest African Air Forces, *The Air Forces in North Africa*, frame 276, 612.719-1, mcflm roll A6009, AFHRA.

37. Memo, Col M. C. Grow to Air Surgeon, HQ AAF, Aug 30, 1943, Subj: Report on Status of Training on Newly Arrived Heavy Bombardment Combat Crews in Oxygen and Safety Equipment, 143.03, mcflm roll A1326, AFHRA.

38. Rprt, to Comdg Gen AAF (attn: Asst Chief of Air Staff OC&R), Apr 13, 1944, Subj: Mission to the Mediterranean and European Theaters of Operations by the Baylor Committee, box 486, 353.01, uncld, AAG, RG 18, NA; Memo, Maj Gen Robert W. Harper to Comdg Gen 3d

Notes

AF, Jul 6, 1944, Subj: Bombardment Training Report, box 485, 353.01, uncld, AAG, RG 18, NA. It should be added that the Fifteenth Air Force flew smaller, more compact formations because it, unlike the Eighth which was subject to heavy German flak, had to fend off strong opposition from Japanese FW 190s and Me 109s.

39. Maj Gen George E. Stratemeyer, Chief of Air Staff, HQ Ofc Inst 25-2, May 26, 1943, Subj: Medical, High Altitude Indoctrination in Low Pressure Chamber, box 67, 353, AAG, RG 18, NA.

40. 2d AF hist, 1943,.

41. Memo, Maj Gen Robert W. Harper to Gen Arnold, May 6 1944, Subj: High Altitude Training for Heavy Bombardment Crews, box 480, 353, uncld, AAG, RG 18, NA.

42. Memo, Maj Gen Robert W. Harper to Comdg Gen 3d AF, May 10, 1944, Subj: Deficiencies in Training, box 480, 353, uncld, AAG, RG 18, NA.

43. Proceedings of a Board of Officers Convened for the Purpose of Investigating Instrument Flying and Instrument Flying Training in the Army Air Forces, box 904, 353, uncld, AAG, RG 18, NA.

44. Tng Intel Rprt 1523, HQ 21st Bom Wg, Recommendations Regarding Training Program, American Theater, Jan 18, 1944, box 137, 319.1, entry 294, RG 18, NA.

45. Daily diaries, Unit Tng Div, Bom Br, Dec 9, 1942, Mar 5 and 25, 1943, and May 13, 1943, box 127, 319.1, entry 294, RG 18, NA.

46. Memo, Brig Gen E. L. Eubank to Comdg Gen 2d AF, Jan 18, 1943, Subj: Bombing Accuracy, box 487, 353.41, uncld, AAG, RG 18, NA; 2d AF hist, 1943.

47. Bishop intvw, July 6-7, 1976, 9.

48. Air Chief Marshal Sir Harry Broadhurst, comment in "Joint RAF/USAF Seminar, Part I: Higher Command Structures and Relationships, 1942-45" in *Air Power History* (Summer 1991): 34.

49. Ltrs, Gen H. H. Arnold to Lt Gen L. J. McNair, Jun 2, 1943; Lt Gen L. J. McNair to Gen H. H. Arnold, Jun 15, 1943, box 488, 353.4, 1942-44, uncld, AAG, RG 18, NA.

50. See Alan F. Wilt, "Allied Cooperation in Sicily and Italy, 1943-1945," 221; Will A. Jacobs, "The Battle for France, 1944," 266, 271; and Joe Gray Taylor, "American Experience in the Southwest Pacific," 303, 329, all in Benjamin Franklin Cooling, ed., *Case Studies in the Development of Close Air Support* (Washington, D.C.: Office of Air Force History, 1990).

51. Minutes, AAF Tng and Gnr Conf, Jul 31 to Aug 1-2, 1944.

52. Daily diary, Unit Tng Div, Bom Br, Oct 17, 1942, box 127, 319.1, entry 294, RG 18, NA; Daily diary, Rqrs Div, Oct 6, 1944, box 245, 319.1, entry 294, RG 18, NA.

53. The purpose of this study is not the debate over the morality and the AAF leadership's views of strategic bombing of civilian populations. For recent works on this subject, see Ronald Schaffer, *Wings of Judgment: American Bombing in World War II* (New York: Oxford University Press, 1985); Michael S. Sherry, *The Rise of American Air Power: The Creation of Armageddon* (New Haven: Yale University Press, 1987); and Richard G. Davis, "Operation 'Thunderclap': The US Army Air Forces and the Bombing of Berlin," *The Journal of Strategic Studies* 14: no. 1 (Mar 1991): 90-111.

54. Wilt, "Allied Cooperation in Sicily and Italy, 1943-1945," in Cooling, ed., *Close Air Support*, 196.

55. *Condensed Analysis of the Ninth Air Force in the European Theater of Operations* (New Imprint, Office of Air Force History, Washington, D.C., 1984).

56. Richard P. Hallion, *Strike from the Sky* (Washington, D.C.: Smithsonian Institution Press, 1989), 173.

57. Memo, Lt Col Charles E. Marion, 303d Bom Gp (H), to Comdg Off 34th Bom Gp (H) (thru: Comdg Gen AAF; attn: Asst Chief of Air Staff, Tng, Unit Tng Div, Bom Br), May 31, 1943, Subj: Correspondence with Units in Training, box 477, 353, uncld, AAG, RG 18, NA.

58. Irene W. McPherson, "Tactical Air Power Symposium," *Air Power History* 37, no. 4 (Winter 1990): 51-55.

Notes

59. Intvw, Lt Gen Arthur C. Agan by Lt Col Vaughn H. Gallagher, Apr 19–22, 1976, 61, 200, K239.0512-900, AFHRA.

60. Intel Rprt, Intel Div, Ofc Chief of Naval Ops, Nov 6, 1943, box 262, 319.1, entry 294, RG 18, NA.

61. Comment 2, Brig Gen Robert W. Harper to Deputy Chief of Air Staff (General Vandenberg), Sep 3, 1943, Subj: Tactical Doctrine, 143.03, mcflm roll A1326, AFHRA.

62. Smith, "Development of AAF and USAF Training," 313–314; [Cable] to Arnold from Brereton, Nov 26, 1942; Memo, Gen Stratemeyer to AAF Comdrs Outside Continental Limits, Subj: High and Medium Altitude Bombing of Maneuvering Vessels [with enclosed memo from Arnold to Admiral King, Nov 27, 1942], box 487, 353.41, uncld, AAG, RG 18, NA.

63. Memos, Lt Col Edgar M. Wittan, 390th Bom Gp (H), to Comdg Off, 306th Bom Gp (H) (thru: HQ AAF; attn: Dir Bom), Apr 9, 1943; Lt Col Claude E. Putnam, Comdg, HQ 306th Bom Gp (H), to Comdg Off, 390th Bom Gp (H) (thru: Comd Chans), May 6, 1943, Subj: Tactical Ideas, box 477, 353, uncld, AAG, RG 18, NA.

64. Memo, Gen H. H. Arnold to Comdg Gen 5th AF, Apr 13, 1943, Subj: Combat Spirit, vol 5, Gen G. C. Kenney Diaries.

65. Erich Maria Remarque, *All Quiet on the Western Front* (New York: Fawcett Crest, 1928), 236.

66. Booklet, *The Air Forces in North Africa*.

67. Edward J. Giering, *B–17 Bomber Crew Diary* (Manhattan, Kans.: Sunflower University Press, 1985), 19.

68. Ardery, *Bomber Pilot*, 54.

69. C. F. Rawnsley and Robert Wright, *Night Fighter* (New York: Henry Holt & Co., 1957), 22.

70. Hynes, *Flights of Passage*, 176, 114.

Chapter 14

1. Kevin Herbert, *Maximum Effort: The B–29's Against Japan* (Manhattan, Kans.: Sunflower University Press, 1983), 21.

2. For a description of preparation and employment of the strategic strike force in the Pacific, see Richard P. Hallion, "Prelude to Armageddon," *Air Power History* (Fall 1995): 39–54.

3. Kit C. Carter and Robert Mueller, comps., *The Army Air Forces in World War II: Combat Chronology, 1941–1945* (Albert F. Simpson Historical Research Center, Air University, and Office of Air Force History, Headquarters USAF, 1973), 5–242.

4. It is noteworthy that in his final report to the Secretary of War Arnold never mentioned *precision* daylight strategic bombardment. He stressed the "strategic theory" of bombardment.

5. AAF Tng Std 20-2-1, Heavy Bombardment Units and Crews, box 915, 353, uncld, AAG, RG 18, NA.

6. Proceedings, Bd Offs to Comdg Gen AAF, Nov 3, 1942, box 475, 353, uncld, AAG, RG 18, NA.

7. 2d AF hists, Dec 1941–Dec 1942 and 1943.

8. Memo, Capt G. R. Dougherty, HQ 2d AF, to Comdg Gen AAF, Jul 14, 1943, Subj: Navigation Personnel, box 907, 353, uncld, AAG, RG 18, NA.

9. 2d AF hist, Dec 1941–Dec 1942.

10. Memo, Lt Col John D. Ryan to Dir Bom, Ofc Chief of Air Corps, Jan 12, 1943, Subj: Report of Bombing Activities in Operational Training Units, box 925, 353.41, uncld, AAG, RG 18, NA.

11. 2d AF hist, 1943.

12. Manpower requirements and allocations are not the subject of this study, but the magnitude of the task of developing reasonable schedules for training huge numbers of men with a chronic shortage of aircraft, coupled with the burning frustration of combat commanders who, daily facing fearsome attrition rates and therefore needed more, more, more, can be

Notes

sensed best from some of the memoirs and oral histories of those who dealt with requirements on both sides of the Atlantic. Lt. Gen. Arthur C. Agan, who served with the Eighth Air Force tactical operations section in both the Mediterranean and European Theaters of Operations, recalled that for some time Eighth Air Force planners framed requirements based on a 10 percent loss rate per month on B-17s and a 15 percent loss for crews, a stupefying 120 percent and 180 percent rate per year. For B-24s, the expectation was even worse: 20 percent attrition for aircraft and 15 percent for crews. Faced with these numbing statistics, Agan was nonetheless sympathetic to the production burden the anticipated loss rate levied on people at home. "You have got to have 15% per month [for B-24s] just to hold your own and here we are trying to build up to 1,500 bombers. The bill you sent back to this country...is so massive that you were really facing a problem back in Washington." Agan intvw, Apr 19–22, 1976, 91–92, 96.

13. Ltr, Gen H. H. Arnold to Gene [Col Eugene H. Beebe, Comdg 308th Bom Gp] [c. Sep 14, 1943], box 488, 353.41, uncld, AAG, RG 18, NA.

14. Memo [fragment], Asst Chief of Air Staff, OC&R, Allocations Br, to Asst Chief of Air Staff, Tng, Sep 1, 1943, Subj: Production of Heavy Bombardment Crews, 143.03, mcflm roll A1326, AFHRA; Memo, Brig Gen Robert W. Harper to Gen H. H. Arnold, Oct 31, 1943, Subj: Training of Four Additional B-24 Groups, box 890, uncld, 353, AAG, RG 18, NA.

15. Wesley Frank Craven and James Lea Cate, eds., *The Army Air Forces in World War II*, vol 3, *Europe: ARGUMENT to V-E Day, January 1944 to May 1945* (Washington, D.C.: Office of Air Force History, 1983), 306.

16. Hist, 1st AF, Tng, Jan–Dec 1944, 420.01–7, mcflm roll A4039, AFHRA.

17. Hist, HQ III Bom Comd, May 1–Dec 31, 1944, 438.01, mcflm roll A4133, AFHRA.

18. Ground Training Program for Combat Crews, Second Air Force, box 76, 353.01, AAG, RG 18, NA; HQ III Bom Comd hist, May 1–Dec 31, 1944; HQ 1st AF, Tng, hist, Jan–Dec 1944.

The 1944 III Bomber Command history provides the following listing of B-17 gun positions under the revised system:

B-17 "A" Crew
Pilot	Pilot's seat (left)
Copilot	Copilot's seat (right)
Armorer	Bendix chin turret
Navigator	Side nose guns
Engineer	Left waist
Radio operator	Right waist
1st Career gnr	Sperry ball turret
2d Career gnr	Sperry upper turret
3d Career gnr	Gunner tail gun

B-17 "B" Crew (changes from above)
Bombardier	Bendix chin turret
Armorer	Sperry ball turret

19. Memo and Annex, HQ AAF to Comdg Gen 2d AF, n.d., Subj: OTU Development Program, box 157, 353, bulky files, 1939–1942, uncld, AAG, RG 18, NA.

20. AAF Tng Std 20–2–1, War Dept, HQ AAF, Heavy Bombardment Units and Crews, Feb 8, 1943, box 69, 353, uncld, AAG, RG 18, NA.

21. The commonly used title shifted from Pilot to First Pilot to Airplane Commander. The April 1945 Training Standard officially dropped the last-mentioned term.

22. Excerpted in Edward Jablonski, *Flying Fortress* (Garden City, N.Y.: Doubleday & Co., 1965), 324.

23. HQ AAF to Comdg Gen 2d AF memo and annex, n.d.; AAF Tng Std 20–2–1, Feb 8, 1943.

24. Jablonski, *Flying Fortress*, 324.

25. Smith, "Development of AAF and USAF Training," 317–319.

26. Memo, 2d Lt Robert J. Clegg, Act Adj Gen HQ 2d AF, to Comdg Gen AAF (attn: Asst Chief of Staff, Tng), Sep 17, 1943, 1403.03, mcflm roll A1326, AFHRA.

27. Smith, "Development of AAF and USAF Training," 339.

28. Tng Intel Rprt 1547, Op Info Bul, Southwest Pacific, "The Bombardier," Dec 14, 1944, box 137, 319.1, entry 294, RG 18, NA.

29. Boeman, *Morotai*, 33.
30. Daily diary, Bom Br, Dec 1, 1942, box 127, 319.1, entry 294, RG 18, NA; AAF Tng Std 20-2-1, Feb 8, 1943; Memo 50-50, HQ 2d AF, Bombardment Training, Evasive Technique, Mar 18, 1943, box 895, 353, uncld, AAG, RG 18, NA.
31. Loftin, *Quest for Performance*, 122.
32. HQ AAF to Comdg Gen 2d AF memo and annex, n.d.; Rprt, HQ 2d AF Tng Conf, Sep 20-22, 1943, box 478, 453, uncld, AAG, RG 18, NA.
33. 1st Ind, HQ 2d AF to Comdg Gen AAF (attn: Air Insp), Sep 8, 1943, box 903, 353, uncld, AAG, RG 18, NA.
34. AAF Tng Std 20-2-1, Feb 8, 1943; Smith, "Development of AAF and USAF Training," 341-345.
35. Telecon, quoted in 2d AF hist, 1944.
36. R&R, Comment No. 1, Chief of Air Staff to Asst Chief of Air Staff, Tng, Jun 9, 1943; Comment No. 2, Asst Chief of Air Staff, Tng, to Chief of Air Staff, Jun 14, 1943, box 895, 353, uncld, AAG, RG 18, NA; Daily diary, Bom Br, Jun 21 and Aug 5, 1943, box 127, 319.1, entry 294, RG 18, NA.
37. Tng Prog, 2d AF, Ground Training Program for Combat Crews, box 76, 353.01, uncld, AAG, RG 18, NA; 1st Ind, HQ 2d AF to Comdg Gen AAF (attn: Brig Gen Robert W. Harper, Asst Chief of Air Staff, Tng), Jul 1, 1943, box 900, uncld, AAG, RG 18, NA.
38. Ardery, *Bomber Pilot*, 44.
39. 2d AF hist, 1944.
40. Proceedings, Bd Offs to Comdg Gen AAF, Nov 3, 1942, box 475, 353, uncld, AAG, RG 18, NA.
41. 4th Ind, HQ 8th AF to Comdg Gen AAF, May 12, 1943, box 904, 353, uncld, AAG, RG 18, NA.
42. Combat rprts, c. Aug 1943 and Apr 15, 1943, boxes 903 and 904, 353, uncld, AAG, RG 18, NA.
43. Smith, "Development of AAF and USAF Training," 321-324; 2d AF hist, 1943. These two sources vary as to the number of model missions.
44. Daily diaries, Unit Tng Div, Bom Br, Mar 27 and Jun 15, 1943, box 127, 319.1, entry 294, RG 18, NA.
45. Daily diary, Unit Tng Div, Bom Br, May 7, 1943, box 127, 319.1, entry 294, RG 18, NA.
46. Boeman, *Morotai*, 34.
47. Memo, Brig Gen U. G. Ent, Comdg Gen 2d AF, to Comdg Gen AAF (attn: Maj Gen R. W. Harper, Asst Chief of Air Staff, Tng), May 30, 1944, box 480, 353, uncld, AAG, RG 18, NA; 2d AF hist, 1944; Daily diary, HQ 4th AF, Jun 27-28, 1944, box 244, 319.1, entry 294, RG 18, NA.
48. Memo, Brig Gen Nathan B. Forrest, Chief of Staff 2d AF, to Comdg Gen AAF, Feb 9, 1943, Subj: Bombing Accuracy, box 487, 353.41, uncld, AAG, RG 18, NA; Daily diary, Bom Br, Aug 5, 1943, box 127, 319.1, entry 294, RG 18, NA.
49. Rprt, Lt Col G. B. Dany, Comdg HQ AAF, Nav Sch, Hondo Army Air Fld, to Comdg Gen AAF (thru: Channels), Jan 27, 1943, Subj: Consolidation of Report of Navigation Officers on Temporary Duty at Salina, Kansas; Topeka, Kansas; Barksdale, Louisiana; Columbia, South Carolina; and Greenville, South Carolina, box 895, 353, uncld, AAG, RG 18, NA.
50. Smith, "Development of AAF and USAF Training," 324-329; 2d AF hist, 1943; Memo, Maj Gen Robert W. Harper to Comdg Gen 4th AF, Apr 14, 1944, Subj: Extract from "Minutes of the General Council Meeting [dated Apr 10, 1944]," box 451, 337, entry 294, RG 18, NA.
51. Daily diary, Unit Tng Div, Bom Br, Jul 12, 1943, box 127, 319.1, entry 294, RG 18, NA.
52. Ltr, Lt Col Ted S. Faulkner, Comdg 333d Bom Gp, to Comdg Off 91st Bom Gp (H), Apr 2, 1943; 1st Ind, Brig Gen Robert W. Harper to Comdg Off 91st Bom Gp (H), Apr 22, 1943, box 890, 353, uncld, AAG, RG 18, NA.
53. 2d AF hist, 1944.
54. Flying Training Guide: Heavy Bombardment, I Bom Comd, Mitchel Fld, N.Y., May 10, 1944, box 76, uncld, 353.01, AAG, RG 18, NA.
55. HQ III Bom Comd hist, May 1-Dec 31, 1944; HQ III Bom Comd hist, Jan

Notes

1–Sep 2, 1945.

56. Flying Training Guide: Heavy Bombardment.

57. 2d AF hist, 1943.

58. Jablonski, *Flying Fortress*, 335.

59. Boeman, *Morotai*, 31.

60. 2d AF hist, 1944.

61. Ground Training Guide: Heavy Bombardment.

62. Ardery, *Bomber Pilot*, 45–46; Boeman, *Morotai*, 29.

63. 2d AF hist, 1943; Memo, Maj J. G. Cozzens to Chief, Ofc Info Svcs, Dec 18, 1944, Subj: Information from AC/AS Offices, copy in Cozzens, *A Time of War*, 195.

64. 1st AF, Tng, hist, Jan–Dec 1944.

65. Burchinal intvw, Apr 11, 1975, 48–49.

66. Ltr, Don Moody to his father, Jul 10, 1944, in Dallas Donald Moody, *Aerial Gunner from Virginia: The Letters of Don Moody to His Family during 1944*, William E. Hemphill, ed., (Richmond: Virginia State University Press, 1950), 176.

67. Boeman, *Morotai*, 26, 30.

68. Carigan, *Ad Lib*, 11.

69. 2d AF hist, 1943; Tng Intel Rprt 1598, Prov Gp Comdr, Training Notes from Overseas Stations, European Theater, Feb 7, 1944, box 137, 319.1, entry 294, RG 18, NA.

70. White, "Combat Crew and Unit Training," 73–86; Fahey, ed., *U.S. Army Aircraft*, 28.

71. Smith, "Development of AAF and USAF Training," 375.

72. For a description of the construction of Wendover Army Air Field, site of training for the 509th Composite Group that dropped the atomic bombs, see Charles G. Hibbard, "Training the Atomic Bomb Group," *Air Power History* (Fall 1995): 24–33.

73. Memo, Brig Gen Robert W. Harper, Asst Chief of Air Staff, Tng, to Asst Chief of Air Staff, Pers, Aug 27, 43, Subj: First Pilots for B–29 Project, 143.03, mcflm roll A1326, AFHRA; Memo, Maj Allen B. Black, AAF Tng Comd, to Comdg Gen AAF (attn: Asst Chief of Air Staff, Tng), Sep 16, 1943, Subj: B–29 Training Program, box 483, 353.01–F, uncld, AAG, RG 18, NA; White, "Combat Crew and Unit Training," 30; 2d AF hist, 1944.

74. Memo, Maj Gen Robert Olds, Comdg Gen 2d AF, to Comdg Gen AAF, Aug 28, 1942, Subj: Crews for B–29 Aircraft, box 476, 353, uncld, AAG, RG 18, NA; Memo, Brig Gen Robert W. Harper, Asst Chief of Air Staff, Tng, to Asst Chief of Air Staff, OC&R, Oct 16, 1943, Subj: Proposed Combat Crew for B–29, 143.03, mcflm roll A1326, AFHRA; 2d AF hist, 1944; Wesley Frank Craven and James Lea Cate, eds., *The Army Air Forces in World War II*, vol 5, *The Pacific: MATTERHORN to Nagasaki, June 1944 to August 1945* (Washington, D.C.: Office of Air Force History, 1983), 55.

75. Daily diary, Rqrs Div, Aug 9, 1944, box 244, 319.1, entry 294, RG 18, NA; Smith, "Development of AAF and USAF Training," 352–353.

76. AAF Tng Std 20–2–2, Special Heavy Bombardment Replacement Units, Crews, and Radar Mechanics, box 915, 353, uncld, AAG, RG 18, NA.

77. 2d AF hist, 1944; Consolidated diaries, HQ 2d AF, Dec 14, 1944, box 240, 319.1, entry 294, RG 18, NA.

78. White, "Combat Crew and Unit Training," 30–31; Memo, Col B. E. Allen, HQ 2d AF, to Comdg Gen AAF (attn: Asst Chief of Air Staff, Tng), Sep 10, 1944, Subj: Theater Indoctrination Training for Very Heavy Units, box 481, 353, uncld, AAG, RG 18, NA; Smith, "Development of AAF and USAF Training," 351.

79. HQ III Bom Comd hist, Jan 1–Sep 2, 1945.

80. Carpenter intvw, Jan 9–12, 1979, 103.

81. 1st AF, Tng, hist, Jan–Dec 1944; Incl, LAB [Low-Altitude Bombing], 111th AAF Base Unit (Search Attack and Staging), May 29, 1944, box 473, 353, uncld, AAG, RG 18, NA; Memo, Brig Gen O. P. Weyland to Chief of Air Staff, [Nov 25, 1943], Subj: Summary of Low Altitude Radar Blind Bombing Projects, box 488, 353.4, 1942–44, uncld, AAG, RG 18, NA; Memo, Brig Gen Robert W. Harper to Comdg Gen 1st AF, Sep 23, 1943, Subj: Training of Combat Crews on

Notes

Electronic Equipment, 143.03, mcflm roll A1326, AFHRA.

82. AAF Tng Std 20-2-2, Special Heavy Bombardment Replacement Units, Crews, and Radar Mechanics, box 915, 353, uncld, AAG, RG 18, NA.

83. Memo, [Gen Gross] to Chief of the Air Staff, Sep 30, 1944, Subj: "Around the Clock" Bombing — B-29 and B-32 Consideration for, box 489, 353.41, 1942-44, uncld, AAG, RG 18, AAG, NA.

84. Burchinal intvw, Apr 11, 1975, 62-63.

85. Jean H. Dubuque, "The Development of the Heavy Bomber, 1918-1944," revised by Robert F. Glockner, USAF Hist Study No. 6, 1951, 90, AFHRA; Smith, "Development of AAF and USAF Training," 254.

86. Hist, AAF Tng Comd, Jan 1, 1939, to V-J Day, Jun 15, 1946, 220.01, mcflm roll 2245, AFHRA.

87. 2d AF hist, 1944.

88. Tng Intel Rprt 2487, "B-29 Gunnery Training," 2 AAF Ops Analysis Sec, Aug 26, 1944, box 137, 319.1, entry 294, RG 18, NA.

89. Smith, "Development of AAF and USAF Training," 375-376.

90. By July 1944, AAF headquarters policy dictated that *all* B-29s be equipped with BTO equipment, and heavy bomber groups receive twelve such aircraft per group. But that latter number potentially could slip, as "under no circumstances" would an attempt to step up heavy bombardment radar programs erode the "A1 priority" of very heavy bombardment. Daily diary, Rqrs Div, Jul 6, 1944, box 244, 319.1, entry 294, RG 18, NA.

91. Robert Buderi, *The Invention That Changed the World* (New York: Simon & Schuster, 1996), 176. Buderi details the personalities and institutions that mobilized civilian science and engineering for wartime purposes. Long-wave radar, developed by military laboratories in consultation with the groundbreaking work of the Radiation Laboratory and Bell Telephone Labs., produced radars for antiaircraft gun direction, airborne interception, and long-range navigation.

92. Tony Devereux, *Messenger Gods of Battle: Radio, Radar, Sonar, the Story of Electronics in War* (London: Brassey's, 1991), 151-159.

93. R&R, Brig Gen H. M. McClelland to Air Force Chief of Air Staff, Dec 7, 1942; R&R, Brig Gen E. L. Eubank, Subj: Information Reference to Dr. Bowles' Conference, Dec 15, 1942, box 446, 337, entry 294, RG 18, NA.

94. Rprt ENG:-M-53-556-470, Mat Cen Engineering Div, Precision Navigation and Blind Bombing, Feb 5, 1943, box 487, 353.41, uncld, AAG, RG 18, NA.

95. Don Baucom, "The Radar War," unpubl. ms., AFHSO; Minutes, Air Tech Sec mtg, SCR-297: Airborne Precision Navigation Equipment, Wed, Dec 30, 1942, box 446, 337, entry 294, RG 18, NA; Lee Kennett, *A History of Strategic Bombing* (New York: Charles Scribner's Sons, 1982), 135, 160.

96. Rprt, Laurance M. Leeds and Dale R. Corson, "Report on the Strategic Bombing of Europe" (prepared for the Tech Comte on Radar Aids to Bombing), Feb 22, 1944, box 488, 353.4, 1942-44, uncld, AAG, RG 18, NA.

97. See 1st AF, Tng, hist, Jan-Dec 1944; Memo, Lt Gen Carl Spaatz to Comdg Gen AAF, [c. Mar 1944], Subj: Requirements for Improved Equipment for Bombing Through Overcast, box 489, 353.41, 1942-44, uncld, AAG, RG 18, NA.

98. R&R, Col William F. McKee, Dep Asst Chief of Air Staff, OC&R, to Asst Chief of Air Staff, Tng, Feb 17, 1944, box 895, 353, AAG, RG 18, NA; Daily diary, AAF Tng Comd, Mar 27, 1944, box 243, 319.1, entry 294, RG 18, NA; Baldwin, "Individual Training of Navigators," 130, 132-134, 228; Memo, Brig Gen Robert W. Harper to Comdg Gen AAF and Comdg Gen 1st AF, Oct 19, 1943, Subj: Training of Pathfinder Force Radar Personnel, box 478, 353, uncld, AAG, RG 18, AAG, NA. For a discussion of the Training Command's radar program, see AAF Tng Comd hist, Jan 1, 1939, to V-J Day, Jun 15, 1946.

99. 2d AF hist, 1944; HQ III Bom Comd hist, Jan 1-Sep 2, 1945.

100. Comments, Lt Col E. S. Allee,

Notes

from Rprt, HQ AAF Tng Comd, Training Conference, Fort Worth, Tex., Jan 10, 11, 12, 1944, box 368, 337, AAG, RG 18, NA.

101. Memo, Edward L. Bowles to Gen Giles, Nov 21, 1944, Subj: Radar Training, box 482, 353, uncld, AAG, RG 18, NA.

102. Baucom, "The Radar War"; Memo, W. B. Shockley, Expert Consultant, Ofc Secy War, to Comdg Gen 2d AF, May 5, 1944, Subj: Recommendations Regarding the VHB Radar Program, box 480, 353, uncld, AAG, RG 18, NA; Ltr, W. B. Shockley to "Eddie" [Dr E. L. Bowles], Nov 29, 1944, box 489, 353.41, 1942–44, uncld, AAG, RG 18, AAG, NA.

103. Daily diary, Rqrs Div, Jun 23, 1944, box 244, 319.1, entry 294, RG 18, NA.

104. Ltr, Maj Gen U. G. Ent to Mac [Brig Gen H. M. McClelland], Jun 26, 1944, box 489, 353.41, 1942–44, uncld, AAG, RG 18, NA.

105. 2d AF hist, 1944; Smith, "Development of AAF and USAF Training," 351.

106. Memo, 1st Lt William O. Smith to Base Tng Off, 552d AAF Base Unit (2d Ftr Gp), Flying Div, Air Tng Comd, New Castle Army Air Base, Wilmington, Del., Sep 22, 1944, Subj: Report on Loran Installation — 2d Air Force, box 482, 353, uncld, AAG, RG 18, NA.

107. Memo, Col Amzi G. Barber, Exec Asst Chief Air Staff, Tng, to Comdg Gen 2d AF, Oct 25, 1944, Subj: Loran Training, box 482, 353, uncld, AAG, RG 18, NA.

108. HQ III Bom Comd hist, Jan 1–Sep 2, 1945.

109. Curtis E. LeMay and Bill Yenne, *Superfortress: The Story of the B–29 and American Air Power* (New York: McGraw-Hill, 1988), 74–75.

110. Ltr, Spaatz, Aug 28, 1942, quoted in 2d AF hist, Dec 1941–Dec 1942.

111. 2d AF hist, Dec 1941–Dec 1942.

112. Tng Intel Rprt 1867, Tech Representative, Sperry Gyroscope Co., Summary of A–5 Equipment Performance in a XV AAF Group, Mediterranean Theater, box 137, 319.1, entry 294, RG 18, NA.

113. Memo, Maj Gen Robert W. Harper to Gen Arnold, May 6, 1944, Subj: High Altitude Training for Heavy Bombardment Crews, box 480, 353, uncld, AAG, RG 18, NA.

114. 2d AF hist, 1944.

115. Memo, Brig Gen Donald Wilson to Col Dean, Dec 20, 1944, Subj: Training of Radar Operators, box 482, 353, uncld, AAG, RG 18, NA; Memos, Asst Chief of Air Staff, Tng, Unit Tng Div, Bom Br, to Asst Chief of Air Staff, OC&R, Very Heavy Bom Proj Off, Mar 27, 1944, Comment No. 1; Lt Col John W. Carpenter III, Dep Chief Bom Br, Unit Tng Div, to Very Heavy Bom Proj Off, Apr 3, 1944, Subj: Weekly Report of VHB Training Problems, box 895, 353, uncld, AAG, RG 18, NA.

116. Memo, Maj Gen Robert W. Harper to Maj Gen H. A. Craig, Jun 29, 1944, Subj: Radar Training Summary, box 480, 353, uncld, AAG, RG 18, NA.

117. Montgomery intvw, Jun 28–30, 1983, 71.

118. Rprt, Leeds and Corson, "Report on the Strategic Bombing of Europe," Feb 22, 1944.

119. Terkel, *"The Good War,"* 200.

120. It is important to note that LeMay was not a maverick who singlehandedly discarded the American strategic doctrine. The Air Staff had in hand studies of incendiary attacks and the susceptibility of Japanese cities to fire bombing well before LeMay began them. Prodded by Arnold and Norstad, Hansell conducted at least one incendiary raid, though from high altitude, so that wind scattered the bombs and the individual blazes did not merge into a fire storm. Yet it may have been Hansell's reluctance to undertake wholesale area bombing that contributed to his removal.

Notes

Chapter 15

1. George C. Kenney, *General Kenney Reports: A Personal History of the Pacific War* (Washington, D.C.: reprint, Office of Air Force History, 1987), 293.
2. Ibid.
3. *Condensed Analysis of the Ninth Air Force in the ETO*, 116.
4. Taylor, "American Experience in the Southwest Pacific," in Cooling, ed., *Close Air Support*, 325–326.
5. Hallion, *Strike from the Sky*, 166, 177.
6. Memo, Brig Gen Robert W. Harper to Asst Chief of Air Staff, OC&R (thru: Asst Chief of Air Staff, Pers), Oct 7, 1943, Subj: Personnel Shortages in the Third Air Force, 143.03, mcflm roll A1326, AFHRA; Memo, Capt R. H. Thom, HQ 3d AF, to Comdg Gen AAF (attn: Asst Chief of Staff, Con Div, Ofc Mgt Con), May 5, 1943, Subj: Replacement Combat Crew Report, box 900, 353, AAG, RG 18, NA.
7. Rprt, Col John W. Egan to Asst Chief of Staff, A-3, Flying Tng Comd, [Jan 2, 1943], Subj: Report on Proficiency Requirements of Graduates of Army Air Force Navigation Schools for the Third Air Force, box 907, 353, uncld, AAG, RG 18, NA.
8. Ltr, Hon Walt Horan, 5th Washington Dist, Hon Hal Holmes, 4th Washington Dist, and Hon Fred Norman, 3d Washington Dist, to Hon Carter Glass, Chmn, Senate Appropriations Comte, May 5, 1943, "Extension of Remarks of Hon Fred Norman of Washington to the House of Representatives, Tuesday, May 11, 1943," *Congressional Record — Appendix*, May 11, 1943, box 172, 353.4, uncld, AAG, RG 18, NA.
9. Memo, Brig Gen William E. Kepner to Comdg Gen AAF, May 6, 1943, Subj: Fourth Air Force Training Exercise No. 6, box 476, 353, uncld, AAG, RG 18, NA.
10. AAF Tng Std 200–1, Combined Air Force Training [Oct 13, 1943], box 915, 353, uncld, AAG, RG 18, NA.
11. Brig Gen Robert W. Harper, Air-Ground Training and Operations, Dec 7, 1943, box 478, 353, uncld, AAG, RG 18, NA.
12. Tom Harmon, B–25 pilot, *Pilots Also Pray*
13. Doolittle intvw, Sep 26, 1970, 40.
14. R&R, with attached material, AFDMR to AFRROM, AFRDB, AFRAD, AFRIT in turn, May 8, 1942, Subj: Intensive Training Program Carried on by Lt Col Doolittle at Eglin Field, box 177, 353.9, AAG, RG 18, NA.
15. AAF Tng Std 20–1–1, Medium Bombardment Units and Crews, Feb 8, 1943, box 69, 353, uncld, AAG, RG 18, NA.
16. Kenney, *Reports*, 21–22.
17. Angelucci, *Encyclopedia of Military Aircraft*, 288.
18. Daily diary, Unit Tng Div, Aug 31, 1942, box 127, 319.1, entry 294, RG 18, NA.
19. Ibid., Mar 16, 1943.
20. AAF Tng Std 90–5, 2-engine Pilot-Specialized Transition, box 915, 353, uncld, AAG, RG 18, NA.
21. Daily diary, Unit Tng Div, Aug 25, 1942, and Feb 16 and Apr 23, 1943, box 127, 319.1, entry 294, RG 18, NA.
22. Rprt, Lt Col G. B. Dany to Comdg Gen AAF (thru: Channels), Jan 27, 1943, Subj: Consolidation of Report of Navigation Officers on Temporary Duty at Salina, Kansas; Topeka, Kansas; Barksdale, Louisiana; Columbia, South Carolina; and Greenville, South Carolina, box 895, 353, uncld, AAG, RG 18, NA.
23. Memo, Lt Col A. J. Bird, Jr., Comdg, 334th Bom Gp (M), to Comdg Gen 3d AF (thru: Channels), Jul 10, 1943, Subj: Narrative Bombing Report for the Fiscal Year Ending Jun 30, 1943, box 385, 353.41, uncld, AAG, RG 18, NA.
24. Memo, Col G. A. McHenry to Comdg Gen 1st AF, Nov 30, 1943, Subj: Difficulties Encountered in Training 454th and 455th Bombardment Groups; 1st Ind, HQ 1st AF to Comdg Gen AAF (attn Asst Chief of Air Staff, Tng), box 888, 353, uncld, AAG, RG 18, NA.
25. Memo, Maj Gen Barney M. Giles, Act Chief of Air Staff, to Comdg Gen 3d AF, May 15, 1943, Subj: Bombing Accu-

Notes

racy, box 488, 353.4, 1942–44, uncld, AAG, RG 18, NA.

26. Memo, Maj Gen Robert W. Harper to Comdg Gen 3d AF, Apr 25, 1944, Subj: Bombardment Training for Medium and Light Bombardment Units, box 895, 353, uncld, AAG, RG 18, NA. It is worth noting once again that, despite the AAF's infatuation with it, the Norden bombsight did not assure pinpoint accuracy.

27. Smith, "Development of AAF and USAF Training," 362–363; HQ III Bom Comd hist, May 1–Dec 31, 1944.

28. R&R, Gen Arnold to Gen Stratemeyer, Oct 13, 1942, AFCAS to Dir Bom, Oct 14, 1942, Brig Gen E. L. Eubank to AFCAS, Oct 17, 1942, Subj: Day and Night Bombing, box 517, 353.4, AAG, RG 18, NA.

29. Memo, Cols D. M. Schlater and O. P. Weyland to Comdg Gens, 2d and 3d AFs, Oct 14, 1942, box 172, 353.4, AAG, RG 18, NA; Gen H. H. Arnold to President, AAF Bd, Sep 26, 1944, Subj: Night Bombing Operations, box 489, 353.41, uncld, AAG, RG 18, NA.

30. Harmon, *Pilots Also Pray*, 50.

31. William Green, *Famous Bombers of the Second World War*, vol 1 (Garden City, N.Y.: Hanover House, 1959), 105–106; HQ III Bom Comd hist, May 1–Dec 31, 1944; Memo, Brig Gen William W. Welsh, Asst Chief of Air Staff, Tng, to Comdg Gen 3d AF, Sep 28, 1944, Subj: Medium and Light Bombardment Crew Training, box 481, 353, uncld, AAG, RG 18, AAG, NA.

32. HQ III Bom Comd hist, May 1–Dec 31, 1944.

33. Ibid., vol 1.

34. Doolittle intvw, Sep 26, 1970, 40.

35. Tng Intel Rprt 2132, South African Air Directorate Bul, "The B–26 as a Combat Plane," Mar 1944, box 137, 319.1, entry 294, RG 18, NA.

36. Memo, Brig Gen J. H. Doolittle to Dir Mil Rqrs (thru: Dir Bom), Jun 22, 1942, Subj: Medium Bombardment; 1st Ind, War Dept, HQ AAF, to Dir Mil Rqrs, Jul 5, 1942, box 157, 353, bulky files, 1939–1942, uncld, AAG, RG 18, NA.

37. Daily diary, Tng Sec, Dec 3, 1942, box 127, 319.1, entry 294, RG 18, NA; Daily diary, Dir Indiv Tng, Feb 19, 1943, box 242, 319.1, entry 294, RG 18, NA; Hist, HQ III Bom Comd, Sep 5, 1941 to May 1, 1944, 438.01, mcflm roll A4133, AFHRA.

38. Daily diary, Bom Br, Jun 22, 1943, box 127, 319.1, entry 294, RG 18, NA.

39. Memo, Maj Charles H. Campbell, Asst AG, HQ 3d AF, to Comdg Gen AAF (attn: AFRDB), Feb 6, 1943, Subj: Medium Bombardment Training, box 473, 353, uncld, AAG, RG 18, NA.

40. Daily diary, AAF Tng Comd, Mar 16, 1944, box 243, 319.1, entry 294, RG 18, NA.

41. Tng Intel Rprt 2132, Mar 1944.

42. Samuel A. Stouffer et al., *The American Soldier: Combat and Its Aftermath*, vol 2 (Princeton, N.J.: Princeton University Press, 1949), 398.

43. HQ III Bom Comd hist, May 1–Dec 31, 1944.

44. Kenney, *Reports*, 55.

45. Maurer, ed., *Air Force Combat Units of World War II*, 29–31.

46. Angelucci, *Encyclopedia of Military Aircraft*, 289; Ray Wagner, *American Combat Planes* (Garden City, N.Y.: Doubleday & Co., 1968), 75, 77, 82.

47. HQ III Bom Comd hist, May 1–Dec 31, 1944.

48. AAF Tng Std 30–3–1, Light and Dive Bombardment Units and Crews, Dec 1, 1942, box 69, 353, uncld, AAG, RG 18, NA. Although light and dive bombardment were linked at this time, the latter only followed the bombardment training pattern until August 1943. Then, units were redesignated fighter-bomber and placed under the III Fighter Command. Dive, or fighter-bomber, training is discussed on page 544, this chapter (Chapter 15).

49. HQ III Bom Comd hist, May 1–Dec 31, 1944.

50. Memo, Capt R. H. Thomas, HQ 3d AF, to Comdg Gen AAF, Jun 11, 1943, Subj: Conference on Light Bombardment Tactics, box 448, 337, entry 294, RG 18, NA. See page 537, this chapter, for mention of the North African experience as it affected medium bomber training.

51. Stouffer et al., *American Soldier*,

vol 2, 398.

52. Memo, Brig Gen William W. Welsh to Comdg Gen 3d AF, Oct 11, 1944, Subj: Specialized A–20 Training, box 482, 353, uncld, AAG, RG 18, NA.

53. HQ III Bom Comd hist, May 1–Dec 31, 1944.

54. Ibid.

55. Hist, 1st AF, Monogr 1, May 1945, First Redeployment: Jan–May 1945, 420.04, mcflm roll A4040, AFHRA.

56. *Yeager: An Autobiography*, 66–67.

57. Barcus intvw, Aug 10–13, 1976, 65–66.

58. Tng Intel Rprt 1526, Fighter Squadrons Summary, European Theater, Dec 28, 1943, box 137, 319.1, entry 294, RG 18, NA.

59. White, "Combat Crew and Unit Training," 52.

60. Barcus intvw, Aug 10–13, 1976, 86–87.

61. Daily diary, Rqrs Div, Oct 13, 1944, box 245, 319.1, entry 294, RG 18, NA.

62. Barcus intvw, Aug 10–13, 1976, 88, 89.

63. Memo, Col Robert W. Harper, Asst Chief of Air Staff, G–3, to Asst Chief of Staff, Ops Div Gen Supp, Aug 28, 1942, Subj: Winter Fighter Training, box 176, 353.9, AAG, RG 18, NA; Partridge intvw, Apr 23–25, 1974, 334–335.

64. Memo, Col Gordon P. Saville, Dir Air Def, to Comdg Off, Ftr Comd School, May 20, 1942, Subj: Mission of the Fighter Command School, box 157, 353, bulky files, 1939–1942, uncld, AAG, RG 18, NA.

65. Lectures, AAF Sch Applied Tactics, "Offensive Fighter Aviation in a Tactical Air Force," Aug 1, 1944, and "Defensive Fighter Aviation, Tactical Air Force: Escort and Cover," Apr 1944, box 379, 352.11, uncld, AAG, RG 18, NA.

66. Colgan, *Fighter-Bomber Pilot*, 37.

67. AAF Tng Std 10–1–1, Dec 1, 1942, and Dec 4, 1943, boxes 69 and 915, 353, uncld, AAG, RG 18, NA.

68. "Fighter Training in the Fourth Air Force, 1942–1945."

69. III Ftr Comd hist, Dec 7, 1941, to Apr 30, 1944.

70. R&R, AFRAD to AFACT (attn: Col Harper), Oct 3, 1942, Subj: Fighter Training Program, box 380, 353, uncld, AAG, RG 18, NA.

71. III Ftr Comd hist, Dec 7, 1941, to Apr 30, 1944.

72. Memo, Brig Gen Gordon P. Saville, Dir Air Def, to Chief of the Air Staff, Comment No. 1, Mar 19, 1943, Subj: Deficiencies in Training of Fighter Pilots, box 910, 353, uncld, AAG, RG 18, NA.

73. Daily diaries, 4th and 1st AFs, Sep and Nov 1944, box 245, 319.1, entry 294, RG 18, NA.

74. *Yeager: An Autobiography*, 50, 67–68; Lecture, c. Aug 1943, box 924, 353.41, uncld, AAG, RG 18, NA.

75. Tng Memo, Noumea, New Caledonia, Oct 2, 1942, Subj: Practical Application of Combat Lessons to Training of Fighter Units, box 911, 353, uncld, AAG, RG 18, NA.

76. Like every other technique, the "standard" four-finger fighter formation was no ironclad rule. Bill Colgan, in his memoir *World War II Fighter-Bomber Pilot*, described the formation flown by the 87th Fighter Squadron, although not her sister squadrons, as "straight line abreast."

77. Memo, Maj Gen Robert W. Harper to Chief of Staff, Jul 25, 1944, Subj: Deficiencies in Fighter RTU Training, box 481, 353, uncld, AAG, RG 18, NA.

78. Joseph W. Noah, *Wings God Gave My Soul* (Annandale, Va.: Charles Baptie Studios, 1974), 59, 120.

79. Tng Intel Rprt 1964, "Hints for Combat Pilots," AAF Redistribution Center Interview, May 30, 1944, box 264, 319.1, entry 294, RG 18, NA.

80. Jacobs, "Battle for France, 1944," in Cooling, ed., *Close Air Support*, 278.

81. Memo, Brig Gen Robert W. Harper to Chief of the Air Staff, Sep 23, 1943, Subj: Fighter and Fighter-Bomber R.T.U. Training, 143.03, mcflm roll A1326, AFHRA.

82. III Ftr Comd hist, Dec 7, 1941, to April 30, 1944.

83. Ibid.

84. Memo, Col Kenneth P. Bergquist, Exec, Asst Chief of Air Staff, OC&R, to

Notes

Comdg Gen, India-Burma Sector, CBI Theater, Jul 25, 1944, box 482, 353, uncld, AAG, RG 18, NA.

85. Memo, Maj Gen Robert W. Harper to Comdg Gen 1st AF and Comdg Gen 2d AF, Sep 11, 1944, Subj: Specialized Theater Training; 2d Ind, HQ 72d Ftr Wg to Comdg Gen, HQ AAF, Sep 22, 1944, box 481, 353, form. cl., AAG, RG 18, NA; Hist, 1st AF, Tng, Jan–Dec 1944, 420.01–7, mcflm roll A4039, AFHRA.

86. III Ftr Comd hist, Dec 7, 1941, to Apr 30, 1944.

87. Minutes, Fixed Gnr Conf, Sep 29–Oct 1, 1943, Washington, D.C., box 449, 337, entry 294, RG 18, NA.

88. Consolidated diaries, HQ 2d AF, Jan 22, 1945, box 76, 319.1, entry 1, RG 18, NA.

89. Wagner, *American Combat Planes*, 213–215.

90. "Fighter Training in the Fourth Air Force, 1942–1945."

91. Rprt, V Ftr Comd Recep Cen to Comdg Gen V Ftr Comd, May 23, 1943, Subj: Status of Training of Replacement Fighter Pilots Prior to Arrival in This Theatre, box 475, 353, uncld, AAG, RG 18, NA.

92. Loftin, *Quest for Performance*, 132.

93. C. F. Rawnsley and Robert Wright, *Night Fighter* (New York: Henry Holt & Co., 1957), 25.

94. *Condensed Analysis of the Ninth Air Force in the ETO*, 126.

95. R&R, Col Gordon P. Saville to AFTSC (thru: AFROM [Col Lowe]), Oct 21, 1942, box 175, 353.9, AAG, RG 18, NA; Garry R. Pape with John M. and Donna Campbell, *Northrop P-61 Black Widow* (Osceola, Wis.: Motorbooks International Publishers & Wholesalers, 1991), 40.

96. Memo, Col L. S. Smith, Dir Indiv Tng, to Comdg Gen AAF Flying Tng Comd, Oct 27, 1942, Subj: Training for Night Fighter Pilots, box 175, 353.9, AAG, RG 18, NA.

97. Memo, Lt Col D. B. Brummel, Dir Night Ftr Div, to Dir Tng Ftr Comd Sch, Sep 1, 1942, Subj: Airplanes, Materiel and Personnel Required and Instruction Accomplished at Fighter Command School for Night Fighter Training, box 160, 353.9, bulky files, 1939–1942, uncld, AAG, RG 18, NA; Memo, Lt Col D. B. Brummel to Dir Tng Ftr Comd Sch, Oct 2, 1942, Subj: Training Report, box 170, 352.9, RG 18, NA; Memo, Col William F. McKee, Dep Asst Chief of Air Staff, OC&R, to Col Libby, Jan 8, 1944, Subj: Notes for General Council Meeting, Report of the Asst Chief of Air Staff, Operations Commitments and Requirements, box 450, 337, entry 294, RG 18, NA.

98. R&R, Asst Chief of Air Staff, OC&R, to Asst Chief of Air Staff, Tng, Apr 26, 1943, Subj: Transfer of Night Fighter Training Activities from AAF School of Applied Tactics, box 477, 353, uncld, AAG, RG 18, NA; 4th Ind, Brig Gen Robert W. Harper to Comdg Gen 4th AF, Sep 16, 1943, 143.03, mcflm roll A1326, AFHRC.

99. Memo, Col Ralph A. Snavely to Comdg Gen 4th AF, May 18, 1944, Subj: P-61 Aircraft for OTU and RTU Training; 1st Ind, HQ 4th AF to Comdg Gen AAF, box 917, 353, uncld, AAG, RG 18, NA.

100. AAF Tng Comd hist, Jan 1, 1939, to V-J Day.

101. "Fighter Training in the Fourth Air Force, 1942–1945."

102. White, "Combat Crew and Unit Training," 73–88.

103. Agan intvw, Apr 19–22, 1976, 83–84.

104. 1st Ind, HQ 7th AF to Comdg Gen AAF (thru: Comdg Gen Pacific Ocean Areas), Oct 2, 1944, box 475, 353, uncld, AAG, RG 18, NA.

105. Ltr, Maj Gen B. K. Yount to Lt Gen H. H. Arnold, Dec 8, 1942, box 902, 353, uncld, AAG, RG 18, NA.

106. Memo, Col Robert M. Caldwell to Col Schneider, May 18, 1944, Subj: Flexible Gunnery Conference; Comment No. 2, Asst Chief of Air Staff, Tng, to The Air Insp, Mar 11, 1944, Subj: Precautions against Collisions in Simulated Attacks on Combat Type Aircraft, box 452, 337, entry 294, RG 18, NA; Rprt, HQ AAF Tng Comd, Training Conference, Fort Worth, Tex., Jan 10, 11, 12,

1944.

107. Memo, Maj Gen Robert W. Harper to Chief of Air Staff, Jul 25, 1944, Subj: Deficiencies in Fighter RTU Training, box 481, 353, uncld, AAG, RG 18, NA.

108. Memo, Brig Gen Robert W. Harper to Gen Perrin, Jul 17, 1943, box 483, 353.01, uncld, AAG, RG 18, NA.

109. Memo, Maj Gen H. A. Craig to Chief of the Air Staff, Aug 26, 1944, Subj: Deficiencies in Fighter Training — Public and Political Considerations, box 481, 353, uncld, AAG, RG 18, NA.

110. Memo, Maj J. G. Cozzens to Chief, Ofc Info Svcs, Dec 30, 1944, Subj: Information from AC/AS Offices, copy in Cozzens, *A Time of War*, 208.

111. III Ftr Comd hist, Dec 7, 1941, to Apr 30, 1944.

112. Memo, Rear Ech HQ VII Bom Comd to Comdg Gen 7th AF, Mar 13, 1944, Subj: Training of Medium Bombardment Crews, box 480, 353, uncld, AAG, RG 18, NA.

113. III Ftr Comd hist, Dec 7, 1941, to Apr 30, 1944.

114. Ltr, Col Leroy A. Rainey to Col T. J. DuBose, May 6, 1944, box 924, 353.41, uncld, AAG, RG 18, NA.

An End and a Beginning

1. Memo, Maj J. G. Cozzens to Chief, Ofc Info Svcs, Dec 16, 1944, Subj: 23 November Momorandum from AC/AS Personnel on Air Crew Training Situation for the Use of Col. McIntyre, Legislative and Liaison Division, copy in Cozzens, *A Time of War*, 193.

2. Memo, Brig Gen Walter Krauss to Comdg Gen AAF, Jan 20, 1944, Subj: Reduction in Training Program; R&R, Asst Chief of Air Staff, Tng, to Mgt Con, Statistical Con Div, Feb 11, 1944, Subj: Reduction in Training Program, box 484, 353.01-H, uncld, AAG, RG 18, NA; R&R, Chief of Air Staff to Asst Chief of Air Staff, Tng [and others], Mar 27, 1944, box 909, 353, uncld, AAG, RG 18, NA; Memo, Maj Gen Robert W. Harper to Chief of Staff G–3, Mar 30, 1944, Subj: Elimination of AAF College Training Program, box 384, 353.01, uncld, AAG, RG 18, NA.

3. Ltr, Maj Gen Ira C. Eaker to Gene [Brig Gen E. L. Eubank], Feb 9, 1943, box 890, 353, uncld, AAG, RG 18, NA.

4. Third Report of the Commanding General of the Army Air Forces to the Secretary of War, Nov 12, 1945.

5. The highly dangerous nature of military air training remained a consistent source of anxiety for the air force throughout its first half century. Accident rates and fatality statistics have been very closely held and, possibly, deliberately kept vague. For example, the 1946 *Air Force Statistical Digest* does not separate washouts from fatalities in its listing of eliminations from air training during World War II. The lack of concrete information as of this writing has resulted in wild guesses from many commentators about the relative numbers of men killed in training versus combat. Unfortunately, the author is unable to provide official figures of training fatalities for the period under discussion.

6. Quoted in Kenneth P. Werrell, "A Case for 'New' Unit History," *Air Power History* 39: no. 1 (Spring 1992): 38-39.

7. Remarks, Lt Gen B. K. Yount, from Rprt, HQ AAF Tng Comd, Training Conference, Fort Worth, Tex., Jan 10, 11, 12, 1944.

Abbreviations Used

AAF	Army Air Forces	cen	center
AAG	Air Adjutant General	chan	channel
AAOS	Artillery Aerial Observation School	chmn	chairman
		cir	circular
acft	aircraft	clas	classification
act	acting	clk	clerk
adj	adjutant	comd	command
admin	administrative	comdg	commanding
AEF	American Expeditionary Forces	comdr	commander
		comdt	commandant
aero	aeronautical, aeronautics	comp	composite
AFHRA	U.S. Air Force Historical Research Agency	comte	committee
		con	control, conversation
AFHSO	U.S. Air Force History Support Office	conf	conference
		Cong	Congress
AI	airborne intercept	corr	correspondence
AIAA	American Institute of Aeronautics and Astronautics	CPTP	Civilian Pilot Training Program
		def	defense
AIC	Aviation Instruction Center	dep	deputy
		dept	department
app	appendix	det	detachment
appl	applied	dir	director, direction
an	annual	disbg	disbursing
arty	artillery	dist	district
asst	assistant	div	division
atch	attachment	docu	document
atk	attack	ech	echelon
att	attache	eqp	equipment
atten	attention	eval	evaluation
avn	aviation	exec	executive
bd	board	exper	experimental
bln	balloon	FAI	Federation Aeronautique Internationale
bmbdr	bombardier		
bom	bomber, bombardment, bombing	fisc	fiscal
		fld	field
bud	budget	flt	flight
bul	bulletin	Ft.	fort
br	branch	ftn	fortification
CAF	Continental Air Forces	ftr	fighter

Abbreviations

GCI	ground-controlled interception	ops	operations
Gen	general	ord	ordnance
GHQ	General Headquarters	org	organization
gnd	ground	OTU	Operational Training Unit
gnr	gunnery, gunner	para	paragraph
GO	general order	pers	personnel
govt	government	pdn	production
gp	group	pres	president
hist	history, historical	prog	program
HR	House Resolution	proj	project
HQ	headquarters	prov	provisional
HUL	Harvard University Library	pur	pursuit
incl	inclose, include	QM	quartermaster
ind	indorsement	R&R	Routing and Record Sheet
indiv	individual	recon	reconnaissance
info	information	reg	regulation
insp	inspector, inspection	res	reserve
inst	instruction	RMA	Reserve Military Aviator
intel	intelligence	ROTC	Reserve Officers Training Corps
LCMD	Library of Congress, Manuscript Division	rprt	report
(LR)	(Long Range)	rqrs	requirements
intvw	interview	rsch	research
ltr	letter	RTU	Replacement Training Unit
(M)	(Medium)	S.C.	Signal Corps
maint	maintenance	sch	school
mat	materiel	sec	section
mcflm	microfilm	secy	secretary
memo	memorandum	sess	session
mgt	management	sig	signal
mil	military	sked	schedule
mtg	meeting	sq	squadron
NA	National Archives	SR	Senate Resolution
NACA	National Advisory Committee for Aeronautics	sta	station
nav	navigator, navigation, navigate	std	standard
		subj	subject
		sup	supply
obsn	observation	supp	support
obsr	observer	supt	superintendent
OC&R	Operations, Commitments, and Requirements	svc	service
		T&O	Training and Operations
		tac	tactics, tactical
ofc	office	tech	technical
off	officer	telecon	telephone conversation
OIC	Officer in Charge	telg	telegraph
op	operational	thru	through
		tng	training

Abbreviations

tnr	trainer	unpubl. ms.	unpublished manuscript
trp	troop	wg	wing
uncld	unclassified	wkly	weekly

Selected Bibliography

Manuscript and Archival Collections

National Archives and Records Administration

RG 18: Records of the Army Air Forces
RG 94: Records of the Adjutant General's Office, 1780s–1917.
RG 111: Records of the Office of the Chief Signal Officer, 1917–1918
RG 120: Records of the American Expeditionary Forces (WWI), 1917–1923
RG 341: Records, Headquarters United States Air Force, 1935–1963

Library of Congress Manuscript Division

Scrapbooks from the Archives of the American Institute of Aeronautics and Astronautics
Personal papers of the following individuals: Frank Maxwell Andrews, Henry H. Arnold, Ira Eaker, Benjamin D. Foulois, Adolphus Greely, Frank Purdy Lahm, Curtis E. LeMay, William Mitchell, Carl Andrew Spaatz, Hoyt S. Vandenberg

Air Force Historical Research Agency

The records, on microfilm at the Air Force Historical Research Agency (AFHRA), Maxwell Air Force Base, Alabama, were used extensively in the study, and cannot be individually cited here. Central to the collection are the Records of the Zone of Interior Commands and Organizations of AAF and USAF (1926–). Also consulted were personal papers, interviews, speeches, reports, etc., that date from the earliest years of Army aviation. The voluminous oral history collection was extraordinarily useful for understanding the individual and sometimes off-the-record aspects of flight training.

Books, Articles, Reports, and Monographs

Some of the materials included cover a longer span of time than the period for which they are listed. Also, the bibliography does not include all sources cited in footnotes in the text.

Bibliography

1907–1917

Hearings

House Committee on Military Affairs. *Aviation School and Training Grounds for the Signal Corps of the United States Army.* 64th Cong., 1st sess. Feb 14, 1916, Doc. 687.

———. *First Heavier-than-Air Flying Machine.* 70th Cong., 1st sess. Apr 27, 1928.

———. *25th Anniversary of First Airplane Flight.* 70th Cong., 2d sess.

———. *Pioneer Aviators: Hearings on H.R. 11273.* 70th Cong., 1st sess. Apr 3, 1928.

Books

Arnold, H. H., and Ira C. Eaker. *This Flying Game.* New York: Funk & Wagnalls Co., 1936.

Bernardo, C. Joseph, and Eugene H. Bacon. *American Military Policy: Its Development Since 1775.* Harrisburg, Pa.: Stackpole, 1955.

Bilstein, Roger E. *Flight in America, 1900–1983.* Baltimore & London: The Johns Hopkins University Press, 1984.

Bonney, Walter T. *The Heritage of Kitty Hawk.* New York: W. W. Norton, 1962.

Cantwell, Gerald T. *Citizen Airmen: A History of the Air Force Reserve, 1946–1994.* Washington, D.C.: Air Force History & Museums Program, 1997.

Casari, Robert B. *Encyclopedia of U.S. Military Aircraft, 1908 to April 6, 1917.* 3 vols. Chillicothe, Ohio: Robert B. Casari, publ., 1970–1972.

Challener, Richard D. *Admirals, Generals, and American Foreign Policy, 1898–1914.* Princeton, N.J.: Princeton University Press, 1973.

Chandler, Charles DeForest, and Frank P. Lahm. *How Our Army Grew Wings.* New York: The Ronald Press Co., 1943.

Crouch, Tom. *The Bishop's Boys.* New York: W. W. Norton, 1989.

Finnegan, John Patrick. *Against the Specter of a Dragon: The Campaign for American Military Preparedness, 1914–1917.* Westport, Conn.: Greenwood Press, 1974.

Foulois, Benjamin D., and Carroll V. Glines. *From the Wright Brothers to the Astronauts: The Memoirs of Major General Benjamin D. Foulois.* New York: McGraw-Hill, 1960.

Hammond, Paul Y. *Organizing for Defense.* Princeton, N.J.: Princeton University Press, 1961.

Hofstadter, Richard. *The Age of Reform.* New York: Vintage Books, 1955.

Howard, Fred. *Wilbur and Orville: A Biography of the Wright Brothers.* New York: Ballantine, 1987.

Johnson, Kenneth M. *Aerial California: An Account of Early Flight in Northern and Southern California, 1849 to World War I.* Los Angeles: Dawson's Book Shop, 1961.

Link, Arthur S. *Woodrow Wilson and the Progressive Era, 1910–1917.* New York: Harper, 1954.

Loening, Grover C. *Our Wings Grow Faster.* Garden City, N.Y.: Doubleday, Doran & Co., 1935.

Lyle, Eugene P., III, and Charles W. Diffin. *The City of Wings: A Narrative of Aviation and Its Swift Development As Seen in San Diego from the Earliest Days of Flying to the Present Time.* San Diego: The City Schools of San Diego, Calif., 1938.

Morison, Elting E. *Turmoil and Tradition: A Study of the Life and Times of Henry L. Stimson.* Boston: Houghton Mifflin Co., 1960.

Nalty, Bernard C., ed. *Winged Shield, Winged Sword: A History of the United States Air Force.* Vol. 1, *1907–1950.* Washington, D.C.: Air Force History & Museums Program, 1997.

Nenninger, Timothy K. *The Leavenworth Schools and the Old Army: Education, Professionalism, and the Officer Corps of the United States Army, 1881–1918.* Westport, Conn.: Greenwood Press, 1978.

Pringle, Henry F. *The Life and Times of William Howard Taft.* 2 vols. New York: Farrar & Rinehart, 1939.

Scott, James Brown. *The Hague Peace Conferences of 1899 and 1907.* Vol. 1, *Conferences;* vol. 2, *Documents.* New York & London: Garland Publishing, 1972.

Spiller, Roger J., ed. *Dictionary of American Military Biography.* 3 vols. Westport, Conn.: Greenwood Press, 1984.

Reports and Pamphlets

Loening, Grover C. *Military Aeroplanes: An Explanatory Consideration of Their Characteristics, Performances, Construction, Maintenance and Operation, for the Use of Aviators.* Prepared for Signal Corps Aviation School, San Diego, California. San Diego: Frye & Smith, 1915.

Scriven, Brig. Gen. G. P. *The Service of Information.* Circular 8. Washington, D.C.: Office of the Chief Signal Officer, 1915.

Squier, George O. *The Present Status of Military Aeronautics.* N.p. 1908.

Swanborough, F. G., and Peter M. Bowers. *United States Military Aircraft Since 1908.* London: Putnam, 1963.

U.S. War Department, War College Division, General Staff Corps, Army War College. *Military Aviation.* Washington, D.C.: Government Printing Office, 1916 [prepared Nov 1915].

Bibliography

Articles and Monographs

During the period leading up to World War I, many aeronautical periodicals very closely followed the activities of military aviation. The articles in these magazines are generally not indexed, and the periodicals such as *Aeronautics, Aviation and Aeronautiacal Engineering,* and *Aerial Age Weekly* must be read issue by issue. The information on military aviation found in these magazines was critical to this study, but the articles are too numerous to be cited individually here.

Brown, Jerold E. "Army-Navy 'Cooperation': The Case of North Island." *Journal of the Council on America's Military Past* 15, no. 3 (Oct 1987): 3–11.

Coffman, Edward M. "The Young Officer in the Old Army." In *The Harmon Memorial Lectures in Military History, 1959–1987,* Lt. Col. Harry R. Borowski, ed. Washington, D.C.: Office of Air Force History, 1988.

Dubina, George N. "Flying with Orville Wright, Montgomery, Alabama, 1910: A Chronology." Historical Study No. 62, Air University, Maxwell AFB, 1974.

Foulois, Benjamin D. "Early Flying Experiences: Why Write a Book? — Part I." *Air Power Historian* 2 (Apr 1955): 17–35.

Hastings, George E. "Notes on the Beginnings of Aeronautics in America." *American Historical Review,* Oct 1919, 68–72.

Hennessy, Juliette A. *The United States Army Air, Arm April 1861 to April 1917.* New imprint. Washington, D.C.: Office of Air Force History, 1985.

Humphreys, Frederic E. "The Wright Flyer and Its Possible Uses in War." *Journal of the United States Artillery* 33 (Mar–Apr 1910): 144–147.

Lahm, Frank P. "Early Flying Experiences." *Air Power Historian* 2 (Jan 1955): 1–10.

———. "Training the Airplane Pilot." *Journal of the Royal Aeronautical Society* 37 (Nov 1933): 916–941.

Milling, Thomas DeW. "Early Flying Experiences." *Air Power Historian* 3 (Jan 1956): 96.

World War I

Official Histories

Maurer, Maurer, ed. *The U.S. Air Service in World War I.* 4 vols. Washington, D.C.: Office of Air Force History, 1978–1979.

Office of the Deputy Chief of Naval Operations (Air), *United States Naval Administration in World War II.* Vol. 13, *Aviation Training, 1911–1939.* Washington, D.C.: n.d.

Bibliography

Raleigh, W., and H. A. Jones. *The War in the Air.* 6 vols. New York: Oxford University Press, 1922–1937.

Wise, S. F. *Canadian Airmen and the First World War: The Official History of the Royal Canadian Air Force.* Vol. 1. Toronto: University of Toronto Press, 1980.

Books

Adams, Briggs K. *The American Spirit: Letters of Briggs Kilburn Adams, Lieutenant of the Royal Flying Corps.* Boston: Atlantic Monthly Press, 1918.

Bingham, Hiram. *An Explorer in the Air Service.* New Haven, Conn.: Yale University Press, 1920.

Brown, Walt, Jr., ed. *An American for Lafayette: The Diaries of E. C. C. Genet, Lafayette Escadrille.* Charlottesville: University Press of Virginia, 1981.

Casari, Robert B. *Encyclopedia of U.S. Military Aircraft: The World War I Production Program.* Chillicothe, Ohio: Robert B. Casari, publ., 1975.

Chamberlin, Clarence D. *Record Flights.* New York: Dorrance & Co., 1928.

Chapman, Victor. *Victor Chapman's Letters from France, with Memoir by John Jay Chapman.* New York: Macmillan, 1917.

Coffman, Edward M. *The War to End All Wars.* New York: Oxford University Press, 1968.

Falls, Cyril. *The Great War.* New York: G. P. Putnam's Sons, 1959.

Fussell, Paul. *The Great War and Modern Memory.* New York: Oxford University Press, 1975.

Gibbs-Smith, Charles H. *Aviation.* London: Her Majesty's Stationery Office, 1985.

Greenhous, Brereton, ed. *A Rattle of Pebbles: The First World War Diaries of Two Canadian Airmen.* Monograph No. 4, Directorate of History, Department of National Defence. Ottawa: Minister of Supply and Services Canada, 1987.

Grider, John M. *War Birds: Diary of an Unknown Aviator.* New York: George H. Doran Co., 1926.

Hall, James Norman, and Charles Bernard Nordhoff, eds. *The Lafayette Flying Corps.* 2 vols. Boston & New York: Houghton Mifflin Co., 1920.

Hallion, Richard B. *Rise of the Fighter Aircraft, 1914–1918.* Annapolis: The Nautical & Aviation Publishing Co. of America, 1984.

Hartney, Harold E. *Up and At 'Em.* Reprint. New York: Arno Press, 1980.

Hudson, James J. *Hostile Skies.* Syracuse, N.Y.: Syracuse University Press, 1968.

Huntington, Samuel F. *The Soldier and the State.* Cambridge, Mass.: Belknap Press, 1957.

Kennan, George. *American Diplomacy, 1900–1950.* Chicago, Ill.: University of Chicago Press, 1951.

Bibliography

Kennedy, David. *Over Here: The First World War and American Society.* New York: Oxford University Press, 1980.

Kennett, Lee. *The First Air War, 1914–1918.* New York: The Free Press, 1991.

———. *A History of Strategic Bombing.* New York: Charles Scribner's Sons, 1982.

Mead, Peter. *The Eye in the Air: History of Air Observation and Reconnaissance for the Army, 1785–1945.* London: Her Majesty's Stationery Office, 1983.

Mitchell, William. *Memoirs of World War I: From Start to Finish of Our Greatest War.* New York: Random House, 1960.

Molter, Bennett A. *Knight of the Air.* New York: D. Appleton & Co., 1918.

Morrow, John H., Jr. *The Great War in the Air: Military Aviation from 1909 to 1921.* Washington, D.C.: Smithsonian Institution Press, 1993.

Patrick, Mason. *The United States in the Air.* Garden City, N.Y.: Doubleday, Doran & Co., 1928.

Pershing, John J. *My Experiences in the World War.* 2 vols. New York: Frederick A. Stokes Co., 1931.

Porter, Harold E. *Aerial Observation: The Airplane Observer, the Balloon Observer, and the Army Corps Pilot.* New York: Harper & Bros., 1921.

Richards, John Francisco, II. *War Diary and Letters of John Francisco Richards II, 1917–1918.* Kansas City, Mo.: George B. Richards, 1925.

Schaffer, Ronald. *America in the Great War: The Rise of the War Welfare State.* New York: Oxford University Press, 1991.

Stokesbury, James L. *A Short History of Air Power.* New York: William Morrow & Co., 1986.

Sweetser, Arthur. *The American Air Service: A Record of Its Problems, Its Difficulties, Its Failures, and Its Achievements.* New York: Appleton, 1919.

Terraine, John. *The First World War, 1914–18.* London: Macmillan, 1965.

———. *To Win a War: 1918, the Year of Victory.* Garden City, N.Y.: Doubleday & Co., 1981.

Thayer, Lucien H. *America's First Eagles: The Official History of the U.S. Air Service, A.E.F. (1917–1918).* San Jose, Calif.: R. James Bender; Mesa, Ariz.: Champlin Fighter Museum Press, 1983.

Todd, Robert M. *Sopwith Camel Fighter Ace.* Falls Church, Va.: Ayjay, 1978.

Toulmin, H. A., Jr. *Air Service, American Expeditionary Force, 1918.* New York: D. Van Nostrand Co., 1927.

Tredrey, F. *Pioneer Pilot.* London: Peter Davies, 1976.

Winter, Denis. *The First of the Few.* Athens, Ga.: The University of Georgia Press, 1982.

Articles and Monographs

Dickey, Philip S., III. "The Liberty Engine 1918–1942." *Smithsonian Annals of Flight* 1 no. 3 (1968).

Levy, Michael H. "Learning from the Ground Up: The Acquisition of Aircrew Training Devices by the Aeronautical Systems Division and its Predecessors." Historical Monograph, Aeronautical Systems Division, USAF Systems Command, Jul 1986.

Snyder, Thomas S. *Chanute Field: The Hum of the Motor Replaced the Song of the Reaper, 1917–1921*. Historical Study, Air Training Command, Chanute Technical Training Center History Office, May 1975.

The World War I Diary of Col. Frank P. Lahm, Air Service, A.E.F. Maxwell AFB, Ala.: Air University, Dec 1970.

Interwar Years

Hearings

House Committee on Military Affairs. *Historical Documents Relating to the Reorganization Plans of the War Department and to the Present National Defense Act*. 69th Cong., 2d sess. Mar 3, 1927.

Books

Angelucci, Enzo. *The Rand McNally Encyclopedia of Military Aircraft, 1914–1980*. New York: The Military Press, 1980.

Bilstein, Roger E. *The American Aerospace Industry*. New York: Twayne, 1996.

Bissell, Clayton. *Brief History of the Air Corps and Its Late Development*. Fort Monroe, Va.: Coast Artillery School Press, 1928.

Bond, Brian. *British Military Policy between the Two World Wars*. Oxford: Clarendon Press, 1980.

Boyne, Walter J. *DeHavilland DH–4: From Flaming Coffin to Living Legend*. Washington, D.C.: Smithsonian Institution Press, 1984.

Byrd, Martha. *Chennault: Giving Wings to the Tiger*. Tuscaloosa: University of Alabama Press, 1987.

———. *Kenneth N. Walker: Airpower's Untempered Crusader*. Maxwell AFB, Ala.: Air University Press, 1997.

Copp, DeWitt S. *A Few Great Captains*. Garden City, N.Y.: Doubleday & Co., 1980.

Fahey, James C. *U.S. Army Aircraft, 1908–1946*. New York: Ships & Aircraft, 1946.

Francillon, Rene J. *McDonnell Douglas Aircraft since 1920*. London: Putnam, 1979.

Frisbee, John L., ed. *Makers of the United States Air Force*. Washington, D.C.: Office of Air Force History, 1987.

Futrell, Robert Frank. *Ideas, Concepts, Doctrine: A History of Basic Thinking*

Bibliography

in the United States Air Force, 1907–1964. Maxwell AFB, Ala.: Air University, 1971.

Harbold, Norris B. *The Log of Air Navigation.* San Antonio, Tex.: Naylor, 1970.

Hewes, James E., Jr. *From Root to MacNamara: Army Organization and Administration, 1900–1963.* Washington, D.C.: U.S. Army Center of Military History, 1975.

Holley, I. B., Jr. *Buying Aircraft: Materiel Procurement for the Army Air Forces.* Washington, D.C.: U.S. Army Center of Military History, 1950.

———. *Ideas and Weapons.* New Haven, Conn.: Yale University Press, 1953.

Hurley, Alfred F. *Billy Mitchell: Crusader for Air Power.* New York: Watts, 1964.

Karsten, Peter, ed. *The Military in America from Colonial Times to the Present.* New York: Free Press, 1985.

Kelly, Lloyd L. *The Pilot Maker.* New York: Grosset & Dunlap, 1970.

Kennedy, Paul. *The Rise and Fall of the Great Powers.* New York: Random House, 1987.

Lay, Beirne, Jr. *I Wanted Wings.* New York: Harper & Bros., 1937.

Leuchtenburg, William E. *The Perils of Prosperity, 1914–1932.* Chicago, Ill.: University of Chicago Press, 1958.

Loftin, Laurence K., Jr., *Quest for Performance: The Evolution of Modern Aircraft.* Washington, D.C.: National Aeronautics & Space Administration, 1985.

Maurer, Maurer. *Aviation in the U.S. Army, 1919–1939.* Washington, D.C.: Office of Air Force History, 1987.

Millis, Walter, ed. *American Military Thought.* Indianapolis, Ind.: Bobbs-Merrill, 1966.

Mitchell, William. *Winged Defense: The Development and Possibilities of Modern Air Power, Economic and Military.* New York: Putnam, 1925.

Murray, Williamson, and Allan R. Millett, eds. *Military Innovation in the Interwar Period.* New York: Cambridge University Press, 1996.

Nelson, Otto L., Jr. *National Security and the General Staff.* Washington, D.C.: Infantry Journal Press, 1946.

Roland, Alex. *Model Research: The National Advisory Committee for Aeronautics, 1915–1958.* 2 vols. Washington, D.C.: Scientific & Technical Information Branch, National Aeronautics & Space Administration, 1985.

Shiner, John F. *Foulois and the U.S. Army Air Corps, 1931–1935.* Washington, D.C.: Office of Air Force History, 1983.

Walker, Lois E., and Shelby E. Wickam. *From Huffman Prairie to the Moon: The History of Wright-Patterson Air Force Base.* Washington, D.C.: Air Force Logistics Command, 1986.

Watson, Mark Skinner. *Chief of Staff: Prewar Plans and Preparations.* Washington, D.C.: Historical Division, United States Army, 1950.

Wiener, Willard. *Two Hundred Thousand Flyers: The Story of the Civilian–AAF Pilot Training Program.* Washington, D.C.: The Infantry Journal, 1945.

Workers of the Writers' Program of the Work Projects Administration in the State of Texas, comp. *Randolph Field: A History and Guide.* New York: The Devin-Adair Co., 1942.

Wright, Monte Duane. *Most Probable Position: A History of Aerial Navigation to 1941.* Lawrence: University Press of Kansas, 1972.

Monographs and Studies

Brown, Jerold E. "Where Eagles Roost: A History of Army Airfields before World War II." Ph.D. diss., Duke University, 1977.

Curtis, Robert I, John Mitchell, and Martin Copp. "Langley Field: The Early Years, 1916–1946." Historical Study, Office of History, Langley AFB, Va., 1977.

Finney, Robert T. *History of the Air Corps Tactical School, 1920–1940.* New imprint. Washington, D.C.: Center for Air Force History, 1992.

Greer, Thomas H. *The Development of Air Doctrine in the Army Air Arm, 1917–1941.* New imprint. Washington, D.C.: Office of Air Force History, 1985.

Killigrew, John W. "The Impact of the Great Depression on the Army, 1929–1936." Ph.D. diss., Indiana University, 1960

Hugh Knerr, "The Vital Era, 1887 1950." Manuscript, Air Force History Support Office, Bolling AFB, D.C.

Layman, Martha E. "Legislation Relating to the Air Corps Personnel and Training Programs, 1907–1939." USAF Historical Study No. 39, Historical Research Agency, Maxwell AFB, Ala., Dec 1945.

McClendon, Robert Earl. *Autonomy of the Air Arm.* New imprint. Washington, D.C.: Air Force History & Museums Program, 1996.

Purtee, Edward O. "The Modification and Development of Training Aircraft for AAF Use, 1918–1945." Historical Study, Air Materiel Command, Wright Field, Ohio, Nov 1946.

Reynolds, Jon A. "Education and Training for High Command: General Hoyt S. Vandenberg's Early Career." Ph.D. diss., Duke University, 1980.

Articles

For the interwar years, an invaluable source of information comes from the *Air Service* and *Air Corps News Letter.* The *News Letter* tracks a variety of activities on air installations, but it is not possible here to cite all the issues that touch upon training.

Finney, Robert T. "Early Air Corps Training and Tactics." *Military Affairs*

Bibliography

(Fall 1956): 154–161.

Greene, Fred M. "The Military View of American National Policy, 1904–1940." *American Historical Review* (Jan 1961): 354–377.

World War II

Official Histories

Craven, Wesley Frank, and James Lea Cate. *The Army Air Forces in World War II*. 7 vols. New imprint. Washington, D.C.: Office of Air Force History, 1983.

Office of the Deputy Chief of Naval Operations (Air). *United States Naval Administration in World War II*. Vol. 14, *Aviation Training, 1940–1945*. Washington, D.C.: n.d.

Books

Ardery, Philip. *Bomber Pilot: A Memoir of World War II*. Lexington: University Press of Kentucky, 1983.

Arnold, H.H. *Global Mission*. N.Y.: Harper & Bros., 1949.

Army Air Forces Statistical Digest, World War II. Office of Statistical Control, Dec 1945.

Bombardier: A History. Paducah, Kentucky: Turner Publishing Co., 1993.

Beck, Alfred M., ed. *With Courage: The U.S. Army Air Forces in World War II*. Washington, D.C.: Air Force History & Museums Program, 1994.

Bodie, Warren M. *The Lockheed P-38 Lightning*. Hiawassee, Ga.: Wideway Publications, 1991.

Boeman, John. *Morotai: A Memoir of War*. Manhattan, Kans.: Sunflower University Press, 1981.

Bond, Charles R., and Terry Anderson. *A Flying Tiger's Diary*. College Station: Texas A&M University Press, 1984.

Boyington, Gregory. *Baa Baa Black Sheep*. New York: Putnam, 1958.

Buderi, Robert. *The Invention That Changed the World*. New York: Simon & Schuster, 1996.

Carigan, William. *Ad Lib: Flying the B-24 Liberator in World War II*. Manhattan, Kans.: Sunflower University Press, 1988.

Carter, Kit C., and Robert Mueller, comps. *Army Air Forces in World War II: Combat Chronology, 1941–1945*. Reprint. Washington, D.C.: Center for Air Force History, 1991.

Chennault, Claire Lee. *Way of a Fighter*. New York: G.P. Putnam's Sons, 1949.

Colgan, Bill. *World War II Fighter-Bomber Pilot*. Blue Ridge Summit, Pa.: Tab Books, 1985.

Bibliography

Cooling, Benjamin Franklin, ed. *Case Studies in the Development of Close Air Support*. Washington, D.C.: Office of Air Force History, 1990.

Copp, DeWitt S. *Forged in Fire*. Garden City, N.Y.: Doubleday & Co., 1982.

Cozzens, James Gould. *A Time of War: Air Force Diaries and Pentagon Memos, 1943–45*. Matthew J. Bruccoli, ed. Columbia, S.C.; Bloomingfield Hills, Mich.: Bruccoli Clark, 1984.

Devereux, Tony. *Messenger Gods of Battle: Radio, Radar, Sonar, the Story of Electronics in War*. London: Brassey's, 1991.

Divine, Robert A. *Roosevelt and World War II*. Baltimore: Penguin Books, 1969.

Dunn, William R. *Fighter Pilot*. Lexington: University Press of Kentucky, 1982.

Fletcher, Eugene. *Mister: The Training of an Aviation Cadet in World War II*. Seattle: University of Washington Press, 1992.

Freeman, Roger A. *The Mighty Eighth: Units, Men and Machines*. London: Macdonald, 1970.

Giering, Edward J. *B–17 Bomber Crew Diary*. Manhattan, Kans.: Sunflower University Press, 1985.

Goodson, James A. *Tumult in the Clouds*. New York: St. Martin's Press, 1983.

Green, William. *Famous Bombers of the Second World War*. Garden City, N.Y.: Hanover House, 1959.

Greenfield, Kent Robert. *American Strategy in World War II: A Reconsideration*. Baltimore: The Johns Hopkins University Press, 1963.

Hallion, Richard P. *Strike from the Sky: The History of Battlefield Air Attack, 1911–1945*. Washington, D.C.: Smithsonian Institution Press, 1989.

Hansell, Haywood S., Jr. *The Air Plan that Defeated Hitler*. Atlanta, Ga.: Higgins-McArthur/Longino & Porter, Inc., 1972.

Harmon, Tom. *Pilots Also Pray*. New York: Thomas Y. Crowell Co., 1944.

Hart, B. H. Liddell. *History of the Second World War*. New York: G.P. Putnam's Sons, 1970.

Hatch, F. J. *The Aerodrome of Democracy: Canada and the British Commonwealth Air Training Plan, 1939–1945*. Monograph No. 1, Directorate of History, Department of National Defence. Ottawa: Minister of Supply and Services Canada, 1983.

Herbert, Kevin. *Maximum Effort: The B–29's Against Japan*. Manhattan, Kans.: Sunflower University Press, 1983.

Hoyle, Martha Byrd. *A World in Flames: A History of WWII*. New York: Atheneum, 1969.

Hoyt, Edwin P. *The Airmen: The Story of American Flyers in World War II*. New York: McGraw-Hill, 1990.

Hynes, Samuel. *Flights of Passage: Reflections of a World War II Aviator*. New York: Frederic C. Beil; Annapolis: Naval Institute Press, 1988.

Jablonski, Edward. *Flying Fortress: The Illustrated Biography of the B–17s*

Bibliography

and the Men Who Flew Them. Garden City, N.Y.: Doubleday & Co., 1965.

Kenney, George C. *General Kenney Reports: A Personal History of the Pacific War.* Reprint. Washington, D.C.: Air Force History & Museums Program, 1997.

LeMay, Curtis E., and Bill Yenne. *Superfortress: The Story of the B-29 and American Air Power.* New York: McGraw-Hill, 1988.

Maurer, Maurer, ed. *Air Force Combat Units of World War II.* Washington, D.C.: Office of Air Force History, 1983.

McFarland, Stephen L. *America's Pursuit of Precision Bombing, 1910-1945.* Washington, D.C.: Smithsonian Institution Press, 1995.

────── and Wesley Phillips Newton. *To Command the Sky: The Battle for Air Superiority Over Germany, 1942-1944.* Washington, D.C.: Smithsonian Institution Press, 1991.

Nalty, Bernard C., ed. *Winged Shield, Winged Sword. A History of the United States Air Force.* Vol 1, *1907-1950.* Washington, D.C.: Air Force History & Museums Program, 1997.

Noah, Joseph W. *Wings God Gave My Soul.* Annandale, Va.: Charles Baptie Studios, 1974.

Overy, R. J. *The Air War, 1939-1945.* New York: Stein & Day, 1980.

Pape, Garry R., with John M. and Donna Campbell. *Northrop P-61 Black Widow.* Oceola, Wisc.: Motorbooks International Books and Wholesalers, 1991.

Parton, James. *"Air Force Spoken Here": General Ira Eaker and the Command of the Air.* Bethesda, Md.: Adler & Adler, 1986

Perret, Geoffrey. *Winged Victory: The Army Air Forces in World War II.* New York: Random House, 1993.

Pisano, Dominick A. *To Fill the Skies with Pilots.* Urbana: University of Illinois Press, 1993.

Ragsdale, Kenneth Baxter. *Wings Over the Mexican Border.* Austin: University of Texas Press, 1984.

Rawnsley, C. F., and Robert Wright. *Night Fighter.* New York: Henry Holt & Co., 1957.

Schaffel, Kenneth. *The Emerging Shield: The Air Force and the Evolution of Continental Air Defense, 1945-1960.* Washington, D.C.: Office of Air Force History, 1991.

Schaffer, Ronald. *Wings of Judgment: American Bombing in World War II.* New York: Oxford University Press, 1985.

Scott, Robert L. *God Is My Copilot.* New York: Charles Scribner's Sons, 1943.

Sherry, Michael S. *The Rise of American Air Power.* New Haven, Conn.: Yale University Press, 1987.

────── . *In the Shadow of War: The United States since the 1930s.* New Haven, Conn.: Yale University Press, 1995.

Stiles, Bert. *Serenade to the Big Bird.* New York: W. W. Norton & Co., 1947.

Tunner, William H. *Over the Hump*. New Imprint. Washington, D.C.: Air Force History & Museums Program, 1998.

Wagner, Ray. *American Combat Planes*. Garden City, N.Y.: Doubleday & Co., 1968.

The War Reports of General of the Army George C. Marshall, General of the Army H. H. Arnold, and Fleet Admiral Ernest J. King. Philadelphia: J. B. Lippincott, 1947.

Watry, Charles A. *Washout! The Aviation Cadet Story*. Carlsbad, Calif.: California Aero Press, 1983.

——— and Duane L. Hall. *Aerial Gunners: The Unknown Aces of World War II*. Carlsbad, Calif.: California Aero Press, 1986.

Weinberg, Gerhard. *A World at Arms: A Global History of World War II*. New York: Cambridge University Press, 1994.

Wiener, Willard. *Two Hundred Thousand Flyers: The Story of the Civilian–AAF Pilot Training Program*. Washington, D.C.: The Infantry Journal, 1945.

Wolk, Herman S. *Planning and Organizing the Postwar Air Force 1943–1947*. Washington, D.C.: Office of Air Force History, 1984.

Yeager, Chuck, and Leo Janos. *Yeager: An Autobiography*. New York: Bantam Books, 1985.

Monographs, Studies, and Articles

Baldwin, Ben R. "Individual Training of Bombardiers." USAF Historical Study No. 5, Historical Research Agency, Maxwell AFB, Ala., May 1944.

———. "Individual Training of Navigators in the AAF." USAF Historical Study No. 27, Historical Research Agency, Maxwell AFB, Ala., Jan 1945.

Boylan, Bernard. "Development of the Long-Range Escort Fighter." USAF Historical Study No. 136, Historical Research Agency, Maxwell AFB, Ala., Sep 1955.

"Combat Crew Rotation, World War II and Korean War." Historical Study, Air University, Maxwell AFB, Ala., Jan 1968.

Dubuque, Jean H., revised by Robert F. Glockner. "The Development of the Heavy Bomber, 1918–1944." USAF Historical Study No. 6, Historical Research Agency, Maxwell AFB, Ala., 1951.

England, J. Merton, and Chauncey E. Sanders, "Legislation Relating to the AAF Training Program, 1939–1945." USAF Historical Study No. 7, Historical Research Agency, Maxwell AFB, Ala., Office, 1946.

Futrell, Robert F. "Command of Observation Aviation: A Study in Control of Tactical Airpower." USAF Historical Study No. 24, Historical Research Agency, Maxwell AFB, Ala., Sep 1956.

Goldman, Martin R. R. "Morale in the AAF in World War II." USAF Historical Study No. 78, Historical Research Agency, Maxwell AFB, Ala., Jun 1953.

Grant, Clement L. "Policies and Procedures Governing Elimination from AAF

Bibliography

Schools, 1939–1945." USAF Historical Study No. 79, Historical Research Agency, Maxwell AFB, Ala., 1952.

Greer, Thomas H. "Preflight Training in the AAF, 1939–1944." USAF Historical Study No. 48, Historical Research Agency, Maxwell AFB, Ala., 1946.

Grow, Natalie. "Procurement of Aircrew Trainees." USAF Historical Study No. 15, Historical Research Agency, Maxwell AFB, Ala., Aug 1944.

Hollon, W. Eugene. "History of Preflight Training in the AAF, 1941–1953." [Title varies.] USAF Historical Study No. 90, Historical Research Agency, Maxwell AFB, Ala., Jun 1953.

Layman, Martha. "Organization of AAF Training Activities 1939–1945." USAF Historical Study No. 53, Historical Research Agency, Maxwell AFB, Ala., Jun 1946.

McMullen, Richard F. "The Role of Civilian Contractors in the Training of Military Pilots." Historical Study No. 2, Air Training Command, Scott AFB, Ill., 1955.

McNarney, Betty Jo. "Pilot Transition to Combat Aircraft." USAF Historical Study No. 18, Historical Research Agency, Maswell AFB, Ala., Sep 1944.

Mooney, Chase C. "Organization of the Army Air Arm 1935–1945." USAF Historical Study No. 10, Historical Research Agency, Maxwell AFB, Ala., Jul 1956.

Nickle, Barry H. "Contract Flying Training in Air Training Command, 1939–1980." Historical Study, Air Training Command, 1981.

———. "Major Changes in Undergraduate Navigator Training, 1940–1980." Historical Study, Air Training Command, 1981.

Purtee, Edward O. "The Modification and Development of Training Aircraft for AAF Use, 1918–1945." Historical Study, Air Materiel Command, Wright Field, Ohio, Nov 1946.

Jones, R. L., revised by Chauncy E. Sanders. "Personnel Problems Relating to AAF Commissioned Officers, 1939–1945." USAF Historical Study No. 11, Historical Research Agency, Maxwell AFB, Ala., 1951.

Simms, Henry. "Flexible Gunnery Training in the AAF." USAF Historical Study No. 31, Historical Research Agency, Maxwell AFB, Ala., Mar 1945.

Smith, Leslie F. "Development of AAF and USAF Training Concepts and Programs, 1941–1952." USAF Historical Study No. 93, Historical Research Agency, Maxwell AFB, Ala., 1953.

Thompson, Robert L. "Initial Selection of Candidates for Pilot, Bombardier, and Navigator Training. USAF Historical Study No. 2, Historical Research Agency, Maxwell AFB, Ala., 1943.

Tibbets, Paul W. "Twenty-Eight Years Ago: Training the 509th for Hiroshima." *Air Force* 56 (Aug 1973): 49–55.

Walters, Raymond. "Individual Training in Aircraft Armament by the AAF, 1939–1945." USAF Historical Study No. 60, Historical Research Agency,

Bibliography

Maxwell AFB, Ala., 1955.

White, Gerald T. "Combat Crew and Unit Training in the AAF, 1939–1945." USAF Historical Study No. 61, Historical Research Agency, Maxwell AFB, Ala., Aug 1949.

Williams, Edwin L., Jr. "History of the Air Force Civilian Training Program, 1941–1951." USAF Historical Study No. 123, Historical Research Agency, Maxwell AFB, Ala., 1956.

———. "Legislative History of the AAF and USAF, 1941–1951." USAF Historical Study No. 84, Historical Research Agency, Maxwell AFB, Ala., Sep 1955.

Index

1-2-3 aircraft control system: 398
Adams, Briggs Kilburn: 130, 183
Aerial Experiment Association: 17
Aero Club of America: 16, 75
Aero Club of San Diego: 57
Aerostation. *See* Balloons.
Agan, Arthur C.: 478
Air Corps, U.S. Army. *See also* Air Service, U.S. Army; Army Air Forces, U.S.; Signal Corps, U.S. Army; Training Center, Air Corps; Interwar era.
  aircraft inventory: 242
  air mail experience: 277
  all-weather proficiency: 268
  autonomy issue: 310
  Baker Board: 277
  balanced forces approach: 559–560
  bombardment: 334, 335, 336–337, 342–347
    accuracy: 355
    curriculum: 342–344
    dive bombing: 358–359
    heavy: 356, 361–365
    level v. dive: 358–359
    light (attack): 356, 359–361
    low-altitude: 359–360
    night v. day: 364
  casualties: 267, 272
  Circular 50-1: 269
  coastal defense mission dispute: 244, 265, 292
  congressional appropriations: 273, 280
  depression, influence of: 241
  expansion: 241–242, 306–307
  Ferry Command: 340–341
  Flying Training Command: 313
  MacArthur influence: 276, 277
  mail delivery experience: 267–268
  mission: 244
  mobilization success: 371–372
  personnel and manpower issues: 241–242, 270, 272
    aviation cadet designation: 334
    career path: 274–275
    Civilian Pilot Training Program graduates: 318
    command staff: 242–243
    enlisted men trainees: 318–319
    expansion forecasts, 1940: 314
    mathematics aptitude: 338
    officer–enlisted issue: 344
    personnel assignment: 279
    pilot ratings and eligibility: 353–354
    pilot shortages: 245–246
    qualification standards, reconnaissance observers: 370
    recruitment: 383
    recruitment and selection: 335–336
    reserve officers: 245, 246
    training pool expansion: 318–319
  planning and organization: 242–246
  public relations emphasis: 272
  public support for: 245
  relief mission, depression: 270
  schools
    Air Corps Training Center: 308
    Gulf Coast Air Corps Training Center: 315
    Langley Field: 263
    Southeast Air Corps Training Center: 315
    Tactical School: 275–276, 298
    technical: 308
    Training Center: 301
    training centers reorganization: 308, 314–315
    Tuskegee Institute: 326
    West Coast Air Corps Training Center: 315
  specialization: 559
  training. *See also* bombardment *above*.
    24-, 41-, 54-, 84-group programs: 309–310, 311, 313
    advanced: 317–318, 328–333
    aircraft. *See* Aircraft, Allied.
    air crews: 334–337
    basic flying: 324–328
    casualties: 320
    civil navigators program: 338–339

Index

copilot: 363
cross training: 336–337, 363–364
fighter, pursuit: 365–367
fire control officer: 363
First Aviation Objective: 309–310, 313, 317
flare landings: 326
formation flying: 327
gunnery: 329, 331, 332, 334, 335, 347–349
high-altitude: 366–367
instructors: 321–322
instrument flying: 333
navigation: 334, 337–342, 354–355, 363
night bombing: 343
night flying: 326–327, 356–357
operational training: 351–358
Pan Am navigation instruction: 338–339, 341
pilot training, individual: 316–320
preflight: 315
primary flying: 320–324
private flying schools: 320–323
reconnaissance: 335, 367–370
schools expansion: 326
Second Aviation Objective: 310
single engine: 331–332
specialization: 307–308, 329, 336, 358
tactical training policy: 307–308
three-phase program: 324–328
Training Regulation 440-15: 244
twin-engine: 329–332
Training and Operations Division: 314–315
Training and War Plans Division: 243
Air Corps Enlisted Reserve: 384
Aircraft industry, U.S.: 134, 300
  aircraft industry production: 309–310, 310, 313, 375, 376, 380
  Bureau of Aircraft Production: 110
  Burgess Company: 52–53
  production capacity, World War I: 204–205
  production capacity deficit: 110
  World War II expansion: 309–310, 313, 375, 376, 380
Aircraft
  A–2: 431
  A–3: 360
  A–3B: 259
  A–11: 290

A–12: 284, 290, 360
A–17: 284, 290
A–17A: 361
A–18: 354, 360
A–20: 368, 413, 522, 535, 537, 541, 545, 546, 551, 552
A–20A: 360, 361
A–20G: 535
A–24: 413, 522, 535, 545
A–26: 413, 535–536
A–31: 413
A–36: 523, 539
AT–6: 328, 332–333, 399–401, 406, 408, 414–415, 444
AT–7: 332, 406, 426
AT–8: 332
AT–9: 415
AT–10: 332, 406, 407
AT–11: 333, 346, 406, 431, 433
AT–17: 406, 407
AT–18: 426
AT–21: 406
AT–23: 412, 444
AT–24: 407, 412
AT–38A: 426
Avro: 128, 159, 167, 173
B–2: 281, 283, 284
B–3A: 259
B–4: 283
B–5: 283
B–5A: 259
B–6: 283
B–10: 292, 332, 354
B–12: 284, 332
B–12A: 284
B–17: 284, 292, 300, 301, 306, 354, 361, 362, 372, 409, 411, 444, 460, 467, 469, 501, 506, 533
B–17D: 491
B–17F: 472
B–17G: 491
B–18: 284, 332, 346, 354, 360, 362, 426
B–24: 361, 372, 411, 444, 460, 467, 472, 486, 497, 498, 501, 503, 509, 533
B–24J: 491, 497
B–25: 406–407, 412, 462, 476, 522, 528, 533, 535
B–25G: 527
B–25J: 531–532
B–26: 412, 444, 462, 533–535
B–29: 372, 413, 460, 461, 503–509

Index

B–32: 372, 460
B–34: 426, 444
B–36: 372
BC–1: 328, 332, 354
BT–2: 255, 256
BT–8: 254–255, 328
BT–9: 255, 325
XBT–12: 255
BT–13: 325, 397, 399
BT–14: 325
BT–15: 325, 399
Baldwin airship: 11
Bleriot: 90, 154
Breese Penguin: 89–90, 154
Breguet 14: 163, 186
Breguet-Renault: 194
Bristol Fighter: 155, 172, 173
Burgess H tractor: 53, 55, 63, 91
Burgess-Wright B: 45
C–33: 354
C–39: 354
C–40: 354
C–60: 426
Curtiss Model D: 38–39
Curtiss Model E: 44, 52, 57, 86
Curtiss Model F: 63
Curtiss Model G: 57, 86
Curtiss Model J: 64
Curtiss Model N: 64
Caproni bomber: 158, 183, 186–187
Caudron G–3/4: 154, 166, 176
DH–4: 103, 176, 177–180, 186, 196, 205, 234, 235, 250
DH–4B: 228, 234, 237
DH–6: 159
Dirigible No. 1: 18
F–5: 415
F–13: 508
Farman: 157, 158, 159, 176
FW 190: 545
GA–1: 234
Handley-Page bomber: 188
JN–2: 64
JN–3: 64
JN–4: 64, 96, 124, 155
JN–4A: 123, 125
JN–4B: 93, 125
JN–4C: 134
JN–4D: 123, 134, 205
JN–4H: 134
JN–6H: 232
Ju 88: 358–359
Keystone B: 259, 268
Keystone LB: 263
MB–2: 235, 263
MB–3A: 231, 232
Martin T/TT trainers: 65, 66
Me 262: 544
Morane-Saulnier: 90, 155, 156, 164
Nieuport: 90, 154–155, 166, 170
Nieuport 17: 154
Nieuport 25: 155
Nieuport 27: 167, 168
Nieuport 28: 171, 192
O–2: 234, 237
O–2H: 234, 250
O–19: 259, 260
O–25: 259, 260
O–46A: 269
O–47B: 368, 369
OA–8: 354
OA–9: 354
P–1: 231, 259
P–6E: 287
P–12: 283, 290, 328, 329
P–12B: 259, 260
P–12D: 259
P–25: 287
P–26: 287, 353, 380
P–35: 287, 353
P–36: 353, 401
P–38: 409, 410, 413, 414, 415, 482, 497, 522, 539, 542
P–39: 353, 366, 404, 413, 444, 464, 539, 544, 552
P–40: 353, 366, 404, 409, 413, 444, 499, 522, 539, 544, 552
P–43: 353
P–47: 482, 522, 539, 540, 542
P–51: 413, 482, 508, 522, 539, 540, 541, 552
P–51B: 473
P–61: 539, 548–552
P–63: 444, 508
P–70: 551
P–322: 409, 410, 414, 415
PT–1: 250
PT–3: 253, 323
PT–11D: 253, 254
PT–13: 254, 323, 393
PT–16: 323
PT–17: 323, 393, 394
PT–18: 393
PT–19: 393
PT–20: 393
PT–21: 393

659

Index

PT–22: 393, 394
PT–23: 393
PT–26: 393
PT–27: 393
PW–8: 232
S–4: 134, 231
SAI: 176
Salmson: 181, 193, 194
S.C. No. 1: 24, 25
S.C. No. 5: 53
S.C. No. 8: 52
S.C. No. 17: 84
S.C. No. 22: 64
"Scout": 52–53
S.E. 5: 172, 173, 231, 232
Sopwith A–2: 176
Sopwith B–2: 176
Sopwith Camel: 155–156, 172
Spad XIII: 163, 166, 167, 231
"Speed Scout": 52–53
Type IV Military: 34, 38–39
White Wing dirigible: 17
Wright B: 28, 29, 44–45, 49, 84
Wright biplane: 8, 18–21
Wright C: 82, 84
TB–24: 444
Waller trainer: 349
Aircraft, pre–World War I era
  classification types, European: 87–88
  seaplanes: 36, 53
  standardization: 47
  tactical aircraft development: 51–53
Air Forces, U.S.
  First Air Force: 378, 454, 458, 459, 463, 540
  Second Air Force: 378, 411, 428–429, 452, 454, 458–459, 460, 463, 465–466, 471, 474, 478, 485, 493, 494, 496, 517, 553
  Third Air Force: 378, 412, 454, 458, 461–462, 499, 506, 517
  Fourth Air Force: 378, 415, 454, 458, 463, 496, 540, 551, 553
  Fifth Air Force: 528, 545
  Seventh Air Force: 554
  Eighth Air Force: 481–482, 509
  Ninth Air Force: 477, 522, 545
  Twelfth Air Force: 507
  Fifteenth Air Force: 469
Air Service, U.S. Army. *See also* Air Corps, U.S. Army; Army Air Forces, U.S.; Signal Corps, U.S. Army; Training Section, Air Service; World War I era.
  Advanced Tactical Section: 214
  aircraft: 231–232
  aircraft inventory decline: 210–211
  air-ground forces coordination: 218–220
  appropriations: 221
  Army Reorganization Act of 1920: 220
  Army Training Regulation 440–15: 208
  attack: 233–234
  Balloon and Airship Section separation: 215–216
  casualties: 221
  Chief of Air Service role: 214
  Civilian Conservation Corps assistance: 210
  coastal defense mission dispute: 208
  command staff: 103, 216
  divisional system: 214–216
  establishment: 103
  expansion program, 1939: 210
  fighter development: 230
  five-year expansion program, 1926: 213
  ground schools training: 112–114
    best 10 percent plan: 120
    Canadian ground schools: 112
    criticism of: 118–119
    equipment shortages: 120
    military training element: 116
    program goals confusion: 117–118
    recruit selection: 114–116
    school rivalries: 118
    schools coordination: 117
    theoretical v. practical instruction: 118–119
    U.S. Schools of Military Aeronautics: 113
  instructional literature: 230, 231, 239
  MacArthur position, 1933: 211
  mission: 205, 206, 219–220
  personnel and manpower issues: 217, 224, 228
    cadet, commissioned officer issue: 138–139
    command staff: 103, 216
    officers recruitment: 220–221
    personnel levels decline: 210–211
    recruitment: 210
    reserve officers: 220–221
    reserves training: 239
    selection and eligibility: 225–226
  Primary and Technical Section: 214
  schools: 222–224

Index

Advanced Flying School: 224, 226–227, 235, 236
Army Air Service School of Application: 274
Observation School: 228
Primary Flying School: 224–226
202 Squadron Program: 131
tactical doctrine and research: 207–208, 209
training: 214, 215
  bombardment: 224, 235–236
  crew: 208
  navigation: 208
  observation: 227–228
  overwater piloting and navigation: 207
  peacetime program: 217–218
  priorities development: 207–208
  pursuit: 224, 229–233, 238
  specialization: 224
  tactical unit: 236–239
  training regulations: 214, 215
Training and Operations Division: 209
Training and Operations Group: 222, 224
Training Regulation 440-15: 219–220
Allen, James: 11, 12, 16, 21, 22, 23, 26, 30, 35, 47, 54
American Defense Society: 92
Anderson, Graeme: 115
Anderson, Samuel: 296
Andrews, Frank M.: 207, 208, 211, 246, 273, 274, 276, 277, 278, 279–280, 282, 286, 289, 292, 299, 300–301, 307, 313
Ardrey, Phillip: 325, 393, 480, 494
Armstrong, Frank A., Jr.: 466
Army, U.S.
  1910 Field Service Regulations: 28
  Air Corps autonomy issue: 310
  Air Corps establishment: 242, 243, 244
  air-ground forces cooperation: 296–298
  Army Reorganization Act of 1920: 220
  coastal defense mission dispute: 206–207, 208, 244, 265, 276, 277
  continental defense: 83
  Four Army Plan: 211
  modernization, 1930s: 310
  protective mobilization program of 1937: 299
  reconnaissance and observation: 295
  specialization: 559
  War Department Victory Program: 307
Army Air Forces, U.S. (AAF). *See also* Air Corps, U.S. Army; Air Service, U.S. Army; Crew and Unit training, AAF; Fighter training, AAF; GHQ Air Force; Signal Corps, U.S. Army; Strategic bombardment training, AAF; World War II era.
accident rate: 380
Air Corps Enlisted Reserve: 384
aircraft, tactical: 522–523
air forces-based training: 378–379, 486–487
Aviation Expansion Program: 376–377
AWPD-1: 381
bombardment units strength: 452
bombing accuracy: 474–475
British and Canadian Organization Training Units: 456
casualties: 479
combat readiness evaluation: 380
Combined Bomber Offensive: 483
cross training: 422
Doolittle raid on Tokyo: 526–527
expansion: 380–381
expansion procurement issues: 378
Field Manual 100-20: 477
fighter-bombers: 545–546
fighter escorts: 473
Flying Training Command role: 378
friendly fire: 472
instructors: 387–388
mission: 467
Model Mission Flight Training study: 494
night fighters: 548–551
night missions: 474
North African experience: 477–478
nuclear era reality: 558
personnel and manpower issues
  Air Corps Enlisted Reserve: 384
  Aviation Cadet Qualifying Examination: 385
  bombardiers: 385, 386, 428, 429
  crew selection: 489
  90-day wonders: 465
  expansion requirements: 377
  local procurement boards: 384
  navigators: 385–386
  pilot classification: 388
  pilot qualifications: 527–528
  pilots: 385
  procurement and classification: 384–385

Index

quota reductions 1944: 557–558
Replacement Training Centers: 386
stanine scores classification: 385
technology, influence of: 377
Troop Basis of January 1942: 375
War Department Circular 59: 375
preflight instruction: 386–387
program coordination and standardization: 379–380
programs decline in 1944: 448
radar, airborne: 509–511
Rainbow 5 plan: 376
Replacement Training Centers: 386
schools
   civilian–military partnership: 390–392
   Combat Crew Training School: 457
   Instrument Instructors School: 398
   Interceptor Command School: 550
   Mira Loma Flight Academy: 392
   Ryan School: 391
   School of Applied Tactics: 541
specialization: 422
tactical forces doctrine: 475–476
training. *See also* Crew and Unit training, AAF; Fighter training, AAF; GHQ Air Force; Strategic bombardment training, AAF.
   accident rate: 403–404, 410
   advanced pilot: 400
   aircraft: 393–394, 397, 399, 400–401, 404
   basic: 395–400
   bombers: 409–413
   casualties: 396
   check pilots, military: 392
   Civilian Pilot Training Program: 390–392
   civilian v. military schools: 395–396
   copilots: 411–412
   dive bombing: 412–413
   elimination rate: 394–395
   fighters: 413–415
   formation flying: 397
   gunnery: 404–405
   instrument flying: 396–399
   observation: 407–408
   preflight: 386–387
   primary phase: 390–395
   program evaluation: 417
   program freeze and decline of 1944: 415–416
   responsibilities confusion: 378
   single engine: 400–405
   standardization: 389
   theater specific: 477–478
   three-stage course: 388–389
   Training Manual 1–210: 393
   transition: 408–410
   twin engine: 389, 405–407
training standards
   1944 Training Standard: 530
   Training Standard 10-1-1: 542
   Training Standard 30-3-1: 536
Army Air Service School of Application: 274
Arnold, Henry H.: 27, 30, 31–32, 36, 45, 47, 48, 58, 62, 66, 67, 77, 87, 136, 278, 280, 283, 285–286, 292, 300, 308, 310, 313, 314, 320, 351, 356, 358, 359, 376, 380, 390, 415, 439, 473, 475, 479, 486, 516, 517, 525, 531, 553, 557
Artillery Aerial Observation School, First: 174–176
Assistant Secretary of War for Air: 243–244
Atkinson, B. M.: 192
Aviation Cadet Qualifying Examination: 385
Aviation Expansion Program: 376–377
Avigation: 264–265
AWPD–1 war plan: 307, 371, 372, 376, 381
AWPD–42 war plan: 376

Baker, Newton D.: 69, 138, 277
Baker Board: 277
Balloons
   Balloon and Airship Section separation: 215–216
   Civil War: 11
   Gordon Bennett International Balloon Race: 15
   White Wing dirigible: 17
Barcus, Glenn O.: 251, 272, 539–540
Barrett, Joseph E.: 12
Baucom, Byrne V.: 230, 233
Bauer, Lawrence J.: 182, 183
Beck, Paul W.: 35–36, 39, 44, 48, 50, 54, 67, 90–91
Bell, Alexander Graham: 17, 94, 96
Bilstein, Roger: 103, 376
Bingham, Hiram: 112–113, 118, 119, 163, 165, 170, 171, 250
Bishop, Cleo: 472

Index

Bliss, Tasker: 35
Board of Academic Schools: 14
Boeman, John: 375, 405, 490, 495, 499, 502
Bolling, Raynal: 147, 148, 149, 184
Bombardment training. *See also under* Air Corps, U.S. Army; Crew and Unit training, AAF.
 aircraft: 431
 Air Service, U.S. Army: 224, 235–236
 AN/APQ–5 blind bombing equipment: 506
 bombardier qualifications: 385, 386
 bombing accuracy: 474–475, 528, 529
 circular error probable: 436
 curriculum: 430–432
 dive bombing: 412–413
 dual: 429–430, 434–436, 450, 511–512
 gunnery: 430–431
 M38A2 practice bombs: 431, 495
 night bombing: 343, 432
 proficiency evaluation: 436–437
 program evaluation: 450
 salvo bombing: 436
 schools: 428, 430
 strategic bombardment: 490–491
 terror bombing: 475
 train bombing: 435
Bombsights: 8, 432–433, 500
 A–5: 516
 D–1: 500
 D–8: 430, 432, 433, 529
 M–1: 343
 M–2: 365
 M–9: 432, 433
 N3A: 529
 Norden: 346, 358, 361, 475, 519
 O–1: 343
 S–1: 432, 433
 Sperry: 346, 500
Bond, Charlie: 329
Bowles, Edward L.: 512–513
Bradley, Follett: 443, 444, 463
Brereton, Lewis H.: 55, 58
Brett, George: 255, 274, 281, 282, 360, 361
Brindley, Oscar A.: 62, 63
British Commonwealth Training Plan: 456
Broadhurst, Harry: 472
Brodie, Bernard: 379
Brummel, Donald B.: 550
Burchinal, David A.: 368, 465, 507

Bureau of Aircraft Production: 110
Burge, Vernon L.: 83
Burgess, W. Starling: 28, 52
Burgess Company: 52–53
Bush, Vannevar: 512
Buzzer classes: 327

Campbell, Douglas: 107
Carberry, Joseph E.: 52, 59, 110, 150
Carigan, Bill: 469
Carpenter, John W.: 328, 329, 354, 355, 356, 361, 506
Chamberlain, J. L.: 121
Chamberlain, Neville: 299
Chandler, Charles DeForest: 11–12, 13, 18, 20, 33, 34, 44, 45, 47, 48, 50, 54, 55, 58, 67, 81, 84, 88, 90, 215
Chandler, Sherman: 55
Chaney, J. E.: 247–248, 249, 274
Chapman, Carleton G.: 84, 85
Chapman Field: 140
Chaute Field: 122
Chennault, Claire: 287
Churchill, Winston S.: 181–182, 299, 474
Ciardi, John: 519–520
Circular 50–1: 269
Circular error probable: 436
Civilian Conservation Corps: 210
Civilian Pilot Training Program: 318, 390–392
Civil War: 11
Clark, Albert P.: 257
Coffyn, Frank: 28, 29, 30
Colgan, Bill: 401, 541
Collier, Robert F.: 28
Combined Bomber Offensive: 483, 522
Commands
 I Bomber Command: 487, 498, 500, 529–530
 I Fighter Command: 463
 III Bomber Command: 462, 487, 498, 506, 508, 512, 514–515, 529, 532, 536–537, 545–546
 III Fighter Command: 463, 542–543, 555
 III Interceptor Command: 366
 III Reconnaissance Command: 415, 552
 IV Fighter Command: 415, 463
 XX Bomber Command: 413, 460
 Ferry Command: 340–341
 Troop Carrier Command: 454
Congress, U.S.
 Air Corps, establishment: 205–206

663

Index

aircraft production expansion: 307
appropriations: 8, 16–17, 75
  Air Service: 221
  Army Air Corps: 273, 280
  pre–World War I era: 16–17, 26, 28, 67
  Signal Corps: 75, 87, 93, 94, 102, 103
  World War II era: 310
Army expansion mandate: 309
Army reform: 72–75
aviation reserve controversy: 92–93
Aviation Section establishment: 9, 72
legislation
  Act of 1914: 75–76
  Air Corps Act: 241
  Army Reorganization Act of 1920: 220
  Aviation Student Act: 318
  House Resolution 448: 73
  Lend Lease Act: 305
  National Defense Act of 1916: 94
  National Defense Act of 1920: 204
  Neutrality Act of 1937: 305
  Overman Act of 1918: 103, 194
  Urgent Deficiency Act: 93, 94
recruitment limits: 220–221
Report on Progress Made in Aeronautics in the Army Since 1913: 90
training centers expansion: 77, 78
Coningham, Arthur: 476
Cook, Orval: 205
Coolidge, Calvin: 242
Cowan, Arthur S.: 27, 58, 59, 60, 62, 64–65
Cozzens, James Gould: 441, 554, 557
Craig, Malin: 274
Crew and Unit training, AAF: 451–454, 521–523, 527–528, 561, 562. *See also* Flying training, Signal Corps; Training Center, Air Corps; Training Section, Air Service.
  accidents: 533, 534
  aircraft: 460, 535–536
  air forces based: 454, 458
  attack tactics: 529
  B-25: 531–532
  B-26: 533–535
  bombardment: 528, 529
    accuracy: 530–531
    heavy: 458–460
    light: 523–524, 525, 535–538
    medium and light: 461–463, 523, 525–535
    skip bombing: 527, 528, 529
    very heavy: 460–461
  British Commonwealth Training Plan: 456
  Combat Crew Training School: 457
  combat readiness issues: 464–467
  copilot: 528
  crew bonding and cohesiveness: 480
  doctrine and tactics: 467–468
    balanced forces concept: 468
    confusion of: 478–479
    Field Manual 100–20: 468, 477
    strategic v. tactical: 468–469
  dual: 528, 529
  fighters: 463, 524
  fighters and bombers, coordination of: 524–525
  flight control equipment, automatic: 471–472
  formation flying: 469
  gunnery: 526, 527, 537–538
  half-truth training: 479
  high-altitude flying: 469–470
  instrument flying: 470–472
  Italian theater experience: 528
  Model Mission Flight Training study: 494
  navigator: 528
  night missions: 531, 534
  Operational Training Units program: 456–457
  pilot qualification: 527–528
  program evaluation: 529–530
  reconnaissance: 523, 524
  Replacement Training Units program: 457–459
  Special Combat Intelligence Memoranda: 452–453
  strafing: 527
  tactics: 537
  theater specialization: 466–467, 532–533
  1944 training Standard: 530–531
  Training Standard No. 30-3-1: 536
Culture, Air Force: 2–3
Curriculum
  acrobatics: 552–553
  advanced flying school: 258–259
  advanced flying training: 139–140, 157–158, 164–165
  aerodynamics response: 34
  aeronautics introduced: 13

Index

B-29: 506
basic flying training: 325-327
blind flying: 256
bombardment training: 132, 142-143, 185-189, 235-236, 342-344, 362, 421-422, 487, 490-491, 526
buzzer classes: 327-328
cross country flying: 27
Curtiss Aviation School: 37-38
electrical communications: 14
engine familiarity deficit: 119
fighter pilots, French: 89-90
fighter pilot training: 289, 540-543, 546-547, 551, 553
fighter training, AAF: 540-543, 546-547
flight simulators: 265-267
flying training stages: 109
formation flying: 133, 289, 327
Foulois influence: 26-28
ground schools: 112-117, 118-119
gunnery training: 349, 422, 440-444, 491, 492-493, 547-548
instrument flying: 265-267, 268-269
instrument landing: 268
map reading: 265
Military Aviation report: 73, 75
navigation training: 289, 327, 339, 341, 421, 422-423, 423, 426, 449
night flying: 326-327, 551
North Island Aviation School: 59-61
observation training: 135
operational training: 363
pilot training: 389-390, 392-394
   1-2-3 aircraft control system: 398
   aircraft identification: 402
   basic: 395-397
   Duckworth full panel aircraft control system: 398
   fighter: 414
   gunnery: 404-405
   heavy bomber: 411
   instrument flying: 396-399
   navigation: 402-403
   observation: 407-408
   primary pilot: 157-158, 164-165, 392-394
   single engine: 402-403
   twin engine: 405
primary flying school: 124-125, 256
program, three-phase: 77-78
pursuit: 133, 229, 366
reconnaissance: 368-370

Signal Corps: 48, 59-61
single engine: 331-332
specialization: 3
strategic bombardment: 489, 493, 498
tactical: 60
tactical unit training: 262-264
Training Center: 250-251
wireless telegraphy: 327-328
Wright school: 31
Curtis, Greely S.: 52
Curtiss, Glenn: 22, 35, 38, 50-51

Dargue, Herbert A.: 69, 84, 85-86, 139, 149, 162, 216, 217, 244, 320
Davidson, Howard: 62-63, 115, 118, 153, 154-155, 169
Davis, Milton F.: 216
Davison, F. Trubee: 206, 242
DoDo Days: 253
Doolittle, James H.: 22, 271, 525
Drum Board: 276, 280
Duckworth, Joseph B.: 398
Duncan, C. E.: 351
Dwyer, Geoffrey J.: 160

Eaker, Ira: 288, 289
Edgar, C. G.: 121, 122
Eisenhower, Dwight D.: 472
Ellyson, Theodore: 35
Ely, H. E.: 261
Emmons, Delos C.: 307, 313
Engines, aircraft: 27, 46
   early: 27, 38-39
   Fiat: 157
   Gnome: 134, 164
   Hispano-Suiza motor: 166
   OX-5 water cooled V-8: 124, 125
   Pratt and Whitney Wasp Junior: 254
   Renault: 63
   Sturtevant four cylinder: 55-56
   Wright Cyclone: 290
   Wright R-975: 325
Ent, Uzal G.: 492, 514
Estes, Howell: 369, 387, 397-398
Estoppey, George: 432
Eubank, E. L.: 427, 472, 558
Exhibition flying: 8

1910 Field Service Regulations: 28
Fairchild, Muir S.: 438
Fechet, James E.: 219, 242-243, 245
Federation Aeronutique Internationale (FAI): 12

665

Index

Fighter training, AAF: 538–539
  accidents: 551–552
  acrobatics: 552–553
  aircraft: 539–540
  air defense mission: 540
  breakaways: 553
  casualties: 548, 553–554, 555
  criticisms of: 554
  curriculum: 540–543, 546–547
  escort: 544–545
  fighter-bomber: 545–546
  formation flying: 544
  gunnery training: 547–548, 551
  gun sight aiming cameras: 547
  night missions: 548–551
  P–38: 548, 552
  P–61: 548–552
  program administration: 542, 543
  program evaluation: 543, 552–555
  radar: 551
  reconnaissance: 552
  replacement training: 542–543
  standards of proficiency: 540
  tactics: 543–544
  theater specialization: 541, 547
  Training Standard 10–1–1: 542
First Aviation Objective: 309–310, 313, 317
Fletcher, Eugene: 392
Flight simulators: 265–267
Flying Fortress Pilot Training Manual: 499
Flying training, Signal Corps: 131, 132–133, 165–166. *See also* Crew and Unit training, AAF; Training Center, Air Corps; Training Section, Air Service.
  202 program: 166
  aircraft. *See* Aircraft, Allied.
  aircraft and equipment shortages: 142, 175–176, 187
  bombardment: 132, 141–142, 142, 181–189, 193
  chain of command issue: 137
  curriculum: 139–140
  Europe: 148–149, 160
    aircraft: 154–156, 159, 164
    bombardment: 158
    British program: 172
    curriculum: 157–158, 164–165
    DH–4 aircraft: 176, 177–180
    dog fighting: 171, 173
    England: 149, 158–160
    equipment shortages: 157
    formation flying: 168, 170
    France: 149–151, 160–162
    Gosport training method: 159
    gunnery: 171
    Italy: 156–158
    Million Dollar Guard: 152
    multiple aircraft types confusion: 162–163
    navigation, cross country: 162
    night flying: 171–172
    organization: 163–164
    program duration: 139
    pursuit: 132–133
    specialization: 191–194, 197
    tailspins: 169–170
    training v. operational aircraft issue: 159–160
    transitional training: 159–160
  gunnery: 132, 140–141, 189–191
  instructors: 142
  night bombardment: 182–183, 186, 187
  observation: 135, 137–138, 140, 173–181, 197
  personnel and manpower issues: 150, 151
  nonflying cadet force: 151–153
  officer procurement: 135, 138–139
  selection and classification: 131–132, 163, 179, 185
  pursuit: 166–172, 197
  schools
    First Artillery Aerial Observation School: 174–176
    Second Aviation Instruction Center: 149, 150, 152
    Third Aviation Instruction Center: 149, 162
    Seventh Aviation Instruction Center: 184, 186
    Eighth Aviation Instruction Center: 156
    French schools: 136–137
  skills and training deficits: 161–162
  United States: 148, 150–151
Flying Training Command: 313
Fog of Peace and war planning: 273
Foulois, Benjamin D.: 11, 17, 19, 22, 23, 25–26, 39, 40, 69, 76, 79, 81–82, 96, 103, 138, 143, 182, 191–192, 207, 235, 241, 243, 252–253, 260–261
Frank, W. H.: 208

Index

Frisbee, John: 327, 331, 392
Fundamental Conceptions of the Air Service: 219

Gardner, A. P.: 92
Garland, William: 337
Geiger, Harold: 55, 58, 80, 86
Genet, Edmond C. C.: 89
Gerstner, Frederick J.: 64
GHQ Air Force: 244
  aircraft. *See also* Aircraft, Allied.
    inventory: 280, 283–284, 298, 300
    overhaul and repair issues: 281, 298
    shortages: 281–282
  Andrews's influence: 277–278
  balloons: 296
  budget and acquisition issue: 279
  command issues: 277, 279
  Drum Board: 276, 280
  dual organization structure issue: 273–274, 279–280
  establishment: 208
  expansion: 308–311
  geographical dispersion: 283
  mission: 277–278
  personnel and manpower issues
    personnel assignment issue: 279
    personnel shortages: 280
    pilot–engineer antagonism: 279–280
    reserve pilots: 280
  schools
    Command and General Staff School: 281
    March Field: 283–285
    Muroc Dry Lake: 285
  side-by-side operations: 283
  three-wing structure: 282–283
  training
    air-ground forces cooperation: 278–279, 291, 296–298
    attack strafing and bombing: 290–291
    blind flying: 289
    bombardment: 291–295
    casualties: 288
    crew: 293–294
    fighters: 287–289
    firing ranges: 285–286
    flying hours ceilings: 290
    gunnery: 285–286
    navigation: 294
    night missions: 295
    reconnaissance and observation: 295–296, 295–298
    specialization: 286–287
    tactical: 279
    wing: 283–285
  Training Directive for 1938–1939: 298
Giles, Barney: 116
Glassford, William A.: 13–14, 16, 69
Glines, C. V.: 398
Goldsborough, W. C.: 298
Goodier, Lewis E., Jr.: 52, 55
Gordon Bennett International Balloon Race: 15
Gorrell, Edgar S.: 182, 197
Gosport training approach: 127–128, 159
Graham, Harry: 55
Greely, Adolphus: 11, 12
Greer, Thomas H.: 219–220
Grider, John M.: 120, 159, 172, 195
Ground Training Guide: 500
Groups
  1st Bombardment Group: 217, 235, 284
  1st Day Bombardment Group: 235
  1st Pursuit Group: 217, 229, 237, 262, 289
  1st Surveillance Group: 217
  2d Bombardment Group: 237, 262, 364
  3d Attack Group: 226, 233, 237, 238, 262, 271, 290
  3d Bombardment Group (Light): 535
  7th Bombardment Group: 283, 295, 364–365
  8th Pursuit Group: 290
  9th Observation Group: 262
  10th School Group: 226
  11th School Group: 224
  17th Attack Group: 284
  17th Bombardment Group: 343
  17th Pursuit Group: 283
  19th Bombardment Group: 294, 339
  21st Airship Group: 273
  23d Composite Group: 365
  25th Bombardment Group: 362
  31st Pursuit Group: 365
  34th Bombardment Group: 485
  39th Bombardment Group: 485
  97th Bombardment Group (Heavy): 425
  305th Bombardment Group: 515
  319th Bombardment Group: 538
  330th Bombardment Group: 502
  354th Fighter Group: 542
  360th Fighter Group: 542
Guard of Honor: 2–3
Gulf Coast Air Corps Training Center: 315

Index

Gunnery training: 404–405
  advanced flying training: 132, 140–141, 171
  aircraft: 433
  aircraft and naval vessel recognition: 441, 442
  air-to-air: 444–445, 508
  angle of reflection: 442–443
  Army Air Corps: 329, 331, 332, 334, 335, 347–349
  B–29: 507–509
  camera guns: 445
  crew training: 526, 527, 537–538
  curriculum: 349, 422, 440–444, 491, 492–493, 547–548
  E–14, Jam Handy simulator: 440, 441
  frangible bullets: 445–446
  Gunnery Branch chart: 129
  gunsight aiming cameras: 547
  instructors: 439–440, 446
  machine gun maintenance: 441–442
  M–6 Bell Adapter: 440
  position firing system: 442–444, 492
  primary flight school: 128–130
  program evaluation: 446, 449–450
  Renshaw System of Identification of Aircraft: 441
  Royal Flying Corps School of Aerial Gunnery: 130
  schools: 438
  simulated fighter attacks: 444–445
  simulators and training devices: 440–441
  specialists ratios: 439
  specialization: 446, 447
  strafing: 527
  strategic bombardment: 492–493, 500, 504, 507–509
  target practice, moving: 441
  turrets: 444, 446–448, 492–493
  Waller trainer: 441

Hall, Leonard R.: 550
Hansell, Haywood S., Jr.: 444, 478, 520
Harbold, N. B.: 337, 341
Harding, Warren G.: 203
Harmon, Tom: 531
Harper, Robert W.: 378, 379, 394, 406–407, 426, 427, 438, 447, 466, 470, 478, 486, 516, 517, 525, 553
Hartney, Harold: 115, 155
Hatfield, Henry: 425
Hay, James: 72

Hazelhurst, Leighton W.: 49, 54
Hazelhurst Field: 122
Heavy Bombing Training Standard: 514
Helicopters: 3
Hennessy, Frederick B.: 49
Herbert, Kevin: 481
Hickam, H. M.: 213
Hitler, Adolf: 299
Hoare, Cuthbert G.: 108
Hodges, James P.: 126
Hoover, Herbert: 206, 209
How Our Army Grew Wings: 90
Humphreys, Frederick: 7, 20, 22, 25, 33
Hunter, Frank O'D. "Monk": 288
Hutton, F. R.: 59
Hynes, Samuel: 480

Instructors: 436–437
Instrument flying: 264–270, 271, 333, 396–399, 470–472
International Aeronautical Congress: 16
Interwar era. *See also* Pre–World War I era; World War I era; World War II era.
  Air Corps, establishment: 205–206
  aircraft industry production expansion: 300
  air power, value of: 204
  appropriations: 209, 236
  armed forces, peacetime: 204
  AWPD–1 war plan for air: 307
  Bolling Mission: 235
  Canadian and British pilot training: 305–306
  civil aviation experience: 205
  civil aviation industry: 210
  Civilian Conservation Corps: 210
  coastal defense mission dispute: 208, 244
  demobilization: 203–204
  depression and economy: 206
  disarmament efforts: 305
  fighter development: 230
  Four Army Plan: 211
  isolationism: 203, 206, 270, 299
  Lend Lease Act: 305
  mission and role uncertainty: 205, 206
  National Defense Act of 1920: 204
  organization and administration: 204
  protective mobilization program of 1937: 299
  research and development: 205
  strategic bomber development: 301

Index

training program, peacetime: 217–218
War Department Victory Program: 307
weaponry, obsolete: 205
World War II looms: 299–300

Janus, Anthony: 43
Japan: 299
Jarrell, Randall: 451
Jervey, Thomas M.: 348
Johnson, Davenport: 334–335

Kelly, G. E .M.: 36, 37, 40, 48
Kenly, William L.: 103, 128, 132, 133
Kennedy, Frank M.: 44, 46, 50, 54, 67
Kennedy, Paul: 203
Kennedy, William L.: 349
Kennett, Lee: 196
Kenney, George C.: 195, 467, 479, 521, 527
Kerwood, Charles W.: 182, 183
Kilbourne, C. E.: 203, 208
Kilner, Walter G.: 155, 161, 162, 177, 186, 190, 227, 244, 280
King, Campbell: 245
Kirtland, Roy C.: 44, 48, 55, 59
Knauss, Walter: 557
Knerr, Hugh: 121, 123, 263, 279–280
Kraus, Walter: 436

Lafayette Escadrille: 89
LaGuardia, Fiorello H.: 157
Lahm, Frank P.: 14–15, 18, 19–20, 22, 23, 25, 33, 58, 83, 84, 90, 94, 197–198, 242, 243, 246
Landon, Truman: 233–234, 264, 290
Lassiter, William: 205
Lee, C. F.: 118
Lee, Robert M.: 477
LeMay, Curtis E.: 506, 507, 515, 520
Lewis, C. S.: 162, 169
Lewis gun: 91
Liddle, Peter: 196
Link trainer: 265–267
Lodge, Henry Cabot: 280, 351
Loening, Grover C.: 33, 34, 50, 62, 64, 67, 169
Love, Moss L.: 83, 85
Lovett, Robert: 415
Lowell, Lawrence A.: 114
Lusitania, sinking of: 92
Lyon, E. B.: 324

MacArthur, Douglas: 207, 210–211, 276, 521
Macready John: 126
Manchu Law: 76
March Field: 123
Marshall, George C.: 306, 358, 370, 467
Martin, Glenn: 65
Martin, H. S.: 236
McConnell, James: 89
McFarland, Stephen: 437
McLeary, Samuel H.: 55
McNair, Leslie J.: 307, 370, 473
Menoher, Charles: 131, 195, 214, 215, 216, 220–221, 227, 239
Mexican intervention: 9–10, 93
Milling, Thomas DeWitt: 30, 31, 32, 45, 47, 48, 49, 55, 76, 80, 82, 84, 91, 107
Million Dollar Guard: 152
Mira Loma Flight Academy: 392
Mitchell, William "Billy": 69, 182, 211, 214, 217, 231
Model Mission Flight Training study: 494
Molineau, Henry S.: 44
Montgomery, Richard: 257, 259, 322, 403–404, 517–518
Moore, Willis L.: 16
Moulthrop, J. R.: 132
Muroc Dry Lake: 285
Mussolini, Benito: 299

Nash, Paul: 101
National Advisory Committee for Aeronautics: 9, 61–62
National Aeroplane Fund: 92
National Guard: 239
National Security League: 92
Navigation training: 208, 264–270, 271, 421, 422–423
  advanced flying training: 162
  Army Air Corps: 334, 337–342, 354–355, 363
  B–29 photo reconnaissance: 508–509
  celestial navigation: 424
  Celestial Navigator Trainer: 500
  crew: 528
  cross country: 162
  curriculum: 422–423, 426, 449
  dead reckoning: 423–424, 425
  dual: 434–435
  Gee medium-range navigational system: 510
  gunnery: 426, 438
  instruments: 425, 426

669

Index

Loran navigation: 424–425
navigators' qualifications: 385–386
Navitrainer G–1: 500
pilots: 402–403
radar: 482
radio navigation: 424
strategic bombardment: 489–490, 500
strategic bombers: 489–490, 490
Navy, U.S., coastal defense mission dispute: 206–207, 208, 244, 265
New York National Guard First Aero Company: 95
Nugent, Dick: 337

Observation and reconnaissance training: 44, 51–52, 79–80, 135, 137–138, 140, 227–228, 407–408
Ocker, Harry B.: 62–63
Ocker Box: 266
Olds, Robert: 468, 472
Operational Training Units (OTUs): 378–379
Orange Plan for Philippines defense: 83
Organization and development
  24-, 41-, 54-, 84-group programs: 309–310, 311, 313
  202 program: 166
  Air Corps: 205–206, 241, 242–246
  Air Force Combat Command establishment: 308
  Air Service establishment: 103
  Air Staff establishment: 308
  Army Air Forces establishment: 375
  Army Reorganization Act of 1920: 220
  Army Training Regulation 440–15: 208
  Assistant Secretary of War for Air: 243–244
  Aviation Expansion Program: 376–377
  Balloon and Airship Section separation: 215–216
  B–29 crew system: 487
  Continental Air Forces: 454
  divisional system: 214–216
  dual organizational structure problem: 273–274, 279–280
  five-year expansion program, 1926: 213
  flight officer designation: 319
  Flying Training Command establishment: 313, 378
  Four Army Plan: 211
  G (Assistant Chiefs of Staff) system: 216
  GHQ Air Force establishment: 244, 273–274, 276–278
  GHQ Air Force three-wing structure: 282–283
  Interwar era: 213–214
  MacArthur Army reform: 276
  Operational Training Units: 378–379, 452, 454, 456–457
  OTU–RTU pattern: 484–486
  Overman Act of 1918 Signal Corps reorganization: 194
  Pre–World War I era
    Act of 1914: 75–76
    Aeronautical Division: 8
    Army reform: 72–75
    Aviation Section established: 72
    command structures: 39–41
    enlisted men pilot training prohibition policy: 83–84
    Enlisted Reserve Corps establishment: 95
    Manchu Law: 76
    Military Aviator rating: 54
    Provisional Aero Company: 39
    Signal Officers' Reserve Corps establishment: 95
    Stimson reccomendations: 75
    tactical squadrons: 58
  program coordination and standardization: 379–380
  protective mobilization program of 1937: 299
  Replacement Training Units: 378–379, 452, 454
  Reserve units establishment: 239
  ROTC establishment: 217
  Training Section establishment: 102–103
Organized Reserve: 239
Overman Act of 1918: 194

Pan Am navigation instruction: 338–339, 341
Park, Joseph D.: 55
Parmalee, Phil: 28, 29
Parrish, Noel: 221, 250, 257, 272, 302
Parsons, Edwin C.: 182
Partridge, Earle: 331
Patrick, Mason M.: 103, 165, 166, 216, 217, 219, 221, 242, 243, 245
Pershing, John J.: 102, 103, 104, 116–117, 134, 143, 147, 150, 172, 182, 216
Philippines deployment: 84–86

Index

Photography: 368
Plan B training system: 248–249
Post, Henry B.: 84–85, 86
Pratt, H. Conger: 127
Pratt, William: 207, 208
Preddy, George: 544, 545
Preparedness movement: 9–10
Preparedness Tournament: 92
Pre–World War I era: 7. *See also* Interwar era; World War I era; World War II era.
  Aero Club of America influence: 16
  aerostation training: 14
  aircraft industry: 28, 103
  aircraft industry–military relationship: 55–56
  aircraft trials, early: 18–21
  Allen early disinterest: 15
  Army reform: 71–73, 72–75
  Aviation section establishment: 9
  balloons: 11–12
  Board of Academic Schools: 14
  congressional appropriations: 16–17, 26, 28, 67
  defense-oriented groups' influence: 92
  Lafayette Escadrille: 89
  European conflict, influence of: 72, 73, 87–88, 92, 101
  foreign aviation, influence of: 15–16
  French aeronautics: 87
  International Aeronautical Congress: 16
  Lusitania sinking influence: 92
  military training, private: 95
  military unpreparedness: 102, 104
  Moore influence: 16
  National Defense Act of 1916: 94
  New York National Guard First Aero Company: 95
  nonmilitary publications influence: 15–16
  Orange Plan for Philippines defense: 83
  Philippines deployment: 84–86
  preparedness movement: 9–10, 95, 97
  Progressive Era: 71–72
  research and development: 9, 13
  specialization: 9
  training, early: 7, 9
  War Department General Order No. 145: 14
  World War I–eve expansion: 102
  Wright brothers' influence: 18, 19
Professionalization: 1, 70
  career path mapping: 274–275
  early: 14
  officer and a gentleman: 117
  pre–World War I era: 8
  1930s: 301
Provisional Airplane Regulations for the Signal Corps: 39
Putnam, Claude E.: 479

Quesada, Elwood R.: 477

Radar. *See under* Technology.
Randolph Field: 252–253
Rawnsley, C. F.: 549
Reber, Samuel: 11–12, 59, 61, 69, 80–81, 88
Reconnaissance and observation: 44, 51–52, 79–80, 335, 367–370, 503, 508–509, 523, 524
Renshaw System of Identification of Aircraft: 441
Replacement Training Centers: 386
Replacement Training Units (RTUs): 378–379
Report on Progress Made in Aeronautics in the Army Since 1913: 90
Research and development
  Aerial Experiment Association: 17
  aircraft development
    B–10: 292
    B–17: 292
    bombers, long-range: 292
    closed cockpit: 288
    engine and airframe design: 210
    P–61: 548–550
    single seaters: 288–289
    strategic bombers: 301
    test pilots: 288
  interwar era: 205, 210
  National Advisory Committee for Aeronautics: 61–62
  pre–World War I era: 9
  1930s: 280–281
  tactical research: 207–208
Reserve Officers Training Corps (ROTC): 239
Rich, C. Perry: 84
Richards, John: 154, 161
Robertson, W. A.: 151, 161
Robins, A. W.: 274
Rockwell Field: 122
Rodgers, John Rogers: 30, 31
Roleur training approach, French: 127
Roosevelt, Franklin D.: 206, 209, 267,

671

Index

299–300, 305, 313, 318, 375
Roosevelt, Philip: 92
Roosevelt, Theodore: 72, 92
Rowe, Josiah P.: 156, 182
Royal Air Force establishment: 187–188
Royal Flying Corps, Canada: 108–110
Royal Flying Corps School of Aerial Gunnery: 130
Royce, Ralph: 76
Ruggles Orientator: 246–247, 248, 265, 267
Russell, F. H.: 33
Ryan School: 391

Saltzman, C. McK.: 20
Schaffer, Ronald: 103
Scriven, George P.: 26, 55, 59, 62, 69, 76–77, 78, 81, 83, 86, 88, 93, 94, 96
Second Aviation Instruction Center: 149, 150, 152
Second Aviation Objective: 310
Selfridge, Thomas E.: 17, 19
Selfridge Field: 122
Sharp, William B.: 73
Sherman, William C.: 48, 58, 214, 217, 218, 219, 227–228
Shockley, William B.: 513
Signal Corps, U.S. Army. *See also* Air Corps, U.S. Army; Air Service, U.S. Army; Army Air Forces, U.S.; Flying training, Signal Corps; Pre–World War Era.
  1910 Field Service Regulations: 28
  administrative issues: 76
  Aero Club of America influence: 75
  Aeronautical Division establishment: 8
  aeronautical engineers: 62
  aircraft: 44, 48, 65, 68
    brakes deficit: 80
    dep control system: 65
    pusher type: 63
    standards and evaluation: 63
    trainer aircraft development: 63–65
    trials: 18–21
    Wright planes removal: 63, 68
  air-ground forces training: 78, 79–81, 81–82, 96
  Aviation Section established: 72
  budget: 67, 78
  Burgess Company training: 55, 56
  casualties: 19, 40, 54, 64, 84
  civilian instructors: 62, 77
  congressional appropriations: 75, 87, 93, 94
  Connecticut joint maneuvers: 79–81
  Curiss-Wright aircraft parity: 49–50
  Curtiss factory training: 55, 56
  Curtiss v. Wright rivalry: 67–68
  decentralization policy: 81–82
  empirical v. theoretical flying: 66
  established: 43–44
  European observation influence: 88–89
  expansion: 53, 77–79
  Experimental and Repair Department: 61, 62
  Experimental Department: 87
  flight performance standards: 12
  foreign aeronautics influence: 87–89
  Fort William McKinley, Manila: 83
  General Orders 79: 57–58
  Lewis gun: 91
  Manchu Law: 76
  mechanics training: 75
  methods and instructional techniques
    application method: 59–60
    balancing machine: 31–32
    catapult launching: 24
    Curtiss v. Wright methods: 37–38
    dual-method trainer: 24
    dual seat, dual control: 44, 45
    flight simulators: 31–32
    Foulois's development of: 26–29
    progressive training sequence: 51
    sawhorse trainer: 31
    self-training v. dual instruction: 37–38
    side-by-side dual controls issue: 44–45
    single pilot: 44
    take-off and landing, early: 24
    turns and stability: 50–51
  Mexican intervention: 9–10, 93
  Military Aviation report: 73, 75
  mission: 8, 11
  night flying: 96
  overseas centers proposal: 81
  overseas deployment: 83–86
  personnel and manpower issues: 30, 54–55
    aviator standards: 61, 62
    cadet requirements: 54
    enlisted men pilot training prohibition policy: 83–84
    morale issues: 66, 67–68
    personnel issues: 58, 67

Index

pilot antagonisms: 67–69
pilot shortages: 93–94
rank v. function conflict: 76
staffing: 12, 44, 55
Wright v. Curtiss factionalism: 46
Philippines deployment: 84–86
pilot proficiency issues: 54–55
pilot training, early: 30–33
Provisional Airplane Regulations: 39
radio communication: 96
reconnaissance and observation: 44, 51–52, 79–80
Report on Progress Made in Aeronautics in the Army Since 1913: 90
safety issues: 47, 50
schools. *See also under* Flying training, Signal Corps.
 accessibility issues: 59
 Augusta, Ga.: 47–50
 civilian instructors: 29, 30–32, 38, 48–49
 College Park, Md.: 22–23, 43, 56, 77
 Curtiss Aviation School: 34–36
 Curtiss relationship: 57, 59
 establishment: 21–22
 Fort Kamehameha, Hawaii: 86
 Fort Leavenworth: 14, 59–60
 Fort Sam Houston: 26–27
 maneuver camp: 40–41
 North Island, Calif.: 56–57, 75, 77
 San Antonio, Tex.: 79
 San Diego, Calif.: 56
 winter camp: 47–50
 Wright brothers instruction: 23–26, 29, 30–34
 Wright brothers training camp: 22
 year-round sites development: 25–26
 technicians: 47
training centers expansion: 95–96
Training Department: 61
training program: 61, 77–78
weather conditions: 19, 22, 27, 41, 46, 48, 50, 58–59, 86
wireless communication, aerial: 79
Signal Officers' Reserve Corps establishment: 95
Simulators: 431
Skip bombing: 527, 528, 529
Smart, Jacob E.: 324
Smith-Barry, Robert: 127, 128
Snavely, Ralph: 551
Sorensen, Edgar P.: 434–435
Soubiran, Robert: 154

Southeast Air Corps Training Center: 315
Spaatz, Carl: 161, 209, 230, 231, 236, 255, 258, 278, 280, 281, 283, 292, 298, 313, 361–362, 467–468, 511
Special Combat Intelligence Memoranda: 452–453
Squadrons
 1st Aero Squadron: 9, 58, 79, 93
 1st Observation Squadron: 262
 1st Pursuit Squadron: 365
 3d Aero Squadron: 96, 136
 5th Observation Squadron: 262
 8th Attack Squadron: 262, 290
 9th Aero Squadron: 193
 11th Bombardment Squadron: 262
 12th Aero Squadron: 193
 13th Attack Squadron: 290
 16th Reconnaissance Squadron: 352
 17th Pursuit Squadron: 262
 20th Bombardment Squadron: 262
 24th Aero Squadron: 193
 24th Bombardment Squadron: 365
 27th Pursuit Squadron: 262
 31st Bombardment Squadron: 295
 34th Attack Squadron: 290
 37th Attack Squadron: 291
 40th School Squadron: 226
 41st School Squadron: 226
 42d School Squadron: 226
 43d School Squadron: 226, 231
 46th School Squadron: 224
 49th Bombardment Squadron: 262–263
 54th Bombardment Squadron: 365
 57th Service Squadron: 262
 59th Service Squadron: 262
 60th Service Squadron: 262
 61st Service Squadron: 262
 63d Service Squadron: 224
 70th Service Squadron: 226
 73d Attack Squadron: 290
 90th Attack Squadron: 262, 290
 91st Aero Squadron: 193
 92d Aero Squadron: 193
 94th Pursuit Squadron: 262
 95th Attack Squadron: 290
 95th Pursuit Squadron: 262
 96th Bombardment Squadron: 262
 97th Observation Squadron: 368
 312th Fighter Squadron: 464
Squier, George O.: 12–13, 14, 16, 20, 26, 88, 108, 112, 114–115
Stiles, Bert: 386
Stimson, Henry L.: 72, 75, 92, 512

Index

Straham, Victor H.: 161
Strategic bombardment training, AAF: 481–484
  accidents: 516
  AN/APQ–5 blind-bombing equipment: 506
  B–17: 501
  B–24: 501–503
  B–29: 482, 487, 488, 517
  B–24, fear of: 501–502
  blind flying: 483
  bombardment: 484–486, 490–491, 500
  bombing accuracy: 496, 519–520
  B–29 very heavy bombardment program: 503–509
  combat readiness based: 506–507
  Combined Bomber Offensive: 483, 522
  copilot: 488–489
  crew composition: 504
  crew selection: 489
  crew system: 487–488
  curriculum: 506
  daylight bombing: 483
  defensive armament: 491
  ditching and bailout procedures: 500
  engine fires, B–29: 517
  fighter escorts: 484
  flight engineer: 504, 505
  Flying Fortress Pilot Training Manual: 499
  formation flying: 483, 499–500
  Ground Training Guide: 500
  gunnery: 492–493, 500, 504, 507–509
  Heavy Bombing Training Standard: 514
  joint exercises: 497
  mission: 499
  navigation: 489–490, 500
  Pacific theater: 505, 507
  pilot responsibility: 488
  program evaluation: 515–520
  radar: 509–512, 514–515, 517
  radio operators: 492
  reconnaissance: 503
  standardized v. flexible approaches: 498
  standards: 505–506
  Training Standard 20-2-1: 488, 492
  unit training, three-phase: 493–497
Stratemeyer, George E.: 322, 436, 470
Stratton, Julius: 513
Sullivan, Mark: 71
Sweet, G. C.: 7, 20, 25
Sweetser, Arthur: 169

Tactical History of the Air Force: 180–181, 193
Tactics and doctrines: 561
  air superiority: 104
  area bombing: 475
  balanced forces concept: 468
  continental defense: 306
  fighter-bomber: 476–477
  formation bombing: 478
  formation flying: 133
  GHQ Air Force: 279
  interwar era refinement: 218–219
  pursuit mission: 287
  schools
    Barksdale Field: 345, 346–347
    School of Applied Tactics: 456
    Tactical School: 275–276
  strategic bombardment: 182–183, 287, 291, 467–468, 473
  tactical forces doctrine: 475–476
  terror bombing: 475
  theater differences: 453–454, 465
  two-ocean war: 306
Taft, William Howard: 18, 72, 83
Taliaferro, W. R.: 52, 59, 60, 64
Technical Training Command: 378
Technology
  B–2, B–3, B–5 driftmeters: 426–427
  bubble sextant: 265
  camera gun: 141
  catapult launchers: 24
  C–1 autopilot: 432, 433, 435, 472
  cockpits, pressurized: 366
  compasses: 265
  E–6B computer: 408, 423
  flight instrumentation: 265
  Hollocombe-Clift airspeed indicator: 114
  K–6 camera: 297
  Lewis gun: 91, 141
  machine guns: 128, 130, 173
  Marlin gun: 140–141
  radar: 410, 475, 476, 482, 509–512
    AN/APG–15: 512
    AN/APN–4: 512
    AN/APN–9: 512
    AN/APQ–5 blind-bombing equipment: 506
    AN/APQ–13 system: 511
    Gee: 510
    H–25: 510, 511
    H2S: 482
    H2X: 482, 510–511

Index

Oboe: 482, 510
SCR–718: 512
SCR–720: 551
radio: 264
tachometer: 66
wheel assembly: 28
wireless telegraphy: 87–88
Wright warping mechanism: 24
Thayer, Lucien H.: 107, 153, 159, 184
Third Aviation Instruction Center: 149
Training Manual 1–210: 393
Todd, Robert: 116
Training and Operations Division: 314–315
Training and War Plans Division: 243
Training Center, Air Corps: 246–252, 270–271. *See also* Crew and Unit training, AAF; Flying training, Signal Corps; Training Section, Air Service.
   advanced flying school: 250, 257–261
      aircraft: 259–260
      curriculum: 258–259
      specialization, delay of: 258–259
   aircraft, basic trainer: 254–255
   aircraft inventory: 246
   Avigation: 264–265
   basic: 250
   budget reduction proposal: 244–246
   casualties: 251
   class size, 1933: 242
   cooperative training: 261
   curriculum: 250–251, 268
   flying hours ceilings: 250–252, 262
   formation flying: 271
   instructors: 251
   instrument flying: 264–270, 271
   Kelly Field: 257–258
   methods and instructional techniques
      flight simulators: 265–267
      Link trainer: 265–267
      Ocker Box: 266
      Plan B training system: 248–249
      Ruggles Orientator: 246–247, 248, 265, 267
   personnel and manpower issues
      admittance policy, overage: 247–248
      all-through system: 249
      candidate selection: 246–249
      collegiality: 261
   primary flying school: 250, 252–257
      aircraft: 253–254
      blind flying: 256
      curriculum: 256–257
      DoDo Days: 253
   program evaluation: 252
   Randolph Field: 252–253
   Ruggles Orientator: 246–247, 248, 265, 267
   stage system: 249
   standardization: 250
   tactical unit training: 261–264
   training centers reorganization: 308, 314
Training Regulation 440–15: 219–220
   characteristics: 293
Training Section, Air Service. *See also* Crew and Unit training, AAF; Flying training, Signal Corps; Training Center, Air Corps.
   administration: 103
   aircraft: 196
   aircraft shortages: 196
   American and European training methods: 168–170
   best 10-percent plan: 120
   Bolling Mission: 182
   cadet statistics: 196
   Camp Taliaferro: 108
   Canadian reciprocal agreement: 108–110
   casualties: 197
   duration of training: 194–195
   establishment and organization: 102–103
   European flight training: 120
   expansion: 107
   First Pursuit Organization and Training Center: 192
   flying branch: 110
   flying fields list: 145
   flying schools chart: 111
   foreign instruction: 108
   graduation quotas: 137–138
   Pershing plan: 143–144
   personnel issues: 110–112
   pilot preparedness: 195
   post–World War I: 194
   primary flight training
      aircraft shortages: 123
      airfields: 121–123
      curriculum: 124–125
      facilities expansion: 121–122
      Gosport training approach, British: 127–128
      graduation standards: 125

Index

gunnery: 128–130
instructors: 126
program goals: 121
Roleur training approach, French: 127
specialization: 130
procurement and supply: 103
school specialization: 110
Training standards
Training Standard 10–1–1: 542
Training Standard 20–2–1: 488, 492
Training Standard 30–3–1: 536
Trenchard, Sir Hugh: 188
Troop Basis of January 1942: 375
Troop Carrier Command: 454
Tunner, William: 270–271
Turpin, Cliff: 30
Tuskegee Institute: 326, 396

Voorhis, Daniel Van: 356

Walker, John C., Jr.: 36, 40
Waller trainer: 441
de Walmer, Baron: 190
Ward, Edward: 12, 13
War Department Circular 59: 375
War Department General Order No. 145: 14
War Department Training Regulation 10–5: 218
War Department Victory Program: 307
Warrior culture: 1
Watry, Charles: 399, 405, 407
Weaver, Walter R.: 348
Weigley, Russell F.: 370–371
Wiener, Willard: 305, 322
Welsh, A. L.: 30, 32, 54
West Coast Air Corps Training Center: 315
Westover, Oscar: 207, 216, 274, 275, 276
Wilbur Wright Field: 122
Wilder, Harold: 181
Wildman, Francis "Doc": 62
Williams, Stanley T.: 119
Wilson, Woodrow: 92
Winder, C.B.: 48
Wings
1st Bombardment Wing: 283
15th Bombardment Wing: 485
17th Bombardment Wing: 485
18th Bombardment Wing: 357
21st Bombardment Wing: 471
58th Bombardment Wing: 460

73d Bombardment Wing: 460
Wood, Leonard: 72, 78, 92, 95
Wood, Robert E.: 48
Woodring, Harry: 292
Woods, Arthur: 119
World War I era: 101, 132, 141–142, 158, 181–189, 193. *See also* Interwar era; Pre–World War I era; World War II era.
aircraft industry, U.S.: 134
Air Service, American Expeditionary Forces command: 103
American front establishment: 104–105
Bolling Mission: 147–148
Bureau of Aircraft Production: 110
casualties: 197
demobilization: 203–204
German offensive, 1918: 165
industrial mobilization: 148
night bombardment: 188–189
Ribot cable of 1917: 196
U.S. Schools of Military Aeronautics: 113
Zepplin attacks: 181–182
World War II era. *See also* Interwar era; Pre–World War I era; World War I era.
aircraft industry production: 309–310, 310, 313, 375, 376, 380
air-ground forces cooperation: 296–298
AWPD–1 war plan: 307, 371, 372, 376, 381
AWPD–42 war plan: 376
Battle of Britain: 305
Civilian Pilot Training Program: 318
Combined Bomber Offensive: 483, 522
congressional appropriations: 310
curriculum specialization: 3
Doolittle raid on Tokyo: 526–527
First Aviation Objective: 309–310, 313, 317
friendly fire: 472
Mediterranean campaign: 522
North African experience: 477–478
Pacific Theater radar: 511–512
Rainbow 5 plan: 376
Tokyo bombing: 520
Tuskegee Institute: 326, 396
Wright, Orville: 18, 19–20, 33, 34, 82
Wright, Wilbur: 7, 18, 19, 22, 23, 24, 25, 26, 31, 50
Wuest, Jacob W. S.: 204

Index

Yeager, Chuck: 388, 543, 545
Yeats, W. B.: 147
Yount, Barton K.: 218, 306, 308, 314, 378, 415, 416, 466, 517, 518, 553, 562

Zoller, V. L.: 473

www.ingramcontent.com/pod-product-compliance
Lightning Source LLC
Chambersburg PA
CBHW060242240426
43673CB00047B/1861